AF716151

A Richer Picture of Mathematics

David E. Rowe

A Richer Picture of Mathematics

The Göttingen Tradition and Beyond

Springer

David E. Rowe
Institut für Mathematik
Johannes Gutenberg-Universität Mainz
Rheinland-Pfalz, Germany

ISBN 978-3-030-09812-4 ISBN 978-3-319-67819-1 (eBook)
https://doi.org/10.1007/978-3-319-67819-1

Softcover re-print of the Hardcover 1st edition 2018

Printed on acid-free paper

This Springer imprint is published by Springer Nature
The registered company is Springer International Publishing AG
The registered company address is: Gewerbestrasse 11, 6330 Cham, Switzerland

For Hilde and Andy

A RICHER PICTURE OF MATHEMATICS

A symposium in honor of David E. Rowe

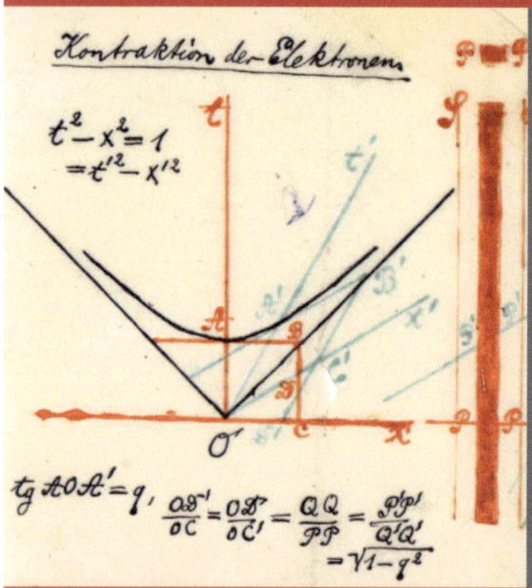

May, 18-20, 2016

Johannes Gutenberg-University,
Institute of Mathematics
Staudinger Weg 9, 55099 Mainz

For further information, visit
www.history.mathematics.uni-mainz.de

Organizer: Tilman Sauer (tsauer@uni-mainz.de)

SPEAKERS:

Tom ARCHIBALD, Vancouver
June BARROW-GREEN, Milton Keynes
Moritz EPPLE, Frankfurt
Tinne HOFF KJELDSEN, Copenhagen
Jesper LÜTZEN, Copenhagen
Walter PURKERT, Bonn
Volker REMMERT, Wuppertal
Erhard SCHOLZ, Wuppertal
Marjorie SENECHAL, Northampton, MA
Reinhard SIEGMUND-SCHULTZE, Kristiansand
Klaus VOLKERT, Wuppertal
Scott WALTER, Nantes

Layout: www.artefont.de

JOHANNES GUTENBERG UNIVERSITÄT MAINZ

JG|U

Foreword

"A Richer Picture of Mathematics" was the title of a symposium held in honor of David Rowe in May 2016 at the University of Mainz, his academic home for 25 years, as a sign of gratitude from his longtime friends and colleagues. The title conveyed the impetus and spirit of David's various and many-faceted contributions to the history of mathematics: Penetrating historical perspective paired with a sensitivity for the relevant broader context of mathematics as a human activity, an immensely broad knowledge and care for historical detail, and attention to remote and sometimes surprising but always revealing cross references.

The title also seems to us to capture particularly well David's regular contributions to the *Mathematical Intelligencer* over the long time of his association with this journal. Taken together, these articles display a surprising (perhaps unintended) coherence and a charming seriousness. They cover topics ranging from ancient Greek mathematics to modern relativistic cosmology. They were written for a broad readership of open-minded and curious but mathematically trained and educated readers. They are collected here, augmented by two other of David's papers and set in context by a foreword and new introductions. This volume is indeed "a richer picture of mathematics."

Northampton, MA
Mainz, Germany

Marjorie Senechal
Tilman Sauer

Preface

This is a book that explores how mathematics was made, focusing on the period from 1870 to 1933 and on a particular German university with a rich history all its own. Relatively little of that mathematics will be explained in these pages, but ample references to relevant studies will appear along the way. My intention is not so much to add to that scholarly literature as to shed light on the social and political contexts in which this intellectual activity took place. Or, to put a twist on a well-known book title, instead of asking, *"what is mathematics?"*, I wish to explore a different question, namely "how was mathematics made?" The reader should note the verb tense here: this is a book about the past, and as a historian, I make no prescriptive claims. Clearly, mathematicians can, and often do, voice their opinions about how mathematics should be made, or about the relative importance of the people and ideas discussed in the pages that follow, but that is not my purpose here. Rather, my aim in presenting this series of vignettes centered on the Göttingen mathematical tradition is to illustrate some important facets of mathematical activity that have received far too little attention in the historical literature. It is my hope that by bundling these stories together a richer picture of mathematical developments emerges, one that will be suggestive for those who might wish to add other dimensions to those described here.

Although the central actors in this book were not always situated in Göttingen, I have attempted to show how each was drawn, so to speak, into that university's intellectual force field. Over the course of time, the strength of that field grew steadily until it eventually came to dominate all others within the German mathematical community. At the same time, Göttingen's leading representatives, acting in concert with the Prussian Ministry of Education, succeeded in transforming this small university town into one of the world's leading mathematical research centers. This larger story has been told, of course, before; what I offer here is a collection of shorter tales, many based on archival and primary sources, while drawing on a wide range of secondary literature. Taken as a whole, these essays point to several new dimensions that have been overlooked in earlier accounts. In particular, I believe they provide a basis for better understanding how Göttingen could have attracted such an extraordinary array of talented individuals. A number of them played central roles in creating a new kind of mathematical culture during the first decades of the twentieth century.

A brief word about the structure of the book: its six parts reflect thematic aspects of the overall story, whereas the individual essays largely recount episodes related to these six themes. These essays were originally written over a period of more than three decades, a circumstance that helps account for their rather uneven character. I have arranged them largely in accordance with chronological developments, but also in order to bring out different thematic elements. While the main threads running through each section are described in the introductions to each, the essays themselves were written as self-standing pieces that can be read out of order or even at random. There is no grand narrative as such; the six introductions mainly have the purpose of drawing out the underlying themes rather than providing a road map that shows how all the pieces fit together. With two exceptions, the essays in this book represent slightly revised versions of earlier ones that were published over the course of three decades in *The Mathematical Intelligencer*, a magazine with an interesting and novel history.

From its inception in the late 1970s, *MI* has served both to inform and to entertain its readers with topical writings pertaining to present-day mathematical culture, but also with an eye directed toward the past. Historical themes and issues can be found in practically every issue of *MI*, which started as a typewritten newsletter, but eventually grew into a full-fledged magazine familiar to mathematicians around the world. While a few of my contributions to *MI* found here were written during the 1980s, most come from the 15 years when I edited the column "Years Ago." Like many other regular features of the magazine, this column has evolved over time. It was first introduced by Jeremy Gray under the heading "50 and 100 Years Ago," a rubric that had long been used in the British journal *Nature*. Jeremy and others later transformed the column into a platform for historical essays and open-ended reflections on mathematical themes from past eras.

My own long association with *MI* started in a small way. During the 1980s, I was a fellow of the Alexander von Humboldt Foundation, which gave me the chance to delve into the rich archival sources in Göttingen, in particular the papers of Felix Klein and David Hilbert. As it happened, John Ewing, who was editor of *MI* at that time, was visiting Göttingen as a guest of the institute's *Sonderforschungsbereich*. So he and I sometimes had occasion to chat about various things, including my historical interests. John clearly recognized the importance of *MI* as a new type of venue for mathematicians seeking to air their opinions about more general concerns. His editorials from those years are still well worth reading today. As for what came later from my pen, I have to admit that this collection exposes me to the charge that I never managed to escape the pull of Göttingen and its famous mathematical tradition. So let me take this opportunity to thank Hans Becker, Helmut Rohlfing, and the many helpful staff members in the *Handschriftenabteilung* of the Göttingen library, past and present, who offered me their kind assistance over the years.

A Richer Picture implies that the reader probably already has a picture of many of the people and topics discussed in this book, as the essays in it were written with such an audience in mind. It should also be remembered that *The Mathematical Intelligencer* is a magazine, not a scholarly journal. Its mission from the beginning has been both to inform and to entertain. So I tried over the years to write about historical things that I happen to know about in such a way as to interest historically minded mathematicians. Some may nevertheless wonder: how could I write about mathematics in Göttingen and say so little about Riemann? Or about relativity, without saying more about Einstein's brilliant ideas? Of course, I am well aware of these and other glaring gaps. This volume does not pretend to give a comprehensive picture of all the many significant themes that run through the Göttingen tradition. In some of the introductions, I have taken the opportunity to address missing parts of the story, and for Part II I added new material based on two essays published elsewhere to fill in parts of Felix Klein's early career. I have also tried to give ample attention to the scholarly literature, including recent publications, for those who might want to learn about matters that go well beyond what one finds in these essays. The visual images of mathematical objects in the book were greatly enhanced by the talents of Oliver Labs, with whom I have worked closely in recent years, a collaboration I have greatly enjoyed. For the cover design, I am grateful to his wife, the graphic designer Tanja Labs, whose talents I and others have long admired.

The title chosen for this book was actually an idea I owe to Tilman Sauer, though it reflects very well my general views as an historian. That outlook can be put succinctly by citing the words of a famous philosopher from the first half of the twentieth century, the Spanish-born man of letters and Harvard don, George Santayana, who wrote many famous aphorisms. One of them – perhaps more appropriate than ever for the present century – maintains that "those who neglect history are condemned to repeat its mistakes" (or words to that effect). No one should want to dispute the wisdom of that saying, but I would offer a kind of inversion that runs like this: "those who think that history repeats itself – whether as tragedy or farce – are condemned to misunderstand it." A corollary to that claim would suggest that engagement with the past is its own reward; we only demean that activity by insisting that we should study history to learn what went wrong or to feel morally superior to those whose mistakes are so self-evident to us today. The essays presented here reflect my own efforts to learn about mathematics in Germany over many years during which I tried to learn from my own mistakes, some of which surface in the pages that follow.

Of course, so much went wrong in Germany that this theme was bound to be inescapable for historians, many of whom regard the Holocaust as the great looming moral problem of the last century. Quite a number of those who have studied this dark chapter in European history have done so exceedingly well. In the introduction to Part V, I briefly describe some of the key events bearing on mathematics in the Weimar and early Nazi eras by drawing on some of that literature. For this volume, I have included among the reference works several books and articles that offer background information on larger developments in German history relevant for understanding what I have written about. Here I would like to mention just two books that I have personally found particularly insightful. The first is *The Germans*, a collection of essays by Gordon A. Craig that first came out in 1982 when I was studying late modern European history at the CUNY Graduate Center. Craig was long considered the dean of American authorities on modern German history, a reputation gained in part from his volume *Germany, 1866–1945* published in the Oxford History of Modern Europe series. His *The Germans*, a kind of sequel, still makes delightful reading today. Particularly relevant are the five chapters entitled: Germans and Jews, Professors and Students, Romantics, Soldiers, and Berlin: Athens on the Spree and City of Crisis.

The second book was another best-seller, *The Pity of It All: A Portrait of the German-Jewish Epoch, 1743–1933,* written by the Israeli journalist and writer Amos Elon. Many historians have doubted that the much-vaunted notion of a German-Jewish symbiosis ever had any chance of succeeding, but none would deny that for many Jews this was a very real dream. Elon begins his account with the arrival of fourteen-year-old Moses Mendelssohn at an entrance gate to Berlin. Mendelssohn's rise to prominence as a philosopher and proponent of the German Enlightenment served as a symbol for German Jewry throughout the nineteenth century. His friend, the playwright Gotthold Ephraim Lessing, immortalized him in *Nathan der Weise*, first performed in Berlin in 1783, three years before Mendelssohn's death. In 1763, he won a prize competition offered by the Berlin Academy for his essay "On Evidence in the Metaphysical Sciences," in which he applied mathematical proofs to metaphysics (Immanuel Kant finished second). Mendelssohn's larger importance for mathematics in Germany was highlighted in the travelling exhibition "Transcending Tradition: Jewish Mathematicians in German-Speaking Academic Culture." This exhibition appeared at several venues around the world, including the Jewish Museum in Berlin, where it opened in July 2016 on the occasion of the VIIth European Congress of Mathematics.

Over the years that these essays were written, I have benefitted from the encouragement of a great many people, both within academia and outside it. Some of their names recur throughout this volume, but I should begin by thanking Leonard Rubin, my mathematical mentor at the University of Oklahoma, who taught me what doing mathematics was all about and then forgave me for leaving the field to do history. The inspiration to make that shift first came from members of the history of science department in Norman: Ken Taylor, Mary Jo Nye, and above all Sabetai Unguru, who opened my eyes to ancient Greek mathematics. Realizing that I would never learn ancient Greek, I tried German, and with my wife's help succeeded, at least to some extent. Our first visit to Germany in the summer of 1980 was what first put an unlikely idea into my head: why not try to study the history of mathematics in Göttingen? That thought soon led to a letter from Joe Dauben, inviting me to study with him in New York. Without his help and support, I'm sure nothing would have come of what was then only a dreamy idea.

To my own amazement, after two years at the CUNY Graduate Center, I was on my way to Göttingen with a fellowship from the Alexander von Humboldt Foundation. My advisor in Germany was Herbert Mehrtens, one of the first historians to explore the dark secrets of mathematics during the Nazi period. We barely saw each other, though, since he was then working at the Technische Universität in Berlin, where he later habilitated. In 1990, he published *Moderne – Sprache – Mathematik*, a book that created quite a stir at the time. Unfortunately, it went out of print some years ago without ever having been translated into English. During my two years as a Humboldt fellow I got to know several other historians of mathematics with whom I've been in contact ever since, among them: Umberto Bottazzini, Jeremy Gray, Jesper Lützen, John McCleary, Walter Purkert, Norbert Schappacher, and Gert Schubring. Afterward my circle of colleagues and friends widened considerably. So along with the above group, I'd like to thank Tom Archibald, June Barrow-Green, Jed Buchwald, Leo Corry, Michael Eckert, José Ferreirós, Livia Giacardi, Catherine Goldstein, Ulf Hashagen, Tom Hawkins, Jens Høyrup, Tinne Kjeldsen, Karen Parshall, Jim Ritter, Tilman Sauer, Erhard Scholz, Reinhard Siegmund-Schultze, Rossana Tazzioli, Renate Tobies, Klaus Volkert, Scott Walter, and several unnamed others who have helped inspire my work.

An especially important friendship for my wife and me was the one that developed with Walter Purkert, who took a leave of absence from the Karl Sudhoff Institute in Leipzig to spend a semester teaching with me at Pace University – not in New York City, but in the idyllic atmosphere of Westchester County. Marty Kotler, who as department chair at Pace was always highly supportive, helped make that invitation possible. Walter's visit also gave John McCleary and me the incentive to co-organize a memorable conference at Vassar College. That event, which brought together several of the aforementioned historians from Europe and North America, took place during a scorching hot week in the summer of 1988. It led to two volumes of essays that John and I published with the help of Klaus Peters, who was then at Academic Press in Boston.

The following summer we were in Göttingen again, but also spent a month in Leipzig, thanks to an invitation arranged by Walter. That was in August 1989, just before the famous Monday demonstrations that began at the Nikolaikirche. The atmosphere in the city was noticeably tense, though some people were very eager to meet and talk with us the moment they realized we were Americans. One could see younger folks standing in long lines, hoping to get visas for Hungary, which had recently opened its border to Austria. In Leipzig, it was easy to watch reports of such things on West German television and, of course, there was much talk about *Glasnost* and *Perestroika*, the sweeping reforms Michael Gorbachev was calling for just as the GDR was preparing to celebrate it fortieth anniversary. His state visit in early October put renewed pressure on the East German government, setting off the October demonstrations that led to the fall of the Berlin Wall in early November. Having gained an outsider's sense of life in the former GDR during the late 1980s, coupled with the experience of living in western Germany during the early 1990s, I am still inclined to view the reunification as a missed opportunity, though the sands of time have by now left that brief window of promise covered beyond all recognition.

In August of 1990, I had the opportunity to speak at a special symposium on the history of mathematics held in Tokyo and organized by Sasaki Chikara along with several other Japanese historians. The larger occasion was the Twenty-First International Congress of Mathematicians that took place in Kyoto one week earlier. This was the first ICM ever held outside the Western hemisphere and a truly memorable event. At the symposium, I spoke about "The Philosophical Views of Klein and Hilbert," whereas Sasaki presented an overview showing how Japanese mathematics had gradually become westernized during the latter half of the nineteenth century. I only realized then that his story and mine were closely related. From a purely intellectual standpoint, the great number theorist Takagi Teiji stands out as the key figure in this regard, and so it was fitting that two of the speakers at the symposium, Takase Masahito and Miyake Katsuya, gave summary accounts of developments in number theory that both preceded and succeeded Takagi's fundamental contributions to the field.[1] Readers who would like to find out how Hilbert enters into this story – in particular by way of his famous Paris problems – can find a stirring account of this in Ben H. Yandell's *The Honors Class: Hilbert's Problems and their Solvers.*

Although we often spent summer months in Göttingen, the thought that I might someday be offered a professorship at a German university rarely crossed my mind, and then only as a remote possibility. So I was both surprised and delighted to receive such an offer from the Johannes Gutenberg University in Mainz, where I began teaching history of mathematics in the spring of 1992. The adjustment wasn't easy for any of us, but probably it was hardest for my wife, Hilde, and our five-year-old son, Andy. We tend to think back on the good times now, but both of them had to deal with a lot of discouraging things. Luckily, we managed to get through difficult episodes together, for which I'm extremely grateful to them. Adversity, too, can make the heart grow fonder.

When I first came to Mainz, I had some doubts about whether this new life was going to work out. The atmosphere in the department was friendly, but very formal and somewhat provincial. I was one of the youngest members of the faculty and the only foreigner. Probably my colleagues were dimly aware that this was not going to be an easy transition for me. I had virtually no experience teaching history of mathematics, nor had I even taken courses at a German university, let alone taught in German. Luckily, the professorship came with a second position, and so I was pleased to bring in Moritz Epple as a post-doc after my first semester. During his years in Mainz, Moritz taught several different types of courses while working on his splendid book, *Die Entstehung der Knotentheorie* (1999). A second stroke of luck came when Volker Remmert replaced Moritz, who went to Bonn on a Heisenberg fellowship. Both of them contributed greatly to the activities of our small group in Mainz, where I soon felt very much at home. Two senior colleagues, Albrecht Pfister and Matthias Kreck, deserve a special note of gratitude for their support during those first years. Among the younger generation that came to Mainz after me, I am especially grateful to Duco van Straten, Volker Bach, Manfred Lehn, Stefan Müller-Stach, and Steffen Fröhlich. Many others in our immediate *Arbeitsgruppe* ought to be named as well, but six who deserve special thanks are Martin Mattheis, Annette Imhausen, Andreas Karachalios, Martina Schneider, Eva Kaufholz-Soldat, and our wonderful secretary, Renate Emerenziani.

Beginning in the mid 1990s, when I was a fellow at MIT's Dibner Institute for the History of Science, I had the opportunity to interact with colleagues at Boston University's Einstein Papers Project. Some of the fruits of those interactions can be found in Part IV of this volume, for which I thank Hubert Goenner, Michel Janssen, Jürgen Renn, Tilman Sauer, Robert Schulmann, and John Stachel. Two historians who deserve special thanks for their support over many years are Reinhard Siegmund-Schultze and Scott Walter, both of whom have been exceedingly generous in sharing their time and expertise. During my tenure as column editor, which ended in 2016, I had the pleasure of working closely with Marjorie Senechal, now editor-in-chief of *MI*. Marjorie was, from the beginning and ever afterward, a great source of support and intellectual stimulation. Her efforts on behalf of the magazine have been tireless and appreciated by everyone involved with its success. As a retirement present, she and Tilman Sauer, my successor in Mainz, came up with the idea of bundling my writings from *MI* into this book. To both of them, as well as to Marc Strauss and Dimana Tzvetkova at Springer-New York, go my heartfelt thanks.

Marjorie has a keen sense for the importance of history and memory in mathematical cultures, particularly when it comes to understanding that mathematics is a human activity: it's about people. A fine example of this can be found in the special thirtieth anniversary issue

[1]Most of the papers from the Tokyo symposium were later published in *The Intersection of History and Mathematics*, ed. Sasaki Chikara, Sugiura Mitsuo, Joseph W. Dauben, Science Networks, vol. 15 (Proceedings of the 1990 Tokyo Symposium on the History of Mathematics), (Basel: Birkhäuser, 1994).

(*MI,* vol. 30(1)) she put together. This contains her interview with Alice and Klaus Peters, who founded *MI* when both were working for Springer. Since *The Mathematical Intelligencer* will soon celebrate its fortieth birthday, those readers with a sense for history will surely appreciate the chance to page through earlier issues of the magazine, which offer so many insights into people, places, and ideas of the recent past. The selection offered here represents only a narrow slice of what one finds in the magazine as a whole, but I am pleased to have had the chance to contribute to its historical dimension.

Mainz, Germany David E. Rowe
August 2017

Contents

Part I

Two Rival Centers: Göttingen vs. Berlin

1 Introduction to Part I

Mathematicians can sometimes be ferociously competitive, making rivalry a central theme in the history of mathematics. Various forms of competitive behavior come easily to mind, but here I am mainly concerned with a specific rivalry between two leading research communities as this evolved during the latter-half of the nineteenth century in Germany. This concern has, in fact, less to do with specific intellectual achievements, important as these were, than with the larger context of professional development that eventually led to a clearly demarcated German mathematical community by the end of that century. Unlike the highly centralized French community, in which Parisian institutions and their members dominated the scene, the German universities were largely autonomous and tended to cultivate knowledge in local settings, some more important than others.

Germany's decentralized university system, coupled with the ethos of *Wissenschaft* that pervaded the Prussian educational reforms, created the preconditions for a new "research imperative" that provided the animus for modern research schools.[1] Throughout most of the nineteenth century, these schools typically operated in local environments, but with time small-scale research groups began to interact within more complex organizational networks, thereby stimulating and altering activity within the localized contexts. Mathematicians have often found ways to communicate and even to collaborate without being in close physical proximity. Nevertheless, *intense* cooperative efforts have normally necessitated an environment where direct, unmediated communication can take place. This kind of atmosphere arose quite naturally in the isolated settings of small German university towns. Such collaborative research presupposes suitable working conditions and, in particular, a critical mass of researchers with similar backgrounds and shared interests. A work group may be composed of peers, but often one of the individuals assumes the role of a "charismatic leader,"[2] typically as the academic mentor to the junior members of the group. This type of arrangement – the modern mathematical research school – has persisted in various forms throughout the nineteenth and twentieth centuries (Rowe 2003).

By the mid nineteenth century, two institutions in Germany had emerged as the leading centers for research mathematics: Göttingen and Berlin. Both were newer universities, founded by royal patrons who looked to the future. Göttingen's Georgia Augusta was first established in 1737 under the auspices of the King of Hanover, better known in the English-speaking world as George II, King of Great Britain and Ireland, and Prince-Elector of the Holy Roman Empire. The Georgia Augusta has often been considered the prototype for the modern university, owing to the fact that its philosophical faculty enjoyed the same status as the other three – theology, law, and medicine – these being traditionally regarded as higher faculties. George II delegated the task of launching this enterprise to the Hanoverian Minister Gerlach Adolph Freiherr von Münchhausen, who as the university's first Kurator made the initial appointments to the faculty. A similar arrangement took place in Prussia in 1810. King Friedrich Wilhelm II, whose country had been overrun and then annexed in large part by Napoleon, empowered the educational reformer and linguist Wilhelm von Humboldt to found Berlin University, renamed Humboldt University in 1949 when it reopened in the German Democratic Republic. Its main building, still located on the avenue Unter den Linden, was originally built during the mid-eighteenth century as a palace for Prince Heinrich, the brother of the previous king. Acting in concert with other scholars, Humboldt conceived

[1]On the preconditions, see Turner (1980); for the character of research schools in the natural sciences, see Servos (1993).

[2]The notion of charismatic leadership was made famous by Max Weber, who however denied that it had a rightful place in academic life since he believed that personal authority had to be based on purely scientific qualifications (*Wissenschaft als Beruf*). Weber's view, however, was clearly idealized; even within the field of mathematics charismatic leaders have often exerted an influence far beyond their own achievements, three noteworthy examples being Weierstrass, Klein, and Hilbert.

D.E. Rowe, *A Richer Picture of Mathematics*, https://doi.org/10.1007/978-3-319-67819-1_1

Berlin's university in the spirit of neohumanism, an ideology in many ways opposed to the utilitarianism associated with French science (Rowe 1998).

When mathematicians think of Göttingen, famous names immediately spring to mind, beginning with Gauss, as well as his immediate successors, Dirichlet and Riemann. These, to be sure, are names to be conjured with, but they are not central players in the larger story told here. Still, they are relevant because their ideas continued to inspire later generations of mathematicians who lived and worked in a radically different intellectual milieu whose leaders strongly identified with this older Göttingen mathematical tradition. Indeed, to a considerable extent these modern representatives of Göttingen mathematics created a near cult-like worship of Gauss and Riemann that enabled them to bask in the reflected light of their predecessors past glory. Thus, in a word, they were engaged in constructing an image of the grand tradition to which they belonged. Before addressing that theme, however, I should first say something about what made this particular university such a fertile environment for mathematical creativity. This requires taking a glance further backward into the history of German higher education as it developed over the course of the eighteenth and nineteenth centuries. During that time important reforms were inaugurated at leading German universities, spurred by the example of Göttingen, where teaching and research came to be combined in a fundamentally new way.

Despite many common features, the German universities did not form a centralized system even after the nation's unification in 1871. They were administered instead by the individual states just as today. Moreover, since a variety of different cultural, historical, and regional factors have shaped their character and traditions, it is difficult to make sweeping generalizations about how they functioned in practice. For this reason, even though these institutions certainly shared a number of important features, they can only really be understood through an examination of their individual histories. Göttingen's *Georgia Augusta* initially served as Hanover's answer to an institutional dilemma posed by the state of Prussia ruled by the Hohenzollern monarchy. Prior to its founding, prospective candidates for the Hanoverian civil service who wished to pursue a higher education were forced to take up studies outside their home state, most notably at the Prussian University in Halle. Soon after its founding, however, young men, often from aristocratic families, thronged to the new university in the town of Göttingen, located in the southern part of Hanover. By mid-century, the Georgia Augusta not only surpassed Halle in scholarly reputation but also in popularity among German students, many of whom spent more time in beer halls than in lecture rooms (Fig. 1.1).

Historically, European universities had been dominated by the sectarian interests vested in their theological faculties. Göttingen marked an abrupt break with this longstanding tradition by *inverting* the status of its theological and philosophical faculties. Its rapid rise during the middle of the eighteenth century owed much to von Münchhausen's acumen and foresight. Up until his death in 1770, he exercised virtually complete control over the university's affairs. Moreover, his liberal policies contrasted sharply with those of Prussia's Frederick William I. The latter banished Germany's leading Leibnizian philosopher, Christian von Wolff, from Halle in 1723 for espousing doctrines inimical to the monarch's strict Pietism. Although Wolff was allowed to return (and even

Fig. 1.1 Göttingen University and library, ca. 1815.

served as Chancellor of Halle's university after Frederick II's ascension to the throne in 1740), Halle never regained its scientific reputation. In the meantime, under von Münchhausen's beneficent leadership, Göttingen shook off the yoke of sectarianism that had restrained free thought at European universities for centuries.

Göttingen's philosophical faculty – which offered the traditional elementary training in the "liberal arts" to students who hoped to study in one of the three "higher faculties"' (theology, medicine, or law) – rapidly developed a reputation for serious scholarship based on a fertile combination of teaching and research. Initially, the vaunted ideals of *Lehr- und Lernfreiheit*, flourished to a remarkable degree in the philology seminars of Johann Matthias Gesner and Christian Gottlob Heyne. Spreading from the humanities, the reform spirit soon took hold in the natural sciences after Göttingen acquired the services of such prominent scholars as the anatomist and botanist, Albrecht von Haller; the physicist, Georg Christoph Lichtenberg; the astronomer, Tobias Mayer; and the mathematician, Abraham Kaestner. Münchhausen's innovative policies thus created the preconditions for a university dedicated to serious scholarly pursuits undertaken with a modicum of academic freedom. Within this setting, several of those who studied or taught in Göttingen went on to make lasting contributions in a number of fields. In mathematics, however, their influence would become legendary.

Göttingen's famous mathematical tradition commenced with the career of Carl Friedrich Gauss, who studied, worked, and taught there for some 50 years. No book about mathematics in Göttingen can afford to neglect Gauss, the famous "Prince of Mathematicians," about whom so much has been written. That being the case, Chap. 2 deals with some mythic aspects of his life in the form of an *Intelligencer* quiz. Much of what we know about Gauss comes from a later time and a different mathematical culture that had largely broken earlier ties with astronomy and physics. Gauss embodied all three disciplines, which made him the perfect symbol for what Felix Klein hoped to achieve in Göttingen: the re-integration of pure and applied mathematics.

Gauss grew up in the city Brunswick (Braunschweig), where he was raised in a modest home as the only child of Gebhard Dietrich and Dorothea Gauss. His parents sent him to the local Volksschule, where his intellectual talent was discovered early. This brought him to the attention of Duke Karl Wilhelm Ferdinand, who financed his further education at the Collegium Carolinum in his native city. There, already as a teenager, Gauss was reading works like Newton's *Principia*, which he was able to purchase in 1794. Thanks to a royal stipend, he then had the opportunity to spend three years pursuing university studies. Years later, when he recounted why he chose to study in Göttingen rather than at the local university in Helmstedt, Gauss gave one simple reason: books (Küssner 1979, 48). Few universities, in fact, could rival the holdings in Göttingen's library. Travelling the 90 km by foot, he arrived there in October 1795 and found what he was looking for. Shortly thereafter, he reported to his former teacher Eberhard A. W. Zimmermann about how he was pouring over volumes of the Proceedings of the Petersburg Academy and other such works.

It was in this manner that the young Gauss first entered the still quiet, intellectually isolated world of the German universities. Soon thereafter, Napoleon would arrive upon the scene, but until then life went on as usual. At first undecided as to whether he should take up mathematics or philology, Gauss chose mathematics after he succeeded in proving that the 17-gon can be constructed with straightedge and compass. On 30 March 1796, he recorded the first entry in his mathematical diary: "The principles upon which the division of the circle depends, and geometrical divisibility of the same into seventeen parts, etc." (Gray 1984). In other words, he had cracked the problem of determining which regular polygons can be constructed by straightedge and compass alone, including the first non-classical case, the 17-gon. These principles led him to explore the theory of cyclotomy, thereby entering the portals that led to the theory of algebraic number fields. Symbolically, this *Tagebuch* entry may be regarded as marking at once the birth of higher mathematics in Germany along with the Göttingen mathematical tradition. By the end of the year, though still not yet 20 years of age, Gauss had already filled his scientific diary with 49 entries for results he had obtained during the preceding nine months! Several would remain secrets he took with him to his grave. In fact, he seems to have told no one about the existence of this famous little booklet documenting his early mathematical interests and findings, so its survival was something akin to a small miracle. Not until 1899 did Paul Stäckel find it among Gauss's "personal papers"; it was then in the possession of a grandson who was living in Hameln.

In 1807, six years after he had correctly predicted the location of the asteroid, Ceres, and published his monumental *Disquisitiones Arithmeticae*, Gauss was appointed professor of astronomy in Göttingen. Another nine years passed before he and his family moved into the wing of the newly built astronomical observatory, the *Sternwarte*, which he would occupy for the remainder of his life. Gauss's own star shone brightly in the mathematical firmament, but there were few others nearby. Moreover, unlike his contemporary, Augustin Cauchy, he felt no compulsion to rush into print, remaining true to his motto: "*pauca sed matura*" (few, but ripe). Even more characteristic of his conservatism, Gauss showed no interest in imparting his research results (or those of other mathematicians) in the courses he taught. Instead, he preferred to confide these only to a handful of friends and peers with whom he carried on an extensive scientific correspondence. Among the larger memoirs that

Fig. 1.2 C. F. Gauss, the Prince of Mathematicians.

he did choose to publish, several were written in Latin. In these respects, Gauss's scholarly orientation was entirely traditional (Fig. 1.2).

Throughout his career, Gauss was mainly known for his accomplishments as an applied mathematician. As a professional astronomer, he corresponded regularly with other leading practitioners, including Wilhelm Olbers (Bremen), Friedrich Wilhelm Bessel (Königsberg), and Christian Schumacher (Altona). Moreover, like several other astronomers, he worked on geodetic surveys. Beyond these standard activities, he also took part in Alexander von Humboldt's project to study global fluctuations in the earth's magnetic field. Humboldt had earlier taken note of large-scale magnetic storms, a phenomenon he hoped to study by coordinating data from a worldwide network of magnetic observatories. In 1828, he managed to win over Gauss for this endeavor when the latter attended a scientific conference in Berlin (the setting for Daniel Kehlmann's novel, *Measuring the World*, my main source for recent mythologizing about Gauss).

This then led to Gauss's famous, though brief collaboration with Wilhelm Weber, who was appointed professor of physics in Göttingen in 1831 at the age of 27. Gauss had previously used spherical harmonic analysis in celestial mechanics, but he now adapted these techniques to geomagnetism. He and Weber could thereby show how to represent the global magnetic field of the Earth by combining observations at many locations. In 1834, Gauss and Weber founded the Göttingen Magnetic Union to coordinate research for a network of European observatories. Humboldt later made this into a truly international undertaking by linking with similar efforts in Britain and Russia. The British set up stations in Greenwich, Dublin, Toronto, St. Helena, Cape of Good Hope, and Tasmania, with the British East India Company adding four more in India and Singapore. Humboldt persuaded the Russian Czar to build observatories across his vast territory, making it possible to draw up worldwide magnetic charts. Gauss and Weber also demonstrated the feasibility of telegraphy by building an instrument that linked Weber's physics laboratory with Gauss's *Sternwarte*. Their collaboration briefly linked mathematics and physics within the Göttingen mathematical tradition, a bond that Klein would later seek to revive during the 1890s. For Gauss and Weber, however, their alliance of interests ended abruptly in the wake of an event that had disastrous repercussions for the whole university.

With the death of William IV in 1837 and Queen Victoria's ascension to the throne, the Hanoverian line in England ended. Ernst August, the Duke of Cumberland and a younger son of George III, thereby became King of Hanover. One of his first acts as monarch was to annul the constitution, replacing it with an older version that preserved the former privileges of the aristocracy. This evoked considerable protest among Göttingen's student body, and seven professors, Weber and the Grimm brothers among them, submitted a formal protest. All seven were removed from their positions, and three were even forced to leave the Kingdom of Hanover altogether. Passions ran high, leading the king to send in troops to maintain order; his actions left a deep scar in the university's collective psyche. Politically, the "Göttingen Seven" incident marked a serious setback for the forces of democracy and reform in Hanover. It also sent a chilling message to Göttingen scholars, who thenceforth realized that their cherished academic freedoms had strictly proscribed limits. Throughout this whole drama, Gauss stood by on the sidelines. After the blow had fallen, he hoped to see Weber (one of the least vocal among the seven) restored to his chair, but to no avail. Another member of this "Göttingen Seven" was the Orientalist Heinrich Ewald, who was married to Gauss's daughter Wilhemine. Ewald and Weber were the only two to be reappointed years later, but by then Gauss was too old to continue his collaboration with Weber. In the meantime, the results of their earlier work had been published by Gauss and his assistant, Carl Wolfgang

Benjamin Goldschmidt, in *Atlas des Erdmagnetismus: nach den Elementen der Theorie entworfen.*

Gauss's reticence and failure to speak out during this critical escapade reflected not only his own deep political conservatism but also his academic roots as well. For even though his ideas broke fresh ground and gave a decisive impulse to several new branches of pure mathematics, outwardly Gauss's career and personality bore many of the typical traits of an eighteenth-century scholar. Like Goethe and von Humboldt, he aspired to the ideal of universal knowledge. He ignored hard and fast distinctions between fields such as mathematics, astronomy, and mechanics, and his research ran the gamut from number theory and algebra to geodesy and electromagnetism. Felix Klein, whose reverence for Gauss bordered on idolatry, regarded him as the culmination of an earlier age, likening him to the crowning peak in a gradually ascending chain of mountains that drops off precipitously, leading to a broad expanse of smaller hills nourished by a steady stream flowing down from on high (Klein 1926, 62).

Looking back to the first decades of the nineteenth century, the German universities still appear very much like a backwater in the world of mathematics. However, that would change by the mid-1820s, thanks in large part to the efforts of two scientists who worked outside the mathematical field. One of these was the building commissioner August Leopold Crelle (1780–1855), a well-known figure in Berlin with broad scientific interests. His name is still remembered today as the founder of the first research journal for mathematics in Germany, *Das Journal für die reine und angewandte Mathematik*, long known simply as Crelle. Founded in 1826, its reputation for publishing outstanding original work was almost immediate, though certainly not in the direction Crelle had originally intended.

Nominally dedicated to both pure and applied mathematics, it quickly became a stronghold for pure research alone, so much so that Crelle's creation was jokingly called the "Journal für die reine unangewandte Mathematik" (pure unapplied mathematics). Its very first volume contained several articles by two brilliant young foreigners, the Norwegian Niels Henrik Abel und the Swiss geometer Jakob Steiner. One of Abel's papers presented his famous result that the general quintic equation cannot be solved by means of an algebraic formula. This landmark result stands at the threshold of Galois theory and thus represents one of the great advances in the history of algebra. Altogether, Abel published six articles and notes in this first volume, Steiner five, and the 22-year-old C. G. J. Jacobi contributed a paper in analysis. Crelle clearly had a sharp eye for young talent; when he was seen in the company of Abel and Steiner strolling around Berlin, local humorists made up the idea of calling the two young men Crelle's sons, Cain und Abel.

The other key talent scout was Wilhelm von Humboldt's brother, Alexander, who discovered Peter Gustav Lejeune-Dirichlet (1805–1859) when both were living in Paris. Dirichlet grew up as the youngest child in a large family from Düren, a town situated between Aachen and Cologne on the left bank of the Rhine. At the time of his birth, this region belonged to the French Empire, but with the 1815 Congress of Vienna, it was granted to Prussia. Already at age twelve, Gustav Dirichlet showed a strong interest in and aptitude for mathematics, so his parents sent him to Bonn, where he attended the Gymnasium for two years. Afterward he went to the Cologne Gymnasium, where he was a pupil of the physicist Georg Simon Ohm of Ohm's Law fame, but left without a graduation certificate. By then, Dirichlet was intent on pursuing a career in mathematics, a decision that led him to Paris in May 1822. There he was hired as a private tutor by General Foy, a lucky turn of events that enabled him to remain in the French capital.

Over the next five years, he studied under or met with many of the era's most illustrious figures, including Fourier, Hachette, Laplace, Lacroix, Legendre, and Poisson. During these years, Dirichlet also spent a good bit of time struggling with Gauss's *Disquisitiones Arithmeticae,* a book that would remain important to him throughout his entire life. His hopes for a career in mathematics took a turn for the better after Alexander von Humboldt came to hear about him through Fourier and Poisson (Fig. 1.3). This was surely no accident. Having spent nearly 18 years in Paris, Humboldt knew nearly every scientist there. In later years, he recalled his sense of duty in keeping a watchful eye out for budding German talents in every field, whether astronomers, physicists, chemists, or mathematicians. His efforts clearly helped set the stage for the flowering of scientific excellence at the Prussian universities. Soon after leaving Paris in 1826, he arranged for Dirichlet's first academic appointment in Breslau by asking Gauss to write a letter of recommendation on his behalf.

This brings us to Chap. 3, which focuses on events from this time and the sharply contrasting personalities of Gauss and Dirichlet. The famous Gaussian motto *pauca sed matura* was often taken as an admonition declaring that one should strive for perfection in mathematical publications, voluntarily holding back new results until they can be presented elegantly. In practice, however, Gauss often took advantage of this austere attitude toward publication by referring to his "works in progress." Thus, when corresponding with potential competitors he made it a point to inform them that he was already familiar with their results: he had just not found time yet to polish them up for publication. Chapter 3 takes up this theme by way of Gauss's early efforts to uncover a number-theoretic law of biquadratic reciprocity together with Dirichlet's parallel discoveries along the same lines.

The latter found dramatic expression in a letter Dirichlet wrote to his mother in late October 1827. I found a typescript

Fig. 1.3 Statue of Alexander von Humboldt at Humboldt University located on Unter den Linden in Berlin.

of a portion of this letter among Klein's papers, as this document was presented in a seminar on the psychology of mathematics (Klein Nachlass 21A). One of the participants was the philosopher Leonard Nelson, whose mother was a granddaughter of Gustav Dirichlet. Gert Schubring had, in fact, tracked down the three parts of Dirichlet's Nachlass, which he described in (Schubring 1984) and (Schubring 1986). The latter article contains interesting information about Nelson's political activities, indicating why he was a likely target for Nazi persecution, even though he had died some years before Hitler assumed power. For this reason Nelson's papers, including the family documents related to Dirichlet, were placed in the Murhard library in Kassel, now part of the Kassel University Library. By far the most important documents for Dirichlet's biography can be found in this portion of the Nachlass, which contains the many letters he wrote to his mother. This includes the original of the letter I translated using the typescript in Klein's papers, which contains some minor inaccuracies. Schubring also pointed out a glaring mistake soon after my essay appeared in print (see his letter to the editor in *Mathematical Intelligencer* 12(1)(1990): 5–6.): the photograph that appears on the cover of *MI* and in my original paper is not a picture of the young Dirichlet, although it was identified as such in the Göttingen collection. I should also add that the story surrounding this dramatic letter and how Dirichlet eventually unlocked the secrets of biquadratic reciprocity – though with no help from Gauss – has recently been retold by Urs Stammbach in (Stammbach 2013).

Little more than a year after he came to Breslau, Dirichlet was on his way to Berlin, again thanks to the influence of Alexander von Humboldt. He would remain in the Prussian capital until 1855, at which time he succeeded Gauss in Göttingen. It was also through Humboldt that Dirichlet met and later married Rebecka Mendelssohn-Bartholdy, the younger sister of the famous composer (Fig. 1.4). Although she was overshadowed as a musician by her brother Felix and older sister Fanny, Rebecka nevertheless maintained many friendships with celebrities and artists of the Romantic period. She continued to cultivate these cultural ties when the couple moved to Göttingen. During his years in Berlin, Dirichlet exerted a strong influence on several talented mathematicians who went on to brilliant careers: Kummer, Eisenstein, Kronecker, Riemann, and Dedekind.

The year 1844 marked a turning point for mathematics in Berlin with the arrival of Carl Gustav Jacob Jacobi (1804–1851) as a salaried member of the Prussian Academy of Sciences (Fig. 1.5). Jacobi, a native of Potsdam, had been teaching in Königsberg alongside the physicist Franz Neumann and the astronomer Friedrich Wilhelm Bessel. Jacobi and Neumann founded the famed mathematical-physical seminar that would later serve as a model for several seminars established at other German universities. These seminars were the principle vehicle for educating future researchers, an innovation that made Germany a magnet for so many young foreign scholars during the latter half of the nineteenth century. Dirichlet and Jacobi got along splendidly, and the latter was soon a regular guest at the musical soirees hosted by Rebecka Dirichlet.

During this period, Berlin stood in the forefront of Jewish emancipation, a major theme in (Elon 2002). Rebecka's grandfather, the philosopher Moses Mendelssohn, had paved the way for the *Haskalah* (Jewish enlightenment) of the eighteenth and nineteenth centuries. In 1763, he was granted the status of *Schutzjude* (Protected Jew) by the King of Prussia, which assured him the right to live in Berlin without being disturbed by local authorities. In keeping with this

Fig. 1.4 Drawings by Wilhelm Hensel of Dirichlet and young Rebecka Mendelssohn-Bartholdy.

Fig. 1.5 Carl Gustav Jacob Jacobi.

family tradition, the Dirichlets were on close terms with other prominent liberal families in Berlin, including Rahel and Karl Varnhagen von Ense. Both supported the political uprisings in 1848 that called for democratic reforms in the German states. After this failed, Rebecka helped two prominent revolutionaries escape after the Prussian army had crushed the last remaining resistance in July 1849. One of those arrested, Gottfried Kinkel, was sentenced to life imprisonment in Spandau, just west of Berlin. His friend Carl Schurz escaped, and a little more than a year later, he managed, partly with the help of Rebecka Dirichlet, to free Kinkel from the Spandau prison. Both became prominent figures in exile, Kinkel as a writer in London, and Schurz as a general in the army of Abraham Lincoln during the Civil War (Lackmann 2007, 244–245).

The third figure in this circle of important Berlin mathematicians contrasted sharply with the other two, who belonged to an intellectual elite and its surrounding salon culture. In fact, Jakob Steiner, one of the most celebrated mathematicians of his time, stood poles apart from German high culture in general (Fig. 1.6). Whereas younger contemporaries, like Jacobi and Abel, produced brilliant work in their early twenties, Steiner was nearly forty before he gained an appointment as extraordinary professor in Berlin.

Fig. 1.6 Jakob Steiner, before he grew a beard.

Afterward he enjoyed a legendary career, attracting throngs of students to his courses on synthetic geometry. Yet he was never promoted to a full professorship (*Ordinariat*), probably because his Swiss-German dialect, gruff personality, and outspoken liberal views on political matters led many of his colleagues to avoid him whenever they could. Not only was Steiner regarded as unpleasant, he was also thought to be uncultivated and far too uncouth to be welcomed into Berlin's elite society.

Born in 1796 near the small Swiss village of Utzenstorf, 25 km north of Bern, Steiner grew up as the youngest of eight children in a farming family. The local school he attended offered its charges only the most rudimentary skills, so that at age 14 Steiner could barely write. Afterward some deep inner urge must have taken hold of him, leading him to pursue a more advanced education. In the spring of 1814, at age 18, he left home to take up studies at Johann Heinrich Pestalozzi's school in Yverdon. There Steiner's genius for geometry was quickly discovered, and soon he was even allowed to teach classes. Four years later, he left this stimulating and supportive atmosphere to take up studies in Heidelberg, the beginning of a long, difficult road to academic success. All his life, Steiner saw himself as a crusader for the teaching methods he learned and practiced at Pestalozzi's school. He gave an early testimonial to this effect in a document that he submitted to the Prussian Ministry of Education on 16 December 1826:

> The method used in Pestalozzi's school, treating the truths of mathematics as objects of independent reflection, led me, as a student there, to seek other grounds for the theorems presented in the courses than those provided by my teachers. Where possible I looked for deeper bases, and I succeeded so often that my teachers preferred my proofs to their own. As a result, after I had been there for a year and a half, it was thought that I could give instruction in mathematics (Lange 1899, 19).

An interesting testimonial concerning Steiner's unusual teaching style, but also his cantankerous personality, comes from the pen of the English geometer Thomas Archer Hirst (1830–1892). After taking his doctorate in Marburg in 1852, Hirst spent the academic year 1852–53 in Berlin, where he attended the lectures of Steiner and Dirichlet. He later became a fixture in the British community as a member of the Royal Society, the British Association for the Advancement of Science, and the London Mathematical Society. In his diary, he recorded this impression of Steiner:

> He is a middle-aged man, of pretty stout proportions, has a long, intellectual face, with beard and moustache, and a fine prominent forehead, hair dark and rather inclining to turn grey. The first thing that strikes you on his face is a dash of care and anxiety almost pain, as if arising from physical suffering. . . . He has rheumatism. All these point to physical nervous weakness. His Geometry is famed for its ingenuity and simplicity. He is an immediate pupil of Pestalozzi: in his youth he was a poor shepherd boy, and now a professor.
>
> His argument is that the simplest way is the best; he tries ever to find out the way Nature herself adopts (not always, however, to be relied upon). Mathematics he defines to be the "science of what is self- evident." . . .
>
> I listened with great interest to [Prof. Riess] talk about Dirichlet, Jacobi and Steiner. He told me fully the relations on which the latter stands with them all, and truly it is unexplainable. Riess says his vulgarity has by them all been slightly borne in consideration of his undoubted genius. But that some time ago without provocation Steiner cut them all. The probable reason is that Steiner, naturally of a testy disposition, which has been increased, too, by bodily illness, feels slighted that he has been 33 years "Ausserordentliche" [Extraordinary] Professor. The reason is clear: firstly he does not know Latin, and that among German professors is held as a necessity: 2nd he is so terribly one-sided on the question of Synthetical Geometry that as an examiner he would not be liked. The more I hear, the more I am determined to see him and study him for myself (Gardner and Wilson 1993, 622–624).

In earlier years, Steiner spoke often with Jacobi about common geometrical interests, and later he did the same with Weierstrass. After his colleague's death in 1863, Weierstrass continued to uphold and honor the Steinerian tradition by teaching his standard course on synthetic geometry (Biermann 1988). It should be noted that Jacobi was no longer living at the time Hirst visited Berlin. Thus, the period when Dirichlet, Jacobi, and Steiner were together in Berlin lasted only seven years, and even these were clearly troubled ones as far as Steiner's relations with his colleagues were concerned. This era in Berlin came to an end in 1855 with the death of Crelle and the departure of Dirichlet for Göttingen, where he assumed the chair in mathematics formerly occupied by Gauss (his position as director of the astronomical observatory, however, remained for many years unfilled).

There followed a brief, but brilliant era for mathematics in Göttingen, now represented by Dirichlet and two talented young researchers – Bernhard Riemann and Richard Dedekind – both of whom profited greatly from nearly daily contacts with their older friend, who was still only fifty. Dirichlet's decision to leave Berlin for a more provincial city was made easier by the fact that he no longer had to put up with teaching at the Military school in the Prussian capital. Rebecka surely would have preferred staying, but saw that her husband had no desire to negotiate better conditions with Prussian ministerial officials. She corresponded regularly with her friends, playing for sympathy by calling the little university town she now lived in "Kuhschnappel," the name of a rural village in Saxony made famous in the novel *Siebenkäs* by Jean Paul (Lackmann 2007, 246). It took her little time, though, to make their new home into a local musical center. After briefly renting an apartment, Dirichlet bought a large house located near the town's center from a colleague (today it serves as student housing (see Fig. 1.7)). Here they lived for the next few years with their two youngest children and Dirichlet's mother. Rebecka opened their new home to upwards of seventy guests at a time and with plenty of gaiety for all. Dedekind, a gifted pianist and cellist, played waltzes here for the dancers. The famed violinist Joseph Joachim performed in the Dirichlet home, and it was during a stay in Göttingen that he was invited by Rebecka to hear a young sopranist named Agathe von Siebold. One year later, she met Joachim's close friend, Johannes Brahms, to whom she was briefly engaged.

Not long after his arrival, Dirichlet was contacted by Johann Karl Kappeler. As President of the Swiss School Board, Kappeler was seeking advice about a suitable candidate for the chair in mathematics at the newly founded Federal Polytechnical School in Zürich (Frei u. Stammbach 1994, 37). Not surprisingly, Dirichlet named both Riemann and Dedekind, which prompted Kappeler to travel to Göttingen so that he could hear lectures by both men! He chose Dedekind, who in 1858 became the first in a long line of eminent German mathematicians to spend part of their careers teaching in Zürich. Incidentally, it was while teaching a course in analysis there that Dedekind realized his inability to prove a standard property of the real numbers. That insight was what led him to invent so-called Dedekind cuts, his famous method for constructing the real numbers; however, he only published this much later in 1872 (Fig. 1.8).

Fig. 1.7 The house at Mühlenstrasse 1 in Göttingen where the Dirichlets lived from 1856–1859.

Riemann enjoyed the strong support of Dirichlet, with whom he had studied in Berlin. On the other hand, his relationship with Gauss, whose career was already nearing its end when Riemann first came to Göttingen in 1846, had been distant by comparison. Gauss, however, did play a role in drawing out some of Riemann's boldest ideas, which he expressed in a manuscript written in 1854 for an auspicious occasion: Riemann's final qualification to join the faculty as a *Privatdozent.* This text would later exert a strong influence on developments in differential geometry, followed by a wave of new interest after Einstein showed its relevance for interpreting gravitational effects as variations of curvature in a space-time manifold. In short, Riemann's lecture came to be regarded as a central document presaging the "relativity revolution," the theme of Part IV. The third essay "Geometry and Physical Space" (Chap. 20) in Part IV touches on some of the novel ideas in Riemann's lecture in connection with Einstein's theory of gravitation. A few brief remarks about the events of 1854 would therefore seem appropriate here.

Fig. 1.8 Richard Dedekind.

In that year, Riemann wrote his younger brother Wilhelm that he had given "hypotheses on the foundations of geometry" as the third topic for his habilitation lecture. Hoping that the faculty would choose one of his first two topics, he learned that Gauss wanted to hear the third. Probably Gauss knew nothing about the direction of Riemann's thoughts just then, but he may have wondered whether the young candidate had something new to say about recent speculations concerning non-Euclidean geometry. Originally, Gauss wanted to postpone the colloquium until August owing to his poor health. Riemann had to pester him so that the final act for his qualification as a *Dozent* could take place on June 10, 1854. The ordeal began at 10:30 and ended at one o'clock, but no protocol survives, which means that nothing is known about what Gauss might have asked during the examination. Writing some 20 years after the event, Dedekind mentions that immediately after the colloquium Gauss spoke to Wilhelm Weber with great excitement about the depth of Riemann's ideas. Whether Riemann himself ever learned of this no one can say, and we can only speculate as to why he never published this monumental essay.

During his lifetime, Riemann's ideas exerted a strong influence on research in real and complex analysis, though his use of topological tools such as Riemann surfaces also met with strong resistance (Bottazzini and Gray 2013, 259–341). This part of his legacy, however, stands in sharp contrast with the speculations on foundations of geometry as set forth in his habilitation lecture; in fact, these ideas exerted no impact at all during his lifetime. His famous lecture only appeared in print after his death when Dedekind managed to locate it among his friend's posthumous papers (Riemann 1868). In all likelihood, Dirichlet would have heard about Riemann's lecture after his arrival in Göttingen, but if so, no trace of such discussions seems to have survived. In fact, personal misfortunes soon intervened that would cut short the careers of both men. Consequently, the university's promising potential as a research center never actually took root. In the summer of 1858, Dirichlet suffered a heart attack from which he at first recovered; but then his wife's sudden death led to his own in May 1859. Riemann was then appointed as his successor, but by this time his own health was already in a precarious state. Suffering from tuberculosis, he spent most of his time convalescing in Italy. When he died there in 1866, the mathematical school in Berlin was in full ascendency.

The year 1866 was a fateful one, not only for Göttingen mathematics but for the course of German political history as well. In early July, the Austrian army was crushed by the Prussians at Königgrätz, the latter fighting under the command of Helmuth von Moltke. His forces were outnumbered, but they had a decisive technical advantage over the Austrians, who had to stand while reloading their traditional rifles. This made them easy targets for the Prussians, since they had fast-firing, breech-loading guns, which they could reload while lying on their stomachs. This battle proved decisive for Bismarck's larger strategy, which aimed to nullify the Austrians before uniting the German states under the dominion of the Prussian monarchy. The big loser in this Seven-Weeks War between Austria and Prussia, though, was the Kingdom of Hanover, which had allied itself with the Austrian side. As a consequence, it disappeared from the map altogether, taken, so to speak, as war booty by the Prussians. Riemann was residing in Göttingen until June, when he left for Italy just as the Hanoverian army was about to be defeated near Langensalza. When he died a month later, the Georgia Augusta was already a Prussian university.

Since 1855, Berlin had been led by the triumvirate of Ernst Eduard Kummer, Karl Weierstrass, and Leopold Kronecker. All three exemplified ideals fully consistent with the neo-humanist tradition that had long animated Berlin's philosophical faculty, which included both the humanities as well as the natural sciences. Thus, their appeal to purism and systematic rigor in mathematics was thoroughly in accordance with the research orientation in other disciplines. This ethos served not only to instill a spirit of high purpose, it also gave clear definition to the special type of training imparted in Berlin, thereby providing graduates of the program with a sense of collective identity. In many cases, the Berlin tradition also gave its members a feeling of self-assuredness that outsiders might envy or even despise, but which surely had to be acknowledged.

Kummer's career had followed in the wake of Dirichlet's: he succeeded him first in Breslau and then again in 1855, when the latter left Berlin for Göttingen. Like Dirichlet, he also married into the Mendelssohn family; Kummer's first wife, Ottilie, was the daughter of Nathan Mendelssohn and a cousin of Rebecka Mendelssohn-Bartholdy, the wife of Dirichlet. Although Berlin's leaders were, each in his own way, advocates of disciplinary purism, they came from very different backgrounds: Kummer was a Protestant Prussian, Weierstrass a Catholic Westphalian, and Kronecker a wealthy Jewish businessman from Silesia. Weierstrass eventually became the most prominent of the three. His lectures drew huge numbers of auditors, few of whom had any chance of understanding what he was saying. That seemed not to matter, though, as word got round that his cycle of lectures on analysis was the very latest word on the subject (Bottazzini and Gray 2013, 343–486). Since he never wrote these up for publication, but rather continuously revised their content during each cycle, it was very difficult to learn Weierstrassian analysis without attending his lectures. (Felix Klein managed to do so through the help of his friend Ludwig Kiepert.)

Chapter 4 sketches the rivalry that developed between Berlin and Göttingen. This conflict arose only after the era when Gauss, Dirichlet, and Riemann lived and taught at the Hanoverian university. One should also take note that none of these three luminaries succeeded in establishing a mathematical school there comparable with the one Jacobi founded in Königsberg. After the passing of Riemann, this older Göttingen tradition gave way to new trends that became common characteristics of *Wissenschaft* in general as academic specialization began to take hold. By the time young Felix Klein first set foot in the town, three years after Riemann's death, the Göttingen of Gauss and Riemann was already the stuff of legends. Klein joined the circle of students who surrounded Alfred Clebsch, a leading exponent of the Königsberg tradition and a staunch opponent of the then dominant mathematical culture in Berlin.

Riemann's demise left Richard Dedekind as the last living representative of the older Göttingen mathematical legacy. Dedekind had studied under Gauss, Dirichlet, and Riemann before becoming a *Privatdozent*. In 1862, after three years at the Zurich Polytechnique, he accepted a professorship at the Polytechnic Institute in his native Brunswick, where higher mathematics played virtually no role in the curriculum. He would remain in Brunswick all his life, shunning any opportunity to assume a more distinguished position at a leading German university. From this unlikely outpost, Dedekind produced work that would inspire others for decades to come, including several editions of Dirichlet's *Vorlesungen über Zahlentheorie*, which he supplemented with new concepts and results that were to prove fundamental for modern algebra and algebraic number theory. Like Riemann, Dedekind was intent on pursuing original ideas that were well ahead of their time. Emmy Noether later paid tribute to him by saying "es steht alles schon bei Dedekind" (everything can be found in Dedekind), an exaggeration that actually reflected her own modesty (Dedekind 1930–32). Such hero worship certainly has its endearing side, particularly when coming from such a towering figure as Noether, whose father was no doubt a kind of role model in this regard. Not only was Max Noether an eminent algebraic geometer, he also studied the works of other leading mathematicians with great care. Moreover, like him, Emmy Noether proved to be a true mathematical scholar, as attested by were work on the Cantor-Dedekind correspondence (Noether and Cavaillès 1937). Similarly, Dedekind made an important scholarly contribution when he joined with Heinrich Weber in preparing the first edition of Riemann's *Werke*, which appeared in 1876.

Yet despite his redoubtable mathematical stature, Dedekind always preferred to work in quiet isolation. His modest, almost reclusive personality made him ill-suited for the task of promoting and sustaining the Göttingen tradition of Dirichlet and Riemann. And so finding a suitable successor for Riemann proved difficult: Dedekind had no desire to leave Brunswick, nor did Leopold Kronecker wish to give up his position in Berlin, where as a member of the Prussian Academy of Sciences he could teach whenever and whatever he wanted.

Eventually the ministry settled on Königsberg-trained Alfred Clebsch, who already headed a small research group in Giessen. After moving to Göttingen in 1868, this fledgling school quickly matured, attracting several promising new talents. That same year, Clebsch joined forces with another Königsberg product, Carl Neumann, in founding a new journal, *Die Mathematische Annalen*. By doing so, they effectively threw down the gauntlet to the Berlin mathematicians, who till then had dominated the scene through Crelle's journal, which since 1855 was edited by Carl Wilhelm Borchardt, a former pupil of Jacobi and close friend of Weierstrass. Clebsch stood on quite bad terms with the Berliners, who took a critical view of his influence, including his use of mixed methods in fields like algebraic geometry and invariant theory (see Chap. 4). Unfortunately for Göttingen, Clebsch's tenure at the Georgia Augusta was brief: in November 1872, he contracted diphtheria, which quickly extinguished his young life. He was only 39 years old when he died.

Following Clebsch's sudden and premature death, Göttingen became just another outpost for the now dominant Berlin school. His professorship was briefly occupied until 1875 by Lazarus Fuchs and afterward by Hermann Amandus Schwarz; both had studied in Berlin and would later become members of its faculty. In the meantime, memories of the

Fig. 1.9 Felix Klein.

older Göttingen tradition began to fade. Although Clebsch's school continued to thrive, supported by *Mathematische Annalen*, it did so through a network of mathematicians situated outside of the Prussian university system. This group often referred to themselves as "southern Germans," despite the fact that Clebsch and Neumann had grown up in Königsberg. Yet whatever their background, they worked at institutions located in the southern German states: Bavaria, Saxony, Baden, Württemberg, et al. Those who had studied with Clebsch in Göttingen had all dispersed. By the 1870s, the former Hanoverian university had quickly established itself as a Prussian institution, but its glory days in mathematics had seemingly come to an end.

The man largely responsible for restoring Göttingen's former luster and who, by so doing, initiated a process that transformed the whole fabric of mathematics at the German universities was Felix Klein, whose early career forms the focus of Part II (Lorey 1916) (Fig. 1.9). As a major power broker in the world of mathematics, Klein developed an extensive network of contacts both within Germany and beyond. It should not be overlooked, however, that his power and influence had not come easily. Indeed, he spent the early part of his career battling the influence of the still stronger network of mathematicians associated with Berlin. This conflict began to take on clear form soon after 1870 with the creation of Bismarck's new empire, dominated by the Prussian crown.

By this time, Weierstrass stood at the height of his powers as the leading analyst of the day. Soon thereafter, however, his reputation and fame would be linked with that of an exotic foreigner who came knocking on his door, a talented young Russian named Sofia Kovalevskaya. She belonged to a generation of idealistic Russian women who broke new ground by pursuing higher education in the West, while others, like Kovalevskaya's older sister Anyuta – who joined the ill-fated Paris Commune in 1871 – left Russia to become political activists. As a woman, Sofia had no chance of being admitted to university lectures in Berlin, but Weierstrass agreed to give her private lessons in his home. She soon became his favorite and most famous student, a story filled with intrigue and rumors that have continued to flourish until this very day. Weierstrass used his connections with Lazarus Fuchs in Göttingen so that Kovalevskaya could be awarded a doctorate *in absentia* in 1874. Eventually she received a professorship in Stockholm, making her a famous pioneering figure for women in higher education.

I first became interested in the older literature on Kovalevskaya through Reinhard Bölling, who prepared the authoritative edition of Weierstrass's letters to her, published in Bölling (1993). The letters she wrote to him were all destroyed by Weierstrass at some point after her premature death in 1891. All, that is, except for one that accidentally escaped the flames: this survived as a partially burnt fragment that Bölling discovered during a visit to the Mittag-Leffler Institute. That sensational find led him to write the essay in Chap. 5, which I translated for *MI* when the two of us spent a month together at the Mittag-Leffler Institute in February 1991. As it happened, our stay fell at the time of the hundredth anniversary of Sofia Kovalevskaya's death, which followed a harrowing trip back to Stockholm (the basis for Alice Monroe's short story "Too much Happiness"). With our Russian colleague, Sergei Demidov, who contacted an orthodox Russian priest, we visited her gravesite on that anniversary, the 10th of February, to commemorate a courageous and gifted woman whose life was cut short, but whose name and fame have grown ever since.

Recalling the career of Kovalevskaya, one can hardly avoid remembering the man who brought her to Sweden in the first place, Gösta Mittag-Leffler. Although a foreigner, his role in the Berlin-Göttingen rivalry can hardly be overstated, as will emerge more clearly in Chap. 11. In Chap. 4 below, he serves as a star witness reporting on this conflict from the mid-1870s, when he first left Sweden to undertake studies in Paris and then Berlin. In a letter to his former mentor, Hjalmar Holmgren, the young Swede wrote that "[i]t seems unlikely that the mathematics of our day can point to anything that can compete with Weierstrass's function theory or Kronecker's algebra." He goes on to provide a very clear account of why the disciples of Weierstrass looked askance at Riemann's approach to complex analysis, but thought even less of the works of Alfred Clebsch, who had died three years before this letter was written.

During the 1880s, Mittag-Leffler would begin to develop an impressive network of allies in both Berlin and Paris. He did so mainly by linking these two centers with the new one he created in Stockholm, where Kovalevskaya had joined him. She, in fact, enjoyed excellent relations with the leading mathematicians in Paris as well. Mittag-Leffler liked to portray himself as an internationalist, a role he could assume rather easily from his outpost in Sweden after 1882 when he launched the new journal *Acta Mathematica*. This formed a new center of power in the world of mathematics, part of a triangle whose vertices were Paris, Stockholm, and Berlin. Mittag-Leffler's network obviously posed a major challenge, not to say threat, to the one Klein had been slowly building, mainly within Germany, as the principal editor of a competing journal, *Mathematische Annalen* (Rowe 2008). These circumstances reshaped the initial rivalry that pitted Berlin against Göttingen, which now took on international dimensions, some of which reverberated up through the First World War and even beyond. It should come, therefore, as no surprise that Klein and Mittag-Leffler saw themselves as leading representatives of two rival camps. How their personal relations played out is one of the main themes in the final, somewhat lengthier Chap. 11 in Part II. As a prelude to that story, a brief word should be said about mathematical schools and Klein's public stance with regard to the Berlin tradition.

The appellation "school" was often used rather loosely by mathematicians to describe informal groups of individuals with a shared academic pedigree and research interests (Rowe 2003). This usage usually reflected certain intellectual affinities or a particular orientation to their subject. Thus, one might speak of the practitioners of Riemannian or Weierstrassian function theory as members of two competing "schools." Yet Riemann never drew more than a handful of students, whereas Weierstrass was the acknowledged leader of the dominant research school of his day (even France's leading analyst, Charles Hermite, was supposed to have said "Weierstrass est notre maître à tous").[3] Felix Klein identified strongly with the Riemannian approach, which he learned mainly through Clebsch, yet neither ever met Riemann personally. Clearly the spheres of activity and influence of Riemann and Weierstrass were radically dissimilar, suggesting that here, as elsewhere, it would be more apt to distinguish between two competing mathematical *traditions* rather than schools, particularly in view of the complex methodological issues involved. Mittag-Leffler's letter speaks mainly about such methodological differences, though with a strong ideological bias. Klein's opinions, although perhaps no less pointed, were generally expressed in more conciliatory language, at least publicly.

In later years, Mittag-Leffler deftly cultivated his longstanding ties with leading mathematicians in Paris and Berlin, whereas Klein exerted great efforts to forge links with the older Göttingen tradition of Gauss and Riemann. Shortly after Gauss's death in 1855, his Nachlass had been divided into two parts, scientific and personal. The "scientific portion" was acquired by the Göttingen Scientific Society, which charged Gauss's former student, Ernst Schering, with ordering these papers for publication in the *Werke*. This project proceeded very slowly, however, a source of frustration for Klein, who had to await Schering's death in 1897 before he could take over control. He thereafter threw himself into it with a passion: between 1898 and 1921, Klein published no fewer than fourteen separate reports on the ongoing Gauss edition. After the discovery of Gauss's *Tagebuch* in 1899, he also prepared a special Festschrift edition. This was published in 1901, together with preliminary commentary, and then republished two years later in *Mathematische Annalen* (Klein 1903).

Klein made similar efforts to recover and promote Riemann's works and mathematical legacy following publication of the first edition of his *Werke* (Riemann 1876). Riemann's *Nachlass*, though far smaller than that of Gauss, posed numerous challenges, as Dedekind, Clebsch, and Heinrich Weber discovered when they worked through it (Scheel 2015). By the 1890s, however, Klein was intent on recovering additional documents, particularly lecture notes taken by some of Riemann's former auditors. This initiative eventually brought to light some 20 sets of lecture notes that were carefully studied by Max Noether and Wilhelm Wirtinger when they prepared the *Nachträge* for the expanded edition of Riemann's Collected Works (Riemann 1902). Scholarly interest in Riemann's original works remained strong throughout the twentieth century. Erich Bessel-Hagen, a protégé of Klein and close friend of Carl Ludwig Siegel, was actively involved in preserving documents relating to Riemann's life and work. In 1932, Siegel published a number of new results on the Riemann zeta-function based on formulas he had found scattered across papers in the Nachlass (Siegel 1932). The most complete edition of Riemann's works with commentaries published to date is (Riemann 1990).

By way of underscoring the vast differences between the Göttingen of Gauss and the atmosphere inculcated there during the era of Klein and David Hilbert (Fig. 1.10), it is worth noting some of the sharply contrasting differences that separate them. Gauss was evidently content to stand on his

[3] Hermite was a Catholic conservative who took a strong interest in mathematics east of the Rhine; see Archibald (2002).

Fig. 1.10 David Hilbert.

laurels, even to the point of keeping his personal discoveries secret. When he did publish, he took care to "remove the scaffolding" from the finished building, preferring that the reader admire the finished product rather than understand how it was constructed. Klein, by contrast, not only put his own, sometimes half-baked ideas on public display, he also wrote about his efforts to uncover the roots of discoveries made by his two famous Göttingen forebears, Gauss and Riemann (Klein 1926). While probing their works, he also celebrated the aura of genius long associated with these two names, often with his own larger interests in mind. For Gauss, teaching was only a burdensome chore, and he restricted his lecture courses to elementary topics in applied mathematics. Moreover, as noted in Chap. 3 on Dirichlet, Gauss confided little about his researches in pure mathematics to his contemporaries, especially to the younger generation of mathematicians who struggled to understand his *Disquisitiones Arithmeticae* and other works in number theory.

In this respect, a direct comparison with another classic work in number theory, David Hilbert's *Zahlbericht*, proves illuminating, particularly when we ask ourselves why these two paradigmatic texts were written and for whom (Schappacher 2005). Whereas Gauss' *Meisterwerk* stares back at us like a sphinx, as if reflecting its author's gaze, Hilbert opens the *Zahlbericht* with an upbeat preface that places his work at the pinnacle of a grand tradition dominated by the prior achievements of German number theorists.[4] Gauss clearly wrote for posterity, not for his fellow man. His *Disquisitiones Arithmeticae* was thus much admired, but most of it was poorly understood until Dirichlet broke its seven seals.[5] The latter's posthumously published *Vorlesungen über Zahlentheorie* provided the first readable textbook on algebraic number theory. Hilbert's *Zahlbericht*, on the other hand, was written as a monograph, not as a textbook.

Recognizing this, he offered a 4-hour elementary lecture course in the winter semester 1897–98, largely to help students ease into the subject. Much of this course consisted of routine examples, chosen in order to illustrate the general theory by calculating results in specific number fields. The following summer, he offered a 2-hour continuation of this course in which he took up two central topics in the theory of algebraic number fields: factorization and reciprocity laws. Right from the beginning, Hilbert alluded to their importance in Gauss's *Disquisitiones Arithmeticae*, while emphasizing that his goal was to show how the classical results could be extended to general number fields. Yet Hilbert also recognized the need for a standard textbook on algebraic number fields, so he encouraged the Göttingen *Privatdozent* Julius Sommer to write such a text (Sommer 1907). By 1899, Hilbert himself had quit the field, but not before leaving others a fertile terrain for future work: no fewer than twelve of his students went on to write dissertations on topics in number theory.

The contrast with Gauss could hardly be greater, and since the present collection of essays aims to shed light on modern mathematics in Göttingen and beyond, the reader may begin to appreciate why the life and times of Carl Friedrich Gauss mainly have a symbolic significance for this volume. By Klein's day, the Prince of Mathematicians was already a legendary figure, but this was a legend still in the making. Clearly, Hilbert's influence on the generation that followed him was incomparably greater than Gauss's had been, owing in large part to the inner dynamics of the Göttingen community Klein and Hilbert helped to create (Reid 1970; Rowe 1989). Gauss and Hilbert were not only very different types of personalities, their sharply contrasting workaday environments and social relations also reflected important differences in their respective local cultures. These contrasts tell us much about the dramatic changes that took place over the course of the nineteenth century.

Clearly many of Gauss's contemporaries viewed him with awe, but his exalted place as a peerless genius in the history of mathematics came later. That image of the celebrated Göttingen mathematician was especially promoted by Felix Klein, beginning in the 1890s, when a virtual cult of worshipers made Gauss into a great German hero. As a universal mathematician, Gauss was celebrated by virtually everyone in Germany who followed in his wake. For the Berliner E.

[4]As noted by Petri and Schappacher, the rhetoric that Hilbert employed in his preface was largely self-serving and provided little information about what was in the work itself (Goldstein et al. 2007, 362–366).

[5]As described by Goldstein and Schappacher, Gauss's *Disquisitiones Arithmeticae* was for 60 years a "book in search of a discipline" (Goldstein et al. 2007, 3–66).

E. Kummer, Gauss was the great model for rigor, just as for Klein he symbolized the highest incarnation of the applied mathematician, alongside Archimedes and Newton. By the time Klein took over the reins of the Gauss edition, however, Göttingen had already begun to eclipse Berlin. Furthermore, no mathematician from the Berlin tradition could match Klein's skills as an orator and propagandist for what he liked to call the Gaussian program, a watchword for his efforts to make Göttingen the dominant center for all facets of mathematical research in Germany.

After Klein's death in 1925, the Berlin Ludwig Bieberbach gave this hero worship a new twist by linking Felix Klein himself with Gauss. For Bieberbach, Gauss and Klein represented a distinctive German style of thought grounded in the fertile soil of their genial race, *das Volk* (Mehrtens 1987). This brilliant intellectual tradition, he argued further, owed nothing to Jacobi and other leading Jewish mathematicians, any more than one could imagine confusing Gauss's works with those of a foreign competitor, like the French analyst Augustin Louis Cauchy. Bieberbach's typology of mathematical styles was crude, to be sure, but then so was the larger Nazi ideology he identified with (see the introduction to Part V). Like Klein, he clearly sought to appropriate Gauss's legacy for his own purposes and to construct his own distinctive narrative.

The fifth and final essay (Chap. 6) takes us back to an earlier time. It comes from another *Intelligencer* quiz, one aimed at placing Kummer within the larger intellectual and spiritual currents of his day. One of the most famous and influential of all Berlin professors was the philosopher Georg Wilhelm Friedrich Hegel, who taught there from 1818 until his death in 1831. His son Karl became a well-known historian in Erlangen, where he edited some of his father's works for publication. Karl Hegel also became Felix Klein's father-in-law when his daughter Anna agreed to marry the young mathematician, then on his way from Erlangen to Munich. Such academic marriages were commonplace back then, but so was the *Zeitgeist* that made Hegel's philosophy so popular along with the more general current of German idealism he represented. It may seem hard to imagine the extent to which this influenced the thought of a mathematician like Kummer, who only came to Berlin a quarter century afterward. It must be remembered, however, that he and his contemporaries were members of a philosophical faculty. What many thus shared was their knowledge of and love for Latin and Greek literature, Luther's Bible, and their devotion to *Wissenschaft*, the life of the mind. Hegel's philosophy, in particular his lectures on the phenomenology of mind and the philosophy of history, taught that human destiny was bound with the quest for Mind to discover itself. That was, indeed, heady stuff in its day.

Klein later identified the Berlin tradition with a general trend toward "arithmetization" in mathematics (Klein 1895/1922). Underlying this was an explicitly stated doctrine that Gauss himself had articulated, namely that the number concept alone was the true font of all firm mathematical knowledge. Kummer, who deeply admired Gauss, was a firm advocate of the view that number was a pure idea grounded in human thought – or as a good Hegelian would say, Mind. Accordingly, other branches of mathematics, in particular geometry and mathematical physics, were to be regarded either as inferior or, at the very least, subordinate to those that had been arithmetized. The broader outlines of that divide – a crucial factor for understanding the ongoing rivalry between Göttingen and Berlin – are sketched in Chap. 4.

Arithmetization was also a step toward professional autonomy. As such, it reflected a quite general tendency to liberate purely mathematical research from applications to other disciplines. Many mathematicians in Germany identified with this trend, which was by no means exclusively associated with the Berlin tradition: two of its leading proponents were Dedekind and Hilbert. Still, the Berlin mathematicians were very much in the forefront of the program to build mathematics on the foundations of the number concept. This program motivated Weierstrass's constructive approach to the foundations of analysis, both for the real and for the complex numbers, by avoiding all appeal to geometrical intuition. Kronecker went even further, making the rational numbers not only the *sine qua non* for all pure mathematics, but also its only legitimate basis. Although these two seminal authorities had fallen into profound disagreement over such issues by the 1880s, they nevertheless shared this general purist ideology, just as did Kummer.

Klein, on the other hand, was an outspoken critic of mathematical purism. In fact, during his youth he cultivated a style of geometrical research that relied heavily on visualization and even occasionally made use of physical models, as discussed in Chap. 7 of Part II. In Germany, model making mainly came to flourish at the higher technical schools, where it was taught to future engineers and architects.[6] An important exception, as described in Chapter 6 and also in Chap. 8, was Kummer in Berlin. By the 1890s, however, Klein had begun to integrate these and other applied methods into the curriculum at Göttingen, thereby making a sharp break with the norms adopted at most other German universities (Rowe 1989). This was but one of the many ways in which Klein reformed and adapted the Göttingen tradition to what he saw as the needs of a new era. The broader contours of this story are by now quite familiar, since much has been written about the careers of Klein and Hilbert, the

[6] In a recently completed dissertation, Anja Sattelmacher has carefully followed how interest in models as well as the lack thereof played out during later periods in Germany (Anja Sattelmacher, *Anschauen, Anfassen, Auffassen. Eine Wissenschaftsgeschichte mathematischer Modelle*, Humboldt Universität Berlin, 2016).

two central figures in the modern Göttingen tradition. Many of the essays in the present volume delve into less familiar aspects of their careers, including special episodes that cast new light on their personalities and influence. Lurking in the background of several such episodes, though, were ongoing rivalries and competition that reflected various tensions within the German mathematical community. In no other case was this more apparent than at its two leading centers, Göttingen and Berlin.

References

Archibald, Thomas. 2002. Charles Hermite and German Mathematics in France. In *Mathematics Unbound: The Evolution of an International Mathematical Research Community, 1800–1945*, ed. K.H. Parshall and A.C. Rice, 123–138. Providence: American Mathematical Society.

Biermann, K.-R. 1988. *Die Mathematik und ihre Dozenten an der Berliner Universität, 1810–1920*. Berlin: Akademie Verlag.

Bölling, Reinhard, ed. 1993. *Der Briefwechsel zwischen Karl Weierstraß und Sofja Kowalewskaja*. Berlin: Akademie Verlag.

Bottazzini, Umberto, and Jeremy Gray. 2013. *Hidden Harmony – Geometric Fantasies: The Rise of Complex Function Theory*. New York: Springer.

Dedekind, Richard. 1930 bis 1932. *Gesammelte mathematische Werke*. Emmy Noether, Robert Fricke, Øystein Ore, Hrsg., Braunschweig: Vieweg, 3 Bände.

Elon, Amos. 2002. *The Pity of it All: A History of Jews in Germany*. New York: Metropolitan Books.

Frei, Günther, and Urs Stammbach. 1994. *Die Mathematiker an den Zürcher Hochschulen*. Basel: Birkhäuser Verlag.

Gardner, Helen J., and Robin J. Wilson. 1993. Thomas Archer Hirst – Mathematician Xtravagant. III. Göttingen and Berlin. *The American Mathematical Monthly* 100 (7): 619–625.

Goldstein, Catherine, Norbert Schappacher, and Joachim Schwermer, eds. 2007. *The Shaping of Arithmetic after C. F. Gauss's. Disquisitiones Arithmeticae*. Heidelberg: Springer.

Gray, Jeremy. 1984. A commentary on Gauss's mathematical diary, 1796–1814, with an English translation. *Expositiones Mathematicae* 2 (2): 97–130.

Klein, Felix. 1895/1922. Über die Arithmetisierung der Mathematik, in (Klein 1921–1923, vol. 2, 232–240).

———. 1903. Gauß' wissenschaftliches Tagebuch 1796–1814. *Mathematische Annalen* 57: 1–34.

———. 1921–1923. *Gesammelte mathematische Abhandlungen*. 3 Bde., Berlin: Julius Springer.

———. 1926. *Vorlesungen über die Entwicklung der Mathematik im 19. Jahrhundert*. von R. Courant u. O. Neugebauer hrsg., Bd. 1, Berlin: Julius Springer.

Küssner, Martha. 1979. *Carl Friedrich Gauss und seine Welt der Bücher*. Göttingen: Musterschmidt.

Lackmann, Thomas. 2007. *Das Glück der Mendelssohns. Geschichte einer deutschen Familie*. Berlin: Aufbau-Verlag.

Lange, Ernst Julius. 1899. *Jacob Steiners Lebensjahre in Berlin, 1821–1863: Nach Personalakten dargestellt*. Berlin: Gärtner.

Lorey, Wilhelm. 1916. *Das Studium der Mathematik an den deutschen Universitäten seit Anfang des 19. Jahrhunderts*, Abhandlungen über den mathematischen Unterricht in Deutschland, Band III. Vol. Heft 9. Leipzig: Verlag B. G. Teubner.

Mehrtens, Herbert. 1987. Ludwig Bieberbach and "Deutsche Mathematik", *Studies in the History of Mathematics*. ed. Esther R. Phillips, MAA Studies in Mathematics, vol. 26, 195–241. Washington: Mathematical Association of America.

Noether, E., and J. Cavaillès, eds. 1937. *Briefwechsel Cantor-Dedekind*. Paris: Hermann.

Nye, Mary Jo, ed. 2003. *The Cambridge History of Science, vol. 5, Modern Physical and Mathematical Sciences*. Cambridge: Cambridge University Press.

Reid, Constance. 1970. *Hilbert*. New York: Springer.

Riemann, Bernhard. 1868. Über die Hypothesen, welche der Geometrie zu Grunde liegen, *Abhandlungen der Königlichen Gesellschaft der Wissenschaften zu Göttingen*, 13: 133–150.

———. 1876. *Bernhard Riemanns Gesammelte Mathematische Werke und wissenschaftlicher Nachlass*. hrsg. von Heinrich Weber mit Richard Dedekind, Leipzig: B. G. Teubner., 2. Auflage 1892.

———. 1902. *Bernhard Riemanns Gesammelte Mathematische Werke und wissenschaftlicher Nachlass*, hrsg. von Heinrich Weber mit Nachträgen hrsg. von Max Noether und Wilhelm Wirtinger, Leipzig, B. G. Teubner.

———. 1990. In *Riemanns Gesammelte Werke*, ed. von Raghavan Narasimhan. Leipzig: Teubner/Springer.

Rowe, David E. 1989. Klein Hilbert, and the Göttingen Mathematical Tradition. *Osiris* 5: 186–213.

———. 1998. Mathematics in Berlin, 1810–1933. In *Mathematics in Berlin*, ed. H.G.W. Begehr et al., 9–26. Basel: Birkhäuser.

———. 2003. Mathematical Schools, Communities, and Networks, in (Nye 2003, 113–132).

———. 2008. Disciplinary Cultures of Mathematical Productivity in Germany, in *Publikationsstrategien einer Disziplin: Mathematik in Kaiserreich und Weimarer Republik*. Volker Remmert u. Ute Schneider, eds. Mainzer Studien zur Buchwissenschaft 19, Wiesbaden: Harrassowitz, 9–51.

Schappacher, Norbert. 2005. David Hilbert, Report on Algebraic Number Fields (*Zahlbericht*) (1897). In *Landmark Writings in Western Mathematics*, ed. Ivor Grattan-Guinness, 700–709. Amsterdam: Elsevier.

Scheel, Katrin, ed. 2015. *Der Briefwechsel Richard Dedekind - Heinrich Weber*. Berlin: De Gruyter.

Schubring, Gert. 1984. Die Promotion von P. G. Lejeune Dirichlet. Biographische Mitteilungen zum Werdegang Dirichlets. *NTM (Schriftenreihe für Geschichte der Naturwissenschaften, Technik und Medizin)* 21 (1): 45–65.

———. 1986. The Three Parts of the Dirichlet. *Nachlass, Historia Mathematica* 13: 52–56.

Servos, John. 1993. Research Schools and their Histories, *Research Schools. Historical Reappraisals*, Gerald L. Geison and Frederic L. Holmes ed., (Osiris, 8, 1993), 3–15.

Siegel. 1932. Über Riemanns Nachlass zur analytischen Zahlentheorie, *Quellen und Studien zur Geschichte der Mathematik, Astronomie und Physik*, Abt. B: Studien 2, (1932), S. 45–80; reprinted in (Siegel 1979) and (Riemann 1990).

———. 1979. *Carl Ludwig Siegel, Gesammelte Abhandlungen*. Bd. 1 ed. Berlin/New York: Springer.

Sommer, Julius. 1907. *Vorlesungen über Zahlentheorie: Einfuhrung in die Theorie der algebraischen Zahlkörper*. Leipzig: Teubner.

Stammbach, Urs. 2013. Gauss und Dirichlet – eine Episode. *DMV-Mitteilungen* 21: 182–188.

Turner, R. Steven 1980. The Prussian Universities and the Concept of Research. *Internationales Archiv für Sozialgeschichte der deutschen Literatur* 5: 68–93.

2 On Gauss and Gaussian Legends: A Quiz

(Mathematical Intelligencer 37(4)(2015): 45–47;
38(4)(2016): 39–45)

For the last few years, students in my history of mathematics course have been required to do a bit of research on the web. Each of them chooses from a list of specially chosen questions designed to make them ponder whether the information they find on standard internet sites is solidly grounded and clearly sourced, or whether subsequent research (pursued in such unlikely places as the local university library) might lead a person to doubt what one reads online. The idea here is not to push for a definitive answer; in many cases, this would be a hopeless undertaking anyway. Instead, I ask students merely to report on what they found and how they went about tracking down the information cited in their short reports. It's all about learning to read sources critically. Here I offer a taste of what my students have to put up with, a set of challenge questions to test the mettle of those *Intelligencer* readers who might care to chase after some trivia.

Some of my queries involve quotes that allegedly came from the mouths or pens of famous figures. Others concern opinions, accomplishments, or animosities attributed to noteworthy mathematicians. Since the students nearly all come from Germany, it should not be surprising that several of the questions concern an enigmatic genius named Carl Friedrich Gauss (1777–1855) (Fig. 2.1), whose legendary life served more recently as part of the canvas for Daniel Kehlmann's historical novel, *Measuring the World.* Kehlmann's Gauss is a pretty nasty guy, obsessed with measurement, just like his counterpart, Alexander von Humboldt (1769–1859). Yet being a mathematician, he stumbles on an uncomfortable idea (somewhat garbled in Kehlmann's telling), to wit, that Euclidean geometry might not be true. Shaken, he slithers off to Königsberg, hoping to gain enlightenment from the great philosopher who claimed that the truths of Euclidean geometry are not grounded in this world; they enjoy a still higher transcendental truth status. Thus, we witness Gauss barging his way into the home of Immanuel Kant, only to meet a senile old man. This brings us to our first query (relatively easy):

1. How do we know that this meeting never took place?

One of the standard legends found in virtually every biography of Gauss tells how he stunned his grade school teacher by quickly adding up a series of natural numbers, a task that was supposed to keep a bunch of little boys busy for a good long time. In older sources, however, one finds no mention of the precise series, whereas later sources typically refer to the numbers $1 + \ldots + 100$. So we might wonder whether there is a trustworthy source for this latter claim, or was it merely an embellishment added later to the standard story? Indeed, we should ask (not too hard):

2. What solid evidence do we have for either version of this tale?

A commonly cited saying attributed to Gauss reflects his aesthetic tastes when it came to mathematical works. He was quoted by a noteworthy authority as having remarked that the observer of an impressive building would surely wish to view it with the scaffolding removed. This statement would also seem consistent with the high esteem with which Gauss held the works of Archimedes and Newton.

3. But who was it that first reported this remark and when was it made (fairly easy)?

Long before then, a famous Norwegian mathematician came to Germany, where he met August Leopold Crelle (1789–1855) in Berlin. It was above all Crelle who discovered the great talent of Niels Henrik Abel (1802–1829) (Fig. 2.2), whose work adorns the first volumes of a new journal founded in 1826, *Das Journal für die reine und angewandte Mathematik,* but usually known simply as "Crelle." Abel had no doubt heard much about the legendary Gauss, then professor of astronomy in Göttingen, and he surely could have tried to visit him there on his way to Paris. But

D.E. Rowe, *A Richer Picture of Mathematics*, https://doi.org/10.1007/978-3-319-67819-1_2

Fig. 2.1 C. F. Gauss, as portrayed by the painter Gottlieb Biermann, who never met Gauss. Biermann painted this famous portrait over three decades after Gauss's death.

Fig. 2.2 Niels Henrik Abel, Crelle's great discovery. This is the only authentic portrait of Abel to survive. It was made by Johan Gorbitz in Paris during Abel's stay there in 1826.

influenced by what he had heard about his character, he chose to avoid passing through Göttingen. If one googles in "Gauss Abel fox" or just goes to the Abel article in Wikipedia, a "famous" quote comes up, according to which Abel once said of Carl Friedrich Gauss's writing style: "He is like the fox who effaces his tracks in the sand with his tail." Clearly, Abel did not say this in English, so one would hope to find an authentic source for this in the original language. Wikipedia merely cites G. F. Simmons, *Calculus Gems*. New York: McGraw Hill, p. 177, but please don't bother with that because Simmons doesn't cite any such source. I have elsewhere found this statement attributed to Jacobi, but let me assure the reader: the statement is authentic, but these attributions are wrong. Tracking this down, however, is not at all easy:

4. Who spoke of Gauss as a sly fox who knew how to erase his tracks and on what occasion was this remark made?

Gauss died a wealthy man, at least by the standards of a university professor. He might have become even wealthier if he had had a knack for patenting some of his nifty inventions. Among these, surely the most famous was the telegraph he and the physicist Wilhelm Weber strung up (Fig. 2.3), a device that enabled them to send messages between Weber's institute, located near the center of the city, and Gauss's observatory, which was built outside the walls of the old town. They began to build it in the spring of 1833 and it remained functional until 1845 when the cables snapped in a storm. By that time Weber was in Leipzig, having lost his professorship in 1837 as one of the Göttingen Seven who protested against the annulment of the Hanoverian constitution. In the meantime, a certain Samuel F. B. Morse had far surpassed his potential competitors, having demonstrated the possibility of sending telegraphic messages over long distances. Many years later, a mathematician who knew Gauss well revealed that his interests in the electromagnetic telegraph had been purely scientific. Gauss thus expressed no regrets that he had neglected to promote the commercial development of telegraphy.

5. Who was this witness and on what occasion did he make Gauss's views known (hard)?

Fig. 2.3 A replica of the telegraph invented by Gauss and Weber.

Fig. 2.4 Gauss's heliotrope used for the triangulation of Hanover as shown on the back of a ten-mark banknote.

Another famous instrument invented by Gauss was the heliotrope (Fig. 2.4), which he used for sighting purposes in geodesy. Kehlmann's Gauss was primarily a geodesist, befitting the title of the novel, but he also had the rich imagination of a troubled genius. So, like the real Gauss, he thought a lot about an "anti-Euclidean" geometry, and he even wrote about it in letters to fellow astronomers. Legend has it that Gauss even found a way to test the possibility that space is curved by measuring the angles in a huge terrestrial triangle he had used for surveying purposes. This reference triangle was the largest ever employed in his day, but using a heliotrope all three of its vertices could be sighted from the roof of Gauss's observatory. From there he could make and control precision measurements, carried out during the years 1818–1826 when he was charged with mapping out the Kingdom of Hanover. Adding a little theory into the mix, he supposedly used this data to probe the geometry of space. That test, however, turned out to be inconclusive; at any rate, he supposedly found no appreciable deviation from the usual angle sum of 180°.

6. But where does this story come from and how believable is it really (disputed)?

The vertices of this famous triangle were located in three remote mountainous locations: Brocken in the Harz, Hohen Hagen near Dransfeld south of Göttingen, and Inselberg in the Thuringian Forest. These three points happen to be separated by considerable distances: 69, 85 and 107 km, respectively. For geodesy, impressive numbers, but not surprisingly, a terrestrial triangle turns out to be far too small for the detection of curvature effects on light rays.

7. Still, if Gauss might have been led to check his figures, what would he have found (nontrivial)?

People in Göttingen certainly believed this story, and not just the guy on the street or those who might have read Kehlmann's novel. Felix Klein (1849–1925) and Karl Schwarzschild (1873–1916) helped to raise funds to erect a Gauss tower on Hohen Hagen, and they did so by commemorating Gauss's early test of non-Euclidean geometry. That fundraiser took place in the same year that David Hilbert (1862–1943) invited a famous mathematician to attend the ground-breaking ceremony for its construction.

8. Who was that mathematician and did he decide to accept Hilbert's invitation (very hard)?

Answers to the Gauss Quiz

In December 2015, I posed eight questions about the near legendary Carl Friedrich Gauss (1777–1855). When dealing with someone as famous as Gauss, it certainly isn't always easy to sort out fact from fiction, but trying to do so can be instructive. So this quiz began with a legend of rather recent vintage. It concerns a certain episode told by Daniel Kehlmann in his entertaining historical novel, *Measuring the World* (Kehlmann 2006). For those unfamiliar with this book, or the movie based on it, just a word about the title. The story begins with Gauss travelling to Berlin to meet the naturalist Alexander von Humboldt. Kehlmann portrays these two heroes of German science as oddball types; for some reason, both of them are obsessed by a need to measure terrestrial things. Beyond that realm, however, Gauss even reflects on certain new possibilities for measuring space itself.

Several of the queries in the quiz touch on Gauss's interest in testing non-Euclidean geometry, beginning with Kehlmann's new approach to this theme. From Gauss's correspondence with friends and colleagues, it has long been known that he had seriously contemplated an alternative to Euclidean geometry. But he also hated public controversy, which probably accounts for why he remained silent, even when others (notably N. Lobachevsky and J. Bolyai) stuck their necks out. Gauss was, in this respect, conservative to the core.

Kehlmann puts a new spin on all this, though, by staging a fictitious meeting between Gauss and an aged philosopher who happened to live in far off Königsberg: an old sage by the name of Immanuel Kant. Picking up with this episode in the novel, we learn that the young Gauss has recently stumbled on the thought that Euclidean geometry might not be true if we measured the world more carefully. Knowing that the great Kant had taught that the theorems of Euclidean geometry are transcendent truths that govern our perceptions of space, and which are therefore independent of all possible measurement results, he sets off for Königsberg hoping to attain enlightenment. Since Gauss was living in Brunswick, this journey of nearly 1000 km would in any event have been an arduous undertaking, lasting perhaps two weeks (one way!). Here Kehlmann's omniscience helps us appreciate Gauss's mental state on arrival:

> When he reached Königsberg Gauss was almost out of his mind with exhaustion, back pain, and boredom. He had no money for an inn, so he went straight to the university and got directions from a stupid-looking porter. Like everyone here, the man spoke a peculiar dialect, the streets looked foreign, the shops had signs that were incomprehensible, and the food in the taverns didn't smell like food. He had never been so far from home.

Kehlmann's Gauss had some difficulty persuading Kant's butler that he had come to consult with the master about a matter of great urgency. Whether persuaded by this or not, the butler reluctantly gave in and escorted this persistent visitor into Kant's private chambers. At this climactic point, the great chance he was waiting for, Gauss could now finally pour his heart out, hoping thereby to gain the wise counsel that brought him all this way:

> In a hushed voice, he made his request: he had ideas he had never been able to share with anyone. For example, it seemed to him that Euclidean space did not, as per the *Critique of Pure Reason,* dictate the form of our perceptions and thus of all possible varieties of experience, but was, rather, a fiction, a beautiful dream. The truth was extremely strange: the proposition that two given parallel lines never touched each other had never been provable, not by Euclid, not by anyone else. But it wasn't at all obvious, as everyone had always assumed. He, Gauss, was thinking that the proposition was false. Perhaps there were no such things as parallels. Perhaps space also made it possible, provided one had a line and a point next to it, to draw infinite numbers of different parallels through this one point. Only one thing was certain: space was folded, bent, and extremely strange.

Many readers will know what happens next, but for those who don't, I'll say no more. Overlooking all the mathematical obscurities in this passage – Kehlmann is just telling a story after all – I raised the question: 1) how do we know that this meeting never took place? Of course, actually proving that a certain event in the distant past did not happen or could not have occurred is next to impossible, so certainly we should not neglect considerations of plausibility here. For example, I am unaware of any evidence that Gauss ever expressed a positive view regarding Kant's famous doctrine regarding the geometry of space. On the contrary, he seems to have held a skeptical view of Kantian epistemology in general. This being the case, to suspend our disbelief would presumably mean swallowing the idea that even Gauss was impressed by Kant's reputation. If so, he might well have been eager to learn his reaction when confronted with a sharply different understanding of the theory of parallels. Let us, then, allow for this possibility.

We can next turn to certain circumstantial and biographical details; these, at any rate, were the type of data I had in mind when I first thought of this question. First, we should note that Kant died on 12 February 1804, which means that Gauss would have had to visit him by early 1804, at the very latest. Thankfully, Kehlmann provides us with just enough information so that we can quite easily zero in on the time of this alleged journey. That's because his Gauss is not only in a desperate state of mind owing to his dark thoughts regarding the strange non-Euclidean character of space. He is probably even more distraught because he has recently fallen madly in love with a lovely young girl named Johanna Osthoff. Like him, she is a Brunswick native from a similar working-class family background. When he proposes to her, though, she rebuffs him with worries about their compatibility. Soon after this, he suddenly decides to set off on his eastward journey in hopes of gaining reassurance from the elderly Kant.

This added romantic dimension works very nicely in the novel, and what is more Kehlmann only needed to stretch the truth a tiny bit when describing the situation as regards his hero's romantic life. Little mystery remains on that score, for the extant letters from this time clearly show that Gauss was head over heels in love with Hannchen Osthoff, whom he eventually took as his wife on 9 October 1805. We also know that he first met her in July 1803. This, of course, left Kehlmann rather little time to conjure up the wedding proposal, which she then casts aside, thereby prompting the heartbroken Gauss to rush off to Königsberg, and still have him arrive there while Kant was yet among the living. After all, he would need two weeks just to make the journey.

But even if this were all theoretically possible, it clearly did not happen because we know that the courtship only moved into high gear after Kant was already dead. This can be discerned from a letter Gauss wrote on 28 June 1804 to Farkas Bolyai, his Hungarian friend from their student days in Göttingen. There he expresses his deep feelings for Johanna, but he also confides that he had only recently dared approach her. Clearly dazzled by her beauty and character, he waited another month before proposing to her in writing. She accepted, and their engagement was announced on 22 November 1804. These letters, together with many more details, can be found in (Biermann 1990). Gauss's correspondence travelled far and wide – some 7000 of his letters are still extant – but he himself never set foot in Königsberg.

The second query concerned one of the best-known stories found in virtually every biography of Gauss. As a youngster, he stunned his grade school teacher by quickly adding up a series of natural numbers almost before the other little boys had a chance to begin calculating (Fig. 2.5). The source for this tale was actually Gauss himself, as we learn from Wolfgang Sartorius von Waltershausen, his friend and first biographer:

> In 1784, after his seventh birthday, the little fellow entered the public school, where elementary subjects were taught, and which was under the supervision of a man named Büttner. . . . Here occurred an incident which [Gauss] often related in old age with amusement and relish. In this class the pupil who first finished his example in arithmetic was to place his slate in the middle of a large table. On top of this, the second placed his slate, and so on. The young Gauss had just entered the class when Büttner gave out a problem for summing an arithmetic series. The problem was barely stated before Gauss threw his slate on the table with the words (in the low Brunswick dialect): "There it lies." While the other pupils continued counting, multiplying, and adding, Büttner, with self-conscious dignity, walked back and forth, occasionally throwing an ironical, pitying glance toward the youngest of the pupils. The boy sat quietly with his task ended, as fully aware as he always was on finishing a task that the problem had been correctly solved and that there could be no other result. At the end of the hour the slates were turned bottom up. That of the young Gauss with one solitary figure lay on top. When Büttner read out the answer, to the surprise of all present that of young Gauss was found to be correct, whereas many of the others were wrong. (Sartorius 1966, 12)

This is surely the earliest written account of this famous story, but I noted that here, as well as in other older sources, one finds no mention of the actual problem Gauss solved so quickly, whereas later versions typically refer to the numbers 1 +. .. + 100. This led me to ask about the evidence underlying either version of this tale.

When I wrote this, I thought it would be a novel question to ask, but since then I have come to learn that others have asked this same question before. For example, it is taken up in *The Tower of Hanoi – Myths and Maths* (Hinz et al. 2013, 35–36). My eyes were really opened, though, by a letter from José Hernández Santiago from Morelia, Mexico. Hernández informed me that Brian Hayes, a senior author for *American Scientist*, had published a

Fig. 2.5 The house at Wilhelmstraße 30 in Brunswick where Gauss was born on 30 April 1777 (Photo from (Cajori 1912)).

thorough search of the literature, citing over one hundred versions of this story. His results can be found online by going to: http://bit-player.org/wp-content/extras/gaussfiles/gauss-snippets.html. In his published article, "Gauss's Day of Reckoning" (Hayes 2006), he wrote that:

> [the] *locus classicus* of the Gauss schoolroom story is a memorial volume published in 1856, just a year after Gauss's death. The author was Wolfgang Sartorius, Baron von Waltershausen, professor of mineralogy and geology at the University of Göttingen, where Gauss spent his entire academic career. As befits a funerary tribute, it is affectionate and laudatory throughout.

Hayes, of course, noticed the discrepancy I asked about, and he seemed to confirm my guess as to where the series came from. Regarding this he wrote: "In the literature I have surveyed, the 1–100 series makes its first appearance in 1938, some 80 years after Sartorius wrote his memoir. The 1–100 example is introduced in a biography of Gauss by Ludwig Bieberbach (a mathematician notorious as the principal instrument of Nazi anti-Semitism in the German mathematical community)."

Hernández informed me, though, that this book (Bieberbach 1938) was not the first place that refers to these numbers. In fact, Hayes has updated his list since its first appearance in (Hayes 2006), so it now contains an obscure pamphlet written in 1906 by one Franz Mathé. There one finds again that the task was to add the first hundred natural numbers. So perhaps Bieberbach took his version of the story from this source, or maybe an even older one. Hayes considered the possibility that: "someone to whom Gauss told the story 'with amusement and relish' [may have] left a record of the occasion. The existence of such a corroborating document cannot be ruled out, but at present there is no evidence for it.. .. If an account from Gauss's lifetime exists, it remains so obscure that it can't have had much influence on other tellers of the tale." (Hayes 2006).

Sartorius von Waltershausen's memorial for Gauss (Sartorius 1966) has served as the principal source for numerous stories relating to the life of his friend (Reich 2012). Given their special relationship, there is good reason to view the information he related as highly reliable, and I exploited it for some of the quiz questions, including the third one. This concerns Gauss's aesthetic views regarding mathematical works, where I asked about the source for his remark that the observer of an impressive building would surely wish to see it with the scaffolding removed. One finds this on page 82 of (Sartorius 1966), followed by a statement that Gauss had come to prize this synthetic style of presentation through his studies of the works of Archimedes and Newton. Sartorius went on to explain that it was these sensibilities that informed the famous Gaussian motto "pauca sed matura" ("few but ripe").

Many of Gauss's readers took less delight in his Spartan writing style, which made it very difficult to penetrate the original line of thought that motivated the finished works. If we were to believe what one finds on Wikipedia, it would seem that Niels Henrik Abel was one of those who expressed such displeasure. Referring to Gauss's formal prose, Abel supposedly said, "He is like the fox who effaces his tracks in the sand with his tail." Since no reliable source for this statement is given, however, my fourth question asked whether this saying was due to Abel or perhaps someone else, and to determine on what occasion the remark was made.

It seemed to me quite odd that such a striking and colorful statement is nowhere to be found in the modern biographical literature, e.g. in (Ore 2008) or (Stubhaug 2000). I, therefore, took the opportunity to pursue this question a few years ago in Kristiansand, Norway, not far from Abel's final resting place, during a visit with Reinhard Siegmund-Schultze. He happened to own the French translation of an older Abel biography by C. A. Bjerknes, and this source, indeed, contains a reference to the mysterious statement about Gauss as a fox, though only in a footnote (Bjerknes 1885, 92) and without attribution. Reinhard then suggested that we contact Henrik Kragh Sørensen, a leading expert on Abel. Henrik quickly solved the puzzle and sent us copies of two articles by Christopher Hansteen (1784–1873), published in March 1862 in the *Illustreret Nyhedsblad.* Hansteen was a well-known geophysicist, astronomer and physicist in Christiania (present-day Oslo) who also happened to be Abel's mentor. His two essays were based on five letters he had received nearly 40 years earlier from Abel when the latter was touring Europe. The first of these five letters contains some interesting remarks apropos Gauss (Fig. 2.6).

After tracking this down with the help of two historians, I was certain that this was one of the hardest of the quiz questions. So I was surprised to learn from José Hernández that he was able to crack this puzzle. He even sent me a scan of the page in *Illustreret Nyhedsblad* where one finds the first published reference to Gauss, the fox. Hernández noted further that this magazine was a Norwegian weekly published in Christiania from 1851 to 1866. Among its contributors were such famous names in Norwegian literature as Ibsen, Bjørnson, and Collett. He also pointed out that passages from Abel's first letter to Hansteen can be found in Stubhaug's biography, where one reads:

> Abel [is remarking] on how all the young mathematicians in Berlin "nearly deify Gauss. He is for them the quintessence of all mathematical [e]xcellence." Abel went on to comment, "It may be granted that he is a great [g]enius, but it is also well-known that he gives rotten [l]ectures. Crelle says that all Gauss writes is an abomination [Gräuel], since it is obscure to the point of being almost impossible to understand." (Stubhaug 2000, 332)

It was in this connection that Hansteen added a footnote containing this statement:

"A German student said about him on this occasion, 'er macht es wie der Fuchs, der wischt mit dem Schwanze

Illustreret Nyhedsblad.

Nr. 9. 2den Marts 1862. 11. Aarg.

Niels Henrik Abel.

Hansteen.

Det er over al Maade, hvor de unge Mathematikere her i Berlin, og som jeg hører overalt i Tydskland, ligesom forgude Gauss. Han er for dem Indbegrebet af al mathematisk Fortræffelighed; men lad være, at han vist nok er et stort Genie, saa er det ligesaa vist, at han har et slet Foredrag. Crelle siger, at alt hvad Gauss skriver er „Gräuel", da det er saa dunkelt, at det næsten ikke er muligt at forstaae det**). G. arbeider nu paa et stort Værk over den physiske Astronomie, hvoraf de tre første Dele ere færdige til Pressen. Der skal deri findes mange nye Ting.

Da jeg var i Hamburg, besøgte jeg Professor Schumacher, som modtog mig meget forekommende; men han var dengang ikke ret frisk. Der gjorde jeg ogsaa Bekjendtskab med Thomas Clausen***), som

*) Formodentlig den samme, som i Humboldts Breve omtales som Medarbeider af Kreuzzeitungs Feuilleton.

**) En Tydsk Student sagde om ham i den Anledning: „er macht es wie der Fuchs, der wischt mit dem Schwanze seine Spuren im Sande aus".

Fig. 2.6 Christopher Hansteen's first article on Abel in *Illustreret Nyhedsblad* with the footnote referring to Gauss as a "mathematical fox".

seinen Spuren im Sande aus'" (Hansteen 2 March 1862). Clearly, then, it was not Abel, but rather a German student whom he had met in Berlin who suggested that the *princeps mathematicorum* resembled "a fox who erases his tracks in the sand with his tail."

The fifth question concerned Gauss's attitude with regard to technological progress in general and the invention of the electromagnetic telegraph in particular (Fig. 2.7). He and his collaborator, Wilhelm Weber, had come up with such a gadget in the early 1830s. Sartorius discusses various plans for its development in the Saxon railroad industry, but economic circumstances caused these to be dropped. Not until the invention of relays, however, was it possible to transmit messages over longer distances, a breakthrough exploited by Samuel F. B. Morse. Gauss expressed the view that science should by all means befriend the practical arts, but should never be a slave to the latter. Perhaps he said this often, but this statement was recorded for posterity in 1877 by his former student, Moritz Abraham Stern (1807–1894), on the occasion of the Gauss centenary celebrations in Göttingen (Stern 1877, 15).

For many decades, Stern was a fixture in the Göttingen mathematical community, having begun teaching there in 1830 as a Privatdozent. As an un-baptized Jew, many were convinced that he had virtually no chances of becoming a full professor, though in 1848 he was finally appointed to an extraordinary professorship. Throughout these years he taught many subjects, but also as a researcher he was generally held in high esteem by his colleagues. Finally, on 30 July 1859, the Hanoverian Ministry of Culture granted

Fig. 2.7 This statue of Gauss sending a message on his telegraph could once be seen at one corner of the Potsdam Bridge in Berlin-Schöneberg. The other three corners were adorned with statues of Röntgen, Siemens, and Helmholtz. All four were erected in 1898 and destroyed during the bombing of Berlin in 1944 (Photo from Cajori 1912).

him the title of *Ordinarius*, the very day that Bernhard Riemann was also appointed full professor.

We come now to the last three quiz questions, which revolve around the measurement of space, as discussed above. Kehlmann's Gauss was an astronomer and geodesist, befitting the title of the novel, but he also had the rich imagination of a troubled genius. Like the real Gauss, he thought a lot about an "anti-Euclidean" geometry, and we know that Gauss wrote about this topic in letters to colleagues and friends. During the years 1818–1826, he was charged with surveying the Kingdom of Hanover, for which purpose he invented the heliotrope to improve the accuracy of sightings. The sixth question concerns a famous test of Euclidean geometry that Gauss was purported to have carried out based on measurements taken for a giant spherical triangle whose vertices were located in three remote mountainous locations: Brocken in the Harz, Hohen Hagen near Dransfeld south of Göttingen, and Inselberg in the Thuringian Forest. These three points happen to be separated by quite considerable distances: 69, 85 and 107 km, respectively, but by using a heliotrope all three of these points could be sighted from Gauss's observatory in Göttingen.

As by now is well known, Sartorius is the source for the story that Gauss used the data from these sightings to test whether the sum of the angles in this giant optical triangle deviated from 180°. The result he obtained was astonishingly small, only two-tenths of 1 s of arc from the Euclidean result (Sartorius 1966, 53). In describing Gauss's views on the foundations of geometry, Sartorius wrote:

> Gauss regarded geometry as a consistent structure only if the theory of parallels is conceded as an axiom and placed at the summit. He was nevertheless convinced that this proposition could not be proved, even though one knew from experience, e.g. from the angles in the triangle Brocken, Hohenhagen. Inselberg that it was approximately correct. (Sartorius 1966, 81).

This answers the first part of question six, but it hardly helps is answering the second, namely, how believable is it that Gauss used the BHI triangle to test the possibility that space might be curved? The historian of physics Arthur Miller disputed this claim in a short paper entitled "On the myth of Gauss's experiment regarding the Euclidean nature of space" (Miller 1972). Miller based his critique on arguments taken from Gauss's work on surface theory – in particular, his discussion of the same triangle in *Disquisitiones generales circa superficies curvas* (1827) – but apparently he was unaware of Sartorius's testimony cited above. Thus, he speculated that this "Gaussian myth" only arose in the wake of Einstein's gravitational theory, which drew interest to Riemann's famous lecture of 1854 on higher-dimensional manifolds with intrinsic curvature, a generalization of Gauss's 1827 theory, which dealt with the intrinsic curvature properties of surfaces embedded in Euclidean 3-space.

Miller's argument was immediately challenged by B. L. van der Waerden in (van der Waerden 1974), but the first detailed examination of the issues at stake came a decade later with the publication of (Breitenberger 1984). More recently, Erhard Scholz has parsed the various issues involved in this debate in (Scholz 2004). The thrust of his argument aims to show that Gauss had all the means at hand to calculate a deviation from 180° due to an assumption that the light rays in space followed geodesics in a spatial geometry of constant curvature.

With this in mind, the seventh question about the results Gauss would have obtained based on his own data can best be answered by referring readers to the detailed analysis in (Scholz 2004). In fact, Breitenberger's paper also confirms that the figure given by Sartorius accords well with Gauss's data, but he argues that Gauss lacked a clear conception of a 3-dimensional non-Euclidean geometry within which he could frame an alternative to the Euclidean theory (Breitenberger 1984, 285). Scholz, on the other hand, cites Gauss's letter to Schumacher from 1831, in which Gauss states, "we know from experiment that the constant k [in anti-Euclidean geometry] must be incredibly large compared with all that we can measure. In Euclid's geometry k is infinite" (Scholz 2004, 24). He thus finds the testimony of Sartorius fully believable and consistent with what Gauss must have found.

Turning the historiographic tables, then, he dismisses what he calls "Miller's myth." For another view, however, one should take into account what Jeremy Gray writes in (Gray 2006) about this and other difficulties associated with Gauss's place in the history of non-Euclidean geometry. As Gray states at the outset, "[the evidence] suggests that Gauss was aware that much needed to be done to Euclid's Elements to make them rigorous, and that the geometrical nature of physical Space was regarded by Gauss as more and more likely to be an empirical matter, but in this his instincts and insights on this occasion were those of a scientist, not a mathematician." (Gray 2006, 60).

Kehlmann's myth only adds another layer of confusion to a story already shrouded in a good deal of mystery. In the meantime, his best-selling novel has given rise to a batch of efforts to separate fact from fiction in his various stories. Unlike novelists, historians obviously have no such freedom; they are bound by documentary evidence. Miller's myth, as Scholz pointed out, was concocted without taking due account of a crucial piece of evidence that Sartorius obtained from Gauss himself. At the very least, the analysis in (Scholz 2004) clearly refutes Miller's claim that Gauss could not have tested the Euclidean hypothesis for spatial geometry.

Fig. 2.8 The old Gauss Tower near Dransfeld (Dransfelder Archiv, http://dransfeld.knobelauflauf.de).

The eighth and final question concerned the Gauss Tower, located on Hohenhagen near Dransfeld (Fig. 2.8). An announcement, published in 1909 by Felix Klein and Karl Schwarzschild, pointed to the significance of the location, which was meant to commemorate Gauss's purported test of non-Euclidean geometry. That same year David Hilbert invited a famous mathematician to attend the groundbreaking ceremony for its construction. This led me to ask who this mathematician was, and to query whether he decided to accept Hilbert's invitation.

This was a difficult question certainly, though Scott Walter has known the answer for a long time. He transcribed the letter Hilbert wrote to Henri Poincaré on 25 February 1925 in which plans for Poincaré's Wolfskehl lectures, to be delivered in April, were discussed. This ends with these remarks: ".. . Finally, on the 30th of April, the birthday of Gauss, the dedication of a Gauss-Tower is planned. This will take place at nearby Dransfeld on the "Hohenhagen" (a vertex of the Gaussian triangle of straight lines for which he observed that the angle sum was π). Your presence on that occasion is urgently wished." (The full letter is reproduced in Chap. 16.)

Surely Hilbert appreciated that Poincaré's own views on the non-empirical nature of this whole problem would have put him in an awkward position on such an occasion. In fact, everyone in Göttingen seems to have heard some version of the story first told by Sartorius von Waltershausen, and none doubted it was true. That meant they believed that Gauss really had used the data from the great BHI triangle to test the empirical bounds within which Euclidean geometry held. So to imagine this scene, we would have to picture Schwarzschild making a speech to celebrate the profundity of this astronomical event, while Poincaré made an odd face, or quietly shook his head in disbelief. Clearly, he had not been invited to Göttingen to explain to his hosts that even the great Gauss had not fully appreciated the nature of the situation. Luckily for Poincaré, the groundbreaking ceremony was postponed, leaving the esteemed French mathematician a chance to return to Paris in peace. The Gauss Tower was completed in 1911 and thereafter it became a local tourist attraction. These excursions ended abruptly in

1963, however, when basalt mining in the area surrounding the tower caused it to collapse.

References

Bieberbach, Ludwig. 1938. *Carl Friedrich Gauß. Ein deutsches Gelehrtenleben*. Berlin: Keil.

Biermann, Kurt-R., ed. 1990. *Carl Friedrich Gauss - Der Fürst der Mathematiker in Briefen und Gesprächen*. Berlin: Urania-Verlag.

Bjerknes, C.A. 1885. *Niels Henrik Abel. En skildring af hans Liv og vitenskapelige Virksomhed*. Stockholm, 1880; French trans.

Breitenberger, Ernst. 1984. Gauss's Geodesy and the Axiom of Parallels. *Archive for History of Exact Sciences* 31: 273–289.

Cajori, Florian. 1912. Notes on Gauss and his American Descendants. *Popular Science Monthly* 81: 105–115.

Gray, Jeremy. 2006. Gauss and Non-Euclidean Geometry. In *Non-Euclidean Geometries: Janós Bolyai Memorial Volume*, ed. András Prékopa and Emil Molnár, 60–80. New York: Springer.

Hansteen, Christian. 1862. Niels Henrik Abel, *Illustreret Nyhedsblad*, 2 March/9 March, 1862.

Hayes, Brian. 2006. Gauss's Day of Reckoning. *American Scientist* 94(3): 200–205.

Hinz, A.M., S. Klavžar, U. Milutinović, and C. Petr. 2013. *The Tower of Hanoi – Myths and Maths*. Birkhäuser: Basel.

Kehlmann, Daniel. 2006. *Measuring the World*. Trans. Carol Brown Janeway. New York: Pantheon Books.

Miller, Arthur. 1972. The Myth of Gauss' Experiment on the Euclidean Nature of Physical Space. *Isis* 63: 345–348.

Ore, Oystein. 2008. *Niels Henrik Abel: Mathematician Extraordinary*. Providence: AMS Chelsea Publishing.

Reich, Karin. 2012. Wolfgang Sartorius von Waltershausen (1809–1876), *Wolfgang Sartorius von Waltershausen, Gauß zum Gedächtnis*. K. Reich, Hrsg., Leipzig: Edition am Gutenbergplatz Leipzig.

Sartorius von Waltershausen, Wolfgang. 1966. Wolfgang: *Gauß zum Gedächtnis,* Leipzig: S. Hirzel, 1856; *Carl Friedrich Gauss: A Memorial*. Trans. Helen Worthington Gauss. Colorado Springs.

Scholz, Erhard. 2004. C. F. Gauß' Präzisionsmessungen terrestrischer Dreiecke und seine Überlegungen zur empirischen Fundierung der Geometrie in den 1820er Jahren. In *Form, Zahl, Ordnung. Studien zur Wissenschafts- und Technikgeschichte. Ivo Schneider zum 65. Geburtstag*, ed. Folkerts Menso, Hashagen Ulf, and Seising Rudolf, 355–380. Stuttgart: Franz Steiner Verlag. http://arxiv.org/math.HO/0409578.

Stern, Moritz Abraham. 1877. *Denkrede auf Carl Friedrich Gauss zur Feier seines hundertjährigen Geburtstages*. Kästner: Göttingen.

Stubhaug, Arild. 2000. *Niels Henrik Abel and his Times*. New York: Springer.

van der Waerden, Bartel L. 1974. Comment II. *Isis* 65: 85.

3 Gauss, Dirichlet, and the Law of Biquadratic Reciprocity

(*Mathematical Intelligencer* 10(2)(1988): 13–25)

Gauss and Dirichlet are two of the most influential figures in the history of number theory. Gauss's monumental *Disquisitiones Arithmeticae,* first published in 1801, synthesized many earlier results and served as a point of departure for the modern approach to the subject (Goldstein et al. 2007). The three principal sections of the book were devoted to the theory of congruences (where Gauss introduced the still standard notation $a \equiv b(\text{mod } m)$), the classical subject of quadratic forms that had been studied by Fermat, Euler, Lagrange, and even Diophantos, and the theory of cyclotomic equations. Although the appearance of this masterpiece did much to establish Gauss's early mathematical fame, the sheer novelty of the work together with its rigidly formal exposition made it difficult for all but the greatest of his contemporaries to appreciate it fully. One of those who did was Lagrange, the aged giant who, along with Euler, stood at the pinnacle of eighteenth-century mathematics. In 1804, he wrote the young protégé of the Duke of Brunswick, "With your *Disquisitiones* you have at once arrayed yourself among the mathematicians of the first rank, and I see that your last section on cyclotomic equations contains the most beautiful analytic discoveries that have been made for a long time." (Wussing 1974, 32) (Fig. 3.1).

Another mathematician who was deeply influenced by Gauss's *Disquisitiones* was Johann Peter Gustav Lejeune Dirichlet, who was born in 1805 in the town of Düren near Aachen in the Rhineland. Although he spent his formative years in Paris studying mathematics with some of the leading French mathematicians of the day, nothing attracted Dirichlet's interest so much as did Gauss's *magnum opus.* In the words of Ernst Eduard Kummer:

> This [work] exercised a much more significant influence on his whole mathematical education and development than his other Paris studies. Rather than having merely read through it once or even several times, his whole life long, over and again, he never stopped repeatedly studying the wealth of deep mathematical thoughts it contained. That is why it was never placed on his bookshelf, but rather always lay on the table at which he was working. One can well imagine the exertion it must have cost him in analyzing this extraordinary work, considering that more than 20 years after it had appeared there was still no one alive at that time who had studied all of it and understood it completely. Even Legendre, who dedicated a large part of his life to higher arithmetic, had to confess in the second edition of his *Thdorie des nombres* that he would have liked to have enriched it with Gauss's results, but the methods of this author, being so peculiar, made it impossible to do so without the greatest digressions or without merely assuming the role of a translator. Dirichlet was the first not only to understand this work completely, but also to have made it accessible to others, in that he made its rigid methods, behind which the deepest thoughts lay hidden, fluid and transparent, and replaced many of the main points by simpler more genetic ones, without compromising the complete rigor of the proofs in the slightest. He was also the first to go beyond it and reveal the rich treasures and still deeper secrets of number theory (Kummer 1897, 315–316).

One of the founders of analytic number theory, Dirichlet presented his first paper in this field in 1837. This contained the famous proof that there are infinitely many primes in every arithmetic progression $an + b$, where a and b are relatively prime. The following year he published the first portion of a two-part article entitled "Recherches sur diverses applications de l'analyse infinitésimale a la théorie des nombres," which introduced what have since become known as Dirichlet series. Although he published relatively little in his lifetime, Dirichlet, unlike Gauss, placed a high value not only on the discovery of new ideas but on their dissemination as well. In the course of his career at Berlin and Göttingen, his students included some of the outstanding figures of the next generation – Eisenstein, Kronecker, Dedekind, and Riemann. His *Vorlesungen über Zahlentheorie,* which only became accessible to the mathematical public in 1863 through the efforts of Richard Dedekind, also exerted a profound influence on number theory and algebra in the late nineteenth century.

In view of these circumstances, the intellectual and personal relationship of Gauss and Dirichlet would seem to possess an importance all its own for the history of mathematics. Nevertheless, this fascinating and significant chapter in nineteenth-century mathematics has never been adequately dealt with by those who have written about either man. In part, this is because, while there is a plethora of literature

D.E. Rowe, *A Richer Picture of Mathematics*, https://doi.org/10.1007/978-3-319-67819-1_3

Fig. 3.1 A lithograph of Gauss made by Siegrfried Detlev Bendixen, published in *Astronomische Nachrichten* in 1828, the year he met Alexander von Humboldt in Berlin.

on Gauss, no one has yet produced even a small biography of Dirichlet. In the English language, practically nothing has been written about his life, work, and extraordinary influence.[1] The following contribution to this subject concentrates on the years 1825–1831, a formative period in Dirichlet's life when he and Gauss were working on closely related problems in the theory of biquadratic residues. Before turning to this story directly, however, a few introductory remarks regarding Gauss's prior arithmetical researches are called for here.

The second entry in Gauss's *Tagebuch* (Gray 1984), dated April 8, 1796, indicates that he had already begun work on quadratic reciprocity during his first year as a student at Göttingen. By June of that year he had obtained two entirely independent proofs of what he called the *Theorema Fundamentale.* Euler had already known this "fundamental theorem," which was later rediscovered by Legendre, but neither of them had given a complete demonstration of its validity (Edwards 1983). For any prime p, an integer a is said to be a quadratic residue mod p if the congruence $x^2 \equiv a \pmod{p}$ has a solution. Gauss's "fundamental theorem," better known as the law of quadratic reciprocity, is then stated most succinctly by means of the Legendre symbol:

$\left(\frac{p}{q}\right) = \begin{matrix} 1 \\ -1 \end{matrix}$, depending on whether p is a quadratic residue mod q, or not.

If p and q are distinct odd primes, then the following reciprocal relationship holds between them:

If p, q are not both $\equiv 3 \pmod{4}$, then $\left(\frac{p}{q}\right) = \left(\frac{q}{p}\right)$, and if $p \equiv q \equiv 3 \pmod{4}$, then $\left(\frac{p}{q}\right) = -\left(\frac{q}{p}\right)$. This relationship may be reformulated as follows:

$$\left(\frac{p}{q}\right) = (-1)^{\frac{p-1}{2}\cdot\frac{q-1}{2}} \left(\frac{q}{p}\right).$$

The prime 2 can also be dealt with as a special case in this theory, since:

$$\left(\frac{2}{p}\right) = (-1)^{\frac{p^2-1}{8}}.$$

The two proofs Gauss discovered in 1796 were first presented to the world in the *Disquisitiones Arithmeticae* (Gauss 1996). The first of these (see § 133) is generally considered unreadable; the second, however, is clearer (see § 262) and makes use of the theory of quadratic forms developed earlier in the book. In 1808 Gauss published two new proofs, and nine years later two more – clearly this was one of his favorite theorems, and he returned to it often for inspiration. In fact, this result prompted him to search for an analogue in the theory of cubic and biquadratic residues. During February of 1807 he made the following entries in his *Tagebuch*: "Began the theory of cubic and biquadratic residues. . . . Further worked out and completed. Proofs thereto are still wanted . . .". Out of this work arose a new proof (Gauss's sixth) of the "fundamental theorem," as he recorded on May 6: "We have discovered a completely new proof of the fundamental theorem based on totally elementary principles."

Following the death of his first wife, whom he loved dearly, on October 11, 1809, Gauss recorded nothing in his diary for more than two years. The next entry, dated February 29, 1812, began: "The preceding catalogue interrupted by misfortunate times resumed a second time . . . ." During the six year period since recording his initial work on higher degree residues, Gauss only once made mention of further progress on this subject. In 1809 he noted, "the theorem for the cubic residue 3 proved with elegant special methods. . ." From this one may surmise that he was still groping with special cases at this time and had not yet proved a general relation for either cubic or biquadratic residues. Of course, by now the number of entries in Gauss's diary had become sparse, so it is dangerous to make conjectures on this basis alone. Indeed, there is nothing in the diary that would prepare us for the following dramatic entry of October 23, 1813:

[1]In the meantime, Jürgen Elstrodt has helped to fill this gap to some extent with Elstrodt (2007), which is available as a pdf online.

Fig. 3.2 The Göttingen Astronomical Observatory where Gauss lived and worked from 1816 to 1855 (Special Exhibition, Niedersächsische Staats- und Universitätsbibliothek Göttingen, 2005. http://webdoc.sub.gwdg.de/ebook/e/2005/gausscd/html/hauptmenue.htm).

> The foundation of the general theory of biquadratic residues, which we have sought for with utmost effort for almost seven years but always unsuccessfully, at last happily discovered the same day on which our son is born. . . . This is the most subtle of all that we have ever accomplished at any time. It is scarcely worthwhile to intermingle it with mention of certain simplifications pertaining to the calculation of parabolic orbits.

These remarks were made just before Gauss's final diary entry, recorded July 9, 1814, when he wrote: "I have made by induction the most important observation that connects the theory of biquadratic residues most elegantly with the lemniscatic functions." Yet, as with so many other pioneering researches, Gauss would publish nothing on biquadratic residues for many years to come. One may speculate on a number of plausible reasons for this; the demands of his astronomical and geodetic studies may have played a role here, for example, or his general reluctance to publish anything that was not in polished form. Perhaps an even more telling reason for Gauss's failure to present this work to the mathematical public was his instinctive conservatism when it came to matters of potential controversy. In this particular case, Gauss's work on biquadratic residues required a bold new approach to number theory in which the so-called Gaussian integers were introduced for the first time. As we shall have occasion to see later, Gauss was well aware that this represented a radical break with the number theory of the past, and he took pains to argue that this daring step was both natural and necessary. His defensive posture in this matter brings to mind how he shunned all controversy regarding the status of Euclidean geometry. He once wrote his friend, the renowned astronomer Bessel, that he was reluctant to publish in this field because his views were bound to evoke the outcries of certain "Boeotians" (Wussing 1974, 57). Gauss kept his word, and even after the younger Bolyai presented arguments for the existence of non-Euclidean geometry in 1831, he maintained his stony silence on this subject (Fig. 3.2).

Gauss had been aware of Dirichlet's work in number theory since 1826, when the young man sent him a copy of an earlier paper asking him if he would read it and then write a letter expressing his judgment of its contents to someone in Berlin. Over the course of the preceding four years Dirichlet had been studying under Fourier and Poisson at the Collège de France and the Faculté des Sciences. By the summer of 1825 Alexander von Humboldt became aware of this young talent and was preparing to use his influence to create a position for him within the Prussian university system. Like Dirichlet, Humboldt was about to return to Prussia after a prolonged stay in Paris, and he was well informed as to the high esteem in which the young Rhinelander was held by such leading French mathematicians as Fourier, Lacroix, and

Poisson. Still, nothing would carry as much weight with the Berlin *Kulturministerium* as a letter of recommendation from Gauss, to whom Humboldt wrote the following lines on May 21, 1826:

> As you know, I cannot pretend to have a serious opinion when it comes to the higher regions of mathematics, but I do know through the great mathematicians that Paris possesses, and especially through my oldest friends Fourier and Poisson, that Herr Dirichlet has by nature the most brilliant talent, that he is progressing along the best Eulerian paths, and that one day Prussia will have in him (he is barely 21 years old!) an outstanding professor and academician. Grant my young friend, whose fortune interests me dearly, the protection of your great name (Biermann 1959, 13).

One week later Dirichlet sent Gauss the above-mentioned letter, along with an offprint of his *Mémoire sur l'impossibilité de quelques équations indeterminées du cinquième degré*. This article had been accepted for publication by the French Academy in July 1825 after receiving a favorable referee's report from Legendre and Lacroix. It dealt with Diophantine equations of the form $x^5 + y^5 = Az^5$. Shortly after it appeared Legendre applied its techniques to prove that $x^5 + y^5 \neq z^5$, which was one of the first breakthroughs on Fermat's last theorem since Euler resolved the case $n = 3$ (Fermat himself proved the case $n = 4$). Even by the conventions of the day, Dirichlet's letter was written in a tone of exaggerated humility and almost embarrassing self-effacement. Of course his youth must be taken into account here, as well as his awareness of Gauss's reputation for being aloof and difficult.[2] Nevertheless, this letter gives a distinct impression of how truly daunted he was by Gauss's stature as a mathematician:

> At the kind recommendation of Baron von Humboldt I take the liberty of flattering myself with the hope that you will read and judge this first work of a young German with kind indulgence. If Your Honor should find it not altogether unworthy of your attention, I would dare to humbly beseech you with the request for permission to write you on occasion and to ask you for some indications that would guide my further scientific efforts. I would regard this permission as the greatest fortune, as the love with which I pursue indeterminate analysis makes me wish for nothing with more longing than that the author of the immortal *Disquisitiones Arithmeticae* might take some part in my efforts. Since I have busied myself primarily with higher arithmetic, I have completely followed my inclination without duly considering how little I may dare to hope of accomplishing anything substantial in this difficult area of mathematics given my restricted aptitude. (Lejeune Dirichlet 1897, 373–374).

Six weeks passed before Gauss got around to writing his former student Johann Friedrich Encke, an astronomer in Berlin who was well connected with the Prussian Ministry of Culture. The "Prince of Mathematicians" was a reclusive sort of figure, and it was not often in the course of his career that he went out of his way to lend a helping hand to younger talents. In this case, however, when Dirichlet's future as a mathematician largely lay in his hands, he was unusually lavish in his praise of him. A few days after receiving Gauss's letter, Encke wrote to the Ministry urging that Prussia create a position for Dirichlet before the French won him for themselves. Speaking of Gauss's high regard for Dirichlet's work, Encke added: "In my eyes what gives this opinion of Gauss such high value is that this unique man has always sharply distinguished between works that are worthy through their diligence and those of true genius, and that so long as I have had the fortune to know him, he has never spoken of anyone with such warmth, however respectfully he may have spoken of the accomplishments of others (Biermann 1959, 14).

Two months later Dirichlet received a letter from Gauss, who excused himself for not having written earlier by saying that he had wanted to await further news from Berlin. Gauss wrote with uncharacteristic charm, the almost fatherly tone of his letter suggesting that Dirichlet's young talent may well have brought back memories of his own youth. He assured him that the prospects for an appointment did indeed look promising, and then went on to express his pleasure that Dirichlet had taken an interest in number theory while recounting his own love for the subject:

> It is all the more pleasing to me that you have a great attachment to that part of mathematics that has always been my favorite field of study, however seldom I have pursued it. I dearly wish you a situation in which you will have as much control over your time and the choice of your work as possible. Immediately after the appearance of my *Disquisitiones,* I myself was very much hindered by other business, and later by my external circumstances, from following my inclinations to the degree that I would have wished. (Lejeune Dirichlet 1897, 375)

He went on to say that he had given up his original plan of publishing a sequel to the *Disquisitiones* and instead would content himself with the publication of an occasional memoir on number theory. His plans called for a three-part study on biquadratic residues, the first of which would appear shortly. "The main materials for the rest," he added, "as well as for the related theory of cubic residues are essentially finished, although little of it has yet been written up adequately" (Lejeune Dirichlet 1897, 375).

By early November *Kultusminister* Altenstein informed Alexander von Humboldt that the state would agree to provide Dirichlet with a minimal salary of 400 *Taler* while he habilitated at Breslau. Humboldt pressed him for 600–700 *Taler* and also to have Dirichlet appointed as *außerordentlicher Professor*. This plea, however, was spurned by the Ministry. All the same, Dirichlet was delighted with this opportunity to commence his career as a mathematician,

[2] A letter written by Dirichlet to his mother expressed surprise that he had been received very cordially by Gauss in Göttingen. After meeting him, Dirichlet apparently had a much more favorable impression of the "Prince of Mathematicians" than he had had beforehand; see Kummer (1897, 341).

and, having received assurances of continued support from Humboldt, he had every reason to be optimistic about his future.

By this time the winter semester was already a month old, so Dirichlet requested permission to stay at home for the duration, proposing that he take up his duties at Breslau in the spring. This was granted, but shortly thereafter Altenstein learned that during his stay in Paris young Gustav had been a private tutor to the children of General Maximilien Foy, the leader of the liberal opposition party in the Chamber of Deputies. By the standards of the day, Altenstein was a reasonably liberal politician, but as K.-R. Biermann points out this was the era when the Carlsbad decrees were being carried out, a period when the "Metternichian reaction made it much too dangerous for Altenstein to place an unknown young man at a Prussian university simply on Humboldt's recommendation (Biermann 1959, 12). He therefore had the Minister of the Interior investigate Dirichlet's character. Luckily his agents in Paris turned up nothing damaging, and subsequently he was allowed to teach at Breslau after all. In the meantime Dirichlet had been awarded a doctorate *honoris causa* by the Bonn faculty, and sometime during the middle of March he began the long journey from his home in the Rhineland to Breslau.

Along the way Dirichlet made stops in Berlin and in Göttingen, where he met Gauss for the first time. In all likelihood it was during the course of this visit that he first became aware of an announcement Gauss had published about two years earlier in the *Göttingische Gelehrte Anzeigen.* This notice conveyed the contents of *Theoria Residuorum Biquadraticorum, Commentatio prima,* the first article alluded to by Gauss in the letter cited above. Its main results were concerned with the biquadratic character of the number 2, which, as remarked earlier, forms a special case in the theory of quadratic residues. Now, in general, for a to be a biquadratic residue mod p (i.e., for $x^4 \equiv a (\text{mod } p)$ to be solvable) it is clearly necessary that a also be a quadratic residue mod p. In the *Commentatio prima* Gauss offered criteria for determining whether or not $a = -1, 2, -2$ will be a biquadratic residue mod p for some given prime p. These, he remarks, are special cases "that allow of being worked out without too much machinery and can serve as preparation for the general theory to be given in the future".[3] The case $a = -1$ was already dealt with in the *Disquisitiones Arithmeticae,* where Gauss showed that -1 is a biquadratic residue mod p if and only if p is of the form $8n + 1$ (thus when $p = 8n + 5$ it is a quadratic but not a biquadratic residue).

For the cases $a = \pm 2$, however, Gauss gave two distinct criteria for deciding this question. Since in the special case $a = 2$, p an odd prime, 2 is a quadratic residue mod p if and only if p is of the form $8k + 1$ or $8k + 7$, it follows that for p of the form $8k + 3$ or $8k + 5$, 2 is a biquadratic *nonresidue* mod p. Gauss, moreover, pointed out that it suffices to consider primes p of the form $4n + 1$, which means that there is no loss of generality in assuming p to be of the form $8k + 1$. Gauss's first criterion is obtained by writing $p = g^2 + 2h^2$, a representation that is both possible and unique. Then -2 is a biquadratic residue mod p if and only if g is of the form $8n + 1$ or $8n + 7$. The second criterion uniquely represents p as $e^2 + f^2$, where e is odd and f even. Since $p = 8k + 1, f/2$ must be even. Then ± 2 is a biquadratic residue mod p if $f = 8m$, but a nonresidue if $f = 8m + 4$.

Whether directly from its author or through some other possible source, it was around this time that Dirichlet became aware of Gauss's announcement, and from this point forth he could not take his mind off it. How he struggled with these ideas and what resulted therefrom are best conveyed by the following excerpts from a letter (containing parts of another letter) Dirichlet wrote to his mother about seven months after he met Gauss in Göttingen. To the best of my knowledge, this is the first time this letter has been referred to in print.[4] The setting is as follows: his first semester as a *Dozent* having come to an end, Dirichlet had just returned to Breslau after spending a brief vacation in Dresden.

Breslau, October 29, 1827

Dearest mother!

Although I have already been back from Dresden 14 days, it was impossible until now for me to answer your two letters of which the first, dated September 11, only arrived the day before my return. You will find the reason in the following passage from a letter to Encke, which Professor Schoeller took with him to Berlin a few days ago. I quote this passage for you not so much to excuse myself as to enable you to judge for yourself whether the oft-mentioned love is serious or merely a jest, whereby it will now be altogether clear to you that (as you yourself said) one who is in love is incapable of proper work[5]:

> "The signs of good-will with which you have favored me give me the courage to trouble vou with a new request. I have been busy for some time with arithmetical investigations that were prompted by an article in the *Göttingische Gelehrte Anzeigen.* In this article *(Gelehrt. Anz.,* April 11, 1825) a till now un-

[3] *Göttingische Gelehrte Anzeigen,* 59. St., April 11, 1825; republished in Gauss (1973, 165–168).

[4] A transcription of the original letter can be found in *Nachlass Klein XXIA,* Niedersächsische Staats- und Universitätsbibliothek Göttingen. The author wishes to thank this institution for permission to publish it here in translation.

[5] These remarks and much of what follows reveal how eager Dirichlet was to persuade his mother that mathematics was a viable career choice for him. Like so many parents of famous mathematicians, Dirichlet's mother and father wanted him to take up something practical and suggested that he study law at a German university. See Kummer (1897, 314).

known work by our Gauss is announced that was submitted to the Society there [die Gesellschaft der Wissenschaften zu Göttingen]. It concerns biquadratic residues and its two main results [the criteria mentioned above] are communicated *without proofs*. The elegance of these theorems set forth in me the urge to prove them, and after a number of vain efforts I succeeded in doing so a short time ago. My proofs appear to be completely different from those of the famous author of the *Disquisitiones arithmeticae*, for his, as they are described in the article, require a series of subtle preliminary investigations, whereas mine consist of a simple application and combination of results that have long been known. The method I have used, moreover, has led me to the discovery and proof of a *large* number of new theorems, some in the theory of *biquadratic* residues, some in related branches of higher arithmetic. I am now eagerly pursuing these things further and hope through my efforts to bring out a worthy piece of work that will attract the attention of mathematicians.[6] As, however, some time must pass before this work is completed, I wish to show that I am already in possession of a suitable method for the treatment of these matters. I have, therefore, written up the main features of this method and take the liberty of sending you this sketch by separate post in a sealed packet with the request that you kindly deposit it with your Academy, etc."

Regarding the above mentioned investigations I experienced a most peculiar fortune. Already in the course of the summer I had made a number of steps that brought me nearer to the goal I sought. Still, there always loomed one difficulty that needed to be overcome before I had the proof of Gauss's theorems. I concentrated on this matter incessantly, not only during my trip to the mountains but also in Dresden, and yet without gaining any real insight (Fig. 3.3). One evening, as I wandered alone on the Elbe bridge (which, by the way, occurred only seldom, as I enjoyed too much being in the company of such kindly people as the Remer family) I had a few ideas that appeared to put me within grasp of the so long and zealously searched-for results. On the gorgeous Brühl terrace I let my thoughts go for several hours (till around ten o'clock), but still I could not see my way to the end of the matter (probably because nonmathematical ideas kept mixing together with the mathematical ones). With very weak hopes I went to bed and was extremely restless until around one o'clock when I finally fell asleep. But then I woke up again around four o'clock, and even awoke the health official, who slept in the same room, by hollering: "I've found it" ["Ich habe es gefunden"]. It took but a moment for me to get up, turn on the light, and, pen in hand, convince myself of its correctness. After this my investigations expanded everyday, and 14 days later I was in a position to send Herr Encke my six-page sketch. I have every reason to expect that this work will accomplish a good deal for my promotion, since Gauss announced his results, which indeed do not contain nearly so much as mine, with a certain pomp. I will certainly not neglect to mention in a tactful manner what he said about the difficulties that must be overcome in carrying out the proofs . . .[7]

The history of mathematics is filled with tales of remarkable discoveries: Poincaré saw the connection between non-Euclidean geometry and the Fuchsian functions in a moment's flash, Klein his *Grenzkreistheorem* in much the same way, and after weeks of intense deliberation Lie discovered the key property of his line-to-sphere transformation while lying awake in bed one morning.[8] On July 10, 1796, Gauss had reason to remember the legendary reaction of Archimedes after discovering his hydrostatic law during a visit to the public baths.[9] It was on that day that Gauss wrote in his diary: "EUREKA. num $= \Delta + \Delta + \Delta$." Evidently he had found a proof that every number can be written as the sum of three triangular numbers, a conjecture first made by Fermat. Still for sheer drama, this "Eureka!" story, culminating with Dirichlet's impetuous cry "Ich habe es gefunden", is practically in a class by itself.

It may at first seem hard to believe that an authentic episode like this, involving a mathematician of Dirichlet's stature, could have become totally forgotten over the years. Yet I doubt that this phenomenon will surprise anyone who has spent a good bit of time sifting through the letters and unpublished papers of some great mathematicians of the nineteenth century. Often such papers are untapped mines of information providing insights into their lives and work that simply cannot be found in published sources. In the case of this particular letter, my translation is based on a typewritten transcription of the original (which is included among Felix Klein's posthumous papers). A portion of the letter also appears in the protocol book for Klein's seminar of 1909–1910 on mathematics and psychology. These notes indicate that the letter was presented to the seminar by the philosopher Leonard Nelson, who was a great-grandson of Dirichlet through his mother's side of the family. Apparently the original letter was at one time in the personal possession of the Nelson family, and it is possible that it is among the

[6]A marginal note reads: "La modestie est. une bien belle chose."

[7]A marginal note further reveals that Dirichlet was eager to leave Breslau at the earliest opportunity: "As much as I am satisfied with my present abode so far as the company is concerned, there is so little here to offer from the scientific side that I will mobilize all the forces I can to bring about mv transfer to Berlin." ("So sehr ich auch mit meinem hiesigen Aufenthalt, insofern vom Umgang die Rede ist, zufrieden bin, so ist doch in wissenschaftlicher Hinsicht hier so wenig zu machen, daß ich alles aufbieten werde, um meine Versetzung nach Berlin zu bewerkstelligen.")

[8]The circumstances surrounding Lie's discovery are described by Felix Klein in Klein (1921, 97); Klein's discovery of the *Grenzkreistheorem* is discussed in Klein (1926, 379). The psychological implications of Poincaré's discoveries are analyzed in Hadamard (1954, 11–15).

[9]The famous story of Archimedes running from the bath comes from Vitruvius's *De architectura*. See Thomas (1968, 36–39).

Fig. 3.3 A view of Dresden not long after the time when Dirichlet stayed there.

collection of letters (now located at the University Library in Kassel) that Dirichlet wrote to his mother.[10]

After discovering this letter, I was naturally curious about the surrounding circumstances and turned to Gauss's announcement in the *Göttingische Gelehrte Anzeigen* to see what it was that may have sparked Dirichlet's interest in the first place. The first sentence confirmed Dirichlet's assertion that Gauss had already submitted his memoir to the Scientific Society on April 5, 1825, although the *Commentatio prima* only appeared in print three years later. Another statement that appears in the introduction must have caught Dirichlet's attention: " . . . the present work is in no way intended as an exhaustive treatment of this rich topic. On the contrary, the development of the general theory, which requires an altogether special extension of the field of higher arithmetic, remains for the most part reserved for a future continuation" (Gauss 1973, 165–166). The "certain pomp" that Dirichlet referred to at the end of his letter to his mother appears in the following passage of Gauss's announcement:

> . . . as so often in higher arithmetic it is not so much the simplicity and beauty of the theorems as the difficulty of the proofs that distinguishes them so remarkably. As soon as one is prompted to conjecture the existence of a connection between the behavior of the number - 2 and the two decompositions of the number p presented here, it is extremely simple to actually discover this connection through induction. Yet already with the first criterion it is not altogether easy to carry out the proof, and with the second the matter lies much deeper, as it is intimately bound with other subtle preliminary investigations, which for their part lead to a remarkable extension of cyclotomic theory. As has often been remarked, this wonderful chain of truths is primarily what gives higher arithmetic its so special attraction. Naturally these proofs themselves cannot be sketched here, and must be read in the monograph itself (*ibid.*).

Dirichlet's simplified proofs of Gauss's criteria regarding the biquadratic character of the number 2 appeared in his article "Recherches sur les diviseurs premier d'une classe de formules du quatrième degré".[11] Regarding this work, Bessel wrote Alexander von Humboldt on April 14, 1828, that no one would have noticed the error had the name Lagrange stood at the top rather than Dirichlet's (Kummer 1897, 323). In addition to to being simpler, Dirichlet's methods also turned out to be more powerful than those employed by Gauss. Not only did he succeed in handling the case $p = 2$, but he also answered the same question for an arbitrary prime p, thereby providing a complete analysis of those primes q for which p is a biquadratic residue mod q. Thus it appeared he was only a step away from establishing a reciprocity law for biquadratic residues analogous to Gauss's "fundamental theorem" in the theory of quadratic residues. It was probably only after Gauss announced his results on this subject three years later that Dirichlet began to realize why his own work

[10]Gert Schubring later confirmed that the original letter, translated here from the transcription, is indeed among the large collection housed in Kassel. See the introduction to Part I above.

[11]Dirichlet's article appeared in Crelle's *Journal für die reine und angewandte Mathematik*, 3(1828), 35–69. H. M. Edwards pointed out to me that the results on the biquadratic character of 2 mod p were already known to Euler. His work on this subject, however, was only published posthumously, and not until 1848!

on biquadratic reciprocity had reached an impasse at this point. Shortly after his article appeared, Dirichlet sent a copy to Gauss along with some remarks as to how he derived the results presented therein:

> . . . As soon as I had familiarized myself somewhat with my professional duties here, I began to busy myself with these things. My exertions remained for some time fruitless, until I succeeded in deriving your second criterion, whose proof by way of the path you have chosen appears to require a number of preliminary investigations, directly from the first. After this fortunate success my work again came to a standstill, and for a long time I could not find a suitable means for establishing the first criterion. Finally, around the beginning of winter I came upon the proof presented in the enclosed article. It is so simple that it seems hard to comprehend that one would not grasp it immediately, just as soon as the proposition that requires proving became known. The continuation of my investigations, which are only partly contained in the article, has led me to a large number of results that one would certainly not have conjectured stood in any connection whatsoever with these matters. In the course of this work, I have encountered numerous remarkable examples of the often wonderful interconnections of arithmetical truths, which you regard as the main reason for the attraction that the investigations of indeterminate analysis afford us (Lejeune Dirichlet 1897, 376–78).

From this letter there would appear to be little doubt that the discovery Dirichlet made while in Dresden was related to the proof of Gauss's first criterion. Apparently, this was the discovery that gave him insight into a whole series of problems that animated his work during this formative period of his career. Surely this matter deserves some attention by experts in the history of number theory, as it may well be an important clue to understanding Dirichlet's intellectual development.

In his letter to Gauss, Dirichlet also mentioned that he was still unable to obtain a copy of the *Commentatio prima,* despite numerous pleas to his bookdealer that he speed the order along. When Gauss wrote back seven weeks later, he mentioned that he would try to send an offprint of the article, but assumed that in the meantime it had arrived in Breslau and that Dirichlet was familiar with its contents. After expressing his pleasure with Dirichlet's latest effort, he added the following cryptic remarks to explain why his proofs differed so much from Dirichlet's own:

> I could have chosen a number of different forms of proof for the theorem there arising; it will not have escaped you, however, why I have preferred the one carried out here, namely primarily because the classification of 2 with respect to those modules for which it is a quadratic nonresidue (under *B* or *D*) must be regarded as an essentially integral part of the theorem for which most of the other forms of proof appear to be inapplicable. I have already had all of the material for this whole investigation in my possession for 23 years, except for the proof of the main theorem (to which that in the *Commentatio prima* is still not be counted), which I have had for about 14 years. I still hope and wish to be able to simplify the proofs of the latter somewhat, and plan to write approximately three works altogether on this subject. I have already made a beginning with the composition of the second, and hope to have it completed fairly soon, so long as the geodetic surveying that I have recently been assigned to again does not cause some delay (Lejeune Dirichlet 1897, 378–80).

Four months later Gauss and Dirichlet met in Berlin, where Alexander von Humboldt had invited them to attend a meeting of the German Association of Scientists and Physicians. Somewhat earlier Dirichlet had been granted a leave of absence from Breslau; he never went back. For the next twenty-seven years he remained in Berlin, where he taught at the military school and as a *Privatdozent* at the University. In 1831, he was made an *außerordentlicher* and finally in 1839 he was promoted to *ordentlicher Professor.* After Gauss's death in 1855 Dirichlet left Berlin for Göttingen, where he spent the last three and one-half years of his life as Gauss's successor.

It would appear unlikely that Dirichlet learned anything more about the nature of Gauss's "main theorem" when he saw him again in Berlin. Considering, moreover, that the *Commentatio prima* gave only the barest of hints as to what lay behind it, one must assume that Dirichlet (as well as other specialists) looked forward with great anticipation to the second of the three installments Gauss had planned. Both Dirichlet and Jacobi continued their investigations of biquadratic residues during these intervening years, and they must have grown rather impatient by the time Gauss's *Commentatio secunda* finally appeared in 1832. As before, it was preceded by a notice in the *Göttingische Gelehrte Anzeigen* describing the work's contents. This announcement was dated April 23, 1831, and in it Gauss indicated that the proofs for his earlier results on the congruence $x^4 \equiv k \pmod{p}$ for $k = \pm 2$ and similar findings for k $= -3, 5, -7$, etc., could not be extended beyond a certain point. He then went on to make his first public pronouncement regarding the status of the so-called Gaussian integers and their place in the theory of biquadratic residues:

> One soon recognizes after this, that one can only break into this rich area of higher arithmetic by completely new paths. The author had already given an indication in the first work that to do so a remarkable extension of the whole field of higher arithmetic was essentially necessary, but without at the time explaining more closely what this consisted of: the present work has the intention of bringing this matter to light. For the true foundation of the theory of biquadratic residues, this is none other than the extension of the field of higher arithmetic, which has otherwise only been extended to the real whole numbers, to also include the imaginaries as well, for these must be granted exactly the same status as the others. As soon as this has once been observed, that theory appears in a completely new light and its results take on a highly surprising simplicity (Gauss 1973, 171).

Gauss also mentioned that this idea had already been familiar to him for many years in another context. Nevertheless, he knew it would appear like a radical step to some, and he was probably hoping to ward off certain "Boeotians" when he wrote the following defense of this bold leap forward:

> Regarding the reality of negative numbers the situation has been clear for some time. It is only the imaginaries – formerly and sometimes still called *impossibles* [unmögliche] – which standing opposite the real numbers still remain less accepted. Merely tolerated, they thus appear more like an in itself contentless sign-game to which one categorically denies any intelligible substrate, without however wanting to disdain the rich tribute that this sign-game affords among the wealth of relationships between the real numbers. For many years the author has studied this highly important part of mathematics from a different standpoint, whereby the imaginary numbers can be used in connection with an object just as well as the negatives. Until now, however, there has been no appropriate occasion for publicly stating this in a definite manner, although the attentive reader can easily find traces of it in the work of 1799 on equations and in the *Preisschrift* on transformations of surfaces (Gauss 1973, 173).

Gauss then went on to argue that much of the confusion surrounding imaginary numbers was merely a matter of terminology. He pointed out that had the quantities $+1$, -1, and $\sqrt{-1}$ been named direct, inverse, and lateral units – an allusion to their geometric properties under multiplication – the metaphysical doubts regarding the status of imaginaries never would have arisen.

Gauss's *Theoria Residuorum Biquadraticorum, Commentatio secunda* turned out to be his last major contribution to the theory of numbers. This paper was also one of the first to introduce and develop arithmetic in an algebraic number field. Much of it was devoted to establishing analogues of familiar number-theoretic ideas or techniques for the Gaussian integers. Thus, for example, § 37 gives the analogue to the fundamental theorem-of arithmetic, showing that every Gaussian integer can be uniquely factored into complex primes. § 51 then presents a version of the Little Theorem of Fermat by utilizing a generalized Euler ø-function. The ring Z[i] of Gaussian integers contains four units – l, –1, i, – i; numbers which differ by a unit multiple are termed associates. This theory is not a simple extension of ordinary arithmetic, since some numbers, like 3, remain prime, whereas others do not: e.g., $5 = (1 + 2i)(1 - 2i)$.

Gauss introduced the norm $N(a + bi) = a^2 + b^2$ and classified the complex prime numbers according to three classes: real primes of the form $4n + 3$, non-real primes whose norm is a prime of the form $4n + 1$, and the special prime $1 + i$ that appears in the factorization of $2 = (1 + i)(1 - i)$. The fact that 2 is not a prime led Gauss to define an odd number to be one that is not divisible by $1 + i$. To simplify the statements of certain results, he further introduced the notion of a primary number. A nonunit z is said to be primary if $z \equiv 1 \pmod{(1 + i)^3}$. In particular, primary numbers are odd, and if w is an arbitrary nonunit odd number then there exists a unique unit u such that uw is primary (for more, see e.g. (Ireland and Rosen 1982)).

Rather than basing his analysis of biquadratic residues on three classes, depending on whether p is a biquadratic residue mod q, a quadratic residue but not a biquadratic residue, or neither, Gauss introduced four classes corresponding to the four units in *Z[i]*. In the *Commentatio prima* he thus used four equivalence classes to analyze the biquadratic character of $z = -1$ and $z = 2$. For a given prime p not dividing z, he showed that these four classes *A, B, C,* and *D* were determined by four congruences, one of which always holds: $z^{p-1/4} \equiv 1, f, f^2, f^3 \pmod{p}$, where f satisfies $f^2 \equiv -1 \pmod{p}$. To generalize to the Gaussian integers, however, he had to discover the following relation: For any complex prime p not dividing z, there exists a $k = 0, 1, 2, 3$ satisfying the congruence $z^{N(p)-1/4} \equiv i^k \pmod{p}$. This enables one to define a generalized Legendre symbol $\left(\frac{z}{p}\right)_4 = i^k$, and by means of this the law of biquadratic reciprocity may then be stated in the following form[12]: If p and q are distinct primary primes, then

$$\left(\frac{p}{q}\right)_4 = (-1)^{\frac{N(p)-1}{4}\frac{N(q)-1}{4}}\left(\frac{q}{p}\right)_4.$$

Gauss's formulation of this remarkable theorem appears in § 67 of the *Commentatio secunda.* It differs somewhat from the version given above, but the latter formulation has the advantage of revealing more clearly the strong analogy between this result and the law of quadratic reciprocity. One of the principal aims of Gauss's third memoir was to present a proof of this theorem, which, in his words, "belongs among the most deeply-hidden truths of higher arithmetic." For reasons that remain unclear, he never published this final article. A sketch of a proof for the fundamental theorem of biquadratic reciprocity was found among his posthumous papers, and this was published 50 years after his death in Volume X of Gauss's *Werke* along with tables he had compiled for computing the first $(N(m) - 1)/4$ powers of primitive roots of the module m. The proof he outlined was based on ideas similar to those utilized in his sixth proof of the "fundamental theorem." Since it has been impossible to date these documents reliably, one cannot be sure whether this proof was original with Gauss or based on other sources. The first proof of the law of biquadratic reciprocity was published by Eisenstein in 1844, although Jacobi had presented a proof somewhat earlier in his lectures at Königsberg.[13]

In September 1838 Dirichlet again wrote to Gauss, forwarding him a copy of the article in which he proved that every arithmetic progression $a + bn$, with a and b relatively prime, contains infinitely many primes. In his letter Dirichlet remarked that he thought there might be some connection

---

[12]For background on this, see Collison (1977), Rieger (1957), and Bachmann (1911).

[13]Eisenstein presented two proofs of the law of biquadratic reciprocity, one in the 1844 paper "Lois de réciprocité" (Eisenstein 1975, 126–140), and another the same year in "Einfacher Beweis und Verallgemeinerung des Fundamentaltheorems für die biquadratischen Reste" (Eisenstein 1975, 141–163).

Fig. 3.4 The grave of Gustav and Rebecka Lejeune-Dirichlet in Göttingen.

between his methods and those alluded to in the closing remarks at the end of the *Disquisitiones,* wherein Gauss indicated that his findings could be used to shed light on a number of areas in analysis. These remarks naturally aroused Dirichlet's curiosity, since the analytic methods he was beginning to introduce in number theory were intimately related to discontinuous functions represented by trigonometric series, and these, as he emphasized to Gauss, were "still completely unexplained at the time the *Disquisitiones* appeared" (Lejeune Dirichlet 1897, 382). He then went on to say: "You will perhaps still remember that more than 10 years ago when I was in Breslau you informed me of a criterion for deciding the question raised at the end of your first work on biquadratic residues (for the case $p = 8n + 5$), mentioning that you had derived it by means of the results announced at the end of the *Disquisitiones Arithmeticae.*" And then Dirichlet proceeded to inform him about how this criterion could easily be derived from certain propositions he had discovered in the meantime.

Gauss's number-theoretic investigations remained a source of inspiration for Dirichlet throughout the remainder of his career. In fact, fate alone prevents us from reading about this inspiration firsthand. In the summer of 1858, while vacationing in Switzerland, Dirichlet was preparing a memorial speech on Gauss that he was scheduled to deliver before the Göttingen Scientific Society. It was during this stay that he suffered a nearly fatal heart attack. At first he slowly began to regain his health, but the following winter his wife Rebecca, the sister of composer Felix Mendelssohn-Bartholdy, passed away unexpectedly. Filled with grief, Dirichlet's condition quickly worsened, and he followed her in death a few months later. They were buried together in Göttingen not far from the gravesite of the Prince of Mathematicians (Fig. 3.4).

Acknowledgments The author would like to thank Joseph W. Dauben and Harold M. Edwards for reading an earlier version of this article and offering a number of helpful suggestions for improving its style and substance.

References

Bachmann, Paul. 1911. Ueber Gauss' Zahlentheoretische Arbeiten, in *Materialen far eine wissenschaftliche Biographie von Gauss*, Heft 1, *Nachrichten der K. Gesellschaft der Wissenschaflen zu G6ttingen, Math.-phys.* Klasse.

Biermann, K.-R. 1959. Johann Peter Gustav Lejeune Dirichlet, Dokumente für sein Leben und Wirken, *Abhandlungen der Deutschen Akademie der Wissenschaften zu Berlin, Klasse fiir Mathematik, Physik und Technik*, Nr. 2, 2–88.

Collison, M.J. 1977. The Origins of the Cubic and Biquadratic Reciprocity Laws. *Archive for History of Exact Sciences* 17: 63–69.

Edwards, Harold M. 1983. Euler and Quadratic Reciprocity. *Mathematics Magazine* 56 (5): 285–291.

Eisenstein, Gotthold. 1975. *Mathematische Werke*. Vol. 1. New York: Chelsea.

Elstrodt, Jürgen. 2007. The Life and Work of Gustav Lejeune Dirichlet (1805–1859). In *Analytic Number Theory: a Tribute to Gauss and Dirichlet, Clay Mathematics Proceedings, 7*, ed. William Duke and Yuri Tschinkel, 1–37.

Gauss, Carl Friedrich. 1966. *Disquisitiones Arithmeticae*. Trans. Arthur A. Clarke. New York & London: Yale University Press.

———. 1973. *Werke*. Bd. II ed. Göttingen: G. Ohms Verlag.

Goldstein, Catherine, Norbert Schappacher, and Joachim Schwermer, eds. 2007. *The Shaping of Arithmetic after C. F. Gauss's. Disquisitiones Arithmeticae*. Heidelberg: Springer.

Gray, J.J. 1984. A Commentary on Gauss's Mathematical Diary, 1796–1814, with an English Translation. *Expositiones Mathematicae* 2: 97–130.

Hadamard, Jacques. 1954. *An Essay on the Psychology of Invention in the Mathematical Field*. New York: Dover, reprint.

Ireland, Kenneth, and Michael Rosen. 1982. *A Classical Introduction to Number Theory, Graduate Texts in Mathematics, 84*. New York: Springer.

Klein, Felix. 1921. *Gesammelte Mathematische Abhandlungen*. Vol. Bd. 1. Berlin: Julius Springer.

———. 1926. *Vorlesungen über die Entwicklung der Mathematik im 19. Jahrhundert, Bd. 1*. Berlin: Julius Springer.

Kummer, E.E. 1897. Gedächtnisrede auf Gustav Peter Lejeune Dirichlet. In *G. Lejeune Dirichlet's Werke*, ed. L. Kronecker and L. Fuchs, vol. 2, 311–344. Berlin: Georg Reimer.

Lejeune Dirichlet, J.P.G. 1889. In *G. Lejeune Dirichlet's Werke*. ed. L. Kronecker, Bd. 1, Berlin: Georg Reimer.

———. 1897. In *G. Lejeune Dirichlet's Werke*. ed. L. Fuchs, Bd. 2, Berlin: Georg Reimer.

Rieger, G.J. 1957. Die Zahlentheorie bei C. F. Gauss. In *C. F. Gauss Gedenkband anlässlich des 100. Todestages am 23. Februar 1955*, ed. Hans Reichardt. Leipzig: Teubner.

Thomas, Ivor. 1968. *Selections Illustrating the History of Greek Mathematics*. Vol. 2. Cambridge, Mass/London: Harvard University Press/William Heinemann.

Wussing, Hans. 1974. *Carl Friedrich Gauss*, Biographien hervorragender Naturwissenschaftler, Techniker und Mediziner, 15, Leipzig: BSB Teubner.

4 Episodes in the Berlin-Göttingen Rivalry, 1870–1930

(*Mathematical Intelligencer* 22(1)(2000): 60–69)

One of the more striking features in the development of higher mathematics at the German universities during the nineteenth century was the prominent role played by various rival centers. Among these, Berlin and Göttingen stood out as the two leading institutions for the study of research-level mathematics. By the 1870s they were attracting an impressive array of aspiring talent not only from within the German states but also from numerous other countries as well.[1] The rivalry between these two dynamos has long been legendary, yet little has been written about the sources of the conflicts that arose or the substantive issues behind them. Here I hope to shed some light on this theme by recalling some episodes that tell us a good deal about the competing forces that animated these two centers. Most of the information I will draw on concerns events from the last three decades of the nineteenth century. But to understand these it will be helpful to begin with a few remarks about the overall development of mathematics in Germany, so I will proceed from the general to the specific. In fact, we can gain an overview of several of the more famous names in German mathematics simply by listing some of the better-known figures who held academic positions in Göttingen or Berlin. As an added bonus, this leads to a very useful tripartite periodization:

Periodization of Mathematics in Göttingen and Berlin

	Göttingen	Berlin
1801–1855	Gauss	Dirichlet
	W. Weber	Steiner
	Stern	Jacobi
1855–1892	Dirichlet	Kummer
	Riemann	Weierstrass
	Clebsch	Kronecker
	Schwarz	Fuchs
	Klein	
1892–1917	Klein	Fuchs
	H. Weber	Schwarz
	Hilbert	Frobenius
	Minkowski	Schottky
	Runge	
	Landau	
	Carathéodory	

The era of E. E. Kummer, Karl Weierstrass, and Leopold Kronecker (Fig. 4.1) – the period from 1855 to 1892 in Berlin mathematics – has been justly regarded as one of the most important chapters in the history of nineteenth-century mathematics.[2] Still, it is difficult from today's perspective to appreciate the degree to which Berlin dominated not only the national but also the international mathematical scene. Berlin's preeminant position derived in part from the prestige of the Prussian universities, which throughout the century did much to cultivate higher mathematics. During the 1860s and 70s practically all the chairs in mathematics at the Prussian universities were occupied by graduates of Berlin, several more of whom also held positions outside Prussia. Berlin's dominance was reinforced by the demise of Göttingen as a major center following Dirichlet's death in 1859 and Riemann's illness, which plagued him throughout most of the 1860s and eventually led to his passing in 1866. Afterward, Richard Dedekind, who spent most of his career in the relative isolation of Brunswick, was the only major figure whose work revealed close ties with this older Göttingen tradition.

By 1870 a rival tradition with roots in Königsberg began to crystallize around Alfred Clebsch, who taught in Göttingen from 1868 to 1872. Together with Carl Neumann, Clebsch founded *Die Mathematischen Annalen*, which served as a counterforce to the Berlin-dominated journal founded by Crelle, edited after 1855 by Carl Wilhelm Borchardt. Other leading representatives of this Königsberg tradition

[1] For the case of North Americans who studied in Göttingen and Berlin, see Parshall and Rowe (1994, Chap. 5).

[2] For an overview, see Rowe (1998a); the definitive study of mathematics at Berlin University is Biermann (1988).

D.E. Rowe, *A Richer Picture of Mathematics*, https://doi.org/10.1007/978-3-319-67819-1_4

Fig. 4.1 Karl Weierstrass and Leopold Kronecker.

during the 1860s and 1870s included Otto Hesse, Heinrich Weber, and Adolf Mayer. Along with Clebsch and Neumann they operated on the periphery of Berlin and its associated Prussian network. These mathematicians had very broad and diverse interests, making it difficult to discern striking intellectual ties. What they shared, in fact, was mainly a sense of being marginalized, and they looked up to Clebsch as their natural leader (Fig. 4.2).

With the founding of the German Empire in 1871, these "Southern" German mathematicians made a bid to found a nationwide organization. Clebsch's unexpected death in November 1872 slowed the momentum that had been building for this plan, but the effort was carried on by Felix Klein and other close associates of the Clebsch school. A meeting took place in Göttingen in 1873, but the turnout was modest and the results disappointing. None of the prominent "Northern" German mathematicians attended – the label "Northern" being a euphemism within the Clebsch school for "Prussian". Klein and his allies soon thereafter gave up this plan, as without the support of the Berliners there could clearly be no meaningful German Mathematical Society.[3] The same situation prevailed 20 years later, and the result would have very likely been the same had not Georg Cantor persuaded Leopold Kronecker, the most powerful and important Berlin mathematician of the 1880s, to throw his support behind the venture.

After 1871, the Franco-Prussian rivalry loomed large in the minds of many German mathematicians. That Berlin should occupy a place analogous to Paris was, for many mathematicians, merely the natural extension of political developments to the intellectual sphere. One need only read some of Kummer's speeches before the Berlin Academy – which as its Perpetual Secretary he was required to deliver on ceremonial occasions like the birthday of Frederick the Great – in order to realize how deeply this celebrated and revered mathematician identified with the world-historical purpose of the Prussian state and its innermost spirit, its *Geist*.[4] I doubt that Hegel himself could have described that mysterious dialectical linkage more eloquently. This was the same Kummer who, in a letter to his young pupil Kronecker written in 1842, urged him to attend Schelling's lectures in

[3]Further details on this early, abortive effort to found a national organization of mathematicians in Germany can be found in Tobies and Rowe (1990, pp. 20–23, 59–72).

[4]Another sterling example from Kummer's Breslau period is his lecture on academic freedom (Kummer 1848) delivered in the midst of the dramatic political events of 1848.

Fig. 4.2 Alfred Clebsch.

Fig. 4.3 Ernst Eduard Kummer.

Berlin, despite the fact that Schelling's brand of idealism failed to grasp the deeper Hegelian truth that "mind and being" were initially one and the same. Schelling, according to Kummer, was the "only world-historical philosopher still living."[5] Kummer held the office of perpetual secretary of the Berlin Academy for 14 years, from 1865–1878, during which time he conducted himself in a manner that won much admiration. His role during Berlin's "golden age of mathematics" bore a strong resemblance to that played by Max Planck after 1900. Indeed, Planck's worldview (about which, see Heilbron (1986)) had much in common with Kummer's belief in the harmony of Prussia's intellectual, spiritual and political life (Fig. 4.3).

Kummer is, of course, mainly remembered today for his daring new theory of ideal numbers, which served as the point of departure for Dedekind's ideal theory; we also think of him in connection with Kummer surfaces, special quartics with 16 nodal points. But since he founded no special school as such, his impact was clearly more diffuse than that of his colleagues Weierstrass and Kronecker. Still, he embodied for many the heart and soul of the Berlin tradition, and like his colleagues he instilled in his students the same sense of lofty ideals – the purism and rigor for which the Berlin style was soon to become famous. Later representatives – Schwarz, Frobenius, Hensel, Landau, and Isaai Schur – saw themselves as exponents of this same Berlin tradition, though they drew their main inspiration from the lecture courses of Weierstrass and Kronecker.

To gain a quick, first-hand glimpse of Berlin mathematics during the 1870s we can hardly do better than follow the description given by Gösta Mittag-Leffler (Fig. 4.4) in a letter written to his former mentor, Hjalmar Holmgren, on 19 February 1875. The young Swede was traveling abroad on a postdoctoral fellowship that had first brought him to Paris. There he met Charles Hermite, a great admirer of German mathematical achievements despite his limited knowledge of the langauge, who told him that every aspiring analyst ought to hear the lectures of Weierstrass.[6] Here is Mittag-Leffler's account of what he encountered in the Prussian capital:

> . . . With regard to the scientific aspect I am very satisfied with my stay in Berlin. Nowhere have I found so much to learn as here. Weierstrass and Kronecker both have the unusual tendency, for Germany, of avoiding publication as much as possible. Weierstrass, as is known, publishes nothing at all, and Kronecker only results without proofs. In their lectures

[5] See the letter from Kummer to Kronecker, 16 January 1842, published in Jahnke et al. (1910, pp. 46–48).

[6] On Mittag-Leffler's career, see Garding (1998, pp. 73–84) and Stubhaug (2010).

Fig. 4.4 Gösta Mittag-Leffler.

they present the results of their researches. It seems unlikely that the mathematics of our day can point to anything that can compete with Weierstrass's function theory or Kronecker's algebra. Weierstrass handles function theory in a two- or three-year cycle of lectures courses, in which, starting from the simplest and clearest foundational ideas, he builds a complete theory of elliptic functions and their applications to Abelian functions, the calculus of variations, etc.[7] What is above all characteristic for his system is that it is completely analytical. He rarely draws on the help of geometry, and when he does so it is only for illustrative purposes. This appears to me an absolute advantage over the school of Riemann as well as that of Clebsch. It may well be that one can build up a completely rigorous function theory by taking the Riemann surfaces as one's point of departure and that the geometrical system of Riemann suffices in order to account for the till now known properties of the Abelian functions. But [Riemann's approach] fails on the one hand when it comes to recovering the properties of the higher-order transcendentals, whereas, on the other hand, it introduces elements into function theory which are in principle altogether foreign. As for the system of Clebsch, this cannot even deliver the simplest properties of the higher-order transcendentals, which is quite natural, since analysis is infinitely more general than is geometry.

Another characteristic of Weierstrass is that he avoids all general definitions and all proofs that concern functions in general. For him a function is identical with a power series, and he deduces everything from these power series. At times this appears to me, however, as an extremely difficult path, and I am not convinced that one does not in general attain the goal more easily by starting, like Cauchy and Liouville, with general definitions that are, of course, completely rigorous. Another distinguishing characteristic of Weierstrass as well as Kronecker is the complete clarity and precision of their proofs. By the same token, both have inherited from Gauss the fear of any kind of metaphysics that might attach to their fundamental mathematical ideas, and this gives a simplicity and naturalness to their deductions, which have hardly been seen heretofore presented so systematically and with the highest degree of precision.

In respect to form, Weierstrass's manner of lecturing lies beneath all criticism, and even the least important French mathematician, were he to deliver such lectures, would be considered completely incompetent as a teacher. If one succeeds, however, after much difficult work, in restoring a lecture course of Weierstrass to the form in which he originally conceived it, then everything appears clear, simple, and systematic. Probably it is this lack of talent which explains why so extremely few of his many students have understood him thoroughly, and why therefore the literature dealing with his direction of research is still so insignificant. This circumstance, however, has not affected the nearly god-like reverence he enjoys in general. Presently there are several young and diligent mathematicians in whom Weierstrass places the highest hopes. At the top of the list as ‚the best pupil that I have ever had"he places the young Russian Countess Sophie v. Kovalevskaya, who recently took her doctorate *in absentia* from the faculty in Göttingen on the basis of two works that will soon appear in Crelle; one on partial differential equations, the other on the rings of Saturn.[8]

Clearly, these views reflect more than just one man's opinion. Indeed, Mittag-Leffler put his finger on an important component of the Berlin-Göttingen rivalry with his claims for the methodological superiority of Weierstrassian analysis over the geometric function theory of Riemann or the mixed methods of Clebsch. Still, what he wrote must be placed in proper perspective. During Riemann's lifetime, the Göttingen mathematician's reputation stood very high in Berlin, and it remained untarnished after his death in 1866. He was elected as a corresponding member of the Berlin Academy in August 1859, which gave him occasion to travel to the Prussian capital the following month. There he was welcomed by the leading Berlin mathematicians – Kummer, Kronecker, Weierstrass, and Borchardt – with open arms, as his friend Dedekind, who accompanied him on this journey, later recalled (Dedekind 1892, p. 554). Weierstrass practically worshiped Riemann, calling him, according to Mittag-Leffler, an "anima candida" like no one else he ever

[7] Mittag-Leffler took Weierstrass's standard course on elliptic functions during the winter semester of 1874–75; he also was one of only three auditors who attended his course that term on differential equations. During the summer semester of 1875, he followed Weierstrass's course on applications of elliptic functions to geometry and mechanics. This information can be found in Nörlund (1927, p. vii), along with the claim that Mittag-Leffler was offered a *Lehrstuhl* in Berlin in 1876. Presumably this story stemmed from Mittag-Leffler himself, and while difficult to refute, its patent implausibilty is so apparent that we may safely regard this as a Scandinavian legend. A similar conclusion is reached in Garding (1998, pp. 75–76).

[8] Quoted from (Frostman 1966, 54–55), (my translation). For a discussion of Kovalevskaya's work, see Cooke (1984).

knew.[9] His colleague Kronecker, to be sure, had a far less flattering opinion of Riemann's successor, Clebsch, but he, too, had already passed from the scene in November 1872. Thus the subterranean rumblings within the German mathematical community so apparent in Mittag-Leffler's letter reflected not so much personal animosities directed toward Riemann and/or Clebsch but rather the way in which their work had become bound up in an ongoing rivalry between Berlin's leading mathematicians and those associated with the "remnants" of the Clebsch school. Within the latter group the most visible figure was its youngest star, an ambitious and controversial fellow named Felix Klein.

The Berlin establishment had gotten a first taste of Klein during the winter semester of 1869–70 when the 20-year-old Rhinelander, who had worked closely with Julius Plücker in Bonn, arrived in the Prussian capital to undertake post-doctoral studies. Like nearly all aspiring young Prussian mathematicians, Klein recognized the importance of making a solid impression in the Berlin seminar run by Kummer and Weierstrass. Before presenting himself as a candidate, therefore, he took about five weeks to write up an impressive paper on a topic in his special field of line geometry.[10] He then submitted the manuscript to Kummer, thereby fulfilling one of the requirements for membership in the seminar. Klein's paper dealt systematically with the images of ruled surfaces induced by a mapping found a short time earlier by his friend, Max Noether (the Noether map sends the lines of a linear complex to points in complex projective 3-space). Some weeks later Kummer returned the manuscript without saying so much as a word about it; Klein apparently soon lost interest in this topic as well as the results he had obtained.[11]

In the meantime, he introduced himself to Weierstrass and Kronecker, though he otherwise kept his distance from their lecture halls. This aloofness, however, did not prevent him from asking Weierstrass for his assistance in helping him cultivate contacts with his advanced students. Klein no doubt made it plain that he could not spare the time it would take to learn Weierstrassian analysis from the ground up; what mattered to him was getting to know the "inner life" of mathematics in Berlin. Presumably not many would have dared to approach Weierstrass in this way, but the latter willingly obliged, suggesting that Klein seek out Ludwig Kiepert's counsel.[12] Their meeting marked the beginning of a lifelong friendship which both Klein and Kiepert came to value, and for good reason: it turned out to be one of the few bridges connecting members of the Berlin and Göttingen "schools."

Klein made other significant contacts in Berlin, but mainly with other outsiders like the Austrian Otto Stolz, from whom he learned the rudiments of non-Euclidean geometry. By far the most significant new friendship, however, was the one Klein made with Sophus Lie, a Nordic giant whose ideas and personality captivated him so completely. As a backdrop to future events, a few words must be said with regard to the Klein-Lie collaboration. Like Klein, Lie was also an expert on Plückerian line geometry, and thus someone Klein knew by reputation beforehand. In fact, Klein's mentor, Clebsch, had already alerted his protégé to the possibility of meeting Lie personally in Berlin, and in October 1869 they greeted each other at a meeting of the Berlin Mathematics Club. Before long they were getting together nearly every day to discuss mathematics.

Since Kummer's seminar theme concerned the geometry of ray systems, a topic intimately connected with line geometry, both Klein and Lie soon emerged as its two stars. Although Lie was still without a doctorate – a circumstance so embarrassing to him that he introduced himself as Dr. Lie anyway[13] – the Norwegian's brilliant new results dazzled Klein, who was six years younger. At the time, Lie's German was far from flawless, so Klein offered to present his work to the members of the Berlin seminar. Kummer expressed admiration for Lie's originality, though he was perhaps less impressed by Klein's presentation. Such recognition was surely appreciated by both young men, and it may have helped intensify their collaboration. This continued with a sojourn in Paris during the spring of 1870 and lasted until Klein's appointment as *Professor Ordinarius* in Erlangen in the fall of 1872. Lie even accompanied Klein when he moved from Göttingen, the whole while discussing with him the ideas that soon appeared in Klein's famous "Erlangen Program." During the years that followed, however, their interests drifted apart, though they continued an avid correspondence.

Returning now to our main theme, the first overt signs of struggle between Klein and Berlin came to light in the early

[9]Mittag-Leffler (1923, p. 191). Mittag-Leffler's remark was undoubtedly the source E. T. Bell drew upon for the title ("Anima Candida") of the chapter on Riemann in his popular, but idiosyncratic *Men of Mathematics*.

[10]The manuscript can be found in Klein Nachlass 13A, Handschriftenabteilung, Niedersächsische Staats- und Universitätsbibliothek Göttingen. According to the dating in Klein's hand at the top, he began to write the paper on 5 September 1869 and completed it on 15 October 1869.

[11]I have found no traces of this original study in Klein's published work, although there are several references to Noether's mapping, which is related to the famous line-sphere map investigated by Sophus Lie soon thereafter. Erich Bessel-Hagen later added a note to the unpublished manuscript relating that, according to Klein, Kummer returned the manuscript to him after a few weeks without any comments and apparently unread ("anscheinend ungelesen").

[12]This story is recounted in Kiepert (1926, p. 62).

[13]Lie was apparently advised by mathematicians in Norway that he should introduce himself this way in Germany, as they assumed he would otherwise not be taken seriously by those he hped to meet, including Göttingen's Alfred Clebsch.

1880s when Klein was Professor of Geometry in Leipzig. Six years after Mittag-Leffler had given his *private* description of how Berlin mathematicians assessed the drawbacks of a geometrically-grounded theory of complex functions, this issue was taken up by Klein in a *public* forum. Klein's remarks were prompted by a priority dispute with the Heidelberg analyst, Lazarus Fuchs, a leading member of the Berlin network (Fuchs assumed a chair in Berlin in 1884). Like so many priority claims in the history of mathematics, the issues at stake here were far more complicated than might first meet the eye. Particularly interesting in this connection were the international dimensions of the conflict, including the part played by Mittag-Leffler, who was then busy plotting plans to launch *Acta Mathematica* (see Chap. 11).

This episode began innocently enough in 1881 when Henri Poincaré published a series of notes in the *Comptes Rendus* of the Paris Academy in which he named a special class of complex functions, namely those invariant under a group with a natural boundary circle, "Fuchsian functions." Klein soon thereafter entered into a semi-friendly correspondence with Poincaré, from which he quickly learned that the young Frenchman was quite unaware of the relevant "geometrical" literature, including Schwarz's work, but especially Klein's own.[14] Before long Poincaré found himself in the middle of a German squabble that he very much would have liked to avoid. Quoting a famous line from Goethe's *Faust*, he wrote Klein that "Name ist Schall und Rauch" ("names are but sound and smoke"). Nevertheless, he found himself forced to defend his own choice of names in print, while hoping he could placate Klein's wrath by naming another class of automorphic functions after the Leipzig mathematician. In the meantime, Klein and Fuchs exchanged sharp polemics, Klein insisting that the whole theory of Poincaré had its roots in Riemann's work, and that Fuchs's contributions failed to grasp the fundamental ideas, which required the notions of group actions on Riemann surfaces (see Chap. 11). Klein's brilliant student, Adolf Hurwitz, apparently enjoyed this feud, especially his mentor's attacks which reminded him of a favorite childhood song: "Fuchs, Du hast die *Funktion* gestohlen / Gieb sie wieder her." Fuchs and the Berlin establishment were, of course, not amused at all, and neither was Klein.

Over the next 10 years, Klein launched a series of efforts, nearly all of them futile, to make inroads against the entrenched power of the Berlin network. In 1886 he finally managed to gain a foothold in Prussia when he was called to Göttingen (the former Hanoverian university had become prussianized in 1866). But on arrival he found that the mathematics program there was firmly in the hands of H.A. Schwarz, Weierstrass's leading disciple. Both Schwarz and his teacher were incensed that Klein had managed to engineer the appointment of a foreigner, namely his friend Sophus Lie, as his successor in Leipzig.[15] Thereafter, both Lie and Klein were scorned by leading Berliners, particularly Georg Frobenius. According to Frobenius, Weierstrass had made it known that Lie's theory would have to be junked and worked out anew from scratch.[16] Klein sought to make a common front with his Leipzig colleagues, Lie and Adolf Mayer, but Lie became increasingly wary of this behind-the-scenes maneuvering. He gradually distanced himself from Klein as he came to take a critical view of the latter's efforts to enhance his power and prestige.

When Kronecker suddenly died in December 1891, Weierstrass could finally retire in peace – they had been archenemies throughout the 1880s – and this led to a whole new era in German mathematics. The deliberations over their successors opened with the clear understanding that Klein would not be taken into consideration. He was vehemently rejected by the search committee, including Weierstrass and Hermann von Helmholtz, who characterized him as a dazzling charlatan. Fuchs, who was still stinging from Klein's attack a decade earlier, merely added that he had nothing against Klein personally, only his pernicious influence on mathematical science.[17] This much being settled, they chose two candidates from the Berlin camp: Schwarz got Weierstrass's chair, and Kronecker's went to Frobenius. In the midst of these negotiations, Klein tried to gain Hurwitz for Göttingen, even though the faculty placed Heinrich Weber first on its list of candidates. Klein was hoping that Friedrich Althoff, the autocratic head of Prussian university affairs, would reach over Weber and appoint Hurwitz, who was second on the list. This plan backfired, though, leaving Klein in a state of despair, largely due to his loss of face in the faculty. Its members thus witnessed how Schwarz once again won his way against Klein, even though this time he was already sitting in Berlin.[18] Göttingen's informal religious policy, which limited the number of Jews on the faculty to one per discipline, may well have been the decisive factor that prevented Hurwitz's appointment (see Chap. 10). Strangely enough, after Hurwitz's death in 1919 the fallacious story circulated that he had turned down the call to Göttingen in 1892 out of a sense of loyalty to the ETH.[19]

As it turned out, the decisive year 1892 was nothing short of a fiasco for Klein. Following his futile efforts on behalf of Hurwitz, the alliance with Lie, whom he wanted to appoint to the board of *Mathematische Annalen*, fell apart completely. Lie had been under stress practically from the

[14] For details, see Gray (2000, pp. 275–315).

[15] For details, see Rowe (1988, pp. 39–40).

[16] See Biermann (1988, p. 215).

[17] See Biermann (1988, p. 305–306).

[18] For details, see Rowe (1986, pp. 433–436).

[19] See Young (1920, p. liii).

moment he came to Leipzig as Klein's successor in 1886. At the same time, he grew increasingly embittered by the way he was treated by his Leipzig colleagues and certain allies of Klein, who regarded him mainly as one of Klein's many subordinates.[20] By late 1893, the whole mathematical world knew about Lie's displeasure when he published a series of nasty remarks in the preface to volume three of his work on transformation groups. To clarify his relationship with Klein he wrote: "I am not a student of Klein's nor is the opposite the case, even if it comes closer to the truth" (Lie 1893, p. 17).

So, Klein had to regroup his forces and try again, something he knew how to do. In retrospect, the decisive turning point came with Hilbert's appointment in December 1894, a goal Klein had long been planning. Mathematically, Klein and Hilbert complemented one another beautifully; moreover, both shared a strong antipathy for the Berlin establishment which they considered narrow and authoritarian. Whereas Klein tried to advance a geometric style of mathematics rooted in the work of Riemann and Clebsch, Hilbert championed an approach to abstract algebra and number theory that was largely inspired by ideas first developed by Dedekind and Kronecker. With regard to foundational issues, on the other hand, Hilbert's ideas clashed directly with the skeptical views Kronecker had championed in Berlin. In a schematic fashion, we may picture Klein and Hilbert as universal mathematicians whose strengths were mainly situated on the right, respectively, left sides in the following hierarchy of mathematical knowledge:

NUMBER	**FIGURE**
Arithmetic	Euclidean Geometry
Algebra	Projective Geometry
________________	
Analysis	Higher Geometry
_ . _ . _ . _ . _ . _ . _ . _	

Analytic & Differential Geometry

Analytical Mechanics   Geometrical Mechanics

This bifurcated scheme, I would argue, fairly accurately portrays in a general way how most mathematicians in Germany saw the various components of their discipline during the late nineteenth century. In effect, what I've pictured here is the "tree of mathematics" turned upside down so that its roots (number and figure) appear at the top. This is meant to reflect, in particular, the high status accorded to pure mathematics, especially number theory and synthetic geometry, by many influential German mathematicians. The Berlin tradition of Kummer, Weierstrass, and Kronecker clearly favored that branch of mathematics derived from the concept of number, but the tradition of synthetic geometry tracing back to Steiner also played a major part in the Berlin vision (Weierstrass thus taught geometry after Steiner's death in 1863 in an effort to sustain the geometrical component of Berlin's curriculum).

Still, as we have seen, pure mathematics, for Weierstrass, mainly meant analysis, and the foundations of analysis derived from the properties of numbers (irrational as well as rational). He thus drew a reasonably sharp line (indicated by _. _. _ above) that excluded geometrical reasonings from real and complex analysis, whereas his colleague Kronecker drew an even sharper line (marked above as _______) that excluded everything below algebra. In other words, Kronecker wished to ban from rigorous, pure mathematics all use of limiting processes and, along with these, the whole realm of mathematics based on the infinitely small. This, of course, was one of the main sources of the conflict between Kronecker and Weierstrass that severely paralyzed Berlin mathematics during the 1880s and beyond, right up until Kronecker's death.

This is not the place to go into details about how Göttingen quickly outstripped Berlin during the years that followed, but we should at least notice that part of this story concerns a very different vision of this ,inverted tree"of mathematics, a vision shared by Klein and Hilbert. As I have argued elsewhere, this common outlook helps explain how they managed to form such a successful partnership in Göttingen despite their apparent differences.[21] Both were acutely aware of the possibilities for establishing a linkage between the two principal branches of the tree. Indeed, both made important contributions toward securing these ties (Klein through his work on projective non-Euclidean geometry; Hilbert with his arithmetical characterization of the continuum which he hoped to anchor by means of a proof of the consistency of its axioms). Noteworthy, beyond these contributions, was their background in and familiarity with invariant theory, a field that was particularly repugnant to Kronecker and which occupied a rather low rung in the purists' hierarchy of knowledge.

From an institutional standpoint, we can easily spot some other glaring contrasts between Göttingen and Berlin during the era 1892 to 1917. Whereas Frobenius and Schwarz largely saw themselves as defenders of Berlin's purist legacy, Klein and Hilbert promoted an open-ended interdisciplinary approach that soon made Göttingen a far more attractive

[20]Lie's difficulties in Leipzig were compounded by a variety of other factors, included jealousy aroused by the publications of Wilhelm Killing on the structure theory of Lie algebras, about which see Rowe (1988, pp. 41–44). For a detailed account of Killing's work, see Hawkins (1982).

[21]For more on the Hilbert-Klein partnership, see Rowe (1989).

center, drawing international talent in droves (Reid 1970). A glimpse of this multi-disciplinary style in Göttingen can be captured merely by looking at some of the appointments Klein pushed through with the support of the Prussian Ministry and the financial assistance of leading industrial concerns: Karl Schwarzschild (astronomy), Emil Wiechert (geophysics), Ludwig Prandtl (hydro- and aerodynamics), and Carl Runge (applied mathematics/ numerical analysis).

Symptomatic of the Göttingen style was an intense interest in physics, both classical and modern. Arnold Sommerfeld, Max Born and Peter Debye interacted closely with Klein, Hilbert, and Minkowski, all of whom were deeply interested in Einstein's relativity theory. Einstein came to Berlin in 1914 on a special appointment that included membership in the Prussian Academy. Just one year later the Göttingen Scientific Society offered him a corresponding membership, elevating him to an external member in 1923. Ironically enough, general relativity was followed far more closely within Göttingen circles – especially after the publication of (Einstein 1917) – as well as by the Dutch community surrounding Paul Ehrenfest, than it was in Einstein's immediate Berlin surroundings. Klein, Hilbert, Einstein, and Weyl were friendly competitors during the period 1915–1919 (see Part IV and Rowe (1998b)). In Göttingen the work of Klein and Hilbert on general relativity was supported by several younger talents, including Emmy Noether, whose famous paper on conservation laws grew out of these efforts (Rowe 1999).

In the case of Hilbert and Einstein, we can also observe a strong affinity in their insistence on the need to uphold international scientific relations and to resist those German nationalists who supported the unity of Germany's military and intellectual interests. Thus, the controversial pacifist and internationalist, Georg Nicolai, enlisted the support of both Hilbert and Einstein for these causes.[22] Scientifically, perhaps the most important link joining Hilbert and Einstein came through one of Hilbert's many doctoral students, an Eastern European Jew named Jacob Grommer. Grommer's name appeared for the first time in Einstein's famous 1917 paper introducing the cosmological constant and his static, spatially-closed model of the universe (Einstein 1917). Soon thereafter Grommer joined Einstein and worked closely with him until 1929 when he apparently left Berlin – the longest collaboration Einstein had with anyone.[23]

Relations between Göttingen and Berlin mathematicians largely normalized after World War I, but they heated up again in 1928 when Brouwer and Bieberbach sought to boycott the Bologna ICM. As is well known, Hilbert, who was then on the brink of death suffering from pernicious anemia, overcame this effort by organizing a delegation of German mathematicians to attend the congress. He prepared a political statement for this occasion, and although these remarks were not recorded in the Congress Proceedings they can be found, scratched in Hilbert's hand, among his unpublished papers. There one can read: "[Bologna Rede]. .. It is a complete misunderstanding to construct differences or even contrasts according to peoples or human races ... mathematics knows no races.. .. For mathematics the entire cultural world is one single land." (For recent reflections on this text, see Siegmund-Schultze (2016).

These views were very different from the ones held earlier by Ernst Eduard Kummer, admittedly during an era when the German mathematical community was still barely forming. More striking, however, is the clash with Bieberbach's vision, which asserted that mathematical style could be directly understood in terms of racial types. That story leads, of course, into the complex and messy problematics of mathematics during the Nazi era and its historical roots – a topic I can only mention here.[24] Nevertheless, I hope these few glimpses into the mathematical life of Germany's two leading research centers has managed to convey a sense of the clashing visions and intense struggles that took place behind the scenes. The setting may be unfamiliar, but the issues of pure vs. applied or national allegiance vs. international cooperation most certainly are not. The present ICM in Berlin represents not only an opportunity for mathematicians to gather in celebrating recent achievements but also to reflect on the role of mathematics and its leading representatives from the not so distant past, drawing whatever lessons these reflections may offer for mathematics today.

Acknowledgments This essay was based on a lecture presented at a special symposium on history of mathematics that took place in August 1998 as part of the Berlin ICM. I wish to thank the symposium organizers, Georgio Israel and Eberhard Knobloch, for inviting me to speak on that occasion.

References

Biermann, Kurt-R. 1988. *Die Mathematik und ihre Dozenten an der Berliner Universität, 1810–1933*. Berlin: Akademie Verlag.

Cooke, Roger. 1984. *The Mathematics of Sonya Kovalevskaya*. New York: Springer.

Dedekind, Richard. 1892. Bernhard Riemann's Lebenslauf. In *Bernhard Riemann's Gesammelte Mathematische Werke*, ed. H. Weber, 2nd ed., 539–558. Leipzig: Teubner.

Einstein, Albert. 1917. Kosmologische Betrachtungen zur allgemeinen Relativitätstheorie, *Sitzungsberichte der Preußischen Akademie der Wissenschaften zu Berlin*, physikalisch-math. Klasse, 1917, 142–152.

Fölsing, Albrecht. 1997. *Albert Einstein: A Biography*. New York: Viking.

[22]On Einstein's alliance with Hilbert, see Fölsing (1997, p. 466).

[23]For more details on this collaboration, see Pais (1982, pp. 487–488).

[24]See the portrayal of Bieberbach's political transformation in Mehrtens (1987).

Frostman, Otto. 1966. Aus dem Briefwechsel von G. Mittag-Leffler. In *Festschrift zur Gedächtnisfeier für Karl Weierstraß, 1815–1965*, ed. H. Behnke and K. Kopfermann, 53–56. Köln: Westdeutscher Verlag.

Garding, Lars. 1998. *Mathematics and Mathematicians: Mathematics in Sweden before 1950*, History of Mathematics. Vol. 13. Providence/London: American Mathematical Society/London Mathematical Society.

Gray, Jeremy J. 2000. *Linear Differential Equations and Group Theory from Riemann to Poincaré*. 2nd ed. Birkhäuser: Basel.

Hawkins, Thomas. 1982. Wilhelm Killing and the Structure of Lie Algebras. *Archive for History of Exact Sciences* 26: 127–192.

Heilbron, John L. 1986. *The Dilemmas of an Upright Man. Max Planck as Spokesman for German Science*. Berkeley: University of California Press.

Jahnke, E., et al., eds. 1910. *Festschrift zur Feier des 100. Geburtstages Eduard Kummers*. Leipzig: Teubner.

Kiepert, Ludwig. 1926. Persönliche Erinnerungen an Karl Weierstrass. *Jahresbericht der Deutschen Mathematiker-Vereinigung* 35: 56–65.

Kummer, E.E. 1848. Über die akademische Freiheit. Eine Rede, gehalten bei der Übernahme des Rektorats der Üniversität Breslau am 15. Oktober 1848. In *Ernst Eduard Kummer, Collected Papers*, ed. A. Weil, vol. II, Berlin: Springer, 1975, pp. 706–716.

Lie, Sophus. 1893. *Theorie der Transformationsgruppen*. Vol. 3. Leipzig: Teubner.

Mehrtens, Herbert. 1987. Ludwig Bieberbach and Deutsche Mathematik. In *Studies in the History of Mathematics*, ed. Esther Phillips, 195–241. Washington: The Mathematical Association of America.

Mittag-Leffler, Gösta. 1923. Weierstrass et Sonja Kowalewsky. *Acta Mathematica* 39: 133–198.

Nörlund, N.E. 1927. G. Mittag-Leffler. *Acta Mathematica* 50: I–XXIII.

Pais, Abraham. 1982. *Subtle is the Lord. The Science and the Life of Albert Einstein*. Oxford: Clarendon Press.

Parshall, Karen, and David E. Rowe. 1994. *The Emergence of the American Mathematical Research Community, 1876–1900. J.J. Sylvester, Felix Klein, and E.H. Moore*, History of Mathematics. Vol. 8. Providence: American Mathematical Society/London Mathematical Society.

Reid, Constance. 1970. *Hilbert*. New York: Springer.

Rowe, David E. 1986. "Jewish Mathematics" at Göttingen in the Era of Felix Klein. *Isis* 77: 422–449.

———. 1988. Der Briefwechsel Sophus Lie-Felix Klein, eine Einsicht in ihre persönlichen und wissenschaftlichen Beziehungen, *NTM. Schriftenreihe für Geschichte der Naturwissenschaften, Technik und Medizin* 25 (1): 37–47.

———. 1989. Klein Hilbert, and the Göttingen Mathematical Tradition, *Science in Germany: The Intersection of Institutional and Intellectual Issues*, Kathryn M. Olesko, ed. (Osiris, 5, 1989). In *189–213*.

———. 1998a. Mathematics in Berlin, 1810–1933. In *Mathematics in Berlin*, ed. H.G.W. Begehr, H. Koch, J. Kramer, N. Schappacher, and E.-J. Thiele, 9–26. Basel: Birkhäuser.

———. 1998b. Einstein in Berlin. In *Mathematics in Berlin*, ed. H.G.W. Begehr, H. Koch, J. Kramer, N. Schappacher, and E.-J. Thiele, 117–125. Basel: Birkhäuser.

———. 1999. The Göttingen Response to General Relativity and Emmy Noether's Theorems. In *The Symbolic Universe. Geometry and Physics, 1890–1930*, ed. Jeremy Gray, 189–234. Oxford: Oxford University Press.

Siegmund-Schultze, Reinhard. 2016. Mathematics Knows No Races: A Political Speech that David Hilbert Planned for the ICM in Bologna in 1928. *Mathematical Intelligencer* 38 (1): 56–66.

Stubhaug, Arild. 2010. *Gösta Mittag-Leffler. A Man of Conviction*. Heidelberg: Springer.

Tobies, Renate, and David E. Rowe, eds. 1990. *Korrespondenz Felix Klein-Adolf Mayer, Teubner Archiv zur Mathematik, Band 14*. Leipzig: Teubner.

Young, W.H. 1920. Adolf Hurwitz. *Proceedings of the London Mathematical Society* 20 (Ser. 2): xlviii–xlvliv.

5 Deine Sonia: A Reading from a Burned Letter by Reinhard Bölling, Translated by D. E. Rowe

(*Mathematical Intelligencer* 14(3)(1992: 24–30)

It was in January 1990. Finally, just two months after the Berlin Wall had fallen, I had the opportunity to spend a few days at the Mittag-Leffler Institute in Djursholm, a small town just northeast of Stockholm. The palatial villa that today houses the Institute is the former home of Gösta Mittag-Leffler (1846–1927), and on entering its doorway I felt as if I had taken a step back into the world in which he lived. For me, the Institute's single greatest attraction lay in its archival holdings, and particularly the extensive correspondence that linked Mittag-Leffler with many of the era's leading mathematicians. A former student of Karl Weierstrass (1815–1897), Mittag-Leffler sought to preserve everything he could get his hands on from the master's estate. Thanks to his efforts, many letters, manuscripts, and other documents associated with Weierstrass have survived today. In Berlin, on the other hand, where Weierstrass passed his entire scientific career, there is no corresponding collection of materials.

I had just completed my work on a new edition of the letters that Weierstrass wrote to Sonya Kovalevskaya (Fig. 5.1), his trusted pupil and friend [cf. (Bölling 1993a); this is a re-edition of (Kochina and Ozhigova 1984), together with a detailed commentary on the biographical and mathematical contents of the letters, published in 1993 by Akademie Verlag (Berlin)]. Several questions remained open, and I hoped that during my visit I might find some new documents that could shed light on these matters, if not resolve them completely. With only three days available to look through the archival material, I had to make use of every second. From dawn to dusk and without a break, I scanned letters and leafed through large quantities of mathematical notes and sketches. Mittag-Leffler had already made a first attempt to put the papers of Weierstrass and Kovalevskaya into some kind of order, but there still remained a good number of boxes and folders among which sheer chaos reigned (cf. also (Grattan-Guinness 1971)). Nevertheless, there was compensation to be found in the adventure of the hunt, and as it turned out I was not to be disappointed.

It was already late afternoon during the second day of my stay. I was just about to examine the contents of a box that contained material from Kovalevskaya's posthumous papers. At first glance, the documents appeared to be exclusively concerned with details surrounding the events of her death and the arrangements that had to be made afterward. I found various bills and receipts in connection with the funeral, others from purchases she had made shortly before her death, and a number of telegrams from friends and acquaintances expressing condolences. Between these, I found some photographs and several pages filled with handwriting. Staring at these, illuminated by the glimmer of the table lamp, I discovered to my surprise a text filled with crossed-out words and lines, and containing in places afterthoughts written in an almost illegible, scrawling hand that looked like Sonya's own. Then I noticed that the text, whatever it was, had been written in German. At first I thought that perhaps this was a sketch for a literary work, but then I noticed some comments about Mittag-Leffler and his lectures.

Thunderstruck, I strained to decipher a few more passages; the words *Dampfschiff* (steamboat), and *Stockholm* flashed by, and then: *Deine arme, kleine [. ..] Schülerin.* Unbelievable as it seemed, this appeared to be a page from the first draft of a letter that Kovalevskaya intended to send to Weierstrass. I knew very well – Mittag-Leffler had reported it – that Weierstrass had burned all the letters he had received from Kovalevskaya soon after her death in February 1891. Thus, these writings of Kovalevskaya are gone forever, but now the thought raced through my mind: could it be that what I held in my hands was a draft of such a letter from which we might gain an impression of what at least one of them actually looked like? With growing excitement and anticipation, I soon found another page which proved to be a continuation of the one I already had before me. After making a first transcription of the text, I could no longer doubt it: this was a fragment from the draft of a letter (shown below) composed by Kovalevskaya approximately two weeks after she first arrived in Stockholm in early

D.E. Rowe, *A Richer Picture of Mathematics*, https://doi.org/10.1007/978-3-319-67819-1_5

Fig. 5.1 Sonya Kovalevskaya (1850–1891).

December 1883. In all likelihood, it is the only remaining document that has survived from the many letters she addressed to Weierstrass. Whether it was accident or design that spared these pages from destruction is a mystery that will probably never be solved. Beyond the flavor of her idiosyncratic style, Kovalevskaya's letter is particularly significant because it was written at a critical turning point in her life. Since there are several good biographies of Kovalevskaya available (e.g., Cooke 1984; Koblitz 1983; Kochina 1981; Leffler 1895), and, moreover, a previous article about her by Ann Hibner Koblitz in *The Mathematical Intelligencer,* (vol. 6(1) (1984), 20–29), I shall restrict myself here to the events in her life directly relevant to the letter under consideration.

Kovalevskaya had just come from St. Petersburg to Stockholm on November 18, 1883 in order to lecture at the newly founded *Högskola* (which later became Stockholm University). Thus, she stood at the very beginning of her academic career. Only a little more than a half year earlier her husband, Vladimir, had taken his own life after his business ventures in connection with an oil company had brought him and his family to the brink of financial disaster from which he could see no way out. Even earlier, during the late 1870s, a series of building projects and real estate investments by the Kovalevskys had seriously depleted their financial resources, including the money Sonya had inherited following the death of her father.Between 1870 and 1874 Kovalevskaya had studied privately with Weierstrass in Berlin, and, on the latter's recommendation, Göttingen University awarded her a doctorate *(in absentia).* She returned to Russia in August 1874. However, her efforts to find a suitable position in her native country proved futile, and so she gradually lost interest in mathematics. Eventually she gave it up altogether, and thereafter she fell out of contact with her teacher and friend, Karl Weierstrass. Three years went by during which he heard nothing more from her. In February 1876 Mittag-Leffler happened to be in St. Petersburg, and during his stay he paid Kovalevskaya a visit. This was their first meeting. Through his former Swedish pupil, Weierstrass again learned something about Sonya, but it was not until August 1878, two months before the birth of her daughter, that she wrote to him directly. Even so, it would take another two years before she began to correspond with him regularly.

During this period, Vladimir was often away from home trying to manage the affairs of his oil company. Sonya sought to persuade her husband to break his ties with the company, which she rightly viewed as a dubious enterprise. She hoped in vain that he would eventually return to his scientific work. Vladimir, who had taken his doctorate from the University of Jena in 1872 and had since made a name for himself through his work in paleontology, received an appointment as *a Dozent* at Moscow University in 1880. But even this proved of little consequence. Her hopes crushed, Sonya's relationship with her husband grew more and more distant. Finally (evidently in the spring of 1882), she wrote him the following mournful words: "Our dispositions are so different; you are at times capable of making me really crazy and I can only come to my senses when I am left alone. If I look at matters soberly, then I find that you are completely right, and it is best for both of us if we live our lives separately" (translated from (Kochina 1981, 103)). Weierstrass spoke of their "irredeemably broken marriage" (letter to Kovalevskaya from 5 August, 1882 ((Bölling 1993a), letter 111, p. 1). And indeed each went his separate way: Vladimir traveled to North America on business, and Sonya stayed in Paris where she joined her friend Maria Jankowska (1850–1909), a Polish socialist and journalist whom she met that year. According to a remark of Charles Hermite (1822–1901), a divorce procedure was planned, but it never came to pass. It was here in Paris that Sonya learned of her husband's suicide. For five days she refused to see anyone or even eat. After that she fell into a coma and the doctor, whom she refused to see earlier, could finally be brought in to treat her.

Following her recovery, she traveled to Berlin in the summer of 1883 in order to meet with Weierstrass. There she discussed with him her work on the mathematical treatment of the refraction of light as well as her plans to begin lectures

Fig. 5.2 Mittag-Leffler Institute, the former home of Gösta Mittag-Leffler.

in Stockholm. Before her arrival, Weierstrass had no idea of the fateful blow that his former student had suffered, but her appearance told him right away that she was in very bad health. Kovalevskaya's desire to take up an academic career did not result from Vladimir's death, rather it had been her intention for some time past. But now the realization of this early dream clearly took on considerable urgency for her. Back in early 1881 when he was associated with the University of Helsingfors, Mittag-Leffler already learned that she would be willing to accept a university appointment. His efforts there, however, proved unsuccessful, not so much because Kovalevskaya was a woman but because she was a Russian. Just before he took up his new position at the Stockholm *Högskola* in September 1881, Mittag-Leffler informed her that he thought the chances would be better at this new institution. His evident interest in the realization of this plan can be seen from a letter he wrote her on June 19, 1881: "I have no doubt that once you are here in Stockholm ours will be one of the first faculties in the world" (Kochina and Ozhigova 1984, 27) (Fig. 5.2).

Weierstrass remained skeptical. He even feared that Mittag-Leffler's efforts on behalf of Kovalevskaya might compromise his own situation in Stockholm. For this reason, Kovalevskaya also thought that it would be best for all concerned if, for the time being, no further steps in this direction were taken, at least not until she found time to complete her ongoing work. Mittag-Leffler, on the other hand, did not wish to postpone plans for her appointment, partly because he was concerned that Stockholm might ultimately lose her to another institution. As he tried to assure her on 25 February 1882, "Don't you think that it is only right that the one who comes first should also have the best chances?" (Kochina and Ozhigova 1984, 32–33).

Kovalevskaya's health improved during the course of her stay in Berlin. It was during this time that the situation regarding her teaching position was finally resolved. In the meantime, she continued to work with success on her paper dealing with the refraction of light. Just before Weierstrass left Berlin for Switzerland (where he spent the winter of 1883–84), she gave him a small *exposé* summarizing the results she had by then obtained. Already in August 1883 he communicated a detailed report on these results to Mittag-Leffler. That month, Kovalevskaya left for Odessa, where she participated in the Seventh Congress of Russian Natural Scientists and Physicians from August 30th to September 9th. On September 3rd, she lectured on the "Integration of the partial differential equations that determine the refraction of light in a transparent crystalline medium." Full of thanks, she wrote to Mittag-Leffler from Odessa, "I am very grateful to the Stockholm *Högskola,* the first among all the European universities to open its doors to me. .. I hope to remain many years and to find there a second home" (Kovalevskaya to Mittag-Leffler, 9 September 1883 (Kochina and Ozhigova 1984, 34)). After the Odessa Congress, she spent some time in Moscow and Petersburg, and from there, as already mentioned, she traveled to Stockholm in mid-November. Originally she planned to spend two or three months once again in Berlin with Weierstrass in order to fill in some of the gaps in her work and to prepare her lectures. But this visit never came about due to Weierstrass's prolonged absence from Berlin.

We now have reached the point when Kovalevskaya composed her first letter to Weierstrass after arriving in Sweden.

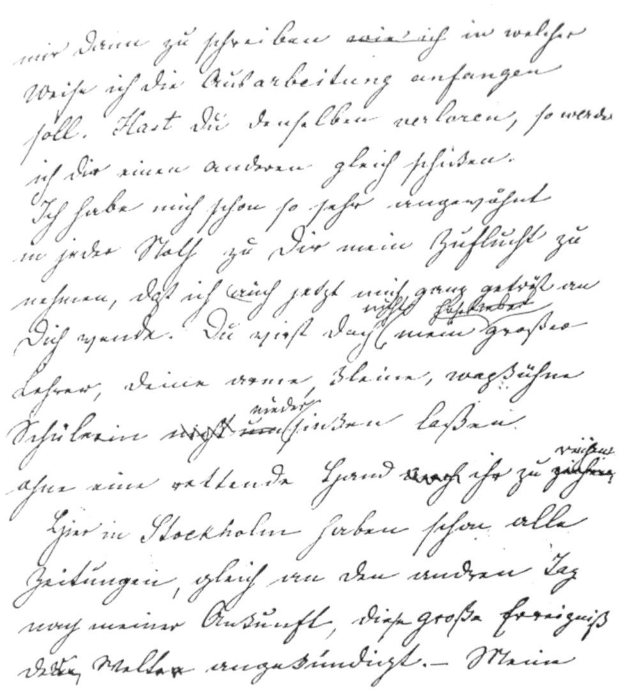

Fig. 5.3 Fragment from Kovalevskaya's letter to Weierstrass.

It should be noted that the English translation of the text that follows gives only a very rough idea of the highly idiosyncratic German of the original. In her biography of Kovalevskaya, Anne Charlotte Leffler (1849–1892), the sister of Gösta Mittag-Leffler, mentioned that although Sonya had studied for several years in Germany, she always spoke a rather broken German (cf. note 7). "She always spoke fluently, always succeeded in expressing what she wanted to say, and in giving an individual stamp to her utterances, however imperfectly she spoke the language she was using. When she had learned Swedish she had nearly forgotten all her German, and when she had been away from Sweden a few months, she spoke Swedish very badly on her return" (Leffler 1895, 59).

The original text is presented as part of a more detailed article in *Historia Mathematica* (see (Bölling 1993b)). The present text of Kovalevskaya's letter (Fig. 5.3) is followed by some brief explanations and commentary on its contents.

Sonya Kovalevskaya to Karl Weierstrass Stockholm, Early December 1883

. . . [the students] are all very talented and have a feel especially for function-theoretic views.[1] Last semester Mittag-Leffler lectured on the theory of analytic functions, and now he has just finished a course on analytic functions of several variables based completely on your lithographed text.[2] Next semester, as I said, he will deal with ordinary differential equations, so that lectures on partial differential equations would appear to tie in naturally with his own. It is true that I regret somewhat that I did not choose from the first to lecture on the calculus of variations, as I own such a good copy [of your lectures] and you would have perhaps allowed me to base my lectures on these.[3] Unfortunately, this thought only occurred to me after it was too late. If, however, you declare yourself dissatisfied with my choice for my first lectures, perhaps I could still take refuge in this subject. But please, be so kind, my dear best friend, and help me by giving me your advice in my distress.[4]

Also in another matter I very much need your support and your help. I wish to turn to the detailed work *[Ausarbeitung]* on my last study,[5] as it is most necessary that it appear this winter in *Acta Mathematica,* and without your help I cannot take a step forward. Would you not be so very kind and read through the small exposé that you took with you and then write me as to how I should begin my work.[6] If you have lost it, I will send you another right away. I have already gotten so used to flying to you in every emergency, so that I again with assurance turn to you. You, my supreme teacher, would not

[1]Following Weierstrass's suggestion, Kovalevskaya's first lectures were to be held before a special group chosen by Mittag-Leffler as a kind of test to determine whether she possessed the requisite talent to be a successful teacher. It is conceivable that she is referring here to this special circle; another possibility, of course, would be that she is writing about those who were attending Mittag-Leffler's lectures.

[2]As a young *Dozent* in Uppsala, Mittag-Leffler received a stipend to study abroad, which he used to study first in Paris and then under Weierstrass in Berlin during 1874–1875. Thereafter, he played a key role in disseminating Weierstrass's mathematics throughout Scandinavia.

[3]In 1882, Weierstrass had a copy made of his lectures on the calculus of variations for his pupil.

[4]Weierstrass approved of Kovalevskaya's choice for her lectures, particularly since in her dissertation she had already done some original work in this field (Theorem of Cauchy-Kovalevskaya). Beyond this, he gave her some hints about which methods and results she should present in detail. Kovalevskaya gave her first lecture on 30 January 1884 (she spoke in German).

[5]During 1883, Kovalevskaya had worked intensively on the mathematical treatment of the refraction of light in a crystalline medium (integration of the Lamé differential equations).

[6]For his vacation on Lake Geneva, Weierstrass took the *exposé* Sonya gave him. He also promised to study it and give her advice, but then he put this off over and again. In her work, Kovalevskaya applied an unpublished method of Weierstrass for the integration of linear partial differential equations with constant coefficients. During the summer of 1884, she was urging him to complete the final version of his work on this subject so that she could submit her own. In a letter of 13 September 1884, he wrote her that he was "horribly tired" and that this had "made him apathetic and filled him with antipathy for all thinking and writing," and he described himself as "mentally exhausted *[gehirnmfide]*" (Bölling 1993a, letter 131, p. 1). He therefore left it to her to use his unpublished work as she wished. Her paper appeared the following year in volume 6 of *Acta Mathematica*.

let your poor little bold[7] student drown without extending a hand to save her.

The day after my arrival here in Stockholm all the papers announced this great world event. My arrival itself did not come off without a rather humorous adventure. The trip from Petersburg to Stockholm was quite arduous for me, since I still could not understand a word of Swedish. As bad luck would have it, the steamship that carried me to Stockholm had but one single man on board who spoke German or any language for that matter other than Swedish. He was quite friendly to me, and in my distress I was happy to take advantage of his company. But since I already knew that several newspapers, both Russian and Swedish, had written about me, I tried to behave as simply and modestly as possible so that he would not guess why I was travelling to Stockholm, and so as not to draw general attention. He appeared very interested to know why I was going to Stockholm so utterly alone and without knowing the language.[8] I don't know exactly what I told him, but he apparently imagined that I was some kind of little governess, alone and unprotected in the world, on my way to Stockholm to take a position there with a family.

As our steamship departed, I had telegraphed Mittag-Leffler [the telegram is pictured here], but the trip went so exceptionally well that we arrived in Stockholm 2–3 h earlier than we had expected. Because of this, Mittag-Leffler was not at the harbor to meet me. Well, you can well imagine my forlorn state, especially since my letters to Mittag-Leffler had always been addressed to the University and I didn't even know his actual address.

Naturally I was relieved when the German-speaking man offered to accompany me to a hotel he evidently knew of, and since there was nothing else I could do under the circumstances, I thankfully accepted his proposal. But then, I suppose in accordance with his impression of me, he took me to a *really* out-of-the-way hotel where no one could understand a foreign word and where the service was altogether [lacking?]. After I had passed enough time that this sad situation became clear to me, I wanted to find the man again and ask him to take me elsewhere, but as I approached the room he had taken, I noticed that there were already a number of other men with him. There were loud voices and laughter coming from his room, and I did not dare to knock. So there was nothing left for me to do but return to my [room?] again and quietly wait several hungry hours in the saddest state in the world.

Only around evening did Mittag-Leffler succeed in learning with [great trouble?] and inquiries what had become of me, and then he came with his whole family, wife and sister, in the most elegant carriage to take me away.[9] . . . to find me in such a sad [state?] . . . .[10] I have seldom seen anything so peculiar as the astonishment of the hotel employees and my travelling companion when they learned that I am such a highly regarded lady. For you must know that Stockholm is the funniest little city in the world in which everything about everybody is known immediately and every little incident takes on the proportions of a world event.[11]

In order to give you an idea of what Stockholm is actually like, I will tell you about how a democratic newspaper announced my arrival the next day approximately as follows: This is not the visit of an insignificant prince or some other distinguished personality that we today have to report to our readers. No, it concerns something completely and incomparably different. The princess of science, Frau Sophie von Kowalevsky, has honored our city with her visit and plans to lecture at our university.[12]

How do you like that? Mittag-Leffler has already gathered a whole collection of newspaper announcements about me, but this one is by far the nicest. My letter has already become so long that I am compelled to bring it to a close. I would like to have told you about the grand soirée that Mittag-Leffler gave in my honor, during which I was introduced to most of the notables of the University here.[13] I liked Herr Nordenskjöld most of all. That is a wonderful man and so

[7]The German reads *wagkühne,* an invention of Kovalevskaya's that probably stands closest to the word *wagemutig.* Anne Charlotte Leffler mentions that Sonya's "German friends used to laugh at the ridiculous and often impossible words she coined. She never allowed herself to be stopped in the flow of her conversation by any such minor considerations as the correct choice of words" (Leffler 1895, 61).

[8]Kovalevskaya afterward made an addition to this sentence and then added about three more words that are impossible to decipher.

[9]Possibly she meant to add two more words to complete the sentence.

[10]Kovalevskaya added this sentence in which approximately four words appear to be missing.

[11]The details given here about her arrival in Stockholm are entirely new. Anne Charlotte's biography merely states: "She arrived from Finland in the evening by boat, and came as a guest to my brother [Mittag-] Leffler's house" (Leffler 1895, 55). According to the recollections of Anne Charlotte, she first greeted Sonya only on the morning of the following day.

[12]On 19 November 1883, the Swedish newspaper *Dagens Nyheter* carried an article about Kovalevskaya entitled "An Important Guest in Stockholm." The report began with the words: "This does not have to do with an insignificant king or prince from some friendly nation but rather a queen from the empire of science." Before giving a few details about the monarch herself, the article pointed out that an earlier published account in another newspaper indicating that the Russian widow Kovalevskaya had come to teach as a "Privatdozent at the Stockholm Högskola" was incorrect, and her teaching activity had been arranged privately with Mittag-Leffler for a specially selected group.

[13]Anne Charlotte recalled this social gathering in these words: "My brother, as soon as she arrived, told her that he wanted to give a *soiree* in order to introduce her to his scientific friends. But she begged him to wait until she could speak Swedish. This seemed to us rather optimistic, but she kept her word. In a fortnight she could speak a little, and during the first winter she had mastered our literature . . ." (Leffler 1895, 58).

simple, and even a democrat as well.[14] Herr, and especially, Frau Gyldén are also very nice people.[15] For the most part, I have been very busy learning the Swedish language and in the last 2 weeks I have made significant progress, although it is indeed difficult, much more so than I had believed. Still, I can already understand what I read and also a little of that which is spoken around me. Naturally I can't yet speak it myself.[16] Tomorrow I must go to an authentic Swedish gathering where Swedish alone is supposed to be spoken! It will be interesting to see how I make out.

I implore you, my dearest friend, to write me very soon. I await the next direct news from you with altogether indescribable anticipation, all the more so, for even though I am [so honored?][17] here, I feel really lonely and estranged. May the spring come quickly so that I can see you again, strong and healthy, in Berlin. My best and warmest greetings to your dear sisters and please do not forget

Your most devoted

Sonia

The question naturally arises whether Weierstrass ever received the letter corresponding to this fragment, and, if so, what kind of reply he gave. As a matter of fact, his answer to Kovalevskaya exists, for on the 27th of December he wrote to her, "Now I will answer your letter from Stockholm (undated) in a suitable fashion. ... . You have thus crossed the Rubicon. I had hardly thought that you would still decide to do so this winter" ((Bölling 1993a), letter 124, pp. 1–2). Weierstrass also gave her the advice she requested regarding her lectures (as mentioned in note 4). He apologized for not going into her work on the refraction of light, saying that he found writing to be difficult, but promised to do so later. Regarding the commotion prompted by her arrival in Stockholm, he wrote, "That so much has been and will continue to be spoken about you is not good, but that is Swedish nature and the good Mittag-Leffler himself is not free of this" (*ibid.*).

Weierstrass's answer also allows us to say something about the contents of the missing first part in the draft of Kovalevskaya's letter. This evidently began by inquiring about the health of her former teacher, because he describes in some detail his present rather unsatisfactory condition. Beyond what is contained in the fragment, it is possible that Kovalevskaya also said something more in the beginning of her letter about the city of Stockholm and the provincial impression it made on her, as Weierstrass replied: "What you write about the *Kleinstädterei* in Stockholm is perfectly understandable to me." He also thanked her for telling him the "charming story about Mittag-Leffler's potato sacks over which we all got a good chuckle." This amusing episode can be reconstructed from the letters that Weierstrass's sisters exchanged with Sonya. A shipload of 30 sacks of potatoes ordered by Mittag-Leffler had arrived in Stockholm. However, Mittag-Leffler had not received a permit to import them. Finally, after many futile discussions, he had to personally consult with the King before he was granted the permit; but in the meantime the ship carrying all his precious potatoes had departed for England.

To close this mini-adventure into the remains of a burned letter, we cite the final words that Sonya's revered teacher wrote to her as she prepared herself for a fresh start in her new life and career ((Bölling 1993a), letter 124, p. 6):

And now a warmest fare-thee-well, my dear friend, and my heartfelt wishes that the New Year may bring you all the best, not in the form of wonderment on the part of Stockholm journalists, but in the satisfaction that only serious striving and successful work can guarantee. To that, best of luck!

Your true friend,

K.W.

Acknowledgments I would like to thank the Institut Mittag-Leffler for the kind permission to publish the materials used in this paper, which was completed together with the translation during my joint stay at the Institut with D. E. Rowe in February 1991 – exactly 100 years after the death of Sonya Kovalevskaya.

[14]The reference is to Eric Nordenskjöld (1832–1901), Swedish geologist, geographer, and Arctic explorer.

[15]This couple was Hugo Gyldén (1841–1896) and his wife Teresa. Gyldén, a Swedish astronomer, was a well-known authority on celestial mechanics. Nordenskjöld and Gyldén later played a vital role in securing Kovalevskaya's appointment to a five-year professorship beginning in June 1884. Following Kovalevskaya's death, Teresa Gyldén cared for Sonya's daughter until she had completed primary school in Sweden, after which she returned to Moscow.

[16]Anne Charlotte reported that during the first weeks of her stay, Sonya did nothing but read Swedish from morning to night (cf. note 13).

[17]Meaning unclear.

References

Bölling, Reinhard, ed. 1993a. *Briefwechsel zwischen Karl Weierstrass und Sofja Kowalewskaja*. Berlin: Akademie-Verlag.

———. 1993b. Zum ersten Mal: Blick in einen Brief Kowalewskajas an Weierstraß. *Historia Mathematica* 20: 126–150.

Cooke, Roger. 1984. *The Mathematics of Sonya Kovalevskaya*. New York/Berlin: Springer.

Grattan-Guinness, Ivor. 1971. Materials for the history of mathematics in the Institut Mittag-Leffler. *Isis* 62 (3): 363–374: (213).

Koblitz, Ann Hibner. 1983. *A Convergence of Lives: Sofia Kovalevskaia: Scientist, Writer, Revolutionary*. Boston/Basel: Birkhäuser.

Kochina, Pelageia Ya. 1981. *Sofya Vasilevna Kovalevskaya*. Moscow: Nauka.

Kochina, Pelageia Ya., and E.P. Ozhigova, eds. 1984. *Perepiska S. V. Kovalevskoy i G. Mittag-Lefflera, Nauchnoye nasledstvo*. Vol. 7. Moscow: Nauka.

Anne Charlotte Leffler. 1895. *Sonya Kovalevsky*. London: Fisher Unwin.

6 Who Linked Hegel's Philosophy with the History of Mathematics?

(*Mathematical Intelligencer* 35(1)(2013): 38–41; 35(4)(2013): 51–55)

Standard histories of mathematics are filled with names, dates, and results, but seldom do we find much attention paid to the contexts in which mathematics was made or past achievements recorded. Yet by widening the net, one can easily retrieve many interesting examples that reveal how mathematicians thought about these matters and much else besides. This column deals with one such person – whose identity readers are hereby challenged to uncover – in order to illustrate in a particularly striking way the potential confluence of mathematical and philosophical ideas. The sources to which I allude below are all in print and readily accessible, so I have reason to hope that these hints will lead readers fairly quickly in the right direction. A reading knowledge of German and at least some tolerance for German philosophical prose will prove useful aids in this endeavor. Those who wish not only to answer the query above, but also to add their own reflections to mine, are invited to forward such musings directly to me.

Let me begin with a sweeping claim that pertains to understanding the rise of higher mathematics at the German universities during the course of the nineteenth century. Virtually all these mathematicians were products of an educational system that stressed classical languages. Furthermore, they taught in *philosophical faculties* (sometimes later divided into two sections, one for humanities, the other for natural sciences and mathematics). One should hardly have to wonder that these mathematicians were accustomed to pondering philosophical questions. Indeed, as members of the *Bildungsbürgertum* – the "educated citizenry" who formed a special elite class within German society – philosophizing in the academic sense of the term was a natural part of their collective cultural identity (Fig. 6.1). Thus, in this respect, our Professor X was hardly exceptional. Indeed, idealist thought in the tradition of Fichte, Hegel, and Schelling held considerable attractions for other contemporary German mathematicians. What makes this particular figure notable, however, is that he took up Hegel's philosophy and applied it directly to the history of mathematical ideas, as I shall briefly indicate below. However he did so – please take note of this – without explicitly mentioning Hegel's name. So how can we conclude that our mystery man was a true-blue follower of Hegel? My contention is that the mere allusion to Hegel's ideas suffices in this case: Professor X was an academic addressing his peers, nearly all of whom certainly knew their Hegel. Indeed, not a few in his audience would have been steeped in the very same *Weltanschauung*. So before proceeding further with his philosophical views, let me first offer a few hints about this individual's life and mathematical career.

Professor X was a leading Prussian mathematician of the nineteenth century. He undertook groundbreaking research on topics in a wide range of fields, spanning analysis, number theory, geometry, and mathematical physics. Several results he obtained along the way still bear his name, though with the passage of time he has been somewhat overshadowed by contemporaries whose reputations rose while his fell. Yet during his lifetime few mathematicians were as famous as this now somewhat forgotten figure. His great hero was Carl Friedrich Gauss, but he also recognized the genius of Gauss's younger contemporary, Augustin Cauchy. For the most part his relations with French mathematicians were stiff and cordial, even though he was once awarded a Grand Prize by the Paris Academie des Sciences. Soon afterward, he made important new discoveries in the field of geometrical optics. Although nearly all his activities were confined to the academic circles in which he moved, many viewed this distinguished scholar as an embodiment of the value system embraced by Germany's intellectual and cultural elite. Widely read, he represented a distinctively German current of idealist philosophical thought most often associated with the writings of Hegel, which brings us to the theme of this column.

Much of the historical literature devoted to mathematics ignores the abundant ties that once bound mathematical with philosophical thought. More recently, philosophers have engaged in historical and critical studies of mathematics, but these normally have rather special aims that historians would label as Whiggish. That term was coined in the wake

D.E. Rowe, *A Richer Picture of Mathematics*, https://doi.org/10.1007/978-3-319-67819-1_6

Fig. 6.1 Wedding photo of Felix and Anna Klein, née Hegel, 1875. She was the daughter of the historian Karl von Hegel and granddaughter of the famous philosopher Georg Wilhelm Friedrich Hegel (1770–1831).

of work by Herbert Butterfield, a historian who argued that the English Whigs tended to look at history as an inevitable march of events leading to the kind of political institutions they themselves held dear. Few would dispute that the Whigs had no lock on this kind of teleological view of history, and even today some seem to believe that the "end of history" lies on the horizon, an era in which a stable system of liberal democracies will rule around the world. One notes, however, that this "enlightened" opinion has yet to take root among the political elite in China, the newest great power of the twenty-first century.

Within the history of science, the term "Whiggish history" came to have a purely pejorative function: it was a warning sign that historians should avoid reconstructing the past as a road paved with "crucial experiments" and "key breakthroughs" that inevitably led to our own present-day understanding of the physical world. Such warnings tend to go unheeded when physicists write about the history of physics or when mathematicians write histories of mathematics. Yet any critical examination of the past must take into account not only the sources available but also how to use them thoughtfully and for what purpose. It should be an axiom for historical studies, whatever their subject, that the researcher engage the material at hand sympathetically, bringing to it a deep intrinsic interest born of *curiosity about the past* rather than preconceived ideas rooted in another time and place.

These general remarks came to mind as I thought about the speeches and writings of Professor X. Not only was he a Prussian patriot and one of the stellar mathematicians of the nineteenth century, he was also a scholar deeply moved by the momentous events of his time, which culminated with the creation of the modern German state in 1871. Steeped in classical learning but attuned to the latest currents of thought, he came to view the history of mankind as a struggle among its peoples to attain ever higher ideals of freedom. His notion of *Freiheit*, however, had little in common with freedom in the tradition of English liberalism, as enshrined in J. S. Mill's famous essay of 1859, *On Liberty.* A decade earlier, this now-half-forgotten Prussian defined with mathematical precision what he meant by the "realization of freedom" in a speech that received local acclaim. The year, to be precise, was 1848, and the ideas were straight out of Hegel, whose followers had by now split into leftist and rightist factions. Our mathematician belonged to neither; he was an "old school" Hegelian. His speech, however, shares certain affinities with another text from 1848, written by two leading Left Hegelians: the *Communist Manifesto.*

Philosophy and politics in the German states were by the mid-nineteenth century thickly entwined with the search for cultural roots that would serve as mainsprings for the construction of a new national identity. As German Romantic writers spun their own versions of exotic folk tales and sagas of medieval chivalry, our mathematician took on the philosopher's task of searching out the deeper meanings of these texts for the German people. No text exercised his imagination more than the *Nibelungenlied*, that newly recovered epic poem that came to be compared with Homer's *Iliad* as the legend around which the German nation might form its cultural identity. Richard Wagner comes immediately to mind here, of course, though his *Ring* cycle was a concoction far removed from the more authentic sources that moved our mathematician turned philosopher. The latter's interpretation of the story also departs from standard readings, including Wagner's: he tells us that Siegfried is *not*, in fact, the true hero of this tragedy but rather it is Hagen, his murderer, who eventually draws our sympathy! (In his recent book, *Shell Shock Cinema: Weimar Culture and the Wounds of War*, Anton Kaes describes how Fritz Lang exploited this same interpretation in the immediate sequel to his *Siegfried,* a film entitled *Kriemhild's Revenge.*)

Who was this mathematician qua philosopher and how did he manage to see the history of mathematics through the lens of Hegel's philosophy? Let us proceed by first recalling the famous philosopher and his time. Georg Wilhelm Friedrich Hegel (1770–1831) had a lively imagination. Struggling to launch his career as a *Privatdozent* in Jena, he felt the hand of history passing through the town just one day prior to the momentous battle in which Napoleon's forces would decimate the beleaguered troops of Prussia's King Friedrich Wilhelm III. Military and political historians would later mark that date, 14 October 1806, as a turning point in modern European history, but Hegel already had a vision of the *Weltgeist* in action when he spotted Napoleon on horseback the day before. Writing to a friend, he could hardly contain his emotions:

> I saw the Emperor – this world-soul – riding out of the city on reconnaissance. It is indeed a wonderful sensation to see such an individual, who, concentrated here at a single point, astride a horse, reaches out over the world and masters it . . . this extraordinary man, whom it is impossible not to admire.

Hegel's philosophy of history drew heavily on the notion of a grand design, one that gradually unfolds via a dialectical struggle over time. Napoleon merely served as the instrument of a higher purpose; he was acting as an agent of a transcendent Spirit, though he had little awareness of this as such. Hegel would later call this the "cunning of Reason" (*die List der Vernunft*), a notion that would, indeed, seem to apply most aptly to this mercurial French hero. Whether Hegel's World Spirit was religious or secular in nature remains a matter of scholarly dispute, but whatever the case may be, it would be hard to overstate the degree to which this *Weltanschauung* held sway within wider intellectual circles over the course of the nineteenth century and beyond. That story, however, begins three years after Napoleon's defeat at Waterloo, when in 1818 Hegel succeeded Johann Gottlieb Fichte as professor of philosophy at Berlin University, a new institution that had been founded under French occupation in 1810.

Fichte and Hegel have often been portrayed as glorifiers of the Prussian state, even though neither was Prussian by birth. Their philosophical systems, to be sure, were deeply indebted to the ideas of the most famous of all Prussian philosophers, Immanuel Kant, a writer who had far deeper affinities with mathematics than did they. Kant was universally regarded as the philosopher who broke with the German tradition of speculative metaphysics associated with Leibniz and Christian Wolff. Our Professor X not only shared this view, he saw Kant as playing for philosophy the same role that Gauss had assumed for mathematics, namely that both elevated their respective intellectual disciplines by introducing new critical standards and principles. Yet he also admired Hegel, a philosopher whose nebulous ideas have long been regarded as antithetical to rational thought. Indeed, from a modern, largely Anglo-American perspective, Hegel seems to represent a retrograde tendency in German philosophical thought, a return to the murky metaphysical tradition of his forbears, though admittedly with some new dialectical twists. But here we would do well to cast aside the "lessons" we have been taught about what went wrong in German history and look at the matter afresh.

As already suggested, Hegel's approach to history was bound together with a peculiar notion of freedom quite foreign to most other Western cultures. His World Spirit, which somehow guides the over-arching course of individual historical events, realizes itself in ever-higher forms of freedom. What this means in concrete terms can perhaps best be understood by quoting a typical passage from the master himself:

> . . . world history is the exhibition of spirit striving to attain knowledge of its own nature. As the germ bears in itself the whole nature of the tree, the taste and shape of its fruit, so also the first traces of Spirit virtually contain the whole of history. Orientals do not yet know that Spirit – Man as such – is free. And because they do not know it, they are not free. They only know that *one* is free; . . .. This one is therefore only a despot, not a free man. The consciousness of freedom first arose among the Greeks, and therefore they were free. But they, and the Romans likewise, only knew that some are free not man as such. This not even Plato and Aristotle knew. For this reason the Greeks not only had slavery, upon which was based their whole life and the maintenance of their splendid liberty . . . Only the Germanic peoples came, through Christianity, to realize that man as man is free and that freedom of Spirit is the very essence of man's nature. This realization first arose in religion, in the innermost region of spirit; but to introduce it in the secular world was a further task which could only be solved and fulfilled by a long and severe effort of civilization.

Many others from this time period were drawn to some such view of human history, but presumably only a few made any serious attempt to use Hegelianism as a master plan for understanding the evolution of mathematical knowledge. Our mathematician did just that, noting that while all early civilizations had a need for practical mathematics, only the Greeks possessed the capacity to produce a pure science of mathematics. True, they took certain things from earlier peoples, as Herodotus duly reported, but nothing that could explain the "Greek miracle." Just as one may assume that the origins of Greek architecture can be traced back to Egypt, no one would claim that the magnificent temples of the Acropolis took their inspiration from Egyptian models, and the same with mathematics: we cannot imagine that the intricate theories of the Greeks owe anything to the Egyptians. Why not? Because (following Hegel) "to accomplish this required a higher degree of intellectual freedom ("geistige Freiheit") that no ancient people other than the Greeks had attained."

As the Greek cultural achievements dissipated with the spread of Hellenism, Indian and Arabic mathematics enter into the picture, offering a new impulse through their work on algebraic problems. Yet their contributions, we are told, fell below the high ideals embodied in the classical texts of Euclid, Archimedes, and Apollonius. Why? Because their works reflected a more limited form of freedom, one that saw the introduction of new methods or tricks for solving algebraic equations, but in which they failed to arrive at a truly symbolic algebra. That step was reserved for the Christianized culture of Western Europe.

The mathematicians of early modern Europe achieved this by "emancipating the idea of magnitude from earlier concepts that were inessential and unfruitful." This breakthrough, however, could easily have led to a sterile formalism had it not been for another new intellectual impulse that came fast on the heels of the first, namely the introduction of the notion of continuous variable magnitudes. By this means "the mathematical constructs that were formerly confined to the sphere of static Being were now elevated to the free region of Becoming where they could at last move about and live." Mathematics, at this point, could embrace and contemplate the realm of the infinitely large and small, and from these investigations emerged the infinitesimal calculus of Newton and Leibniz.

One might note that this Hegelian picture of the history of mathematics is, in nearly all respects, thoroughly familiar; it is also largely outdated and, of course, wholly Eurocentric. No surprise, after all Professor X came from an intellectual world that glorified the Greeks almost like no other before or since. His picture of mathematical events after the discovery of the calculus can be gleaned from numerous texts, but instead of adding further layers to this story let me end with an anecdote taken from the writings of another contemporary German mathematician, a tale that bears on the role of historical allegory during this time. The theme concerns the traditional rivalry between French and Prussian mathematicians, as both seek to bestow honor and glory on their respective countries: "After France was with good fortune defeated on the field of battle, we [the mathematicians] have continued to fight on in the higher regions of thought, like the shadows that fought on in the saga of the Battle of the Huns, and we have brought forth many glorious scientific victories, supported by the Spirit of the Holy Alliance, to which Prussia belongs. And so we may acclaim that we are no longer stand second in the mathematical sciences."

Who wrote these words and to whom were they addressed? To what historical event does the saga of the Battle of the Huns (*Hunnenschlacht*) refer?

Answers to the Hegel Quiz

Earlier this year, I challenged readers to identify a certain well-known German mathematician who saw the history of mathematics through the lens of Hegel's philosophy. Regrettably, the response was close to nil, which might have something to do with the nature of the question, or the fact that it was pitched at those with an ability to read German, or perhaps I threw out the wrong kind of bait for this particular fishing expedition. Who, after all, reads Hegel any more, especially in the English-speaking world? Bertrand Russell once did, of course, but he also warned us that German Idealism tended to soften the mind and harden the heart. The adherents of Prussian militarism, glorified by Hegelian metaphysics, certainly did cast a long, dark shadow over European history. To the extent that Hegel's philosophy of history is known at all today, it has largely been repudiated as part of a reactionary tradition that culminated in the infamous German ideology of the Nazi state.

Yet surely we ought to look at events from so long ago more soberly and without filtering everything that occurred during the nineteenth century through the lens of the twentieth. The idea for this quiz came, in fact, from a reading course I conducted last year, which centered heavily on the speeches and popular writings of our mysterious Professor X, a representative figure of his day. I described him as "one of the stellar mathematicians of the nineteenth century, a scholar deeply moved by the momentous events of his time, which culminated with the creation of the modern state of Germany in 1871. Steeped in classical learning but attuned to the latest currents of thought, he came to view the history of mankind as a struggle among its peoples to attain ever higher ideals of freedom." That sounds like Hegel, but I also cautioned that the famous philosopher's name does *not appear explicitly* in the writings of our mystery mathematician. It was my contention that the mere allusion to Hegel's ideas sufficed, since "Professor X was an academic addressing his peers, nearly all of whom certainly knew their Hegel. Indeed, not a few in his audience would have been steeped in the very same *Weltanschauung.*" Perhaps this lack of an explicit textual reference discouraged certain history buffs, fact-finders who might have liked to activate their search engines in order to troll for this Hegelianized mathematical fish.

I did receive amusing feedback concerning Hegel from the Copenhagen historian of mathematics, Jesper Lützen, who also correctly surmised that the Prussian patriot I had described must have been Ernst Eduard Kummer (1810–1893). Lützen was tipped off by my remark that Professor X "made important new discoveries in the field of geometrical optics." Regarding Hegel's writings, he pointed out to me that Otto Neugebauer liked to entertain his Danish colleague, Asger Aaboe, by sending him little gems, like this remarkably murky (and seemingly untranslatable) pronouncement from Hegel about the essence of time. Neugebauer to Aaboe:

> "Guess who said that:
> Die Zeit, als die negative Einheit des Außersichseins, ist gleichfalls ein schlechthin Abstraktes, Ideelles. - Sie ist das Sein, das, indem es ist, nicht ist, und indem es nicht ist, ist.[1]
> Skidegodt (Danish for: "damn good")"

Kummer's Hegelian Orientation

To learn about Kummer's more sympathetic appreciation of Hegel, one need not search far and wide. In 1975 André Weil paid homage to Kummer by publishing his *Collected Papers* (Kummer 1975) in two volumes. Weil was especially interested in the papers in Volume I, containing Kummer's highly original contributions to number theory, about which he, as editor, wrote a lengthy introduction. Volume II is more of a hodgepodge that includes Kummer's early work in analysis, followed by his once celebrated investigations of ray systems and their caustic surfaces, including the famous quartic surface that bears his name. This second volume also contains a few book reviews as well as numerous speeches, some of which tell us a good deal about Kummer's public life, including his political and philosophical views. Among these, one finds his moving tribute to the memory of Dirichlet, his esteemed predecessor at the University of Berlin.

Weil clearly had no great interest in the documents gathered in Volume II. His introduction was at once brief – filling less than a page – and flippant, noting merely that Kummer's early papers on function theory dealt with special cases and were written before "Riemann and Weierstrass renovated the subject." Weil showed slightly more appreciation for Kummer's old-fashioned work in algebraic geometry, but he could not resist adding that these papers, too, were written "before Riemann's far-reaching ideas had had time to exert their influence." What he apparently admired most about these works was not their content but rather the author's tenacity, or as he put it Kummer's "amazing ability to cut his way through dense thickets of algebraic or analytic brushwood with frequently inadequate conceptual equipment." As for the reviews and speeches, these were merely of "historical interest," but still "it has been not thought proper to omit them."

Historians can be truly thankful to André Weil for that decision, since these texts offer a most vivid picture of the cultural and political world Kummer moved in. Here we find his speech from October 1848 as the newly elected Rector at the University of Breslau, delivered during the midst of fervent revolutionary activity (Kummer 1975, 706–716). After receiving numerous requests for copies of the text from those who heard him read it, Kummer agreed to have this speech printed in Breslau and distributed locally. His theme, fittingly enough for the year 1848, was academic freedom, but like a true mathematician he began by defining his terms! In order to pursue rational discourse on freedom in the special context of academic life, he explained to his audience, one must first establish a clear conception of what human freedom means in general terms. This highly political speech can be seen as Hegelian through and through, yet Kummer's eloquent text attains a level of clarity probably nowhere to be found in Hegel's own writings.

In 1855 Kummer left Breslau for Berlin, where he replaced Dirichlet, who had just succeeded Gauss in Göttingen. This year marks the beginning of the "golden age" of Berlin mathematics, which flourished for three decades under Kummer, Weierstrass, and Kronecker (Biermann 1988, 79–152). Kummer stood at the center of it all; indeed, he was seen as the very embodiment of Berlin mathematics, even if his two colleagues may be better remembered today. From 1863 to 1878 he was perpetual secretary of the physics-mathematics section of the Berlin Academy, which meant he was often called upon to deliver public speeches on ceremonial occasions. One of the most illuminating of these was a speech honoring King Wilhelm on his birthday, 22 March, 1866 (Kummer 1975, 775–788). This took place just a few months before the decisive war with Austria and her allies that would pave the way for German unification under Prussia.

In this speech, Kummer drew on the theme of German loyalty as exemplified in the *Niebelungensage,* a spiritual ideal he saw at work in modern Prussia under the benevolent guidance of its king. Distancing himself from the democratic strivings of the 1848 movement, he criticizes such causes as misguided partisan politics unworthy of the moral ideals of the German people and their united destiny. No need to invoke Bismarck's name here, any more than Hegel's, and yet this was all just a prelude to his main theme. Speaking in the name of the Prussian Academy, Kummer would now turn to consider the social and political conditions under which creative scientific thought can best flourish. Characteristically, he chooses not to discuss this matter in an abstract form but rather by viewing this problem from an

[1]Georg Wilhelm Friedrich Hegel, *Enzyklopädie der philosophischen Wissenschaften in Grundrisse,* Zweiter Teil: Die Naturphilosophie,1. Abteilung: Die Mechanik, A. Raum und Zeit, b. Die Zeit, § 258.

historical perspective, focusing on the case of mathematics in order to illustrate the underlying general principles.

Kummer's master plan here is highly Hegelian, beginning with the "Greek miracle," which presumes, of course, that Greek cultural achievements were clearly superior to those of the Egyptians or any other older civilizations. Following Hegel, the reasoning behind this was both obvious and clear: Greek civilization achieved a higher level of achievement than ever before, but "to accomplish this required a higher degree of intellectual freedom ("geistige Freiheit") that no ancient people other than the Greeks had attained." If Indian and Arabic mathematicians demonstrated new methods for solving algebraic problems, their achievements, according to Kummer, still fell well below those of Euclid, Archimedes, and Apollonius. After all, Islamic mathematics only mimicked the Greek form of rigorous argument. The new quantitative methods thus reflected a more limited form of freedom, hence they could not lead to a truly symbolic algebra, a step reserved for the Christianized culture of Western Europe.

This culture, in turn, could only advance by "emancipating the idea of magnitude from earlier concepts," by introducing the notion of continuous variable magnitudes. By this means "the mathematical constructs that were formerly confined to the sphere of static Being were now elevated to the free region of Becoming where they could at last move about and live." Kummer's next historical stage thus leads to the infinitesimal calculus of Newton and Leibniz. This sweeping vision of the history of mathematics as a dialectical development in which different, ever higher forms of freedom are realized, captures the very essence of Hegel's philosophy. Indeed, we have it on the authority of his lifelong friend Leopold Kronecker that "Kummer was a thoroughgoing Hegelian" (Boniface and Schappacher 2001, 223).

Interested readers can find many similar motifs in other speeches as well as in the reviews collected in Volume II of Kummer's *Collected Papers.* In one of those reviews he dealt at some length with the traditional rivalry between French and Prussian mathematicians, a passionate affair in which many on both sides were determined to bestow honor and glory on their respective countries (Kummer 1975, 695–705). Kummer likened these two nations' mathematical luminaries with stars in a great intellectual firmament, each glowing with a certain discernible intensity. With some pride he took note of the fact that France could only boast of having a single star of the first magnitude, namely Cauchy, whereas Germany had three: Gauss, Dirichlet, and Jacobi. Still, he was quick to point out that the French could count more stars of the second and third orders than the Germans, and indeed, French numerical superiority increased the deeper one looked into this sky. He imagined going all the way out to telescopic stars of the sixteenth order, claiming that by passing through this scale one would see that French mathematicians enjoyed an ever greater quantitative advantage over their German counterparts and that this disproportion increased as the stars grew dimmer. One must assume that French readers would have found this mathematical metaphor less than flattering, but it surely accords well with contemporary mid-century assessments of the relative strengths of these two rival communities. As usual, Kummer appealed to the natural course of history for an explanation: because the period of intellectual flowering had begun earlier in France, this country's development was already showing clear signs of maturity; Germany, by contrast, had only begun to attain a high mathematical culture, though its ascent was rapid. Kummer's generation was thus catching up fast.

This review, written in Breslau in 1846, also contains the suggestive quote that I included in the quiz. The passage reads: "[a]fter France was with good fortune defeated on the field of battle, we [the mathematicians] have continued to fight on in the higher regions of thought, like the shadows that fought on in the saga of the Battle of the Huns, and we have brought forth many glorious scientific victories, supported by the Spirit of the Holy Alliance, to which Prussia belongs. And so we may acclaim no longer to stand second in the mathematical sciences" (Kummer 1975, 695). Kummer cited this passage with decisive approval, which led me to ask: who wrote these words and to whom were they addressed? Furthermore, to what historical event does the saga of the Battle of the Huns (*Hunnenschlacht*) refer?

This passage comes, in fact, from the preface to the work under review, a collection of mathematical writings dedicated to King Friedrich Wilhelm IV of Prussia. The author was none other than Carl G. J. Jacobi, who owed special thanks to the king for enabling him to leave Königsberg and spend the year 1844 in Italy in order to regain his health. Jacobi, who had a deep interest in classical culture, took full advantage of this opportunity, visiting archaeological sites in and around Rome almost daily. The *Hunnenschlacht,* better known in English as the Battle of the Catalaunian Fields, was fought in 451 AD. This famous battle pitted the armies of Attila the Hun against a coalition of Roman and Visigoth forces. According to legend, the fighting was so ferocious that the dead warriors continued to fight each other in the sky as they rose to Heaven.

Steiner's Roman Surface

During his convalescence in Rome, Jacobi spent a considerable amount of time with his future colleague in Berlin, Peter Gustav Lejeune Dirichlet, and his wife Rebecka, sister of the composer Felix Mendelsohn-Bartholdy. Dirichlet's colleague, the Swiss geometer Jakob Steiner, also joined this party. During most evenings, Steiner and Jacobi would relax while discussing various geometrical problems, no doubt occasionally struggling to understand one another. Steiner had considerable respect for analytical methods, though scarcely any ability when it came to applying them, whereas Jacobi was, of course, a virtuoso when it came to wielding such instruments. What they talked about we will surely never know, but history records that it was during his stay that Steiner discovered the famous quartic surface, often called the Roman surface, that became a celebrated object of study after his death. In truth, though, the historical record documenthing Steiner's discovery is exceedingly thin. So to tell this story properly we should flash forward to the year 1863 and return once again to Berlin.

Only a few months after Steiner's death on April 1, 1863, Kummer presented a paper to the Berlin Academy (Kummer 1863) involving several new results concerning quartic surfaces that contain families of conics (see Chap 8 for computer graphics of these various quartics). By far the most interesting of his findings concerned an exotic quartic surface whose tangent planes cut the surface in pairs of conics. This real quartic is a rational surface containing three double lines that meet in a triple point; after passing through the surface these singular lines emerge as six isolated line segments that run off to infinity. Kummer had thus found a truly striking geometrical phenomenon, but he soon came to realize that this part of his paper was not altogether new. For in the course of writing up his results, Weierstrass informed him that Steiner, their recently deceased colleague, had stumbled upon this strange type of quartic surface many years earlier. Both knew how eccentric this self-taught Swiss genius had been, so perhaps it came as no surprise to them that Steiner never published his discovery; in fact, he nearly took this secret with him when he went to his grave.

Kummer's paper, submitted to the Academy on July 16, was followed by a short elaborative note written by Weierstrass, who explained the general idea behind Steiner's purely synthetic construction (Weierstrass 1863). Weierstrass mentioned that he felt compelled to write this up from memory since he thought it unlikely that Steiner had left behind any written account of this himself. Moreover, neither he nor Kummer indicated anything about Steiner's trip to Rome as the occasion for this discovery. At the time of Steiner's death, Weierstrass was nearly the only one in Berlin who enjoyed friendly relations with him, as he had become an increasingly embittered old man who likened himself with a "burnt-out volcano" (Biermann 1988, 84). As a Gymnasium student in Paderborn, Weierstrass had studied the early works of Steiner in *Crelle's Journal*, and as colleagues in Berlin he had promised the feisty Swiss that he would continue the tradition of offering courses in synthetic geometry once Steiner could no longer do so. He kept that promise; during the decade that followed Steiner's death Weierstrass taught courses in geometry no fewer than seven times.

Later in the year 1863, Kummer submitted a short note to the Berlin Academy on the Steiner surface, accompanied by a plaster model illustrating the case: $y^2z^2 + z^2x^2 + x^2y^2 - 2cxyz = 0$. Somewhat ironically, his discovery of the Steiner surface actually preceded his more celebrated work on so-called Kummer surfaces – quartics with 16 double points and 16 double planes – which began shortly after this (Kummer 1975, 418–432). In fact, all his work on quartic surfaces during the 1860s was closely tied to his studies of ray systems in geometrical optics, in particular researches of William Rowan Hamilton (see Chap. 8). This context being part of an older physical tradition, it is quite misleading to speak of Kummer's papers in Volume II as contributions to algebraic geometry, much less to wonder – as André Weil did – about how Kummer could have found these surfaces without the help of theta functions (Kummer 1975, vii). Clearly his discoveries were guided by earlier intensive investigations of the Fresnel wave surface, which, as Kummer himself noted, represents a "Kummer surface," but with only four *real* double points (the remaining 12 being imaginary). As for the "Steiner surface," Kummer's initial paper led to a flurry of other related work by Schroeter, Clebsch, Cremona, Lie, and others. After two decades, Steiner's name was securely attached to this object, even though Weierstrass's memory was the only known link connecting Kummer's work with what Steiner had earlier conceived.

But with the early 1880s came another strange turn of events. As the editor of Steiner's *Collected Works*, Weierstrass stumbled upon a brief note in the geometer's papers that described his original construction. Except for a few details, Weierstrass' recollection from 1863 now proved to be quite accurate after all. So in 1882 he decided to include this sketchy note as an appendix to the second volume of Steiner's *Werke* (Steiner 1882, 723–724), adding to it his own now much elaborated recollection of his discussions with Steiner about this surface (Steiner 1882, 741–742). Here, for the first time, we read when and where Steiner had originally found this surface, namely in 1844 while in Rome; indeed, we learn that it was for this reason that he liked to refer to this quartic as his "Roman surface."

Weierstrass further relates that Steiner was truly tortured by the question of whether his construction actually led to a quartic surface. He suspected, in fact, that the Roman surface might be a sextic that contained an imaginary component as a "ghost image." It was not until around one year before his death that he approached Weierstrass about this, asking him to clarify the matter analytically. Weierstrass then relates how he was able to do this rather easily, producing a rational parameterization of the surface by homogeneous quadratic polynomials. Still, Steiner made no further use of this information, which only makes this story all the more puzzling.

In the year of Steiner's death, *Crelle* published a short obituary for Steiner written by the era's leading analytic geometer, Otto Hesse (Hesse 1863). Hesse's tribute to the "leading geometer of his day" evokes the same sense of mystery, combined with awe, but also regret. Referring to Steiner's later period, he saw this as marked by his struggle with the imaginary, or as Steiner liked to say, his quest to seek out those "ghosts" that hide their truths in a strange geometrical netherworld. As Hesse clearly saw, this was a fight he could not win, especially given the constraints of his purely synthetic approach to geometry. Still, he was clearly awed by Steiner's daring vision, likening his numerous unproved results with the many puzzles Fermat bequeathed to the mathematical world. Still, in Steiner's case the sense of tragic loss prevailed, as he spent his final years feeling like a dormant volcano. One would like to imagine him in happier times, when he was sharing his latest mathematical discoveries with Jacobi in Rome.

Addendum: Plane Sections of a Steiner Surface (Computer Graphics by Oliver Labs)

Kummer's model of the Steiner surface (Fig. 6.2) exhibits the case

$$y^2z^2 + z^2x^2 + x^2y^2 - 2cxyz = 0,$$

a quartic with four double tangent planes that touch the surface along four circles. The surface also has three singular lines running through it that intersect in a triple point.

Here the three double lines are the coordinate axes and the triple point is (0,0,0). Taking the plane z = 0, we see that this cuts the surface along $x^2y^2 = 0$, which represents the coordinate axes x = 0, y = 0 each counted twice. This shows that the coordinate planes are tangents to the surface that touch it along the two respective coordinate axes.

Plane sections of the Steiner surface are rational quartic curves, hence genus $p = 0$, and so contain three singular points. These occur where the plane intersects the three singular lines. Figures 6.3 and 6.4 illustrate typical cases:

Fig. 6.2 Kummer's model of a Steiner surface (Photo from Gerd Fischer (ed.): *Mathematische Modelle. Aus den Sammlungen von Universitäten und Museen*. 2. Bde. Braunschweig: Vieweg 1986. Courtesy of Gerd Fischer).

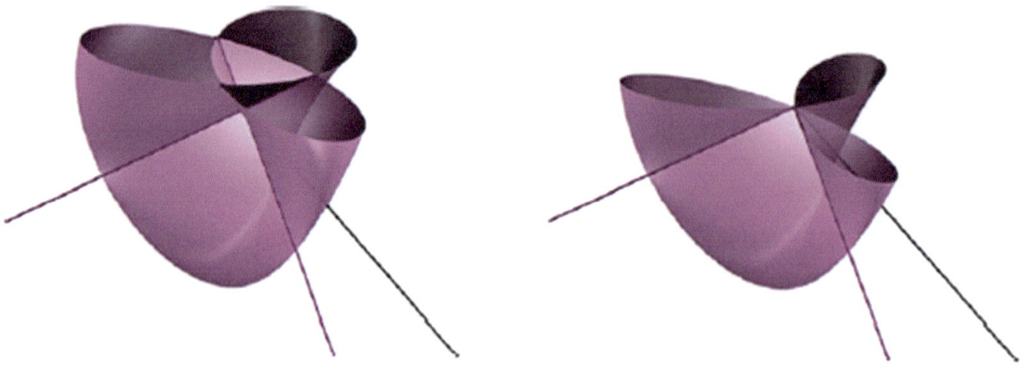

Fig. 6.3 (*Left*) quartic curve with 3 ordinary double points; (*Right*) quartic with a triple point.

Fig. 6.4 (*Left*) quartic with 3 cusps; (*Right*) quartic with 3 isolated double points.

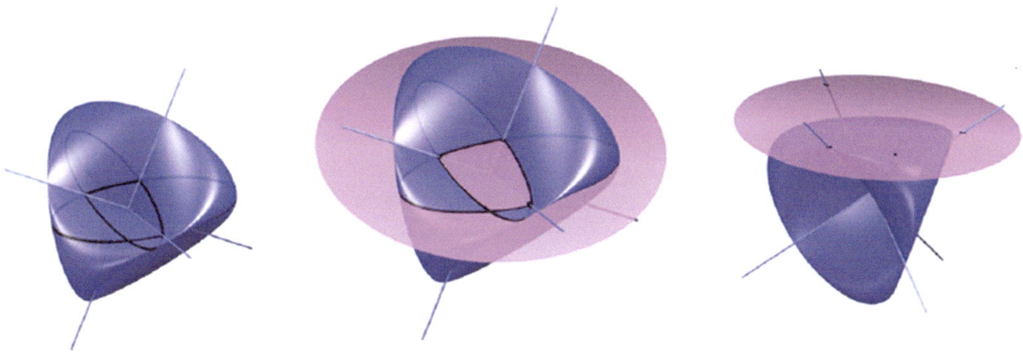

Fig. 6.5 (*Left*) a pair of real conics on the surface; (*Middle*) a pair of real conics in the tangent plane to the surface; (*Right*) a pair of imaginary conics with 4 real points of intersection.

In the case of tangent planes (Fig. 6.5), the quartic splits into a pair of conics whose four points of intersection are the point of tangency and the 3 points where the tangent plane meets the 3 singular lines of the surface. As a special exception, this pair of conics collapses into a single double conic in the case of 4 exceptional double planes.

References

Biermann, K.-R. 1988. *Die Mathematik und ihre Dozenten an der Berliner Universität 1810–1933*. Berlin: Akademie-Verlag.

Boniface, Jacqueline, and Norbert Schappacher, eds. 2001. Sur le concept de nombre en mathématique. Cours inédit de Leopold Kronecker à Berlin (1891). *Revue d'histoire des mathématiques* 7 (2): 207–275.

Hesse, Otto. 1863. Jakob Steiner. *Journal für die reine und angewandte Mathematik* 62: 199–200.

Kummer, E.E., 1863. Über die Flächen vierten Grades, auf welchen Schaaren von Kegelschnitten liegen, *Monatsberichte der Königlich Preußischen Akademie der Wissenschaften zu Berlin,* 16. Juli 1863, 324–336; (Reprinted in (Kummer 1975, 405–416).)

———. 1975. In *Ernst Eduard Kummer, Collected Papers*, ed. A. Weil, vol. II. Berlin: Springer.

Steiner, Jakob. 1882. In *Jakob Steiners Gesammelte Werke*, ed. K. Weierstrass, Bd. II ed. Berlin: Reimer.

Weierstrass, Karl. 1863. Bemerkung zum Vorstehende, *Monatsberichte der Königlich Preußischen Akademie der Wissenschaften zu Berlin,* 16. Juli 1863, 337–338.

Part II

The Young Felix Klein

7 Introduction to Part II

Felix Klein's name has long been identified with Göttingen mathematics, and rightly so, since he was the principal architect of its modernized community. In assuming that role, he also spared no effort in portraying himself as the apotheosis of that celebrated mathematical tradition, which came to full fruition after the turn of the century. His reputation as a great mathematician by then securely established, Klein focused his attention on various large-scale projects, leaving to others – notably David Hilbert and Hermann Minkowski – the responsibility of training younger researchers. Although by then only in his fifties, Klein already seemed an *eminence grise* to that flock of ambitious young men who encountered him. He was remembered by that generation as an aloof Olympian, an image that lived on through Richard Courant and Max Born, two central figures in the Göttingen community during the Weimar era who lost their positions when the Nazis came to power (see Part V). By the end of their 12-year reign, few of those who had known the young Felix Klein were still among the living.

In the essays that follow, I recall various episodes from Klein's largely overlooked, but highly relevant early career. Doing so helps to throw the traditional Göttingen – Berlin rivalry, as sketched in Part I, into sharper relief. Here it should be borne in mind that Klein saw his whole life as a heroic struggle, one that began as a fight for recognition. No one appreciated that dimension of his personality better than Hilbert, who struck this very chord in a speech honoring Klein on his sixtieth birthday (see Chap. 16 in Part III):

> . . . [R]ight from the beginning, you stressed the general role of geometric intuition, placed it in the foreground, and cultivated it through drawings and models, and by emphasizing the physical, kinematical, and mechanical aspects of mathematical thinking. Riemann was the name that stood on your flag and under whose sign you have been victorious right down the line – victorious against opponents because of the correctness of your ideas, which brought you support from altogether unexpected quarters, from Minkowski, for example, who continually utilized geometric intuition as an arithmetical method.

A closer look at Klein's early publications actually shows that this Riemannian thread entered in a quite special way, at least initially. Up until the late 1870s, his work focused mainly on problems concerning the visualization of the real and imaginary parts of geometrical objects defined by equations over the complex numbers (about which see Chap. 11). Only a small sampling of these ideas can be touched on here, but hopefully enough to show that Klein's vaunted visual style was always a work-in-progress.

Today, only a few experts have any real knowledge of what Klein accomplished in this direction, even though he promoted geometric intuition all his life. Yet a century ago, his name and fame continued to shine on brightly. Indeed, hardly anyone could claim comparable success when it came to refashioning themselves and their accomplishments. During the years immediately following the collapse of the German empire, Klein was afforded the luxury of being supported by a number of young mathematicians, who as paid assistants helped him in preparing the three volumes of his annotated collected papers (Klein 1921–1923). In many ways, this arrangement mimicked an earlier one that enabled Klein to churn out written texts for his lectures courses, a task that placed heavy burdens on those called upon to produce them. Still, the stakes were clearly far higher for those who assisted him in annotating his *Gesammelte Mathematische Abhandlungen*. What is more, the financial burden for this last grand endeavor proved to be at least as great as the intellectual one, considering that the final volume came out almost at the peak of hyperinflation in the Weimar Republic.

Little more than a decade later, Springer published another set of three volumes of collected papers under similarly inauspicious circumstances, this time for David Hilbert (Hilbert 1932–1935). Ironically, the resulting publication was a haphazard effort that fell far below the standard set by the edition of Klein's work, despite the fact that Hilbert had produced an impressive number of significant papers, far more than his senior colleague had written. Part of the reason for this disparity can be traced to two fundamentally different attitudes when it came to collective mathematical

D.E. Rowe, *A Richer Picture of Mathematics*, https://doi.org/10.1007/978-3-319-67819-1_7

knowledge. Klein was constantly looking backward, not only in his own works but also to those of past heroes, like Gauss and Riemann. Hilbert, by contrast, was intent on shaping the future course of mathematical research. He had little desire to return to his earlier work, much less produce the kind of extensive annotations Klein wrote, commentaries that, however, helped to bring many older papers back to life.

Some aspects of Klein's early career are quite familiar even today, in particular his collaboration with Sophus Lie that culminated with the publication of Klein's "Erlanger Programm" (Klein 1872b, 1893). Readers already familiar with his more famous exploits as a young man may well wonder why these receive such short shrift here. Part of the justification for omitting topics like the prehistory of transformation groups is that these developments from the years 1869 to 1872 have been described elsewhere, for example in (Hawkins 1989; Hawkins 2000; Rowe 1989b). Most of the themes I touch on here, by contrast, are less well known, though they deserve to be portrayed as part of a richer picture of the period. Many fall under the general category of "anschauliche Geometrie," the heading Klein chose for the first 15 papers in volume two of (Klein 1921–1923). Part of this story centers on Klein's fascination with physical models, which served as important tools for his geometrical researches. That interest went back to his student days, when he worked closely with the geometer-physicist Julius Plücker. Nevertheless, geometrical models first took on real importance for him during his years as a post-doc in Göttingen, as described in the opening essay below. To gain a sense of how Klein thought about these things, though, a few words about his researches in line geometry are necessary.

Klein's academic training as a mathematician was in many respects unusual (Rowe 1989a). As a student in Bonn, he took courses that spanned the entire curriculum in the natural sciences (Schubring 1989). At age 17, he was already assisting Plücker in preparing experiments for his physics courses. Plücker held a joint chair in mathematics and experimental physics, but by this time his research interests had swung back to geometry after a nearly 20-year hiatus from the field. Working at Plücker's side for nearly three years, Klein became an expert in the new field of line geometry. The lines in space form a 4-dimensional manifold, so the fundamental objects in line geometry are certain submanifolds defined by equations in line coordinates: line complexes and congruences of lines. Klein was still a student when his mentor died in May 1868. He afterward completed his dissertation, written under the supervision of Rudolf Lipschitz, while taking on the responsibility for completing the second part of Plücker's study on line geometry, *Neue Geometrie des Raumes, gegründet auf die Betrachtung der geraden Linie als Raumelement* (Plücker 1868, 1869). This work laid the foundations for the analytic treatment of line geometry, but it also opened the way for visualizing the structure of general quadratic line complexes, the field in which Klein undertook his earliest research (Rowe 2017).

Quadratic complexes are easily defined in analytical terms, once a suitable system of coordinates is found for describing the lines in space. Since these form a 4-dimensional manifold, four independent parameters are required. An algebraic line complex is then given by an equation in line coordinates, which will determine a 3-parameter family of lines. A simple, though highly degenerate example of a second-degree complex comes from the tangent lines to a quadric surface. In the general case, however, the structure will be far too complicated to describe using simple geometrical language. Plücker found a way to visualize the local structure of a quadratic complex, but doing so led to many complications and a huge number of possible cases. Never one to shrink from such a challenge, he began designing models in order to study some of the principal types of quartic surfaces enveloped by certain subfamilies of lines that lie within a given quadratic complex. These models are briefly described in Chap. 8, which focuses primarily on the quartic surfaces that Kummer studied around the same time.

Like Plücker, Kummer also used models to illustrate some of the special types of surfaces he first brought to light in the 1860s, one of these being Steiner's Roman surface (discussed in Chap. 6). Indeed, Kummer's work on quartic surfaces fit squarely within Klein's early research interests, since special types of quartics play a major role in the theory of quadratic line complexes. Plücker had studied the cases he called complex surfaces, which he used to visualize the local properties of a quadratic line complex Ω. His idea was simply to fix a line ℓ in space and then look at all the lines in Ω that happen to intersect ℓ. These will form a congruence of lines enveloping a quartic surface that has ℓ as a double line along with other isolated singularities.

The inspiration behind much of this research came from geometrical optics, a lively topic during the nineteenth century when many new phenomena were first discovered. Plücker was well aware that the lines in Ω that meet ℓ form a special type of ray system and that these rays envelope a caustic surface. In the mid 1860s, Kummer studied ray systems that envelope quartics with 16 singular points and planes, later called Kummer surfaces (Kummer 1864). Plücker's complex surfaces, on the other hand, possess eight singular points and planes plus the double line ℓ (so every plane passing through ℓ intersects it twice). Neither Plücker nor Kummer, however, paid the slightest attention to the other's work, despite the fact that Kummer surfaces were known to play a key role in line geometry. This circumstance was particularly odd since their findings were so closely related. Plücker showed, in fact, how a quadratic line complex determines a so-called singular surface, given by the points (planes) for which the cone (line conic) degenerates into two pencils of lines. Furthermore, he knew that this surface will be a Kummer quartic in the generic case. Yet it was left to young Felix Klein to put all these various pieces together, a task evidently well perfectly to his special talents as a geometer (Klein 1870; Hudson 1905/1990).

One can easily imagine that Klein's life-long conflict with leading representatives of the Berlin school traces back to the mutual animosity that Plücker shared with an earlier generation of Berlin figures, in particular Steiner. Still, Klein could never quite understand why Kummer seemed so aloof to him (Klein 1921–1923, 2: 51). Klein's editorial work on Plücker's monograph also brought him into close contact with Alfred Clebsch, who had only recently assumed Riemann's long vacant chair in Göttingen. A product of the Königsberg school, Clebsch was strongly influenced by the Jacobi tradition as represented by the analyst Friedrich Richelot, but especially by the geometer Otto Hesse. The latter led him to investigate works on algebraic geometry written by the British trio: Arthur Cayley, James Joseph Sylvester, and George Salmon. Clebsch succeeded in combining their ideas with a brilliant new interpretation of Riemann's function theory (Klein et al. 1874). His breakthrough came in 1863 with a lengthy paper entitled "Über die Anwendung der Abelschen Functionen in der Geometrie," a work that Igor Shafarevich once characterized as marking the birth cry of modern algebraic geometry (Shafarevich 1983, 136).

In the 1860s, Clebsch founded a fledgling school at Giessen that specialized in algebraic geometry and invariant theory. This group expanded after his arrival in Gottingen in 1868. Among its leading figures were Paul Gordan, Alexander Brill, Max Noether, Ferdinand Lindemann, Aurel Voss, and Jacob Lüroth, all of whom played important supporting roles during the years when Klein's career was in its ascendancy. Soon after moving to Gottingen, Clebsch joined Leipzig's Carl Neumann in founding *Mathematische Annalen*, which was long regarded as the unofficial organ for publications of the Clebsch school and its allies. After a brief study tour in Berlin and Paris, Klein returned to Gottingen, where he habilitated in January 1871. Thereafter, he quickly gravitated to the center of Clebsch's inner circle (Tobies and Rowe 1990).

Clebsch and Klein had many common interests, which included the use of physical models to explore complex mathematical structures in three-space. Both undertook pioneering studies of cubic surfaces, while drawing on the brilliant work of the Swiss mathematician Ludwig Schläfli. This theory was launched around 1850, after Cayley and Salmon proved that a general cubic over the complex numbers contains exactly 27 straight lines. These lines form an intricate structure in space with highly complex incidence properties: they are cut out by 45 tritangent planes that determine 135 points, ten of which lie on each of the 27 lines (Henderson 1911). This structure remained quite unclear, however, until Schläfli discovered that these 27 lines could be constructed by starting with 12 lines that form a so-called double-six (Schläfli 1858, 1863). These consist of two sets of six mutually skew lines that meet all but one line in the opposite family, thereby forming an incidence configuration of this type (Fig. 7.1):

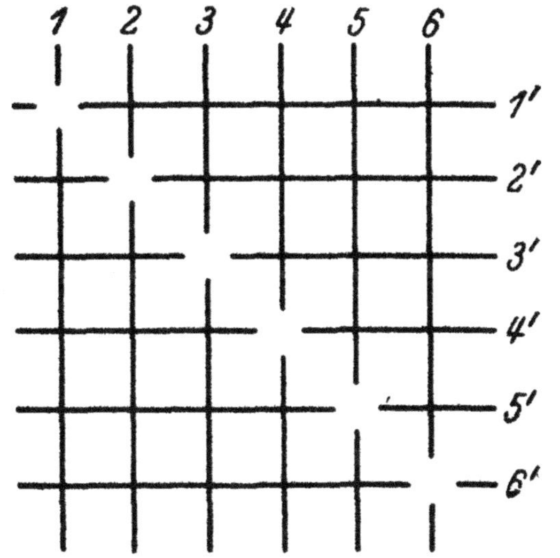

Fig. 7.1 Incidence structure of the lines in a Schläfli double-six.

Schläfli showed that by taking pairs of lines in a double-six that meet, one gets 15 planes, each of which cuts out a new line on the cubic surface, thereby accounting for all 27 lines. The double-six forms a (12,5; 30,2) subconfiguration within the configuration (27,10; 135,2) for all 27 lines and 135 points of the cubic surface. Moreover, these lines can all be real, which is not the case for the 45 tritangent planes. In fact, Schläfli showed that there are just five possibilities for the number of real lines and tritangents, namely: (27,15), (15,15), (7,5), (3,13), (3,7).

Clebsch was eager to see what such a surface looked like, and so he asked his former colleague in Karlsruhe, Christian Wiener, to build a model illustrating the line structure on a general cubic (Fig. 7.2). Wiener did even better: in 1869 he offered to make plaster copies of his model for 50 Gulden, though this was hardly a modest price. Apparently, the idea of casting the original model came to him after he received a number of inquiries from colleagues. Wiener's model was thereafter seen and admired by several geometers, but unfortunately, none of the copies he made seems to have survived. All that remains of it are stereoscopic photos found in the collection of slides at the Göttingen Mathematics Institute. This collection, which dates back to around 1900 when Friedrich Schilling taught geometry in Göttingen, was recently catalogued by Anja Sattelmacher, an historian studying the various uses of mathematical artifacts.

As can be see above, these slides convey rather little about Wiener's model. Luckily, however, they came with a text

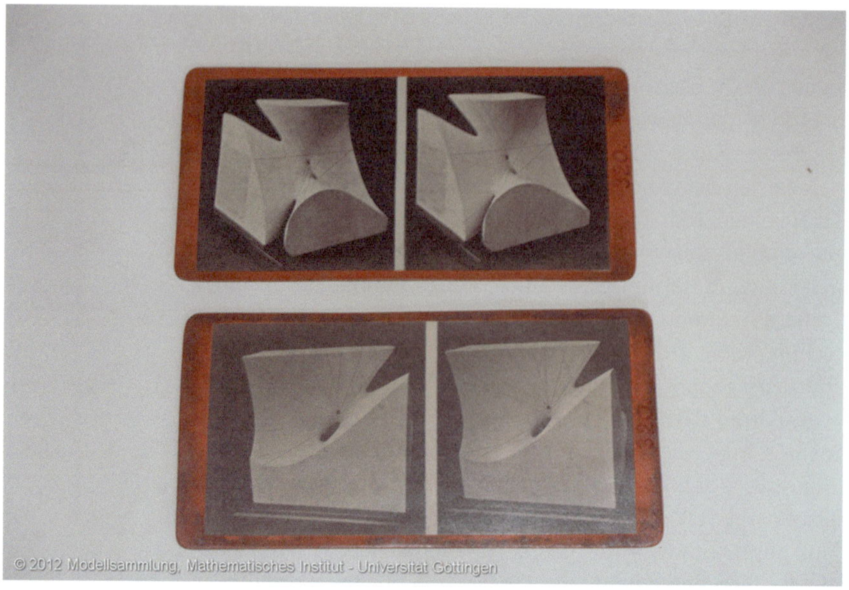

Fig. 7.2 Two stereoscopic images of Wiener's model for a general cubic surface (Courtesy of the Mathematisches Institut – Universität Göttingen).

Fig. 7.3 Rudolf Alfred Clebsch (1833–1872) (Courtesy of the Mathematisches Institut – Universität Göttingen).

in which he gave a detailed account of the mathematical principles underlying its construction (Wiener 1869). Wiener began by choosing four points on a given line that meets five other mutually skew lines. He then took triples of points on these five lines to obtain 19 points. These data then suffice for the construction of a Schläfli double-six, from which he could then derive all 27 lines on the cubic. Wiener's main idea was to fix one of these 27 lines and use this as an axis to sweep out the entire surface. Passing planes through the fixed line ℓ, these will cut the surface in non-degenerate conics, except for the tritangent planes, which he avoids. Since the remaining 26 lines intersect ℓ in ten fixed points, the other 16 lines determine 16 variable points that lie on the conics in each plane. Since five points suffice for constructing a conic, Wiener could calculate the coordinates of enough data points to construct these curves very accurately. He then built a skeletal frame for the model out of 16 planar strips of carton hung together at equal angles around an axis representing the line ℓ. After carefully drawing the conics in each strip, he cut away the portion outside them. To ensure that the lines on the surface were discernible, he bounded the solid figure by a cube that was large enough to indicate the direction of all 27 lines.

Wiener's model attracted much attention at first, but after 1872 interest in it gradually waned when Clebsch produced a far more striking model of his own (Fig. 7.3). More precisely, in that year he asked his student Adolf Weiler to design a model of a so-called diagonal surface. This special cubic surface takes its name from a property connected with the form of its equation. In fact, a generic quaternary cubic form can always be written as a sum of five cubes of linear forms, a result first conjectured by J. J. Sylvester and then proved

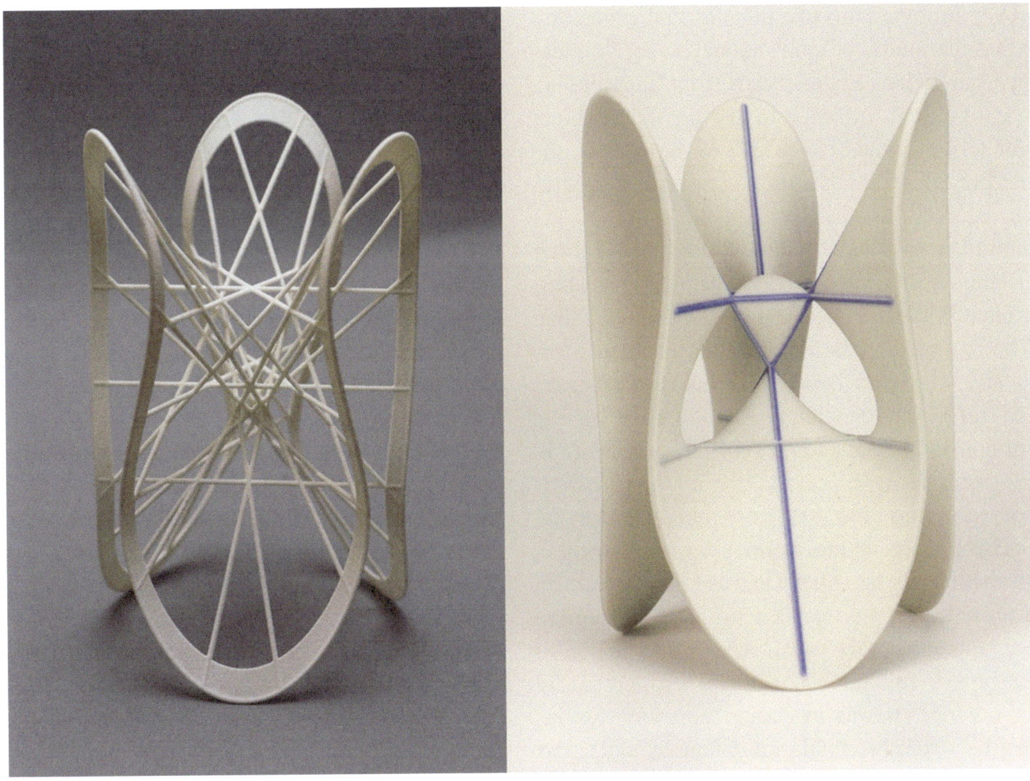

Fig. 7.4 Models by Oliver Labs of the special configuration of 27 lines on a diagonal surface and on Klein's cubic with 4 singular points, where $4 \times 6 = 24$ lines form the edges of a tetrahedron (Courtesy of MO-Labs. www.mo-labs.com).

by Clebsch in (Clebsch 1866, 1871). The union of the five planes obtained by letting these linear forms vanish is called the Sylvester pentahedron associated with the cubic. Its five faces are tritangents and the three lines in each plane are just the diagonals of the complete quadrilateral cut by the other four faces of the pentahedron. The 27 lines thus arise from a double-six and the 15 diagonal lines that lie on the faces of the Sylvester pentahedron. Unlike a general cubic, though, a diagonal surface has a special incidence structure due to its symmetry. Ten of the 45 tritangents meet the surface in concurrent lines that pass through so-called Eckhardt points. It can be shown, in fact, that ten is the maximal number of such Eckhardt points. Some can be clearly seen in the model shown below that was produced by Oliver Labs using the technology of 3D printing (Fig. 7.4).

Klein was working closely with Clebsch in Göttingen when both were studying the diagonal surface. In fact, at the very meeting of the Göttingen Scientific Society when Clebsch unveiled Weiler's model, Klein presented a related plaster model with four singular points, the maximum possible for a cubic surface (Fischer 1986). These were situated symmetrically to form the vertices of a tetrahedron. Like the diagonal surface, all 27 lines are real, but here 24 of them have collapsed into the six edges of the tetrahedron, each counted four times (as shown in the model by Labs above). Klein boldly asserted that by starting with this highly degenerate cubic one could obtain all possible types of cubics by means of suitable deformations that remove the singularities. This required either splitting the surface at a singular point or else enlarging the surface to make it disappear; these were delicate procedures and it would take many years before the results were completely justified (Klein 1873b/1922). Klein only had intuitive continuity arguments at his disposal, but his inspiration gradually gave rise to what later became known as modern deformation theory.

Even earlier than this, Klein made a discovery that did much to establish his later fame. In two papers entitled "On the so-called Non-Euclidean Geometry" (Klein 1871, 1873a) he showed that from the "higher standpoint" of projective geometry one could view Euclidean and non-Euclidean geometry as special cases of a projective space to which one adjoins a conic section (for the plane) or quadric surface (in three-space). His proof depended on a construction due to von Staudt in which projective coordinates are introduced independent of a metric, an approach that met with strong skepticism in some quarters. Ironically, one of the skeptics was the prominent English mathematician Arthur Cayley from whom Klein adapted the so-called "Cayley metric" to the case of non-Euclidean geometries. To counter such misgivings, Klein elaborated on his original argument, but found to his dismay that Cayley, Robert Ball, and others persisted in believing his reasoning still contained a vicious circle. Klein later attributed this to an intellectual generation gap, arguing that the inertia of conventional scientific ideas often prevented otherwise capable minds from transcending certain tacit assumptions in order to see an old problem in a new light (Klein 1926, 152–153). As he saw it, the anomalies that present themselves to one scientific generation are often regarded as only apparent difficulties to the next. Once viewed from a higher plane of discourse, former contradictions often resolve themselves quite naturally. Thus, for example, once Plücker recognized that lines rather than points could be regarded as the fundamental geometric entities, he was able to clarify many of the once-mysterious phenomena of projective geometry. Klein's work on the foundations of non-Euclidean geometries was undertaken in a similar spirit.

Recognition, not only for this work but also for his fundamental achievements in line geometry, was not long in coming. Clebsch certainly regarded Klein as a mathematician of unusual promise, and it was thanks to his recommendation that Klein was appointed to a vacant chair in Erlangen in 1872; he was now a full professor at the extraordinarily young age of only 23. It was in conjunction with his inauguration that he wrote his single most famous work, das "Erlanger Programm" (1872/1893). This had the form of an extended essay that Klein conceived in cooperation with his friend and collaborator, the Norwegian Sophus Lie.

Klein and Lie were both exceedingly ambitious men with a deep longing for acclaim, a theme that recurs often in their correspondence. They sought such recognition in Paris in the spring of 1870, at which time they met several of the era's leading mathematicians. Lie later returned to Paris in 1882 and wrote to Klein from the French capital. His three letters, translated in Chap. 10, provide a glimpse of how several Parisians saw the mathematical world a dozen years later. These letters also allude to the competition between Klein and Henri Poincaré, including the French reaction to Klein's squabble with Lazarus Fuchs. These matters are taken up in the much longer closing essay (Chap. 11), which focuses on events from the period 1881–82 (see also Gray 2000).

Lie reported further that Poincaré's first extensive memoir would soon appear in Mittag-Leffler's new journal, *Acta Mathematica.* Mittag-Leffler's accomplishments as an editor make for a suggestive comparison with Klein's, a juxtaposition that forms one of the principal themes in Chap. 11. Another related topic concerns the belated publication of Poincaré's letters to Klein from the early 1880s. These letters became the flash point for new tensions between Klein and Mittag-Leffler during the First World War. As noted in the introduction to Part I, Mittag-Leffler was intent on cultivating friendships with leading mathematicians in Paris as well as in Berlin, a strategy that threatened to forge a triangle of power that might marginalize Klein's *Mathematische Annalen* with its network centered in Göttingen.

The young Felix Klein is mainly remembered today as the author of the "Erlangen Program," composed and published in 1872. So it comes as a surprise to read Lie's letter to Klein in which he relates a full decade later that Poincaré had never heard of it. Afterward, in fact, Poincaré read it with interest and even suggested to Mittag-Leffler that he would welcome a French translation in *Acta Mathematica.* No doubt, Mittag-Leffler thought this would only weaken his position vis-à-vis Berlin, so he declined. Lie later urged Klein to consider reprinting the original German text in *Mathematische Annalen*, but during the early 1880s Klein was reluctant to do so. When he changed his mind 10 years later, Lie was no longer happy about the new wave of interest in the "Erlangen Program," fearing that its visionary ideas would be conflated with his own pioneering investigations in the new field of continuous groups. One can easily understand why: Klein's "Erlangen Program" had, in fact, very little to do with Lie's research program, as Thomas Hawkins first pointed out in (Hawkins 1984).

What Klein emphasized instead amounted to a new framework for understanding different methods in geometry. Traditional geometrical investigations had long centered on the properties of concrete figures – special curves in the plane or surfaces in space – rather than sorting out different types of properties and studying figures that possess them. Projective geometry focused attention on those properties of figures that remained invariant under projections, as for example the degree of an algebraic curve or surface. From this vantage point, the theory of conic sections and quadric surfaces emerged in a new light. Klein's "Erlanger Programm" represented a far-reaching generalization of this trend by shifting the focus away from geometrical objects to the various types of transformation groups that can act on them. Thus, instead of exclusively focusing on familiar entities like conic sections or other objects invariant under projective transformations, Klein pointed to numerous examples involving other groups of transformations that act on a so-called manifold, by which he meant a parameterized space whose elements can be arbitrary objects (e.g. line complexes, as can be seen from (Klein 1872a)). By means of these concepts, Klein hoped to unite various disparate studies undertaken by Cayley, Lie, Hermann Grassmann, William Rowan Hamilton, and others. His claim was that from a "higher standpoint" all geometrical research could be seen as the study of the invariants associated with a given transformation group.

This was the philosophical message Klein sought to convey in his "Erlanger Programm," but a closer reading also reveals the larger disciplinary agenda he had in mind. His dissatisfaction with the fragmentation and disarray that characterized the field of geometry in Germany during this period was clearly expressed in a note on the "separation of present-day geometry into disciplines." There he commented: "If.

.. one observes how mathematical physicists continually ignore the advantages that a merely modest development of projective thinking could provide in many cases, or, on the other hand, how projective geometers leave untouched the rich treasure of mathematical truths that has been uncovered in the theory of surface curvature, then one must regard the present situation with respect to geometrical knowledge as truly wanting and, one hopes, only a passing one." (Klein 1872b/1893, 491).

Like other classics in the history of mathematics – Riemann's *Habilitationsvortrag* comes immediately to mind – Klein's "Erlanger Programm" began in obscurity; only later did contemporary mathematicians gradually become aware of its significance. One notes another parallel with Riemann's text as well. For although Klein chose a different topic for his inaugural address to the Erlangen faculty (Rowe 1983), the written "Programm" that accompanied his lecture was also intended for a broad audience. So what prevented Klein from following up on these ideas? And what accounts for the sudden resurgence of interest in this work during the early 1890s? While there is no simple answer to the first question, an important factor was surely the unexpected death of Clebsch, an event that had profound consequences for Klein's work in the years that followed. As for the second question, one of the main sources of interest came from Italy, where Corrado Segre took up the theory of quadratic line complexes again (Rowe 2017). Segre's engagement with Klein's work led not only to an ongoing correspondence with him but also to the idea of producing an Italian translation of the "Erlanger Programm" (Luciano and Roero 2012). Segre's student, Gino Fano, published that translation in 1890, after which others soon followed in English, French, and other languages. For a fuller understanding of the text, one should also read Klein's lectures from this period, *Höhere Geometrie* (Klein 1892–93), which were intended as a vehicle for explaining the ideas he and Lie had set forth 20 years earlier. A recent fresh look at Klein's "Erlanger Programm" from the point of view of geometrical equations can be found in (Lê 2015).

Although Klein referred to Grassmann's work more than once in the pages of his "Erlanger Programm," his enthusiasm for these bold new ideas had its limits. Indeed, the second essay (Chap. 9) shows how the young Klein distanced himself from the views of Victor Schlegel, whom he regarded as a radical proselytizer for the Grassmannian cause. Original thinkers never have an easy time winning recognition from their contemporaries, to be sure, though Hermann Grassmann presents a striking example of a mathematician who only attained fame posthumously. Why his ideas failed to take root had much to do with contemporary research trends and fashions, but surely the number of experts who can today claim to have a deep knowledge of Grassmann's original works is very small indeed. To a lesser degree, the same could be said even for Klein's mathematical papers, though he managed to keep many of his more programmatic ideas in circulation, not least through the efforts of an impressive army of students and *Mitarbeiter.* As a young professor, though, he had to fight for recognition, just like anyone else, and this often meant drawing lines and attacking the works of rivals.

Klein had hoped to discuss the bold new ideas in his "Erlanger Programm" with Clebsch soon after its publication, but this was not to be. In November 1872, his second mentor was suddenly stricken by diphtheria and died within days, thereby ending a brilliant and still promising career. As in Riemann's case, death came even before he had reached his fortieth birthday; it would take Klein some time to get over the shock. Although a number of Clebsch's talented students migrated to Erlangen from Göttingen, Klein was hardly in a position to build up a comparable school at this small Bavarian university where he never had more than seven students enrolled in his courses. The most important of these was Ferdinand Lindemann, later to become famous for proving that π is a transcendental number (see the introduction to Part III and Rowe (2015)). Under Klein's supervision, Lindemann edited Clebsch's lectures on geometry and wrote a dissertation in which he applied Klein's approach to non-Euclidean geometry to the problem of describing infinitesimal motions in rigid body mechanics (Ziegler 1985).

Klein left Erlangen in 1875, when he succeeded Otto Hesse at the Technische Hochschule in Munich, then located near the campus of the university (Tobies 1992). As an indication of the contrast with his former position, he was now teaching analytic geometry to over 200 students. Nor was he lacking for young talent: over the next five years his seminar and advanced classes were attended by Adolf Hurwitz, Walther von Dyck, Karl Rohn, Carl Runge, Max Planck, along with two young Italians, Luigi Bianchi, and Gregorio Ricci-Curbastro. It was during these six years in Munich that Klein's gifts as a teacher began to unfold, and it was here that he found a way to integrate model making into his teaching activity. His highly visual style was quite idiosyncratic, but he drew on it heavily to make impressive advances in algebraic geometry, Galois theory, and complex analysis.

As noted above, Chap. 8 begins by discussing Klein's early interest in geometrical models going back to his student days with Plücker in Bonn. During that time, such models largely served as artifacts for research, a trend that persisted in Germany until roughly 1880. In recounting that shift to a more didactical emphasis in model making, I turn to his Munich years when Klein taught alongside Alexander Brill. Together they founded a laboratory in which students learned how to construct such models, several of which soon became part of the collection marketed by Brill's brother over the next 20 years. The Darmstadt firm of Ludwig Brill was the first German company to sell geometric models during an era when Klein and others promoted visual and intuitive aspects of mathematics education. By the early twentieth century, models from the collections of Brill and the company later founded by Martin Schilling were a staple at universities and other institutions throughout Europe and the United States.

In 1880 Klein became the first occupant of a new chair in geometry established by the Saxon Ministry of Education in Leipzig (Tobies 1981). He wasted little time making his presence felt, beginning with the renovation of a large auditorium soon to be equipped with facilities suitable for mathematics instruction on a large scale. Klein also made plans for founding a mathematics seminar, library, and a collection of models (Fig. 7.5). He would later introduce similar innovations at the Georgia Augusta in cooperation with the Prussian government. In Leipzig, his presence clearly exerted a strong influence on the mathematics program, one indicator being the number of doctorates granted. Throughout the 1870s, Leipzig produced only nine doctoral students, as compared with 29 in Berlin and a high of 60 in Göttingen. During Klein's six-year tenure, by contrast, Leipzig produced 36 doctorates, more than half of whom took their degrees under him. Even more telling on a national scale were the number of post-doctoral posts created during these years. The Leipzig faculty accepted no fewer than five post-doctoral theses (*Habilitationsschriften*), a total far exceeding that of any other university. Moreover, five of Leipzig's six *Privatdozenten* – Walther von Dyck, Friedrich Schur, Karl Rohn, Eduard Study, and Friedrich Engel – were closely associated with Klein and his school.

By this time, Klein's influence seriously rivaled that of the leading Berlin mathematicians. Having assumed Clebsch's role as de facto editor-in-chief of *Mathematische Annalen*, he soon succeeded in transforming this fledgling enterprise into one of the leading mathematics journal in the world, surpassing the Berlin-dominated *Journal für die reine und angewandte Mathematik* (Crelle's Journal). One of the keys to this success was Klein's knack for promoting the works of mathematicians who were either estranged from or stood outside the mainstream influence of the Berlin school: two prominent examples being Georg Cantor and David Hilbert. There had been little love lost between Plücker and the synthetic geometer Jakob Steiner, and relations between Clebsch and his Berlin contemporaries were little better. When Lindemann was called to an Ordinariat at Königsberg in 1883, Klein congratulated him by saying: "Your appointment. .. benefits all of us, as it is a victory for our principles. Just remember how 10 years ago an authoritative figure said that he would never agree to have a 'Clebschian' appointed in Prussia." For several years after Clebsch's death, his former students felt virtually ostracized from the German mathematical community. The Annalen relied heavily on contributions from them and their allies in order to survive, thereby acting as a natural counterpoise to Crelle's Journal throughout the 1870s and 1880s.

This polarization within the German mathematical community can easily be documented by comparing the respective lists of leading contributors to these two journals from 1869 to 1901 (Tobies and Rowe 1990). Those who published in *Mathematische Annalen* often specialized in complex analysis, particularly the theory of elliptic, hyperelliptic, and Abelian functions as well as the two specialties of the Clebsch school: algebraic geometry and invariant theory. The vast field of real analysis was also well represented, as was group theory. Aside from the publications of the two Leipzig mathematicians Adolf Mayer and Carl Neumann, relatively little work was done in differential equations and mathematical physics, and only toward the end of the century, when Hurwitz and Hilbert were publishing, was there anything of substance in number theory at all. This pattern contrasts sharply with that of Crelle's Journal. During the 1870s and 1880s, the Berlin mathematicians Kummer and Kronecker, both first-class number theorists, were among the more influential members of its editorial board, and throughout those decades, algebraic geometry and invariant theory rarely appeared in the pages of their journal. Of the 53 mathematicians who published five or more articles in Crelle's Journal from 1869 to 1901, only six – Cantor, Hurwitz, Paul du Bois-Reymond, Leo Koenigsberger, Rudolf Sturm, and Heinrich Weber – published regularly in *Mathematische Annalen*. What is more, many of the leading figures associated with one or the other of these two factions never wrote a single paper for the rival journal.

Still, Klein's years in Leipzig were hardly just a series of success stories, as can be seen in Chap. 11. This describes the fateful events that culminated with Klein's collapse from overwork in the autumn of 1882. Health problems were nothing new for him, but this was a severe breakdown, hastened by an ongoing competition with young Henri Poincaré, who in 1881 began publishing the outlines of a new theory of automorphic functions (Gray 2013, 207–246). After seeing one of Poincaré's preliminary announcements in the *Comptes Rendus*, Klein struck up a correspondence with him. Poincaré had come to this topic by reading the work of Lazarus Fuchs, a leading representative of the Berlin school. Fuchs had already clashed with Klein before this, though Poincaré had little inkling of these intra-German rivalries. Having honored Fuchs by naming a new class of functions after him, he soon learned that Klein was profoundly unhappy with this appellation. After scolding Poincaré privately, he issued a public protest, forcing Poincaré to defend his choice of terminology. In the meantime, the Frenchman came to realize that he had unwittingly walked straight into a veritable hornets' nest.

As these events unfolded, Poincaré received periodic reports from a young graduate of the École Normale, Georges

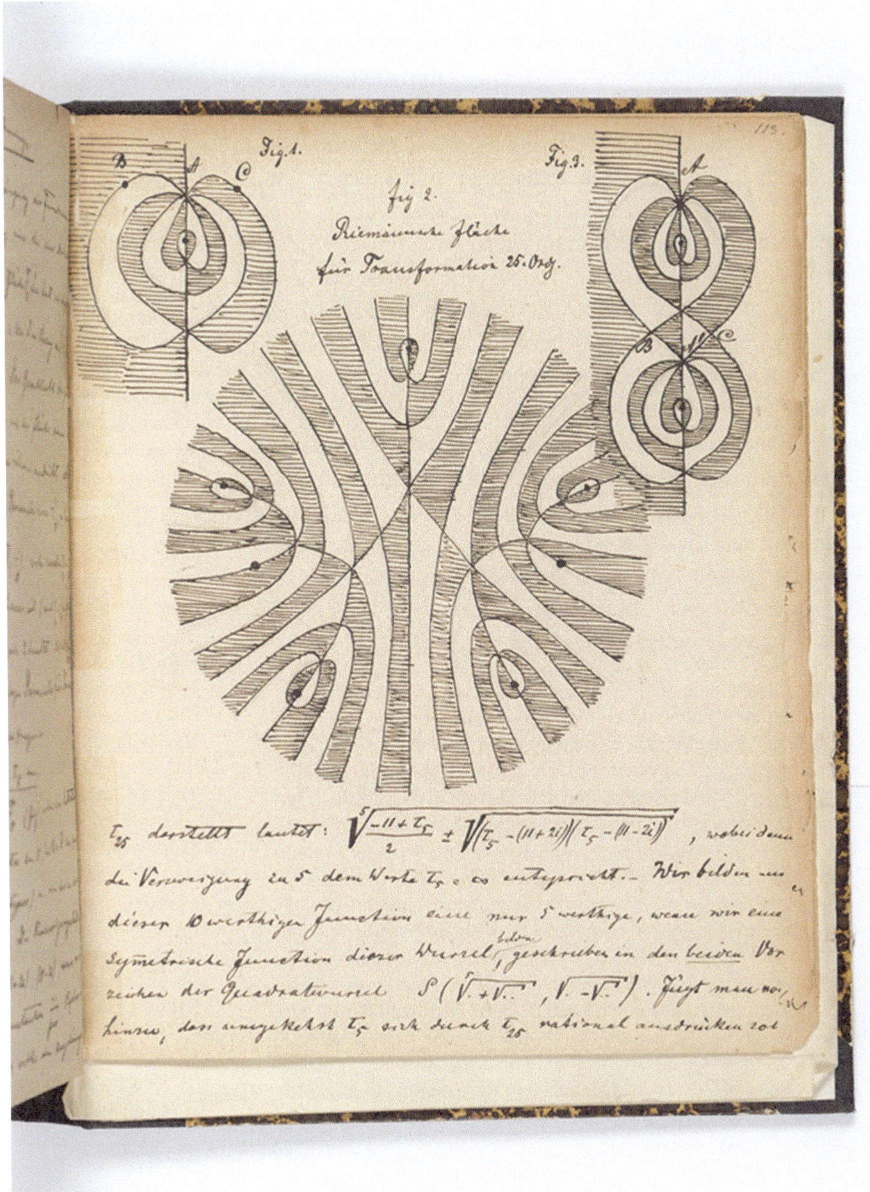

Fig. 7.5 A page from Klein's protocol books covering topics presented in his seminars. (Courtesy of the Mathematisches Institut – Universität Göttingen).

Brunel, who happened to be studying with Klein in Leipzig. Brunel's letters shed fresh light on the discussions behind the scenes and in Klein's seminar, where the main topic was Riemannian function theory. Klein's teaching activity served as background for his booklet, *Über Riemann's Theorie der algebraischen Functionen und ihrer Integrale.eine Ergänzung der gewöhnlichen Darstellungen* (Klein 1882). Here, as elsewhere, Klein adopted a genetic approach to mathematical ideas; in general, he was convinced that the road to discovery was more important than any purely formal argument, however elegant, concocted after the fact. In this booklet, for example, he chose to overlook the technical problems associated with Riemann's use of the Dirichlet principle in order to explore the original underpinnings of his theory. It was Klein's contention that Riemann's insights were motivated by studying current flows on closed surfaces, which he then interpreted as the real and imaginary parts of a complex potential function. Klein was for some time convinced that he had rediscovered Riemann's original inspiration, but when he wrote to the latter's former student, Friedrich Prym, he realized that he had probably been mistaken. In fact, Klein's conception was more general than the conventional view, according to which a Riemann surface was a branched covering of the complex plane or the Riemannian sphere. Hermann Weyl later gave this Kleinian approach a modernized form in his classical monograph, *Die Idee der Riemannschen Fläche* (Weyl 1913).

Poincaré was clearly most unhappy about the way Klein chose to drag him into his ongoing battle with Fuchs. Citing a famous line from Goethe's *Faust*, he made his opinion clear:

this debate over names was sterile, just a lot of "sound and smoke." Apparently this incident left no deep scars, as both men later developed an amicable relationship (see Chap. 16). Following this initial tiff, their correspondence turned to the problem of formulating and proving a general uniformization theorem that would serve as a capstone to this theory (Paul de Saint-Gervais 2010). Working under great stress, Klein came up with such a theorem, while sketching a strategy for proving it. But during the course of this work his health collapsed completely; for nearly two years he was plagued by lethargy and depression. Afterward, he gradually realized that his truly creative years as a mathematician were over.

There followed a transition phase in Klein's career. After the publication of his *Vorlesungen über das Ikosaeder* (Klein 1884), teaching became the primary outlet for his mathematical activity. In addition to his many German admirers, he began to attract an impressive number of foreigners. One of his many fervent American disciples, Frank Nelson Cole, later gave this vivid description of the orderly manner in which Klein conducted his research seminars:

> [His] management of the *Seminar* has always been exceptionally efficient, even among the German models. It is Klein's custom to distribute among his students certain portions of the broader field in which he himself is engaged, to be investigated thoroughly under his personal guidance and to be presented in final shape at one of the weekly meetings. An appointment to this work means the closest scientific intimacy with Klein, a daily or even more frequent conference, in which the student receives generously the benefit of the scholar's broad experience and fertility of resource, and is spurred and urged on with unrelenting energy to the full measure of his powers. When the several papers have been presented, the result is a symmetric theory to which each investigation has contributed its part. Each member of the *Seminar* profits by the others' points of view. It is a united attack from many sides of the same field. In this way, a strong community of interest is maintained in the *Seminar*, in addition to the pleasure afforded by genuine creative work. (Cole 1893, 107)

In his lecture courses, Klein usually just sketched the broad outlines of a theory while offering suggestions as to how it might best be carried out. The task of doing so he left to his apprentices, such as Robert Fricke, who came to Leipzig in 1884. Soon he became Klein's closest *Mitarbeiter* and the workhorse behind four giant co-authored volumes on automorphic and elliptic modular functions. Klein continued this mode of teaching during his first years in Gottingen. When he arrived there in 1886, mathematics enrollments throughout Germany had begun to drop off precipitously. Over the next seven years, his advanced classes were attended almost entirely by foreigners, especially Americans. Graduate education in the United States was then still in its infancy, but the first generation to promote it found in Klein their ideal teacher. No fewer than six of his students went on to become presidents of the American Mathematical Society (Parshall and Rowe 1994).

References

Clebsch, Alfred. 1866. Die Geometrie auf den Flächen dritter Ordnung. *Journal für die reine und angewandte Mathematik* LXV: 359–380.

———. 1871. Über die Anwendung der quadratischen Substitution auf die Gleichungen 5ten Grades und die geometrische Theorie des ebenen Fünfseits. *Mathematische Annalen* 4: 284–345.

Cole, Frank Nelson. 1893. Klein's Modular Functions. *Bulletin of the American Mathematical Society* 2: 105–120.

Fischer, Gerd, ed. 1986. *Mathematische Modelle*. 2 Bde. ed. Berlin: Akademie Verlag.

Gray, Jeremy. 2000. *Linear Differential Equations and Group Theory from Riemann to Poincaré*. 2nd ed. Boston: Birkhäuser.

———. 2013. *Henri Poincaré: A Scientific Biography*. Princeton: Princeton University Press.

Hawkins, Thomas. 1984. The *Erlanger Programm* of Felix Klein: Reflections on its place in the history of mathematics. *Historia Mathematica* 11: 442–470.

———. 1989. Line Geometry, Differential Equations, and the Birth of Lie's Theory of Groups, in (Rowe and McCleary 1989, 274–327).

———. 2000. *Emergence of the Theory of Lie Groups: An Essay in the History of Mathematics 1869–1926. New York: Springer.*

Henderson, Archibald. 1911. *The Twenty-Seven Lines upon the Cubic Surface*. New York: Hafner.

Hilbert, David. 1932–1935. *Gesammelte Abhandlungen*. 3 Bde. ed. Berlin: Springer.

Hudson, Ronald W. H. T. 1905/1990. *Kummer's Quartic Surface*. Cambridge: Cambridge University Press, 1905; reprinted in 1990.

Klein, Felix. 1870. Zur Theorie der Liniencomplexe des ersten und zweiten Grades. *Mathematische Annalen* 2: 198–228.

———. 1871. Ueber die *sogenannte Nicht-Euklidische* Geometrie. *Mathematische Annalen* 4: 573–625.

———. 1872a. Über Liniengeometrie und metrische Geometrie. *Mathematische Annalen* 5: 257–277.

———. 1872b/1893. *Vergleichende Betrachtungen über neuere geometrische Forschungen*. Erlangen: Verlag von Andreas Deichert, 1872; (Reprinted in *Mathematische Annalen*, 43(1893): 63–100; in (Klein 1921–1023, 2:. 460–497).)

———. 1873a. Ueber die *sogenannte Nicht-Euklidische* Geometrie. *Mathematische Annalen* 6: 112–145.

———. 1873b/1922. Über Flächen dritter Ordnung, 6(1873): 551–581; in (Klein 1921–1923, 2: 11–62).

———. 1882. *Über Riemann's Theorie der algebraischen Functionen und ihrer Integrale. Eine Ergänzung der gewöhnlichen Darstellungen*. Leipzig: Teubner. in (Klein 1921–1923, 3: 499–573).

———. 1884. *Vorlesungen über das Ikosaeder und die Auflösung der Gleichungen vom fünften Grade*. Leipzig: Teubner.

———. 1921–1923. *Gesammelte mathematische Abhandlungen*. 3 Bde. ed. Berlin: Julius Springer.

———. 1926. *Vorlesungen über die Entwicklung der Mathematik im 19. Jahrhundert*. Bd. 1, R. Courant u. O. Neugebauer, Hrsg., Berlin: Julius Springer.

Klein, Felix, et al. 1874. Rudolf Friedrich Alfred Clebsch, Versuch einer Darlegung und Würdigung seiner wissenschaftlichen Leistungen. *Mathematische Annalen* 7: 1–55.

Kummer, Ernst Eduard. 1864. Über die Flächen vierten Grades mit sechzehn singulären Punkten. *Monatsberichte der Akademie der Wissenschaften zu Berlin*: 246–260.

Lê, François. 2015. "Geometrical Equations": Forgotten Premises of Felix Klein's Erlanger Programm. *Historia Mathematica* 42: 315–342.

Luciano, Erika, and Clara Silvia Roero. 2012. From Turin to Göttingen: Dialogues and Correspondence (1879–1923). *Bollettino di Storia delle Scienze Matematiche* 32 (1): 7–232.

Olesko. 1989. Science in Germany: The Intersection of Institutional and Intellectual Issues, Kathryn M. Olesko, ed., *Osiris* (2nd Series), 5, Chicago: University of Chicago Press.

Parshall, Karen H., and David E. Rowe. 1994. *The Emergence of the American Mathematical Research Community 1876–1900: J. J. Sylvester, Felix Klein and E. H. Moore*. Providence/London: AMS/LMS History of Mathematics 8.

Paul de Saint-Gervais, Henri. 2010. *Uniformisation des surfaces de Riemann. Retour sur un théorème centenaire*. Lyon: ENS Éditions.

Plücker, Julius. 1868. *Neue Geometrie des Raumes gegründet auf die Betrachtung der geraden Linie als Raumelement, Erste Abteilung*. Leipzig: Teubner.

———. 1869. In *Neue Geometrie des Raumes gegründet auf die Betrachtung der geraden Linie als Raumelement, Zweite Abteilung*, ed. von F. Klein. Leipzig: Teubner.

Rowe, David E. 1983. A Forgotten Chapter in the History of Felix Klein's "Erlanger Programm". *Historia Mathematica* 10: 448–454.

———. 1989a. Klein, Hilbert, and the Göttingen Mathematical Tradition, in (Olesko 1989,186–213).

———. 1989b. Klein, Lie, and the Geometric Background of the Erlangen Program, in (Rowe and McCleary 1989, 1: 209–273).

———. 2015. Historical Events in the Background of Hilbert's Seventh Paris Problem. In *A Delicate Balance: Global Perspectives on Innovation and Tradition in the History of Mathematics, A Festschrift in Honor of Joseph W. Dauben*, Trends in the History of Science, ed. D.E. Rowe and W.-S. Horng, 211–244. Basel: Springer.

———. 2017. Segre, Klein, and the Theory of Quadratic Line Complexes. In *From Classical to Modern Algebraic Geometry: Corrado Segre's Mastership and Legacy*, Trends in the History of Science, ed. G. Casnati et al., 243–263. Basel: Springer.

Rowe, David E., and John McCleary. 1989. In *The History of Modern Mathematics*, ed. David E. Rowe and John McCleary, vol. 2. Boston: Academic Press.

Schläfli, Ludwig. 1858. An attempt to determine the twenty-seven lines upon a surface of the third order, and to divide such surfaces into species in reference to the reality of the lines upon the surface. *Quarterly Journal for pure and applied Mathematics* II: 55–66. 110–220.

———. 1863. On the Distribution of Surfaces of the Third Order into Species, in Reference to the Presence or Absence of Singular Points and the Reality of their Lines. *Philosophical Transactions of the London Royal Society* CLIII: 193–241.

Schubring, Gert. 1989. The Rise and Decline of the Bonn Naturwissenschaften Seminar, in (Olesko 1989, 56–93).

Shafarevich, I.R. 1983. Zum 150. Geburtstag von Alfred Clebsch. *Mathematische Annalen* 266: 135–140.

Tobies, Renate. 1981. *Felix Klein, Biographien hervorragender Naturwissenschaftler, Techniker und Mediziner, 50*. Leipzig: Teubner.

———. 1992. Felix Klein in Erlangen und München. In *Amphora: Festschrift für Hans Wussing zu seinem 65. Geburtstag*, 751–772. Basel: Birkhäuser.

Tobies, Renate, and David E. Rowe. 1990. *Korrespondenz Felix Klein – Adolf Mayer*, Teubner – Archiv zur Mathematik. Vol. 14. Leipzig: Teubner.

Weyl, Hermann. 1913. *Die Idee der Riemannschen Fläche*. Leipzig: Teubner.

Wiener, Christian. 1869. *Stereoscopische Photographien des Modells einer Fläche dritter Ordnung mit 27 reellen Geraden. Mit erläuterndem Texte*. Leipzig, Teubner.

Ziegler, Renatus. 1985. *Die Geschichte der geometrischen Mechanik im 19. Jahrhundert: Eine historisch-systematische Untersuchung von Möbius und Plücker bis zu Klein und Lindemann*. Stuttgart: Franz Steiner Verlag.

8 Models as Research Tools: Plücker, Klein, and Kummer Surfaces

(*Mathematische Semesterberichte*, 60(1) (2013), 1–24)

In the late summer of 1869, 20-year-old Felix Klein made his way to Berlin, where he planned to attend the renowned seminar founded by Ernst Eduard Kummer and Karl Weierstrass. Klein had already taken his doctorate in Bonn and he would soon be recognized as a leading expert on line geometry, a new approach to 3-space launched by his mentor in Bonn, Julius Plücker. Just before Plücker died in 1868, he entrusted Klein to complete the classic monograph, *Neue Geometrie des Raumes gegründet auf die Betrachtung der geraden Linie als Raumelement*. Overall responsibility for this project fell to Alfred Clebsch in Göttingen, which was how Klein first came to the prestigious *Georgia Augusta*. There he was quickly drawn into the orbital field surrounding this stellar figure, the co-founder of a new journal, *Die Mathematische Annalen* (Shafarevich 1983). It was during this first brief stay in Göttingen that Klein struck up his lifelong friendship with Max Noether. Their initial encounter also bore immediate mathematical fruit; both found mappings that transformed the 3-parameter system of lines in a linear complex to the points in 3-space, an approach to line geometry that would soon have far-reaching consequences (Klein 1921, 89).

Although Klein found this new-found environment highly convivial, he was also eager to travel. So, after only eight months and against the advice of his new mentor – Clebsch had long ago burned his bridges to the leading mathematicians in Berlin – he left Göttingen for the Prussian capital (Klein 1923, 8). No doubt his personal ambition to see the "world of mathematics" heavily influenced this decision; after Berlin, he planned to visit Paris and then England, possibly Italy as well. Still, he had special reasons to visit Berlin first. For in the meantime, he had been probing deeper into Plücker's theory of quadratic line complexes, and he was beginning to realize its close links with Kummer's work in geometrical optics, an older field of research that had taken on new life in the 1860s. So what better place to delve more deeply into these matters than in Kummer's own seminar?

Klein knew that admission to the Berlin seminar required candidates to submit a manuscript demonstrating their ability to undertake independent mathematical research (Biermann 1988, 279–281). How stringently Kummer and Weierstrass chose to implement this requirement would have been difficult for an outsider to know, but Klein took no chances. Although he surely would have been admitted on the basis of the work he had already completed, he saw this as a golden opportunity to impress Kummer right from the start. Besides, he had an excellent topic in mind for just this purpose: Noether's newly discovered mapping of a linear complex onto the points of projective 3-space. This mapping is bijective except for a single line in the complex, which corresponds to a conic in 3-space. Klein thus began an extensive study of the images of different types of ruled surfaces under this transformation, a project that occupied his attention from 5 September to 15 October 1869, shortly before the winter semester began. Clearly his main objective was to awaken Kummer's interest in these new results. So one can imagine how he felt when a short time later Kummer simply returned the manuscript to him without any comment at all; apparently he barely bothered to glance through it! (this was Klein's impression, according to a note attached to the manuscript in (Klein NL, 13A).)

For someone as ambitious and gregarious as Klein, such an experience must have been downright demoralizing (Fig. 8.1). Many years later he still felt puzzled that he had never managed to establish a closer relationship with Kummer. Part of the reason may have been temperamental, but deeper differences seem obvious in retrospect. Kummer was an old-fashioned, serious-minded Prussian who thought of Gauss as the embodiment of the highest ideals in mathematics (Biermann 1988, 81–82). Like his colleagues, Weierstrass and Kronecker, he valued nothing higher than clarity and rigor. As a mathematician, Klein came nowhere near this ideal. Somewhat like Plücker but with far greater breadth,

D.E. Rowe, *A Richer Picture of Mathematics*, https://doi.org/10.1007/978-3-319-67819-1_8

Fig. 8.1 Christian Felix Klein (1849–1925).

he relied heavily on intuition, bold analogies, and hunches in his work. Little wonder that Kummer gently chided him to "emulate Gauss more" (Klein NL 22J, S.1), a remark that surely did nothing to ingratiate Klein. Private conversations with Weierstrass and Kronecker proved more rewarding, but these encounters were too seldom to have any lasting importance. Klein's general reaction, especially in conversations with younger mathematicians, was disdain for their narrow-minded opinions and intellectual arrogance. His overall disillusionment with mathematics in Berlin shines through clearly in letters he wrote at this time both to his mother as well as to his friend Max Noether ((Rowe 1989) and (Klein NL 12, 523–526)).

Clearly Klein's disappointment also had much to do with the contrast between the vibrant atmosphere surrounding Clebsch in Göttingen and the far more formal ambiance he experienced in Berlin. Kummer had no advanced students at this time, so Klein naturally felt starved for intellectual nourishment. Luckily, he soon met another attendee in the seminar with interests similar to his own. They first ran into each other at a meeting of the Mathematics Club, where a gigantic foreigner introduced himself as Sophus Lie; he had just arrived from Christiania, the former name for Oslo. This odd pair hit it off immediately, and soon Klein had all the mathematical stimulation he needed (Rowe 1989).

Like his famous countryman, Niels Henrik Abel, who visited the Prussian capital a half-century earlier, Lie had just begun a Continental tour that would eventually take him to Paris and then Rome. Nothing could have suited Klein better; for he soon realized that Lie was a genius waiting to be discovered. Their tumultuous friendship would mark both men for the rest of their lives. Klein complained to Noether that no one in Berlin showed the slightest interest in the kind of geometry practiced by Plücker or Clebsch; Weierstrassian analysis was all the rage and Klein, out of a sense of inner opposition, simply refused to attend his lectures, a decision he would later come to regret (Klein 1923). In the meantime, he spent long hours talking to Lie, who was deeply immersed in research on curves associated with a special type of quadratic line complex formed by the lines that meet the faces of a tetrahedron in a fixed cross ratio. The properties of such tetrahedral complexes and their curve systems would continue to occupy Lie and Klein for well over a year (Rowe 1989).

Following a short break in early spring, they renewed their conversations in Paris. Over the next few months they strolled through the streets of the city, but they also had the opportunity to meet a number of its leading mathematicians. Among them, the most important by far was Gaston Darboux, who quickly directed their attention to the field of sphere geometry, a French specialty. Soon Klein and Lie were eagerly pursuing links between Plücker's line complexes and their analogues, 3-parameter families of spheres in 3-space. By early July, Lie had made one of his most spectacular discoveries: his line-to-sphere transformation. This variant of the Noether mapping induces a contact transformation between the lines of one space and the spheres of another. Darboux was startled and deeply impressed by the virtuosity with which Lie handled this new mapping and its properties (Rowe 1989).

Unfortunately for Lie and Klein, the Franco-Prussian War intervened; plans to visit Italy or the British Isles would have to wait. When war broke out in July 1870, Klein quickly scrambled to the Gare du Nord to take the first train headed back home (his recollections of this and the events that followed during the war were recorded in (Klein NL 22L, 2)). Rejected for military service, he joined an ambulance crew of emergency workers who assisted the men in battle. The work was sheer drudgery, even dangerous at times, especially due to exposure to diseases. Klein, in fact, came down with typhoid fever and had to be sent back to his parents' home in Düsseldorf. It took him a good month to recover. Lie, who loved to wander in the mountains of his native land, decided to leave Paris for Rome – on foot! He got as far as Fontainebleau, where sentries took him in for questioning (Engel 1900, 35–36). Discovering that he was carrying letters from Klein, written in what seemed a strange German code language, he was arrested after an inconclusive interrogation.

The French authorities suspected that "Komplex" was the code name for his German contact, and that "Linien" und "Kugeln" alluded to secret military formations. Lie tried to convince them otherwise, but his mathematical oration got no further than something like "let X, Y, Z be the usual axes of a coordinate system in 3-space," after which the bewildered Frenchmen concluded that if Lie turned out not to be a spy, then he was probably crazy and even potentially dangerous (Klein NL 22G). So they locked him up, and after several weeks had passed Darboux came to his rescue, enabling Lie to pursue his original plan. He again set off for Rome, this time by train.

On his return trip back to Norway, he stopped in Düsseldorf to spend a few days with Klein, who was still slowly recovering. In the meantime, the latter had decided his next move would be to Göttingen, this time to work alongside Clebsch as a *Privatdozent*. To qualify, he had to deliver the customary lecture before the philosophical faculty. That took place in January 1871, just as Prussia's princes and military officers were about to celebrate the founding of the new German Reich in the Palace of Versailles. His topic: a model of a Plücker quartic surface that he had designed and built with the help of a friend. Six months earlier he had made use of a similar model of a Kummer quartic in order to study the *Gestalt* and singularities of its asymptotic curves. The latter, as Lie had discovered, were algebraic of degree 16, a result he deduced directly from his line-to-sphere mapping.

Models as Artefacts for Discovery

It was only five years earlier that Ernst Eduard Kummer, the senior Berlin mathematician, had uncovered the remarkable quartics since known as Kummer surfaces, a discovery that eventually opened the way to numerous subsequent investigations. This era, in fact, witnessed several such discoveries in algebraic geometry. In rapid- fire fashion, mathematicians now began to study the rich variety of surfaces that lie beyond the realm of quadrics, which had already been studied and classified by Euler. An early breakthrough came around 1850 when Cayley and Salmon discovered that non-singular cubic surfaces contain 27 lines that form a complicated spatial configuration (Dolgachev 2004).

It is important to recall that geometry at this time was still strongly tied to the study of figures in 3-space. Leading mathematicians continued to think of geometrical objects as idealizations drawn from the world in which we live. Thus, an algebraic surface was not seen as some special case of an abstract concept, like Riemann's notion of a manifold, but rather as the mathematical counterpart of an entity that one could imagine as situated in real physical space.Seen from this vantage point, it should come as no surprise that Klein and his teachers were intent on gaining a visual image of various new-fangled mathematical objects, an interest that prompted several contemporary geometers to build various types of models in order to study their properties in greater detail.

These geometric investigations were pursued almost like botanical studies in which the geometer went about collecting various specimens associated with certain classes of equations. These then had to be classified according to some larger scheme, certainly less ambitious than the Linnaean framework, but nevertheless with a similar purpose in mind. By the mid-nineteenth century, determinants and invariant theory had begun to provide invaluable tools for studying objects in the projective plane or 3-space. Bezout's theorem made it possible to exploit the invariant degree of algebraic curves and surfaces by moving to the complex domain, thereby posing a new challenge: how to interpret the imaginary elements that lie outside the realm of *real* 3-space, the arena of actual interest for the geometer. Special incidence configurations – like Hesse's inflection point configuration for cubic curves, or Schläfli's double-six in connection with the 27 lines of a cubic surface – also helped identify key features.

Still, the wealth of material was immense, making it difficult to sort out the relevant structures and the number of possibilities for each. While some geometers pursued taxonomical studies, others began to seek out underlying organic principles to identify the morphology of the objects under investigation. Special archetypes were usually identified in order to identify structural differences, etc. This style of research had many variants, but for roughly three decades, from 1850 to 1880, many leading mathematicians all across Europe were exploring this new terrain almost like scientific progeny of the naturalist Alexander von Humboldt touring South America with virgin eyes. Among the leading explorers were Cayley and Salmon in Britain, Cremona in Italy, Schläfli in Switzerland, Clebsch and H. A. Schwarz in Germany, and Lie in Norway.

During this period, model-making went hand in hand with this cutting edge research. In Bonn, Klein assisted Plücker in designing around 30 different models that displayed select features of a certain class of quartic surfaces linked to quadratic line complexes. Klein emphasized the connection between these mathematical models and Plücker's earlier research in physics, especially his famous experiments on electrical discharges in rarefied gases (Müller 2006). Plücker carried these out with the assistance of Heinrich Geissler, famous for his invention of the glass tubes that bear his name. In both of these research fields, Plücker was drawn

Fig. 8.2 Julius Plücker (1801–1868).

Fig. 8.3 Plücker's model made for the London Mathematical Society of an equatorial surface with eight real singular points lying in pairs on two sets of parallel lines. (Photos of this and other Plücker models can be seen at http://www.lms.ac.uk/content/plucker-collection).

to describe complex, never-before-seen spatial phenomena (Clebsch 1871).

Though little appreciated in Germany, Plücker (Fig. 8.2) had an excellent reputation as an experimental physicist in England, where his work was championed by Michael Faraday, a physicist who thought in pictures, not formulae. Klein later recalled how Plücker once told him that Faraday had given him the initial impetus to build models illustrating different types of the so-called complex surfaces he unveiled as the centrepiece of his new line geometry (Klein 1922, 7). Faraday was by no means the only one in England to take an interest in these exotic spatial artefacts. Thomas Archer Hirst, who had studied under Jakob Steiner in Berlin, was another enthusiast for Plücker's models. In 1866, Plücker delivered a well-received lecture at a meeting in Nottingham in which he employed a group of his models (Cayley 1871). Hirst was intent on acquiring copies, and so Plücker afterward donated a set of these made in boxwood to the London Mathematical Society (the correspondence between Plücker and T. A. Hirst can be found at http://www.lms.ac.uk/content/plucker-collection).They can still be seen today on display at the headquarters of the LMS or just by going online (Fig. 8.3).

Not long after Plücker's death, Klein designed four additional models to show the main types of real singularities that can arise in complex surfaces (Klein 1922, 7–10). These models were, like the original Plücker models, produced and sold by the firm of Johann Eigel Sohn, located in Cologne. Originally made in zinc, they are far heavier than the more familiar plaster models built afterward. For his *Habilitation* lecture in Göttingen, held in January 1871, Klein presented one of these models in order to convey some newly discovered properties of special curves associated with complex surfaces. Later the next year, he and Clebsch presented two new models of cubic surfaces to the Göttingen Scientific Society. It was on this occasion that Clebsch unveiled his famous diagonal surface which illustrates the intricate configuration formed by the 27 lines that lie on a non-singular cubic (Fischer 1986, Kommentarband, 7–14). Klein's model, built by his doctoral student Adolf Weiler, illustrated a cubic surface with four real singular points, the maximum possible. These two models represent the two extremes in what later became known as the Rodenberg series, a set of 26 models designed and published by Carl Rodenberg in 1881 (Schilling 1911).

Equally famous were the special quartics studied by Kummer in the mid 1860s (Kummer 1975, 418–432). These objects exhibit a special configuration of 16 singular points and 16 singular planes in space that soon led to a plethora of investigations connected with the properties of these so-called Kummer surfaces. Each plane passes through six of the 16 singular points, which are constrained to lie on a conic, forming a (16, 6) configuration. Such surfaces, like the Fresnel wave surface, are self-dual, hence of the fourth order and class. This means that the lines in space intersect the surface in four (real or imaginary) points, whereas four tangent planes will pass through any given line. Kummer noted that these dual singularities are the maximum possible for a quartic since the class of a surface with d singular points is given by $k = n(n-1)^2 - 2d = 4(3)^2 - 2(16) = 4$.

A Context for Discovery: Geometric Optics

Kummer surfaces are familiar to algebraic geometers today for the special role they play within the larger setting of K3 surfaces. When in 1958 André Weil coined this term as a shorthand for Kummer, Kähler, and Kodaira, he certainly had nothing in mind like the plaster models that Kummer built in Berlin shortly after Klein came to study there. As editor of Kummer's *Collected Papers,* Weil wrote in praise of his "amazing ability to cut his way through dense thickets of algebraic or analytic brushwood, with frequently inadequate intellectual equipment, while never losing sight of the main goal." (Kummer 1975, vii). Weil certainly took a serious interest in Kummer's work on number theory – see his lengthy introduction to volume 1 – but I doubt he spent much time looking over the more than 200 pages in volume 2 that span the period from 1860 to 1878. These papers deal with ray optics, atmospheric refraction, and include a lengthy study on algebraic ray systems (1864), which led directly to the study of Kummer surfaces as special types of caustics.

All these papers fall under the rubric of geometrical optics, a field that earlier saw considerable activity in France, where Monge and his pupils often studied geometrical problems closely connected with experimental physics. Augustin Fresnel, whose wave theory of light established a new paradigm for optical research, was able to account successfully for various phenomena related to polarization. This included the longstanding mystery of double refraction in certain crystals, such as Iceland spar. Fresnel's explanation for this led him to consider the wave fronts formed by light entering biaxial crystals, which are anisotropic media (Knörrer 1986). He described this wave surface in the early 1820s, and soon afterward Ampère gave a precise mathematical description of it as a quartic surface with four singular points. By the early 1830s this exotic surface caught the attention of W. R. Hamilton, who exploited its properties to make a startling new prediction: conical refraction, which was quickly confirmed experimentally by his colleague Humphrey Lloyd. Fresnel was no longer alive, but his theory finally began to win adherents even in Britain, thanks in part to Hamilton's work on the wave surface (Hankins 1980).

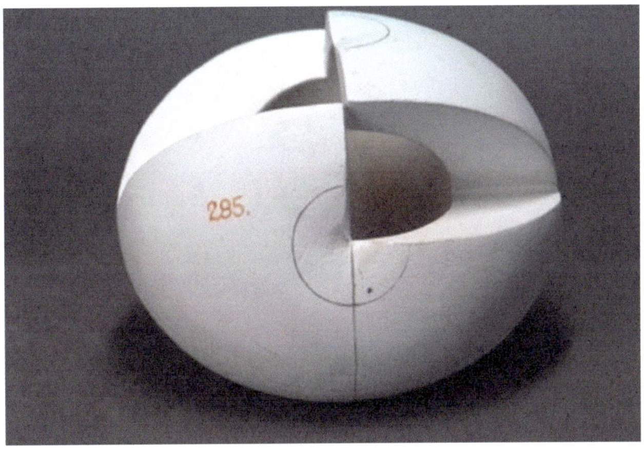

Fig. 8.4 A model of the Fresnel wave surface showing the circle of tangency for one of its four double planes. The point in the interior represents one its four singular points at which the two sheets of the quartic surface come together. (From the Collection of Mathematical Models at the Göttingen Institute of Mathematics).

This historical background had a direct relevance for Kummer's research in the 1860s. Indeed, his quartic surfaces can be seen as a natural generalization of the famous wave surface of Fresnel. As mentioned, Kummer surfaces have 16 singular points, but in the case of a Fresnel wave surface only four of these points are real. At those four singular points the tangent lines form cones rather than tangent planes. Furthermore, associated with these points are four singular planes, which touch the surface along circles. These planes appear in symmetric pairs each of which is perpendicular to the two principal optical axes of the crystal. The normal lines to these planes, directed inward from the centres of the four circles, intersect the surface at the four singular points. Hamilton likened this geometric structure with that of a piece of fruit on a kitchen table: the Fresnel surface touches one of its singular planes along a circle "somewhat as a plum can be laid on a table so as to touch and rest . . . [there] on a whole circle and has in the interior of the circular space a sort of conical cusp" (Hankins 1980, 88) (Fig. 8.4).

So when Weil wrote that Kummer surfaces were discovered "in connection with problems in line geometry in three-dimensional projective space," he not only overlooks this context but he also misuses the term line geometry, which deals mainly with the study of line complexes, 3-parameter families of lines, rather than the 2-parameter ray systems studied by Kummer (Fig. 8.5). Here is the full quotation from 1975: "To many mathematicians of the present generation, it may come as a surprise to learn that Kummer discovered that family of surfaces in connection with problems in line geometry in three-dimensional projective space, entirely without the powerful tool provided by theta functions" (Kummer 1975, vii). Presumably Weil knew that Kummer

Fig. 8.5 Ernst Eduard Kummer (1810–1893).

Fig. 8.6 Klein designed this zinc model of a Kummer surface in 1871. The vertices of the five interior tetrahedra correspond to the 16 conical points of the surface. Attached to these are three tetrahedra that pass through infinity, so these appear only partially as pairs of opposite pieces that attach to one vertex for each of the four tetrahedra that surround the one in the centre. (From the Collection of Mathematical Models at the Göttingen Institute of Mathematics).

surfaces were deeply tied to line geometry, but if he had read Kummer's papers carefully he would have quickly realized that Kummer never made that connection.

In fact, it was Klein who first pointed this out in a fundamental paper (Klein 1921, 53–80) written earlier that same summer of 1869, just before he departed for Berlin. There he introduced Kleinian coordinates and showed that for a general quadratic line complex – in fact, for a whole 1-parameter family of such complexes – the associated singularity surface will be a quartic with precisely the same configuration of singular points and planes found by Kummer in 1864. Klein cited Kummer's paper, of course, and noted further how to derive the special case of a Fresnel wave surface. Furthermore, he took note of the model in metal wiring that Kummer had designed to illustrate the shape of such a special quartic, one in which all 16 conical points are real and visible within a bounded portion of space. Klein would soon design a new model for this type of surface to illustrate the *Gestalt* of the singular surface for a generic quadratic complex (Fig. 8.6). He also designed three other models related to this general case. These represent three principal types of Plücker complex surfaces, chosen to highlight the organic principle underlying all four cases (Klein 1922, 6–7).

Almost certainly, Klein spoke about these findings in Kummer's seminar, where the main topic was ray systems. Yet if Kummer took note of these results, he never found occasion to refer to them in print. Indeed, one finds no hint in his papers on ray systems that he was aware of recent research on Plückerian line geometry and its striking connections with the 2-parameter systems of lines studied in ray optics. One of Klein's main contributions, in fact, was to demonstrate the substantial links between Plücker's theory of quadratic line complexes and Kummer surfaces. Yet Kummer ignored this fundamentally new insight, even though he continued to work on the latter topic until at least 1878 (contrary to Weil's claim that he had lost interest in it by then). By this time a number of mathematicians had shown that Kummer surfaces could be dealt with analytically by means of theta functions. Already in 1871 Klein indicated the possibility of parameterizing Kummer surfaces by means of genus two hyperelliptic integrals, a line of attack later pursued by his student, Karl Rohn, in the late 1870s (Klein 1921, 52).

Kummer's Seven Models

In the meantime, Kummer found an elegant and surprisingly simple description for a 2-parameter system of quartic surfaces in which the real points lie within a bounded portion of space. This system included as a special case the so-called Roman surface of Jacob Steiner, which Kummer had independently re-discovered in the mid 1860s (Weierstrass

1903, 180). To give a vivid picture of the organic connections underlying the various types of singularities that can arise, he designed seven plaster models, presented to the Prussian Academy at a meeting held on 20 June 1872 (Kummer 1975, 575–586). All seven exhibit tetrahedral symmetry, as reflected in the form of their quartic equation:

$$\varphi^2 = \lambda \, p \, q \, r \, s = 0,$$

where $\varphi = 0$ is the equation for a family of concentric spheres with parameter μ:

$$\varphi = x^2 + y^2 + z^2 - \mu \, k^2.$$

The right side of the equation represents the four planes of a regular tetrahedron given by:

$$\begin{aligned} p &= z - k + \sqrt{2}\,x &&= 0, \\ q &= z - k - \sqrt{2}\,x &&= 0, \\ r &= z + k + \sqrt{2}\,y &&= 0, \\ s &= z + k - \sqrt{2}\,x &&= 0. \end{aligned}$$

For a fixed value of k this leads to a 2-parameter system of equations:

$$\left(x^2+y^2+z^2-\mu\,k^2\right)^2=\lambda\,\left(\left(z-k\right)^2-2x^2\right)\right)\,\left(\left(z+k\right)^2-2y^2\right)\right)=0.$$

Kummer restricted the parameterλ to the values $-3 \leq \lambda \leq 1$ so that the real portion of the surface will be bounded; otherwise it extends to infinity. For $\lambda = 0$ the quartic collapses to a double sphere, so its shape alters drastically in passing over this boundary value. The cases where $\mu < 1$ hold little interest since these surfaces will have no real singularities due to the circumstance that the radii of the spheres are too small to intersect with the edges of the tetrahedron. When $\mu = 1$, however, the sphere just touches the six edges, leading to six real bi-planar singular points, as illustrated by Kummer's first model[1] (Fig. 8.7).

By holding $\mu = 1$ fixed and letting λ increase until it approaches $\lambda = 1$, Kummer obtained a second model with six singular points, but where the osculating planes are imaginary so that only a real line passes through these conical points (Fig. 8.8).

Kummer next passes to the special case where $\mu = \lambda = 1$ by letting λ attain unity at which stage three concurrent single lines form, meeting in a triple point. This is the famous Steiner surface of order four and class three. Kummer discovered that all its tangent planes form pairs of conics (Fig. 8.9).

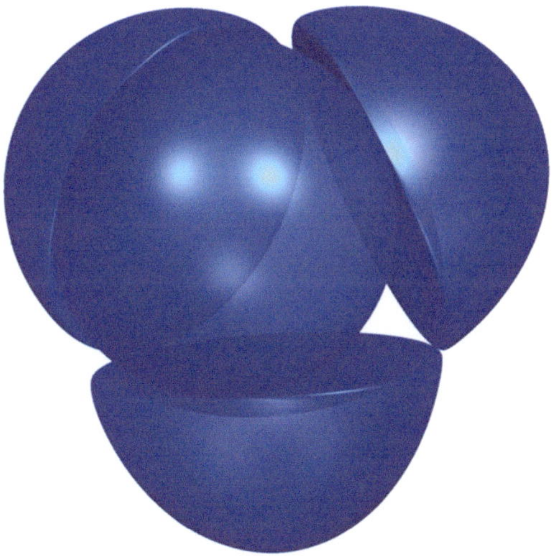

Fig. 8.7 Model I, $\mu = 1, \quad \lambda = -\frac{1}{8}, \quad k = 40\ mm.$

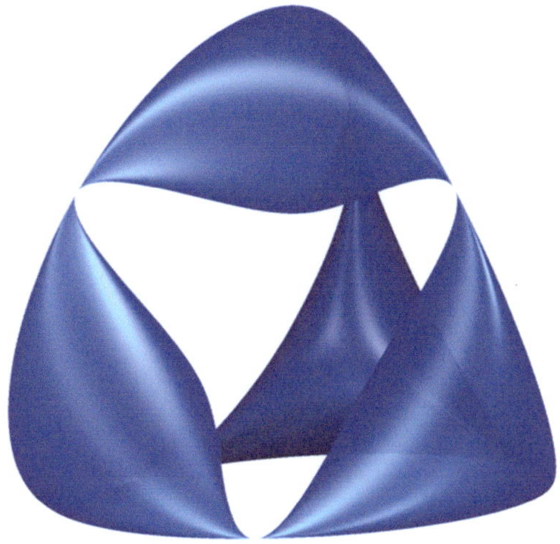

Fig. 8.8 Model II, $\mu = 1, \quad \lambda = \frac{9}{10}, \quad k = 50\ mm.$

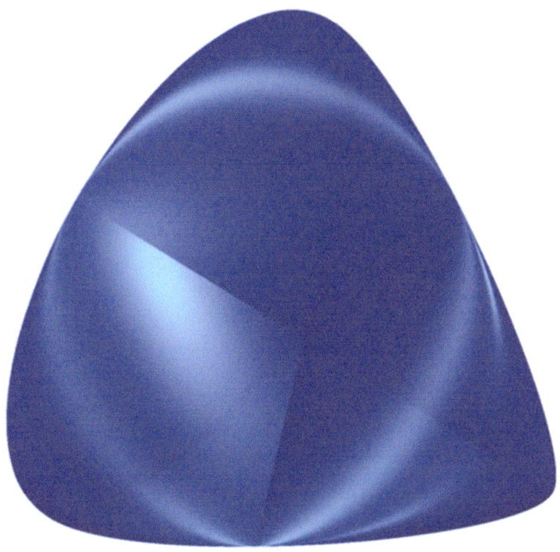

Fig. 8.9 Model III (Steiner surface), $\mu = 1, \quad \lambda = 1, \quad k = 50\ mm.$

[1]The images that follow were created by Oliver Labs, whose expertise in computer graphics greatly enhanced the visual quality of this essay. For further discussion of these models along with lovely black-and-white photos of most of them, see Fischer (1986).

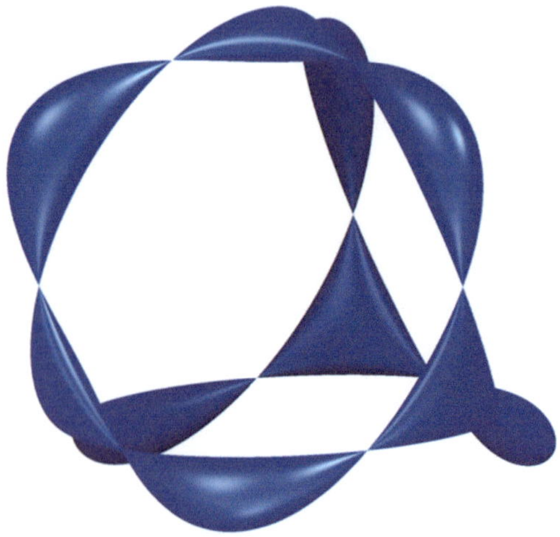

Fig. 8.10 Model IV, $\mu = \frac{4}{3}, \quad \lambda = \frac{1}{2}, \quad k = 50\ mm.$

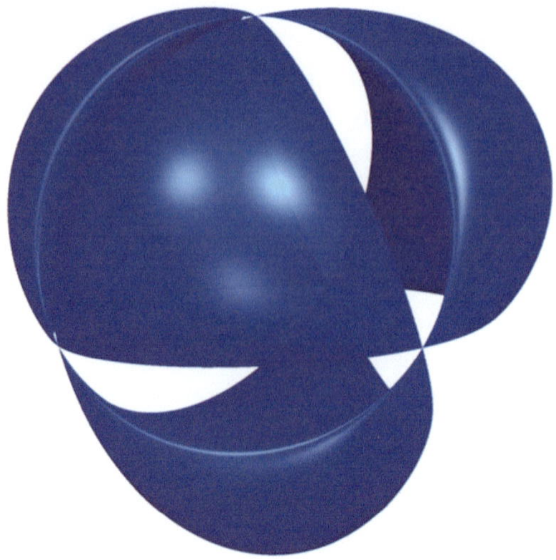

Fig. 8.12 Model VI, $\mu = 3, \quad \lambda = -\frac{1}{8}, \quad k = 30\ mm.$

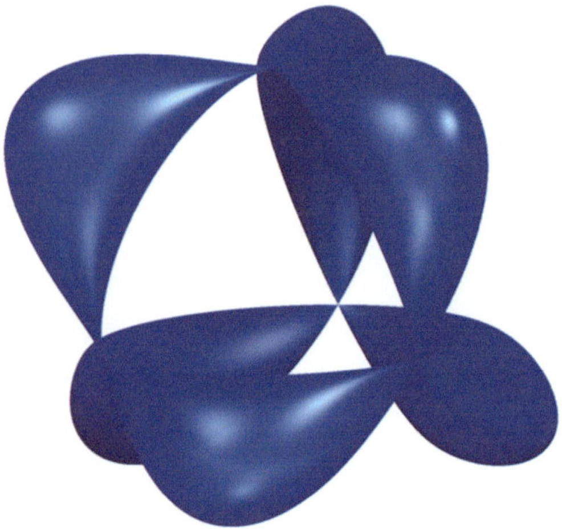

Fig. 8.11 Model V, $\mu = 3, \quad \lambda = \frac{1}{2}, \quad k = 25\ mm.$

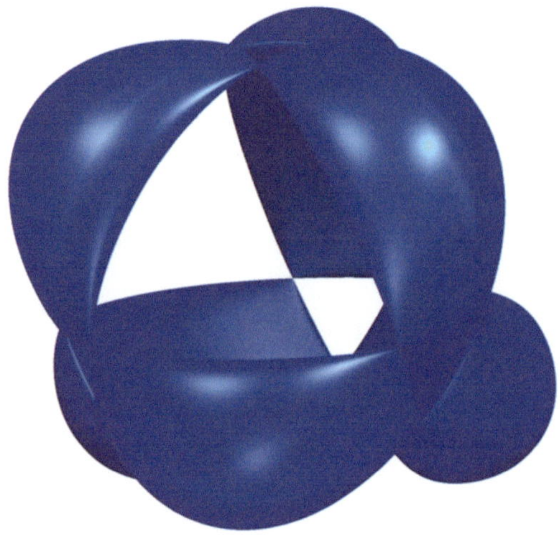

Fig. 8.13 Model VII, $\mu = 9, \quad \lambda = \frac{1}{4}, \quad k = 18\ mm.$

When $\mu > 1$ 12 singular points appear (except when $\mu = 3$), as illustrated by the next model, where the sphere intersects the six edges of the tetrahedron in pairs of points (Fig. 8.10):

In the exceptional case,$\mu = 3$, the sphere passes through the vertices of the tetrahedron so that only four singularities arise (Fig. 8.11):

Passing over the critical value $\lambda = 0$ (the double sphere) to a negative λvalue leads to an "inversion" of the surface (Fig. 8.12):

Finally, if $\mu > 3$ the sphere will intersect the extended edges of the tetrahedron in paris of points, leading to the 12 singularities in model 7 (Fig. 8.13):

Kummer further noted that the wire model he had presented to the academy six years earlier (Kummer 1975, 425–426) corresponded to the condition $\lambda = \frac{3\mu-1}{3-\mu}$. In the special case when $\lambda = \mu = 1$ this will be a Steiner surface, but passing to $\lambda > 1$, $\mu > 1$ the surfaces are no longer bounded, whereas all 16 double points and planes will now be real. In closing, Kummer also mentioned the model in zinc that Klein had recently published for just such a surface.

Over the next 12 years, these seven plaster models could only be seen by visiting the Berlin mathematics seminar. Surely a number of mathematicians knew of their existence from Kummer's publication, but very few had any idea of what they actually looked like. Not until the 1880s did these models leave their local confines in Berlin, after which they quickly became familiar objects to mathematicians around the world. How that came to pass reflects an important shift in model making that took place during the intervening period, which takes us to the next chapter in our story.

Models in Standardized Production

The production of mathematical models took on a far more public face in the late 1870s. This began in 1875 with Klein's appointment as professor of mathematics at the Technische Hochschule in Munich, where he succeeded the Otto Hesse. There Klein was joined by Alexander Brill, who had studied under Clebsch in Giessen in the 1860s and afterward formed a close working relationship with Max Noether. Brill's background and training, however, were highly unusual, though ideally suited for model making (Finsterwalder 1936). His uncle, Christian Wiener, gave him early lessons and the incentive to take up mathematics. Wiener, who taught for many years at the TH Karlsruhe, was the first to construct a plaster model showing the 27 lines on a cubic surface. In fact, it was Clebsch who urged him to do so. Wiener's young nephew, Alexander Brill, thus grew up with a hands-on appreciation of higher mathematics. Yet Brill chose first to study architecture in Darmstadt before gravitating to Giessen to work with Clebsch. He thus had the opportunity to develop the kinds of skills in technical drawing and model design that would later bear fruit during his nine-year-tenure at the TH Munich (Fig. 8.14).

Fig. 8.14 Alexander Brill (1842–1935).

Unlike most other such institutions, the Munich TH trained teaching candidates as well as engineers and architects. Klein exploited this opportunity by obtaining funds for a new mathematics institute with special seminar and laboratory rooms exclusively for the students pursuing this program (Tobies 1992). From the beginning, Klein's aim was to combine courses in higher mathematics with the kinds of practical instruction in drawing and modelling that had long been cultivated at the leading polytechnical institutes. He had tried to do something similar during his brief tenure in Erlangen, but met with little success. Klein found that the typical university student had far too little experience with technical drawing, so he was pleased to have a chance to pursue this *anschauliche* approach to geometry with students who already had the requisite skills. Perhaps even more crucial for the success of this venture was the appointment of Brill, who represented precisely the kind of fusion of practical and theoretical interests Klein hoped to achieve.

Alexander Brill, like his uncle Christian Wiener, was a native of Darmstadt, where the family owned a publishing company (Finsterwalder 1936). His brother, Ludwig, took over the business from his father, a circumstance that turned out to have considerable import for model-making in Germany. For just two years after they began their collaboration in Munich, Brill and Klein struck on the plan to market several of the models from their Munich collection through the Darmstadt firm of L. Brill. In the meantime, their modest workshop had become a beehive of activity, especially when it came to producing plaster models for visualizing the new objects geometers had recently brought to light in their research. Klein was especially interested in these, whereas the bulk of the work done under Brill's supervision had a more immediately didactical aim.

When Brill and his brother launched this commercial venture in 1877, they probably had no inkling of just how successful it would turn out to be. From the beginning, the models were produced and sold in series, the first being a modestly priced set of carton models illustrating the seven types of quadric surfaces. Fittingly enough, the inspiration behind these came about when Brill and Klein first met in 1873 at a mathematical conference held in Göttingen. Klein, as one of the principal organisers, had arranged that various models would be put on display. One of these was designed by Olaus Henrici, a former student of Clebsch who taught at University College, London. This showed how the contour of an elliptical paraboloid can be constructed by splicing together an array of half-circles. Klein presented this model at the meeting, and soon afterward Brill realized that Henrici's method could be adapted to build any of the seven types of quadric surfaces (Brill 1889) (Fig. 8.15).

Among the earliest plaster models designed by the students of Brill and Klein, the following examples may be taken as representative:

Fig. 8.15 Carton models of quadric surfaces designed by Alexander Brill, 1877 (From the Collection of Mathematical Models at Marburg University).

- A surface of rotation with its geodesic and asymptotic curves – J. Bacharach
- A caustic surface derived an elliptic paraboloid – L.Schleiermacher
- An ellipsoid with its geodesic curves – K. Rohn
- A cubic surface with 4 conical points and its asymptotic curves – J. Bacharach
- Four rotational surfaces of constant mean curvature with geodesics – A. v. Braunmühl
- Two surfaces of constant negative curvature with geodesics – W. Dyck

Although the motivation behind many of these models was primarily didactical, some cases involved new results of potential interest to researchers. In such cases, fairly elaborate explanations were sometimes necessary. These could be found in brochures written by the respective designers, many of whom provided detailed mathematical descriptions without which even an educated observer would surely remain baffled. This literature accompanied the models upon purchase, but was otherwise unavailable to a wider public. By 1899, when Ludwig Brill sold the publication rights for his collection of models to the firm of Martin Schilling in Halle, the demand for them had not yet reached its peak; nor had the number of different models available for purchase flattened out. Indeed, a dozen years later M. Schilling, now located in Leipzig, put out a catalogue listing nearly 400 models available for delivery (Schilling 1911). Some years earlier, after repeated requests for the technical brochures that came with the models, Schilling also published the entire collection for the first 23 series of models (Schilling 1904). Many of these are an invaluable resource for understanding the mathematical meaning of the objects on display.

By this time, the era of mathematical model-making had long since entered an era of standardized production. While the use of models and other visual aids took on a growing importance in mathematics education, they rarely played a role any longer as artefacts for research. Indeed, their demise as tools for research was already under way when the Brill firm began to market them. A glance backward reveals just how successful this venture proved to be. The models pictured below, many of them familiar from collections scattered around the world, were already available for purchase ca. 1880 (Fig. 8.16).

When he arrived at the TH Munich, Klein was eager to take up research on Kummer/Plücker surfaces again. He brought with him the zinc models he had designed back in 1871, but these he found unsatisfactory in certain respects. So he was pleased to enlist the support of a student, Karl Rohn, who built three new models under Klein's watchful eye. All three were prepared in plaster of Paris and marketed in the second series of models offered by L. Brill. In the catalogues they are identified as Kummer surfaces, but Rohn emphasized that these models illustrate the passage from a general Kummer surface to the special cases that arise in Plücker's theory of quadratic line complexes (Rohn 1877). So the latter two models represent a general Plücker complex surface with eight real singular points together with a degenerate Plücker surface for which only four such points remain. The former case appears in the middle of the third row of the advertisement above. A more aesthetically pleasing image is reproduced below (Fig. 8.17).

Geometry and model building continued to play a major role in mathematical pedagogy at the TH Munich for decades after the departure of Klein and Brill. Two key figures thereby were Walther von Dyck and Sebastian Finsterwalder,

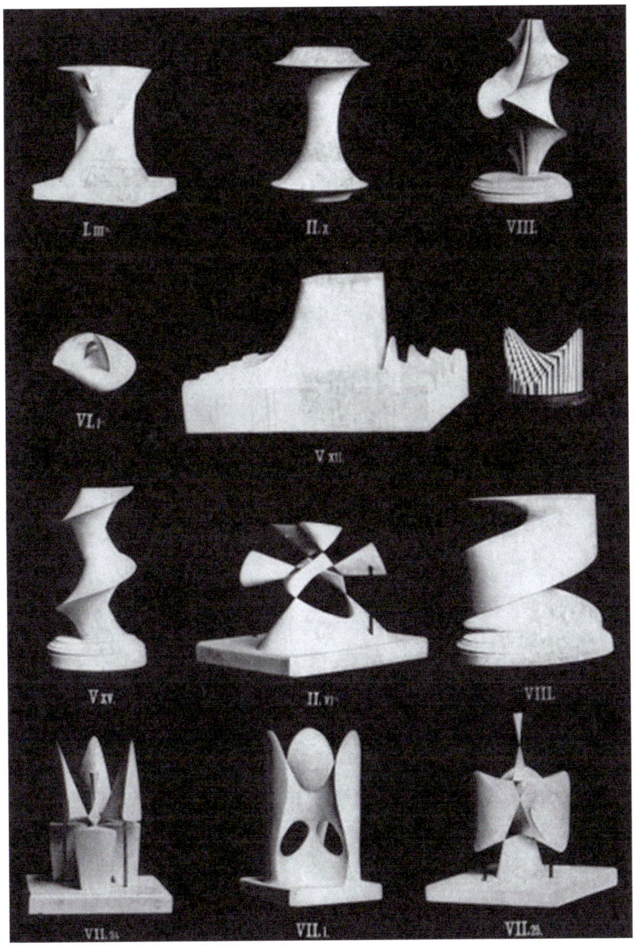

Fig. 8.16 An early advertisement made for the firm L. Brill illustrating some of its more popular models. From left to right: surface of curvature centres for a one-sheeted hyperboloid, rotational surface of constant negative curvature, Kuen surface, Fresnel wave surface, Jacobi amplitude function, hyperbolic paraboloid, Dini surface, Kummer surface with 8 real nodes, Meusnier minimal surface, Hessian surface with 14 singular points, Clebsch diagonal surface, Hessian surface with 13 singular points.

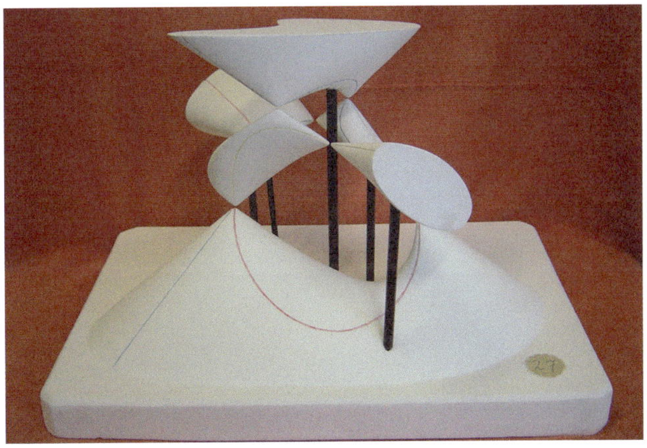

Fig. 8.17 Karl Rohn's model of a Plücker complex surface showing eight real singular points, the remaining eight lie in pairs on a double line (in this case imaginary) (Mathematics Institute, University of Groningen) All three of the models designed by Rohn can be viewed at http://www.math.rug.nl/models/serie_2.html.

both of whom were skilled in drawing and design. In 1880 Felix Klein left Munich to accept a new chair as Professor of Geometry in Leipzig. Four years later Alexander Brill took a professorship in Tübingen, where he spent the remainder of his long career. During his final year at the TH Munich, however, he had the opportunity to make a new acquisition for the collection of models marketed by his brother in Darmstadt, namely copies of the original models of quartic surfaces with tetrahedral symmetry designed by Kummer in 1872.

How this came about can be surmised from the correspondence published in the appendix to this paper. The original impetus came once again from Klein, who was interested in acquiring copies of the Kummer models for his collection at the new mathematics institute in Leipzig. He thus wrote to Kummer in early 1883, asking him for permission to use the originals for this purpose. The latter responded affirmatively in a letter from the 4th of March, expressing his pleasure that Klein wished to use them as didactical aids. He advised, however, that Klein contact Weierstrass to make the necessary arrangements owing to the circumstance that he, Kummer, was stepping down as co-director of the Berlin Seminar as part of his transition into full retirement.

Not long afterward, Brill visited Kummer in Berlin and learned from him about Klein's request. Brill also learned that Kummer had some reservations about the technical facilities available in Leipzig for preparing and producing the castings for his models. Brill then told him about his longstanding working relationship with an expert in this domain, Joseph Kreittmayr, who prepared castings of the antique statuary in the Bavarian National Museum. Little wonder that the models in the collection of L. Brill had such aesthetic appeal! (Figs. 8.18, 8.19).

So Weierstrass sent the Kummer models directly to Brill in Munich. On their arrival the latter was delighted by their elegance. He was also pleased to gain permission to reprint the earlier article Kummer published with the Prussian Academy in 1872 when he first presented his seven models. These famous models now became the ninth in the series of L. Brill, though they could just as easily have fallen into oblivion like many others from this time. In mathematics, success sometimes depends on circumstances; in this case at least, it came with a knack for marketing the right artefacts at the optimal time.

Fig. 8.18 Kummer's Model IV as copied for the Brill Collection, Seires IX, n. 4.

Fig. 8.19 Kummer's Model V as copied for the Brill Collection, Seires IX, n. 5.

Appendix[2]

Auszug aus einem Brief von Ernst Eduard Kummer an Felix Klein:

Berlin d. 4te März 1883

Geehrter Herr Kollege!

Die Modelle zu den Flächen vierten Grades, welche ich früher selbst angefertigt habe, und welche Sie im mathematischen Seminar der Universität gesehen haben, scheinen auch mir recht nützlich um das Gebiet der geometrischen Anschauungen für den Studierenden zu erweitern und manche Besonderheiten klar und bestimmt anschaulich zu machen. Ich kann es danach nur mit Freude und mit Dank anerkennen, wenn Sie sich der Mühe unterziehen wollen dieselben vervielfältigen zu lassen. . . .

. . .

Leben Sie recht wohl,

Ihr ergebenen

E. E. Kummer

Drei Briefe von Alexander Brill an Karl Weierstrass:

München, 12. April 1883

Hochverehrter Herr Professor!

Als ich bei meiner jüngsten Anwesenheit in Berlin Herrn Professor Kummer besuchte, wurde ich von ihm zu einer brieflichen Anfrage an Sie veranlasst, um deren freundliche Aufnahme ich ganz ergebenst bitten wollte.

Herr Kummer wurde vor einiger Zeit von Herrn Klein in Leipzig darum angegangen, seine in dem mathematischen Seminar der Universität Berlin befindlichen Modelle ihm zur Entnahme von Abgüssen auf kurze Zeit zu überlassen. Er hatte ihm dies zugesagt, mit dem Beifügen, dass, da er von der Verwaltung des Inventars zu Ostern zurücktreten würde, Klein sich an Sie mit der Bitte um Übersendung der Modelle wenden möchte.

Herr Kummer äusserte, indem er mir von dem Vorstehenden Mitheilung machte, das Bedenken, ob ich wohl in Leipzig ein geschickter Gipsformator fände, dem man die an den Modellen nothwendigen [-2-] vorgängigen Reparaturen überlassen könnte. Ich erlaubte mir den Vorschlag zu machen, diese Arbeit, sowie die Herstellung von Abgüssen, in München von dem Formator des k[öniglichen] b[ayrischen] Nationalmuseums Herrn J.[oseph] Kreittmayr, einem mir als geschickt bekannten

[2] The letter from Kummer to Klein is found in Klein's posthumous papers, Cod. Ms. F. Klein, Niedersächsische Staats- und Universitätsbibliothek Göttingen. The three letters from Brill to Weierstrass are located in the Weierstrass Nachlass, Geheimes Staatsarchiv Preußischer Kulturbesitz, Berlin. My thanks go to Eva Kaufholz-Soldat for alerting me to these letters and for her help in preparing the transcriptions.

Mann, unter meiner Aufsicht vornehmen zu lassen, sowie weiterhin gestatten zu wollen, dass die Modelle in Copien weiteren Kreisen zugänglich gemacht würden. Die Herstellung und Vertreibung der Copien könnte etwa der Buchhandlung von L. Brill in Darmstadt, die einen ausgedehnten Verlag von mathematischen Modellen bereits besitzt, durch meine Vermittlung übertragen werden. Herr Kummer erklärte sich mit diesen Vorschlägen einverstanden und empfahl mir, nach vorgängiger Verständigung mit Herrn Klein, den Sachverhalt Ihnen darzulegen.

Diesem Wunsche nachkommend erlaube ich mir, nachdem mir Herr Klein die weiteren Schritte zu Ihnen überlassen hat, an Sie, hochverehrter Herr Professor, die ergebenste Anfrage zu richten, ob es möglich ist, die fraglichen Modelle, die ich bei meiner Unbekanntschaft mit denselben nicht näher bezeichnen kann, zu verschicken? [sic] und im bejahenden Falle die Bitte beizufügen, dass dieselben an meine Adresse, hiesige technische Hochschule, versandt werden. [-3-] Die sorgsame baldige Rücksendung der Originale sowie je eines Abgusses an das mathematische Seminar der Universität Berlin wäre meine Sorge, die Kosten der Verpackung und Versendung, sowie alle besonderen Ausgaben würden der Buchhandlung zur Last fallen.

Ist die Veröffentlichung der Modelle, etwa so wie ich sie vorstehend darzulegen mir erlaubte, möglich, so zweifle ich nicht, dass Erzeugnisse aus der Hand Kummer's, die in die Zeit zurückreichen, wo das Interesse an gestaltlichen Untersuchungen noch eine äusserst geringes war, der allgemeinsten Theilnahme sicher sind, und ich würde es mir zur hohen Ehre anrechnen, bei der Publication dieser Modelle mitthätig gewesen zu sein.

Ihre Gesundheit, hochgeehrter Herr Professor, ist hoffentlich wieder im Stand, und das Unwohlsein, das Sie nicht hinderte, eine so inhaltsreiche Unterredung mir freundlichst zu gewähren, wieder beseitigt.

Gestatten Sie mir zum Schluss meinen vorläufigen Dank für die Bemühung auszusprechen, die aus dieser Angelegenheit Ihnen erwächst, sowie die Versicherung ausgezeichneter Hochachtung, mit welcher zeichnet,

Ihr ergebenster
Alexander Brill

München, 19. Mai 1883

Hochverehrter Herr Professor!

Ihre hochgeschätzten Zeilen sowie die Kiste mit Modellen habe ich erhalten und beeile mich, meinen verbindlichen Dank für die Güte und Bereitwilligkeit zu sagen, mit der Sie meinem Wunsche entsprochen haben. Die Modelle sind im besten Zustande angekommen, bis auf eines der im inneren Kästchen befindlichen, das eine unbedenkliche leicht reparierbare Beschädigung erhalten hat, wenn es dieselbe nicht schon früher besass.

Die Modelle haben, was den Inhalt und was die Eleganz der Ausführung angeht, alle meine Erwartungen übertroffen, und ich werde zufrieden sein, wenn [-2-] die Feinheit und Genauigkeit der Originale an den Copien überhaupt noch zur Erscheinung kommt.

Die Ausführungen in den Berliner Monatsberichten, auf die Sie mich hinzuweisen die Güte hatten, würden im Abdruck sich vortrefflich als Beilage zu der Serie eignen, und ich werde seitens der Buchhandlung die Erlaubniss zum Abdruck von Herrn Professor Kummer zu erwirken suchen.

Die Modelle sind inzwischen bereits in den Händen des Formators, doch wird die Herstellung genauer Copien einige Zeit in Anspruch nehmen.

Bis dahin bleibe ich mit dem wiederholten Ausdruck lebhaften Dankes und der Versicherung ausgezeichneter Hochachtung

Ihr
ergebenster
A. Brill

München, 15. Juli 1883

Hochverehrter Herr Professor!

Nachdem die Schwierigkeiten, die sich dem Abgusse der Kummer'schen Modelle entgegen gestellt hatten, überwunden sind, wird die Rücksendung der mir gütigst überlassenen Originale an Ihre Adresse in diesen Tagen erfolgen. Der Formator hat diesselben bestmöglich reparirt [sic] und fügt der Sendung eine Copie der Serie bei, die von der Buchhandlung dem mathematischen Seminar Ihrer Universität zur Verfügung gestellt wird.

Mit dieser Anzeige verbinde ich den wiederholten Ausdruck verbindlichen Dankes für die Güte, mit der Sie, hochverehrter Herr Professor, die Mühwaltung der Versendung übernahmen, sowohl im Namen des mathematischen Instituts der hiesigen technischen Hochschule, das um eine werthvolle Serie bereichert wurde, wie im Namen der Buchhandlung, welche die Copien weiteren Kreisen zugänglich machen wird.

Was den Abgüssen, etwa an Eleganz der äußeren Erscheinung abgeht, das ersetzen sie, wie ich glaube, durch Solidität [-2-] der Construction, indem in jedes Gipsmodell als Gerüste das Drahtgestell der vier Doppelberührkreise eingesetzt ist.

Mit dem Wunsche, dass die Sendung in gutem Zustande ankommen möge zeichnet mit vielen Empfehlungen und dem Ausdruck ausgezeichneter Hochachtung

ergebenst
A. Brill

References

Archival Source: Nachlass Felix Klein (Klein NL), Handschriften und Seltene Drucke, Niedersächsische Staats- und Universitätsbibliothek, Göttingen.

Biermann, Kurt R. 1988. *Die Mathematik und ihre Dozenten an der Berliner Universität 1810–1933 – Stationen auf dem Wege eines mathematischen Zentrums von Weltgeltung*. Berlin: Akademie-Verlag.

Brill, Alexander, 1889. Über die Modellsammlung des mathematischen Seminars der Universität Tübingen, Druckfassung eines Vortrages vom 7.11.1887, *Mathematisch-Naturwissenschaftliche Mitteilungen*, Tübingen, Band II, 69–80.

Cayley, Arthur. 1871. On Plücker's Models of certain Quartic Surfaces. *Proceedings of the London Mathematical Society* 3: 281–285.

Clebsch, Alfred. 1871. Zum Gedächtnis an Julius Plücker. *Abhandlungen der Königlichen Gesellschaft der Wissenschaften in Göttingen* 16: 1–40.

Dolgachev, Igor. 2004. Luigi Cremona and Cubic Surfaces, in arXiv: math/\penalty\z@0408283 [math.AG].

Engel, Friedrich. 1900. Sophus Lie. *Jahresbericht der Deutschen Mathematiker-Vereingung 8*: 30–46.

Finsterwalder, Sebastian. 1936. Alexander v. Brill: Ein Lebensbild. *Mathematische Annalen* 112: 653–663.

Fischer, Gerd, ed. 1986. *Mathematische Modelle, 2 Bde*. Berlin: Akademie Verlag.

Hankins, Thomas. 1980. *Sir William Rowan Hamilton*. Baltimore: Johns Hopkins University Press.

Klein, Felix. 1921. *Gesammelte Mathematische Abhandlungen*. Bd. 1 ed. Berlin: Springer.

———. 1922. *Gesammelte Mathematische Abhandlungen*. Bd. 2 ed. Berlin: Springer.

———. 1923. Göttinger Professoren (Lebensbilder von eigener Hand): Felix Klein. *Mitteilungen des Universitätsbundes Göttingen* 5: 11–36.

Knörrer, Horst. 1986. Die Fresnelsche Wellenfläche. In *Arithmetik und Geometrie*, Mathematische Miniaturen, Bd. 3 ed., 115–141. Basel: Birkhäuser.

Kummer, E.E. 1975. In *Ernst Eduard Kummer, Collected Papers*, ed. A. Weil, vol. 2. Berlin: Springer.

Müller, Falk. 2006. Purifying Objects, Breeding Tools: Observational and Experimental Strategies in Nineteenth-Century Gas Discharge Research, in *Max-Planck-Institut für Wissenschaftsgeschichte*, Preprint 318.

Rohn, Carl. 1877. Drei Modelle der Kummer'schen Fläche, *Mathematische Modelle angefertigt im mathematischen Institut des k. Polytechnikums zu München*. Wiederabgedruckt in [Schilling 1904].

Rowe, David E. 1989. Klein, Lie, and the Geometric Background of the Erlangen Program. In *The History of Modern Mathematics: Ideas and their Reception*, ed. D.E. Rowe and J. McCleary, vol. 1, 209–273. Boston: Academic Press.

Schilling, Martin, ed. 1904. *Mathematische Abhandlungen aus dem Verlage Mathematischer Modelle von Martin Schilling*. Halle: M. Schilling.

———, ed. 1911. *Katalog mathematischer Modelle für den höheren mathematischen Unterricht*. Leipzig: M. Schilling.

Shafarevich, Igor. 1983. Zum 150. Geburtstag von Alfred Clebsch. *Mathematische Annalen* 266: 135–140.

Tobies, Renate. 1992. Felix Klein in Erlangen und München. In *Amphora. Festschrift für Hans Wussing zu seinem 65. Geburtstag*, 751–772. Basel: Birkhäuser.

Weierstrass, Karl. 1903. *Mathematische Werke*. Vol. Bd. 3. Berlin: Mayer & Müller.

9 Debating Grassmann's Mathematics: Schlegel vs. Klein

(*Mathematical Intelligencer* 32(1)(2010): 41–48)

Mathematical fame can be a fickle thing, little more enduring than its mundane counterparts, success and recognition. Sometimes it sticks, but for odd or obscure reasons. Take the case of a largely forgotten figure named Victor Schlegel (1843–1905): googling for "Schlegel diagrams" immediately brings up scads of colored graphics depicting plane projections of 4-dimensional polyhedra. None that I found, however, could compare with the figures that appear in a little-known paper (Schlegel 1883). It seems these figures are aptly named, but *how and when* they came to be called Schlegel diagrams remains a mystery (see also Schlegel 1886). In fact, clicking through Wikipedia, MacTutor, and their progeny for Victor Schlegel turns up nothing; nor does he appear in standard compendia, like the *Lexikon bedeutender Mathematiker*.

Schlegel was not the first to study the properties of 4-dimensioanl polytopes. In fact, his study cites the work of a young American mathematician, Washington Irving Stringham, who was one of the first to take up systematic investigations in this field. In fact, the diagrams he first published in the *American Journal of Mathematics* (see Fig. 35.3) have recently enjoyed a certain vogue: they figure prominently in a study of how such exotic mathematical models entered into modern art by Linda Henderson (Henderson 1983). Even more recently the artwork of Stringham and Schlegel have assumed center stage in Tony Robbin's survey of a similar terrain (Robbin 2006). This book also contains some biographical information on Stringham, but it offers no hints regarding the identity of the man whose name came to be synonymous with Schlegel diagrams.

Nevertheless, during his lifetime Victor Schlegel was a well-known mathematician, though not so much for his contributions to the study of figures in 4-space. He was mainly recognized by his contemporaries as a leading proponent of Hermann Günther Grassmann's ideas and life's work. Indeed, Schlegel was in an excellent position to write about this subject, having taught alongside Grassmann at the Stettin Gymnasium from 1866–68. Afterward he went on to publish over 25 works dealing with Grassmannian ideas, perhaps the most valuable for the historian being his biographical essay (Schlegel 1878) and his retrospective article (Schlegel 1896). Schlegel's biography covers all facets of Grassmann's far-ranging scholarly life, from theology and philology to mathematics and politics, presenting one of the most vivid personal portraits of this struggling genius ever produced. Friedrich Engel, author of the definitive Grassmann biography (Engel 1911) published in volume 3 of the collected works, went out of his way to praise Schlegel's biography as well as his numerous efforts to promote interest in Grassmann and his work. Yet Engel also distanced himself from what he viewed as Schlegel's one-sided hero worship, so typical among Grassmann's closest followers. As we shall see, in taking this critical stance Engel was by no means alone.

Grassmann's *Ausdehnungslehre*

It has often been observed that Grassmann's mathematics was not widely appreciated during his lifetime (Fig. 9.1). Although awareness of his achievements had begun to spread by the early 1870s, few in Germany appear to have been well acquainted with either the original 1844 edition of his *Ausdehnungslehre* or the mathematically more accessible revision of 1862 (Crowe 1967, 54–95). Among Grassmannians, the first edition was the true *Ausdehnungslehre,* a work of audacious and daring vision. Victor Schlegel described it at length in his Grassmann biography, claiming that it occupied a singular place in the history of mathematics:

> Such heights of mathematical abstraction like those reached in the *Ausdehnungslehre* had never before been attained. Like Pallas from the head of Zeus, it sprang suddenly to life, full and ready, leaping over a generation in the course of mathematical developments, it stood as a new science there, and today, 33 years later, it remains new and, unfortunately, for many just as incomprehensible as before (Schlegel 1878, 19–20).

No one would claim that this book was an easy read, but sheer bad luck also had something to do with its weak reception. As Grassmann himself noted in the introduction

D.E. Rowe, *A Richer Picture of Mathematics*, https://doi.org/10.1007/978-3-319-67819-1_9

Fig. 9.1 Hermann Günther Grassmann.

to the second edition of the original *Ausdehnungslehre*, his ideas might have become better known had not two distinguished voices passed on before having a chance to be heard (Grassmann 1894–1911, I.1, 18–19). He was referring to Hermann Hankel (1839–1873) and Alfred Clebsch (1833–1872), both of whom had drawn attention to the importance of Grassmann's ideas before they abruptly died in the early 1870s (Tobies 1996). Hankel presented Grassmann's theory in (Hankel 1867), whereas Clebsch paid tribute to the same in his obituary for Julius Plücker (Clebsch 1872). Afterward, no one of comparable stature arose to champion Grassmann's cause, whereas some of those who did so tended to be seen as fanatics with a narrow, sectarian agenda. It also seems that Grassmann's own efforts to highlight the significance of his *Ausdehnungslehre* during the last years of his life mainly served to reinforce this tendency. As Michael Crowe has described, these circumstances contributed to the tensions between a zealous band of Grassmannians and an equally committed group who promoted Hamilton's calculus of quaternions (Crowe 1967). Their dispute raged into the 1890s, but with the emergence of the new vector analysis, which drew on both systems, it largely subsided.

Leaving this international conflict aside, the present essay will focus on earlier events within Germany, in particular those connected with Victor Schlegel's role in promoting Grassmann's work. In the early 1870s, Schlegel's efforts were resisted by a leading member of Clebsch's school, young Felix Klein, who was then intent on promoting the ideas in his newly hatched "Erlangen Program." As a self-appointed spokesman for mainstream mathematicians, Klein was eager, on the one hand, to adapt Grassmann's concepts to his own new vision, while, on the other, asserting his own authority with regard to the relevance of these ideas for future prospects in geometrical research. Before turning to these matters directly, however, a few general observations should be made regarding Grassmann's career and the mathematical world of his day.

Mathematics at University and Gynmasium

The belated recognition of Hermann Grassmann's importance and stature led some of his closer followers to view him as a kind of martyr figure, a man who had been forced to toil away his life as a school teacher in Stettin and whose brilliant genius only began to be appreciated after his death in 1877. Of course, no one today would dispute that Grassmann was a man of extraordinary gifts and impressive accomplishments. Nevertheless, his isolated situation was hardly unique; nor was his the most striking example of a creative genius whose work failed to win swift acclaim. For a more balanced assessment, one must bear in mind the times and culture in which he lived, an era when professional research in pure mathematics was still in its infancy (Klein 1926, 181–182).

In 1852 Grassmann succeeded his father, Justus Günther Grassmann (1779–1852), as *Oberlehrer* at the Stettin Gymnasium (Fig. 9.2). He apparently took no great pleasure in his duties there, for he longed instead to become a university professor. This circumstance has often been seen as the crux of Grassmann's dilemma: for he never gained such a post, leading many of his latter-day followers to conclude that the German mathematical establishment failed to appreciate the merits of his new ideas and methods. Had they recognized his genius, so ran the argument, surely he would then have taken his place among Germany's mathematical elite with the opportunity to spread his ideas through a close band of intellectual disciples. Maybe, but leaving such hypothetical conclusions aside, we might begin by noting that the market for research mathematicians in Germany ca. 1850 was close to infinitesimally small. Moreover, however justified Grassmann's desire to gain a university professorship may seem today, the fact that it remained unfulfilled can hardly be considered a qualification for martyrdom. For in the 1850s, Prussian *Oberlehrer* were generally treated with considerable respect and deference, if not by their pupils then at least by their peers. What is more, they often consorted with members from the upper echelons of local society.

Indeed, Grassmann himself was the product of a Prussian culture that not only honored, even revered its teachers, but also attached extraordinarily high value to scholarly productivity. As such, he was in many ways a representative figure in an era when the pursuit of higher learning was

Fig. 9.2 Marienstiftsgymnasium in Stettin.

almost taken for granted; it lay at the very heart of Germany's neohumanist tradition (Turner 1971). In Prussian secondary schools this research ethos was cultivated by and transmitted through numerous scholars with considerable élan. During Grassmann's day the gulf separating those who taught at the universities and their counterparts in the Gymnasien was not nearly as wide as it would later become at the end of the century. Thus, both academically and socially, an *Oberlehrer* was only one remove from a university professor, and those who held the position were referred to by their proper title: Herr Professor.

One should also note that gymnasium professors like Grassmann played a major part in Germany's swift ascent in the world of mathematics, which until around 1830 had been totally dominated by France. Ernst Eduard Kummer (1810–1893) and Karl Weierstrass (1815–1897), the two most influential mathematics teachers in Germany during the 1860s and 1870s, had both begun their careers teaching at secondary schools. Other distinguished mathematicians like Hermann Schubert (1848–1911), inventor of the Schubert calculus, spent their entire professional careers teaching young charges elementary mathematics. Some highly significant, non-mathematical findings took place in these settings. Thus, Leopold Kronecker (1823–1891) was discovered by Kummer when he taught at the Gymansium in Liegnitz, whereas Schubert befriended young Adolf Hurwitz (1859–1919), his most gifted pupil, in Hildesheim. These particularly striking examples attest to a pattern of truly impressive quality. Indeed, it is safe to say that had it not been for the high standards and intense dedication demonstrated by scores of now forgotten *Oberlehrer*, Germany's sudden ascent in mathematics during the second half of the nineteenth century would have been unthinkable.

As these remarks imply, the scholarly ideals that animated professors at the Prussian Gymnasien did not differ markedly from the research ethos at institutions like Berlin University, where Grassmann studied theology, philosophy, and philology in the late 1820s. This background could hardly be considered ideal preparation for someone who aspired to a career in mathematics at the university level, and yet it was by no means atypical for this time. Both Gauss and Jacobi studied classical philology before ultimately turning to mathematics, and like Grassmann, they were both self taught. Only very few students in Germany seriously contemplated pursuing a career as a research mathematician, and for two obvious reasons. First, there were simply too few positions available, and second, one faced a long, arduous struggle that posed considerable financial hardships. Little wonder that even during the late 1800s some of the most promising talents, like David Hilbert (1862–1943), opted to take the Staatsexamen, which qualified a person to teach in the secondary schools, rather than risk having nothing to fall back on later. Most of those who did choose to go on, by habilitating at a German university, came from families with fairly substantial means. Habilitation gave one the right to teach as a *Privatdozent* and to collect fees from the students, but nothing more. The hardships such an unsalaried position imposed discouraged many from even pursuing a professorship. Thus the fact that Grassmann never attained this more exalted position – although he was given serious consideration for a position at Greifswald in 1847 – can hardly be considered unusual or surprising.

Professionalisation and Patterns of Reception

With regard to the slow diffusion of Grassmann's ideas within wider mathematical circles, it should also be noted that this type of transmission pattern was closer to the norm rather than the exception throughout most of the nineteenth century. Even when the ideas issued from highly esteemed individuals, like Gauss or Riemann, it often took many decades before important new breakthroughs could be absorbed and understood. As a general rule, the more novel or "revolutionary" the ideas (e.g., Galois's theory of algebraic equations), the longer it took to assimilate them. Yet within Germany's decentralized institutional structures, even relatively routine transmission of information was not easy as there were few publication outlets and even fewer possibilities for meeting face to face. Before the founding of the *Mathematische Annalen* in 1869, Crelle's *Journal für die reine und angewandte Mathematik* represented the only specialized mathematics journal in Germany, and it was not until 1890 that German mathematicians managed to establish a national organization, the Deutsche Mathematiker-Vereinigung, which provided a forum for meetings and other professional activities. Thus, prior to 1890 most university mathematicians conducted their research in rather closed and isolated settings.

Among the leading geometers of this period, Julius Plücker (1801–1868) worked in quiet solitude in Bonn, August Ferdinand Möbius (1790–1868) was employed as an astronomer in Leipzig, and Karl von Staudt (1798–1867) taught for many years in the solitude of Erlangen. Much of Möbius's mathematical work only became more widely accessible in the 1880s, and it took several decades before Staudt's fundamental contributions to the foundations of projective geometry became widely known. Thus, Klein was unaware in 1872 of the extent to which the ideas in his Erlangen Program had already been anticipated by Möbius until the mid-1880s when he co-edited the latter's collected works (see Klein 1921–23, vol. 1, p. 497).

No doubt Grassmann faced even graver difficulties than either Möbius or Staudt, especially due to the esoteric philosophical notions that pervaded his early work, ideas closely akin to Schleiermacher's thought (Lewis 1977). Even those who were motivated enough to penetrate into the misty realm of ideas set forth in Grassmann's 1844 edition of the Ausdehnungslehre usually came away from it feeling like they had entered into a strange new world. One of the few foreigners who undertook the journey was William Rowan Hamilton (1805–1865), whose own work on quaternions had been partly inspired by German idealism, in particular Kant's notion that our conception of number derives from an intuition of pure time (Hankins 1980).[1] Hamilton, who began reading Grassmann's book in 1852, undoubtedly possessed the proper philosophical temperament (and linguistic abilities) needed to tackle Grassmann's work. Yet his principal concern appears to have been to reassure himself that he alone had discovered the quaternions. Still, he expressed considerable admiration for Grassmann's originality, an opinion that was also shared in Italy by Luigi Cremona (1830–1903) and Giusto Bellavitis (1803–1880).

In Germany it seems fair to say that Grassmann's work failed to attract serious and sustained interest among leading contemporary mathematicians, though not so much for lack of sympathy with his goals but rather because of the manner in which he approached them. The elderly Carl Friedrich Gauss wrote Grassmann that he had no time to study his book, but in glancing through it he was reminded of his own longstanding interest in complex numbers and the metaphysical views he had expressed in his 1831 note on this topic (see Chap. 3). But he also admitted that he had published nothing like Grassmann's calculus for spatial quantities (Schlegel 1878, 22–23). Many expressed bewilderment when faced with the convoluted formulations found in the *Ausdehnungslehre*. Particularly telling was Möbius's reaction, since his *Baryzentrische Calcül* of 1827 revealed strong affinities with Grassmann's fundamental conceptions. Grassmann even visited Möbius after completing work on his *Ausdehnungslehre*, but when he asked the latter to write a review of the book Möbius answered:

> I reply that I was sincerely pleased to have come to meet in you a kindred spirit, but our kinship relates only to mathematics, not to philosophy. As I remember telling you in person, I am a stranger to the area of philosophic speculation. The philosophic element in your excellent work, which lies at the basis of the mathematical element, I am not prepared to appreciate in the correct manner or even to understand properly. Of this I have become sufficiently aware in the course of numerous attempts to study your work without interruption; in each case however I have been stopped by the great philosophic generality (A. F. Möbius to H. G. Grassmann, 2 February 1845; (Schlegel 1878, 23)).

Despite this mild rebuff, Möbius's sympathy was apparent; he clearly understood and appreciated Grassmann's book, but he still declined to review it, recommending instead the philosopher Moritz Drobisch (who also declined). Möbius proved more helpful, however, in suggesting to Grassmann that he compete for a prize announced the previous year by the Jablanowski Society in Leipzig. This scientific society wished to receive a work which would

[1] Kant's ideas in this respect exerted a similarly strong influence on Brouwer, for which see Freydberg (2009).

develop the "geometrical calculus" Leibniz had sketched in 1679 in a letter to Huygens, first published in 1833 (Schlegel 1878, 28–29). Taking up Möbius's advice, Grassmann submitted his "Geometrische Analyse, geknüpft an die von Leibniz erfundene geometrische Charakteristik "(Grassmann 1894–1911, I.1, 321–398). This essay won the prize (as the only entry submitted) in 1846 and was published the following year along with a commentary by Möbius (Grassmann 1894–1911, I.1, 613–633.). Like the *Ausdehnungslehre*, however, it failed to elicit sustained interest, despite favorable comments from some leading European mathematicians.

Somewhat better known were a number of articles Grassmann published in Crelle's Journal between 1842 and 1856. During these years, this highly regarded journal carried 14 papers by Grassmann, most of them concerned with methods for generating algebraic curves and surfaces. Although these results were both elegant and easy to comprehend, few geometers appear to have followed up on them. Nevertheless, many of these papers were clearly read and appreciated (Tobies 1996, 123–124). Felix Klein later emphasized that Grassmann's methods – unlike those of the celebrated geometer Jacob Steiner – enable one to construct all possible algebraic curves synthetically, a truly important breakthrough (Klein 1926, 180–181).

After 1856, Grassmann's mathematical productivity slackened, but certainly not his zeal for propagating what he had achieved. He presented his Ausdehnungslehre again in 1862, but in strictly Euclidean fashion and without reference to its philosophical roots. Yet even in this form, the unfamiliarity of Grassmann's terminology and the algorithms he employed posed major obstacles for his readers, most of whom gave up before they had gotten very far into the book. Thus, like the 1844 edition, this account, too, failed to awaken substantial interest, and soon after its appearance Grassmann turned to other endeavors, most notably philological studies, discouraged by the indifference of professional mathematicians.

Promoting Grassmannian Mathematics at the Gymnasium

In the meantime, however, he and his brother, Robert, had turned to a different audience, hoping for a more enthusiastic response from mathematics teachers at the Gymnasien. Thus, in 1860 they published the first of a projected three volumes on elementary mathematics which were to serve as modern textbooks for the secondary schools (Grassmann 1860). Yet the response of the mathematics teachers proved just as discouraging as that of the research mathematicians. Whatever they might have thought about the mathematical merits of this new, fully rigorous approach to arithmetic, the immediate consensus of opinion indicated that the presentation of it in this book was altogether inappropriate for teaching purposes (Engel 1911, 225–228).

Despite this setback, Grassmann was convinced that his ideas could be and one day would be incorporated into the standard mathematical curriculum at German secondary schools; and others eventually came to share that conviction. The most important new convert was Victor Schlegel, Grassmann's colleague at the Stettin Gymnasium from 1866 to 1868. Afterward Schlegel accepted a position as Oberlehrer at the Gymnasium in Waren, a small town in Mecklenburg. It was there that he began an intensive study of Grassmann's *Ausdehnungslehre*, an experience that prompted him to write a textbook entitled *System der Raumlehre*, which he dedicated to the master. Part I (Schlegel 1872) dealt with elementary plane geometry, whereas Part II (Schlegel 1875) presented a Grassmannian version of those portions of algebraic geometry closely connected with invariant theory.

By virtue of this two-part study, but especially his biography of Grassmann published in 1878, one year after the latter's death, Schlegel emerged as the leading spokesman for the Grassmannian cause. Like his hero, he hoped to show that the *Ausdehnungslehre* offered more than just another useful tool for the research mathematician. Indeed, he saw Grassmann's methods as applicable to mathematics at nearly every level, and consequently as holding the key to a much needed reform of the mathematics curriculum in the secondary schools. Thus, following the master's lead, Schlegel's *System der Raumlehre* had two principal objectives: to show the importance of Grassmannian ideas for pedagogical purposes, and to indicate the superiority of vector methods over techniques based on coordinate systems in presenting the main results of geometry, particularly projective geometry. He pursued the first of these goals in Part 1, which dealt with the basic theorems of school geometry, whereas the second objective was relegated to Part II, which only appeared three years later.

In the introduction to Part I, however, Schlegel not only made the two-fold purpose of his study more than clear, he went on to attack both research mathematicians and pedagogues for ignoring Grassmann's work. Adopting a strident tone, he suggested that Grassmann's *Ausdehnungslehre* of 1844 had simply been too radical an advance forward for a "generation that viewed imaginary quantities as impossible and that shook its head in dismay at non-Euclidean geometry" (Schlegel 1872, vi.). But this was not the only reason Schlegel gave to explain the indifferent reaction to Grassmann's theory: he also intimated that the playing field for mathematicians in Germany was anything but fair, and that Grassmann's ideas had largely been neglected because he had not held a university chair, a circumstance that had prevented him from establishing a school for the promulgation of his ideas. Whereas younger mathematicians, as Schlegel observed, "had been preoccupied with the elaboration of the theories set forth by Jacobi, Dirichlet, Steiner, Möbius, Plücker, Hesse, and others, all of whom gathered a circle of active pupils around them, fate failed to grant the author of the *Ausdehnungslehre* a similar influence" (Schlegel 1872, vi.).

Schlegel seemed to imply further that Grassmann could claim priority with regard to various methods usually ascribed to others. For, having ignored Grassmann's work, other mathematicians (he referred to Otto Hesse explicitly) had in the meantime developed alternative techniques very close to Grassmann's, but without fully realizing the advantages of the latter's coordinate-free, n-dimensional mode of representation. Still, priority claims were the least of Schlegel's concerns. Far more pressing, from his point of view, was the need to clarify the foundations for the treatment of extensive magnitudes in mathematics. In this connection, Schlegel mentioned Martin Ohm's *Versuch eines vollkommen consequenzen Systems der Mathematik* as about the only noteworthy recent effort to clarify the foundations of mathematics.[2] Thus, he decried the state of arithmetic and algebra, calling it a "conglomeration of loosely connected rules for calculation" supplemented by "a collection of arbitrary foundational principles, more or less vague explanations, and geometric artifices" (Schlegel 1872, xi.). Rather than developing a sound basis for their analytic methods, he claimed that contemporary mathematicians tended to invent new symbolisms on an ad hoc basis, creating a Babel-like cacophony of unintelligible languages. This type of criticism, in fact, had long been hurled at analytic geometers, leading many to conclude that the only legitimate approach to geometry was the purely synthetic one.

Schlegel went on to argue that this weakness had exerted an adverse effect not only on higher mathematics but also on school mathematics as well. Indeed, he contended that such chaos inevitably reinforced the widespread view of mathematics as a kind of "black magic" accessible only to those with a predisposition for its abstruse formalisms. Schlegel further asserted that the neglect of foundations had led to the widely acknowledged lack of interest in mathematics in the schools. Echoing the views expressed in the preface of Hermann and Robert Grassmann's textbook, he proclaimed that the standard mathematics curriculum contained nothing whatsoever that could serve as a basis for a truly "scientific method." Thus his goal in presenting Part I of *System der Raumlehre* was to demonstrate how Grassmann's ideas could provide just such a foundation.

While Schlegel's claims were, no doubt, exaggerated and his remedy far-fetched, his book did offer some fresh perspectives that might well have borne fruit. Certainly it appeared at a propitious point in time, as by now the sorry state of mathematics instruction in the Gymnasien had become a matter of considerable concern. In fact, Schlegel could even call attention to an editorial statement in the newly founded *Zeitschrift für mathematischen und naturwissenschftlichen Unterricht* indicating that the journal had come out strongly in favor of a rational reform of the teaching methods in geometry (Schlegel 1872, xii). But he also realized that any significant reform was bound to face strong resistance from those who had a vested interest in the status quo. Rather than mincing words, he confronted the conservative opposition head-on, stating that he was well aware that his book would not please "those who regard the present state of elementary mathematics as satisfactory" (Schlegel 1872, xiii). Drawing battle lines in advance, he appealed not to these readers, but to all those capable "of'an *unbiased* consideration of new viewpoints" (Schlegel 1872, xiii).

Klein's Critique

Schlegel's book appears not to have engendered the kind of response its author had hoped for; nor did it garner accolades from the German mathematical community. Friedrich Engel later wrote about it rather dismissively, asserting that "Schlegel was not the man to put the old Grassmannian wine into new vessels" (Engel 1911, 324). Three years after its publication, however, the book was reviewed at considerable length by Felix Klein in *Jahrbuch über die Fortschritte der Mathematik* (Klein 1875). Klein, then a fast-rising star in German mathematics, had in 1872 been appointed full professor in Erlangen at just 23 years of age. It was this circumstance which led him to compose his famous "Erlangen Program" (Klein 1872) in which he sought to establish a unified view of geometrical research based on the study of structures left invariant under various transformation groups.

Ironically, but perhaps not by chance alone, Klein's unusually lengthy review appeared in the *Jahrbuch* alongside a favorable review of his "Erlangen Program" written by his friend, the Austrian mathematician Otto Stolz. At first sight this would seem like an unlikely pairing since the two works were clearly written for different audiences and with quite different goals in mind. Nevertheless, both dealt with geometry in a programmatic manner, and as such a comparison of their approaches offers certain suggestive insights. In particular, this helps to shed light on the dynamics surrounding the reception of Grassmannian ideas in Germany during the early 1870s. Indeed, a comparison of their overall orientations reveals a good deal about the clashing agendas of research mathematicians at the universities, on the one hand, and Grassmann's followers, most of whom were Gymansium teachers, on the other.

Klein had been exposed to Grassmann's *Ausdehnungslehre* already in 1869 as a member of Clebsch's school. He had studied both the 1844 and 1862 editions (the latter was richer

[2] On Ohm, see Bekemeier (1987) and Schubring (1981).

mathematically, the former more suggestive philosophically) and he clearly held Grassmann's ideas in high regard, referring to both works prominently in his Erlangen Program (Klein 1872; 478, 480, 483, 489, 492). Like Schlegel, he also favored a reform of the mathematics curriculum, and indeed the whole philosophy behind mathematics instruction at the Gymnasien (Rowe 1985). His reaction to Schlegel's book, however, was decidedly negative, though not so much due to its substance as its tone and tendency. Thus, while he found the proofs of familiar results, like the Pythagorean Theorem, by means of Grassmannian methods interesting, he also noted that these proofs often required extensive calculations. This, Klein wryly observed, stood in flat contradiction to the book's stated objective. He could have added that such excessive formalism was precisely what Schegel had criticized in the work of contemporary analytic geometers, a criticism that Klein and other members of the Clebsch school had heard often enough.

Aside from this particular point, however, Klein largely ignored the pedagogical issues that Schlegel's book sought to address, preferring instead to comment on matters that he personally found more interesting. In fact, his account suggests that he was far less interested in the book under review than in the convenient opportunity it offered him to expound on the merits and shortcomings of Grassmann's mathematics as a whole, particularly from the perspective of his own "Erlanger Program." Thus, he pointed out the limitations of Grassmannian conceptions for projective geometry, noting that neither Grassmann nor Schlegel introduced imaginary elements in their treatments of projective constructions. He also called attention to their general neglect of key aspects of the theory, such as cross ratios, polar curves, etc. After remarking on these drawbacks, Klein went on to say:

> Perhaps instead of presenting Grassmann's conceptions as such, and only in elementary form, it would have been more important to show how these connect with and compare to similar directions that research has taken afterward, independent of Grassmann. So far as the present book is concerned, the principal accomplishments of Grassmann are essentially three. First, it is due to him that formal algebra gained an unsuspected depth, as he showed how to grasp the essence of the operations of addition and multiplication in a much more general way than had been done before him. In this regard, Grassmann stands alongside the English investigators, such as Hamilton. Furthermore, he was the first to develop the theory of higher-dimensional manifolds, which contains as special cases the theory of space and, especially, the theory of linear (projective) manifolds. Finally, he opened a new, wide-reaching field of research through his methods for generating all algebraic structures by means of linear mechanisms (Klein 1875, 233–234).

This praise for Grassmann's achievements was followed by two critical observations. First, with regard to invariant theory, Klein found that Grassmann's methods – with the exception of the mixed product, which had not been fully incorporated into algebra – offered nothing essentially new. "The two forms of representation [Grassmann's and those employed by invariant theorists] are, at bottom, barely distinguishable and even the formulas that express them are often identical" (Klein 1875, 235). This opinion was later repeated at considerable length by Eduard Study in his commentary on Grassmann's last mathematical works; see Grassmann (1894–1911, II.1, pp. 431–433). Second, with respect to Grassmann's notion of n-dimensional manifolds, Klein noted that, from the present-day standpoint, this marked only a beginning. Grassmannian manifolds were "only the direct generalization to higher dimensions of ordinary space with its positional and metrical properties, whereas Riemann's investigations opened a much more general line of inquiry that has since been extended significantly in various directions" (Klein 1875, 235).

In his "Erlangen Program" Klein had sought to integrate a wide spectrum of geometrical research by placing the concepts of manifold, transformation group, and the associated group invariants at the center stage. This offered a bold new conceptual framework, but its design was clearly dominated by the particular field of research Klein knew best, namely projective geometry and the related theory of algebraic invariants of the projective group. Remarkably enough, the theory of affine spaces was not even mentioned in the "Erlangen Program," a clear indication that Grassmannian ideas played a peripheral role in Klein's overall conception of geometry at this time. Klein later openly admitted that this omission stemmed from the predominance of projective geometry in his background (Klein 1921–23, 320–321). This limitation along with Klein's strong predilection for projective methods must, therefore, be taken into account when reading these reflections on the significance of Grassmann's mathematics for geometrical research in the year 1872.

Following these comparative remarks, Klein summed up his opinion of Schlegel's book in a single sentence, offering this decidedly negative verdict: "If, like the author, one presents Grassmann's ideas divorced from such comparisons, it is this reviewer's opinion that the reader will tend to be repelled rather than attracted by them; he will be forced to accept that Grassmann's methods are absolutely superior, and such a claim always contains something improbable about it" (Klein 1875, 235).

Schlegel's Response

Schlegel must have been dismayed that one of Clebsch's closest associates could have written such an unsympathetic, self-serving review of Part I of his book. Not only had Klein largely ignored the core issues at stake, namely the efficacy of Grassmannian methods for the presentation of elementary geometry, but the whole thrust of his final remarks seemed to say: why didn't the author write the kind of book I would have written instead of this one? Clearly, Klein had a very different mathematical agenda than the one Schlegel wished to pursue. The whole aim and tendency of his "Erlangen Program" was to unite disparate strands of geometrical research within a broad conceptual framework. As noted above, in setting forth this new scheme Klein drew his main inspiration from projective geometry (with Galois theory serving as a loose structural analogy). Moreover, several passages in the "Erlangen Program" criticized tendencies toward disciplinary fragmentation and methodological purism, trends Klein spent his whole life fighting (Klein 1872, 461, 490–491).

Felix Klein was not one to back down from a fight, but when it came to Grassmann's followers he only rarely felt inclined to take issue with them. His true enemies sat in Berlin, where elitism and purism reigned supreme. He had no desire to waste his time scolding the Grassmannians, a powerless faction whose members were without prominent standing within the German mathematical community. Shortly after Klein's review appeared, Schlegel took the opportunity to respond in the introduction to Part II of his *System der Raumlehre*. He began by addressing the criticism voiced in Klein's closing *judicium* in which Schlegel was taken to task for having failed to draw comparisons between Grassmann's results and those obtained by others. After noting where this had been done in Part II, Schlegel curtly pointed out that with respect to Part I this criticism was vacuous, "since this part only comprises theories of elementary geometry" (Schlegel 1875, viii). As for the assertion that his book forced the reader to accept Grassmann's methods as absolutely superior, Schlegel replied straightforwardly that, in his opinion, Grassmann's methods offered "the shortest and easiest approach to the results of ancient and modern geometry and algebra" (Schlegel 1875, viii). But he also emphasized that the reader stood under no compulsion to accept this view; all he asked of the mathematical public was that they study the material presented in the book before drawing their own conclusions. Schlegel further added that the issue as to whether Grassmann's *Ausdehnungslehre* was merely of historical interest, perhaps "with the added regret, after a comparison with other methods, that it had not exerted its influence earlier," or whether it was "also at present worthy of further development and capable of being usefully employed in scientific advancements, that is a question that cannot be answered in three lines; nor can it be answered only on the basis of my merely introductory writings, but rather alone by a thorough study of Grassmann's original works" (Schlegel 1875, viii).

Klein apparently did not bother to reply to these comments, at least not in print; nor did Part II of Schlegel's *System der Raumlehre* ever receive a review in *Jahrbuch über die Fortschritte der Mathematik*, despite the fact that its mathematical content went far beyond that contained in Part I. Needless to say, neither party in this dispute succeeded in persuading the other side to alter their views, and in any case it seems unlikely that any sort of compromise could have been reached. Klein later played a major role in engineering the appointment of Friedrich Engel as editor of Grassmann's *Gesammelte Mathematische und Physikalische Werke*, a disappointing turn of events for Schlegel. Grassmann's leading disciple made his displeasure known by noting that his own participation in the Grassmann edition had been restricted to the preparation of a bibliography (Schlegel 1896, 3). In fact, Engel later adopted a position with respect to "fanatical Grassmannians" very similar to Klein's own (Engel 1910, 12–13).

There appears to be no extant evidence indicating how Grassmann himself responded to Klein's review of 1872. But if his last mathematical works serve as any indication, he apparently felt very strongly that his methods and ideas, as presented by Schlegel, had, once again, failed to receive a fair hearing (Grassmann 1874, 1877). Grassmann tried to recoup some loses in these final works, but his efforts probably only resulted in a hardening of opinion. In 1904 Engel wrote in this regard that these last works reveal "a striking imbalance between what is actually accomplished and the claims Grassmann made in them. This, however, has not restrained Grassmann's unconditional admirers from praising even these works far above their merits" (Grassmann 1894–1911, II.1, vi). Eduard Study sounded a similar critical note in his commentary in (Grassmann 1894–1911, II.1, 431–435).

Grassmann chose to publish these last papers in *Mathematische Annalen*, the journal associated with Alfred Clebsch and his school. Clearly he considered Clebsch to have been a strong and sympathetic supporter; his premature death in November 1872 was thus seen as a serious blow to the cause. Given these circumstances, it seems only plausible to assume that Grassmann felt stung by the critical remarks published by Clebsch's leading disciple three years later. These events of the mid 1870s surely only widened the gulf that already separated the Gymnasium teachers in Grassmann's camp from those who taught at the German universities. As such, they seem to have marked a turning point in the tense relations between Grassmann's growing contingent of followers and influential members within Germany's still fragmented and often divisive mathematical community. Thus by the time of the master's death in 1877, the myth of his martyrdom was already firmly in place, shaped and cultivated by true believers who felt increasingly marginalized by professional mathematicians at German universities, many of whom adopted Grassmannian ideas in their own research.

References

Bekemeier, Bernd. 1987. *Martin Ohm (1792–1872): Universitätsmathematik und Schulmathematik in der neuhumanistischen Bildungsreform*, Studien zur Wissenschafts-, Sozial- und Bildungsgeschichte der Mathematik. Bd. 4 ed. Göttingen: Vandenhoeck & Ruprecht.

Clebsch, Alfred. 1872. Zum Gedächtnis an Julius Plücker, *Abhandlungen der Königlichen Gesellschaft der Wissenschaften zu Göttingen*, vol. 16.

Crowe, Michael J. 1967. *A History of Vector Analysis*. Notre Dame: University of Notre Dame Press.

Engel, Friedrich. 1910. Hermann Grassmann. *Jahresbericht der Deutschen Mathematiker-Vereinigung* 19: 1–13.

———. 1911. Grassmanns Leben, in (Grassmann 1894–1911, III.2).

Freydberg, Bernard. 2009. Brouwer's Intuitionism vis à vis Kant's Intuition and Imagination. *Mathematical Intelligencer* 31 (4): 28–36.

Grassmann, Hermann. 1860. *Lehrbuch der Arithmetik für höhere Lehranstalten*. Berlin: Enslin.

———. 1874. Die neuere Algebra und die Ausdehnungslehre. *Mathematische Annalen* 7: 538–548; Reprinted in (Grassmann 1894–1911, II.1, 256–267).

———. 1877. Der Ort der Hamilton'schen Quaternionen in der Ausdehnungslehre. *Mathematische Annalen* 12: 375–386; Reprinted in (Grassmann 1894–1911, II.1, 268–282).

———. 1894–1911. *Hermann Grassmanns Gesammelte Mathematische und Physikalische Werke*. Friedrich Engel ed., vol. 3. in 6 pts., Leipzig: Teubner.

Hankel, Hermann. 1867. *Theorie der complexen Zahlensysteme*. Leipzig: Teubner.

Hankins, Thomas. 1980. *Sir William Rowan Hamilton*. Baltimore: Johns Hopkins University Press.

Henderson, Linda Dalrymple. 1983. *The Fourth Dimension and Non-Euclidean Geometry in Modern Art*. Princeton: Princeton University Press.

Klein, Felix. 1872. *Vergleichende Betrachtungen über neuere geometrische Forschungen*. Erlangen: Deichert; Reprinted in (Klein 1921–23, I: 460–497).

———. 1875. Review of Victor Schlegel, *System der Raumlehre, Erster Teil, Jahrbuch über die Fortschritte der Mathematik*, Jahrgang 1872, Berlin: Georg Reimer, 231–235.

———. 1921–1923. *Gesammelte Mathematische Abhandlungen*. Vol. 3. Berlin: Springer.

———. 1926. *Vorlesungen über die Entwicklung der Mathematik im 19. Jahrhundert*. Vol. 1. Berlin, Springer.

Lewis, Albert C. 1977. H. Grassmann's 1844 Ausdehnungslehre and Schleiermacher's Dialektik. *Annals of Science* 34: 103–162.

Robbin, Tony. 2006. *Shadows of Reality: The Fourth Dimension in Relativity, Cubism, and Modern Thought*. New Haven: Yale University Press.

Rowe, David E. 1985. Felix Klein's Erlanger Antrittsrede: A Transcription with English Translation and Commentary. *Historia Mathematica* 12: 123–141.

Schlegel, Victor. 1872. *System der Raumlehre, nach den Prinzipien der Grassmannschen Ausdehnungslehre und als Einleitung in dieselbe dargestellt, Erster Teil: Geometrie*. Leipzig: Teubner.

———. 1875. *System der Raumlehre, nach den Prinzipien der Grassmannschen Ausdehnungslehre und als Einleitung in dieselbe dargestellt, Zweiter Teil: Die Elemente der modernen Geometrie und Algebra*. Leipzig: Teubner.

———. 1878. *Hermann Grassmann: Sein Leben und seine Werke*. Leipzig: Brockhaus.

———. 1883. Theorie der homogen zusammengesetzten Raumgebilde, *Nova Acta Leopoldina Carolinium*. (Verhandlungen der Kaiserlichen Leopoldinisch-Carolinischen Deutschen Akademie der Naturforscher), Band XLIV, Nr. 4.

———. 1886. *Über Projectionsmodelle der regelmässigen vierdimensionalen Körper*. Waren.

———. 1896. Die Grassmannsche Ausdehnungslehre. Ein Beitrag zur Geschichte der Mathematik in den letzten fünfzig Jahren. *Zeitschrift für Mathematik und Physik* 41 (1–21): 41–59.

Schubring, Gert. 1981. The Conception of Pure Mathematics as an Instrument in the Professionalization of Mathematics. In *Social History of Mathematics*, ed. H. Mehrtens, H. Bos, and I. Schneider, 111–134. Birkhäuser: Basel.

Tobies, Renate. 1996. The Reception of Grassmann's Mathematical Achievements by A. Clebsch and his School, in (Schubring 1996, 117–130).

Turner, R. Steven. 1971. The Growth of Professional Research in Prussia, 1818–1848-Causes and Contexts. *Historical Studies in the Physical Sciences* 3: 137–182.

10 Three Letters from Sophus Lie to Felix Klein on Mathematics in Paris

(*Mathematical Intelligencer* 7(2)(1985): 74–77)

Sophus Lie and Felix Klein first met in 1869 as students in Berlin. They soon became daily companions and spent the spring of 1870 together in Paris where they met the French mathematicians Michel Chasles, Gaston Darboux, and Camille Jordan. Jordan had just published his classic *Traité des substitutions,* and the two foreigners read it avidly. Mathematics has not been the same since, for it has often been said – and not altogether unjustly – that from this moment on they made group theory their common property: Lie taking the continuous groups and Klein those that were discontinuous. It should not be overlooked, on the other hand, that this observation was first made by Klein himself in the preface to his *Lectures on the Icosahedron* (Klein 1884, iv), making this an effective piece of propaganda for a particular view of their early work in geometry.

This liaison in Paris was relatively short-lived, however, as the Franco-Prussian War broke out forcing Klein to return to Germany, while Lie decided to set out on foot for Italy. This ill-fated idea cost him 4 weeks in prison; he was arrested just south of Paris in Fontainebleau, where guards found that he was carrying some letters written by Klein in what appeared to them to be a secret code! Lie had to wait until Darboux could set the matter straight with the authorities. He then went directly to Italy and a short time later returned to Norway (Figs. 10.1 and 10.2).

In 1872 Lie returned to the continent and stayed with Klein in Göttingen and Erlangen, where together they discussed the key ideas that developed into the latter's Erlangen Program (Stubhaug 2002), (Yaglom 1989). After this, however, their work began to drift apart, although they continued to stay in close contact with one another. Lie was working in almost total isolation until 1884 when Klein and his Leipzig colleague Adolf Mayer sent their student, Friedrich Engel, to Norway for nine months to serve as his assistant. Engel ended up dedicating his life to Lie's work, first by co-authoring the three-volume *Theorie der Transformationsgruppen,* and then, after Lie's death, editing his *Collected Works.* Lie's situation changed dramatically in 1886, when Klein succeeded in bringing him to Leipzig as heir to his position as Professor of Geometry, a maneuver that infuriated Weierstrass and his student H. A. Schwarz.

During his many years of isolation, however, Lie often thought about making a second trip to Paris with Klein. Over and again this idea pops up in their correspondence, and in 1882 Lie actually took the plunge and tried to persuade Klein to join him. Unfortunately, his timing could not have been worse. Klein was in the thick of his competition with Henri Poincaré to find a proof for the uniformization theorems involving automorphic functions (Gray 2013), and at the same time he was experiencing serious difficulties with his health. He completed the lengthy study, entitled "Neue Beiträge zur Riemannschen Funktionentheorie," in early October, shortly before the semester began. But he was only able to meet his classes on an irregular basis before finally giving them over to his assistant Walther von Dyck. After this, it took him more than a year before he was able to fully recover and work again on a steady, if more moderate, basis.

It was during the first few months following this breakdown that Lie wrote Klein the following letters from Paris.[1] These provide an intimate glimpse of his impressions on meeting that astonishing array of mathematical talent that frequented the *Académie des sciences* (Fig. 10.3). Henri Poincaré had only arrived in the capital city about a year earlier and, as Lie relates, he had already completed the first of the five famous articles on automorphic functions that he eventually published in Mittag-Leffler's *Acta Mathematica.* One year before Lie's visit, Klein had published certain objections to Poincaré's having named the functions with a boundary circle after Lazarus Fuchs – the so-called "fonctions fuchsiennes." Echoes from that controversy still rang loudly, even though the mathematicians Lie spoke with were careful when broaching this topic.

[1]The original letters are located in the Klein *Nachlass* housed at the Niedersächsische Staats- and Universitätsbibliothek, Göttingen. The author would like to thank that institution for permission to publish them in translation here.

D.E. Rowe, *A Richer Picture of Mathematics*, https://doi.org/10.1007/978-3-319-67819-1_10

Fig. 10.1 Sophus Lie.

Fig. 10.2 Felix Klein.

One should bear in mind when reading these letters that Lie was anything but an objective observer; he was, rather, a highly temperamental and one-sided advocate of his own work and ideas, and thus he tended to evaluate other mathematicians according to the degree to which they happened to find his work praiseworthy. Even before he arrived in Paris, he had a whole agenda of business he wanted to square with the mathematicians there. For example, as he wrote Klein in June 1882: "During my stay in Paris I will make it clear to Halphen that my theory of *differential invariants* embraces not just general linear groups but *arbitrary transformation groups*. The concept of *differential invariant* belongs to *the two of us* (Klein and myself). His propositions are in part obvious consequences of a rather important body of simple corollaries to propositions concerning infinitesimal transformations that I published long ago."

Yet despite Lie's lack of objectivity, his remarks pertaining to the animosity between German and French mathematicians were not something he merely invented. The nationalist sentiment among leading mathematicians in both countries was much stronger than any counterbalancing ideal of international scientific cooperation. At the same time, Gösta Mittag-Leffler, who is sometimes portrayed as having played a sort of mediating role between Paris and Berlin, appears to have often aggravated the situation more than helped it with his high-handed political maneuvering. And while the leading mathematicians of France and Germany usually had high respect for one another, they nearly always viewed their foreign counterparts as rivals rather than friends.

Paris, October 1882

Dear Klein,

Having just arrived in Paris, I send you these few lines for the time being. First my warm thanks to you and your wife for the wonderful days in Leipzig. You must also extend my greetings to Mayer.

The trip to Switzerland went well except that the high tides on Lake Constance forced me to miss the afternoon meeting I had planned with the Zurich mathematicians. But considering the circumstances, I had good fortune in Switzerland. Not much rain and fog, although the snow and avalanches hindered my way quite a bit.

Fig. 10.3 Four figures who loomed large in Lie's letters to Klein.

In the Academy I met with Halphen, Darboux, Poincaré, Levy, and Stephanos, all of whom were very obliging. In any case my geometrical works are better known in Paris than in Germany. I hope to find an open ear for transformation groups as well. I also spoke with Tchebicheff in the Academy – a remarkable old gentleman who is as lively as a youngster. He spoke about a variety of topics and has published many things that we had drawn from Abel's posthumous papers. He also spoke about first-order differential equations whose multiplier can be exhibited because one knows an infinitesimal transformation.

Poincaré already has page-proofs of his work for Mittag-Leffler. As I understood him to say, Weierstrass's Note has delayed the appearance of the first volume. I will ask him more about it, especially when it appears. Hermite disappeared from the Academy just as I entered. As soon as I become better oriented I'll write you more. I say, however, like Abel: It is most difficult to gain the intimacy of the French, etc.

With heartfelt greetings,

Sophus Lie

This letter is already somewhat old. Soon more. .. I am quite satisfied with my stay in Paris.

Dear Klein,

I must write you something about the situation in Paris even though I don't have that much of interest to communicate. Concerning Poincaré's work for Mittag-Leffler, it is still undecided when the first volume will appear. Even the name of the journal has yet to be decided. Leffler has recently been corresponding with Hermite about this. In all probability it will be at least a month before something appears.

I have spoken now with Hermite about all kinds of things. He has a very amiable nature, but I still don't know how much of it is genuine. People here say it is *certain* that he can't read a word of German, which indeed explains a number of things. The most remarkable thing he said was the following (which I communicate to you in confidence): Mittag-Leffler told him that the German mathematicians hate the French mathematicians. Nor did he want to hear anything of my protests against this. That is certainly strong. He was eager to hear about friction between German mathematicians, whereas he described the situation in Paris as idyllic in this regard. Probably it is no better in Paris than in Germany.

Naturally, he praised Weierstrass a great deal, although he also spoke a lot about how difficult it is to follow his presentations. He considers Fuchs to be the German mathematician who presents his theories most clearly. He spoke with marked recognition about your results, although in such a vague way that I can't tell how much about them he actually knows. .. It is true, he said, that Fuchs is Klein's *bête noire* . . . I answered that you gladly recognized Fuchs's accomplishments, but that you were of the opinion that, in France, too much was ascribed to him that others, for example Riemann, had done earlier. "But we know that well" was his reply again.

Picard is Hermite's son-in-law and he makes a very intelligent impression naturally.

Poincaré mentioned on one occasion that all of mathematics was a matter of groups. I told him about your [Erlangen] Program which he did not know. Halphen, Darboux, and Stephanos spoke with the highest praise about you. Until now I have spoken very little with Jordan, whose mother died recently.

Fig. 10.4 Henri Poincaré.

... In the meantime I have spoken at length with Jordan. He finds your investigations difficult to understand. Poincaré said that at first it was hard for him to read your work, but that now it goes very easily. A number of mathematicians, such as Darboux and Jordan, say that you make great demands on the reader in that you often do not supply proofs. I am trying to report on this as correctly as possible (Figs. 10.4 and 10.5).

So far as my own things go, I am more or less satisfied. Darboux has studied my work with remarkable thoroughness. This is good insofar as he has given gradually more lectures on my theories at the Sorbonne, for example on line and sphere geometry, contact transformations, and first-order partial differential equations. The trouble is that he continually plunders my work. He makes inessential changes and then publishes these without mentioning my name. Now he is starting on the surfaces of constant curvature. I must therefore rework my papers from Christiania [present day Oslo] for the *Mathematische Annalen* just as soon as possible.

Halphen is very friendly. He knows something of my minimal surfaces, which interest him a great deal. I am trying to explain to him the relationships between his differential invariants and my integration theories. Moreover, I have spoken in detail about this with Hermite, Jordan, Poincaré, Picard, and Tchebicheff. Tchebicheff and Jordan gave me energetic encouragement.

Poincaré comprehended me clearly, and I believe Picard as well. On the other hand, I fear that Hermite's recognition was only pleasant phrases. Hermite is a man of the world in conversation, but he appears now to be just a *little* old.

Bonnet is colossally fat; he is very friendly and knows my geometrical work fairly well as it interests him greatly.

Fig. 10.5 Gaston Darboux.

I did not look for Bertrand. To Bouquet I expressed my thanks because he and Briot have paid wonderful tribute to Abel. Hermite's quotations regarding Abel and Jacobi in his course at the Sorbonne are remarkably inaccurate. He even says that Euler had already accomplished the inversion of the integrals. That is a remarkable misunderstanding. As for the rest, whenever Abel had priority he says that Abel and Jacobi made their discoveries at the same time.

Mannheim is friendly as always. He is really a good fellow and warns me constantly about Darboux, for which he really has good reason. But I must speak with Darboux, as he is the one who understands me the best. And for the pleasure I must pay something. In any case he is promoting mathematical science. Levy is a good mathematician who knows my integration theories a little. In the mathematical society I must try to do some advertising for my integration theories.

I am now calculating for *every* group in the plane the associated differential invariants. It has been going easily so far, and I am giving a detailed integration theory for these differential invariants. These are forming themselves very prettily, although in principle they are only examples of the new theory. I will use them in a large work that I will publish first in Christiania. I think I wrote you already that Darboux praises Bäcklund, which pleases me.

Greet Mayer and your wife.

Sophus Lie

Dear Klein:

... I have a bit of bad conscience that I told you about Mittag-Leffler's remark to Hermite. Having passed through Hermite, it could well have obtained too strong a coloring. Mittag-Leffler is, I believe, not only an outstanding mathematician and excellent socializer, but also basically a *good* person. By the same token, he is by all means a much too intriguing diplomat to appeal altogether to such an undiplomatic person as myself. It appears that his journal is going well now. I am afraid, however, that he will set out in all too grand a style with it. If he begins to get tired, I fear it will decay. But I conjecture that in the next few years everything will go well. The time-frame was certainly well chosen, as the discoveries of the last years are without question epoch-making in the history of mathematics.

Tell me when you get a chance whether Fuchs has answered your last remarks and, if so, where. I have no doubts that the mathematical world will give the essential work you've done prior to Poincaré's discoveries its just due. With all of your students you have an army that represents a mighty force. ...

With heartfelt greetings,

Sophus Lie

References

Gray, Jeremy. 2013. *Poincaré: A Scientific Biography*. Princeton: Princeton University Press.

Klein, Felix. 1884. *Vorlesungen über das Ikosaeder und die Auflösung der Gleichungen vom fünften Grade*. Leipzig: Teubner.

Stubhaug, Arild. 2002. *The mathematician Sophus Lie: It was the audacity of my thinking*. Heidelberg: Springer.

Yaglom, I.M. 1989. *Felix Klein and Sophus Lie: Evolution of the idea of symmetry in the nineteenth century, translated by Sergei Sossinsky*. Basel: Birkhäuser.

11 Klein, Mittag-Leffler, and the Klein-Poincaré Correspondence of 1881–1882

(*Amphora. Festschrift for Hans Wussing*, 1992, 598–618)*

If a modern-day Plutarch were to set out to write the "Parallel lives" of some famous modern-day mathematicians, he could hardly do better than to begin with the German, Felix Klein (1849–1925), and the Swede, Gösta Mittag-Leffler (1846–1927). Both lived in an age ripe with possibilities for the mathematics profession and, like few of their contemporaries, they seized upon these new opportunities whenever and however they arose. Even when their chances for success looked dismal, they forged ahead, winning over the skeptics as they did so. Although accomplished and prolific researchers (Klein's work has even enjoyed the appellation "great"), they owed much of their success to their talents as lecturers. Indeed, as teachers they exerted a strong influence on the younger generation of mathematicians in their respective countries. Klein's German students included such prominent figures as Ferdinand Lindemann, Walther von Dyck, Adolf Hurwitz, Robert Fricke, Philipp Furtwängler, and Arnold Sommerfeld. His influence on North American mathematicians was, if anything, even stronger, as will be briefly described in this essay. Since Stockholm's *Högskola*, founded in 1878, could hardly compete with the much older universities where Klein taught, most notably Leipzig and Göttingen, Mittag-Leffler was clearly not in a position to draw large numbers of doctoral students. Nevertheless, he attracted several, four of whom left their mark on modern mathematics: Edvard Phragmén, Ivar Bendixson, Helge von Koch, and Ivar Fredholm (Stubhaug 2010).

Beyond their pedagogical gifts, Klein and Mittag-Leffler held influential positions that served to enhance their respective roles in the European mathematical community. Both served as editors-in-chief of two of the leading mathematics journals of the day: *Mathematische Annalen* and *Acta Mathematica*. In what follows, I describe some pivotal events from the years 1881–1882 when Klein reached the zenith of his mathematical achievements and Mittag-Leffler succeeded in attaching his new journal to a rising mathematical star, Henri Poincaré.[1] It was during this period that Klein took up Riemannian ideas with a passion, though I also describe his older interest in so-called projective Riemann surfaces. After considering some highlights connected with the brief competition between Klein and Poincaré as reflected in their correspondence from this period, I turn to look at some of the cat and mouse games played by Klein and Mittag-Leffler in the wake of this famous duel. These not only shed significant light on their personal relationship but also say much about their respective styles in cultivating scientific relations in an atmosphere rife with tension fed by the rivalries between leading mathematical camps. Finally, I take up some background events that surrounded the publication of the Klein-Poincaré correspondence after World War I. Mittag-Leffler's handling of this matter reveals a side of his character usually overlooked in accounts of his role as a mediator between the mathematicians of France and Germany.

Launching *Acta Mathematica*

Mathematische Annalen had been founded by Alfred Clebsch and Carl Neumann in 1868. After the untimely death of Clebsch in 1872, Klein gradually took control of the reins, assisted by the Leipzig mathematician, Adolf Mayer. Realizing that the journal's success depended on the support of mathematicians outside the narrow confines of the Clebsch school, Klein developed a broad network of contacts both within Germany and abroad. Among Scandinavian mathematicians, for example, the *Annalen* published the work of Klein's close friend, Sophus Lie, and numerous articles by the Swedish mathematician, Albert Viktor Bäcklund.[2] By the time of Klein's appointment in 1880 as Professor of Geometry in Leipzig, *Mathematische Annalen* had

*This essay, based on (Rowe 1992), has been supplmented with additional material on Klein's mathematics and his influence as a teacher.

[1]On Mittag-Leffler's relationship with Poincaré, see their correspondence in Nabonnand (1999).

[2]See the table of major contributors to the *Annalen* in Tobies and Rowe (1990, 38–39).

D.E. Rowe, *A Richer Picture of Mathematics*, https://doi.org/10.1007/978-3-319-67819-1_11

overtaken the *Journal für die reine und angewandte Mathematik* to become the leading mathematics periodical in Germany.

Whereas Klein essentially inherited the editorship of his journal by filling the vacuum formed by Clebsch's premature death, Mittag-Leffler had to create *Acta Mathematica* out of thin air, as it were. His success came like a bolt of lightning from the blue, and the way he achieved it reflected not only his imagination and daring but also his unusual talent for scientific diplomacy. Mittag-Leffler had taken his doctorate at Uppsala in 1872 before travelling to Paris, where he studied with Charles Hermite. On the latter's advice, he then went on to Berlin to hear Karl Weierstrass's lectures. Although nearly 60 years old, Weierstrass was now entering the period of his real fame and his ideas exerted a profound influence on Mittag-Leffler (see Chap. 4). As it happened, the young Swede arrived in Berlin in late 1874, just after Weierstrass's favorite student, Sonja Kovalevskaya, had returned to Russia. Kovalevskaya had been studying privately with him during the previous five years. Mittag-Leffler heard, of course, about the talented Russian woman from Weierstrass, and in February 1876 he had the opportunity to meet her during the course of a trip he made to St. Petersburg.

In 1877, Mittag-Leffler accepted a professorship in Helsingfors. He taught there until 1881, when he accepted a position at Stockholm's newly-founded *Högskola*. Even while in Finland, Mittag-Leffler had begun a determined effort to find a position for Kovalevskaya in Scandinavia. Never one to take "no" for an answer, he pressed his case using a mixture of flattery and charm. At one point, he even wrote Kovalevskaya: "I have no doubt that once you are here in Stockholm ours will be one of the first faculties of the world" (Kochina and Ozhigova 1984, 27). By the autumn of 1883, she had joined him at the *Högskola* in Stockholm, an affiliation she maintained up until her premature death in 1891. Her arrival came just as Mittag-Leffler war preparing to publish the third volume of his new journal, *Acta Mathematica*, (an issue that contained Poincaré's *Mémoire sur les groups kleinéens*), and thereafter Kovalevskaya took an active part in its editorial affairs.

Although not generally known, the initial impulse for founding a Scandinavian mathematics journal had not come from Mittag-Leffler, nor from Hermite and Weierstrass, as Mittag-Leffler once suggested (Domar 1982, 3). In fact, the idea came from Sophus Lie. In the spring of 1881, Lie met with Mittag-Leffler in Stockholm and suggested that the latter should consider editing a journal that would showcase the work of Scandinavian mathematicians. At first they planned the venture together, but Mittag-Leffler quickly made it his own. With his many contacts abroad, particularly in Paris and Berlin, he began to conceive of the journal in more international terms, however. Although he sensed a splendid opportunity to capitalize on his good relations with both Hermite and Weierstrass, he also recognized the need to proceed with great delicacy if he hoped to receive significant support from both the French and German mathematical communities (Fig. 11.1).

One factor he had to contend with concerned the *Journal für reine und angewandte Mathematik*, first founded by Leopold Crelle and later edited by Carl Wilhelm Borchardt. The latter's death in 1880 left this prestigious periodical in the uncertain hands of Weierstrass and Kronecker. Given these circumstances, Mittag-Leffler had good reason to believe that the Berlin mathematicians would not greet his plans with enthusiasm. In the meantime, Hermite had kept him well abreast of mathematical developments in Paris and, in particular, the work being done by his talented young pupils: Paul Appell, Émile Picard, and Poincaré. Hermite spoke about Poincaré's current discoveries in function theory in the most glowing terms, and Mittag-Leffler, remembering what Abel's work had accomplished for *Crelle's Journal*, soon realized that if he could launch his journal by publishing Poincaré's work its future would be assured (Domar 1982, 5). Poincaré was then composing his lengthy memoir, *Théorie des groupes fuchsiens*, which he planned to submit to the *Journal de l'école polytechnique*. Drawing on his full powers of persuasion, Mittag-Leffler convinced Poincaré not only to submit this paper to him but also four other lengthy articles he planned to write on the theory of automorphic functions (Poincaré 1928). At the same time, Mittag-Leffler also kept Hermite informed of his plans for the new journal, obtaining his support as well.

Still, Mittag-Leffler had two difficult hurdles to clear. First, he needed to find donors to finance the printing costs and other expenses, and, second, he needed a strategy to win over the Berlin mathematicians. He decided on a daring course, not without intrigue. Rather than ask for his mentor's blessing, he opted not to inform Weierstrass or any of his confidants in Germany in order to present them with a *fait accompli*. Together with Carl Johan Malmsten, a former mathematics professor in Uppsala and close ally of Mittag-Leffler, he requested an audience with King Oscar II, who agreed to donate the modest sum of 1500 crowns and his royal name to the cause. Afterward, Malmsten and Mittag-Leffler received the king's permission to invoke his name in requesting the support of German mathematicians for the undertaking. Malmsten then wrote to Kummer, Kronecker, Weierstrass, and Schering (all former recipients of a Swedish royal order) to break the news and convey the king's interest in it (Domar 1982, 6–7). The plan worked like a charm, as all of them agreed to support the new journal, which was to be called *Acta Mathematica Eruditorum*. At the last minute the word "Eruditorum" was dropped, perhaps to avoid confusion with the older journal *Acta Eruditorum* (Domar 1982, 8). Almost overnight, Mittag-Leffler had launched a major international mathematical publication.

Fig. 11.1 Mittag-Leffler posing in front of the hearth in his mansion.

Klein's Projective Riemann Surfaces

Whereas Mittag-Leffler identified strongly with the Weierstrassian approach to complex analysis, Klein championed competing Riemannian ideas, which the Berlin school rejected (see Chap. 4). Yet in certain respects, Felix Klein's most original work had an idiosyncratic flavor owing to the fact that nearly everything he touched in mathematics – and this included many different things – always turned into geometry. Visual aspects abound in his works, though he particularly emphasized the role of visualization in the set of papers he bundled together for the opening section of volume 2 of his *Gesammelte Mathematische Abhandlungen* (Klein 1922). These were his contributions to "Anschauliche Geometrie." Most of these papers were written during the mid-1870s, the period immediately following his collaboration with Lie, which effectively ended with the publication of Klein's "Erlangen Program" (Klein 1872). Afterward, Klein pursued algebraic problems via a geometric approach to Galois theory, and this work brought him into the arena of Riemann's theory of complex-valued functions. At the center of his interests, though, stood a special concern: how to visualize the properties of algebraic curves and surfaces as objects embedded in space? Lie had developed a "theory of the imaginary" with a similar objective, and this had helped him to launch his early career as a geometer. Later, though, he made little effort to distinguish the real and imaginary elements that together define a geometric object, having in the meantime gained confidence that he could reason intuitively when working with the complex-valued entities. Klein, by contrast, never lost interest in sorting out these two cases.

One of his most novel ideas in approaching this problem was a new type of Riemann surface, sometimes called projective Riemann surfaces since their construction takes place within the context of algebraic curves defined over the complex projective plane (Klein 1874; Klein 1876a; Klein 1876b). Such curves, given by implicit equations with real coefficients, correspond analytically to surfaces in a 4-space. The same holds for conventional Riemann surfaces, which cannot literally be visualized in 3-space owing to the virtual intersections of their leaves. So Klein sought a way to represent these curves in 3-space, but in such a way that the *Gestalt* of their real part would be preserved. In effect, he would build the imaginary part around this, once he had established the relation between the two.

This meant proceeding by concrete examples, starting with the simplest non-trivial case, a degree-two curve, the real part of which can be visualized as an ellipse in the real projective plane Π (Klein 1874). Klein's idea was to take advantage of duality, which interchanges the order and class of a curve. (The class is just the number of tangents to the curve through an arbitrary point, which is invariant if real and imaginary tangents are counted together.) The dual of a given point curve thus produces an image of the curve as a locus of tangent lines. In the case of an ellipse, the dual is just another ellipse, since the order n and class k both equal two. Now the Riemann surface of a degree two curve is just a topological sphere, a surface of genus zero, so Klein wanted to construct such an object in 3-space around this given class curve. Since its equation has real coefficients, its imaginary tangents will appear in conjugate pairs, which Klein visualized using a continuity argument. Imagine the plane Π of the curve as embedded in 3-space. This plane is partitioned by the ellipse into two regions: in the exterior there are two real tangents to the ellipse through a given point, so no imaginary elements arise, whereas in the interior we have two imaginary tangents. Fixing a point P in the interior, Klein imagines two equidistant points Q_1, Q_2 on the normal to Π through P, one point above Π, the other below it. Moving P throughout the interior of the ellipse, one ensures that Q_1, Q_2 vary continuously and in such a way that the distances shrink toward zero whenever P approaches the real part of the curve. This produces a topological sphere in 3-space.

This example is far too simple to be compelling, in part because this construction only produces interesting visualizations when singularities enter the picture. Furthermore, Klein was mainly interested in whole classes of curves and the *changes in Gestalt* that take place in passing from one class to another. He used such deformation techniques to study the properties of real cubic surfaces in (Klein 1873), taking up the classification of cubics made by Ludwig Schläfli in (Schläfli 1863). Schläfli's scheme began with a double-six of lines from which one can easily obtain the other 15 that make up the 27 lines on a cubic surface. In some cases, all 27 are real, but there are many others with fewer. Klein started with a very special case, a cubic with four singular points that determine a tetrahedron. Here 24 of the 27 lines have collapsed into the six edges of the tetrahedron, each counted four times. Klein showed how, starting with this special case, one could utilize two basic types of deformation processes to desingularize the surfaces step-by-step, obtaining all possible types. He thereby described the space of all cubic surfaces as a 19-dimensional *connected manifold.*

Numerous other geometers were exploring similar terrain in 3-space for the first time. A noteworthy result in this connection was published by Carl Geiser in the very first volume of *Mathematische Annalen*. In (Geiser 1869) he established a striking connection between the 27 lines on a cubic surface and the 28 bitangents to a quartic curve. This argument is simple and very general, since it starts from an arbitrary point P on a cubic surface and an arbitrary plane Π in space. The tangential cone at P has degree six and decomposes into the tangential plane T_P and a ruled surface F_4. The latter meets Π in a quartic curve $F_4 \cap \Pi = C_4$, whereas the plane T_P intersects Π in one of the bitangents to C_4. Geiser then shows that the other 27 bitangents are the projections of the lines on the cubic surface from the point

Fig. 11.2 This modern account of Klein's quartic, (Levy 1989), can be downloaded at http://library.msri.org/books/Book35/contents.html.

P mapped into Π. Klein's friend, the Danish geometer H. G. Zeuthen, made an extensive study of plane quartic curves and their double tangents (Zeuthen 1874), also published in *Mathematische Annalen.* These were among the many researches Klein could build on for his own work.

One of the most famous objects he discovered a few years later came to be called the Klein quartic (Klein 1879). Jeremy Gray wrote about the discovery and some of its many properties in one of the early issues of *The Mathematical intelligencer* (Gray 1982). Klein's quartic has also been celebrated at MSRI in Berkeley, where "The Eightfold Way" – a sculpture made by Helaman Ferguson – is on prominent display (see Levy (1989)) (Fig. 11.2).

Many mathematicians will immediately associate the Klein quartic with a famous picture that shows how its automorphism group G_{168} acts on the complex curve by moving the 168 triangles (in shaded and unshaded pairs that remain invariant) around the figure. Under this action, generic points have orbits of 168 points, but the vertices have smaller orbits consisting of 24, 56, and 84 points, corresponding to special points on the quartic curve as well as branch points of the Riemann surface. Klein called these a, b and c points, respectively, and by using this information he calculated the genus to be:

$$p = \frac{1}{2}(2 - 2 \cdot 168 + 56 \cdot 2 + 84 \cdot 1 + 24 \cdot 6) = 3.$$

To obtain this closed surface of genus 3, Klein made the identifications of sides specified in Fig. 11.3, taking due note that there are two types of vertices that need to be aligned. As a genus 3 curve, the Klein quartic has no point singularities, and the number of tangential singularities follow directly from the Plücker formulas. Indeed, the a, b and c points correspond, respectively, to the tangential singularities of the curve: its 24 inflection points, 56 bitangential points, and 84 sextatic points, where special conics have six-fold contact with the quartic. Klein could also identify how many of each type were real by noting that the real part of the curve was one of the 28 symmetry lines that run through the center of the figure (Fig. 11.3).

This gives a splendid picture of the full complex curve, but what about its real part? Klein wrote about that, too, but he left the task of constructing the associated projective Riemann surface to one of his doctoral students, the American Mellen W. Haskell. His dissertation (Haskell 1890) was published in the *American Journal of Mathematics*, which J.J. Sylvester had founded during his tenure at Johns Hopkins University. Haskell made a precise calculation of the points on the real quartic curve by using theta series to calculate its parameterized coordinates. This enabled him to give a beautifully precise picture of the symmetries of the curve, including its various tangential singularities. Six of the 24 inflection tangents are real yielding three pairs of a point, separated by a c point, but lying between the two b points determined by a bitangent line. Beyond these three real bitangents there is a fourth: the line at infinity is an isolated bitangent that touches the curve in the two imaginary points I, J, the circular points at infinity. Note that the six b points lie on a circle, which therefore contains eight b points since every circle passes through I, J. This conforms with a theorem of Plücker according to which eight of the 56 b points of a quartic curve lie on a conic (Fig. 11.4).

Dualizing this curve turns tangential singularities into point singularities, which means inflection tangents become cusps and bitangents will pass over into double points. These will alternate according to the pattern above, whereas the isolated bitangent at infinity will become an isolated double point at the origin. Haskell thus came up with this figure (Fig. 11.5):

This real class curve served as the skeleton for the projective Riemann surface Haskell next constructed following Klein's general procedure. He noted that for points in the triangle whose vertices are double points the four tangents to the curve are all imaginary, whereas for points inside the other three triangles the tangents are all real. In the exterior region of the curve, exactly two tangents are real and two imaginary. So these results imply that the genus three surface can be built by letting the four leaves in the center collapse to two at the boundary where the three outer triangles become holes in the object so formed. This forgets the isolated double point in the center, however, which forms a special type of singularity; there the four imaginary tangents pass over into two pairs.

In fact, Klein had years before pointed out that such isolated singularities yield branch points for the projective Riemann surface. Haskell thus had to make a careful investigation of the structure of the four leaves that hang together at the center of the figure. There are two identical large leaves and two small ones, which are also identical. In his paper, Haskell drew one of each type, the larger leaf containing an image of the real part of the curve. The letters S, T, U refer to respective elements of the group G_{168} that can be related to simple transformations of the modular figure in the upper half-plane (see Gray (1982)). The group is generated by the first two elements: the parallel translation S and the reflection T, whereas $U = TS^2TS^4TS^2$. The elements T and U are of order two and three, respectively, and together generate a six element subgroup: $1,\ U,\ U^2,\ T,\ TU,\ TU^2$. Here then are the constituent parts Haskell needed to build the projective Riemann surface for Klein's quartic curve (Figs. 11.6 and 11.7):

Klein's Influence on American Mathematics

Haskell was one of a large group of New Englanders who took up graduate studies in Germany during the 1880s, most of whom chose Felix Klein as their mentor (Parshall and Rowe 1994). Haskell's dissertation on the Klein quartic was quickly forgotten; nevertheless, it shows the extent to which he steeped himself in the complexities of this rather idiosyncratic mathematical style. Little wonder that Klein turned to him when he sought an English translator for his "Erlangen Program," which was published in (Klein 1893). Although Klein founded no school of American mathematics comparable to the one that Sylvester established at Johns Hopkins, his influence on the emergent mathematical research community in the United States proved far more pervasive.

After Sylvester left Baltimore in 1883 to accept the Savilian Chair for Geometry at Oxford, his position at Hopkins was offered to Klein, who was then in his mid-thirties. Sylvester, who was already approaching seventy, hoped Klein would succeed him. He therefore wrote him to explain his resignation as well as the enormous challenge

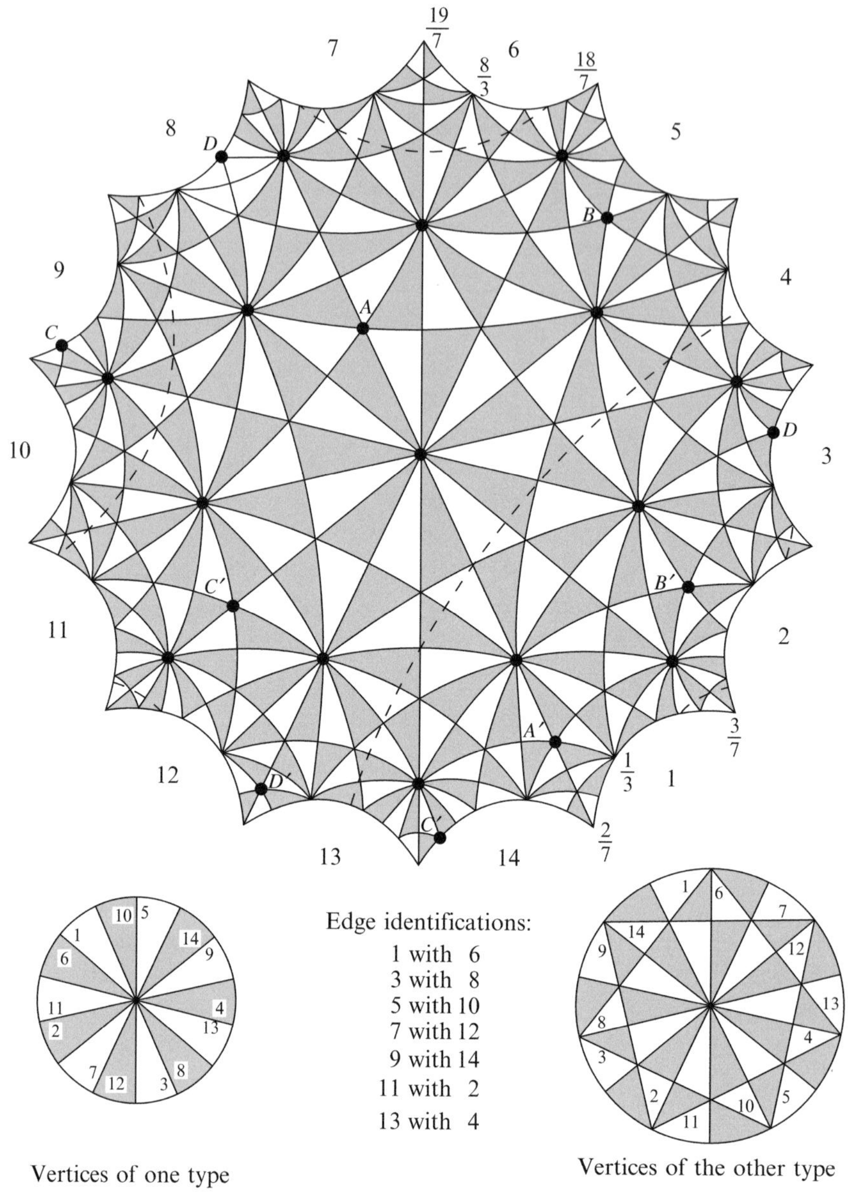

Fig. 11.3 Klein's quartic split open (based on Table I in (Haskell 1890)).

the position afforded. "I did not consider," he admitted, "that my mathematical erudition was sufficiently extensive nor the rigour of my mental constitution adequate to keep me abreast of the continually advancing tide of mathematical progress to that extent which ought to be expected from one on whom practically rests the responsibility of directing and molding the mathematical education of 55 millions of one of the most intellectual races of men upon the face of the earth" (James Joseph Sylvester to Felix Klein, 17 January, 1884, Klein Nachlass, SUB). Klein was seriously tempted by the Hopkins' offer, but decided in the end to remain in Germany. Nevertheless, from his outposts in Leipzig and, beginning in 1886, Göttingen, he managed to exert a remarkably strong influence on the next generation of American mathematicians.

Ten years after Sylvester had departed for England his impact on mathematics in the United States had largely dissipated. During the decade from 1883 to 1893, aspiring American mathematicians not only looked to Germany as the standard-bearer but also often traveled there to obtain the kind of higher-level training. No fewer than six of those who studied under Klein went on to become Presidents of the American Mathematical Society, and thirteen served as Vice-Presidents. During the last quarter of the nineteenth century, American mathematicians thus came increasingly under the sway of research trends that had developed in Germany, and Klein served as the principal conduit for the transmission of German mathematics and its associated ideals into the United States. (For details, see Parshall and Rowe (1994, Chap. 5)).

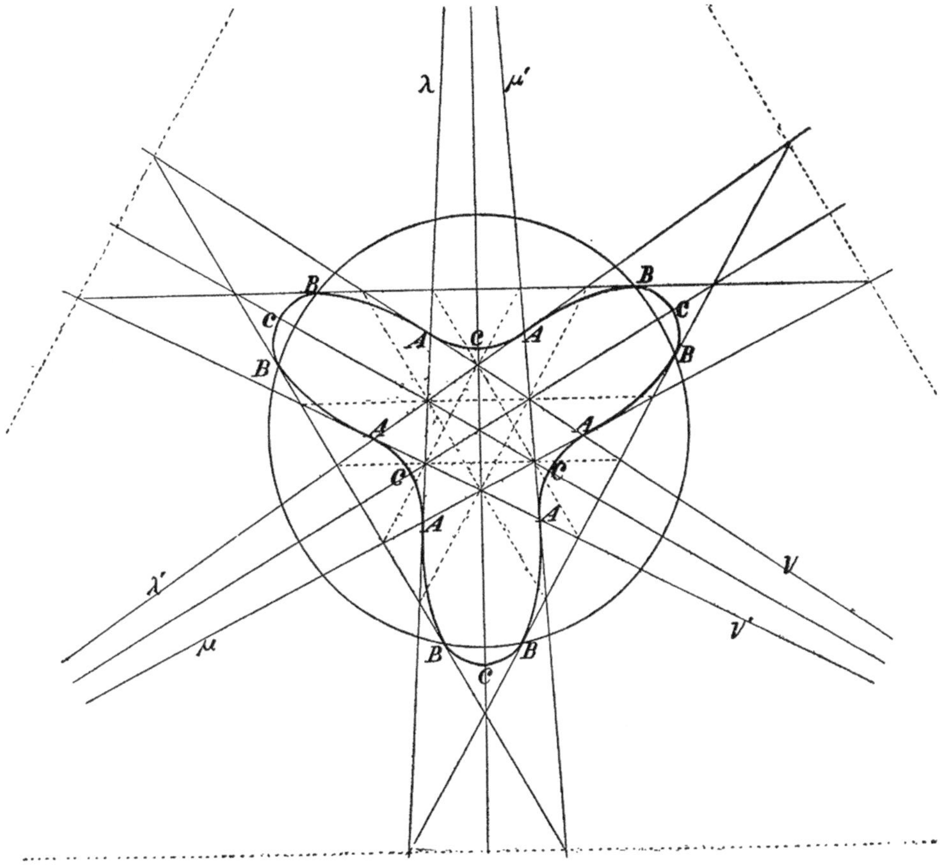

Fig. 11.4 The real part of Klein's quartic (Haskell 1890, 18).

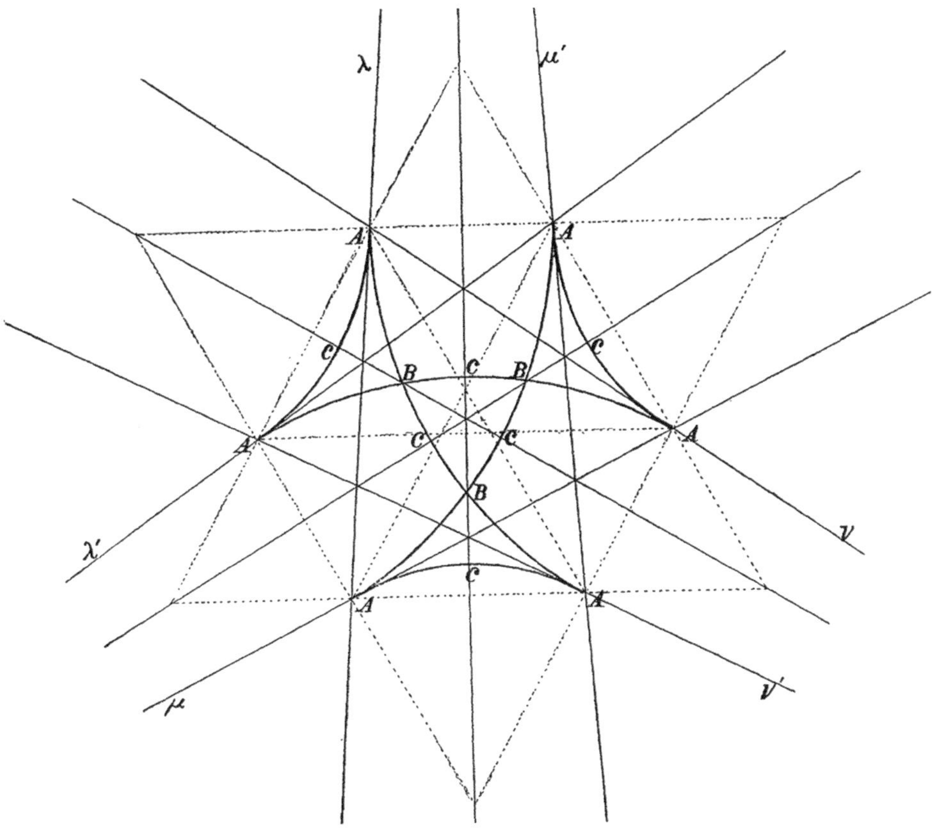

Fig. 11.5 The dual to the quartic in Fig. 11.4 (Haskell 1890, 28).

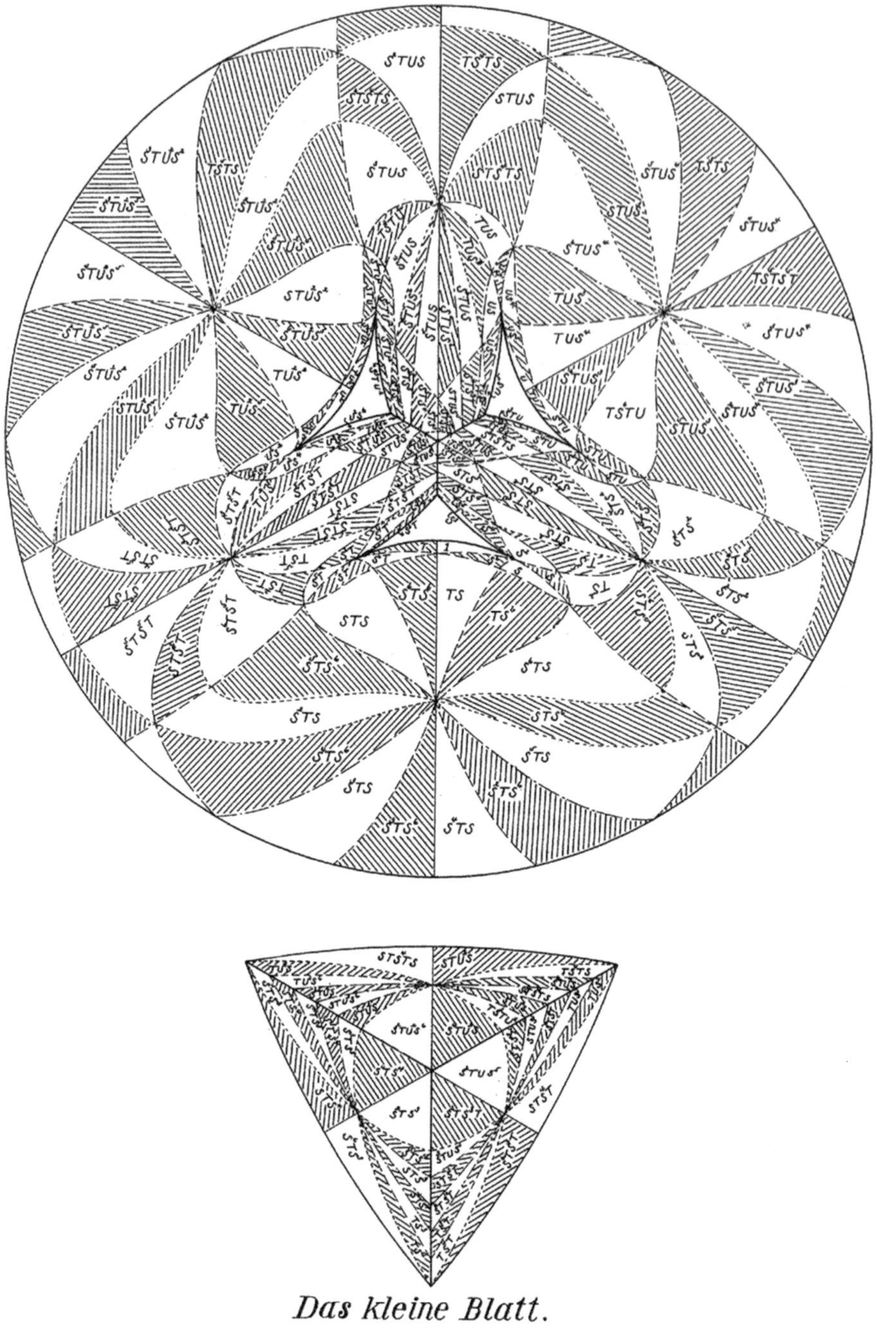

Fig. 11.6 Two of the four leaves of the projective Riemann surface (Haskell 1890, Tafel II).

One of the first Americans to study with Klein was Frank Nelson Cole, who arrived in Leipzig in 1884 on a fellowship from Harvard. After a year abroad, he returned to finish his doctoral degree and shared his newly attained knowledge in the form of two courses based on material from Klein's lectures. These attracted several members of the Harvard faculty as well as two undergraduates, William Fogg Osgood and Maxime Bôcher, both of whom would later become fixtures at Harvard. Cole's courses evidently inspired them to study in Göttingen, where Osgood attended Klein's three-semester course on Abelian functions. He was one of only six students and five of these six happened to be Americans.

Osgood opted to complete the entire course, but his compatriot, Harry W. Tyler, decided to move on to Erlangen. While there, Tyler wrote letters to Osgood detailing his impressions of Erlangen's mathematical atmosphere as compared with the far livelier one in Göttingen (Harry W. Tyler to William Fogg Osgood, 20 February and 1 March, 1889, Osgood Nachlass, SUB, Göttingen). Like Osgood, Tyler found Klein's geometric style appealing, but he also recognized its lack of rigor. In any case, he had little trouble convincing his friend to leave Göttingen and join him in Erlangen, where both subsequently earned their degrees: Tyler under Germany's then leading invariant theorist, Paul

Fig. 11.7 Mellen Woodman Haskell studied under Klein in Göttingen from 1886 to 1890. Afterward he joined the faculty of the University of California, Berkeley, where he succeeded Irving Stringham as chair in 1909. Haskell remained chair of the Berkeley mathematics department until his retirement in 1933.

Gordan, and Osgood under Max Noether. On returning to the United States, Tyler took a position at the Massachusetts Institute of Technology (MIT), where he eventually became departmental chair, whereas Osgood returned to Harvard and taught there for the next 40 years.

By the end of the 1880s, Klein's interests were shifting away from function theory toward other topics, including mathematical physics. He took up potential theory, pursuing an approach that had briefly occupied him during his early Leipzig years. Inspired by his reading of Thomson and Tait's *Treatise on Natural Philosophy*, Klein sketched a broad geometrical framework within which one could embrace all the usual coordinate systems of potential theory as special cases. He lectured on these ideas during the winter semester of 1889–1890, a course attended by Maxime Bôcher. Afterward Klein wrote up the philosophy behind his theory of Lamé functions in (Klein 1890), leaving it to Bôcher to work out the details. Although he had studied potential theory earlier under William E. Byerly and Benjamin O. Peirce at Harvard, Bôcher hardly brought with him all the various tools he would need to tackle Klein's theory, which required a mastery of the theory of elementary divisors, boundary value problems in partial differential equations, Lamé polynomials, Lamé products, and more. Indeed, the problems involved proved far too complicated for an exhaustive treatment within the limited space of the prize-winning doctoral dissertation he completed in 1891. Bôcher therefore continued to work on this topic for three more years following his return to a faculty position at Harvard. His efforts led to the classic 1894 volume, *Über die Reihenentwicklungen der Potentialtheorie*, a masterpiece of mathematical exposition that represents the high watermark of American achievement under the tutelage of Felix Klein.

While Cambridge, Massachusetts clearly formed one important focus for Klein's influence on American mathematics, another such locale was Wesleyan College in the southern New England town of Middletown, Connecticut. There, the astronomer John Monroe Van Vleck taught three undergraduates who went on to take doctoral degrees under Felix Klein in Göttingen: Henry Seely White, Frederick Shenstone Woods, and his own son, Edward Burr Van Vleck (father of the Nobel laureate in physics, John Van Vleck). Henry White had originally planned to study under Sophus Lie and Eduard Study in Leipzig (White 1944).[3] During his first semester in Leipzig, however, White learned from a friend about Klein's lectures. So he left for Göttingen in the summer of 1888, joining Osgood, Tyler, Haskell, and Princeton's Henry Dallas Thompson as the fifth American taking Klein's course on Abelian functions. White later recalled that "Klein expected hard work, and soon had in succession Haskell, Tyler, Osgood, and myself working up the official *Heft* or record of his lectures, always kept for reference in the mathematical *Lesezimmer*. This gave the fortunate student extra tuition, since what Klein gave in one day's lecture (2 h) must be edited and elaborated and submitted for Klein's own correction and revision within 48 h" (White 1944, 23–24). After taking his degree in 1891, White taught briefly at Clark University before moving on to Northwestern and finally Vassar College. A leading American geometer, he enjoyed the distinction of serving as the ninth President of the AMS.

Woods and Van Vleck arrived in Göttingen somewhat later, so after Klein's interests had shifted to potential theory and differential equations arising in mathematical physics. Woods studied differential geometry, and his dissertation,

[3]In fact, Lie's lectures on the theory of transformation groups steadily attracted Americans from the time he succeeded Klein at Leipzig in 1886 right up until 1898 when Lie returned to his native Norway. At least twenty Americans – including Leonard Dickson and G. A. Miller – attended his lectures during this 12-year period, and six (James Morris Page, Hans Blichfeldt, Edgar O. Lovett, Charles Leonard Bouton, David Andrew Rothrock, and John van Elten Westfall) wrote doctoral dissertations under him.

submitted in 1895, dealt with pseudo-minimal surfaces. Along with Harry Tyler, he played an important role in upgrading the quality of the mathematics program at MIT. The Woods and Bailey calculus text (Woods and Bailey 1907–1909) also exerted considerable influence nationwide. Van Vleck spent five semesters in Göttingen, during which he took courses from nearly every mathematician and scientist on the faculty. He completed his dissertation on analytic expansions in continued fractions under Klein (Van Vleck 1894). Although he was not a prolific mathematician, Van Vleck's work on this topic and in other areas of analysis was regarded highly by his peers (Birkhoff 1938, 295. After graduation, he taught briefly at Wisconsin, then from 1895 to 1906 at his alma mater, Wesleyan, and then finally in Wisconsin again, where he spent the remainder of his career.

Two of Klein's German protégés, Oskar Bolza and Heinrich Maschke, also promoted research mathematics in the US after joining the faculty at the newly founded University of Chicago, which opened its doors in 1892. As postdoctoral students, Bolza and Maschke had studied together with Klein during the academic year 1886–87. Since neither felt encouraged by the prospects for climbing the German academic ladder, both eventually opted to seek employment in America. Thus in April 1888, Bolza arrived in Baltimore armed with nothing more than a letter of introduction from Klein. Soon afterward, Simon Newcomb wrote Klein in very negative terms about Bolza's chances: "I never advise a foreign scientific investigator to come to this country, but always tell him that the difficulties in the way of immediate success are the same that a foreigner would encounter in any other country. . . . We have indeed several hundred so-called colleges; but I doubt . . . if one half of the professors of mathematics in them could tell what a determinant is. All they want in their professors is an elementary knowledge of the branches they teach and the practical ability to manage a class of boys, among whom many will be unruly" (Simon Newcomb to Felix Klein, 23 April 1888, Klein Nachlass XI, SUB).

Bolza taught for just one semester at Hopkins as an unpaid "Reader in Mathematics before accepting a promising three-year appointment at Clark University, which was just opening in 1889. Clark employed several strong mathematicians during this period, including William Story, Joseph de Perott, Henry Taber, and Henry White. Unfortunately, Clark's President, G. Stanley Hall, its benefactor, Jonas Clark, and its trustees had difficulty sorting out their respective responsibilities, leading to discontent within the faculty. When the newly named President of the University of Chicago, William Rainey Harper, decided to visit the Clark campus, he easily managed to attract most of Clark's outstanding scholars, including Bolza, to Chicago. In the meantime, Maschke had also immigrated to the United States, where he had taken a job as an electrical engineer. Bolza managed to work out a deal with Harper and the new acting head of the Chicago Mathematics Department, Eliakim Hastings Moore, whereby both he and Maschke obtained appointments.

Thus, in the fall of 1892, when the University of Chicago first opened, these two protégés of Felix Klein constituted two-thirds of its mathematical faculty. This triumvirate can be seen in the back row of this group photo, taken one year later when Klein attended the Congress of Mathematics held in connection with the World's Columbian Exposition in Chicago. Immediately afterward, he spent two weeks at Northwestern University, where he delivered the Evanston Colloquium Lectures (Fig. 11.8). Klein's stay in the Chicago area was only part of a whirlwind trip to the United States in 1893. This event effectively marked the end of his career as a master teacher, a role he gladly relinquished to Hilbert starting with the latter's arrival in Göttingen in 1895.

Klein's Leipzig Seminar

We now backtrack to the years when Klein's rivalry with the Berlin mathematicians first began (see Chap. 4). The 15-year period that preceded Hilbert's call to Göttingen was one in which he achieved a great deal, but also suffered a number of serious defeats. It began with his appointment to a new professorship in geometry created in Leipzig, where he spent six fruitful, but also personally demanding years. Riemannian ideas had by now come to play a central role in his work (including Klein's new "projective" Riemann surfaces). These had been crucial for his research in algebraic geometry, but also for his studies of modular functions and algebraic equations. In Leipzig, he turned to investigate the foundations of Riemann's theory of algebraic functions by considering fluid flows on closed surfaces, an approach he made famous in his booklet *Über Riemanns Theorie der algebraischen Funktionen und ihrer Integrale* (Klein 1882). Klein's innovative approach was later adopted by Hermann Weyl in his influential book, *Die Idee der Riemannschen Fläche* (Weyl 1913).[4]

[4]Weyl's preface, written in sparkling German prose, is worth quoting in full: "Die Riemannsche Fläche ist ein unentbehrlicher *sachlicher* Bestandteil der Theorie, sie ist geradezu deren Fundament. Sie ist auch nicht etwas, was a posteriori mehr oder minder künstlich aus den analytischen Funktionen herausdestilliert wird, sondern muß durchaus als das prius betrachtet werden, als der Mutterboden, auf dem die Funktionen allererst wachsen und gedeihen können. Es ist freilich zuzugeben, daß Riemann selbst die wahre Verhältnis der Funktionen zur Riemannschen Fläche durch die Form seiner Darstellung etwas verschleiert hat—vielleicht nur, weil er seinen Zeitgenossen allzu fremdartige Vorstellungen nicht zumuten wollte; dies Verhältnis auch dadurch verschleiert hat, daß er nur von jenen mehrblättrigen, mit einzelnen Windungspunkten über der Ebene sich ausbreitenden Überlagerungsflächen spricht, an welche man noch heute in erster Linie denkt, wenn von Riemannschen Flächen die Rede ist, und sich nicht der (erst später von Klein zu durchsichtiger Klarheit entwickelten)

Fig. 11.8 Felix Klein (front center) at the mini-congress held in 1893 during the Chicago World's Fair. Immediately behind him sits John Monroe Van Vleck;flanking Klein stand two former students: Henry White (left) and Harry Tyler (right). Sitting to the right in the back row are (l. to r.): Oskar Bolza, E.H. Moore, and Heinrich Maschke (Parshall and Rowe 1994, 316).

During the academic year 1880–1881, Klein's first in Leipzig, he began to immerse himself in this topic. Full of enthusiasm and big plans, he surprised his colleagues by launching his career with a two-semester course on Riemannian function theory that attracted 75 auditors. Alongside this course, he offered a seminar on various topics in geometry and complex function theory. This drew a number of advanced students, including Adolf Hurwitz and Walter Dyck, who had followed Klein from Munich to Leipzig. There were also a few participants from foreign countries, including Giuseppe Veronese, then an assistant under Luigi Cremona in Rome, and Washington Irving Stringham, a recent graduate of Johns Hopkins University, where he took his doctorate under J. J. Sylvester. These two foreigners presented their own recent research in Klein's seminar, whereas a young Frenchman named Georges Brunel reported on topics he was just learning. Although Brunel remained in Leipzig for only one academic year, this proved to be a decisive period, not only for him but also for Klein and above all for Brunel's young compatriot, Henri Poincaré, as well. As it turned out, the lives of all three would become intertwined in a dramatic way; for it was during this summer semester that Klein and Poincaré began their famous correspondence, which documents an important chapter in the early history of automorphic functions.

Brunel had attended schools in Abbeville, Lille, and Paris before entering the École normale supérieure in 1877, where Paul Appell and Émile Picard were a few years ahead of him. Moreover, Brunel was one of the first normaliens to continue his studies in Germany. Shortly before Christmas 1880, he spoke in Klein's seminar about Riemann's approach to the genus of surfaces and its role in the theory of algebraic curves. Clearly, Klein suggested this topic along with some relevant literature that Brunel used in preparing his talk. Not long afterward, in late January 1881, Brunel spoke for a second time on a related topic. On this occasion, he began with topological ideas in Riemann's theory of Abelian functions before turning to their generalization to higher dimensions in the hands of Enrico Betti. He also described earlier related work by the Göttingen physicist and mathematician Johann Benedict Listing, a pioneering figure in the history of topological studies. One can easily recognize Klein's interests here, including his fascination with the older Göttingen tradition. In fact, Betti and Riemann first met in Göttingen, though their friendship grew far closer during Riemann's final years in northern Italy.

Brunel had already spent a good deal of time in Leipzig before he wrote to Poincaré, and he made it plain that his stay abroad had not been easy (G. Brunel to H. Poincaré, 22 June 1881, Archives Henri-Poincaré, Nancy). At the same time, he clearly felt a deep urge to serve his country while behaving

---

allgemeineren Vorstellung bediente, als deren Charakteristikum man dieses nennen kann: daß in ihr die Beziehung zu der Ebene einer unabhängigen komplexen Veränderlichen, sowie überhaupt die Beziehung zum dreidimensionalen Punktraum grundsätzlich gelöst ist. Und doch ist darüber kein Zweifel möglich, daß erst in der Kleinschen Auffassung die Grundgedanken Riemanns in ihrer natürlichen Einfachheit, ihrer lebendigen und durchschlagenden Kraft voll zur Geltung kommen. Auf dieser Überzeugung basiert die vorliegende Schrift" (Weyl 1913, iv–v).

properly as a guest in a foreign land. As he explained to Poincaré, he came to Leipzig hoping to learn what German mathematicians had to teach the French, a view altogether consonant with that taken by Hermite, who urged his pupils to follow the new currents of research pursued on the other side of the Rhine. In Brunel's case, he clearly found himself in the right place at the right time, for Klein was intent on breaking new ground as professor of geometry in Leipzig. During the summer semester, when Klein's lectures moved into the heart of Riemann's theory, nearly all the talks in his seminar were concerned with topics closely tied to his course. The one striking exception was Brunel's presentation, which dealt with Georg Cantor's theory of point sets, highlighting his proof that the algebraic numbers constitute a countable subset within the uncountable infinite numbers of the continuum. Besides several works by Cantor, Brunel also cited papers by several other German authors, including Jakob Lüroth, Eugen Netto, and Enno Jürgens. He also cited the paper in which Joseph Liouville gave his classical method for constructing transcendental numbers. These new studies of the real number continuum constitute a second new field of research that had not yet made deep inroads into French mathematics, making Georges Brunel a likely envoy for this mission.

In fact, when he returned to Paris the next year he took a position as agrégé-préparateur de mathématique at the École normale. Then followed an appointment as chargé du cours de mécanique at the École des Sciences in Algiers, where he completed his doctoral thesis on a topic in complex analysis. Finally, in 1884 Brunel obtained the chair for pure mathematics in the faculty of sciences at Bordeaux. His predecessor, Jules Hoüel, was perhaps the leading authority on non-Euclidean geometry in France, having translated works by both Lobachevsky and Bolyai. Brunel remained in this position until his early death at the age of 43. His colleague, the distinguished physicist, historian and philosopher of science, Pierre Duhem, wrote a lengthy obituary in which he noted that Brunel had published 97 works during his short lifetime (Duhem 1902).

During the course of this summer semester of 1881, Klein came across three notes that Poincaré had published earlier that year under the title *Sur les fonctions fuchsiennes*. With them, Poincaré had begun to lay the groundwork for a comprehensive new theory of complex-valued functions, which remain invariant under an infinite discontinuous group of linear fractional transformations. Klein immediately wrote the young Frenchman on 12 June to inform him of his own previous work in this connection, a letter that marks the beginning of their famous correspondence, which lasted until September of the following year.[5] At the time Klein initially contacted him, Poincaré was a little-known 27-year-old mathematician living in Caen. Although he had studied Fuchs' work on the problem of determining when a linear differential equation possesses algebraic solutions, Poincaré had relatively little knowledge of Klein's research or that of other German function theorists. Through Klein, however, he quickly became aware of such fundamental contributions as Schwarz's triangle-functions, Klein's own work on elliptic modular functions, and even some of the deeper aspects of Riemann surface theory. All this, Poincaré managed to assimilate without any apparent effort.[6]

Already in his second letter to Poincaré, written on 19 June 1881, Klein informed him that Georges Brunel had been studying in Leipzig and that he would be able to give Poincaré details about Klein's research program. More specifically, he mentioned a manuscript from a course Klein had taught two years earlier in Munich, adding that he would discuss this with Brunel so that the latter could inform Poincaré of its contents. It was in this same letter that Klein also objected to the name the young Frenchman had given to automorphic functions with a natural limiting circle. Poincaré had dubbed these "fonctions fuchsiennes" in honor of Lazarus Fuchs, a choice Klein simply could not accept. This refusal soured the mood in the letters that followed, and Klein's persistence eventually produced a public debate in the pages of *Mathematische Annalen*. The four letters from the summer of 1881 that Brunel sent to Poincaré provide a vivid impression of how he felt compelled to take the French side in this conflict while striving to give an objective account of what he saw and heard. As an eyewitness to Klein's initial anger over this whole affair, his accounts shed a good deal of light on the emotional dimensions of what Poincaré would later call "un débat stérile pour la science" (Poincaré to Klein, 4 April 1882).

While Klein's letters proved a fertile source of information for Poincaré, the tone with which the German mathematician imparted his knowledge and insights must have grated on the nerves of his younger rival at times. In particular, Klein raised a mighty fuss over Poincaré's decision to name an important class of functions after Fuchs, a leading exponent of the Berlin school who was then teaching in Heidelberg. Already in his second letter to Poincaré, Klein flatly rejected this terminology ("[d]ie Bennenung 'fonctions fuchsiennes' weise ich zurück ..."), thereby initiating a dispute that would have major repercussions (Klein to Poincaré, 19 June 1881, (Klein 1923, 590–593)). After pressing him on this point, Poincaré admitted that he would have chosen a different name for the functions had he then known of Schwarz's work (Poincaré to Klein, 27 June 1881, (Klein 1923, 594)). Clearly, he had no ulterior motive in

[5]The letters were first published in Klein (1923, 587–621), and then immediately afterward in *Acta Mathematica* 39 (1923), 94–132.

[6]For a discussion of this work and the Klein-Poincaré correspondence, see Gray (2000).

invoking Fuchs' name. In fact, he had even obtained Fuchs' permission to do so, making any kind of public retraction out of the question. Still, he was blissfully unaware not only of Klein's work but also the methodological divide that stood between it and that of Fuchs. Poincaré soon discovered that his generosity had unleashed a swarm of controversy among German mathematicians whose work he barely knew.

Klein had longstanding relations with a number of leading French mathematicians, including Gaston Darboux and Camille Jordan, dating back to the spring of 1870 when he and Sophus Lie were in Paris together. In writing to Poincaré, he recalled that trip and expressed his interest in extending his contacts in France to leading analysts, especially Hermite and Picard. As he confided to Poincaré, he had thought about striking up a scientific correspondence with Hermite, but inability to express himself in the French language had inhibited him from doing so; Hermite, unlike Poincaré, had a quite limited knowledge of German. For his part, Hermite expressed delight when he learned that Klein and Poincaré were corresponding; he had high regard for Klein's works relating to modular equations and the general quintic (C. Hermite to H. Poincaré, 30 June 1881, Archives Henri-Poincaré, Nancy). At this time, Hermite apparently had no idea that Klein had criticized Poincaré for choosing the name "fonctions fuchsiennes."

Klein was sincerely interested in establishing stronger relations with leading French analysts, but he had a quite different and much stronger motivation for writing to Poincaré. More than anything, he was deeply concerned that his own recent work – much of which had been published in *Mathematische Annalen* – might be overlooked by other mathematicians, particularly those in Hermite's circle. Considering the strong ties between the Parisian and Berlin mathematicians, Klein had good reason to fret over this. His unwillingness to accept the newly baptized *fonctions fuchsiennes* as belonging to Fuchs clearly reflects the larger rivalry within the German mathematical community. This divide was already clearly visible in 1868 when Alfred Clebsch and Carl Neumann founded *Mathematische Annalen* as a counterweight to the Berlin-dominated *Journal für die reine und angewandte Mathematik*, established already in 1826 by A.L. Crelle. The Annalen, now led by Klein, was strongly supported by those who felt estranged from the Berlin mathematicians and their extensive network of influence within the Prussian universities (see Chap. 4 and the introduction to Part I).

Georges Brunel provided Poincaré with a very frank assessment of how Klein saw these matters. In particular, Klein "complained that the young Frenchmen did not know what had been published in Germany, and that the existence of the Mathematische Annalen was probably not known in France" (G. Brunel to H. Poincaré, 22 June 1881, Archives Henri-Poincaré, Nancy). Brunel felt very annoyed by these remarks, but held his tongue. He wrote Poincaré how he was tempted to respond that even in Berlin Klein's journal was viewed as wholly problematic. As for French ignorance of the works in Crelle, where did he think Poincaré had read the works of Fuchs? Still, most of his comments about Klein's behavior suggest that Brunel held him in high respect despite his Prussian manner; indeed, his view of Klein was far more favorable than his general opinion of the Germans he had encountered.

Brunel reported further how Klein wished his French counterparts – Poincaré, Appell and Picard – would not just announce their results in the *Comptes Rendus*, but also write about their methods. Such a wish Brunel found quite odd, since Klein clearly knew there was no space available for anything more than brief announcements. Yet, for Klein, the issue of competing methods was a vital point, one that Brunel passed on to Poincaré: "The fundamental idea is due to Riemann, and to Schwarz goes the merit of applying Riemann's idea. Later, I myself worked in this same direction and in my courses at the Munich *Polytechnikum* I introduced some results that are the basis of the work of Mr. Poincaré. As for Mr. Fuchs, who once wanted to deal with similar issues, he only succeeded in showing us that he understood nothing of these things." (*Ibid.*)

Thus, Klein's primary motivation for writing to Poincaré in the first place was surely to make him aware of various related work that he and others in Germany had been pursuing in the realm of Riemannian function theory, particularly complex functions that remain invariant under groups of linear fractional transformations. Once informed of these studies, Poincaré immediately acknowledged their importance and relevance, even though his own independent research had arisen from quite another direction. He had been inspired by the work of Fuchs, who had found a wide class of differential equations that admit algebraic solutions. These studies, in turn, had grown out of longstanding interests in hypergeometric functions, a topic taken up in papers by Gauss, Kummer, and Riemann (Gray 2000, Chap. 1).

Somewhat later, beginning in the 1890s, Klein and others began to unearth unpublished documents that showed how both Gauss and Riemann had anticipated many results that were discovered by others independently. Thus, reconstructing the early history of the theory of automorphic functions became a matter of special interest to Klein, as can readily be seen from his lectures on nineteenth-century mathematics (Klein 1926), delivered during the First World War. Klein's picture of these developments clearly exalts the work of his heroes, Gauss and Riemann, while somewhat marginalizing the line of ideas that led to Poincaré's great works, the five papers he published during the years 1882 to 1884 in *Acta Mathematica*.[7]

The rivalry that first brought Klein and Poincaré together lasted little more than a year, but this was a period of great significance for both men. For Poincaré, it saw him launch the general theory of automorphic functions, one of his most stunning achievements and a major turning point for complex analysis. Klein had already approached this terrain some years earlier, though from the special vantage point of elliptic modular functions, a topic he pursued by combining group-theoretic methods with Riemannian function theory. As co-editor of *Mathematische Annalen*, he was also keenly aware of other related work, particularly studies made by other German mathematicians. Poincaré had very limited familiarity with most of this literature, as quickly became apparent when he wrote to Klein. In a letter Klein wrote on 2 July 1881, he ended by wondering whether Poincaré was even aware that Riemann's theta-functions depended on the *3p-3* parameters given by the modules of an algebraic curve of genus *p*.[8] Here and elsewhere in these first letters, Klein imparted a good deal of information, but with an authoritative tone that betrayed his annoyance with Poincaré, who seemed to be rushing ahead to publish new things without taking time to study the literature. In fact, Klein had little inkling at this time of what Poincaré had already discovered on his own.

Poincaré's Breakthrough

Already a year before Klein wrote to him, Poincaré had struck up a scientific correspondence with Lazarus Fuchs, whose work on linear differential equations had long interested Poincaré's teacher, Charles Hermite.[9] Sparked by Fuchs' investigations, Poincaré made the first discoveries that enabled him to create the theory of automorphic functions (Gray 2000). By June 1880, he had come to recognize the key importance of hyperbolic geometry for understanding the special class of functions defined on the interior of a circle, beginning with a special case generated by a circular-arc triangle. At that time he wrote that "I propose to call such a function fuchsian . . . The fuchsian functions are to the geometry of Lobachevsky that which the doubly periodic functions are to that of Euclid" (Poincaré 1997, 12). A month later, in a letter to Fuchs, he described how one could start with a suitable polygon rather than a triangle to generate "les fonctions fuchsiennes."

Poincaré kept Hermite informed about his correspondence with Fuchs, though his former teacher freely admitted that he could not form any opinion about Poincaré's appeal to non-Euclidean geometry as a way to explore questions in analysis and number theory (Hermite to Poincaré, 20 July 1880, Archives Henri-Poincaré). One might easily imagine that Hermite's resistance to such reasoning was a factor that later led Poincaré to suppress the important role of hyperbolic geometry when he began to write up his researches the following year. On the other hand, his very first letter to Klein from 15 June 1881 emphatically highlighted this very connection: "I know how well versed you are in knowledge of non-Euclidean geometry, which is the veritable key to the problems which occupy us" (Poincaré to Klein, 15 June 1881, (Klein 1923, 590)).

Many years later, Poincaré recalled the unusual circumstances that led him to this discovery:

> Just at this time I left Caen ... to take part in a geological expedition organized by the École des Mines. The incidents surrounding the journey made me forget my mathematical work. Having arrived at Coutanees, we boarded an omnibus bound for some place or other. At the moment when I put my foot on the step, the idea came to me, without anything in my previous thoughts seeming to have paved the way for it, that the transformations I had used to define the Fuchsian functions were identical with those of non-Euclidean geometry. I did not verify this idea. I did not have time to do so, as, upon taking my seat in the bus, I resumed a conversation already begun, but I was entirely certain at once. On returning to Caen, to soothe my conscience, I verified the result at my leisure (Poincaré 1908, 51–52).

What Poincaré saw as he stepped into the bus was the connection between the motions of certain circular-arc polygons inside a disc and the figures that arise in a now familiar model of plane hyperbolic geometry. This discovery took place in June of 1880, so exactly one year before his correspondence with Klein began, which accounts for Poincaré's opening claim, namely that non-Euclidean geometry was "the veritable key to the problems which occupies us." He evidently made that remark believing that Klein probably already recognized this same connection before him. Klein affirmed in his response that the analogy with non-Euclidean geometry was familiar to him, but if so, then he surely failed to take full advantage of it.

Brunel met several times privately with Klein and acted as a kind of mediator between him and Poincaré, whom he did not know personally. Thus, he introduced himself in his first letter as a fellow comrade of two older *normaliens*, Paul Appell and Émile Picard, whom Poincaré knew well. Together with Klein, Brunel examined notes from Klein's earlier course on the theory of modular functions, which he had taught in Munich. However, they found only traces

[7]One should add, however, that Klein had originally planned to discuss Poincaré's work at length, but then broke off this plan in 1916 in order to engage with fast-breaking developments in the wake of Einstein's work on general relativity; see Klein (1921, 553–612).

[8]Poincaré likely did not know this, but he explained in his reply from 5 July that he merely needed an upper bound for the argument in his paper, which he easily derived as $4p + 4$.

[9]The Fuchs-Poincaré correspondence was published by G. Mittag-Leffler in *Acta Mathematica*, 38(1921): 175–187; see the discussion in the introduction to Poincaré (1997, 7–11).

of ideas Klein claimed he already had at that time. As an explanation for this, Klein told Brunel: "I did not want to overburden my students, but I already had the ideas that we are talking about" (G. Brunel to H. Poincaré, 22 June 1881, Archives Henri Poincaré, Nancy).

At their next meeting, though, Klein expressed his annoyance at reading about certain ideas in the *Comptes Rendus*, which Brunel took to mean that he was lamenting the fact that he had not published his own general ideas earlier. The young Frenchman realized that Poincaré, who was then in Caen, had no ready access to foreign literature, so he assured him that he would send anything he could get his hands on in Leipzig. He summed up his patriotic feelings in this tense situation with these words:

> As Frenchmen it is our duty to fight the Germans by all possible means, but fairly. By which I mean that we must forthrightly acknowledge what they have accomplished, but we must also not attribute everything to them. If in his theory of modular functions Mr Klein has already published certain special results in the theory of Fuchsian functions, I find it only fair that you should render him justice, as you say. If he has not gone further, so much the worse for him! (G. Brunel to H. Poincaré, 22 June 1881).

Considering how Klein had done pioneering work showing the connection between non-Euclidean spaces and projective geometry, it may seem a little surprising that he failed to exploit that link. Gray offers one possible explanation for this, namely Klein's tendency to rely on projective thinking (Gray 2000, 299–300). On the other hand, Poincaré only published his famous model for non-Euclidean geometry more than a year later. In fact, it first appeared in his paper of July 1881, entitled *Sur les groups kleinéens*, although not in the more familiar two-dimensional form but as a three-dimensional model. In the two-dimensional model, one studies the reflections of circular-arc polygons whose edges, when extended, intersect a fixed circle – the *Grenzkreis* – at right angles. These motions can be regarded as isometries in a hyperbolic space. Klein only introduced various metrics (parabolic, hyperbolic, and elliptic) a bit later in his programmatic memoir, *Neue Beiträge zur Riemannschen Funktionentheorie* (Klein 1883, 179–180); there, but only in a footnote, he alludes to the connection with Poincaré's use of non-Euclidean geometry.

"Name ist Schall und Rauch"

The squabble over names quickly became complicated. Klein knew from Friedrich Schottky about other functions invariant under a discontinuous group of transformations but which fail to possess a limit circle. When he pointed out such functions to Poincaré, the latter took the liberty of naming these *fonctions kleinéennes*. This failed to mollify Klein, who expressed astonishment when he learned that his name had been attached to a class of functions he had never written about. So he insisted on using the "old-fashioned" terminology of functions invariant under a group of linear transformations instead of the words "fuchsiennes" and "kleinéennes" (Klein to Poincaré, 9 July 1881, (Klein 1923, 601)). He only introduced the term "automorphic functions" somewhat later, in (Klein 1890).

In late 1881, Klein invited Poincaré to write a summary of his recent results for publication in *Mathematische Annalen*. The latter agreed and sent Klein a paper bearing the "old-fashioned" title: *Sur les fonctions uniformes qui se reproduisent par des substitutions linéaires*. Before this article went to press, however, Klein forewarned Poincaré that he had appended a note to it explaining his objections to the terminology the latter had adopted (Klein to Poincaré, 13 January 1881, (Klein 1923, 606–606)). In this note, he characterized such names as "premature" for the following reasons[10]: ". . . On the one hand, all the investigations that Herr Schwarz and I have published point in the same direction as the field of 'fonctions fuchsiennes', about which Herr Fuchs has never published anything. On the other hand, I have published nothing on the more general functions to which Herr Poincaré has attached my name; I merely brought the existence of these functions to his attention."

Klein then underscored that all such geometrically motivated investigations, as opposed to the more analytical studies of the solutions of linear differential equations, should be regarded as rooted in Riemannian conceptions (Gray 2013, 233). Regarding Poincaré's work, he went even further, asserting that his studies could be seen as developing the general function-theoretic program that Riemann had first set out in his doctoral dissertation. In short, Klein argued that Poincaré's work fell squarely within the Riemannian approach to function theory, a geometric style to which he himself identified completely. Since he saw no trace of this in the work of Lazarus Fuchs, he felt certain that Poincaré had simply blundered: Fuchs was a conventional analyst whose work clearly reflected his roots in the tradition of the Berlin school. From Klein's point of view, Poincaré had naively allowed Fuchs, who had no grasp of Riemannian methods, to misappropriate results that belonged to geometric function theory. Schwarz, on the other hand, clearly did deserve recognition for his publications in this direction, so Klein's polemic could have been read as taking sides in the personal conflict between Fuchs and Schwarz, who were not on good terms.

The tone of Klein's remarks was polite, but forceful, whereas Brunel's letters make plain how furious he was that Poincaré had innocently bestowed this laurel wreath on Fuchs. Clearly, the importance he attached to this matter went far beyond the bounds of a conventional priority

[10]Klein's note appears in Poincaré (1928, 104–105).

dispute. True, he was concerned that his own work would receive sufficient acclaim, but, as his final remarks revealed, the overriding issue hinged on whether the mathematical community would regard the burgeoning research in this field as an outgrowth of Weierstrassian analysis or as falling within the province of the Riemannian tradition. Klein had by now become deeply committed to the notion that Riemann's work represented the principal source of that peculiar brand of "anschauliche" mathematics that Klein had made all his own. Indeed, he hoped somehow to unlock the "spirit" behind Riemann's mathematics, the motive that inspired his exposition of Riemannian function theory in terms of current flows on surfaces. Fuchs, on the other hand, stood in the mainstream of the Berlin tradition, and Klein could not afford to stand idly on the sidelines while the Berlin – Paris axis heaped unearned plaudits on each other's work. As the foremost representative of the Clebschian tradition, he knew that recognition in the world of German mathematics seldom came easily, particularly if one's field of research or style of exposition ran counter to current fashion.

Poincaré, of course, had little understanding of and even less interest in these internal conflicts among the Germans. For him, these just seemed like petty disputes over names. While he raised no immediate objections to Klein's provocative commentary, he did request that he be given the opportunity to respond in the pages of *Mathematische Annalen*. These exchanges proceeded politely enough, but in the meantime, Fuchs entered the fray by publishing a note in which he sought to rebut Klein's position. In a letter to Poincaré, Klein characterized Fuchs' argument as "completely amiss". As for Poincaré's own views, as expressed in the note he published in *Mathematische Annalen*, Klein cut abruptly to the heart of the matter (Klein to Poincaré, 3 April 1882, (Klein 1923, 609)). In his opinion, Poincaré would never have suggested Fuchs' name had he known the relevant literature from the beginning. Nor could he accept, so to speak as compensation (Entschädigung), Poincaré's friendly offer of the "fonctions kleinéennes", since this name, too, carried a false historical implication.

Though surely annoyed by these remarks, Poincaré remained unruffled. He denied that "Entschädigung" had motivated his decision to coin the term "fonctions kleinéennes", but he also expressed relief that they could put this controversy over a mere name behind them and thus move on to more substantive concerns. "Name ist Schall und Rauch" he wrote, quoting a famous line from Goethe's *Faust*.[11] Adolf Hurwitz, on the other hand, took some delight in Klein's polemic *contra* Fuchs, which reminded him of a popular children's song with a new twist on the stolen goose[12]:

> Fuchs, Du hast die *Funktion* gestohlen
> Nun, gieb sie wieder her ...
> (Fox, you stole that [goose]
> Now give her back to me ...)

Within the German mathematical community, on the other hand, the fallout from this affair had significance consequences for Klein's future relations with mathematicians allied with the Berlin school. Not surprisingly, Lazarus Fuchs was most unhappy about this whole turn of events, and in a letter to Mittag-Leffler, he gave vent to his anger (Fuchs to Mittag-Leffler, 27 December 1882, Mittag-Leffler Correspondence, Institut Mittag-Leffler):

Klein's most recent accomplishment with respect to me is so clumsy and held in such a tone, as luckily has not hitherto been custom among mathematicians. It would be so easy to expose his sophistries and arrogance, but I regard it as beneath my dignity to do so. Any expert will see the distortion of the truth behind all these machinations. I will leave it to people of Klein's caliber if they wish to influence the opinions of those who understand nothing of these matters.

In the spring of 1882, Klein planned to spend part of his vacation working on a new presentation of the existence theorems for algebraic functions on Riemann surfaces as a sequel to his earlier booklet. He traveled to the North Sea island of Norderney, but wrote to Adolf Hurwitz in a despondent mood, complaining about how he seemed to lack a definite goal for his work. He concluded by saying "... actually the Poincaré material should give me enough to do, but I don't have the slightest desire to go into it" (Klein to Hurwitz, 14 March 1882, Mathematiker-Archiv, SUB, Göttingen). Perhaps even worse for Klein's vacations plans, he found himself trapped on the island in stormy weather that brought on violent asthmatic attacks. After 8 days of misery, he decided to leave for Düsseldorf the following morning. Unable to fall asleep, he sat up on a sofa contemplating the 14-sided figure he had discovered three years earlier in connection with his work on the seventh-degree modular equation. Suddenly he recognized its broader significance for algebraic curves of higher genus: he saw the basic ingredients for the *Grenzkreistheorem* (Klein 1923, 584). This result built upon Poincaré's earlier achievements: for the latter had already shown how circular arc-polygons, like Klein's 14-sided figure, can be used to generate Fuchsian groups that tessellate the hyperbolic plane. Moreover, as noted above, Poincaré's key insight had also come to him in a sudden flash of inspiration.

By March of 1882, when Klein discovered the *Grenzkreistheorem*, Poincaré's work had already made clear the central importance of non-Euclidean geometry for the theory of automorphic functions. A few days after the fateful sleepless night he spent on Norderney, Klein wrote a short note announcing the result and had proof sheets sent to

[11] Poincaré to Klein, 4 April 1882, (Klein 1923, 611); the motto comes from *Faust*, Part I, line 3457.

[12] Hurwitz to Klein, Klein Nachlass IX, SUB Göttingen.

Poincaré, Schwarz, and Hurwitz. Initially, Schwarz had doubts about the counting of certain constants, but after overcoming these misgivings, he suggested an important simplification of Klein's argument by means of a universal covering space. Klein passed this information on to Poincaré in a letter of 14 May, and the latter eventually employed it in his own work on uniformization theory.

In studying the uniformization of algebraic curves by means of automorphic functions, Klein's overall strategy involved an attempt to establish a certain one-to-one correspondence between two higher-dimensional manifolds. Before he could hope to accomplish this, however, he needed assurance that the dimensions of the two manifolds actually agreed (thus the concern about proper counting of constants). One of these two manifolds, the space of algebraic curves of genus $p > 1$, clearly had complex-dimension $3p - 3$, though Schwarz had disagreed with the assertion Klein made in his booklet claiming that these form a connected manifold. Klein, in fact, continually emphasized the importance of this fact, later calling it "the root of the fundamental theorem" (Klein 1923, 579–580).

After his breakthrough on the North Sea, Klein apparently found little difficulty formulating an even more general theorem, which he already outlined to Poincaré in a letter of May 7. This "fundamental theorem" subsumed the two uniformization theorems he had announced earlier, the *Rückkehrschnitt-Theorem* and the *Grenzkreistheorem*, as special cases. A week later, Klein wrote Poincaré again, this time giving more information about his "proof" and emphasizing that the correspondence he had constructed between the two higher-dimensional manifolds had to be analytic as well (Klein to Poincaré, 14 May 1882, (Klein 1923, 615)). Thus, by mid-May, Klein was struggling to thrash out some of the insuperable difficulties associated with proving this general uniformization theorem. Over the next several months, he delivered a series of lectures on this subject in the Leipzig Mathematical Seminar. These lectures were first transcribed and edited by Eduard Study, and then reworked by Klein during his fall vacation in Tabarz, a small village just west of Gotha. The finished product, Klein's lengthy article "Neue Beiträge zur Riemannschen Funktionentheorie" (Klein 1883), was completed in October of 1882. Klein circulated offprints of it in late November shortly before the appearance of the first of Poincaré's five lengthy articles in *Acta Mathematica*.

The central idea behind Klein's argument in (Klein 1883) turned out to be reasonably sound, but impossible to carry through.[13] These early uniformization theorems would only be proved much later with the help of new topological techniques. Klein had assumed that the correspondence between his two manifolds was analytic in nature. As Erhard Scholz has pointed out, however, this assumption proved to be a dead end. The key ideas came from Brouwer's subsequent work, particularly his proof of the topological invariance of dimension, which laid the groundwork for the final breakthrough by Poincaré and Koebe some 25 years later. These same results of Brouwer, not accidentally, also proved essential new tools for Weyl, who shortly thereafter succeeded in placing the theory of Riemann surfaces on a firm, rigorous foundation.

Although his *Neue Beiträge* was surely one of the more impressive papers Klein ever wrote, he paid a very heavy price for his exertions. No doubt, his sense of "duty" and a deeply ingrained work ethic drove him on. Later in life, he blamed his physicians for having proffered faulty advice, such as cold baths and gymnastics, when they should have been paying more attention to his digestive disorders.[14] Nor did he have any luck following their suggestions for a change of climate. Whether he took in the salty air of the North Sea or the gentle breeze of a mountain village like Tabarz, his asthma always seemed to get worse rather than better. Still, in the final analysis, Klein had no one but himself to blame for the ensuing "breakdown". His collapse merely came as the climactic episode in a long-term process in which he continually pushed himself to the edge of his physical capacity for work. Despite his delicate constitution, he failed over and again to heed the warning signs of intermittent poor health. In Leipzig, he had even taken on numerous new teaching and administrative responsibilities.

When he returned from his vacation in the autumn of 1882, the geometer Friedrich Schur found him so exhausted he was surprised to learn that he had not applied for additional time off. Still, he went on teaching a course on applications of calculus to geometry for another month before allowing his assistant, Walther von Dyck, to take over for him. Klein continued to conduct meetings of his seminar on Abelian functions, all the while assuming that his health would improve. It did, but only very slowly. In the meantime, he felt forced to sit back and watch as Poincaré filled the pages of *Acta Mathematica* with his remarkable memoirs. Having sketched some of the circumstances surrounding the Klein-Poincaré correspondence, we can take a closer look at Klein's relationship with Acta's editor, Gösta Mittag-Leffler.

On Cultivating Scientific Relations

Klein and Mittag-Leffler had been moving in very different scientific circles throughout their early careers, so they knew

[13] See the discussion in Scholz (1980, 205–222).

[14] Was ist die Ursache der Krankheit? Doch die Überanstrengung durch vielseitige Tätigkeit bei intensivstem math. Denken. Statt mit Ruhe machen es die Aerzte mit Anregung: Kalte Bäder, Turnen in Aussicht, Verdauung unberücksichtigt"; Klein's notes ‚Vorläufiges über Leipzig", p. 2, Klein Nachlass XXII L, SUB, Göttingen.

one another more by reputation than through personal contacts. One opportunity to meet did arise in July of 1880, when Klein was just finishing out his final semester in Munich and Mittag-Leffler paid him a short visit there. Klein was always on the lookout for new allies and scientific stimulation, and so he looked forward to the visit with enthusiasm (Klein to Mittag-Leffler, 4 July 1880, Mittag-Leffler Papers, Institut Mittag-Leffler): "... With these lines, I would like to express how precious your visit would be for me, not just for a few hours, but for a few days. You stand in the middle of current mathematics and you can give me an essential revitalization, as I have hardly in the last years come out of the narrow circle of my closest acquaintances."

At this time, Klein and Mittag-Leffler appear to have been on cordial terms, and when the first issue of *Acta Mathematica* appeared Klein received a copy along with a friendly letter (Mittag-Leffler to Klein, 15 December 1882, Klein Nachlass X, SUB, Göttingen):

> I have the honor of hereby sending you the first issue of my journal. In the future, I will have each issue sent to you as soon as it appears. I hope you will be so good as to send me your journal in exchange. I will also take the liberty of sending you offprints of those papers of special interest to you as these appear. I thank you most heartily for the paper you recently sent me. I plan to study it carefully during the four weeks of vacation that I now have.

Klein's reply, written during the immediate aftermath of his collapse from overwork and exhaustion, certainly reflected a willingness to cooperate with his new competitor. He may have felt uneasy about the implications of a prospective Paris-Stockholm-Berlin axis within the European mathematical community, but if so, he certainly did not let on. Perhaps he even sympathized with Mittag-Leffler's efforts to facilitate the restoration of international scientific cooperation between France and Germany. On 30 December 1882, Klein wrote (Klein to Mittag-Leffler, 30 December 1882, Mittag-Leffler Papers, Institut Mittag-Leffler):

> You have anticipated me with your suggestion of an exchange of *Acta Mathematica* and our *Annalen*. Today I'm pleased to be able to inform you that Teubner has agreed to this without reservation and that they have sent you the recently published first issue of volume XXI! The second issue (which ... contains my paper) will soon follow. Up until now, I only saw Poincaré's first paper, which of course interested me greatly, and I'm very curious to read the sequels. As for me, I won't for the time being do any further work in this direction because of other tasks that require attention but also because I urgently need a certain period of rest. But later I plan to take up these investigations again in connection with ideas developed by you and Cantor. For even though I primarily strive to orient myself as broadly as possible in the fields of mathematics, the center of my research efforts will for years to come lie in the area of function theory....

From this exchange of letters, one might assume that Mittag-Leffler and Klein had managed to overcome their territorial instincts and the natural rivalry that existed between their two journals. Klein's letter, a typical example of how he dealt with a major figure from the opposition, suggests no reservations about cooperating with Mittag-Leffler's new venture; indeed, his words imply that he was more than happy to do so. Klein genuinely enjoyed cultivating scientific relations with whomever he could, and he nearly always evinced a direct and open manner with mathematical colleagues whether friends, allies, or members of a rival camp.

In contrast with Klein's straightforward style, Mittag-Leffler's conduct in his scientific relationships exuded the charm and sophistication of a born diplomat. In his letters, one often finds him using flowery phrases and a good deal of flattery to win someone over. Since he also had a penchant for gossip, some of which he manufactured and disseminated himself, this style occasionally backfired and turned former friends and potential allies against him. A prominent example was Sophus Lie, who served on the editorial board of *Acta Mathematica* but later became embittered by the manner in which Mittag-Leffler ignored his counsel. Leopold Kronecker, never one to take an offense lightly, broke off relations with Mittag-Leffler when he felt that the latter had betrayed his trust (Kronecker to Mittag-Leffler, 29 July 1885, Mittag-Leffler Papers, Institut Mittag-Leffler).

Probably the most significant instance of this kind involved Georg Cantor. Cantor enjoyed a cordial relationship with Felix Klein and published several papers in *Mathematische Annalen*, the most significant of these being his six-part study *Über unendliche, lineare Punktmannigfaltigkeiten*. Mittag-Leffler hoped to win Cantor over to his side, so he offered to publish French translations of Cantor's work in *Acta Mathematica*. He also tried to sow doubts in Cantor's mind about Klein's interest in his work by writing:

> ... allow me to add a few a few remarks as your true friend and *only for you*. I believe that the philosophical part of your work will excite a great deal of interest in Germany, but I don't believe that the mathematical part will do the same. Aside from Weierstrass and perhaps Kronecker, who however has essentially little interest in these questions and will hardly share your opinions, there are no mathematicians in Germany who possess the fine sense for difficult mathematical investigations needed in order for a correct understanding of your works. For example, a few years ago Klein told me – this is naturally strictly between us – that he couldn't see what purpose all that served. Among Weierstrass's pupils probably Schottky is the only one that will have some understanding for your work. I can very easily imagine how our mutual friend Schwarz will scold you.
>
> But in France things stand differently. There, at this moment, a very lively movement in the mathematical world is taking place. Poincaré, Picard, Appell ... Goursat, and Halphen, just to name the most prominent of Hermite's pupils, are all extremely gifted men with indeed much sense for the finest mathematical researches. Poincaré, in my view at least, is a genius of the first rank. All of them, as well as Hermite himself, will take the liveliest interest in your discoveries, precisely because they now need such investigations as they have encountered difficulties in their beautiful function-theoretical works that can only be overcome through your works. (Mittag-Leffler to Cantor, 10 January 1883, quoted from (Meschkowski 1983, 242–243).

Mittag-Leffler went on to say that, with the exception of Appell, all the mathematicians in Hermite's circle had a poor command of German. This assertion, while applicable to Hermite himself, was certainly incorrect when it came to Poincaré and Picard. At any rate, Mittag-Leffler's plea proved successful, and later that year he published no fewer than eight French translations of articles by Cantor. The reaction of the French mathematicians, however, was precisely the opposite of the one Mittag-Leffler had predicted. Poincaré, for example, thought that Cantor's "numbers in the second, and especially in the third, number-class have the appearance of being form without substance, something repugnant to the French mind" (Poincaré to Mittag-Leffler, 5 March 1883). Hermite, writing on behalf of Appell, Picard, and Poincaré, stated flatly: "[t]he impression Cantor's memoirs make on us is disastrous. Reading them seems, to all of us, to be an utter torture ..." (Hermite to Mittag-Leffler, 13 April 1883, (Dugac 1984, 209); see the discussion in Moore (1989)).

Mittag-Leffler had clearly overplayed his hand, and his machinations, in this case, were to have serious consequences. In early 1885, Cantor sent a short communication to Mittag-Leffler for publication in *Acta Mathematica*. Mittag-Leffler advised him not to put these ideas into print, as they would, in his opinion, "damage [Cantor's] reputation among mathematicians" (Mittag-Leffler to Cantor, 9 March 1885, quoted in (Grattan-Guinness 1970, 102)). Cantor never got over this experience, which marked the effective end of his career as a creative mathematician. Joseph Dauben wrote about this: "Cantor was deeply hurt by Mittag-Leffler's rejection of his latest research. More than his polemic with Kronecker, more than his nervous breakdown or the trouble he was having in finding a proof for his continuum hypothesis, Mittag-Leffler's suggestion that he not print his new article in the *Acta Mathematica* was the most devastating" (Dauben 1979, 138).

In 1883, the same year that Mittag-Leffler published several French translations of Cantor's work, Poincaré wrote to him suggesting that *Acta Mathematica* publish a French translation of Klein's *Erlanger Programm* (Klein 1872). Only one year before this, when Sophus Lie was visiting Paris, Klein's Norwegian friend reported to him that Poincaré had never read this now classic work.[15] In fact, as Thomas Hawkins first showed, relatively few mathematicians had encountered it before the early 1890s. Only then was it reprinted, but also published in a number of different foreign languages, beginning with the Italian translation by Gino Fano in 1890 (Hawkins 1984). Poincaré's letter began:

> I am sending you a pamphlet by Mr. Klein. [Cyparissos] Stephanos thought that a translation of this little-known booklet would be of interest to readers of Acta and he has offered to translate it. I also believe that this pamphlet, though written in a somewhat obscure style, contains ideas that are very justified and unfortunately not very widespread. I thus believe that giving these ideas some publicity would be useful to many geometers, and especially to the French (H. Poincaré to G. Mittag-Leffler, 14 August 1883, Archives Henri-Poincaré, Nancy).

In response, Mittag-Leffler only sent the following mysterious statement: "I am very grateful for Mr. Stephanos' kind offer to translate into French the work you sent me by Mr. Klein. It will, however, be very difficult for me to publish this translation." (G. Mittag-Leffler to H. Poincaré, 25 August 1883, Archives Henri-Poincaré, Nancy). Klein soon learned about how Mittag-Leffler had aborted these plans to publish his *Erlanger Programm* in *Acta*. Its editor then received a curious and rather stiffly worded letter from Klein alluding to this, but also carrying the insinuation that Mittag-Leffler no longer wished to publish translations of any works whatsoever (Klein to Mittag-Leffler, 8 October 1883, Mittag-Leffler Papers, Institut Mittag-Leffler):

> After a long absence from Leipzig, I notice that in the meantime no issues of Acta Mathematica have arrived, whereas based on the off-prints I have received I would expect that a few have since appeared. The last issue I received was 1.4. Proceeding on the assumption that some kind of irregularity has occurred, I kindly ask you to arrange for its elimination. Recently, as far as I know, Mr. Poincaré wrote you at the instigation of Mr. Stephanos about my Erlanger Programm (1872). I was gratified by the idea of the Parisian mathematicians that a translation might be published in Acta because I am interested that the general ideas underlying my program be disseminated as widely as possible. Now, however, having heard that for the time being you do not wish to publish any translations, I gladly accede to this. Would you please take my acceptance as proof that I would like as much as possible to turn the natural competition between Acta and the Annalen into a form of cooperation.

Since no reply from Mittag-Leffler is preserved among Klein's papers, all that can be said with certainty in this connection is that Klein never published any of his works in *Acta Mathematica*. A French translation of his *Erlanger Programm* only appeared some 10 years later. A few months later, Klein was consulting with Lie about the possibility of republishing some of their earlier work in the pages of *Mathematische Annalen*. Lie was eager to cooperate, but he also imparted this advice: "If you intend to let your older works appear successively in *Mathematische Annalen*, then don't you want to publicize your [Erlanger] *Programmschrift* there? It is surely your most important work from the period 1872, and it would now be better understood than at that time." (Lie to Klein, 20 January 1884, Klein Nachlass X, 695, SUB Göttingen). For reasons that remain obscure, however, Klein chose not to follow his friend's counsel.

[15]See Chap. 10: Three Letters from Sophus Lie to Felix Klein on Mathematics in Paris.

Delayed Publication of the Klein-Poincaré Correspondence

The cooperation Klein had hoped for between *Acta Mathematica* and *Mathematische Annalen* would never materialize; over the ensuing years, he corresponded very little with Mittag-Leffler. Klein wrote him once in 1885 to complain again that he had not received the latest issues of *Acta* (Klein to Mittag-Leffler, 21 June 1885, Mittag-Leffler Papers, Institut Mittag-Leffler). After he accepted a chair in Göttingen in 1886, Klein and Mittag-Leffler appear to have had even less contact with one another. In fact, their next significant exchange of letters did not take place until late 1900 when Mittag-Leffler wrote to ask Klein to contribute a scientific autobiography for a new series he was planning. He had in mind to publish a collection of such articles, similar to the French *Notices sur mes travaux scientifiques* composed by leading mathematicians. Not surprisingly, he hoped to begin the series with Poincaré, but he also assured Klein:

> I have so far only asked a very small number of the most important mathematicians for their contribution. I must also watch out not to get too much material all at once. My idea has been greeted with approval by all, and some writers have already begun to work. I'll start with Poincaré; he works very quickly and will probably be finished soon. A commission of Parisian mathematicians has been formed to help Hermite, who is evidently already too old to complete this unaided. My request of you is that you not only give me the support that you always provide in such ample measure to all matters useful to our science, but also, and above all, that you write your own autobiography as quickly as possible. I would be delighted to publish it in the same volume as Poincaré's. I imagine that your autobiography would comprise 100 to 200 pages in the format of Acta. (Mittag-Leffler to Klein, 13 December 1900, Klein Nachlass X, SUB Göttingen)

After thinking it over, Klein declined this invitation, pleading a lack of time (Klein to Mittag-Leffler, 4 January 1901, Mittag-Leffler Papers, Institut Mittag-Leffler):

> I've been thinking a lot about your proposal. I can see that in such a way very interesting material would come together; I have always read the Notices of the French scholars with particular enjoyment. Unfortunately, for the foreseeable future, I am not in a position to satisfy your wishes. I can imagine that in later years I might reprint all my works (but then, incidentally, I would make it my goal to add supplementary comments about the works of others). With such an arrangement, the type of collection you wish for would result more or less automatically. For the time being, however, I will long be fully occupied with work of a different kind. I won't even mention how much I have to do for the Encyclopedia: my main work remains my teaching and caring for the development of mathematics here. That other mathematicians, in particular Poincaré, can move with such ease from one problem to another I admire very much, but I cannot imitate it. Thank you for your offer and all the best wishes for the success of your undertaking.

Mittag-Leffler's ambitious project never really got off the ground. Nevertheless, he did receive Poincaré's contribution in 1901, which he later returned to the author. It was recovered from Poincaré's papers at the time of his death in 1912, and finally published under the title *Analyse des travaux scientifiques de Henri Poincaré* in 1921 in a special issue of *Acta* dedicated to Poincaré's memory (Fig. 11.9). This volume contains articles by such luminaries as Jacques Hadamard, H.A. Lorentz, and Max Planck, and excerpts from letters Poincaré exchanged with Mittag-Leffler and Lazarus Fuchs. Noticeably absent from this collection, on the other hand, were the letters from his correspondence with Klein during the early 1880s. The reason for this forms part of a complex story that is worth telling in some detail.

Mittag-Leffler, in fact, had been planning this special volume of *Acta* for many years, but a variety of circumstances, in particular the outbreak of World War I, had delayed its publication. Already back in November 1913, he wrote to Klein asking for permission to publish his correspondence with Poincaré in such a volume (Mittag-Leffler to Klein, 19 November 1913, Klein Nachlass X, SUB, Göttingen):

> I am busy preparing a volume of my Acta, which will be devoted entirely to the memory of Henri Poincaré. There are two periods in the course of his scientific career where I was involved more than otherwise. The first was the period during which what you call automorphic functions were created; the second was when the prize essay on the three-body problem was produced. I will endeavor to place all documents on the table at once quite impartially, so that the present-day mathematical public will be able to comprehend everything as it really was. It is better to do so now, when several who played a role during these epochs in the history of the mathematical sciences are still living, than for this to happen later. In any case, the many letters and oral communications that stem from these periods belong to the history of science and they will certainly be published sooner or later.
>
> You belong to the first period, the period when the automorphic functions were created, in a very special way.
>
> My suggestion is, therefore, that you allow me to publish all the letters you exchanged with Poincaré during that time. Through Poincaré, I have been familiar with your letters to him for many years. There are hardly any copies of Poincaré's letters, but, of course, you have kept all the originals. All the letters between Fuchs and Poincaré are at my disposal, and they will appear in the relevant volume of Acta. Beyond these, I will also publish letters I received relating to these questions, where I will omit such expressions, but only such, as might perhaps offend still living authors.

At this time, Klein was no longer lecturing in Göttingen. He had spent most of the previous year at a health resort in the Harz Mountains. On returning, he decided to retire in the spring of 1913. He answered Mittag-Leffler's letter as follows (Draft of Klein to Mittag-Leffler, 25 November 1913, Klein-Nachlass X, SUB, Göttingen):

> With regard to the letters addressed to me by Poincaré in 1881–82, I received an inquiry long ago from Pierre Boutroux.[16] I

[16]Boutroux was the nephew of Poincaré. He knew Klein from having spent some weeks in Göttingen; see Chap. 16: Poincaré Week in Göttingen, 22–28 April 1909.

Fig. 11.9 Henri Poincaré (1854–1912).

could only give him a very inadequate answer, which I repeat here: that I placed these letters together in the autumn of 1905, when we conferred the Bolyai Prize on Poincaré and I was collecting material for a report on his work, and now I cannot find them again. A systematic search is useless, as I have already applied all possible care to this. Following the example of similar cases (or a theorem of Poincaré), one must expect that they will one day reappear by themselves, and I will gladly make them available to you as well as to P. Boutroux.

I agree in principle with the printing of my letters to Poincaré. I would only ask that you send me the proofs so I can see how they look and control the details.

Perhaps I should remark why my correspondence with Poincaré suddenly broke off in the summer or autumn of 1882. The cause: my serious sickness due to overwork, which left me for a long time unable to work and from which I recovered only very gradually.

Klein apparently had the letters from Poincaré in his immediate possession as late as 1907. In 1905, as a member of the commission responsible for the first awarding of the Bolyai Prize, Klein led a seminar that reviewed the principal mathematical achievements of Poincaré and Hilbert, the only two serious contenders for the award.[17] Klein presented the Poincaré letters on this occasion, but he may well have referred to them afterward, as this seminar constituted the first of four that he conducted along with Hilbert and Minkowski between 1905 and 1907. One of the main topics they considered in all four seminars concerned Klein's work on linear differential equations and automorphic functions, and on several occasions, Klein spoke on this subject.

Shortly after Mittag-Leffler wrote to him regarding the Poincaré letters, Niels Erik Nörlund contacted Klein with the same request (Nörlund to Klein, 20 December 1913, Klein Nachlass X, SUB, Göttingen). Nörlund, then a professor of mathematics in Lund and an expert in function theory, had studied in Göttingen and Paris between 1910 and 1912. The Paris Academy of Sciences had recently appointed him as editor of Poincaré's works on automorphic functions and their applications. Unfortunately, Klein could only repeat what he had already told Boutroux and Mittag-Leffler (Draft of Klein to Nörlund, 8 January 1914, Klein-Nachlass X, SUB, Göttingen):

It is, of course, in my own interest to clarify the relationship between Poincaré and me regarding the history of the discovery of automorphic functions and their principal theorems as much as possible. You will find a first attempt to do so on the basis of the printed materials in the new report by R[obert] Fricke in volume II of the Mathematical Encyclopedia.

As for the correspondence between Poincaré and me, there is a peculiar difficulty, about which I already wrote to Emile Boutroux[18] when he contacted me about half a year ago regarding an edition of Poincaré's writings. I removed the letters Poincaré wrote to me during the years 1881–82 from my other scientific correspondence for a seminar on automorphic functions, which I held together with Hilbert and Minkowski, and now I can no longer find them!

The letters remained missing throughout the war years, a situation that must have displeased everyone concerned. Scientific relations between Germany and France were by now at an all-time low. As one of the ninety-three representatives of German higher learning who had allowed their names to appear on the notorious appeal "An die Kulturwelt" at the beginning of the war, Klein had been summarily dismissed from the French Academy. Meanwhile, Mittag-Leffler, still trying to play the role of intermediary between Paris and Berlin, was busy spreading nasty rumors and innuendo. At

[17]Extensive material from this seminar can be found in the Klein-Nachlass XXII K, SUB, Göttingen.

[18]Emile Boutroux was the husband of Poincaré's sister, Aline. Klein apparently was contacted either by Boutroux or his son, Pierre, a mathematician who had spent time in Göttingen.

the very end of the war, he wrote the following in a letter to Max Planck:

> On this occasion, I cannot help but tell you a very unpleasant story. There existed a very detailed correspondence between Felix Klein and Henri Poincaré about what are called "automorphic functions" in Germany or "fonctions fuchsiennes et kleinéennes" in France. The letters of Klein were all in Poincaré's estate. They are, from a purely scientific point of view, very weak, and Klein, who pushes Poincaré from top to bottom, making claims that cannot possibly be supported. There are no copies of Poincaré's letters to Klein. Klein, who is usually very careful when it comes to preserving his very extensive correspondence, can no longer find the Poincaré letters. They have disappeared completely. I do not doubt that he has no real fault here, but how can one explain something like this to those who are not very friendly toward German science? That a man like Klein could allow letters of such high scientific importance as Poincaré's to just disappear is especially unfortunate when one considers that the publication of these letters would undoubtedly prove unpleasant for Klein's scientific reputation. Madame P [oincaré] reported to me that P [oincaré] told her several times: "I do not want to do anything that could be harmful to Klein," but I find it only right that he should personally receive the necessary clarifications. (Mittag-Leffler to Planck, 30 March 1919, quoted from (Dauben 1980, 278–279).

In another letter written earlier to Klein's colleague and former Weierstrass pupil, Carl Runge, Mittag-Leffler argued that the republication of this correspondence in *Acta Mathematica* could help build a bridge between the severely divided scientific communities in France and Germany (Dauben 1980, 285). Klein learned through Runge that Mittag-Leffler had been spreading accusations against him, and this led him to drop a hint of his displeasure about this in a letter to Nörlund. The latter, having assumed that Klein's remark had been directed against him, reacted with surprise and dismay: "There is a remark in your letter that astonished me very much and which has hurt me. In the matter of Poincaré's letters, I have never entertained such thoughts as you suggest. They are very far from my mind, and I have not the slightest reason for thinking so. If you or anyone else in Göttingen has gotten a different impression, I am very sorry about that." (Nörlund to Klein, 17 March 1919, Klein Nachlass X, SUB, Göttingen). Klein added a marginal note to this letter that makes it clear he understood the true circumstances: "But Mittag-Leffler wrote this to Runge directly! (namely, that I deliberately kept secret the P. letters)."

The precise circumstances that led to the recovery of the lost letters remain unclear; probably this story will never be clarified completely. One might guess that Klein made a determined search for them as he began preparations for the third and final volume of his *Gesammelte mathematische Abhandlungen*. Therein he devoted considerable attention to the "prehistory" of automorphic functions, the nature of his own contributions, and the issues that arose in his correspondence with Poincaré. Klein published those letters in that volume, but he also agreed to send copies to Mittag-Leffler so that the latter could also publish all the letters in a special issue of *Acta Mathematica*; both publications appeared in 1923. By that time, of course, nothing remained of the goodwill between French and German mathematicians that had made such a fruitful exchange of ideas possible.

Archival Sources

Klein Nachlass, Niedersächsische Staats- und Universitätsbibliothek, Göttingen (SUB).
Osgood Nachlass, SUB, Göttingen.
Mathematiker-Archiv, SUB, Göttingen.
Henri Poincaré Papers, Archives Henri-Poincaré, Université de Lorraine, Nancy.
Correspondence, Mittag-Leffler, Institut Mittag-Leffler, Djursholm, Sweden.

References

Birkhoff, George David. 1938. Fifty Years of American Mathematics. In *Semicentennial Addresses of the American Mathematical Society*, ed. C. Raymond. Archibald: American Mathematical Society.

Dauben, Joseph W. 1979. *Georg Cantor: His Mathematics and Philosophy of the Infinite*. Cambridge, Mass.: Harvard University Press.

———. 1980. Mathematics and World War I: The International Diplomacy of G. H. Hardy and Gösta Mittag-Leffler as Reflected in their Personal Correspondence. *Historia Mathematica* 7: 261–288.

Domar, Yngve. 1982. On the Foundation of Acta Mathematica. *Acta Mathematica* 148: 3–8.

Dugac, Pierre, ed. 1984. Lettres de Charles Hermite à Gösta Mittag-Leffler (1874–1883), *Cahiers du Séminaire d'Histoire des Mathématiques* 5: 49–285.

Duhem, Pierre. 1902. Notice sur la vie et les travaux de Georges Brunel (1856–1900). *Mémoires de la Société des sciences physiques et naturelles de Bordeaux* 2 (6): 1–89.

Geiser, Carl. 1869. Ueber die Doppeltangenten einer ebenen Curve vierten Grades. *Mathematische Annalen* 1: 129–138.

Grattan-Guinness, Ivor. 1970. An unpublished Paper by Georg Cantor: Principien einer Theorie der Ordnungstypen. Erste Mittheilung. *Acta Mathematica* 124: 65–107.

Gray, Jeremy. 1982. From the History of a Simple Group. *The Mathematical Intelligencer* 4 (2): 59–67.

———. 2000. *Linear Differential Equations and Group Theory from Riemann to Poincaré*. 2nd ed. Boston: Birkhäuser.

———. 2013. *Poincaré: A Scientific Biography*. Princeton: Princeton University Press.

Haskell, Mellen W. 1890. Ueber die zu der Curve $\lambda^3\mu + \mu^3\nu + \nu^3\lambda = 0$ im projectiven Sinne gehörende mehrfache Ueberdeckung der Ebene. *American Journal of Mathematics* 13: 1–52.

Hawkins, Thomas. 1984. The *Erlanger Programm* of Felix Klein: Reflections on its place in the History of Mathematics. *Histoira Mathematica* 11 (4): 442–470.

Klein, Felix. 1872. *Vergleichende Bertrachtungen über neuere geometrische Forschungen*. Erlangen: Deichert. Reprinted in *Mathematische Annalen* 43 (1893): 63–100 and in (Klein 1921, 460–97).

———. 1873. Über Flächen dritter Ordnung. *Mathematische Annalen* 6: 551–581.

———. 1874. Über eine neue Art der Riemann'schen Flächen. *Mathematische Annalen* 7: 558–566.

———. 1876a. Über den Verlauf der Abel'schen Integrale bei den Curven vierten Grades. *Mathematische Annalen* 10: 365–397.

———. 1876b. Über eine neue Art von Riemann'schen Flächen. *Mathematische Annalen* 10: 398–416.

———. 1879. Über die Transformationen siebenter Ordnung der elliptischen Funktionen. *Mathematische Annalen* 14: 428–471; Reprinted in (Klein 1923, 90–136).

———. 1882. *Ueber Riemanns Theorie der algebraischen Funktionen und ihrer Integrale*. Teubner: Leipzig; Reprinted in (Klein 1923, 499–573).

———. 1883. Neue Beiträge zur Riemann'schen Functionentheorie. *Mathematische Annalen* 21: 141–218.

———. (1890). Zur Theorie der allgemeinen Laméschen Funktionen, *Nachrichten der Gesellschaft der Wissenschaften zu Göttingen, Mathematisch-Physicalische Abteilung* (1890): 85–95, Reprinted in (Klein 1922, 540–549).

———. 1893. A Comparative Review of Recent Researches in Geometry," English translation of (Klein 1872) by M. W. Haskell". *Bulletin of the American Mathematical Society* 2: 215–249.

———. 1921. *Gesammelte Mathematische Abhandlungen*. Bd. 1 ed. Berlin: Springer.

———. 1922. *Gesammelte Mathematische Abhandlungen*. Bd. 2 ed. Berlin: Springer.

———. 1923. *Gesammelte Mathematische Abhandlungen*. Bd. 3 ed. Berlin: Springer.

———. 1926. *Vorlesungen über die Entwicklung der Mathematik im 19. Jahrhundert*. Bd. 1 ed. Berlin: Springer.

Kochina, P.Ya., and E.P. Ozhigova, eds. 1984. *Perepiska S. V. Kovaleskoy i G. Mittag-Leffler, Nachnoye nasledstvo, vol. 7*. Moscow: Nauka.

Levy, Silvio, ed. 1989. *The Eightfold Way: The Beauty of Klein's Quartic Curve*. Berkeley: Mathematical Sciences Research Institute Publications.

Meschkowski, Herbert. 1983. *Georg Cantor. Leben, Werk und Wirkung*. Mannheim: Bibliographisches Institut.

Moore, Gregory H. 1989. Towards a History of Cantor's Continuum Problem. In *The History of Modern Mathematics*, ed. David E. Rowe and John McCleary, vol. 1, 79–122. Boston: Academic Press.

Nabonnand, Philippe, ed. 1999. *La Correspondance entre Henri Poincaré et Gösta Mittag-Leffler*. Birkhäuser: Basel.

Parshall, Karen, and David Rowe. 1994. *The Emergence of the American Mathematical Research Community, 1876–1900. J.J. Sylvester, Felix Klein, and E.H. Moore*, AMS/LMS History of Mathematics Series. Vol. 8. Providence: American Mathematical Society.

Poincaré, Henri. 1908. *Science et méthode*. Paris: Flammarion.

———. 1928. In *Oeuvres de Henri Poincaré*, ed. N.E. Nörlund, vol. 2. Paris: Gauthier-Villars.

———. 1997. In *Three Supplements on Fuchsian Functions*, ed. Jeremy J. Gray and Scott A. Walter. Berlin: Akademie-Verlag.

Rowe, David E. 1992. Klein, Mittag-Leffler, and the Klein-Poincaré Correspondence of 1881–1882. In *Amphora. Festschrift für Hans Wussing*, 598–618. Basel: Birkhäuser.

Schläfli, Ludwig. 1863. On the Distribution of Surfaces of the Third Order into Species, in reference to the presence or absence of Singular Points and the reality of their Lines. *Philosophical Transactions of the Royal Society* 153: 193–241.

Scholz, Erhard. 1980. *Geschichte des Mannigfaltigkeitsbegriffs von Riemann bis Poincaré*. Basel: Birkhäuser.

Stubhaug, Arild. 2010. *Gösta Mittag-Leffler, A Man of Conviction*. New York: Springer.

Tobies, Renate, and David E. Rowe, eds. 1990. *Korrespondenz Felix Klein – Adolf Mayer, Teubner Archiv zur Mathematik, 14*. Leipzig: Teubner.

Van Vleck, Edward Burr. 1894. Zur Kettenbruchentwickelung hyperelliptischer und ähnlicher Integrale. *American Journal of Mathematics* 16: 1–91.

Weyl, Hermann. 1913. *Die Idee der Riemannschen Fläche*. Leipzig: Teubner.

White, Henry Seely. 1944. Autobiographical Memoir of Henry Seely White (1861–1943). *Biographical Memoirs of the National Academy of Sciences* 25: 16–33.

Woods, Frederick S., and Frederick H. Bailey. 1907–1909. *Course in Mathematics for Students of Engineering and Applied Science*. Vol. 2. Boston: Ginn and Co..

Zeuthen, H.G. 1874. Sur les différentes formes des courbes plane du quatrième ordre. *Mathematische Annalen* 7: 410–432.

Part III

David Hilbert Steps Onstage

12 Introduction to Part III

With the exceptions of Newton, Gauss, and Sofia Kovalevskaya, probably no mathematician's biography is better known than the charmed career of David Hilbert.[1] Considering his striking personality and pervasive influence on twentieth century mathematics, this is hardly surprising. Yet most of what has been written about him, aside from analyses of his work, concerns his later life in Göttingen.[2] Thus, the essays in this part of *A Richer Picture*, much like those in Part II on the young Felix Klein, aim to throw fresh light on the earlier years well before he attained fame and glory. Like any ordinary mortal, Hilbert had first to win recognition from senior mathematicians in Germany, most importantly from Klein, who happened to be intent on cultivating his ties with Königsberg through his former student, Ferdinand Lindemann. The latter replaced Hilbert's first mentor, Heinrich Weber, in 1883, a turn of events with significant long-term consequences.

During the decade Hilbert spent with Lindemann, he learned a great deal about number theory and much else besides, though mainly from Lindemann's younger colleague, Adolf Hurwitz. He also learned much through conversations with his friend Hermann Minkowski, who was one year younger. Both Minkowski and Hurwitz had studied briefly in Berlin, whereas Hilbert rarely left Königsberg. Through Minkowski Hilbert gained insight into the mysteries of Kronecker's number-theoretic investigations, whereas Hurwitz introduced him to Weierstrass's still very new approach to complex function theory. As Hilbert later fondly recalled, these early interactions greatly widened his mathematical horizons (Hilbert 1971, 78). This talented trio loved to talk about mathematics and compete with each other in a friendly way. In other respects, though, their personalities differed markedly. Hurwitz and Minkowski were playful spirits who approached mathematics much like artists, whereas Hilbert tackled mathematical problems from a more systematic point of view, hence the importance of axiomatization in much of his later work (Corry 2004). Hilbert was also, far more than his two friends, full of burning ambition.

The fabled friendship that later developed between Hilbert and Minkowski first took root during their student days together in Königsberg, a period of great significance for what came afterward. Memories of this important chapter in Hilbert's life later receded, however, overshadowed by the tumultuous events of the Göttingen years together with the triumphs and tragedies of the new century. Chapter 13 aims to bring these Königsberg years back into focus by way of some telling anecdotes. In these, Paul Gordan, long known as the "King of Invariants," plays a major role, though not the traditional one familiar elsewhere in the secondary literature.[3] Here Gordan emerges as a positive influence, namely as the expert who helped direct Hilbert's ideas rather than criticizing them (that came only afterward). This provides a more convincing picture of Hilbert's entry into the field of algebraic invariant theory, one that casts doubt on the retrospective accounts given later by others, including Hilbert himself (Fig. 12.1).

One can easily pinpoint the dramatic turning point that launched Hilbert's fame, which came quite suddenly. The setting was the Second International Congress of Mathematicians held in Paris and the event Hilbert's lecture on "Mathematical Problems" (Hilbert 1901), surely the most influential address ever given by a mathematician. In Chap. 15, I describe some of the events leading up to it, drawing particularly on Minkowski's letters to Hilbert, which shed much light on the background to this Paris speech and the keen interest Minkowski took in helping shape it. There one reads how Hilbert's friend steered him away from the idea of clashing swords with Poincaré, whose views contrasted sharply with Hilbert's own. Such a clash did come about

[1] Three standard sources are Blumenthal (1935), Reid (1970), and Weyl (1944).

[2] For reflections on three important studies written by German historians of science, see Rowe (1997).

[3] For a recent detailed examination of the myths that later arose due to Gordan's critique of Hilbert's work on invariant theory, see McLarty (2012).

D.E. Rowe, *A Richer Picture of Mathematics*, https://doi.org/10.1007/978-3-319-67819-1_12

Fig. 12.1 David Hilbert, ca. 1885.

Fig. 12.2 Hermann Minkowski, ca. 1885.

later, however, as recounted in Chap. 16, which deals with the events of "Poincaré Week" in Göttingen, when the famous French mathematician was invited to deliver the first series of Wohlskehl lectures.

Poincaré's visit came only months after Minkowski's tragic death in January of 1909, an event that left Hilbert devastated. His closest friend had lived just a stone's throw away and he often stopped by for a chat. In Göttingen they continued where they had left off in Königsberg by taking regular *Spaziergänge* that gave them many opportunities to talk about mathematics. Max Born later recalled how as a student he was invited on an excursion with Hilbert, Minkowski, and their wives to the "Plesse," a castle ruin not far from the town. This gathering made a deep impression on him, especially the witty conversation between the two friends, as they talked about the week's events, both academic and political. Born had heard plenty of chatter about topics like these in his father's circle, but never before had he witnessed such a whirlwind discussion, the "kind of critical and ironical dissection of ideas and conceptions which Hilbert produced when properly stimulated. And Minkowski obviously enjoyed stimulating him, until he went in his rage far beyond his own convictions – then Frau Hilbert stopped him with a sharp 'aber David!' whereupon he often went to the opposite extreme, praising some ridiculous thing he had just condemned" (Born 1978, 83).

Minkowski's mathematical talent had attracted attention long before Hilbert made any great impression on his teachers. Soon after he entered the university in April 1880 – two months before celebrating his sixteenth birthday – Minkowski was drawing praises from Königsberg's senior professor, Heinrich Weber, who wrote Richard Dedekind one year later about this new star on the mathematical horizon. That same year the French Academy announced the following problem for its grand prize competition: to give a proof of Eisenstein's formula for the number of representations of a number as a sum of five squares. Minkowski not only solved this problem, he did so by first classifying all the various types of quadratic forms with integral coefficients that can arise. When he submitted his memoir to the academy, he had yet to celebrate his eighteenth birthday. Its motto – "Rien n'est beau que le vrai; le vrai seul est aimable" – comes from a famous poem by Nicolas Boileau, showing that Minkowski had some familiarity with classical French literature. Unfortunately, however, the he wrote the text itself in German, which meant that his entry might easily have been disqualified on formal grounds. Instead, the jurors convinced themselves that such an important contribution had to receive the Grand Prize, so they awarded it jointly to Minkowski and H. J. S. Smith in 1883 (Fig. 12.2).

Hilbert later devoted considerable space to these events in his obituary for his friend (Hilbert 1910). There he registered his own astonishment at seeing Minkowski's mastery of algebraic methods (the theory of elementary divisors) and transcendental techniques (Dirichlet series and Gaussian sums), ideas that went far beyond the standard repertoire of mathematical knowledge at this time; yet they were needed in order to develop the theory of quadratic forms in its full generality. Hilbert also took note of the fact that not everyone in Paris had been happy with the academy's decision, which led to some uproar in the chauvinistic press. However, Joseph Bertrand and Camille Jordan quickly rebuffed this politicized attack, and the encouragement of the latter led to a warm correspondence between him and the budding young German talent (Hilbert 1910, 447).

Minkowski spent altogether five semesters in Königsberg and three in Berlin. After graduating in 1884, he left Königsberg for Bonn in order to habilitate there. In the meantime, Hilbert remained in the city of his birth to take his doctorate, pass the *Staatsexamen* that qualified him to teach in secondary schools, and then habilitate in 1886. After this, the two friends often wrote about their prospects for gaining a professorship, a goal that eluded both until 1892. As discussed in Chap. 4, 1892 was *the* decisive year for mathematics at the Prussian universities. Kronecker's unexpected death and the long-awaited retirement of Weierstrass marked the end of a great era for mathematics in Berlin; it also set off a chain reaction of appointments elsewhere, including Göttingen. This began when Klein's Göttingen colleague, H. A. Schwarz, accepted the chair formerly occupied by Weierstrass. Lazarus Fuchs was not all happy to have Schwarz as a colleague, but Weierstrass managed to overcome his opposition by noting that the only realistic alternative for the Ministry would be Klein. That settled the matter, as the Berlin faculty was intent on passing over Klein entirely, hardly surprising in view of the partisan conflicts that had long marked his career. For his part, Klein saw Schwarz's departure as a golden opportunity to advance his own plans. He had long been hoping to modernize teaching and research in Göttingen by exploiting its long tradition of excellence in the mathematical sciences, dating back to the time of Gauss.

As a first step toward achieving this, Klein hoped to convince the faculty that they should appoint his star pupil, Adolf Hurwitz, as Schwarz's successor. Much to his dismay, this initial plan backfired: the faculty chose Heinrich Weber rather than Hurwitz. Anti-Semitism was a pervasive obstacle throughout the German universities, and Klein surmised that this might have been decisive in killing Hurwitz's chances, a matter I take up in detail in Chap. 14. When Hilbert learned that Hurwitz had been passed over in Göttingen, he wrote to Klein, "es wär zu schön gewesen" ("it would have been just too wonderful"). Hilbert was thinking of himself as much as Hurwitz, since he would have had an excellent chance to move into Hurwitz's associate professorship. Now "everything looked bleak and hopeless" (*trost- und aussichtslos*), and he predicted that the century would be over before four chairs in mathematics would become vacant in Prussia again (Hilbert/Klein 1985, 80). This prediction, however, turned out to be totally wrong; the chain reaction had just begun.

Minkowski, who like Hurwitz was Jewish, might have had even more reason to be concerned about his future career. Yet he saw the German academic world with far more sanguine eyes than Hilbert, whom he constantly praised as the coming mathematician of their generation. Just one week after Hilbert wrote in despair to Klein, Minkowski had a splendid chance to sing Hilbert's praises again. During a trip to Berlin, he arranged a meeting with Friedrich Althoff, the all-powerful ministerial official responsible for academic appointments at the Prussian universities (Brocke 1980). An autocratic personality, Althoff had a well-deserved reputation for promoting excellence while running roughshod over the interests of faculties and individual professors. He was like the Dean of a small college who knows everyone on the faculty, but with one significant difference: Althoff's watchful eye ranged over the entire system of higher education in Prussia. One way he managed to oversee this vast academic terrain was by relying on information he gathered through a network of informal contacts, his personal "Vertrauensleute." Critics of this approach viewed these "trusted people" with great suspicion, and some thought of them as unscrupulous informants, practically akin to academic spies. Klein attempted to curry favor with Althoff, but he felt the latter had left him in the lurch by failing to support Hurwitz's candidacy. This episode temporarily soured their relationship, but Klein soon got over it, and with time he became one of Althoff's most trusted and favored advisors.

Just after this meeting in Berlin, Minkowski dashed off a postcard to Hilbert with some exciting news: Althoff told him to expect a wave of new appointments in mathematics that would soon bring both of them appointments as associate professors! Their days of misery as *Privatdozenten* were about to end (Rüdenberg and Zassenhaus 1973, 46). Althoff named two others as well – Victor Eberhard and Eduard Study – and, indeed, all four were promoted soon thereafter. This began with Hilbert's appointment in Königsberg, followed by Minkowski's in Bonn. Study gained the latter's chair in 1894, after Minkowski moved into Hilbert's post. Eberhard, a blind geometer who had long been stuck in Königsberg, was finally appointed associate professor in Halle.

Althoff had also prepared well for the next round of negotiations for full professorships, beginning with Kronecker's chair in Berlin. Georg Frobenius, a Berlin product who taught at the ETH in Zurich, had been offered that position (on Frobenius's work, see Hawkins 2013). Still, he vacillated between accepting this chair and the one Schwarz had just vacated. Klein welcomed him to visit Göttingen, but soon after their meeting, Frobenius decided he would prefer to return to the Prussian capital. This created a vacancy at the ETH, which had long served as a waiting room (or "Wartesaal erster Klasse") for mathematicians from the Prussian university system.[4] This time Hurwitz got the call, and Hilbert stood first in line to move up the academic ladder. All this took place within two short months. Hilbert's mood now changed completely after learning that Lindemann promised to recommend him as Hurwitz's successor. Minkowski thought that Hurwitz might try bringing Hilbert to Zürich after Friedrich Schottky left the ETH to take a position in Marburg. In posing that query to Hilbert, Minkowski commented that he regarded his own chances of joining Hurwitz as next to nil, given that they were both Jews ("allein durch die Gleichheit der Konfession" Rüdenberg and Zassenhaus 1973, 48). Later, after Minkowski had risen to a full professorship in Königsberg, even that obstacle no longer stood in his way; in 1896 he joined Hurwitz on the ETH faculty.

All had gone smoothly enough, but by the time Klein wrote to congratulate Hilbert in September 1892, he could also update him on another series of events. For that past July, Klein had been offered the professorship in Munich formerly held by Ludwig Seidel. Klein immediately informed Althoff of this, and the latter wasted no time in making a generous counter-offer to ensure that the Bavarian Ministry of Education would not be able to entice him to leave Göttingen. This all took place rather quickly, so what Hilbert now learned was that Klein had indeed declined and, furthermore, that Lindemann was next in line for the chair in Munich (Hilbert/Klein 1985, 82). Those negotiations would stretch into the next year, but soon afterward Hilbert assumed Lindemann's chair and, as a final triumph, he even succeeded in having Minkowski appointed to his former position so that they were together once again, if only briefly. Althoff played along wonderfully, though obviously no one, not even Klein, would have imagined that both Hilbert and Minkowski would eventually become his colleagues in Göttingen.

Thus 1892 marked a true turning point both for the German mathematical community as well as for its two rising stars, Hilbert and Minkowski. Just two years after the founding of the *Deutsche Mathematiker-Vereinigung* (*DMV*), both had managed to attain associate professorships. Having launched their careers at this propitious moment, they stood poised to make their marks on this small, but growing community with its new national society. In the meantime, Hurwitz stood by in the "waiting room" at the ETH, hoping to receive a call from one of the Prussian universities. As was

Fig. 12.3 Adolf Hurwitz.

customary for young professors, Hurwitz and Hilbert wasted no time in arranging their domestic lives. Thus, in June, just before departing from Königsberg, Hurwitz married Ida Samuel, the daughter of a professor of medicine. Around the same time, Hilbert and his fiancé, Käthe Jerosch, fixed their marriage plans for the coming October. Minkowski was in no great hurry; he married Auguste Adler in 1896, just before joining Hurwitz in Zürich.

Although Hilbert had studied under Weber and Lindemann in Königsberg, his true mentor had been Hurwitz, an important factor that helped propel him into Klein's circle. Hurwitz and Hilbert would meet each other on a nearly daily basis to discuss mathematical matters, at times accompanied by Minkowski. This was an experience Hilbert would remember all his life, as both he and Minkowski were stunned by the breadth of Hurwitz's knowledge. All three were quite close in age and together they represented the cream of the coming generation. Only two of them, however, would attain a full professorship in Germany, despite Klein's efforts to bring Hurwitz to Göttingen in 1892. That story forms a central part of Chap. 14, where I describe how Hurwitz's candidacy was handicapped because of his Jewish background (Rowe 1986) (Fig. 12.3).

[4]Among the prominent German mathematicians who taught at the ETH before obtaining a position in Prussia were: Dedekind, Christoffel, H. A. Schwarz, H. Weber, Frobenius, Schottky, and Minkowski.

The "Jewish question" loomed large at most of the universities in Germany, and at some Jewish candidates had virtually no chances at all (Bergmann et al. 2012). Minkowski's call in 1902 was a highly exceptional situation; for many aspiring young Jews, his alliance with Hilbert was a clear sign that the doors of opportunity swung wide open in Göttingen. For Hurwitz, on the other hand, his great chance to obtain a full professorship in Prussia had come and gone. Stranded in Zürich, he surely felt embittered never to receive an offer from a German university, despite his merits and achievements. For Klein, the larger import of this defeat meant he could regroup and start all over again. Now that the generational change in the Prussian capital was complete, he could strike out on his own, thereby intensifying the long-standing rivalry between Berlin and Göttingen. Klein was greatly relieved when H. A. Schwarz departed for Berlin, and he got along well with his new colleague, Heinrich Weber. A first-class mathematician with roots in the Königsberg tradition, Weber offered support without personal interference, an attitude altogether different from that taken by Schwarz. Still, Klein kept his eyes on Hilbert as the most promising younger hope for rejuvenating mathematics in Göttingen. In view of this, it is worth describing their scientific relations during this short intervening period in some detail.

An important background event in this respect concerns Hilbert's *Doktorvater*, Ferdinand Lindemann, who had been Klein's pupil at Erlangen in the early 1870s. In 1882, Lindemann solved one of the most famous problems in the history of mathematics – the impossibility of squaring the circle – by proving that π is a transcendental number (Rowe 2015). This came roughly 10 years after Charles Hermite famously proved the transcendence of e. Lindemann was personally acquainted with Hermite, and he certainly knew his views as expressed in Borchardt's Journal: "I will not venture to search for a demonstration for the transcendence of π. If others attempt that enterprise, no one will be happier than me should they succeed, but believe me, my friend, it will cost them some real effort" (Hermite 1873, 342). Lindemann's announcement was consequently greeted with much fanfare and excitement, once a consensus was reached that his argument was indeed sound. In fact, Weierstrass played a major part both in assessing Lindemann's proof as well as in streamlining his argument.

We now pick up this story 10 years later. Shortly after his marriage and honeymoon, Hilbert sent Klein a short note for publication in the *Göttinger Nachrichten*. In his accompanying letter, he explained that this note gave a new and far simpler proof of the transcendence of e and π, one that evaded all the thornier technicalities in the papers of Hermite and Weierstrass (Hilbert/Klein 1985, 85–86). Hilbert made no mention at all of Lindemann's paper, which he presumably regarded as obsolete. Klein was clearly elated – "das ist ja wunderschön" – and he assured Hilbert that he would be honored to present his note at the next meeting of the Göttingen Scientific Society (Hilbert/Klein 1985, 86–87). It takes little imagination to understand Klein's delight, as Hilbert showed that what Hermite, Lindemann, and finally Weierstrass had managed to prove using rather elaborate arguments could actually be deduced in just four pages!

Hilbert also wrote to Hurwitz, who quickly saw how to take still another step toward further simplification: in proving the transcendence of e, he was able to dispense with the integrals used in Hilbert's argument (Hurwitz 1893). In one of his letters to Hurwitz, Hilbert confided that his new proof was inspired by reading a paper by Stieltjes on the transcendence of e; this was (Stieltjes 1890), a note that Hurwitz later reviewed for *Jahrbuch über die Fortschritte der Mathematik*. It was hardly longer than Hilbert's note and appeared in the widely read *Comptes Rendus* of the French Academy. Yet Hilbert failed to cite it; in fact, his note contains no references to the literature at all. Minkowski, a far more generous spirit than his friend, wrote Hilbert shortly after reading his paper to congratulate him once again (Rüdenberg and Zassenhaus 1973, 50). He could only think of Euler's reaction, after reading a striking result published by the young Lagrange: "penitus obstupui, quum hoc mihi nunciaretur" ("absolutely amazed when I saw that"). In his customary manner, Minkowski poured out encouragement, predicting that Hilbert's other works would from now on gain even more attention. He even pictured Hermite's frame of mind when reading this paper (for even though the Frenchman could not read German very well, the entire paper consisted merely of strings of mathematical equations, all very familiar to him). "Knowing the old fellow as I do," he wrote, "it would not surprise me if you soon heard from him about the joy he took in being able to experience this." A little later, Minkowski showed Hilbert's note to his senior colleague in Bonn, Rudolf Lipschitz, who returned the paper with just a single word of praise: "meisterhaft" (masterful) (Rüdenberg and Zassenhaus 1973, 52).

Klein wondered whether Hilbert might wish to publish a more extended version of this note on the exponential function in *Mathematische Annalen*, especially since Klein was eager to publicize Hilbert's new proof there (Hilbert/Klein 1985, 86). Nevertheless, Hilbert declined; instead of taking the trouble to give a more complete presentation with references, he simply wrote back to say that he had nothing more to add. Still, he was pleased to have received a message from Paul Gordan, who had expressed his appreciation (Hilbert/Klein 1985, 88).Yet more was to follow, as the immediate reactions of both Hurwitz and Gordan led to two more short notes in the *Göttinger Nachrichten*. Klein then bundled them together with Hilbert's paper for publication in *Mathematische Annalen* (Hilbert 1893; Hurwitz 1893; Gordan 1893). These striking simplifications of the earlier arguments now signaled that the proofs of transcendence for

e and π had reached the stage where they could appear as standard textbook material.

Felix Klein, who had guided the negotiations that followed Lindemann's original breakthrough 10 years earlier, once again stood at the center of this new round of events. No one was in a better position to exploit this opportunity, and he wasted no time finding ways to popularize this knowledge within the larger world of educated mathematicians. Popularizing mathematics in the context of the 1890s meant something quite different than it does today. In the present instance, what Klein had in mind was a novel way to discuss recent mathematical research, while foregoing virtually all the necessary technical details. This was popularization for intellectual elites; its aim was not to reach the proverbial "man on the street" but rather a mathematically educated audience. By this time, Klein's lecture courses had already begun to evolve in a similar direction. While others in Germany continued to emulate Weierstrassian rigor, he settled for short-circuited arguments and vague remarks that could hardly pass muster if one were claiming a "real proof." In Berlin, where many regarded Klein as a charlatan, Frobenius emerged as the foremost critic of his machinations. Yet even Göttingen's Richard Courant had to concede that Klein's works were full of gaping holes; he once likened him to a visionary pilot who lacked the ability to land his plane (Courant 1925, 772). Still, Klein longed for a chance to speak more informally about a whole range of topics in current mathematical research that reflected his own personal interests and tastes. By looking backward, he wanted to give a broader picture, one that largely evaded the underlying technical problems. This, of course, was the very opposite of Hilbert's approach in his Paris lecture delivered seven years later. Still, the outward circumstances were strikingly similar.

Klein's chance came from abroad in 1893 when he received an invitation from the mathematicians in the Chicago metropolitan area. Three of them – Henry S. White, Oskar Bolza, and Heinrich Maschke – had, in fact, been his former students and/or protégés. They now cordially invited him to attend the Mathematical Congress that would convene later that summer. By today's standards, this gathering was more like a small workshop, one of many such scientific congresses that were held in conjunction with the World's Columbian Exposition, an extravaganza that would attract millions (Parshall and Rowe 1994, Chap. 7). Klein arrived as an official emissary of his state government, which sponsored a major exhibition to celebrate the intellectual tradition of the Prussian universities. His former student, Walther Dyck, organized the mathematics section, which was noteworthy for its large collection of mathematical models, about which Klein gave a demonstration (Parshall and Rowe 1994, 304–309). He also brought with him a set of papers written for the occasion by some of Germany's leading mathematicians, including a survey of research results in invariant theory offered by Hilbert, later published as (Hilbert 1896). Immediately after the congress adjourned, Klein took advantage of this opportunity to deliver a series of "popular lectures," in the sense described above, held over a two-week span at Northwestern University. Another German, Alexander Ziwet from the University of Michigan, wrote up these talks using Klein's notes along with those he had taken. Soon afterward Klein's Evanston Colloquium Lectures (Klein 1894/1911) were already in print.

It was in the seventh of these 12 lectures that Klein spoke about the recent simplifications in proving the transcendence of e and π (Klein 1894, 51–57). Most of this talk was devoted to outlining Hilbert's proof, but Klein also alluded to the related contributions of Hurwitz and Gordan. He began by offering a kind of mini-history of work on transcendental numbers over the past two decades, beginning with Hermite's paper on the exponential function. Although he mentioned the contributions of several others, he passed over the fundamental paper by Weierstrass from 1885, while highlighting Lindemann's earlier breakthrough, which was published, of course, in Klein's journal. Even so, the thrust of his message suggested that all this now belonged to the past. In the meantime, Hilbert had dispensed with all the former technical difficulties, making the whole matter far easier to grasp. In short, Klein was convinced that in the near future mathematicians would encounter few difficulties when teaching these methods of proof to their students.

These events from 1892 to 1893 illustrate how skillfully Klein beat the drums to advertise Hilbert's achievements. This promotional activity, in fact, fit a larger pattern of efforts to make Hilbert his new star. A central pillar supporting Klein's role as a power broker in German mathematics stemmed from his editorial leadership with *Mathematische Annalen*. Thus, one of his main concerns was to win over the younger generation of talented mathematicians in order to ensure the future success of his journal. Not surprisingly, he saw Hilbert as the key figure in this transition process. Back in 1894, shortly before he brought him to Göttingen, Klein sought to have Hilbert appointed to the editorial board, only to be rebuffed by some of its "old-timers" who, as he explained to Hilbert, wanted "to preserve the *Annalen* as the organ of a specific school" (Hilbert/Klein 1985, 110–111).

The resistance came from Paul Gordan and Max Noether, both of whom had a stake in maintaining the journal's tradition as an outlet for publications of the Clebschian school and its allies. Klein, on the other hand, had been opposed to this narrower approach from the beginning, since his editorial policy aimed to attract contributions from all quarters and in nearly every field of research (Tobies and Rowe 1990). As he wrote to Hilbert during these negotiations, ". . . [Heinrich] Weber and I want to see the Annalen keep pace as much as possible with modern advances in mathematics by welcoming the newest work from wherever it may come." (*ibid.*) Klein could have succeeded in gaining Hilbert's appointment to the editorial board at this time, but only by striking up a deal that he found unacceptable, namely by expanding the board to include even more elderly "Clebschians," in particular, Alexander Brill and (possibly) Aurel Voss. Rather than antagonize his long-time allies or compromise his own position, Klein opted to wait them out instead, and pleaded for patience from Hilbert. "[I]t is clear to the opposing party," he assured him, "that we will prevail with our approach, as the balance of power can only change in our favor. I ask of you only that in the meantime you will not become disloyal to us." (*ibid.*)

To seal this informal pact, Klein arranged to have Teubner send free issues of *Mathematische Annalen* to Professor Hilbert in Königsberg. Hilbert's response, written just two months before his sudden call to Göttingen, must have been very gratifying for Klein:

> It is for me a great honor that all members of the editorial board of the Mathematische Annalen have agreed in principle to my appointment on the board, and I am even more pleased that you and Professor Weber have attached importance to my joining the board. I am very satisfied already with this success, and I assure you that I am ready at any time to work in promoting the growth of this journal in the event that I should be asked to do so in the future; I would then pursue this task by following precisely the guidelines which you indicated in your letter and with which I am in total agreement (Hilbert/Klein 1985, 113).

Soon afterward, Weber left Göttingen for Strassburg, after which Klein quickly engineered Hilbert's appointment, which began in the spring of 1895. Letters from Käthe Hilbert to her relatives in Königsberg reveal that her husband was delighted with the atmosphere he found in Göttingen (Hilbert Nachlass 777). In later years, he would have good reason to feel differently. Hilbert's first biographer, Otto Blumenthal (Fig. 12.4), witnessed his arrival and later gave a vivid portrait of his former mentor in the late 1890s when he was not yet a legendary figure (Blumenthal 1935, 399–400). Hilbert lacked Klein's elegance and oratorical skills; in fact, Blumenthal described his lectures as colorless and without flair. Still, many came to appreciate the originality and clarity of his presentations, and above all the effort Hilbert made to reach his audience. As Blumenthal put it, "he lectured for the students, not for himself."

By 1897, just before the First International Congress of Mathematicians convened in Zurich, Klein had paved the way for Hilbert to join the Annalen's editorial board. His name appeared on the title page of Band 50 the following year. Beginning with volume 55, Hilbert was promoted to the main editorial board, replacing Adolf Mayer; he thereby became the journal's *de facto* editor-in-chief, though Klein retained his position at the request of Noether and Gordan, since both thought he was needed as a moderating influence on Hilbert. In the meantime, Klein and Alfred Ackermann-Teubner developed a strategy for re-organizing the wide

Fig. 12.4 Otto Blumenthal, ca. 1900.

array of research journals published by the Leipzig firm of B.G. Teubner (Tobies 1987/1987). At Klein's urging, Carl Runge was named coeditor of the *Zeitschrift für Mathematik und Physik,* which was to serve as the primary vehicle for publications in applied mathematics. Klein was also instrumental in founding a new journal devoted to mathematical pedagogy, the *Archiv der Mathematik und Physik*, and in establishing new guidelines for *Bibliotheca Mathematica* as well as for the *Jahresbericht der Deutschen Mathematiker-Vereinigung*. Thus, during the phase when Hilbert, Germany's leading *Fachmathematiker*, took over the reins of *Mathematische Annalen* – the leading journal for pure mathematics – Klein, the *Wissenschaftspolitiker*, was busy behind the scenes orchestrating a whole system of publications covering virtually every facet of mathematical activity in Germany.

Göttingen's preeminent position vis-à-vis Berlin and other universities was bound to intensify traditional rivalries. These tensions were further aggravated by Klein's aggressive style and omnipresent influence, especially within extra-scientific circles. As a trusted confidant of Althoff, the kingpin of the Prussian university system, Klein's opinions carried considerable weight within the Ministry of Culture, and no one doubted that he could make or break the career of a young mathematician. Moreover, no other German mathematician had such an extensive web of contacts with leading industrialists and businessmen from firms like Krupp, Bayer, Siemans & Halske, AEG, and Norddeutscher Lloyd, companies that had pumped considerable amounts of money into the new research institutes founded by the Göttingen Association for the Promotion of Applied Physics and Mathematics. This organization, the first of its kind in Germany at the time of its founding in 1898, was the brainchild of Felix Klein. Meanwhile, he sought to strengthen applied mathematics by means of reforms in both secondary and higher education throughout Prussia (Schubring 1989).

By the turn of the century, Göttingen stood poised to surpass Berlin and become the leading center for mathematical research in Germany. In 1901, Karl Schwarzschild was appointed director of the astronomical observatory, and one year later Minkowski assumed a new professorship created by Althoff to ensure that Hilbert would remain in Göttingen. Two years after Minkowski's arrival, Carl Runge came to fill the first chair in applied mathematics at a German university (Richenhagen 1985). Runge thereby joined two other Göttingen figures with strong ties to applications: Ludwig Prandtl, a pioneer in aero- and hydrodynamical research (Eckert 2016), and the geophysicist Emil Wiechert. These appointments were part of Althoff's policy for establishing disciplinary *Schwerpunkte* within the Prussian university system. Yet even the Ministry itself seems to have been taken aback by this sudden expansion. Shortly after Runge and Prandtl joined the Göttingen faculty, Klein received a letter from one of Althoff's lieutenants, who wrote: "[a]fter Minkowski's appointment, Hilbert said, 'now we are invincible.' But you continue right on becoming ever more invincible, so that I am really curious what proposals you will come up with next. But we are prepared for anything as nothing can surprise us anymore" (G. Elster to Klein, 18 July, 1904, Klein Nachlass XXII L, SUB, Göttingen).

The photo in Fig. 12.5 – taken in the summer of 1902, only months before Minkowski's arrival – shows Felix Klein presiding over the Göttingen Mathematical Society. The seating arrangement reflects more than just the need to have the tall men stand in back. Here Hilbert affirms his position as Klein's "right-hand man," counterbalanced on the left by Klein's star applied mathematician, Schwarzschild. Taking up the wings in the front row are two of Göttingen's more ambitious younger men, Max Abraham and Ernst Zermelo, the former an expert in electrodynamics, the latter soon to become famous for his work on set theory. In back are two of Hilbert's doctoral students, Otto Blumenthal (a close firend of Schwarzschild) and Georg Hamel, alongside Conrad Müller, who took his doctorate under Klein. Klein's attention seems to be riveted on Grace Chisholm Young, the charming Englishwoman who took her doctorate under him in 1895. She and her husband, the mathematician W. H. Young, in fact resided in Göttingen for several years.

Among the foreigners present was the Japanese mathematician Takuji Yoshiye, a good friend of the number-theorist Teiji Takagi, who first studied in Berlin before joining his friend at the Georgia Augusta (Sasaki 2002). Yoshiye spent several semesters in Göttingen, but by the time this photo was taken Takagi had already returned to Japan. Before they came to Germany, both had been students of Rikitaro Fujisawa in Tokyo. The latter had come to Germany during the 1880s when he was deeply impressed by the idealism of Strassburg's Elwin Bruno Christoffel. This was a new university founded by the Prussians when Alsace-Lorraine was annexed into the German Reich. On returning to Tokyo, Fujisawa modeled his seminar on those he had encountered abroad, thereby spearheading a new style of teaching in Japan (Sasaki 1994, 183–184; Sasaki 2002, 238–240). In 1900, he visited Europe to attend the Paris ICM, stopping in Göttingen briefly along the way. His former pupils, Yoshiye and Takagi, were pleased to see him until Fujisawa requested that they take him to visit Gauss's grave (Honda 1975, 153). This request caused some consternation, since neither of them knew where it was located, though the gravesite was only a few minutes' walk away from the *Sternwarte*.

Takagi later reported his amazement over the differences between these two leading mathematical centers: "I was much astonished by the striking contrast in the atmospheres of the mathematics departments of Göttingen and of Berlin. In the former, once a week a meeting was held, and in attendance was a group of brilliant youths from all over the world, as if here were the center of the mathematical world." (Honda 1975, 153–154). Takagi found nothing at all like this in Berlin. There he attended lecture courses offered by Georg Frobenius on algebra, H.A. Schwarz on function theory, Lazarus Fuchs on differential equations, and Kurt Hensel on number theory. Hensel was then a Privatdozent in Berlin, but in 1901 he obtained a professorship in Marburg, where he spent the remainder of his career. The Berlin analysts, Schwarz and Fuchs, were both well over the usual retirement age in Japan, whereas Frobenius was still going strong. Takagi recalled that Fuchs' course turned out to be nothing more than a recapitulation of a paper he had published in Crelle back in 1867! When Fuchs tried to approach an irregular point on the boundary of the circle of convergence, he found to his embarrassment that he had not reckoned with another kind of barrier: he lacked enough space on the blackboard in order to draw the corresponding picture. (Honda 1975, 151).

When he arrived in Göttingen in 1900, Takagi hoped to study number theory with Hilbert. He already knew his *Zahlbericht*, which he had studied in Tokyo (Sasaki 2002, 240). He must have been quite discouraged to learn that

Fig. 12.5 The Göttingen Mathematical Society in the summer of 1902. First row (*left to right*): Max Abraham, Friedrich Schilling, David Hilbert, Felix Klein, Karl Schwarzschild, Grace Chisholm Young, Diestel, Ernst Zermelo. Second row: Fanla, Hansen, Hans Müller, Dawney, Erhard Schmidt, Takuji Yoshiye, Saul Epsteen, Hermann Fleischer, Felix Bernstein. Third row: Otto Blumenthal, Georg Hamel, Conrad Müller. Schwarzschild *Nachlass,* Niedersächsische Staats- und Universitätsbibliothek Göttingen.

Hilbert's research interests had shifted to other fields. Still, they met occasionally, though Hilbert was surely more than a little skeptical when Takagi told him he was working on the core of his twelfth Paris problem: Kronecker's *Jugendtraum.* Takagi tried not to get discouraged, but he felt as though mathematics in Japan was 50 years behind what he witnessed in Germany. Still, after three semesters in Göttingen he thought he had been able to catch up for the most part. In fact, after returning to Tokyo he published his solution of the problem he spoke about with Hilbert.[5] He then joined Yoshiye as a member of the mathematics department at the Imperial University in Tokyo.

As the Göttingen Mathematical Society grew, eminent guests and foreign visitors often came to give talks, an opportunity and challenge few took lightly. The severity of the competition, along with a Prussian sense of order and proper place, made the atmosphere anything but relaxed. Dirk Struik remembered the sarcasm in Göttingen as "a world apart from the courteous atmosphere in Rome," where he had spent the previous year working with Levi-Civita (see Chap. 31). In some cases, the assessment of a young mathematician's performance could make or break his career. Some budding talents, like Norbert Wiener and Max Born, were scarred for life by the daunting experience of facing the hypercritical audiences that gathered at the weekly meetings of the Göttingen Society. Born gave this vivid account of a typical scene as a prelude to describing his own disastrous first lecture before this group:

> The whole atmosphere of that learned Society was neither pleasant nor encouraging. At a long table parallel to the blackboard were seated the most formidable mathematicians, mathematical physicists and astronomers of Germany: Klein, Hilbert, Landau (Minkowski's successor), Runge Voigt, Wiechert, Prandtl, Schwarzschild, often strengthened by guests, German or European celebrities. The younger members and less important guests were seated at two long tables at right angles to the "high table". This younger crowd, if not as famous as the "Mandarins", was yet no less critical and perhaps more conceited: [Ernst] Zermelo, [Max] Abraham, the Müllers [Hans and Conrad], [Otto] Toeplitz, [Ernst] Hellinger, . . . and many newcomers: Gustav Herglotz, Alfred Haar, Hermann Weyl, Paul Köbe, and others. Books were piled on the green cloth of the tables; at the beginning of the meeting Klein gave a short account of his impressions of some of these new publications and then circulated them. So everybody soon had a book in his hand and paid very little attention to the speaker, and what attention he gave was mostly in the way of objection and criticism. There was no friendly listening nor a vote of thanks at the end. It was extremely difficult to catch

[5]For a delightful account of Takagi's later career and central role as the founder of modern class field theory, see Yandell (2002, 219–230).

Fig. 12.6 Das Auditorienhaus stood at the hub of mathematical activity in Göttingen from 1865 to 1930.

> the attention of the audience, to create a spell of interest, and scarcely possible to arouse enthusiasm. (Born 1978, 134–135)

Göttingen's dynamic community (Fig. 12.6) filled the local atmosphere with all kinds of new ideas, though this circumstance often led to additional animosity. Mathematicians who liked to talk about some of their "good ideas" had to be careful about who was listening. Some discovered that such thoughts, once "airborne," might have a way of proliferating in the minds of others, and in Göttingen there were always some who were more attune to the pursuit of new possibilities than in tracing the lineage of a newfound result, Paul Koebe being one such person. He and others were always quick to snatch up a new technique or insight and run with it, a process people in Göttingen liked to call "nostrification." After Max Abraham warned him about this, Einstein later came to feel that Hilbert had engaged in nostrifying his gravitational theory. Their encounter took place in the early summer of 1915 when Einstein came to Göttingen to lecture about his new theory (see Chap. 16).

Hans Lewy recalled how prevalent this phenomenon was during the era of Hilbert and Courant (Part V):

> It was a common failing of Göttingen people that they were not very conscientious in attributing. That was true for almost all of them – Hilbert included. But when you look back at papers by some of the great heroes of mathematics, you very often find that they are careless in these matters. That greater care is taken now is, I think, due to the fact that jobs depend to a more explicit degree on the credit that a person is given. It is undoubtedly true that the group in Göttingen was careless about studying what other people had done and attributing their results to them, but I think this must be seen against a background of less care. The Göttingers had a facetious expression for the process of making someone else's idea one's own. They called it "nostrification." There were many levels of the process: "conscious nostrification" – "unconscious nostrification" – even "self-nostrification." This last occurred when [a mathematician] came up with a marvelous new idea which he later discovered had already appeared in earlier work of his own. (Reid 1976, 120–121)

These remarks reflect the belated perspective of a younger member of the Göttingen community during the 1920s. They suggest how the intensely competitive atmosphere had much to do with accepted norms of behavior among the mathematicians, and he apparently recognized that Hilbert's rather cavalier attitude reinforced the free-for-all feeling that prevailed in many of its circles.

Meanwhile, Klein took on the responsibility for numerous administrative matters. For once he was convinced that Hilbert was safely in the saddle, he could turn to the task of making Göttingen "more invincible than ever," and many of the resources necessary for doing so were already close at hand. He revamped the mathematics curriculum, reorganized the Göttingen Scientific Society, directed the archival work on Gauss's posthumous papers, and offered special vacation courses for gymnasium teachers. By the mid-1890s, enrollment in mathematics courses at the German universities was beginning to rise again. In Göttingen the number of mathematics and science students had dropped from a high of around 240 in 1882 to ninety ten years later; but by 1900 it had climbed back to 300, and from there it continued to grow, reaching nearly 800 by 1914. According to Klein's own estimates, between 10% and 15% of these students were at a sufficiently advanced level to profit fully from the unique combination of mathematical resources available in Göttingen (Lorey 1916, 21–22, 349). These included an impressive collection of mathematical models and instruments situated next to one of the finest mathematics libraries in the world. Known familiarly as *das Lesezimmer*, this library was a *Präsenzbibliothek*, a non-circulating collection open only to members of the Göttingen mathematical community. It was designed not only to facilitate easy reference to current literature but also to stimulate informal contacts.

The relationship between Klein and Hilbert, as it evolved going into the new century, was not without its troubled aspects. Hilbert surely had not anticipated the scope of Klein's organizational plans, nor the demands these would place on his senior colleague's time and energy. Without doubt, he had imagined a somewhat more conventional partnership with reasonably symmetric roles and, most important of all, considerable opportunity for discussing mathematical matters of mutual interest. Instead, he found himself thrust into the role that had dominated Klein's attention and time during his first 10 years in Göttingen, that of the "master teacher." Klein had long wanted to scale back his teaching and abdicate his position as head of the "Göttingen school," and he might have doubted whether Hilbert was capable of shouldering these responsibilities. Clearly, filling Klein's shoes amounted to a daunting task. For more than a decade, he had been the leading mathematics teacher in Germany, whereas during that time Hilbert had failed to promote a single doctoral student at Königsberg. Moreover, unlike Klein, he was hardly a polished lecturer, though he did possess a similarly prodigious wealth of mathematical knowledge and a fertile imagination that he brought to bear on a wide variety of mathematical problems. As it turned out, Hilbert emerged to become not only a great mathematician but also one of the most influential teachers in the history of mathematics.

It has been customary to regard Klein and Hilbert as representatives of two very different mathematical traditions: one rooted in nineteenth-century geometry and emphasizing the role of intuition and imagination (*Anschauung*), the other, with essentially modernist tendencies that stressed axiomatics and rigor. There are good reasons for making this distinction, to be sure. Still, it overlooks certain important similarities while failing to account for how two such diametrically opposed thinkers could have formed such an effective alliance. For theirs was no mere marriage of convenience, but rather a partnership based on a shared understanding of mathematics as a multi-faceted but fundamentally unified body of knowledge (Rowe 1994). It is true that Klein, like Poincaré, saw the burgeoning interest in abstract structures and axiomatics around the turn of the century as a potential threat to the lifeblood of mathematics, whereas Hilbert had none of these reservations.

Nevertheless, it would be mistaken to think that Klein had no appreciation for axiomatic thinking, or that Hilbert

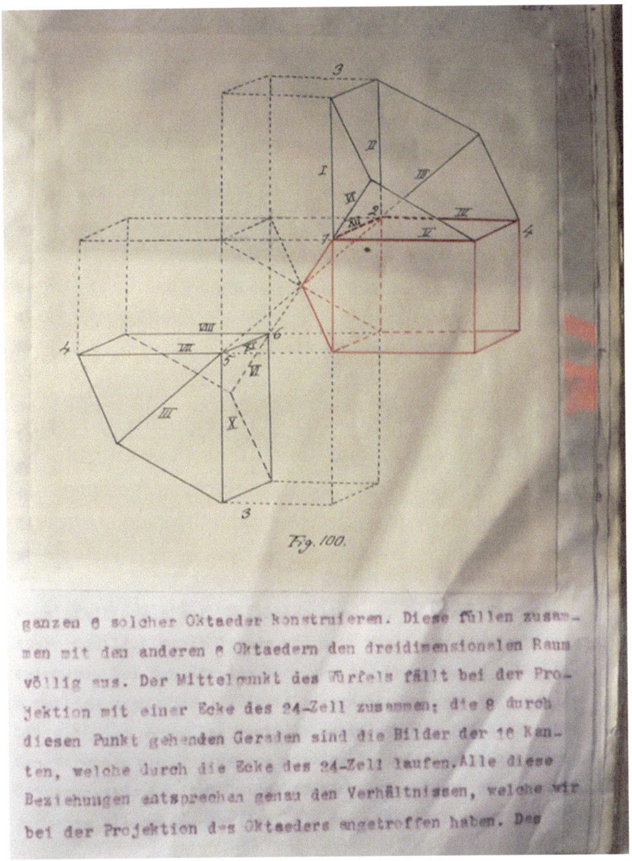
Fig. 100.

ganzen 6 solcher Oktaeder konstruieren. Diese füllen zusammen mit den anderen 6 Oktaedern den dreidimensionalen Raum völlig aus. Der Mittelpunkt des Würfels fällt bei der Projektion mit einer Ecke des 24-Zell zusammen; die 8 durch diesen Punkt gehenden Geraden sind die Bilder der 16 Kanten, welche durch die Ecke des 24-Zell laufen. Alle diese Beziehungen entsprechen genau den Verhältnissen, welche wir bei der Projektion des Oktaeders angetroffen haben. Das

Fig. 12.7 A page from the original *Ausarbeitung* of Hilbert's lecture course "Anschauliche Geometrie" prepared by Walther Rosemann (Hilbert Nachlass 563).

was a reductionist who only valued formal proofs. Indeed, Hilbert's popular lecture course on "anschauliche Geometrie," (Fig. 12.7) published under the same title in (Hilbert u. Cohn-Vossen 1932), reveals that he was a true master of the Kleinian style. These words from its preface might easily have been penned by Klein himself:

> In mathematics, as in all scientific research, we encounter two tendencies: the tendency toward abstraction – which seeks to extract the logical elements from diverse material and bring this together systematically – and the other tendency toward intuition [*Anschaulichkeit*] that begins instead with the lively comprehension of objects and their substantial [*inhaltliche*] interrelations. . . . The many-sidedness of geometry and its connections with the most diverse branches of mathematics enable us in this way [namely, through the anschauliche approach] to gain an overview of mathematics in its entirety and an impression of the abundance of its problems and the rich thought they contain. (Hilbert u. Cohn-Vossen 1932, v)

Klein, long since retired when Hilbert began teaching this course, surely felt very gratified that his colleague chose to illuminate a theme so dear to his own heart. In fact, Hilbert taught this course – both before and after Klein's death in 1925 – no fewer than four times. Walther Rosemann was his assistant when in the winter semester of 1920–21 he taught "anschauliche Geometrie" for the first time. Already then, Rosemann produced a typescript *Ausarbeitung* of well over 400 pages (Hilbert Nachlass 563). We can be sure that Klein was familiar with this text, possibly even during the time it was being written. For, as pointed out in the introduction to Part II, he and his assistants were then hard at work on volume two of Klein's collected works, in which the first 15 papers appear under the rubric "anschauliche Geometrie." That volume was published in 1922, only a few months after Hilbert's sixtieth birthday, which fell on the 23 January. Klein no doubt still had vivid memories of Hilbert's gracious speech honoring him on his own sixtieth birthday (see Chap. 16). For Hilbert's sixtieth, he returned the favor by presenting him with a bound copy of his own contributions to "anschauliche Geometrie," together with a personal dedication that would later adorn the second volume of Klein's collected works.

References

Bergmann, Birgit, Moritz Epple, and Ruti Ungar, eds. 2012. *Transcending Tradition: Jewish Mathematicians in German-Speaking Academic Culture*. Heidelberg: Springer.

Blumenthal, Otto. 1935. Lebensgeschichte, in (Hilbert 1932–1935, 3: 388–429).

Born, Max. 1978. *My Life. Recollections of a Nobel Laureate*. New York: Charles Scribner's Sons.

Brocke, Bernhard vom. 1980. Hochschul- und Wissenschaftspolitik in Preußen und im Deutschen Kaiserreich, 1882–1907: das "System Althoff,". In *Bildungspolitik in Preußen zur Zeit des Kaiserreichs*, ed. Peter Baumgart, 9–118. Stuttgart: Klett-Cotta.

Corry, Leo. 2004. *David Hilbert and the Axiomatization of Physics (1898–1918): From Grundlagen der Geometrie to Grundlagen der Physik*. Dordrecht: Kluwer.

Courant, Richard. 1925. Felix Klein. *Die Naturwissenschaften* 13: 765–771.

Eckert, Michael. 2016. *Ludwig Prandtl: Strömungsforscher und Wissenschaftsmanager: Ein unverstellter Blick auf sein Leben*. Heidelberg: Springer.

Gordan, Paul. 1893. Transcendenz von e und π. *Mathematische Annalen* 43: 222–224.

Hawkins, Thomas. 2013. *The Mathematics of Frobenius in Context: A Journey through 18th to 20th Century Mathematics*. New York: Springer.

Hermite, Charles. 1873. Extrait d'une lettre de Mr. Ch. Hermite à Mr. Borchardt. *Journal für die reine und angewandte Mathematik* 76: 342–344.

Hilbert, David. 1893. Über die Transcendenz der Zahlen e und π. *Mathematische Annalen* 43: 216–219.

———. 1896. Über die Theorie der algebraischen Invarianten. In *Mathematical Papers Read at the International Mathematical Congress Chicago 1893*, 116–124. New York: Macmillan; Reprinted in (Hilbert 1932–1935, 1: 376–383).

———. 1901. Mathematische Probleme, *Archiv für Mathematik und Physik* 1(1901), 44–63 and 213–237; Reprinted in (Hilbert 1932–1935, 3: 290–329); Mathematical Problems: Lecture delivered before the International Congress of Mathematicians at Paris in 1900. Trans. Mary F. Winston. (1902), *Bulletin of the American Mathematical Society*, 8: 437-479.

———. 1910. Hermann Minkowski. *Mathematische Annalen* 68: 445–471; Reprinted in (Hilbert 1932–1935, 3: 339–364).

———. 1932–1935. *Gesammelte Abhandlungen*. 3 Bde ed. Berlin: Springer.

———. 1971. Über meine Tätigkeit in Göttingen. In *Hilbert Gedenkband*, ed. Kurt Reidemeister, 78–84. Berlin: Springer.

Hilbert, David und Stefan Cohn-Vossen. 1932/1952. *Anschauliche Geometrie*. Berlin: Springer, 1932; *Geometry and the Imagination*. Trans. Peter Nemenyi. New York: Chelsea, 1952.

Hilbert/Klein. 1985. In *Der Briefwechsel David Hilbert – Felix Klein: (1886–1918)*, ed. von Günther Frei. Göttingen: Vandenhoeck & Ruprecht.

Honda, Kin-ya. 1975. Teiji Takagi: A Biography. *Commentarii mathematica Universitatis Sancti Pauli* 24: 141–167.

Hurwitz, Adolf. 1893. Beweis der Transcendenz der Zahl e. *Mathematische Annalen* 43: 220–221.

Klein, Felix. 1894/1911. *The Evanston Colloquium. Lectures on Mathematics*. New York: Macmillan; Reprinted 1911.

Lorey, Wilhelm. 1916. *Das Studium der Mathematik an den deutschen Universitäten seit Anfang des 19. Jahrhunderts*, Abhandlungen über den mathematischen Unterricht in Deutschland, Band III, Heft 9, Leipzig: Verlag B. G. Teubner.

McLarty, Colin. 2012. Hilbert on theology and its discontents: the origin myth of modern mathematics. In *Circles Disturbed: the Interplay of Mathematics and Narrative*, ed. A. Doxiadis and B. Mazur, 105–129. Princeton: Princeton University Press.

Parshall, Karen, and David Rowe. 1994. In *The Emergence of the American Mathematical Research Community. AMS/LMS History of Mathematics Series*, ed. J.J. Sylvester, Felix Klein, and E.H. Moore, vol. 8, *187–1900*. Providence: American Mathematical Society.

Reid, Constance. 1970. *Hilbert*. New York: Springer.

———. 1976. *Courant in Göttingen and New York: the Story of an Improbable Mathematician*. New York: Springer.

Richenhagen, Gottfried. 1985. *Carl Runge (1856–1927): Von der reinen Mathematik zur Numerik*. Gottingen: Vandenhoeck & Ruprecht.

Rowe, David E. 1986. "Jewish Mathematics" at Göttingen in the Era of Felix Klein. *Isis* 77: 422–449.

———. 1994, The Philosophical Views of Klein and Hilbert, in (Sasaki, Mitsuo, Dauben 1994), 187–202.

———. 1997. "Perspective on Hilbert": *Moderne – Sprache – Mathematik. Eine Geschichte des Streits um die Grandlagen der Disziplin und des Subjekts formaler Systeme*, by Herbert Mehrtens; *Hilbert-programm und Kritische Philosophie*, by Volker Peckhaus; *Über die Entstehung von David Hilberts "Grundlagen der Geometrie"*, by Michael-Markus Toepell, *Perspectives on Science*, 5(1997): 533–570.

———. 2015. Historical Events in the Background of Hilbert's Seventh Paris Problem. In *A Delicate Balance: Global Perspectives on Innovation and Tradition in the History of Mathematics, A Festschrift in Honor of Joseph W. Dauben*, Trends in the History of Science, ed. D.E. Rowe and W.-S. Horng, 211–244. Basel: Birkhäuser.

Rowe, David E., and John McCleary, eds. 1989. *The History of Modern Mathematics*. Vol. 2. Boston: Academic Press.

Rüdenberg, L., and H. Zassenhaus, eds. 1973. *Hermann Minkowski, Briefe an David Hilbert*. New York: Springer.

Sasaki, Chikara. 1994. The Adoption of Western Mathematics in Meiji Japan, 1853–1903, in (Sasaki, Mitsuo, Dauben 1994), 65–186.

———. 2002. The Emergence of the Japanese Mathematical Community in the Modern Western Style, 1855–1945. In *Mathematics Unbound: The Evolution of an International Mathematical Research Community, 1800–1945*, ed. K.H. Parshall and A.C. Rice, 229–252. Providence: American Mathematical Society.

Sasaki, Chikara, Sugiura Mitsuo, Joseph W. Dauben, eds. 1994. *The Intersection of History and Mathematics*. Science Networks, vol. 15 (Proceedings of the 1990 Tokyo Symposium on the History of Mathematics), Basel: Birkhäuser.

Schubring, Gert. 1989. Pure and Applied Mathematics in Divergent Institutional Settings in Germany: the Role and Impact of Felix Klein, in (Rowe and McCleary 1989, 2: 171–220).

Stieltjes, T. J. 1890. Sur la fonction exponentielle, *Comptes rendus hebdomadaires des séances de l'Académie des sciences*, vol. CX, 267–270.

Tobies, Renate. 1987/1987. Zu Veränderungen im deutschen mathematischen Zeitschriftenwesen um die Wende vom 19. zum 20. Jahrhundert, Teil I," *NTM: Schriftenreihe für Geschichte der Naturwissenschaften, Technik und Medizin,* 23(2) (1986): 19–33; Teil II, 24(1) (1987): 31–49.

Tobies, Renate, and David E. Rowe. 1990. *Korrespondenz Felix Klein – Adolf Mayer*, Teubner – Archiv zur Mathematik. Vol. 14. Leipzig: Teubner.

Weyl, Hermann. 1944. David Hilbert and his Mathematical Work. *Bulletin of the American Mathematical Society* 50: 612–654.

Yandell, Ben H. 2002. *The Honors Class. Hilbert's Problems and their Solvers*. Natick: A K Peters.

13 Hilbert's Early Career

(*Mathematical Intelligencer* 25(2)(2003): 44–50, 27(1)(2005): 77–82)

From Königsberg to Göttingen

David Hilbert's remarkable career falls into two clearly distinct periods: the quiet Königsberg phase, which spanned the period from his birth on 23 January 1862 to that of his full maturity as one of Germany's leading mathematicians, followed by the tumultuous Göttingen years. The latter began with his appointment in Göttingen in 1895 and ended with his death on 14 February 1943 when Nazi Germany had already entered its death throes. It would be difficult to exaggerate the contrast between these two phases, just as it remains difficult to picture life in Germany before the onset of the two world wars that so decisively shaped the course of twentieth century history.[1]

The East Prussian city of Hilbert's youth stood for solidity, integrity, and a harsh, no frills lifestyle rooted in the Protestant work ethic. Hilbert identified with Königsberg's social ambience all his life. In Göttingen, his students and friendly admirers loved to mimic his distinctive East Prussian accent with its unfamiliar twang that set him apart from those who spoke High German in the manner of well-educated Hanoverians. Neither did Hilbert make any particular effort to hide his own preference for Königsberg over Göttingen; he treasured no honor higher than the one the city of his birth bestowed on him in 1930 when he was made an honorary citizen of his home city (Reid 1970).

For many years Königsberg had been the most important Prussian outpost in Eastern Europe. Situated on the Pregel, only a short distance inland from the Baltic Sea, Königsberg lay at the heart of a vibrant trade network that linked it with St. Petersburg to the east and Hamburg to the west. But by the end of the nineteenth century the German empire had become a bustling industrial power whose economic strength stemmed from the steel and chemical firms of the Rhineland. The vast agrarian plains east of Berlin, mainly lands owned by Junker aristocrats – men like Otto von Bismarck and others who dominated the officer corps of the Prussian army – held an increasingly marginal importance for the economy of the Second Reich. Königsberg retained its importance as a regional center, but the surrounding terrain to which it belonged was part of a semi-feudal world with an uncertain future.

Academic life in Germany had a strong local coloring throughout most of the nineteenth century, and the relationship between town and gown was quite different at Königsberg than in the quaint atmosphere of Göttingen. Still, both cities were proud of their distinguished universities, each of which developed strong traditions in mathematics and the natural sciences. The "Albertina" in Königsberg was founded by Duke Albrecht of Brandenburg in 1544 and soon thereafter it became a Protestant bastion fending off the Counterreformation. Its later fame began in the eighteenth century when the great philosopher Immanuel Kant graced its faculty. The younger Georgia Augusta in Göttingen, as the product of a more enlightened Kantian age, managed to leap ahead of the Albertina by attracting a number of leading scholars right from the start. Its reputation in mathematics, astronomy, and physics rose sharply at the beginning of the nineteenth century after acquiring the services of Carl Friedrich Gauss, who from 1807 to 1855 represented all three of these fields. Gauss's works exerted a deep influence on Dirichlet, Jacobi, Eisenstein, and Kummer, the first generation in Germany to pursue pure mathematics, but especially number theory, with an extraordinary passion. The impressive achievements of these figures influenced Hilbert as well, though less directly. He was more profoundly affected by the works of Dedekind, Kronecker, and his first mentor, Heinrich Weber.

Hilbert's professional mathematical career began officially in Königsberg in 1886 when he joined the faculty as a *Privatdozent*, an unsalaried position that left those with meager personal resources in a precarious situation. In Hilbert's case, it took another six years before he became

[1] Hermann Weyl broke Hilbert's career into five phases based on his evolving mathematical interests in Weyl (1944), but this breakdown obscures some of the most striking features of his bipartite career.

D.E. Rowe, *A Richer Picture of Mathematics*, https://doi.org/10.1007/978-3-319-67819-1_13

a Professor *Extraordinarius* in 1892, and then, only a year afterward, he was elevated to *Ordinarius* at the age of 31. It was a long, hard climb for Hilbert, but along the way he acquired an enormous breadth of mathematical knowledge that served him well during the second phase of his career. This began when he was called to Göttingen in 1895 as the successor of his former teacher, Heinrich Weber, who had accepted a chair in Strassburg.

For Hilbert, this meant leaving behind the familiar environment he had enjoyed throughout his student years and beyond – an atmosphere of quiet solitude conducive to deep mathematical contemplation – to enter another kind of academic world, one only just taking form and to which Hilbert was called upon to contribute in an extraordinary way. During the next 20 years he would become the era's most influential mathematician, attracting scores of talented disciples from around the world. As a measure of the contrast between these two phases of his career, in Königsberg he never had a single doctoral student, whereas in Göttingen he supervised the dissertations of no fewer than 60 aspiring mathematicians between the time of his arrival and the outbreak of World War I. That averages out to better than three a year, an impressive figure even today; in those days, it was simply unheard of. A good number of these Hilbert students went on to enjoy distinguished careers, and a few even left a deep imprint on twentieth-century mathematics, among them Hermann Weyl, Richard Courant, Erhard Schmidt, and Erich Hecke.

The Göttingen years were thus the period of Hilbert's real fame and glory, and so it is only natural that they have received far more attention than the Königsberg period. These more remote events in time and space have been diminished all the more because the Hilbert so many knew – the deeply dedicated teacher whose self-confidence and sarcastic wit set the tone for Göttingen's whole mathematical community after the turn of the century – had already become a legend by the time he reached his fortieth birthday. The young Hilbert, the Königsberg *Dozent* who worked for nine long years in virtual isolation, publishing only a small fraction of the mathematical ideas that occupied his attention, this Hilbert was never well known and was therefore easily forgotten after he had stepped to the podium at the Paris ICM in 1900 to speak about "mathematical problems." Thus, to gain an appreciation of the Hilbert phenomenon in full, one must factor into account the academic atmosphere and intellectual climate in which he moved while in Königsberg. This period not only had a deep significance for him personally but it also laid the foundations for the dramatic success he later enjoyed in Göttingen. For, in many respects, Hilbert's triumph there was a matter of reaping the fruits of the 15 years he spent as a student and aspiring academic in the city of his birth.

Hilbert in Königsberg

No university played a greater part in promoting Germany's sudden ascendance in the world of mathematics than did Königberg's Albertina (Fig. 13.1). Its fame in the natural sciences was launched during the 1830s by the physicist Franz Neumann and the brilliant analyst Carl Gustav Jacob Jacobi. Following Jacobi's departure for Berlin in 1844, his successor and leading disciple, Friedrich Richelot, continued to promote the ideals of the master. Largely through his efforts, the dynamic Königsberg tradition in mathematical physics, analysis, and analytical geometry made its influence felt on nearly every university in Germany. Hilbert's initial exposure to higher mathematics came through two representatives of the Jacobian tradition in analysis. In the winter semester of 1880–81, he studied analytical geometry with Georg Rosenhain and differential calculus with Louis Saalschütz. Both were natives of Königsberg who went on to take their doctorates under Jacobi and Richelot. Rosenhain helped pioneer the theory of so-called "ultraelliptic functions," which generalized the elliptic functions first introduced by Niels Henrik Abel and Jacobi. This field of research, soon vastly extended by the works of Karl Weierstrass and Bernhard Riemann, would become a central pillar of complex analysis throughout the entire century.

After a brief exposure to classical analysis, Hilbert spent his second semester in Heidelberg, where he took courses from Lazarus Fuchs on integral calculus and the theory of invariants. He then returned to Königsberg, and remained for the next four years until the completion of his studies. Hilbert thus had plenty of opportunities to learn mathematics from world-class researchers. Still, it seems the first to exert a significant influence on him was the multi-talented Heinrich Weber, who held the chair in mathematics at Königsberg from 1875 to 1883. Weber was born and raised in Heidelberg, where he earned his doctorate in 1863. During this period, however, his teachers included the physicist Gustav Kirchhoff and the geometer Otto Hesse, former students of Neumann and Jacobi who transplanted the Königsberg tradition in the southern German soil of Baden. Following the advice of his Heidelberg teachers, Weber spent the next three years as a post-doctoral student in Königsberg, where he and a number of other gifted young mathematicians and physicists worked under the supervision of Neumann and Richelot. Nine years later, he succeeded Richelot as professor of mathematics at Königsberg, a logical enough choice given Weber's strong ties with the Königsberg mathematical tradition.

At the time Hilbert first met him, Weber had just completed a paper with Richard Dedekind entitled "Theorie der algebraischen Funktionen einer Veränderlichen," which they published a bit later in *Crelle's Journal* in 1882. This work has come to be regarded as a landmark in the history of

Fig. 13.1 The Albertina in Königsberg with its Paradeplatz, where Hilbert and Hurwitz met nearly daily to take walks together.

mathematics, as it presents all the tools necessary in order to give a purely arithmetical approach to the main theorems in Riemann's theory of algebraic functions. Jean Dieudonné wrote that this classic paper of Dedekind and Weber revealed a "remarkable originality, which in all of the history of algebraic geometry is only scarcely surpassed by that of Riemann" (Dieudonné 1985, 29). Strong praise indeed! A new insight gained from this approach – and even more apparent in the closely related work of Kronecker – came with the realization that a deep, but clear analogy underlay the structure of algebraic number fields and the fields of meromorphic functions associated with Riemann surfaces. Hilbert's later work in number theory, as well as that of his students Erich Hecke, Otto Blumenthal, and Rudolf Fueter, was strongly motivated by this analogy. A clear sign of this can be seen from one of the most important of Hilbert's 23 problems, the twelfth, which had to do with Abelian extensions of algebraic number fields (Schappacher 1998). Its aim was to explore the full implications of the analogy between number fields and fields of functions first studied by Kronecker and Hilbert's teacher, Heinrich Weber. Hilbert came to regard this prospect as "one of the deepest and most profound problems of number theory and function theory."

Very few details are known about Hilbert's personal contacts with Heinrich Weber, but we do know that the young Königsberger attended Weber's courses on number theory and on the theory of elliptic functions. He also participated in a seminar Weber directed that dealt with invariant theory. The latter field soon became Hilbert's primary area of expertise, although his interest in algebraic number theory and algebraic functions continued unabated.

When Weber left Königsberg in 1883, he was succeeded by Ferdinand Lindemann, a geometer who had studied under Alfred Clebsch, but also with the latter's youngest disciple, Felix Klein. These names conjure up a host of new connections, personal as well as intellectual, many of which proved important for Hilbert's subsequent career. Clebsch had studied in Königsberg under Franz Neumann and Jacobi's pupil, Otto Hesse; he also befriended Neumann's son Carl, who became a distinguished mathematical physicist, first in Tübingen and then Leipzig. A major event that helped solidify Königsberg's mathematical legacy took place in 1869 when Clebsch and Carl Neumann co-founded *Mathematische Annalen*, which under the editorship of Klein, and then Hilbert, came to be regarded as the premier mathematics journal in the world. Clebsch thus played a leading role in propagating Königsberg's distinguished mathematical tradition. After his sudden death at age 39, his former student Lindemann prepared the publication of Clebsch's lectures on projective algrbraic geometry, a project supervised by Klein.

After his arrival in 1883, Lindemann thus provided a vital link between Königsberg and Klein, who was located in Leipzig from 1880 to 1885 and afterward in Göttingen. Although his influence on Hilbert was by no means deep, Lindemann did act as a catalyst for what became a surprisingly lively, though small, mathematical group at the Albertina. Far more decisive for Hilbert's development, however, were his contacts with two other mathematicians: Hermann Minkowski, a fellow student, and Lindemann's younger colleague, Adolf Hurwitz, who was only three years older than Hilbert. In the years ahead, these three brilliant talents would soon become the leading stars of their generation in Germany. It was through Minkowski and

Hurwitz that Hilbert learned how to "talk mathematics," the art of conversing about the essential ideas and problems that characterize a particular mathematical theory. Yet each of the three had his own distinctive style and personality.

Unlike Hilbert, whose mathematical gifts took long to ripen, Minkowski was a child prodigy who completed the curriculum at the Altstätisches Gymnasium in Königsberg by the age of 15. Two years later he submitted a paper to the Académie des Sciences in Paris in response to a prize announcement related to the problem of expressing a positive integer as the sum of five squares. Although it was written in German, contrary to the stipulations of the Academy, Minkowski's paper was awarded the Grand Prix des Sciences Mathématiques, creating a minor sensation in the world of mathematics. A year later, Minkowski began his studies at the *Albertina* alongside Hilbert. His principal teachers were Heinrich Weber and the physicist, Woldemar Voigt. After five semesters, however, he left Königsberg to study in Berlin, where he attended the lectures of such luminaries as Weierstrass, Kummer, Helmholtz, Kirchhoff, and, above all, Leopold Kronecker. He then returned to Königsberg, passed his doctoral exams, and submitted a dissertation that dealt with the theory of quadratic forms; it was published soon thereafter in volume 7 of *Acta Mathematica*.

Minkowski was a sharp-witted, though outwardly shy young man, with a sarcastic sense of humor that matched Hilbert's own. His relationship with Adolf Hurwitz, who came to Königsberg in the spring of 1884 to assume a newly created *Extraordinariat*, was far more formal, befitting that of professor and student, despite their closeness in age. Nevertheless, it was the quiet Hurwitz who exerted the most decisive influence on Hilbert's mathematical sensibilities, in particular his longing for universal breadth of knowledge. Adolf Hurwitz grew up in Hildesheim, where his father was a merchant. Like Minkowski, he came from a Jewish family and, like him too, his talent for mathematics was apparent from an early age. His Gymnasium teacher was Hermann Schubert, creator of the Schubert calculus (the topic of Hilbert's 15th problem), and one of the leading experts in the difficult field of enumerative geometry (see Kleiman 1976). Hurwitz absorbed his teacher's work like a sponge, and by the time he was only 17 they co-published a paper in the *Göttinger Nachrichten* that dealt with Chasles' theory of characteristics. It was Schubert who advised Hurwitz to study under Klein, who was then teaching at the Technische Hochschule in Munich.

Hurwitz's father had strong reservations about the wisdom of pursuing an academic career, which was hard enough for a family with limited means, but particularly difficult for Jews (see Chap. 14). Most German universities were reluctant to appoint Jewish scholars, and in some fields Jews had virtually no chance of obtaining a professorship. Hurwitz also had health problems, and this was worrisome as well. Felix Klein, already a powerful figure in German mathematics by 1880, was aware of these concerns. In that year, he wrote Hurwitz's father: "Above all, I want to stress that among the totality of young people with whom I have up until now worked there was not one who in specifically mathematical talent could measure up to your son. From now on your son will enjoy a brilliant scientific career, which is all the more certain because his gifts are combined with endearing personality traits" (Rowe 1986, 432). Adolf Hurwitz was, in fact, a charming young man. He also happened to be a talented pianist, who loved to make music with friends. Years later in Zürich, Albert Einstein joined the Hurwitz family in such festivities.

Hurwitz went on to become Klein's star pupil, but he also studied in Berlin before habilitating in Göttingen. During the 1860s and 70s, Berlin held far more attraction for aspiring mathematicians than any other university in Germany. Led by Kummer, Weierstrass, and Kronecker, the Berliners not only drew much young talent they also virtually monopolized university appointments in Prussia. In Hurwitz's case, this sojourn afforded him the opportunity to broaden his knowledge in complex analysis. He had already mastered the theory of Riemann surfaces, a staple of Klein's school, so he now immersed himself in the competing approach – based on the analytic extension of power series representations – that had been developed by Weierstrass and his students. Still more interesting for Hurwitz than Weierstrass's lectures were those of Kronecker. As he related in a letter to his friend Luigi Bianchi, he came to Berlin mainly in order to attend Kronecker's lectures on number theory (Hurwitz to Bianchi, 20 March 1882, in Bianchi (1959), 80). Minkowski had also studied under Kronecker during the early 1880s and, like Hurwitz, he too came away deeply impressed by the elderly master's achievements. At the same time, both Minkowski and Hurwitz found Kronecker's personality much less appealing than his mathematics. In his letter to Bianchi, Hurwitz called him "that great, but very vain mathematician," and Minkowski's letters to Hilbert from the late 1880s and early 1890s are strewn with similar remarks (see Minkowski 1973, 64, 80, 107). In all likelihood, Hilbert had a similar opinion of Kronecker, who was generally seen as a controversial figure.

When he arrived in Königsberg in 1884, Hurwitz was at the height of his powers. He opened up whole new mathematical vistas to Hilbert, who looked up to him with admiration, mixed with a tinge of envy. He later said about him that "Minkowski and I were totally overwhelmed by his knowledge, and we never thought we would ever come that far" (Blumenthal 1935, 390). Hurwitz's appointment had been engineered by Ferdinand Lindemann, Hilbert's future *Doktorvater*, who was called to Königsberg one year earlier after winning international acclaim for his proof that π is a transcendental number, thereby finally establishing the impossibility of squaring the circle. This odd tale has some interesting ties with Hilbert's life story as well (as described above in the introduction).

The young Hilbert was rather fond of Lindemann, even though he hardly viewed him with the kind of awe he felt for Hurwitz's mathematical abilities. Still, he clearly appreciated how his mentor's success in solving a very famous problem had been pivotal for his career. He also eventually came to realize that the truly difficult work had been accomplished by Charles Hermite some 10 years earlier when he proved the transcendence of e. Hilbert would later show that the transcendence of π follows fairly easily once Hermite's fundamental methods are called into play. Still, many claims to fame in the history of mathematics involved good luck, and in this sense Lindemann's famous breakthrough was in no way exceptional. His subsequent career, however, was not nearly so lucky: it was marked by a series of bogus proofs that purported to solve Fermat's famous last conjecture. Both Hurwitz and Minkowski expressed something close to shock at the elementary blunders Lindemann committed along the way; Hilbert, on the other hand, seems to have done his best to avert his eyes from these embarrassments.

Lindemann's memoirs (Lindemann 1971) contain many interesting anecdotes about mathematical life in Königsberg during the 10 years he taught there, and these provide some helpful clues about the role of past tradition. In fact, Lindemann himself was reluctant to leave his professorship in the comfortable environs of Freiburg im Breisgau, but he overcame these doubts after a colleague told him that he simply could not turn down the chair that had once been held by Jacobi. Königsberg's isolated location certainly contributed to the university's stagnation during the 1860s and 70s. But judging from Lindemann's recollections, it would seem that many professors felt inclined to follow in the footsteps of their famous predecessors. Not surprisingly, this attitude had the effect of stifling new innovative ideas. At any rate, when Lindemann came to Königsberg in 1883 he found the facilities woefully inadequate. The mathematics seminar library, which consisted of little more than the 20 or so volumes of the *Mathematische Annalen* that had appeared since 1869, was housed in the university's detention quarters (the *Karzer*), along with some of Königsberg's rowdier students. After several years of futile negotiations, Lindemann managed to obtain a separate room for the mathematics seminar, although the mathematicians were forced to share this with a professor of medicine who used it to store pharmaceutical preparations. Only with Hilbert's appointment as Lindemann's successor did the mathematics seminar and library receive its own quarters, taking over a room on the top floor of the university building that had formerly been occupied by one Secretary Lorkowsky.

In 1883 Königsberg had no special lecture halls for mathematics. Jacobi and his successors had delivered their lectures at a podium next to which stood two blackboards on wooden easels of the same type used in the schools of that day. Lindemann put forth a motion that the mathematics lectures be held in a special room in which a large slate blackboard was to be installed. But the faculty voted this proposal down, pointing out that mathematics had been taught successfully in Königsberg for some 50 years without any such modern devices. The faculty also pointed out to Lindemann that rooms could not be reserved for particular subjects as they were allocated according to the seniority of the professors who requested them. Soon after this, Lindemann decided to go directly to the university's *Kurator*, and not long afterward a modern blackboard was installed in one of the rooms.

Yet Lindemann also recalled how impressed he was by some of the older students in Königsberg who had studied under Heinrich Weber. He even admitted that some of them had already surpassed him in their knowledge of mathematics. Both of Lindemann's teachers, Clebsch and Klein, had excelled in attracting small groups of dedicated disciples, and Lindemann did his best to follow suit. Once a week, he held a colloquium in his home, which he organized together with Hurwitz. The two most active students at these meetings were Hilbert and Emil Wiechert, who later became an outstanding geophysicist and a colleague of Hilbert's in Göttingen. Minkowski also took part in these colloquia. After the meetings, which were sometimes attended by old-timers like Rosenhain, the group would gather in a restaurant for dinner, talking about mathematics the whole time.

On other occasions, Hilbert would join his cohorts on a mathematical walk along the so-called *Paradeplatz* before the university building. One day he ran into Lindemann there and proceeded to tell him about some results he had obtained on continued fractions. Hilbert wondered if this work might prove acceptable for his dissertation (Fig. 13.2). Immediately thereafter, Lindemann went home only to discover that Hilbert's main findings had already been established by Jacobi. The result of this encounter was that Lindemann suggested another dissertation topic to him, one concerned with the invariant-theoretic properties of spherical harmonics. He knew that Hilbert had studied invariant theory in a seminar with Weber, but he probably did not anticipate that his pupil would strike out in a direction different from the methods he himself knew best. Lindemann soon came to realize that the complicated symbolic calculus of Siegfried Aronhold and Alfred Clebsch had little appeal for Hilbert, just as it was quite foreign to Heinrich Weber. Lindemann also recalled how he later met Weber, and when he told him about Hilbert's dissertation research the latter responded in surprise: he would have saved such a beautiful theme for himself rather than assigning it to a student (Lindemann 1971, 91). As it turned out, the topic pointed Hilbert in a direction that would dominate his interests for the next eight

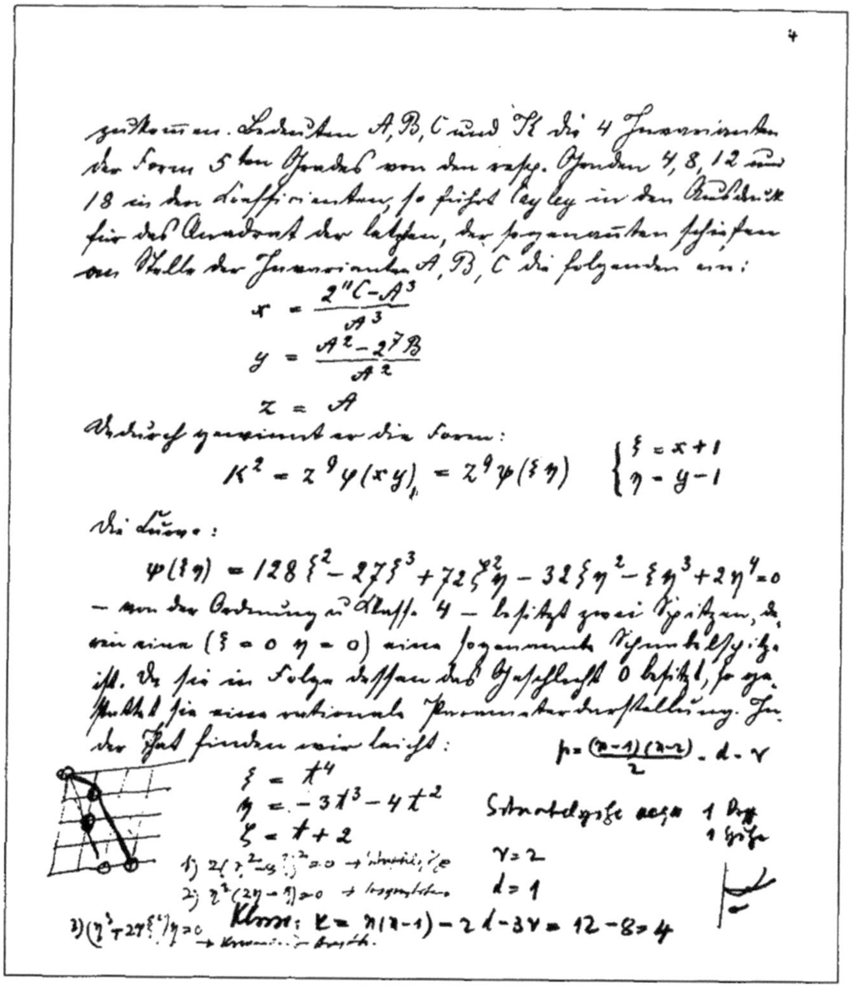

Fig. 13.2 A page from the protocol for the Königsberg mathematical colloquium documenting a presentation by Hilbert on 19 May 1884. The topic was the reality of the roots of quintic equations as treated by Sylvester, Cayley, and others (Teilnachlass von Ferdinand Lindemann, Universität Würzburg).

years, during which time he made his name as one of the world's leading authorities on invariant theory.

Entering Klein's Circle

After taking his doctorate in 1885, Hilbert spent one last summer in Königsberg rounding out his education. Like many other aspiring mathematicians in Germany, he took and passed the *Staatsexamen*, which ensured that he could apply for a teaching position at a Gymnasium in the event that his plans to pursue a university career failed to fructify. During this final semester, he attended Lindemann's geometry course, which dealt with Plücker's line geometry and Lie's sphere geometry, and he also followed Hurwitz's lectures on modular functions. Then he departed for Leipzig, where he was welcomed by Hurwitz's former mentor, Felix Klein.

Hilbert spent the winter semester of 1885–86 working under Klein, who led an active group of young researchers. One of these, Eduard Study, shared Hilbert's interests in algebraic geometry and invariant theory. Study, however, favored the symbolic methods of Aronhold and Clebsch, whereas Hilbert was experimenting with other formalisms at this time. Whether for this reason or simply because he found Study a bit overbearing in his opinions about mathematics, a rather tense relationship developed between them. For his part, Study may well have sensed that Klein was more than a little interested in this newcomer passing through from Königsberg. His own future dealings with Klein were often unpleasant ones, and he was not alone in feeling that Klein treated him unfairly.

Recognizing Hilbert's talent, Klein quickly took him under his wing while confiding to him the latest events of interest in Leipzig. First among these was Klein's own imminent departure for Göttingen and his efforts to appoint Sophus Lie as his successor. After considerable squabbling, Klein was able to inform Hilbert at year's end that Lie had received an official offer from the Saxon Ministry and had accepted the same. This unofficial news was conveyed by Hilbert to Hurwitz in a letter that continued: "New Year's Eve I was invited to Prof. Klein's and found myself in very small but exclusive company consisting of Prof. Klein, his wife, the Prague *Privatdozent* Dr. Georg Pick, and me. We drank a New Year's punch with great pleasure and chattered about all conceivable and inconceivable things, amusing ourselves wonderfully. Prof. Klein tried hard to convince me to spend the next semester studying in Paris. He described Paris as a beehive of activity, especially among the young mathematicians, and thought that in view of this a period of study there would be most stimulating and profitable for me."[2] Klein even joked that Hilbert should try to befriend Henri Poincaré and join him in drinking a toast to "mathematical brotherhood."

For Felix Klein, memories remained fresh of the Parisian mathematical scene he and Lie had encountered in the spring of 1870. Even fresher, to be sure, was the painful experience of watching Poincaré sail by him during the early 1880s when both were working intensively on the theory of automorphic functions (see Gray 2000). But if Klein had special reasons for counseling Hilbert to visit Paris, such advice would hardly have been rare. Indeed, Lindemann had also studied in Paris during the mid 1870s. This gave him the opportunity to learn directly from the source about Hermite's proof of the transcendence of e, a result the French mathematician regarded as among his most important findings (Lindemann 1971, 70). Lindemann clearly thought so, too, since it led him to pursue the closely connected problem of proving the transcendence of π, the solution to which made him famous overnight.

Hilbert's journey to Paris turned out to be considerably less eventful than those undertaken earlier by Klein and Lindemann. Joining his rival invariant-theorist, Study, whom Klein had also persuaded to make the trip, Hilbert got to meet most of the mathematicians he had hoped to see, including Poincaré, Émile Picard, and Paul Appell. Still, difficulties with the language prevented Hilbert from breaking the ice with many, including Poincaré. Political tensions between France and Germany had not lessened noticeably during these years, but Hermite, the leader of a burgeoning school in analysis, took a strong internationalist stance in an effort to cultivate better relations among mathematicians. Hermite had attended the ceremonies in 1877 that were held in Göttingen honoring the centenary of the birth of Gauss, and he never tired of extolling the achievements of leading German mathematicians. Lindemann recalled how Hermite had warmly urged him to pledge himself to the noble cause of promoting the international brotherhood of mathematicians, and Klein may well have had this incident in mind when he urged Hilbert to do the same when he met Poincaré.

Hermite had long been the primary conduit within France for German work in pure mathematics, and despite his advanced age – he was then 64 – it was he who took the strongest interest in the two young visitors. Hilbert found him absolutely fascinating and delightful. He reported to Klein how Hermite had told both him and Study about his reciprocity theorem for binary invariants, encouraging them to try to extend it to ternary forms. He also told them about his ongoing correspondence with Sylvester, who was struggling to find an independent proof of Gordan's theorem (David Hilbert to Felix Klein, 21 April 1886 (Frei 1985, 9)). Klein kept urging Hilbert to seek out as many contacts as possible, but time was short and Hilbert had to worry about his first priority, namely completing his *Habilitationsschrift* so that he could qualify for a position as a *Privatdozent* in Königsberg. Klein, who could roll more into a day than most people would attempt in a week, saw no reason why others should not live by the same insane schedule. Even before Hilbert had departed from Paris, Klein was already scolding him for not taking full advantage of what the city had to offer. "Hold before your eyes," he wrote his young protégé, "that the opportunity you have now will never come again." That proved to be true only for the time being. He would have his second chance in 1900, but in the meantime a subdued rivalry developed between him and Poincaré, the era's leading mathematician. Hilbert returned to the quiet environs of Königsberg, submitted his *Habilitationsschrift* in the summer of 1886, and began a six-year long-period as *Privatdozent.*

The following spring, Minkowski departed for Bonn where he, too, began teaching as a *Privatdozent.* Thereafter, Hurwitz became Hilbert's primary source of intellectual stimulation. Probably the most memorable moments for Hilbert, when he thought back on the years from 1886 to 1892, were the almost daily walks he undertook with Hurwitz. He recalled how together they wandered through nearly every corner of mathematics, with his friend and former teacher acting as guide. No doubt they often discussed ideas that Hilbert was working on in the context of his lecture courses. Königsberg was just about the last place one could find mathematics students during the late 1880s and early 1890s, and Hilbert's courses were often attended by no more than two or three auditors, sometimes even fewer! He complained about these circumstances occasionally, but never seemed to be really bothered by the situation. His goal, after all, was clear enough: he wanted to become a truly universal mathematician. Indeed, his lecture courses, supplemented by nearly daily discussions with Hurwitz, served as vehicles for that purpose, and they spanned practically every area of higher mathematics of the day: from invariant theory, number theory, and analytic, projective, algebraic, and differential geometry to Galois theory, potential theory, differential equations, function theory, and even hydrodynamics. During his entire nine years on the Königsberg faculty he rarely lectured on any subject more than once. Surely Hilbert read a great deal, but more important to him still were the opportunities to talk about mathematics. Thus, right from the very start, the power of the

[2] Hilbert to Hurwitz, 2 January 1886, Mathematiker-Archiv, Niedersächsische Staats- und Universitätsbibliothek Göttingen.

spoken work that stimulates the mathematical imagination played a central role in his work. Without it, his phenomenal success in Göttingen simply would not have been possible.

Encounters with Allies and Rivals

> *It seems to me that the mathematicians of today understand each other far too little and that they do not take an intense enough interest in one another. They also seem to know – so far as I can judge – too little of our classical authors (Klassiker); many, moreover, spend much effort working on dead ends.* – David Hilbert to Felix Klein, 24 July 1890 (Frei 1985, 68).

Probably no mathematician has been quoted more often than Hilbert, whose opinions and witty remarks long ago entered the realm of mathematical lore along with tales of his legendary feats. Fame gave him a captive audience, but as the quotation above illustrates, even before he attained such high standing Hilbert never had any difficulty expressing his views. When he wrote these words, in fact, he had just completed a ground-breaking paper on invariant theory (Hilbert 1890), the first in an impressive string of achievements that would vault him to the top of his profession. Initially, Hilbert made his name as an expert on invariant theory, but his reputation as a universal mathematician grew as he left his mark on one field after another. Yet these achievements alone can hardly account for his singular place in the history of mathematics, as was recognized long ago by his intimates (Weyl 1932; Blumenthal 1935). Those who belonged to Hilbert's inner-circle during his first two decades in Göttingen invariably pointed to the impact of his personality, which clearly transcended the ideas found between the covers of his collected works (see Weyl 1944; Reid 1970). Hilbert's name and thoughts of fame mingled in the minds of many hopeful young mathematicians who felt inspired to tackle one of the 23 "Hilbert problems," some of which hardly deserve that name; several were well known before, but he dusted them off and presented them anew at the Paris ICM in 1900. Suddenly they acquired a fascination all their own. As Ben Yandell puts it in his delightful survey, *The Honors Class*, "solving one of Hilbert's problems has been the romantic dream of many a mathematician" (Yandell 2002, 3).[3]

Hilbert's ability to inspire was clearly central to Göttingen's success, even if only a part. His leadership style fostered what I have characterized as a new type of oral culture, a highly competitive mathematical community in which the spoken word often carried more weight than did information conveyed in written texts (see Rowe 2003, 2004). Hilbert was an unusually social creature: outspoken, ambitious, eccentric, and above all full of passion for his calling. Moreover, he was a man of action with no patience for hollow words. Thus, when in July 1890 he conveyed to Klein the rather harsh views cited in the opening quotation, he was not merely bemoaning the lack of communal camaraderie among Germany's mathematicians; on the contrary, he was expressing his hope that these circumstances would soon change. For plans were just then underway to found a national organization of German mathematicians, the *Deutsche Mathematiker-Vereinigung*, and Hilbert was delighted to learn that Klein would be present for the inaugural meeting which took place a few months later in Bremen (Fig. 13.3). Both knew that much was at stake; as Hilbert expressed it: "I believe that closer personal contact between mathematicians would, in fact, be very desirable for our science" (Hilbert to Klein, 24 July 1890 (Frei 1985, 68)). Soon after his arrival in Göttingen in 1895, Hilbert put this philosophy into practice. At the same time, his optimism and self-confidence spilled over and inspired nearly all the young people who entered his circle.

Hilbert's impact on modern mathematics has been so pervasive that it takes a true leap of historical imagination to picture him as a young man struggling to find his way. Still, many of the seeds of later success were planted in his youth, just as several episodes from his early career have since become familiar parts of the Hilbert legend. The preceding pages describe the quiet early years he spent in Königsberg, where he befriended two young mathematicians who influenced him more than any others, Adolf Hurwitz and Hermann Minkowski. By 1885 Hilbert emerged as one of Felix Klein's most promising protégés. In this role, he traveled to Paris in order to meet the younger generation of French mathematicians, especially Henri Poincaré, reporting back all the while to Klein, who avidly awaited news about the Parisian mathematical scene. Here I pick up the story at the point when Hilbert returned from this first trip to Paris. Afterward he had several important encounters with other leading mathematicians in Germany. These not only shed considerable light on the contexts that motivated much of his work, they also reveal how he positioned himself within the fast-changing German mathematical community to which he belonged.

Returning from Paris

After a rather uneventful and disappointing stay in Paris during the spring of 1886, Hilbert began his long journey home. Stopping first in Göttingen, he learned something about the contemporary Berlin scene when he met with Hermann Amandus Schwarz, the senior mathematician on the faculty. Schwarz had long been one of the closest of Karl Weierstrass's many adoring pupils in Berlin; yet much had changed since the days when Charles Hermite advised young

[3]It was Hilbert's star pupil, Hermann Weyl, who called those who actually succeeded the "honors class"; he also wrote that "no mathematician of equal stature has risen from our generation" (Weyl 1944, 130).

Fig. 13.3 Founding of the Deutsche Mathematiker-Vereinigung (DMV), Bremen, September 1890. Klein (sitting) is fifth from the *right*; Cantor fourth from the *left*. Hilbert and Minkowski are second and third from the *left* in the back row.

Gösta Mittag-Leffler to leave Paris and go to hear the lectures of Weierstrass, "the master of us all" (see Rowe 1998). During the 1860s and 70s, the Berliners had dominated mathematics throughout Prussia, with the single exception of Königsberg, which remained an enclave for those with close ties to the Clebsch school and its journal, *Mathematische Annalen* (see Chap. 4). However, following E. E. Kummer's departure in 1883, the once harmonious atmosphere he had cultivated as Berlin's senior mathematician gave way to acrimony. Weierstrass, old, frail, and decrepit, refused to retire for fear of losing all influence to Leopold Kronecker, who remained amazingly energetic and prolific despite his more than 60 years.

Presumably Hilbert heard something about this situation from Schwarz, who would have conveyed the crux of the matter as seen from Weierstrass' perspective (see Biermann 1988, 137–139). If so, Hilbert would have learned that relations between Weierstrass and Kronecker had deteriorated mainly because of the latter's dogmatic views, in particular his sharp criticism of Weierstrass' approach to the foundations of analysis. Only a few months after Hilbert passed through Berlin, Kronecker delivered a speech in which he uttered his most famous phrase "God made the natural numbers; all else is the work of man" ("Die ganzen Zahlen hat der liebe Gott gemacht, alles andere ist Menschenwerk") (Weber 1893, 19). Kronecker had never made a secret of his views on foundations, but by the mid-1880s he was propounding these with missionary zeal. No one was more taken aback by this than H. A. Schwarz, to whom Kronecker wrote one year earlier:

> If enough years and power remain, I will show the mathematical world that not only geometry but also arithmetic can point the way for analysis – and certainly with more rigor. If I don't do it, then those who come after me will, and they will also recognize the invalidity of all the procedures with which the so-called analysis now operates (Biermann 1988, 138).

Weierstrass wrote to Schwarz around this same time claiming that Kronecker had transferred his former antipathy for Georg Cantor's work to his own. And in another letter, written to Sofia Kovalevskaya, he characterized the issue dividing them as rooted in matters of mathematical ontology: "whereas I assert that a so-called irrational number has a real existence like anything else in the world of thought, according to Kronecker it is an axiom that there are only equations between whole numbers" (Weierstrass to Kovalevskaya, 24 March, 1885, quoted in Biermann (1988, 137)). Whether or not Schwarz alluded to this rivalry when he spoke to Hilbert, he definitely did warn him to expect a cold reception when he presented himself to Kronecker (Hilbert to Klein, 9 July 1886, in Frei (1985, 15)). Instead, however, Hilbert was greeted in Berlin with open arms, and his initial reaction to Kronecker was for the most part positive.

Back in his native Königsberg, Hilbert reported to the ever-curious Klein about these and other recent events. He had just completed all requirements for the *Habilitation* except for the last, an inaugural lecture to be delivered in the main auditorium of the *Albertina*. His chosen theme was a propitious one: recent progress in the theory of invariants. Hilbert was pleased to be back in Königsberg as a *Privatdozent*, even though this meant he was far removed from mathematicians at other German universities. To compensate for this isolation, he was planning to tour various mathematical centers the following year when he hoped to meet Professors Gordan and Noether in Erlangen. Although he had to postpone this trip until the spring of 1888, this venture eventually turned out to be far more fruitful than his earlier journey to Paris. What is more, it helped him solidify his relationship with Klein, who always urged young mathematicians to cultivate personal contacts with fellow researchers both at home and abroad (see for example Hashagen (2003, 105–116, 149–162)). Within Klein's network, the Erlangen mathematicians, Paul Gordan and Max Noether, played particularly important roles. The latter was Germany's foremost algebraic geometer in the tradition of Alfred Clebsch, whereas the former was an old-fashioned algorist who loved to talk mathematics.

Felix Klein knew from first hand experience how stimulating collaboration with Paul Gordan could be. During the late 1870s, when Klein taught at the Technical University in Munich, he took advantage of every chance he got to meet with his erstwhile Erlangen colleague, who was himself enormously impressed by Klein's fertile geometrical imagination. Gordan was widely regarded as Germany's premier authority on algebraic invariant theory, the field that would dominate Hilbert's attention for the next five years. His principal claim to fame was Gordan's Theorem, which he proved in 1868. This states that the complete system of invariants of a binary form can always be expressed in terms of a finite number of such invariants. In 1856 Arthur Cayley had proved this for binary forms of degree 3 and 4, but Gordan was able to use the symbolical method introduced by Siegfried Aronhold to obtain the general result. During his stay in Paris, Hilbert had briefly reported to Klein about these matters (Hilbert to Klein, 21 April 1886, in Frei (1985, 9)).There he learned from Charles Hermite about J. J. Sylvester's recent efforts to prove Gordan's Theorem using the original British techniques he and Cayley had developed for invariant theory. Hilbert thus became aware of the fact that the elderly Sylvester was still trying to get back into this race (see Parshall 1989). Presumably he felt that progress was unlikely to come from this old-fashioned line of attack, but neither had the symbolical methods employed by German algebraists produced any substantial new results since Gordan first unveiled his theorem (Fig. 13.4).

Fig. 13.4 David Hilbert, from the time when he was mainly known as an expert on invariant theory.

Over the next two years Hilbert had ample time to master the various competing techniques of Continental and British invariant theorists. After becoming a *Privatdozent* in Königsberg, he was free to develop his own research program, and his inaugural lecture on recent research in invariant theory clearly indicates the general direction in which he was moving. Still, there were no signs that he was already on the path toward a major breakthrough. Indeed, tucked away in Königsberg, it seems unlikely that he even saw the need to strike out on in an entirely new direction in order to make substantial progress beyond Gordan's original finiteness theorem. That goal, nevertheless, was clearly in the back of his mind when he set off in March 1888 on a tour of several leading mathematical centers in Germany, including Berlin, Leipzig, and Göttingen. During the course of a month, he spoke with some 20 mathematicians from whom he gained a stimulating overview of current research interests throughout the country. Although we can only capture glimpses of these encounters, a number of impressions can be gained from Hilbert's letters to Klein, as well as from notes he took of his conversations with various colleagues.[4]

A Second Encounter with Kronecker

In Berlin, he once again met with Kronecker, who on two separate occasions afforded him a lengthy account of his general views on mathematics and much else related to Hilbert's own research. A gregarious, outspoken man, the elderly Kronecker still exuded considerable intensity, so Hilbert learned a great deal from him during the 4 hours they spent together. Reporting to Klein, he described the Berlin mathematician's opinions as "original, if also somewhat derogatory" (Hilbert to Klein, 16 March 1888 (Frei 1985, 38)). Hilbert told Kronecker about a paper he had just written on certain positive definite forms that cannot be represented as a sum of squares. Kronecker replied that he, too, had encountered forms that cannot be so represented, but he admitted that he did not know Hilbert's main theorem, which dealt with three cases in which a sum of squares representation is, indeed, always possible ("Bericht über meine Reise," Hilbert Nachlass 741).

A noteworthy feature of this paper, (Hilbert 1888a), is that Hilbert actually credits Kronecker with having introduced the general principal behind his investigation. This work lies at the roots of Hilbert's seventeenth Paris problem, which also played an important role in connection with his research on foundations of geometry. Interestingly enough, a second Hilbert problem, the sixteenth, also crept into his conversations with Kronecker. This topic concerns the possible topological configurations among the components of a real algebraic curve. The maximal number of such components had been established by Axel Harnack, a student of Klein's, in a celebrated theorem from 1876 (for a summary of subsequent results, see Yandell 2002, 276–278). Kronecker assured Hilbert that his own theory of characteristics, as presented in a paper from 1878, enabled one to answer all questions of this type, clearly an overly optimistic assessment.

Whatever he may have thought about Kronecker's "priority claims," Hilbert stood up and took notice when his host voiced some sharp views about the significance of invariant theory. Kronecker dismissed the whole theory of formal invariants as a topic that would die out just as surely as had happened with the work of Hindenburg's combinatorial school (the latter had flourished in Leipzig at the beginning of the nineteenth century, but by the 1880s it had already entered the dustbin of history). The only true invariants, in Kronecker's view, were not the "literal" ones based on algebraic forms but rather numbers, such as the signature of a quadratic form (Sylvester's theorem, the algebraist's version of conservation of inertia). He then proceeded to wax forth over foundational issues, beginning with the assertion that "equality" only has meaning in relation to whole numbers and ratios of whole numbers. Everything beyond this, all irrational quantities, must be represented either implicitly by a finite formula (e.g., $x^2 = 5$), or by means of approximations. Using these notions, he told Hilbert, one can establish a firm foundation for analysis that avoided the Weierstrassian notions of equality and continuity. He further decried the confusion that often resulted when mathematicians treated the implicitly given irrational quantity (say $x = \sqrt{5}$) as equivalent to some sequence of rational numbers that serve as an approximation for it.

Not surprisingly, Hilbert took fairly extensive notes when Kronecker began expounding these highly unorthodox views ("Bericht über meine Reise"). But he also jotted down a brief comment made by Weierstrass that sheds considerable light on the differences between these two mathematical personalities. Thus, when Hilbert visited Weierstrass shortly afterward he informed the distinguished analyst about Kronecker's comments regarding invariant theory, including his colleague's prediction that the whole field would soon be forgotten just like the work of the Leipzig combinatorial school. Weierstrass responded by sounding a gentle warning to those who might wish to prophesy the future of a mathematical theory: "Not everything of the combinatorial school has perished," he said, "and much of invariant theory will pass away, too, but not from it alone. For from everything the essential must first gradually crystallize, and it is neither possible nor is it our duty to decide in advance what is significant; nor should such considerations cause us to demur in investigating such invariant-theoretic questions deeply" ("Bericht über meine Reise").

[4]"Bericht über meine Reise vom 9ten März bis 7ten April 1888," Hilbert Nachlass 741, Handschriftenabteilung, Niedersächsische Staats- und Universitätsbibliothek Göttingen.

These words, with their almost fatalistic ring, probably left little impression on the young mathematician who recorded them. For Hilbert's intellectual outlook was filled with a buoyant optimism that left no room for resignation. He may not have enjoyed Kronecker's *braggadacio*, but he was clearly far more receptive to his passionate vision than to Weierstrass' much more subdued outlook. Moreover, mathematically he was far closer to the algebraist than to the analyst. Even in his later work in analysis, Hilbert showed that his principal strength as a mathematician stemmed from his mastery of the techniques of higher algebra (see Toeplitz 1922). True, Klein and Hurwitz had drawn his attention to Weierstrass's theory of periodic complex-valued functions, about which he spoke in his *Habilitationsvortrag* shortly after returning to Königsberg from Paris. Nevertheless, Kronecker's algebraic researches clearly lay much closer to his heart. Soon after their encounter in Berlin, Hilbert would enter Leopold Kronecker's principal research domain, the theory of algebraic number fields. The latter's sudden death in 1891 may well have emboldened him to reconstruct this entire theory six years later in his "*Zahlbericht*." As Hermann Weyl later emphasized, Hilbert's ambivalence with respect to Kronecker's legacy emerged as a major theme throughout the course of his career (Weyl 1944, 613).

Like Hilbert, his two closest mathematical friends, Hurwitz and Minkowski, also held ambivalent views when it came to Kronecker. No doubt these were colored by their mutual desire to step beyond the lengthy shadows that he and Richard Dedekind, the other leading algebraist of the older generation, had cast. Since Dedekind had long before withdrawn to his native Brunswick, a city well off the beaten path of Germany's flowering mathematical community, it was only natural that this trio from Königsberg came to regard the powerful and opinionated Berlin mathematician as their single most imposing rival. In later years, Hilbert developed a deep antipathy toward Kronecker's philosophical views, and he did not hesitate to criticize these before public audiences (see Hilbert 1922). Yet during the early stages of his career such misgivings – if he had any – remained very much in the background. Indeed, all of Hilbert's work on invariant theory was deeply influenced by Kronecker's approach to algebra.

Hilbert's encounters in the spring of 1888 with Berlin's two senior mathematicians left a deep and lasting impression.[5] Based on the notes he took from these conversations, he must have felt particularly aroused by Kronecker's critical views with regard to invariant theory, for he surely found no solace in Weierstrass' stoic advice.

Primed for action and out to conquer, Hilbert could never have contemplated devoting his whole life to a theory that might later be judged as having no intrinsic significance. Whichever problems he chose to work on – even those he merely thought about but never really tried to solve – he always considered these as constituting important mathematics. What makes a problem or a theory important? Probably Hilbert carried that question within him for a long time; one need only read the text of his famous Paris address to see how compelling his views on the significance of mathematical thought could be. From the vantage point of his early, still formative years, we can begin to grasp how his larger views about the character and significance of mathematical ideas fell into place. A few strands of that story emerge from the discussions he engaged in during this whirlwind tour through leading outposts of the German mathematical community.

Tackling Gordan's Problem

From Berlin, Hilbert went on to Leipzig, where he finally got the chance to meet face-to-face with Paul Gordan. Despite their mathematical differences the two hit it off splendidly, as both loved nothing more than to talk about mathematics. Hermann Weyl once described Gordan as "a queer fellow, impulsive and one-sided" with "something of the air of the eternal 'Bursche' of the 1848 type about him – an air of dressing gown, beer and tobacco, relieved however by a keen sense of humor and a strong dash of wit . . . . A great walker and talker – he liked that kind of walk to which frequent stops at a beer-garden or a café belong" (Weyl 1935, 203). Having heard about Hilbert's talents, Gordan was eager to make the young man's acquaintance, so much so that he wished to remain *incognito* while in Leipzig to take full advantage of the opportunity (Hilbert to Klein, 16 March 1888 (Frei 1985, 38)).

Although originally an expert on Abelian functions, Gordan had long since focused his attention almost exclusively on the theory of algebraic invariants. This field of research traces back to a fundamental paper published by George Boole in 1841, as it was this work that inspired young Arthur Cayley to take up the topic in earnest (Parshall 1989, 160–166). Following an initial plunge into the field, Cayley joined forces with another professional lawyer who became his lifelong friend, J. J. Sylvester. Together they effectively launched invariant theory as a specialized field of research. Much of its standard terminology was introduced by Sylvester in a major paper from 1853. Thus, for a given binary form $f(x,y)$, a homogeneous polynomial J in the coefficients of f left fixed by all linear substitutions (up to a fixed power of the determinant of the substitution) is called an invariant of the form f. In 1868 Gordan showed that for *any* binary form one can always construct m invariants such that every other invariant can be expressed algebraically in terms of these m basis elements. Indeed, he proved this held generally for homogeneous polynomials J in the coefficients and variables

[5] He recalled this trip when he spoke about his life on his seventieth birthday; see Reid (1970, 202).

Fig. 13.5 Paul Gordan, the "King of Invariants".

of *f(x,y)* with the same invariance property (Sylvester called such an expression *J* a *concomitant* of the given form, but the term *covariant* soon became standard).

In 1856 Cayley published the first finiteness results for binary forms, but in the course of doing so he committed a major blunder by arguing that the number of irreducible invariants was necessarily infinite for forms of degree five and higher (Parshall 1989, 167–179). Paul Gordan was the first to show that Cayley's conclusion was incorrect (Fig. 13.5). More importantly, in the course of doing so he proved his finite basis theorem for binary forms of arbitrary degree by showing how to construct a complete system of invariants and covariants. Two years later, he was able to extend this result to any finite system of binary forms. His proofs of these key theorems, as later presented in (Gordan 1885/1887), were purely algebraic and constructive in nature. They were also impressively complicated, so that subsequent attempts, including Gordan's own, to extend his theorem to ternary forms had produced only rather meager results. In her dissertation from 1907, written under Gordan, Emmy Noether dealt with the case of fourth-degree ternary forms.

Only scant historical evidence has survived relating to Hilbert's first encounter with Gordan, but it is enough to reconstruct a plausible picture of what occurred. Gordan may have been a fairly old dog, but this does not mean he was averse to learning some new tricks. So even though he and Hilbert had divergent views about many things, they nevertheless understood each other well (Hilbert to Klein, 16 March 1888 (Frei 1985, 38)). Their conversations soon focused on finiteness results, in particular a fairly recent proof of Gordan's finiteness theorem for systems of binary forms published by Franz Mertens in *Crelle's Journal* (Mertens 1887). This paper broke new ground. For unlike Gordan's proof, which was based on the symbolic calculus of Clebsch and Aronhold, Mertens's proof was *not* strictly constructive. Gordan and Hilbert apparently discussed it in considerable detail, and Hilbert immediately set about trying to improve Mertens's proof, which employed a rather complicated induction argument on the degree of the forms. After spending a good week together with Gordan, he was delighted to report to Klein that: "with the stimulating help of Prof. Gordan an infinite series of thought vibrations has been generated within me, and in particular, so we believe, I have a wonderfully short and pointed proof for the finiteness of binary systems of forms" (Hilbert to Klein, 21 March 1888 (Frei 1985, 39)).

Hilbert had caught fire. A week later, when he met with Klein in Göttingen, he had already put the finishing touches on the new, streamlined proof he had found. This paper (Hilbert 1888b) was the first in a landslide of contributions to algebraic invariant theory that would literally turn the subject upside down. Between 1888 and 1890 Hilbert pursued this theme relentlessly, but with a new methodological twist which he combined with the formal algorithmic techniques employed by Gordan. Beginning with a series of three short notes sent to Klein for publication in the *Göttinger Nachrichten* (Hilbert 1888c, 1889a, b), he began to unveil general methods for proving finiteness relations for general systems of algebraic forms, invariants being only a quite special case, though the one of principal interest. With these general methods, combined with the algorithmic techniques developed by his predecessors, Hilbert was able to extend Gordan's finiteness theorem, which was restricted to systems of binary forms over the real or complex numbers, to forms in any number of variables and with coefficients in an arbitrary field.

By the time this first flurry of activity came to an end, Hilbert had shown how these finiteness theorems for invariant theory could be derived from general properties of systems of algebraic forms. Writing to Klein in 1890, he described his culminating paper (Hilbert 1890) as a unified approach to a whole series of algebraic problems (Hilbert to Klein, 15 February 1890 (Frei 1985, 61)). He might have added that his techniques borrowed heavily from Leopold Kronecker's work on algebraic forms. Yet from a broader methodological standpoint, Hilbert's approach clearly broke with Kronecker's constructive principles. For Hilbert's foray into the realm of algebraic forms revealed the power of pure existence arguments: he showed that out of sheer logical necessity a finite basis must exist for the system of invariants associated with any algebraic form or system of forms. Hilbert found his way forward by noticing the following general result, known today as Hilbert's basis theorem for polynomial ideals. It appears as Theorem I in (Hilbert 1888c). This states that for any sequence of algebraic forms in n variables $\varphi_1, \varphi_2, \varphi_3, \ldots$ there exists an index m such that all the forms of the sequence can be written in terms of the first m forms, that is

$$\varphi = \alpha_1\varphi_1 + \alpha_2\varphi_2 + \cdots + \alpha_m\varphi_m,$$

where the α_i are appropriate n-ary forms. Thus, the forms $\varphi_1, \varphi_2, \ldots, \varphi_m$ serve as a basis for the entire system. By appealing to Theorem I and drawing on Mertens's procedure for generating systems of invariants, Hilbert proved that such systems were always finitely generated. There was, however, a small snag.

Hilbert attempted to prove Theorem I by first noting that it held for small n. He then introduced a still more general Theorem II, from which he could prove Theorem I by induction on the number of variables. If this sounds confusing, a number of contemporary readers had a similar reaction, including a few who expressed their misgivings to Hilbert about the validity of his proof. Paul Gordan, however, was not one of them. According to Hilbert, to the best of his recollection he and Gordan had only discussed the proof of Theorem II during their meeting in Leipzig (Hilbert to Klein, 3 March 1890 (Frei 1985, 64)). As it turned out Hilbert's Theorem II, as originally formulated in Hilbert (1888c), is false.[6] Moreover, since it was conceived from the beginning as a lemma for the proof of Theorem I, Hilbert dropped Theorem II in his definitive paper (Hilbert 1890) and gave a new proof of Theorem I. Nevertheless, the latter remained controversial, as we shall soon see.

Today we recognize in Hilbert's Theorem I a central fact of ideal theory, namely, that every ideal of a polynomial ring is finitely generated. Thirty years later, Emmy Noether incorporated Hilbert's Theorem I (from Hilbert 1890) as well as his *Nullstellensatz* (from Hilbert 1893) into an abstract theory of ideals (see Gilmer 1981). Her classic paper "Idealtheorie in Ringbereichen" (Noether 1921) is nearly as readable today as it was when she wrote it. The same cannot be said, however, for Hilbert's papers (for English translations, see Hilbert 1978). Not that these are badly written; they simply reflect a far less familiar mathematical context.

In Hilbert (1889a, 28), Hilbert hinted that much of the inspiration for his terminology and techniques came from Kronecker's theory of module systems. When he wrote this he knew very well that Kronecker held very negative views about invariant theory, making it highly improbable that the era's leading algebraist would view Hilbert's adaptation of his ideas with approval. Indeed, Kronecker had made it plain to Hilbert that, in his view, the only invariants of interest were the numerical invariants associated with systems of algebraic equations. Still, Hilbert quickly recognized the fertility of Kronecker's conceptions for invariant theory. Yet while acknowledging his debt to the Berlin algebraist, he parted company with him by adopting a radically non-constructive approach. Ironically, the initial impulse to do so apparently came from his conversations with Gordan. Thus, with his early work on invariant theory Hilbert sowed some of the seeds that would eventually flower into his modernist vision for mathematics thereby preparing the way for the dramatic foundations debates of the 1920s (see Hesseling 2003).

Mathematics as Theology

Kronecker seems to have simply ignored Hilbert's dramatic breakthrough, but others closer to the field of invariant theory obviously could not afford to do so. Paul Gordan, who had initially supported Hilbert's work enthusiastically, now began to express misgivings about this new and, for him, all too ethereal approach to invariant theory. His views soon made the rounds at the coffee tables and beer gardens where mathematicians liked to gather, and more or less everyone must have heard what Gordan probably said on more than one occasion: Hilbert's approach to invariant theory was "theology not mathematics" (Weyl 1944, 140). Emmy Noether's father, Max, certainly heard this pronouncement from Gordan himself (Fig. 13.6).[7]

No doubt many got a chuckle out of this epithet at the time, but in fact a serious conflict briefly reared its head in Febraury 1890 when Hilbert submitted his definitive paper (Hilbert 1890) for publication in *Mathematische Annalen*. Klein was overjoyed when he received it and wrote back to Hilbert a day later: "I do not doubt that this is the most important work on general algebra that the *Annalen* has ever published" (Klein to Hilbert, 18 Feb. 1890, in Frei (1985, 65)). He then sent the manuscript to Gordan, the *Annalen's* house expert on invariant theory, asking him to report on it.

Since Klein had already heard some of Gordan's misgivings about Hilbert's methods in private conversations, he may well have anticipated a negative reaction. If so, he certainly got one. The cantankerous Gordan forcefully voiced

[6] See the editorial note in Hilbert (1933, 177).

[7] The earliest published reference to Gordan's remark—"das ist keine Mathematik, das ist Theologie"—appears to be Noether (1914, 18). Clearly, however, Minkowski, Klein, and many others were well aware of Gordan's opposition at the time it surfaced. Later mythologizing is discussed in McLarty (2012).

Fig. 13.6 Max Noether, Gordan's colleague in Erlangen.

his objections, aiming directly at Hilbert's presentation of Theorem I, which Gordan claimed fell short of even the most modest standards for a mathematical proof. "The problem lies not with the form," he wrote Klein, " . . . but rather lies much deeper. Hilbert has scorned to present his thoughts following formal rules; he thinks it suffices that no one contradict his proof, then everything will be in order. . . he is content to think that the importance and correctness of his propositions suffice. That might be the case for the first version, but for a comprehensive work for the *Annalen* this is insufficient." (Gordan to Klein, 24 Feb. 1890, in Frei (1985, 65))

Klein forwarded Gordan's report to Hurwitz in Königsberg, who then discussed its contents with Hilbert. After that the sparks really began to fly. Clearly irked by Gordan's refusal to recognize the soundness of his arguments, Hilbert promptly dashed off a fierce rebuttal to Klein. He began by reminding him that Theorem I was by no means new; he had, in fact, come up with it some 18 months earlier and had afterward published a first proof in the *Göttinger Nachrichten* (Hilbert 1888c). He then proceeded to describe the events that had prompted him to give a new proof in the manuscript now under scrutiny.

This came about after he had spoken with numerous mathematicians about his key theorem; he had also carried on correspondence with Cayley and Eugen Netto, who wanted him to clarify certain points in the proof. Taking these various reactions into account, Hilbert had prepared a revised proof which he tested out in his lecture course the previous semester. Afterward he spoke with one of the auditors in order to convince himself that the argument as presented had actually been understood. Having reassured himself that this new proof was indeed clear and understandable, he wrote it up for Hilbert (1890). Hilbert then concluded this recitation of the relevant prehistory by saying that these facts clearly refuted the *ad hominem* side of Gordan's attack, namely his insinuation that Hilbert's new proof of Theorem I was not meant to be understood and that he was merely content so long as no one could contradict the argument.

Regarding what he took to be the substantive part of Gordan's critique, Hilbert stated that this consisted mainly of "a series of very commendable, but completely general rules for the composition of mathematical papers" (Hilbert to Klein, 3 March 1890 (Frei 1985, 64)). The only specific criticisms Gordan made were, in Hilbert's opinion, plainly incomprehensible: "If Professor Gordan succeeds in proving my Theorem I by means of an 'ordering of all forms' and by passing from 'simpler to more complicated forms', then this would just be another proof, and I would be pleased if this proof were simpler than mine, provided that each individual step is as compelling and as tightly fastened" (*ibid.*). Hilbert then ended this remarkable *repartee* with an implied threat: either his paper would be printed just as he wrote it or he would withdraw his manuscript from publication in the *Annalen* "I am not prepared," he intoned, "to alter or delete anything, and regarding this paper, I say with all modesty, that this is my last word so long as no definite and irrefutable objection against my reasoning is raised" (*ibid.*).

Klein was certainly not accustomed to receiving letters like this one, especially not from a young *Privatdozent*. Yet however impressed he may have been by Hilbert's self-assurance and pluck, he also wanted to preserve his longstanding alliance with Gordan. Moreover, in view of his older friend's irascibility, Klein knew he had to handle this squabble delicately before it became a full-blown crisis. So Hilbert received no immediate reply, as Klein wanted to wait until he could confer with Gordan personally. Over a month passed, with no word from Göttingen about the fate of a paper that Klein had originally characterized as one of the most important ever to appear in the pages of *Mathematische Annalen*. Then, in early April, Gordan came to Göttingen to "negotiate" with Klein about these matters, which clearly weighed heavily on the Erlangen mathematician's heart. In order to facilitate the process, Klein also asked Hurwitz to join them, knowing that Hilbert's trusted friend would do his best to help restore harmony.

Gordan spent eight days in Göttingen, following which Klein wrote Hilbert a brief letter summarizing the results of their "negotiations." He began by reassuring him that Gordan's opinions were by no means as uniformly negative as Hilbert had assumed. "His general opinion," Klein noted, "is entirely respectful, and would exceed your every wish"

(Klein to Hilbert, 14 April 1890, in Frei (1985, 66)). To this he merely added that Hurwitz would be able to tell him more about the results of their meeting. But then he attached a postscript that contained the message Hilbert had been waiting to hear: Gordan's criticisms would have no bearing on the present paper and should be construed merely as guidelines for future work!

Thus, Hilbert got what he had demanded; his decisive paper appeared in the *Annalen* exactly as he had written it. Gordan surely lost face, but at least he had been given the opportunity to vent his views. In short, Klein's diplomatic maneuvering carried the day. Gordan knew, of course, that his negotiating partner was a man who had little patient for methodological nit-picking. He also knew that when push came to shove, Klein consistently valued youthful vitality over age and experience. Hilbert represented the wave of the future, and while this conflict, in and of itself, had no immediate ramifications, it clearly foreshadowed a highly significant restructuring of the power constellations that had dominated German mathematics since the late 1860s.

A Final *Tour de Force*

If Hilbert was scornful of Gordan's editorial pronouncements, this does not mean that he failed to see the larger issue at stake. His general basis theorem proved that for algebraic forms in any number of variables there always *exists* a finite collection of irreducible invariants, but his methods of proof were of no help when it actually came to constructing such a basis. Hilbert obviously realized that if he could develop a new proof based on arguments that were, in principle, constructive in nature, then this would completely vitiate Gordan's criticisms. Two years later, he unveiled just such an argument, one that he had in fact been seeking for a long time. In an elated letter to Klein, he described this latest breakthrough which allowed him to bypass the controversial Theorem I completely. He further noted that although this route to his finiteness theorems was more complicated, it carried a major new payoff, namely "the determination of an upper bound for the degree and weights of the invariants of a basis system" (Hilbert to Klein, 5 January 1892 (Frei 1985, 77)).

When Hermann Minkowski, who was then in Bonn, heard about Hilbert's latest triumph, he fired off a witty letter congratulating his friend back in Königsberg:

> I had long ago thought that it could only be a matter of time before you finished off the old invariant theory to the point where there would hardly be an i left to dot. But it really gives me joy that it all went so quickly and that everything was so surprisingly simple, and I congratulate you on your success. Now that you've even discovered smokeless gunpowder with your last theorem, after Theorem I caused only Gordan's eyes to sting anymore, it really is a good time to decimate the fortresses of the robber-knights [i.e., specialists in invariant theory] – [Georg Emil] Stroh, Gordan, [Kyparisos] Stephanos, and whoever they all are – who held up the individual traveling invariants and locked them in their dungeons, as there is a danger that new life will never sprout from these ruins again. (Minkowski 1973, 45).

Minkowski's opinions were a constant source of inspiration for Hilbert, so he surely took these witty remarks to heart. Indeed, this letter may well mark the beginning of one of the most enduring of all myths associated with Hilbert's exploits, namely that he single-handedly killed off the till then flourishing field of invariant theory. As Hans Freudenthal later put it: "never has a blooming mathematical theory withered away so suddenly" (Freudenthal 1981, 389).

Hilbert published his new results in another triad of papers for the *Göttinger Nachrichten* (Hilbert 1891, 1892a, b). Nine months later he completed the manuscript of his second classic paper on invariant theory (Hilbert 1893).[8] He sent this along with a diplomatically worded letter to Klein, noting that he had taken pains to ensure that the presentation followed the general guidelines Prof. Gordan had recommended (Hilbert to Klein, 29 September 1892 (Frei 1985, 85)). Then, in a short postscript, Hilbert added: "I have read and thought through the manuscript carefully again, and must confess that I am very satisfied with this paper" (*ibid.*).

Klein reassured him that "Gordan had made his peace with the newest developments," and emphasized that doing so "wasn't easy for him, and for that reason should be seen as much to his credit" (Klein to Hilbert, 7 January 1893, in (Frei 1985, 86)). As evidence of Gordan's change of heart, Klein mentioned his forthcoming paper entitled simply "Über einen Satz von Hilbert" (Gordan 1892). The *Satz* in question was, of course, Hilbert's Theorem I, which really had caused Gordan's eyes to sting, but *not* because he doubted its validity. Nor did he ever doubt that Hilbert's proof was correct; it was simply incomprehensible in Gordan's opinion. As he put it to Klein back in 1890: "I can only learn something that is as clear to me as the rules of the multiplication table" (Gordan to Klein, 24 Feb. 1890, in (Frei 1985, p 65)).

Hilbert had claimed that he would welcome a simpler proof of Theorem I from Gordan, and here the elderly algorist delivered in a gracious manner. He began by characterizing Hilbert's proof as "entirely correct" (Gordan 1892, 132), but went on to say that he had nevertheless noticed a gap since Hilbert's argument merely proved the existence of a finite basis without examining the properties of the basis elements. He further noted that his own proof relied essentially on Hilbert's strategy of applying the ideas of Kronecker, Dedekind, and Weber to invariant theory (Gordan

[8] For an English translation of this and other works by Hilbert, see Hilbert (1978).

1892, 133). Probably only a few of those who saw this conciliatory contribution by the "King of Invariants" were aware of the earlier maneuvering that had taken place behind the scenes. Nor were many likely to have anticipated that Gordan's throne would soon resemble a museum piece.[9]

Not surprisingly, Hilbert put methodological issues at the very forefront of (Hilbert 1893), his final contribution to invariant theory. Here he called attention to the fact that his earlier results failed to give any idea of how a finite basis for a system of invariants could actually be constructed. Moreover, he noted that these methods could not even help in finding an upper bound for the number of such invariants for a given form or system of forms (Hilbert 1933, 319). To show how these drawbacks could be overcome, Hilbert thus adopted an even more general standpoint than the one he had taken before. He described the guiding idea of this culminating paper as invariant theory treated merely as a special case of the general theory of algebraic function fields. This viewpoint was inspired to a considerable extent by the earlier work of Kronecker and Dedekind, although Hilbert mentioned this connection only obliquely in the introduction where he underscored the close analogy with algebraic number fields (Hilbert 1933, 287).

Hilbert's introduction also contains other interesting features. In it, he set down five fundamental principles which could serve as the foundations of invariant theory. The first four of these he regarded as the "elementary propositions of invariant theory," whereas the existence of a finite basis (or in Hilbert's terminology a "full invariant system") constituted the fifth principle. This highly abstract formulation would, of course, later come to typify much of Hilbert's work in nearly all branches of mathematics. Indeed, the only thing missing from what was to become standard Hilbertian jargon was an explicit appeal to the axiomatic method. Immediately after presenting these five propositions, he wrote that they "prompt the question, which of these properties are conditioned by the others and which can stand apart from one another in a function system." He then mentioned an example that demonstrated the independence of property 4 from properties 2, 3, and 5. These findings were incidental to the main thrust of Hilbert's paper, but they reveal nevertheless how axiomatic ideas had already entered into his early work on algebra.[10]

On 29 September 1892, the day he sent off the manuscript of (Hilbert 1893), Hilbert wrote to Minkowski: "I shall now definitely leave the field of invariants and turn to number theory" (Blumenthal 1935, 395). This transition was a natural one, given that his final work on invariant theory was essentially an application of concepts from the theory of algebraic number fields. One year later, Hilbert and Minkowski were charged with the task of writing a report on number theory to be published by the Deutsche Mathematiker-Vereinigung. Minkowski eventually dropped out of the project, but he continued to offer his friend advice as Hilbert struggled with his single most ambitious work, "Die Theorie der algebraischen Zahlkörper," better know simply as the *Zahlbericht*.[11]

Killing Off a Mathematical Theory

Thus by 1893 Hilbert's active involvement with invariant theory had ended. In that year he wrote (Hilbert 1896), a survey article written in response to a request from Felix Klein, who presented it along with several other papers at the Mathematical Congress held in Chicago in 1893 as part of the World's Columbian Exposition. Hilbert's account offers an interesting participant's history of the classical theory of invariants. At the time he wrote it, invariant theory was a staple research field within the fledgling mathematical community in the United States, which first began to spread its wings under the tutelage of J. J. Sylvester at Johns Hopkins (see Parshall and Rowe 1994). Hilbert briefly alluded to the contributions of Cayley and Sylvester in his brief survey, describing these as characteristic for the "naive period" in the history of a special field like algebraic invariant theory. This stage, he added, was soon superseded by a "formal period," whose leading figures were his own direct predecessors, Alfred Clebsch and Paul Gordan. A mature mathematical theory, Hilbert went on, typically culminates in a third, "critical period," and his account made it clear that he alone was to be regarded as having inaugurated this stage of research. What better time to quit the field? Hilbert realized very well that many aspects of invariant theory had only begun to unfold, but after 1893 he was content to point others in possibly fruitful directions for further research, such as the one indicated in his fourteenth Paris problem. Although he did offer a one-semester course on invariant theory in 1897 (see Hilbert 1993), by this time his eyes were already focused on other fields and new challenges.

Mathematicians are constantly looking ahead, not backward, and by 1893 probably no one gave much thought to the events of five years earlier. Over time, Hilbert's decisive encounter with Gordan in Leipzig was reduced to a minor episode at the outset of his *Siegeszug* through invariant

[9]Gordan later presented a streamlined proof of Hilbert's Theorem I in a lecture at the 1899 meeting of the DMV in Munich. Hilbert was present on that occasion (see *Jahresbericht der Deutschen Mathematiker-Vereinigung* 8(1899), 180. Gordan wrote up this proof soon thereafter for Gordan (1899).

[10]For a detailed examination of Hilbert's work on the axiomatization of physics, see Corry (2004).

[11]Reprinted in (Hilbert 1932, 63–363). For a brief account of the work and its historical reception, see the introduction by Franz Lemmermeyer and Norbert Schappacher to the English edition (Hilbert 1998, xxiii–xxxvi).

theory. Otto Blumenthal, Hilbert's first biographer, even got the city wrong, claiming that Hilbert went to Erlangen to visit Gordan in the spring of 1888 (Blumenthal 1935, 394). By then 40 years had passed, and presumably no one, not even Hilbert, remembered what had happened. Yet his own characterization of this encounter could not be more telling: it had been thanks to Gordan's "stimulating help" that he left Leipzig with "an infinite series of thought vibrations" running through his brain. Scant though the evidence may be, it strongly suggests that the week he spent with the "King of Invariants" gave Hilbert the initial impulse that put him on his way. Back in Königsberg, he adopted several of Gordan's techniques in his subsequent work. Numerous citations reveal that he was thoroughly familiar with Gordan's *opus*, especially the two volumes of his lectures edited by Georg Kerschenstein (Gordan 1885/1887). That work, the springboard for many of Hilbert's discoveries, was by 1893 practically obsolete, though no comparable compendium would appear to take its place.

His friend Hermann Minkowski saw that this presented a certain dilemma: it was all very well to blow up the castles of those robber knights of invariant theory, so long as something more useful could be built on their now barren terrain. Minkowski thus expressed the hope that Hilbert would some day show the mathematical world what the new buildings might look like. In the same humorous vein, he kidded him that it would be best if Hilbert wrote his own monograph on the new modernized theory of invariants rather than waiting to find another Kerschenstein, who would likely leave behind too many "cherry pits" (misspelled by Minkowski as "Kerschensteine") (Minkowski 1973, 45).

Hilbert did neither,[12] leaving the theory of invariants to languish on its own while the original Gordan-Kerschenstein volumes gathered dust in local libraries. Invariant theory thus entered the annals of mathematics, its history already sketched by the man who wrote its epitaph. To the younger generation, Paul Gordan would mainly be remembered as the old algorist who had once declared Hilbert's modern methods "theology." Now that he and his mathematical regime had been deposed from power, classical invariant theory was declared a dead subject, one of those "dead ends" ("tote Stränge") that Hilbert had decried in his letter to Klein from 1890.[13] Although Leopold Kronecker had predicted this very outcome, he would hardly have approved of the executioner's methods. Yet ironically, it was Hilbert's decision to move on to "greener pastures" – even more than the wealth of new perspectives his work had opened – that hastened the fulfillment of Kronecker's prophecy.

[12]Perhaps the closest he came to fulfilling Minkowski's wish was the *Ausarbeitung* of his 1897 lecture course, now available in English translation in Hilbert (1993).

[13]Historical verdicts with regard to the sudden demise of invariant theory have varied considerably (see Fisher 1966; Parshall 1989). The merits of classical invariant theory were later debated in print by Eduard Study and Hermann Weyl. For an excellent account of this and other subsequent developments in algebra, see Hawkins (2000).

References

Bianchi, Luigi. 1959. *Opere*. Vol. XI. Rome: Edizioni Cremonese.

Biermann, Kurt-R. 1988. *Die Mathematik und ihre Dozenten an der Berliner Universität, 1810–1933*. Berlin: Akademie-Verlag.

Blumenthal, Otto. 1935. Lebensgeschichte, in [Hilbert 1935, pp. 388–429].

Corry, Leo. 2004. *Hilbert and the Axiomatization of Physics (1898–1918): From "Grundlagen der Geometrie" to "Grundlagen der Physik"*, to appear in Archimedes: New Studies in the History and Philosophy of Science and Technology, Kluwer Academic.

Dieudonne, Jean. 1985. *History of Algebraic Geometry*. Trans. Judith D. Sally. Monterey: Wadsworth.

Fisher, Charles S. 1966. The Death of a Mathematical theory: A Study in the Sociology of Knowledge. *Archive for History of exact Sciences* 3: 137–159.

Frei, Günther. 1985. *Der Briefwechsel David Hilbert–Felix Klein (1886–1918)*, Arbeiten aus der Niedersächsischen Staats- und Universitätsbibliothek Göttingen, Bd. 19, Göttingen: Vandenhoeck & Ruprecht.

Freudenthal, Hans. 1981. David Hilbert. In *Dictionary of Scientific Biography*, ed. Charles C. Gillispie, vol. 16, 388–395.

Gilmer, Robert. 1981. Commutative Ring Theory. In *Emmy Noether. A Tribute to her Life and Work*, ed. James W. Brewer and Martha K. Smith, 131–143. New York: Marcel Dekker.

Gordan, Paul. 1885/1887. In *Vorlesungen über Invariantentheorie*, ed. Georg Kerschensteiner, vol. 2. Leipzig: Teubner.

———. 1892. Über einen Satz von Hilbert. *Mathematische Annalen* 42: 132–142.

———. 1899. Neuer Beweis des Hilbertschen Satzes über homogene Funktionen. *Nachrichten der Gesellschaft der Wissenschaften zu Göttingen*: 240–242.

Gray, Jeremy. 2000. *Linear Differential Equations and Group Theory from Riemann to Poincaré*. 2nd ed. Boston: Birkhäuser.

Hashagen, Ulf. 2003. *Walther von Dyck (1856–1934). Mathematik, Technik und Wissenschaftsorganisation an der TH München, Boethius*. Vol. Band 47. Stuttgart: Franz Steiner.

Hawkins, Thomas. 2000. *Emergence of the Theory of Lie Groups. An Essay in the History of Mathematics, 1869–1926*. New York: Springer.

Hesseling, Dennis E. 2003. *Gnomes in the Fog. The Reception of Brouwer's Intuitionism in the 1920's*. Birkhäuser: Basel.

Hilbert, David. 1888a. Über die Darstellung definiter Formen als Summe von Formenquadraten, *Mathematische Annalen* 32 (1888), 342–350; Reprinted in [Hilbert 1933, 154–161].

———. 1888b. Über die Endlichkeit des Invariantensystems für binäre Grundformen, *Mathematische Annalen* 33 (1889), 223–226; Reprinted in [Hilbert 1933, 162–164].

———. 1888c. Zur Theorie der algebraischen Gebilde I, *Nachrichten der Gesellschaft der Wissenschaften zu Göttingen* 1888, 450–457; reprinted in [Hilbert 1933, 176–183].

———. 1889a. Zur Theorie der algebraischen Gebilde II, *Nachrichten der Gesellschaft der Wissenschaften zu Göttingen* 1889, 25–34; Reprinted in [Hilbert 1933, 184–191].

———. 1889b. Zur Theorie der algebraischen Gebilde III, *Nachrichten der Gesellschaft der Wissenschaften zu Göttingen* 1889, 423–430; Reprinted in [Hilbert 1933, 192–198].

———. 1890. Über die Theorie der algebraischen Formen. *Mathematische Annalen* 36: 473–534.

———. 1891. Über die Theorie der algebraischen Invarianten I. *Nachrichten der Gesellschaft der Wissenschaften zu Göttingen*: 232–242.

———. 1892a. Über die Theorie der algebraischen Invarianten II. *Nachrichten der Gesellschaft der Wissenschaften zu Göttingen*: 6–16.

———. 1892b. Über die Theorie der algebraischen Invarianten III. *Nachrichten der Gesellschaft der Wissenschaften zu Göttingen*: 439–449.

———. 1893. Über die vollen Invariantensysteme, *Mathematische Annalen* 42 (1893), 313–373; reprinted in [Hilbert 1933, 287–344].

———. 1896. Über die Theorie der algebraischen Invarianten,' *Mathematical Papers Read at the International Mathematical Congress Chicago 1893*, New York: Macmillan, 1896, 116–124; reprinted in [Hilbert 1933, 376–383].

———. 1922. Neubegründung der Mathematik, 1922. Erste Mitteilung,' *Abhandlungen aus dem Mathematischen Seminar der Hamburgischen Universität,* 1: 157–177; Reprinted in [Hilbert 1935, 157–177].

———. 1932. *Gesammelte Abhandlungen*, Bd. 1, Berlin: Springer.

———. 1933. *Gesammelte Abhandlungen*, Bd. 2, Berlin: Springer.

———. 1935. *Gesammelte Abhandlungen*, Bd. 3, Berlin: Springer.

———. 1978. *Hilbert's Invariant Theory Papers*. Trans. Michael Ackermann. in Lie Groups: History, Frontiers, and Applications, Robert Hermann, ed. Brookline: Math Sci Press.

———. 1993. *Theory of Algebraic Invariants*. Trans. Reinhard C. Lauenbacher. Cambridge: Cambridge University Press.

———. 1998. *The Theory of Algebraic Number Fields*. Trans. Iain T. Adamson. New York: Springer.

Kleiman, Steven. 1976. Rigorous Foundation of Schubert's Enumerative Calculus, in [Browder 1976], 445–482.

Lindemann, Ferdinand. 1971. *Lebenserinnerungen*. München.

McLarty, Colin. 2012. Hilbert on theology and its discontents: the origin myth of modern mathematics. In *Circles Disturbed: the Interplay of Mathematics and Narrative*, ed. A. Doxiadis and B. Mazur, 105–129. Princeton: Princeton University Press.

Mertens, Franz. 1887. *Journal für die reine und angewandte Mathematik* 100: 223–230.

Minkowski. 1973. *Hermann Minkowski, Briefe an David Hilbert*. Hg. L. Rüdenberg und H. Zassenhaus, New York: Springer.

Noether, Max. 1914. Paul Gordan. *Mathematische Annalen* 75: 1–41.

Noether, Emmy. 1921. Idealtheorie in Ringbereichen. *Mathematische Annalen* 83: 24–66.

Parshall, Karen H. 1989. Toward a History of Nineteenth-Century Invariant Theory. In *The History of Modern Mathematics: Volume 1: Ideas and their Reception*, ed. David E. Rowe and John McCleary, 157–206. Boston: Academic Press.

Parshall, Karen H., and David E. Rowe. 1994. *The Emergence of the American Mathematical Research Community, 1876–1900. J.J. Sylvester, Felix Klein, and E.H. Moore*, History of Mathematics. Vol. 8. Providence: American Mathematical Society.

Reid, Constance. 1970. *Hilbert*. New York: Springer.

Rowe, David E. 1986. "Jewish Mathematics" at Göttingen in the Era of Felix Klein. *Isis 77*: 422–449.

———. 1998. Mathematics in Berlin, 1810–1933. In *Mathematics in Berlin*, ed. H.G.W. Begehr, H. Koch, J. Kramer, N. Schappacher, and E.-J. Thiele, 9–26. Basel: Birkhäuser.

———. 2003. Mathematical Schools, Communities, and Networks. In *Cambridge History of Science, vol. 5, Modern Physical and Mathematical Sciences*, ed. Mary Jo Nye, 113–132. Cambridge: Cambridge University Press.

———. 2004. Making Mathematics in an Oral Culture: Göttingen in the Era of Klein and Hilbert. *Science in Context* 17 (1/2): 85–129.

Schappacher, Norbert. 1998. On the History of Hilbert's Twelfth Problem. A Comedy of Errors. In Matériaux pour l'histoire des mathématiques au XXe siècle. Actes du colloque à la mémoire de Jean Dieudonné (Nice, 1996). *Séminaires et Congrès (Société Mathématique de France)* 3: 243–273.

Toeplitz, Otto. 1922. Der Algebraiker Hilbert. *Die Naturwissenschaften* 10: 73–77.

Weber, Heinrich. 1893. Leopold Kronecker. *Jahresbericht der Deutschen Mathematiker-Vereinigung* 2: 5–13.

Weyl, Hermann. 1932. Zu David Hilberts siebzigsten Geburtstag. *Die Naturwissenschaften* 20: 57–58; Reprinted in [Weyl 1968, vol. 3, 346–347].

———. 1935. Emmy Noether. *Scripta Mathematica* 3: 201–220; Reprinted in [Weyl 1968, vol. 3, 435–444].

———. 1944. David Hilbert and his Mathematical Work. *Bulletin of the American Mathematical Society* 50: 612–654; reprinted in [Weyl 1968, vol. 4, 130–172].

———. 1968. In *Gesammelte Abhandlungen*, ed. K. Chandrasekharan, vol. 4. Berlin: Springer.

Yandell, Ben H. 2002. *The Honors Class. Hilbert's Problems and their Solvers*. Natick: A K Peters.

Klein, Hurwitz, and the "Jewish Question" in German Academia

14

(*Mathematical Intelligencer* 29(2)(2007): 18–30)

Mathematicians love to tell stories about people they once knew or perhaps only heard about. If the story happens to sound believable, others are apt to repeat it, possibly embellishing on the original tale. Such mathematical folklore occasionally finds its way into print, and once it does, readers are apt to take such stories at face value, lending them additional credibility. Occasionally, though, alleged facts come under scrutiny, and established stories are exposed as fiction. Yet even when someone comes along with decisive evidence refuting an earlier account it can easily happen that the original story just refuses to die.

The case I have in mind here concerns a version of events in the career of Adolf Hurwitz, who taught as an associate professor in Königsberg from 1884 until 1892, when he received a full professorship at the ETH in Zurich, succeeding Georg Frobenius. A rendition of what supposedly transpired at that time can be found on the internet, courtesy of MacTutor:

> Hurwitz remained at Zurich for the rest of his life . . . [but] not because he had not been offered in chair in Germany. Schwarz, who was professor at Göttingen, succeeded Weierstrass by accepting his professorship in Berlin in 1892. Göttingen approached Hurwitz and offered him the vacant chair only weeks after he had accepted the Zurich chair, but he turned down the offer. This must have been a remarkably hard decision for Hurwitz since at that time a chair at a leading German university such as Göttingen would have been much more prestigious to any German than a chair in Switzerland. However Hurwitz was an extremely loyal person, and having given his word that he would accept the Zurich position he would not renege on his promise.[1]

An earlier version of this story was told by Hurwitz's former student, Ernst Meissner, in a memorial speech delivered after his teacher's death in 1919. The text of that speech was later reprinted in volume one of Hurwitz's *Werke*, published in 1932, and thereafter it became a standard source for biographical information on his life (Meissner 1932). About 20 years ago, I wrote an article on various forms of anti-Semitism in German mathematics, part of which dealt with the case of Adolf Hurwitz (Rowe 1986, 433–435). As a point of record, I noted then that Meissner's account of the events of 1892 was obviously untrue. Presumably he was merely embellishing on a story he had once heard, perhaps even from Hurwitz himself. Unfortunately, Meissner's version of the events obscures a far more poignant story, one that deserves to be widely known. In retelling it here, I hope not only to correct the record but also to suggest some larger themes relating to the vulnerability of German Jews who pursued academic careers in mathematics.

In writing about Jewish mathematicians in Germany during the Second Empire, one faces the difficult issue of deciding what should count as "Jewish."[2] Many Jews during this period were no longer practicing their religion, and in such cases they were often inclined to convert to Christianity in order to better their professional prospects. Candidates of "Mosaic confession" were at a distinct disadvantage nearly everywhere; at several universities in the heavily Catholic southern states they had no chances at all. On the other hand, many viewed baptized Jews as pseudo-Christians, though overt signs of discrimination were comparatively rare. Academic anti-Semitism took on subtle guises and the various forms of de facto discrimination against those who were seen as Jewish can in most cases only be discerned through private communications. Thus, Karl Weierstrass confided to Sofia Kovalevskaya that his nemesis, Leopold Kronecker, "shares the shortcoming one finds in many intelligent people, especially those of Semitic stock: he does not possess sufficient fantasy (intuition I would prefer to say). And it is true, a mathematician who is not something of a poet

[1]MacTutor History of Mathematics Archive, http://www-history.mcs.st-and.ac.uk/

[2]The position taken here on Jewish identity follows the one adopted for the exhibition "Jüdische Mathematiker in der deutschsprachigen akademischen Kultur" held in Bonn in July 2006 in conjunction with the annual meeting of the Deutsche Mathematiker-Vereinigung (see Bergmann et al. (2012).

D.E. Rowe, *A Richer Picture of Mathematics*, https://doi.org/10.1007/978-3-319-67819-1_14

will never be a complete mathematician."[3] Such stereotypes were commonplace during this period and went hand in hand with naïve attitudes or prejudices about race and ethnicity. For the present purposes, I will overlook the complicated issues of Jewish identity as these affected German Jews who felt torn by the conflicting pressures they faced. Jewish mathematicians came from a variety of backgrounds and held widely divergent views about Judaism and their identification with it. What they shared, however, was a sense of standing on the outside, of being perceived as somehow alien, and these perceptions could make all the difference.

Klein's Most Talented Student

Adolf Hurwitz (1859–1919) was a *Wunderkind.* The youngest of three sons of a Hildesheim manufacturer (Fig. 14.1), he already made a name for himself as a teenager. Young Adolf's teacher at the Hildesheim Realgymnasium discovered his mathematical talent early on. This teacher's rather exotic name was Hermann Cäsar Hannibal Schubert (1848–1911), the inventor of the Schubert calculus in enumerative geometry. Although an internationally recognized mathematician, Schubert spent his entire career teaching in secondary schools, first in Hildesheim and later in Hamburg. Not that this was so rare: H. G. Grassmann was also a Gymnasium teacher, and even Weierstrass taught "school mathematics" for over a decade before he made the leap to an institution of higher education, the Institute of Trade in Berlin. It took another eight years before Weierstrass gained a professorial appointment at Berlin University in 1864.

In 1874, at age 26, Schubert won the Gold Medal of the Royal Danish Academy of Sciences for his work extending the theory of characteristics to cubic space curves.[4] Around this time Hurwitz was spending Sundays at Schubert's home where he was given private lessons. It did not take long before the youngster had absorbed the main techniques of enumerative geometry and soon he collaborated with Schubert in writing a paper on Michel Chasles' characteristic formula, an enumerative method devised by the French geometer in 1864 to count the number of curves satisfying certain algebraic conditions within a one-parameter family of conics.

Fig. 14.1 Solomon Hurwitz with his three sons, from l. to r.: Max, Adolf, and Julius. This photo was taken on 1 August, 1862, the year the boys' mother died (Courtesy of ETH Bibliothek Zürich).

Chasles was the grand old man of French projective geometry and a connoisseur of classical Greek mathematics. He thus appreciated the fact that the new field of enumerative geometry had classical roots that went back to the famous problem of Apollonius: to construct a circle tangent to three given circles. (With the aid of conic sections, the solution is simple, but the ancients demanded a solution by means of straight edge and compass alone.) In enumerative geometry this traditional type of problem gets a new twist; the question is no longer how to construct a required figure but rather how many different solutions does the problem have. In the case of three circles in the plane, the answer (no surprise) is 8 (at most). Schubert took up the analogous problem in 3-space: how many spheres are tangent to four given spheres? (answer, again no surprise: at most 16). Such questions are easy to pose, but can be very hard to solve. Jakob Steiner claimed that the number of conics tangent to five given conics in the plane was 7776; that's a lot, in fact way too many. Using his theory of characteristics, Michel Chasles showed that the correct answer was a mere 3264. Much later, Hurwitz would return to this field with fundamental work on the most

[3] Weierstrass to Kovalevskaya, 27 August 1883; Weierstrass pointed to other examples: Abel vs. Jacobi, and Riemann as opposed to Eisenstein and Rosenhain in this letter first made public by Gösta Mittag-Leffler, "Une page de la vie de Weierstrass," *Compte Rendu du Deuxième Congrès International des Mathématiciens.* Paris: Gauthier-Villars, 1902, p. 149.

[4] The problem was posed by Chasles' former student, H. G. Zeuthen, a leading authority on enumerative geometry; see Zeuthen (1914).

Fig. 14.2 Adolf Hurwitz as a teenager. Photo courtesy of ETH Bibliothek Zürich.

general algebraic correspondences on Riemann surfaces, generalizing Chasles' correspondence principle (Fig. 14.2).

After graduating from the Realgymnasium, Hurwitz hoped to study mathematics at the university, but his father was less than enthused with this plan. Solomon Hurwitz was a practical man with limited means, and he surely knew that career opportunities in mathematics were few and far between, particularly for unbatized Jews. Schubert, however, persuaded him to reconsider. He also told him about a friend of his, a young geometer at the Munich Institute of Technology who was very eager to attract young talent; his name was Christian Felix Klein.

A *Wunderkind* of another kind, Klein burst upon the scene in the late 1860s as a pupil of Julius Plücker, who taught in Bonn. Plücker was a talented experimental physicist as well as mathematician, but his work received far more recognition in England than within his native Germany. After Plücker's death in 1868, Klein gravitated into the circle surrounding Göttingen's Alfred Clebsch whom he greatly admired. Still, he was itching to see the mathematical world, which in those days centered around two cities: Berlin and Paris. He was visiting the French capitol in July 1870 when the Franco-Prussian War broke out, and had to make a make a wild dash to the train station in order to get back home. After a brief stint helping the invading troops, he rejoined Clebsch in Göttingen. His mentor then paved the way for his appointment as full professor in the sleepy university town of Erlangen. For this occasion Klein prepared his famous "Erlangen Program," spelling out the role of transformation groups and their invariants for geometrical investigations. The author was barely 23 years old, but he had the benefit of consulting with the brilliant Norwegian Sophus Lie, his older friend and collaborator. Soon thereafter, in November 1872, Clebsch suddenly died. Most of his students in Göttingen decided to join Klein in Erlangen, and they soon found that he knew how to keep them busy.

One of these Clebsch pupils, Ferdinand Lindemann, was given the task of editing the master's lectures. The first edition of this volume, which came to be known as Clebsch-Lindemann, appeared in 1876; at least that was the official date of publication. Apparently the book came out earlier, as Lindemann sent a copy to the Göttingen mathematician, M. A. Stern, who wrote back on 27 November 1875 thanking him: "I have till now only looked at the book very superficially, but I believe I can say that the presentation is very illuminating, perhaps more than it would have been under Clebsch's own editorship. For despite all appreciation for the richness of his intellect, one must admit that his presentations were often quite obscure."[5] Moritz Abraham Stern, who had studied under Gauss, was evidently still very sharp when he wrote this at 68 years of age; indeed, he would not retire for another 10 years. When he did finally take his leave in 1885, his chair was assumed by Felix Klein. As we shall see, Stern, Klein, and Lindemann were all destined to play a major part in Hurwitz's career.

Klein also maintained close ties with two older members from Clebsch's circle, Paul Gordan and Max Noether, both of whom – like Stern – were Jewish. In 1874 Klein had the opportunity to fill a new post in Erlangen as associate professor. He consulted with M. A. Stern about this, and Gordan soon thereafter accepted.[6] Klein's second choice was Noether, leaving little doubt that it was on his recommendation that Max Noether assumed Gordan's chair when Klein moved on to Munich in 1875. Gordan was then promoted to Klein's vacant full professorship, thereby giving Erlangen the distinction of having not one, but two Jewish mathematicians. Both were recognized as eminent authorities in their respective fields, invariant theory (Gordan) and

[5]M. A. Stern to Ferdinand Lindemann, 27 November 1875, Lindemann Teilnachlass, Universität Würzburg.

[6]Felix Klein to M. A. Stern, 26 July 1874, Klein Nachlass, Niedersächsische Staats- und Universitätsbibliothek, Göttingen.

algebraic geometry (Noether), and yet for the remainder of their careers they would never receive an offer from another university.

In the summer of 1877, following Schubert's advice, Adolf Hurwitz joined a small group of talented students at the polytechnic school in Munich. Among these were a number of auditors from the nearby university, including Walther von Dyck and Max Planck. These two belonged to an informal group of mathematics students that gathered periodically in Munich pubs. At Klein's urging they joined with six others in May 1877 to form the Munich Mathematics Club, a formal organization that survived for decades afterward (Hashagen 2003, 63–68). Since the founding members enjoyed the social status associated with being enrolled at a university, they were keen to maintain their traditional class privileges in the newly founded club. Thus students at the polytechnic were allowed to join, but only as associate members. Hurwitz entered with this junior status soon after it was founded. At the Munich polytechnic Klein and his colleague Alexander Brill, another former member of the Clebsch circle, held a seminar for those who wished to pursue mathematical research. The modest young newcomer no doubt made a strong impression on both of them, and he would later go on to become Klein's star pupil.

Yet, like Klein earlier, Hurwitz clearly had greater ambitions, and left Munich for Berlin after only one semester. At this time Berlin held far more attraction for aspiring young mathematicians than any other university in Germany. Led by Kummer, Weierstrass, and Kronecker, the Berliners made their influence felt in more ways than one; insiders realized that they virtually monopolised university appointments throughout Prussia. They had thereby succeeded in keeping the "southern German" mathematicians, particularly those associated with the Clebsch group and its journal, *Mathematische Annalen*, out of the Prussian system. For Hurwitz, who was far too young to be concerned about these professional rivalries, Berlin offered the opportunity to broaden his knowledge of complex analysis. Already versed in the geometric approach based on Riemann surfaces, as practiced by Klein, he now took up Weierstrass's theory, which relied on constructive methods based on analytic extensions of local power series representations.

Karl Weierstrass never wrote a textbook on function theory, though many must have wished that he had, for he was anything but a brilliant lecturer. Nevertheless, his courses attracted huge numbers of students who struggled to understand the master's message. This was imparted in a cycle of five lecture courses delivered every four semesters, beginning with an introduction to the theory of analytic functions (6 hours) followed by another 6-h course on elliptic functions. During the third semester, Weierstrass generally offered two 4-h courses, one on applications of elliptic functions and another on the calculus of variations. He then topped off the cycle with a 6-h course on Abelian functions. Although this pattern was fairly consistent, he revised the content of the courses each time he taught them, so that students never emerged with a truly canonical set of lecture notes. Hurwitz was a bit unlucky; when he arrived in Berlin for the winter semester of 1877–78 Weierstrass was lecturing on Abelian functions, the most difficult topic in the cycle. He decided to attend anyway, and dutifully wrote up his notes for the course. Probably he confided to his former teacher that he found the subject very difficult, as Schubert consoled him with the thought that probably no more than six members of the class could follow Weierstrass' presentations.[7]

Hurwitz otherwise felt at home in Berlin, where he befriended Carl Runge, who along with Planck had freshly arrived from Munich. The latter would soon profit from the physics courses taught by such luminaries as Hermann von Helmholtz and Gustav Kirchhoff, whereas Runge was primarily interested in Weierstrass' cycle of lectures. Since he planned to spend several semesters in Berlin, however, he did not plunge in immediately with Hurwitz, but rather joined him when the master started the new cycle during the summer semester of 1878. During the following winter semester they attended Weierstrass' course on elliptic functions, which left a profound impression on Hurwitz.[8]

Runge became a leading expert in applying Weierstrassian techniques without imbibing the orthodox views of the Berlin school. In this connection, his friendship with Hurwitz may well have been a factor. Privately, Hurwitz conceded to Runge that he favored the geometric approach of Cauchy and Riemann over the purely analytical style cultivated by the Weierstrassians.[9] A few years later, he pointed out to Runge that one could simplify the proof of the Runge approximation theorem by using the Cauchy integral formula. Gottfried Richenhagen conjectures that this insight may have contributed to Runge's growing disillusionment with the methodological strictures demanded by Weierstrass and his followers (Richenhagen 1985, 60–66).

Berlin offered another great attraction for Hurwitz; complementing Weierstrass' more methodical lectures were those of his brilliant younger colleague, Leopold Kronecker. In a letter to his friend Luigi Bianchi written during his second stay in Berlin, Hurwitz related that he had returned primarily in order to attend Kronecker's lectures on number theory.[10] In the same letter, he called the controversial Berlin

[7]H. C. H. Schubert to Adolf Hurwitz, 8 December 1877, cited in Hashagen (2003, 106).

[8]Hurwitz wrote up lecture notes for all three of the Weierstrass courses he attended. These Ausarbeitungen are numbers 112, 113, and 115 in his Nachlass at the ETH.

[9]Adolf Hurwitz to Carl Runge, 14 May 1879, cited in Richenhagen (1985, 62).

[10]Hurwitz to Bianchi, 20 March 1882, in Luigi Bianchi. *Opere*, vol. XI, Rome: Edizioni Cremonese, 1959, p. 80.

algebraist "that great, but very vain mathematician"; clearly he found Kronecker's personality less appealing than his mathematics.

After three initial semesters in Berlin, Hurwitz returned to Munich for the summer semester of 1879. At first he felt lonely, and wrote Runge that he had a "great longing" to be back in Berlin.[11] He arrived just in time to make the acquaintance of a student from Pisa, Gregorio Ricci-Cubastro, who would later become famous as the inventor of the Ricci calculus. Ricci returned to Italy after that summer, but the following semester Hurwitz met another Italian, Luigi Bianchi, with whom he struck up a warm friendship (Hashagen 2003, 89–90). This was a critically important time for both young men, particularly for Bianchi, who was able to complete his dissertation under Klein's direction in August 1880. Bianchi's young friend only learned about this by letter, however, as Hurwitz was back home in Hildesheim recovering from a near physical collapse.

This was a recurrent theme, as Hurwitz suffered serious health problems throughout his life; with advancing age these became a matter of deep concern. His father also continued to have strong reservations about the wisdom of pursuing an academic career, in part due to anti-Semitic attitudes in Germany. Klein saw matters otherwise. He reassured his star pupil about his future prospects, but counselled him to take the semester off in order to regain his strength. He also wrote to Hurwitz's father with the same advice:

> Above all I want to stress that among the totality of young people with whom I have up until now worked there was not one who in specifically mathematical talent could measure up to your son. From now on your son will enjoy a brilliant scientific career, which is all the more certain because his gifts are combined with endearing personality traits. The only dangerous point is his health. Your son probably already long ago weakened himself through overwork in his studies. . . . Let me close with the assurance that no one will be happier than I when your son's health . . . fully returns. I need his intensive collaboration for my latest mathematical investigations.[12]

Hurwitz was, indeed, a charming, modest, and unusually warm individual. He was also a talented pianist, who loved to gather with friends and family for music-making. Years later in Zurich, Albert Einstein was a regular participant in such festivities and became a close friend of the Hurwitz clan. At the time this letter was written, Klein already knew that he would be offered a new chair in geometry at Leipzig University. He may even have confided this information to Hurwitz so that he could prepare to submit his doctoral dissertation in Leipzig rather than Munich.

At any rate, the following semester Hurwitz was one of 89 students who attended Klein's lecture course on geometric function theory, the first of several courses in which Klein promoted a Riemannian alternative to the theory taught by Weierstrass in Berlin. Hurwitz's dissertation, published in 1881 in *Mathematische Annalen*, used Kleinian techniques to develop a new foundation for modular functions. He applied these results to a classical problem in number theory: the classification of class numbers of binary quadratic forms with negative discriminant. At age 22 he had already demonstrated his mastery of several disciplines within pure mathematics.

Hurwitz in Göttingen and Königsberg

As a graduate from a Realgymnasium, Hurwitz lacked a strong background in Latin and Greek, the mainstays of the curricula at the humanistic Gymnasien. This posed no difficulty for pursuing his doctorate in mathematics at Leipzig, but the next hurdle presented a real problem. To pursue an academic career meant applying to habilitate, and the Leipzig regulations required applicants to be graduates of a humanistic Gymnasium. Klein must have realized that this stipulation, together with Hurwitz's Jewish confession, virtually ruled out any chance of his habilitating in Leipzig. In a nearly simultaneous case involving Walther von Dyck, who also attended a Realgymnasium, he faced strong opposition within the Leipzig faculty. In the end, he barely managed to push through Dyck's candidacy (Hashagen 2003, 118–121), but in Hurwitz's case he did not even try. Luckily, another option presented itself.

Klein still had good connections in Göttingen, going back to the days when Clebsch acted as his protégé there. Under the latter's watchful eye, Klein had habilitated in 1871, which gave him the opportunity to befriend M. A. Stern, a senior member of the faculty. Stern had been a fixture in Göttingen for many decades. Years later, Klein recalled how Stern told him that the young Bernhard Riemann "sang like a canary" before his tragic illness. Perhaps Klein thought about Hurwitz in similar terms. Certainly he knew that Stern, the first unbaptized Jew to become a full professor at a German university, would offer his talented pupil the kind of support and friendship he needed. So after a second brief stay in Berlin, Adolf Hurwitz became a *Privatdozent* in Göttingen in 1882.

Little has been written about Moritz Abraham Stern's career, which in many ways typifies the hardships faced by Jewish academics.[13] He was born in 1807 in Frankfurt am Main, where the ghetto had long housed the largest Jewish community in Germany. There he was educated at home by

[11]Adolf Hurwitz to Carl Runge, 14 May 1879; cited in Hashagen (2003, 90).

[12]Felix Klein to Solomon Hurwitz, 10 May 1880, Mathematiker-Archiv, Niedersächsische Staats- und Universitätsbibliothek, Göttingen.

[13]The discussion of Stern's career below is based on Küssner (1982).

his grandfather and father who, hoping he would become a rabbi, arranged special lessons in Latin, Greek, Chaldaic, and Syriac. In 1826 he entered Heidelberg University to study mathematics, but just one year later he transferred to Göttingen, where he spent the rest of his long career. From beginning to end he was an active member of the local Jewish community, and during his student days he wrote home in Hebrew. In 1829 Stern was awarded a doctorate with distinction, and the following year he was appointed as a *Privatdozent*, an unsalaried position. To eke out a living he translated Poisson's textbook on mechanics and published two popular works on astronomy. Many years later, in 1838 the Hanoverian Ministry agreed to pay him a modest annual salary of 150 taler, but noted his special status, "as a Jew, Stern cannot become a professor." Nevertheless, he was nominated by the faculty in 1840 for a minor post. Five years later the ministry deigned to answer, "as a Jew it is completely out of the question." In the wake of the political events of 1848, however, Stern finally received an appointment as an associate professor, and throughout the 1850s he was given small salary increases. Finally, in 1859 he was made a full professor.

Stern's career was in one sense unique. During this time, the only real possibility open to Jews who wished to pursue academic careers was conversion to Christianity, as C. G. J. Jacobi had done. Between 1848 and the founding of the Second Empire in 1871, practicing Jews were granted various legal rights within Germany: they were henceforth free to live in cities, move about, attend universities, and take part in civic and political affairs. Yet these gains came at a heavy price, namely the advent of modern anti-Semitism during the 1870s. The wild speculation that took place after the founding of the Reich was followed by a severe financial crash two years later, leading to a depression that lasted some 20 years. Predictably, Jewish banking interests were blamed for this and every other evil brought about by capitalism and swift modernization. At the universities, the historian Heinrich Treitschke published a widely read pamphlet entitled "A Word about our Jewry" containing the infamous phrase "The Jews are our misfortune." This message helped mobilize anti-Semitic elements to form the new Association of German Students in 1880.

Raucous student groups like these may never have crossed Hurwitz's path during his brief stay in Göttingen, but he surely must have realized that the attitude toward Jews had become more openly hostile in some quarters. Whatever notice he may have taken of this, though, he went about his business undistracted. During these two years, Hurwitz published a series of interesting results in function theory. One of these papers confirmed a conjecture of Weierstrass, namely that a single-valued function in n variables that can be locally represented as a rational function has a global representation as a rational function. Hurwitz gave an elegant proof of this claim by means of a newly found result of Georg Cantor, who showed that the continuum was an uncountable set. Fifteen years later, at the First International Congress of Mathematicians, which was held in Zurich, he would deliver a lecture on Cantor's controversial *Mengenlehre* in connection with its applications to problems of analysis.

During the brief time that Hurwitz spent as a *Privatdozent* in Göttingen, the Clebsch school finally gained a foothold in Prussia when Ferdinand Lindemann was called to Königsberg. This came about shortly after 1882, when Lindemann succeeded in proving that π is a transcendental number, thereby finally establishing the impossibility of squaring the circle. One year later he was appointed to succeed Heinrich Weber. Lindemann was particularly impressed by some of the advanced students he inherited from Weber (Lindemann 1971), in particular David Hilbert and Hermann Minkowski. The former was a native son of Königsberg, the latter a new arrival: he was born in Russia to Lewin and Rahel Minkowski, who settled in Königsberg as naturalized Prussian citizens when their son was eight years old. Soon after Lindemann met Hilbert and Minkowski, he learned that the Ministry of Education planned to create a new associate professorship, the position that would eventually go to Hurwitz. (For a survey of the atmosphere in mathematics at the Albertina in the years that followed, see Chap. 13) During the ensuing years, these three brilliant upstarts would become the leading stars of their generation in Germany. Yet only two would attain full professorships in Germany, for reasons that take us into the complex realm of power and authority in German academia over a century ago.

Playing the Game of "Mathematical Chairs"

As described in Chap. 4, the year 1892 marked a decisive turning point for mathematics in Berlin, where a series of events took place that reverberated throughout the German universities and beyond. With Leopold Kronecker's sudden death on 29 December 1891 and Karl Weierstrass' subsequent retirement, the "golden age" of Berlin mathematics came to an end, leaving behind a power vacuum of major proportions. The issue of succession also brought about some interesting behind-the-scenes discussions when a committee of Berlin faculty members met on 22 January 1892 to propose candidates to fill these two vacancies. For although the committee members aired a variety of views on the subject, on one point they were unanimous: under no circumstances would they countenance the candidacy of Felix Klein.

The historian Kurt R. Biermann brought the following excerpts from the committee's protocol to light:

> Weierstrass: Klein dabbles more. A dazzler. [Klein nascht mehr. Blender] . . .

> Helmholtz: Kronecker spoke very disparagingly of Klein. He regarded him as a schemer [Faiseur]. . . .
> Fuchs: I have nothing against his personal qualities, only his pernicious manner when it comes to scientific questions. (Biermann 1988, 205–206)

As the mathematicians on the committee knew, Lazarus Fuchs had been involved in a public dispute with Klein after the latter openly criticized Henri Poincaré for naming a certain class of automorphic functions after Fuchs (see Chap. 11). Weierstrass, now an ailing old man, had been hoping for a long time to install Klein's Göttingen colleague, Hermann Amandus Schwarz, as his successor. Kronecker had opposed this plan, causing Weierstrass to postpone his retirement. Now that his nemesis was gone, he merely had to overcome Fuchs's evident aversion to Schwarz. Fuchs tried to finesse this issue by suggesting that he and Schwarz were too close scientifically. But Weierstrass easily dismissed this claim, demanded that a good analyst be nominated with the potential to draw students, and asserted that only two candidates met this criterion: Klein and Schwarz. Fuchs feebly tried to nullify this argument, noting that he and not Weierstrass was the one who would have to get along with Schwarz. But then he admitted that Schwarz was a much better choice than Klein, at which point the rest of the committee fell in line.

In the faculty's official petition to the Ministry from 8 February, the Berlin consensus on Klein was stated even more sharply:

> Above all else, however, account must be taken that the appointed successor be suitable for continuing the generations-long tradition of our university to lead students in serious and selfless probing of mathematical problems. For this reason, personalities like Professor Felix Klein (born 1849) must be set aside, although opinions among scholars about his scientific achievements are very divided, since the entire effect of his writings and teaching stand in opposition to the tradition of our university as characterized above (Biermann 1988, 307–308).

These views reflect far more than just personal aversion to Klein and his style of mathematics. As the still young leader of a large group of mathematicians formerly associated with Alfred Clebsch, Felix Klein represented a threat to Berlin's hegemony within the German mathematical community, particularly within the Prussian university system. Göttingen, where Clebsch had taught until his unexpected death in 1872, afterward fell under Berlin's influence through the appointment of Weierstrass' star pupil, Schwarz. However, as noted already, Klein still had considerable residual influence in Göttingen, and in 1886, against strong opposition from Schwarz, he managed to obtain a chair there. Thereafter, leading representatives of the Berlin school did their best to keep Klein boxed in. He, in turn, hoped to gain the support of Friedrich Althoff, the powerful ministerial official who oversaw appointments in the Prussian educational system. This conflict set the stage for the events that now followed.

Rather than looking toward the challenges of the future, the Berlin mathematicians were mainly content to keep their own house in order. This conservative outlook was apparent from the start. Thus, it came as no surprise when the appointment committee decided to bring two Berlin products back into the fold by nominating Frobenius to fill Kronecker's chair and Schwarz to succeed Weierstrass. These recommendations were quickly approved by the Prussian Ministry, meaning Althoff, at which point the game of mathematical chairs could begin.

One month later Klein wrote to Adolf Hurwitz to convey his plans for filling Schwarz's vacant post:

> Althoff was here for three days and has decided on the calls to Berlin . . . [Concerning Schwarz's replacement] you will probably have guessed that I want to recommend you and Hilbert as the only two who, together with me, are in a position to assure Göttingen a place of scientific distinction. . . . Naturally, I will name you first and Hilbert behind you.[14]

Klein had in recent years been watching Hilbert's development closely. Whereas most observers regarded him as an expert on invariant theory, Klein realized that his interests went far beyond this field. In a report to Althoff, written in October 1890, he described Hilbert as "the rising man" in mathematics.[15] Still, as a mere *Privatdozent* in Königsberg he had no chance of leaping to the chair in Göttingen once occupied by Dirichlet and Riemann. As for Hurwitz, in his letter Klein expressed three issues of concern:

> There are, however, a series of difficulties associated with your appointment . . . First, there is the problem of your health. . . . Secondly, there is the much subtler difficulty that you are, not only personally but also in your scientific style, much closer to me than is Hilbert. Your coming here could therefore perhaps give our Göttingen mathematics a too one-sided character. There is thirdly – I must touch on it, as repugnant as the matter is to me, and knowing full well your justified sensitivity to this – the Jewish question. Not that your call as such would present difficulties; these I would be able to overcome. The problem is that we already have [Arthur] Schoenflies, for whom I would like to create a firm position as salaried *Extraordinarius*. And having you and Schoenflies together is something I will not get past either the faculty or the Minister.[16]

As this final remark implies, the Göttingen Philosophical Faculty had an unofficial policy aimed at restricting the number of Jewish *Dozenten* within a given discipline. Klein evidently raised this issue directly with Althoff, who apparently informed him of the ministry's position on this matter. The Prussian government clearly saw no reason to intercede in such cases of de facto discrimination against Jews. So Klein faced an impasse. Shortly before the faculty convened

[14]Felix Klein to Adolf Hurwitz, 28 February, 1892, Mathematiker-Archiv, Niedersächsische Staats- und Universitätsbibliothek, Göttingen.

[15]Felix Klein to Friedrich Althoff, 23 October, 1890, Rep. 92 Althoff B No. 92, fols. 76–77, Geheimes Staatsarchiv, Berlin.

[16]Felix Klein to Adolf Hurwitz, 28 February, 1892.

to deal with Schwarz's successor, he wrote to Althoff that, in view of the anti-Semitism within the faculty, he would be willing to "sacrifice" Schönflies in order to bring Hurwitz to Göttingen.[17]

Two weeks later, on 17 March 1892, Klein wrote to Hurwitz again, this time to inform him that he was now the only serious contender for Schwarz's position. It had been impossible even to get Hilbert's name on the list of candidates since he was still a *Privatdozent*.[18] This assessment of Hurwitz's chances, however, did not reflect the true state of affairs. Klein's rivals, Schwarz and Ernst Schering, pushed for a list with: (1) Heinrich Weber, (2) Ferdinand Lindemann, and (3) Georg Hettner, whereas Klein favored (1) Hurwitz, (2) Hilbert, and (3) Friedrich Schottky. After a long and intense debate, the Göttingen Philosophical Faculty compromised on (1) Weber, (2) Hurwitz, and (3) Schottky. These negotiations took place during the week of 6–10 March, that is one week *before* Klein wrote Hurwitz about his chances.[19] Thus, Klein was apparently banking on the autocratic Althoff, knowing full well his reputation for ignoring faculties' wishes. Presumably Klein had received some kind of assurances from him regarding this appointment; otherwise Klein's optimistic letter to Hurwitz would be inexplicable. Clearly he expected Althoff to pass over Weber and choose Hurwitz instead.

This plan might have been realized possibly had not an unexpected circumstance diverted Klein from his original objective. Frobenius, who had not yet accepted the offer from Berlin, began to contemplate whether he might not prefer Göttingen instead. Delighted by this turn of events, Klein assured Althoff that Frobenius would be welcomed with open arms by the Göttingen faculty. He even admitted in a tactless letter to Hurwitz that, had he known there was any possibility of winning Frobenius, he would have placed his name first on his list of nominees.[20] Sensing that his long-term strategy of "divide and conquer" was finally about to pay off, Klein rejoiced at the prospect of stealing away from Berlin the algebraist generally regarded as the natural heir to Kronecker's chair.[21] Not only would this solve the second concern he had confided to Hurwitz – since Frobenius' style clearly complemented his own – but it would also solve the problem posed by the third issue, the "Jewish question," opening the way for Schoenflies' appointment.

Klein invited Frobenius to visit him in Göttingen on 20 March, but the meeting evidently went badly. Klein recalled that Frobenius was exhausted from overwork during the stay. Not surprisingly, he opted for Berlin, and immediately afterward Althoff offered the vacant post in Göttingen to Heinrich Weber. Klein was furious when he learned that Althoff had chosen to honor the faculty's wishes rather than his own, although it is not implausible that his opportunistic fence-jumping may have ultimately ruined Hurwitz's chances. Another possibility is that Hurwitz was the victim on an "anti-Semitic backlash" within the Prussian Ministry, a suspicion Klein himself raised. For this reason, he wrote Hurwitz, he considered the latter's "chances of succeeding Weber in Marburg or obtaining a call anywhere else in Prussia as unfavorable."[22] This assessment turned out to be prophetic, although Hurwitz did receive an offer shortly thereafter from the ETH in Zurich to succeed Frobenius, which he accepted. Little did he know that he would spend the remainder of his career and never receive another offer from a Prussian university. Hurwitz's departure gave Hilbert the chance to move into this position in Königsberg, thereby ending his days as an impoverished lecturer. Minkowski, still in Bonn, was also elevated to associate professor at this time.

Whether or not Klein really believed that anti-Semitism had quashed Hurwitz's candidacy for Göttingen, he was clearly disappointed. However, his older friend and colleague, Paul Gordan, who presumably knew something about anti-Semitism in the German universities from firsthand experience, had a very different perspective of this matter, and wrote Klein accordingly:

> I am sorry to hear that you were not called to Berlin, as your all-embracing spirit would have brought order to the mathematical relationships in Germany. . . It was just that you recommended Hurwitz for Göttingen; Hurwitz deserves this distinction. That your recommendation did not go through, however, is a fortune for which you cannot thank God enough. What would you have had with Hurwitz in Göttingen? You would have taken on the complete responsibility for this Jew; every real or apparent mistake by Hurwitz would have fallen on your head, and all his utterances in the Faculty and Senate would have been regarded as influenced by you. Hurwitz would have been considered nothing more than an appendage of Klein.[23]

Paul Gordan to Felix Klein, 16 April, 1892

„Daß Sie nicht nach Berlin gekommen sind thut mir leid, bei Ihrem umfassenden Geist hätten Sie Ordnung in den wissenschaftlichen Verhältnisse Deutschlands gebracht. Aber für Sie ist es ein Glück. Daß Sie Hurwitz in Göttingen

[17]Felix Klein to Friedrich Althoff, 7 March, 1892, Rep. 92 Althoff AI No. 84, Bl. 21–22, Geheimes Staatsarchiv, Berlin.

[18]Felix Klein to Adolf Hurwitz, 17 March, 1892, Mathematiker-Archiv, Niedersächsische Staats- und Universitätsbibliothek, Göttingen.

[19]Klein Nachlass 22 L Personalia, S. 5, Niedersächsische Staats- und Universitätsbibliothek, Göttingen.

[20]Felix Klein to Adolf Hurwitz, 23 March, 1892, Mathematiker-Archiv, Niedersächsische Staats- und Universitätsbibliothek, Göttingen.

[21]Felix Klein to Friedrich Althoff, 21 March, 1892, Rep. 92 Althoff AI No. 84, Bl. 27–28, Geheimes Staatsarchiv, Berlin.

[22]Felix Klein to Adolf Hurwitz, 7 April, 11 April 1892, Mathematiker-Archiv, Niedersächsische Staats- und Universitätsbibliothek, Göttingen.

[23]Paul Gordan to Felix Klein, 16 April, 1892, Klein Nachlass, Niedersächsische Staats- und Universitätsbibliothek, Göttingen.

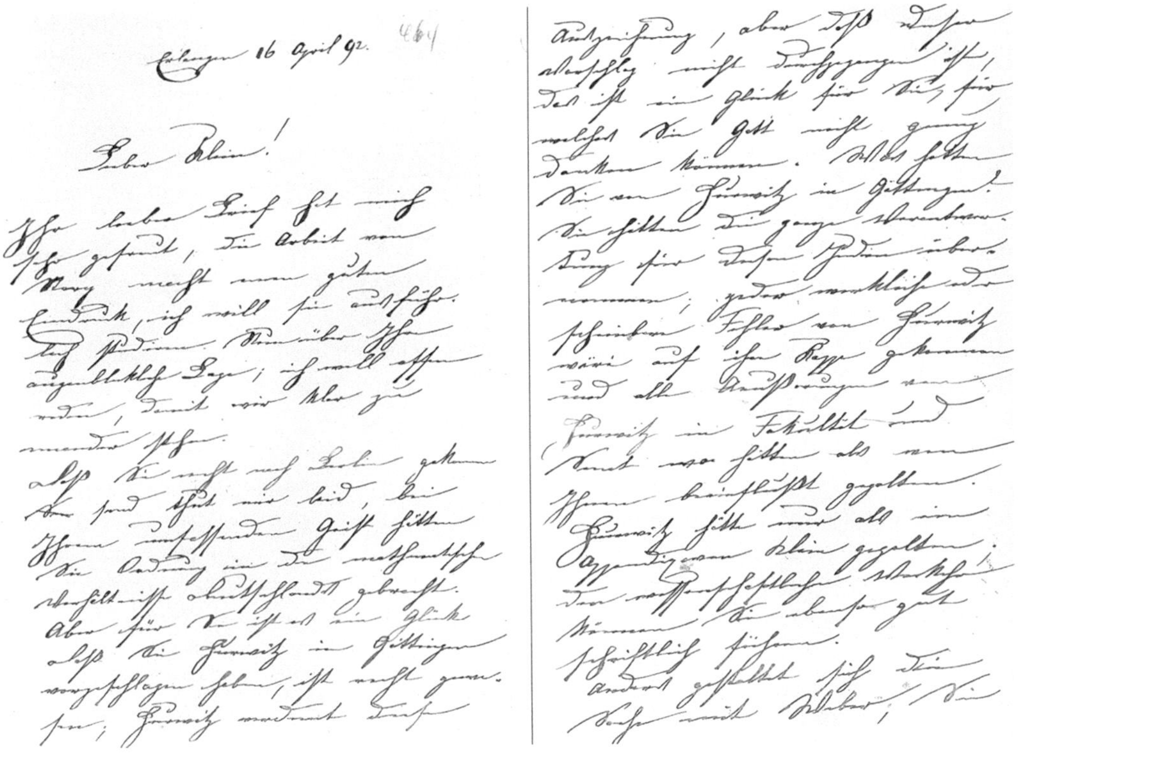

Fig. 14.3 Paul Gordan's letter to Felix Klein from 16 April 1892, commenting on Klein's dashed plans to have Hurwitz appointed to the vacant professorship in Göttingen. A partial transcription appears in the text. Courtesy of Niedersächsische Staats- und Universitätsbibliothek, Göttingen.

vorgeschlagen haben, ist recht gewesen; Hurwitz verdient diese Auszeichnung, aber daß diesen Vorschlag nicht durchgegangen ist, das ist ein Glück für Sie, für welches Sie Gott nicht genug danken können. Was hätten Sie von Hurwitz in Göttingen? Sie hätten die ganze Verantwortung für diesen Juden übernommen; jeder merkliche oder scheinbare Fehler von Hurwitz wäre auf Ihre Kappe gekommen und alle Aeußerungen von Hurwitz in Fakultät und Senat hätten als von Ihnen beeinflusst gegolten. Hurwitz hätte nur als ein Appendix von Klein gegolten." (Fig. 14.3)

Although disappointed, Klein was not one to take such a defeat lying down. He fired off an angry letter to Althoff complaining about the loss of face he had suffered in the Göttingen faculty; he had fought tenaciously for Hurwitz's cause, only to have Weber, the candidate proposed by his opponents, called instead. This turn of events, he asserted,

> . . . could only be somewhat remedied by having Schoenflies named *Extraordinarius*. On the one hand, it is known that I have been working on this appointment for years, on the other that my efforts have only met with resistance, so that I only dispensed from doing so as Hurwitz's call stood in question. Should Schoenflies now be passed over, this impression [i.e. of Klein's impotence in academic affairs] will become a virtual certainty. I would then be forced to advise young mathematicians not to turn to me if they hope to make further advancements in Prussia.[24]

Shortly after this letter was written, Arthur Schönflies was appointed ausserordentlicher Professor in Göttingen, where for the next seven years he taught descriptive geometry and strengthened the University's offerings in applied mathematics.

As it turned out, Klein got along very well with Heinrich Weber, a gifted and highly versatile mathematician whose research interests complemented Klein's. Shortly after Weber's arrival, he and Klein co-founded the Göttingen Mathematical Society, which thereafter played an instrumental role in forging the kind of tightly-knit mathematical community that Klein had long hoped to lead.

Just two months after Weber accepted the call to Göttingen, Klein got a tremendous break, though not from Althoff, but rather from outside Prussia. This came in the form of a very attractive offer from the Bavarian Ministry of Culture, which hoped to induce Klein to accept the chair at Munich University vacated by Ludwig Seidel's retirement. Even before Klein had the official offer in hand, he took up negotiations with Althoff, who matched the Bavarian conditions right down the line, thereby enabling the rejuvenated Göttingen mathematician to decline graciously.[25] In the process of doing so, he advised the Bavarian Ministry regarding other potential candidates, including Ferdinand

[24]Felix Klein to Friedrich Althoff, 10 April, 1892, Rep. 92 Althoff AI No. 84, Bl. 32–34, Geheimes Staatsarchiv, Berlin.

[25]For details, see Toepell (1996, 208–212, 370–378).

Lindemann, who was ultimately appointed.

In the meantime, Hilbert stood waiting in the wings, making every effort to signal his loyalty to Klein. Less than a year after succeeding Hurwitz, he informed Klein that Althoff was going to recommend his appointment as Lindemann's successor in Königsberg.[26] At this point, Hilbert and Klein began trading notes on suitable candidates for the associate professorship that Hurwitz had occupied only a year earlier. Throughout these deliberations, Hilbert indicated that he saw no chance of acquiring Minkowski, in part because his friend had excellent chances of moving up in Bonn. Time ticked on. Four months later, Althoff requested that Hilbert come to Berlin to discuss the matter with him personally, probably because of various other complications with personnel. Hilbert turned to Klein for advice, only to be told that Althoff was totally unpredictable, but Klein also thought that Minkowski's candidature was unrealistic. Six weeks later he congratulated Hilbert on pulling it off all the same (Frei 1985, 104–106). The issue of religious faith never arose, perhaps because the position was not a full professorship (*Ordinariat*). At any rate, the fact that Hilbert got to name Minkowski as his successor was a clear sign of things to come.

Heinrich Weber was not the young dynamo that the empire-builder Klein hoped to find, and Klein now became more determined than ever to bring Hilbert to Göttingen. When Weber accepted a chair in Strasbourg, he quickly seized the opportunity by notifying Hilbert in a letter from 6 December 1894:

> . . . Perhaps you do not yet know that Weber is going to Strassburg. This very evening the faculty will meet, and although I cannot know ahead of time what the commission will recommend, I still wish to inform you that I will make every effort to see that no one other than you is called here. You are the man whom I need as my scientific complement: due to the direction of your work, the power of your mathematical thinking, and the fact that you now stand in the middle of your productive career. I am counting on you to give a new inner strength to the mathematical school here, which has grown and, as it appears, will continue to grow a great deal further, and perhaps you will exert a rejuvenating influence on me as well. . . . I cannot know whether I will prevail in the faculty, even less so whether the recommendation we make will ultimately be followed in Berlin. But this one thing you must promise me, even today: that you will not decline the call if it comes to you! (Frei 1985, 115).

It was the chance of a lifetime, and Hilbert knew it. Klein no longer showed any interest in pursuing Hurwitz's candidacy. As *Dekan* of the Faculty, he wrote up the list of recommendations himself and it contained just two names, rather than the usual three (or more): Hilbert and Minkowski.[27] Some of Klein's colleagues criticized his choice, presuming that he wanted to appoint an easy-going, amenable (*bequem*) younger man. To this he replied: "Ich berufe mir den allerunbequemsten—I want the most difficult of all" (Blumenthal 1935, 399). Klein clearly knew what he wanted as well as what he was getting, and, for his part, Hilbert gladly joined forces with him.

The "Jewish Question" Reconsidered

Many German intellectuals disdained, and even more feared the autocratic Friedrich Althoff, who could make or break academic career at whim (Kayser 1996, 169–177). Hilbert, on the other hand, could only sing his praises, and for good reason: everything went his way. Less than two weeks after he received Klein's letter, he met with Althoff again, this time to seal the Göttingen deal. They quickly reached agreement, whereupon Althoff asked Hilbert whom he would like to have as his successor, practically inviting him to name Minkowski. When Hilbert obliged, Althoff then asked who he might want to fill Minkowski's associate professorship (Frei 1985, 117–118). Whatever its drawbacks, Althoff's style was surely a model of bureaucratic efficiency.

Soon after this conversation, Hilbert wrote Lindemann informing him that he had recommended Minkowski for the Königsberg chair. He received the following illuminating reply, sent on New Year's Day in 1895:

> It also would appear to me most natural to have Minkowski appointed as your successor. If Althoff is actually of the same opinion, then it really does not matter who else you place on the list. Still it seems to me possible that he will collide against [Minkowski's] Judaism; at any rate, he said to me once (in connection with Hurwitz's appointment in Königsberg), that there would be no misgivings about appointed a Jew to an associate professorship, but that the matter would be otherwise were the appointment to be a full professor's position. Of course, Althoff changes his views, too![28]

Althoff's personal views were probably beside the point. It would seem that Minkowski, unlike Hurwitz earlier, was in the right place at the right time. Thus the "Jewish question" never reared its head and Althoff simply followed Hilbert's recommendation. In the meantime, mathematics in Berlin languished. Schwarz largely gave up research, while Frobenius became increasingly frustrated as he saw how Klein

[26] David Hilbert to Felix Klein, 13 August 1893 (Frei 1985, 96).

[27] Klein's list of 13 December, 1894 can be found in the Personalakten Hilbert, Universitätsarchiv Göttingen.

[28] Ferdinand Lindemann to David Hilbert, 1 January 1895, Hilbert Nachlass, Niedersächsische Staats- und Universitätsbibliothek, Göttingen.

Fig. 14.4 Making Music in Zürich. During his early struggles with general relativity, Einstein often liked to relax in the home of his colleague Adolf Hurwitz, shown here pretending to conduct his daughter Lisi and their physicist friend as they play a violin duet. (Pólya 1987, 24).

managed to ingratiate himself with Althoff and the Prussian Ministry of Culture.

Minkowski remained only a little more than a year in Königsberg before joining Hurwitz in Zurich. In his letters to Hilbert (Minkowski 1973, 86–154) he provides many details about the times he and Hurwitz spent together. A critical test then came in 1902 when Hilbert was nominated to succeed Lazarus Fuchs in Berlin. No mathematician had ever before declined a formal offer from the leading German university, and Klein knew it would not be easy for Hilbert to refuse either. He urgently appealed to Althoff for support, and the latter obliged by creating a new *Ordinariat* in mathematics at Göttingen out of thin air, as it were, and appointing Minkowski to fill it. This unprecedented action was all the more daring considering that Minkowski's appointment also overturned the unwritten policy of the Göttingen Philosophical Faculty that restricted the proliferation of Jewish *Dozenten* within a given discipline. Arnold Schoenflies was by now gone, but in the meantime Karl Schwarzschild had been appointed Professor of Astronomy in 1901.

Seven years later came Minkowski's death from appendicitis so that his chair had to be filled anew. The Göttingen faculty deliberated over three possible candidates: Adolf Hurwitz, Otto Blumenthal, and Edmund Landau. All three were Jewish and all three ended up sharing first place on the faculty's *Berufungsliste,* but there the similarities end. Hurwitz had longstanding friendships with Klein and Hilbert, and Blumenthal, now managing editor of *Mathematische Annalen,* had been Hilbert's first doctoral student. Landau, on the other hand, was not only an outsider from Berlin, he was a proud, independent-minded, rich and self-confident Berlin Jew whose mathematics and teaching exuded the very purism that Klein had opposed for so long. Unlike the notoriously difficult and conceited Landau, both Hurwitz and Blumenthal were known for their *Liebenswürdigkeit,* or kind dispositions. In other words, they were the kinds of self-effacing "good Jews" that liberal-minded Germans tended to like, as opposed to the "others" – those "haughty Berlin Jews" and the *nouveau riche,* who (so the thinking went) had the audacity to flout established social convention by trying to hobnob with Prussian high society.

According to a story that Richard Courant liked to tell, the faculty's decision to pass over Hurwitz and Blumenthal was swayed by a statement made by Klein, who supposedly said something like this: "Landau is very disagreeable, very difficult to get along with. But we, being such a group as we are here, it is better that we have a man who is not easy."[29] Whether true or not, this story not only fits Klein but it also captures an important feature of the Göttingen community he wanted to build. Adolf Hurwitz lacked the ruthless arrogance that went with the territory. True, Hurwitz regretted the fact that Althoff saw no place for him in the Prussian university system, and he even queried Hilbert about this in 1904, but those who knew him in Zurich (see the tributes to Hurwitz by Max Born and George Pólya) also reported that his burning love for mathematics never died. His passion for music never ended either (Fig. 14.4).

[29]Reid (1970, 118); however, the surrounding information in Reid's account about those mathematicians who were considered to succeed Minkowski conflicts with the documentary evidence, which makes it difficult to place great confidence in the anecdote cited. See my account in Chap. 28.

Appendix: Two Tributes to Adolf Hurwitz

Max Born Recalling Hurwitz as a Teacher

At the end of the winter semester [1902–03] I again decided to spend the summer [studying] at another university in order to widen my views on science and life. It was only natural that I should consider Zurich, where, in addition to the Cantonal University, there was the *Eidegnössische Technische Hochschule*, an important school of science and engineering. My friend [Otto] Toeplitz approved my choice of Zurich as a mathematician of great renown, Hurwitz, lived there.. .. I have to confess that I went only to two mathematical courses, one (4 h a week) by Hurwitz on elliptic functions, the other (2 h) by [Heinrich] Burkhardt on Fourier analysis.

Hurwitz was a tiny man with the emaciated face of an ascetic in which burned two unnaturally large eyes. He was ailing and very frail. But his lectures were brilliant, perhaps the most perfect I have ever heard. The course was the continuation of another, on analytic functions, which I had not attended; I therefore had some difficulty in following and had to work hard, reading many books. Once when I missed a point in a lecture I went to Hurwitz afterwards and asked for a private explanation. He invited me and another student from Breslau, Kynast,. .. to his house and gave us a series of private lectures on some chapters of the theory of functions of complex variables, in particular on Mittag-Leffler's theorem, which I still consider as one of the most impressive experiences of my student life. I carefully worked out the whole course, including these private appendices, and my notebook was used by Courant when he, many years later and after Hurwitz' death, published his well—known book. .. the so-called Courant-Hurwitz (Born 1978, 72).

George Pólya on Hurwitz as a Colleague

Hurwitz had great mathematical breadth, as much as was possible in his time. He had learned algebra and number theory from Kummer and Kronecker, complex variable from Klein and Weierstrass. It was Hurwitz who arranged for me my first appointment at the ETH (The Swiss Federal Institute of Technology). From the time of my appointment there in 1914 until his death in 1919, I was in constant touch with him. We had a special way we worked. I would visit him and we would sit in his study and talk mathematics—seldom anything else—until he finished his cigar. Then we would go for a walk continuing the mathematical discussion. His health was not too good so when we walked it had to be on level ground, not always easy in the hilly part of Zürich, and if we went uphill, we walked very slowly. I wrote a joint paper with Hurwitz. In fact, it is a paper of mine and a paper of his, linked in a poetic form of correspondence. My connection with Hurwitz was deeper and my debt to him greater than to any other colleague. I played a large role in editing his collected works (Pólya 1987, 25).

References

Bergmann, Birgit, Moritz Epple, and Ruti Ungar, eds. 2012. *Transcending Tradition: Jewish Mathematicians in German-Speaking Academic Culture*. Heidelberg: Springer.

Biermann, Kurt-R. 1988. *Die Mathematik und ihre Dozenten an der Berliner Universität, 1810–1933*. Berlin: Akademie-Verlag.

Blumenthal, Otto. 1935. Lebensgeschichte. In *David Hilbert Gesammelte Abhandlungen*, Bd. 3 ed., 388–429. Berlin: Springer.

Born, Max. 1978. *My Life. Recollections of a Nobel Laureate*. New York: Charles Scribner's Sons.

Frei, Günther, Hrsg. 1985. *Der Briefwechsel David Hilbert—Felix Klein (1886–1918)*, Arbeiten aus der Niedersächsischen Staats- und Universitätsbibliothek Göttingen, Bd. 19, Göttingen: Vandenhoeck & Ruprecht.

Hashagen, Ulf. 2003. *Walther von Dyck (1856–1934)*. Stuttgart: Franz Steiner Verlag.

Hurwitz, Adolf. 1932. *Mathematische Werke*, vol. 2, Basel: Birkhäuser.

Kayser, Heinrich. 1996. *Erinnerungen aus meinem Leben*. München: Institut für Geschichte der Naturwissenschaften München.

Küssner, Martha. 1982. Carl Wolfgang Benjamin Goldschmidt und Moritz Abraham Stern, zwei Gaußschüler jüdischer Herkunft. *Mitteilungen der Gauß-Gesellschaft* 19: 37–62.

Lindemann, Ferdinand. 1971. *Lebenserinnerungen*. Munich.

Meissner, Ernst. 1932. Gedächtnisrede auf Adolf Hurwitz, in (Hurwitz 1932, xxi–xxiv).

Minkowski, Hermann. 1973. In *Briefe an David Hilbert*, ed. Lily Rüdenberg and Hans Zassenhaus. New York: Springer.

Pólya, George. 1987. In *The Pólya Picture Album: Ecounters of a Mathematician*, ed. G.L. Alexanderson. Birkhäuser: Boston.

Reid, Constance. 1970. *Hilbert*. New York: Springer.

Richenhagen, Gottfried. 1985. *Carl Runge (1856–1927): von der reinen Mathematik zur Numerik*. Göttingen: Vandenhoek & Ruprecht.

Rowe, David E. 1986. "Jewish Mathematics" at Göttingen in the Era of Felix Klein. *Isis* 77: 422–449.

Toepell, Michael. 1996. *Mathematiker und Mathematik an der Universität München. 500 Jahre Lehre und Forschung, Algorismus, Heft 19*. München: Institut für Geschichte der Naturwissenschaften.

Zeuthen, H.G. 1914. *Lehrbuch der abzählenden Methoden der Geometrie*. Leipzig: Teubner.

15 On the Background to Hilbert's Paris Lecture "Mathematical Problems"

(*Mathematical Intelligencer* 29(2)(2007): 18–30)

Much has been written about the famous lecture on "Mathematical Problems" (Hilbert 1901) that David Hilbert delivered at the Second International Congress of Mathematicians, which took place in Paris during the summer of 1900 (Alexandrov 1979; Browder 1976). Not that the event itself evoked such great interest, nor have many writers paid particularly close attention to what Hilbert had to say on that occasion. What mattered – both for the text and the larger context – came afterward. Mathematicians remember ICM II and Hilbert's role in it for just one reason: this was the occasion when he unveiled a famous list of 23 problems, a challenge to those who wished to make names for themselves in the coming century (Gray 2000). These "Hilbert Problems" and "their solvers" have long served as a central theme around which numerous stories have been written (Yandell 2002; Rowe 2004a). They have also served as a convenient peg for describing important mathematical developments of the twentieth century (Struik 1987). Yet relatively little has been written about the events that led up to Hilbert's lecture or the larger themes he set forth in the main body of his text. With this in mind, the present essay aims to address these less familiar parts of the story by sketching some of the relevant historical and mathematical background.

In accounts of Hilbert's life, his Paris lecture has rightly been seen as marking *the* great turning point in his spectacular career (Blumenthal 1935; Reid 1970). That career began quietly enough in Hilbert's native Königsberg where he emerged as an expert on algebraic invariants; the famous finiteness theorems for arbitrary systems of invariants stem from the late 80s and early 90s (see Chap. 13). Then, beginning around 1893, he broadened his terrain by taking in the theory of algebraic number fields, work that culminated around 1899, two years after the publication of his *Zahlbericht* (Hilbert 1998). In the meantime, Felix Klein managed to arrange Hilbert's appointment to the most prestigious chair in Göttingen, where he taught from 1895 until his retirement in 1930. Göttingen's subsequent success had everything to do with Hilbert's intriguing role within the context of a remarkable research community (Rowe 2004b). Indeed, when Klein first brought him to Göttingen he did so because he regarded Hilbert as the foremost *pure* mathematician of his generation, and hence the ideal person with whom to counter Berlin's traditional strength. No one could have anticipated – with the possible exception of Hilbert's good friend, Hermann Minkowski – how thoroughly Göttingen would come to dominate German mathematics after 1900 (on the background to the Berlin-Göttingen rivalry, see Chap. 4).

During the 1880s, when Hilbert began his studies in his native Königsberg, it was no easy matter for mathematicians in Germany to meet with one another to discuss their work or just to gossip about the latest news in the profession. The only forum at that time for regular formal gatherings was the annual meeting of the Society of German Natural Scientists and Physicians, an umbrella organization which included astronomers and mathematicians. Most of these conferences, however, were only sparsely attended by leading representatives of the mathematical sciences. The scientific programs seldom featured as many as ten lectures on mathematics, few of them delivered by distinguished researchers.[1] It was at the 1889 meeting of the *Naturforscher* held in Heidelberg that Georg Cantor seized the initiative that would lead to the founding of a national organization for mathematicians in Germany, the *Deutsche Mathematiker-Vereinigung* (DMV), which was formally established the next year in Bremen. Hilbert, along with Klein and Minkowski, were among its charter members (Dauben 1979; Hashagen 2000).

Cantor's seemingly innocuous proposal was by no means novel, but earlier efforts to form such a society had floundered due to factionalism. A long-standing rift divided the "southern" mathematicians from those who taught at the Prussian universities, and none of the more influential Berlin mathematicians saw any need for a national organization. During the early 1870s, Felix Klein had been one of the most active proponents of such a plan, and this earlier failure

[1] A complete list of the lectures delivered between 1843 and 1890 can be found in Tobies and Volkert (1998, 227–248).

D.E. Rowe, *A Richer Picture of Mathematics*, https://doi.org/10.1007/978-3-319-67819-1_15

caused him to doubt whether Cantor could really persuade the old guard, particularly Kronecker and Weierstrass, to go along with this revived plan. Rivalries between leading centers like Göttingen and Berlin, but also the provincial attitudes typical of competing schools at the highly autonomous German universities, had to be put aside in order to realize this enterprise. After a difficult start, the young DMV emerged a decade later as a viable organization representing the larger interests of the German mathematical community. In describing the historical and biographical contexts that led up to Hilbert's Paris lecture, the importance of the DMV as a vehicle that enabled him to publicize his mathematical ideas and larger vision will become very apparent. Indeed, Hilbert's immense power and prestige after 1900 clearly reflected a large-scale transformation, one that created new arenas and platforms for mathematicians with novel messages.

From Königsberg to Göttingen

During Hilbert's Königsberg years the themes of power and polemics only rarely surfaced explicitly. Like most other provincial German mathematicians, he worked in relative solitude, apart from occasional trips to attend meetings or to visit colleagues at other institutions. Nevertheless, Hilbert was acutely aware of Königsberg's importance as the only Prussian university that stood outside the hegemony of Berlin, particularly during the 1880s when Leopold Kronecker emerged as Germany's single most influential mathematician. Only after Klein's appointment in Göttingen in 1886 could the younger generation, including Hilbert, begin to detect that a major shift in power relations in Germany was underway (Rowe 1989).

With Kronecker's death in 1891 and the retirement of Weierstrass soon thereafter, Berlin's "golden age" came to an end. After H. A. Schwarz succeeded Weierstrass, Klein tried to seize the initiative in Göttingen by appointing either Adolf Hurwitz or Hilbert to replace Schwarz. Klein had already signalled to Friedrich Althoff, the influential Prussian Minister for Education, back in 1890 that Hilbert was the "rising man" among Germany's younger mathematicians, and in a letter to Hurwitz he wrote that only he and Hilbert were capable of "ensuring Göttingen a place of scientific distinction" (Rowe 1989, 196–197). As it turned out, Klein got neither: Hilbert, still a *Privatdozent*, was considered too junior to assume the chair once held by Gauss and Riemann, whereas Hurwitz faced a far more difficult hurdle: he was a Jew (see Chap. 14).

After a series of complex negotiations, the Göttingen chair finally went to Hilbert's former teacher, Heinrich Weber. Klein got along well with him, but Weber was not the young dynamo that the empire-builder Klein hoped to win for Göttingen. In the meantime, Hilbert stood waiting in the wings, making every effort to signal his patience as well as his loyalty to Klein. The payoffs were not long in coming. When Hurwitz assumed a position at the ETH in Zurich in 1892, Hilbert was appointed to succeed him as *Extraordinarius*, and within a year he was appointed full professor in Königsberg in the wake of Lindemann's appointment to an attractive post in Munich. As an even clearer sign of the times, Hilbert got to name Minkowski as his successor, thereby bringing the two back together in Königsberg, if only very briefly.

By this time, Klein was more determined than ever to bring Hilbert to Göttingen, and so when his colleague Heinrich Weber accepted a chair at Strasbourg he quickly seized the opportunity by pushing Hilbert's nomination through the faculty without any resistance. As *Dekan*, he wrote up the list of recommendations himself; it contained just two names: Hilbert and Minkowski.[2]

Algebraic Number Fields

When he arrived in Göttingen in the spring of 1895, Hilbert was already hard at work on a new project. This involved writing a report on number theory for the DMV, a task that he and Minkowski had been charged with producing. The two friends agreed to divide their labors so that Hilbert would concentrate on algebraic number theory,[3] whereas Minkowski could pick and choose from other topics in the field. By early 1896, however, Minkowski had still not completed his main project at the time, namely his *Geometrie der Zahlen*, the book with which he vaulted himself into the pantheon of great names in the history of number theory. If Minkowski's masterpiece was slow in coming, Charles Hermite, for one, clearly appreciated its contents when he wrote: "I believe I see the promised land" (Hilbert 1935, 348). What Minkowski had seen amounted to a path linking number theory to convex figures in the plane, a surprising confluence of ideas that caused a major stir in Göttingen. Hilbert drew further attention to it in a letter to Klein, published in *Mathematische Annalen* already in 1895. Five years later he would reformulate the purely geometrical issues at stake when he formulated his fourth Paris problem.

By 1896, Minkowski had reconciled himself to the fact that he could not in good conscience continue with the report he and Hilbert had originally planned. He therefore withdrew from his part in the project, but offered Hilbert

[2] Klein's list of 13 December, 1894 can be found in the Personalakten Hilbert, Universitätsarchiv Göttingen.

[3] Norbert Schappacher pointed out to me that the term "algebraic number theory" was first coined in the twentieth century and therefore arose in the wake of Hilbert's work; for further commentary and analysis, see Schappacher (2005) and Goldstein and Schappacher (2007, 88–90).

his assistance as proof reader and critic for the portion his friend would soon bring to a close. And when the published work – which came to be known as the *Zahlbericht* – arrived in Minkowski's hands in May 1896, he wrote to congratulate Hilbert, "now that the time has finally come, after the many years of labor, when your report will become the common property of all mathematicians, and I do not doubt that in the near future you yourself will be counted among the great classical figures of number theory" (Minkowski 1973, 100). That piece of mathematical prophecy would, indeed, soon come true, no doubt partly because Hilbert's style had a great appeal for the younger generation (Schappacher 2005).

Käthe Hilbert, whose handwriting was far more legible than her husband's, diligently copied out the entire manuscript. After many months filled with the tedium of reworking the text and copy-editing the page proofs, the Hilberts could finally relax in the spring of 1897 after the last corrections had been sent off to the printers. Minkowski, who had since joined Hurwitz at the ETH in Zürich, must have felt nearly as relieved as they did when it was finally finished. He and Hurwitz both scolded Hilbert, however, for neglecting to thank his wife in the preface (a flaw Hilbert remedied in the final page proofs).

Hermann Weyl later recalled his enthralment when he first came to Göttingen and met Hilbert; he simply had to read this man's great monograph, and he did so at the first opportunity. Weyl, whose masterful control of the German language was probably unmatched by any other mathematician, regarded the preface of Hilbert's *Zahlbericht* as one of the great works of prose in mathematical literature (Weyl 1932, 57). Here, in a series of brief allusions to the historical development of the discipline of number theory, Hilbert succeeded in capturing those aesthetic qualities and epistemological characteristics that he regarded as the true center around which all pure mathematics revolved. After explaining the five-part layout of his *Zahlbericht,* Hilbert summarized the architectonic features of algebraic number theory in the following passage:

> The theory of number fields is like a building of wonderful beauty and harmony; the most richly endowed part of this edifice appears to me the theory of abelian and relative abelian fields, a field opened up by Kummer, in his work on higher reciprocity laws, and by Kronecker, in his studies on complex multiplication and elliptic functions. The deep insights into this theory which the works of these two mathematicians offer show us at the same time that in this field an abundance of the most precious treasures still lies concealed, beckoning with rich rewards for the researcher who recognises their value and lovingly practices the art of winning them (Hilbert 1932, 67).

These words were surely still in Weyl's mind when he remembered Hilbert as that "Pied Piper" whose enchanting melodies "seduced so many rats to follow him into the deep river of mathematics" (Weyl 1944, 614).

Hilbert had seemingly found a field of research ideally suited to his temperament and intellectual talent. For, as he noted, unlike other branches of mathematics, which required a lengthy gestation period before they could attain secure foundations, number theory possessed the necessary clarity almost from the very beginning. Nevertheless, higher arithmetic and the theory of algebraic number fields, in particular, had only emerged as a systematic science during the past hundred years or so. Hilbert recalled how Gauss and his followers, Jacobi and Dirichlet, were enchanted by the surprising connections they discovered between number-theoretic questions and algebraic problems, particularly those dealing with cyclotomic equations. The explanation, Hilbert went on, had by now become altogether clear, since algebraic numbers and the Galois theory of equations were both rooted in the theory of algebraic number fields. Thus what was once a mystery to Gauss, Jacobi, and Dirichlet could now be understood – thanks to the efforts of Kummer, Dedekind, and Kronecker – as a self-evident structural feature common to the problems stemming from these two disciplines.

Hilbert saw this as an instance of a far more general pattern of development found over and again in pure mathematics. Indeed, this was a historical lesson he felt the modern mathematician needed to appreciate, namely, that one should not be surprised by seemingly fortuitous interconnections between what might look like unrelated mathematical problems. Rather, one should come to expect that subsequent investigations would eventually reveal the true relations governing such mysterious phenomena once they are understood at a much deeper level. This was a vision of mathematics inspired by the great cathedral of higher arithmetic and algebra whose earlier architects were Gauss, Abel, and Galois. For Hilbert, however, this massive structure extended still further. Just as he noted the integrative power of algebraic number fields for algebra and number theory, so was he drawn to the strong interplay found between number theory and the theory of functions of a single complex variable.

Moreover, he highlighted certain structural analogies between number fields and function fields while emphasizing the importance of analytical formulas like the Riemann zeta-function for the distribution of primes. Within analytic number theory, questions concerning transcendence beckon for attention, as in Hilbert's own strikingly simplified proof demonstrating that π is a transcendental number, hence the impossibility of squaring the circle. His argument made decisive use of the exponential function, which itself can be introduced from the standpoint of invariant theory, since $e^{2\pi iz}$ can be viewed as an invariant for the functional equation $f(z+1) = f(z)$. Yet, here again, Hilbert was following a well-worn historical path. The theory of the exponential function led directly to the theory of cyclotomic equations, a field Gauss cultivated in solving another classical problem bequeathed by the ancient Greeks, namely the determination of all possible constructible regular polygons. All this and more hung together in a beautiful tapestry of pure ideas.

Taken together, Hilbert's remarks outlined ideas he would return to in his Paris lecture, where the very same themes re-emerge in the more precise form of four specific problems – the 7th, 8th, 9th, and 12th – which lay at the heart of his vision not only of number theory but of pure mathematics *in toto*. We need only cite a passage from the preface of the *Zahlbericht* to give a clear impression of the programmatic importance these ideas held for Hilbert:

> Thus we see how far arithmetic, the Queen of mathematics, has conquered broad areas of algebra and function theory to become their leader. The reason that this did not happen sooner and has not yet developed more extensively seems to me to lie in this, that number theory has only in recent years become known in its maturity. . . .Nowadays the erratic progress characteristic of the earliest stages of development of a subject has been replaced by steady and continuous progress through the systematic construction of the theory of algebraic number fields. The conclusion, if I am not mistaken, is that above all the modern development of pure mathematics takes place under the banner of number: the definitions given by Dedekind and Kronecker of the concept of number lead to an arithmetization of function theory and serve to realize the principle that, even in function theory, a fact can be regarded as proven only when in the last instance it has been reduced to relations between rational integers (Hilbert 1932, 65–66).[4]

Here, in a most striking way, we see Hilbert flirting with Kronecker's ideas, which clearly fit in well with the overall vision he sought to promote. Nevertheless, the conceptual approach taken by Dedekind appealed to him even more. Thus, far from following the constructivist tendencies that had motivated the work of Gauss, Kummer, and Kronecker, Hilbert preferred to bypass direct computations whenever he could. This philosophy, too, can be found in the preface, particularly when Hilbert described the culminating fifth section of the *Zahlbericht*, about which he wrote:

> . . . the fifth part develops the theory of those fields which Kummer took as a basis for his researches into higher reciprocity laws and which on this account I have named after him. It is clear that the theory of these Kummer fields represents the highest peak attained today on the mountain of our knowledge of arithmetic; from it we look out on the wide panorama of the whole explored domain since almost all essential ideas and concepts of field theory, at least in a special setting, find an application in the proof of the higher reciprocity laws. I have tried to avoid Kummer's elaborate computational machinery, so that here, too, Riemann's principle may be realized and the proofs completed not by calculations but purely by ideas (Hilbert 1932, 66–67).

These remarks resounded with overtones from the long-standing rivalry that pitted Göttingen against the once-dominant Berlin establishment. Hilbert, now firmly situated at Klein's side in Göttingen, was sounding a message that he and Klein never tired of in their effort to marginalize Berlin, namely the superiority of the Göttingen tradition, which they linked with the glorious names of Gauss, Dirichlet, and Riemann. In the *Zahlbericht*, Hilbert kept his propagandistic remarks to a minimum, but his correspondence with Minkowski clearly reveals how little he thought of Kummer's methods (about which, see Lemmermeyer and Schappacher (1998)). Still, the tone throughout his preface remained highly reverential. He extolled the achievements of Gauss, who, inspired by his work on biquadratic reciprocity, extended arithmetic to the Gaussian integers (see Chap. 3). He praised Kummer, Kronecker, and Dedekind, the creators of modern algebraic number theory, as well as Dirichlet, Riemann, and Jacobi, who established deep connections between complex function theory and the theory of numbers.

Yet if Hilbert's rhetoric often suggested that he was building on the work of his famous predecessors, especially Kummer and Kronecker, insiders must have realized that his main goal was to show how such earlier achievements could be subsumed within the framework of an entirely new theory. No one could have understood Hilbert's true intentions better than Minkowski, who after reading the page proofs of the *Zahlbericht* remarked: "I am extraordinarily pleased with your report in its terse and yet complete form. It will certainly generally meet with great acclaim, and push the works of Kronecker and Dedekind very much into the background" (Minkowski 1973, 79–80).

In his *Zahlbericht*, Hilbert sought to recast the work of his predecessors in such a way that central features of the theory became more evident. He introduced the norm residue symbol, which gave the higher reciprocity laws a more elegant formulation. At the same time, he developed a systematic theory of Galois number fields in which the arithmetic properties of a Galois field are studied by means of the homomorphic images of its Galois group. Hilbert's number-theoretic investigations reached their climax in 1898–99 with the publication of two major papers Hilbert (1932, 370–482) and Hilbert (1932, 483–509). These studies, particularly the first, served as a catalyst for the investigations of a number of leading number theorists during the next three decades (see Hasse (1932)).

Foundations of Geometry

Thus, by 1899 Hilbert had firmly established his reputation as the era's leading authority in both invariant theory and the theory of algebraic number fields, two formerly distinct disciplines that had now been brought together through his work. But then came a most unexpected turn of events. Otto Blumenthal later recalled the buzzing chatter among the students when they read Hilbert's announcement for a course on "Grundlagen der Euklidischen Geometrie" (Hilbert 2004, 185–406) that he would offer during the winter semester of

[4]For an interesting discussion of how Klein and Hilbert put new twists on the term "arithmetization," see Petri and Schappacher (2007, 362–366).

1898–99 (Blumenthal 1935, 402). He and the older students who had been accompanying Hilbert on weekly walks had never heard him talk about geometry, only number fields. Little did they realize that Hilbert had been contemplating the foundations of geometry ever since his years as a *Privatdozent* in Königsberg (see the texts in Hilbert (2004) and the discussion in Toepell (1986)).

The lecture course they attended that semester surprised them even more, for in it Hilbert sought to lay out the fundamental structures underlying Euclidean geometry as no one had ever done before. Hilbert revised this material the following spring, following a request from Klein, in order to present his "Grundlagen der Geometrie" in June 1899 as part of a *Festschrift* commemorating the unveiling of the Gauss-Weber monument in Göttingen. His study bore a motto taken from Kant's *Critique of Pure Reason:* "all human knowledge begins with intuitions (*Anschauungen*), proceeds from there to concepts (*Begriffen*), and ends with ideas" (Hilbert 1899, 1). This genetic characterisation of the process that leads to man's acquisition of sure and certain knowledge aptly reflected Hilbert's own views, particularly with regard to mathematical knowledge. In geometry, the ultimate ideas took the form of axioms, first principles that he placed in five distinct groups. Thus, in Hilbert's system one finds axioms of incidence, order, congruence, a parallel axiom, and axioms of continuity. These groupings, while of no significance for the formal theory itself, served as conceptual signposts identifying those informal elements that were rooted in primitive insights (*Anschauungen*). Hilbert then set himself the task not simply of investigating one special type of geometry; he wanted to explore various types of geometries, which can be obtained by assuming the validity of some, but not all of the axioms.

While remembered today primarily because of its sleek and sophisticated handling of these axioms, it should not be overlooked that the agenda behind this work came from arithmetic and *not* synthetic geometry per se. Hilbert's central concern, in fact, was to build bridges from synthetic to analytic geometry, a task he accomplished without importing arithmetical systems into his geometrical axioms. Thus, coordinate geometry had to be achieved by appeal to a separate axiom system for such number systems. Within geometry proper, Hilbert used two key theorems of synthetic projective geometry, those of Pascal and Desargues, and adapted these as tools for proving the arithmetical properties of two corresponding types of segment arithmetics. By so doing, he emphasised how both of these segment arithmetics could be introduced without any recourse to the axioms of continuity. This meant, in particular, that they also held for non-Archimedean geometries, an innovation introduced by Giuseppe Veronese that was hotly debated in Germany during the 1890s.

Whereas the notion of building geometry on a system of axioms was as old as Euclid's *Elements*, the possibility of developing number systems axiomatically was scarcely heard of in 1899. Even bolder was the idea of setting the two systems of axioms alongside one another in order to weave them together. Hilbert's axioms for what he called "complex number systems" enabled him to introduce algebraic number fields by means of properties that ran parallel to those found in his five groups of geometric axioms. In the original *Festschrift* edition of *Grundlagen der Geometrie* (Hilbert 1899), however, he chose to avoid the thorny problem of arithmetizing the continuum, the case that led to geometry over the field of real numbers, or what Hilbert referred to as Cartesian geometry. This problem, however, soon became the focus of his concern, and already in June of 1899, shortly after the Minkowskis paid a visit to the Hilberts in Göttingen, Hilbert's friend alluded to the $18 = 17 + 1$ axioms of arithmetic that they had recently discussed during their visit (Minkowski 1973, 116–117). This arithmetical system, whose properties were set forth in four groups of axioms, was designed to characterise the real numbers as a complete Archimedean ordered field.

In dealing with more elementary Euclidean geometries, Hilbert had relied on the consistency of the axioms for elementary arithmetic, which were independent of the trickier topological problems associated with the continuum of real numbers. Not even Kronecker had doubted that the operations of ordinary arithmetic were beyond all logical scruples. Kronecker's stringent position, which had much in common with the intuitionistic principles later expounded by L. E. J. Brouwer, amounted to a rejection of all parts of mathematics that could not be rigorously deduced from arithmetic principles. As can be seen from the *Zahlbericht*, Hilbert shared with Kronecker a strong belief in arithmetization, but he was also confident that nearly all branches of geometry could be so treated. In *Grundlagen der Geometrie* he showed that – so long as one avoided all stipulations with regard to the infinitely small (so-called continuity conditions) – one could prove the consistency of the weaker forms of Euclidean geometry by appealing to models based on the arithmetic of countable number fields. Since these arithmetical systems were presumably sound, so reasoned Hilbert, the same must be true of the corresponding geometries.

On the other hand, when it came to Cartesian geometry over the real numbers (which are uncountable) appeal to such a model was far more problematic, since continuity assumptions were in this case unavoidable. Nevertheless, Hilbert thought he had found a way to finesse the issue by invoking his 18th axiom, the notorious "completeness" property of the real numbers. Despite its name, this 18th axiom (*Vollständigkeitsaxiom*) asserted nothing whatsoever about the topological completeness of the real numbers. On the contrary, it simply stipulated the impossibility of extending beyond any set of objects for which the first 17 axioms hold, namely those that obtain for ordered Archimedean fields. Completeness in Hilbert's sense thus has one immediate payoff: it automatically guarantees that only one model could ever satisfy all the conditions of the system. Not surprisingly, the notoriety of this completeness axiom had much to do with the existential nature of the assertion it contains. Thus, the fabled issue of mathematical existence, which had already reared its sphinx-like head in Hilbert's work on invariant theory, once again made its presence felt.

Hilbert's argument for the logical consistency of Cartesian geometry ultimately hinged on knowing that the axiom system he had presented for the real numbers was itself consistent. To address this concern, Hilbert proposed to push his argument one step further. Since the logical consistency of the axioms for Cartesian geometry had been shown to follow from the consistency of the axioms for the arithmetic of the real numbers, it remained to provide a proof that the latter system was itself consistent. In September 1899, Hilbert called attention to this very problem in a lecture entitled "Über den Zahlbegriff," (Hilbert 1900a), which he delivered in Munich at the annual meeting of the DMV. There, he claimed that to prove this – and thereby establish the existence of the real numbers in the sense of Georg Cantor's set theory – required only a suitable modification of known methods of argument" (Hilbert 1900a, 184). One year later, in presenting his famous second Paris problem, he would repeat this conviction.

The upshot of all this was simple and crucial: by showing that one could rigorously arithmetize Cartesian geometry it followed that this discipline stood on the same logical footing as other branches of mathematics, such as analysis. Hilbert thereby sought to allay the long-standing suspicions harbored against classical geometry. The Berlin mathematicians had long treated geometrical arguments as foreign to the rigorous standards demanded by algebra and analysis (Biermann 1988, 103–112). Diagrams and other aids to the imagination were considered taboo. Hilbert's aim was to dispel this orthodoxy by showing that the foundations of geometry were every bit as rigorous as those based on the properties of number systems.

Perhaps more important still, his work would lead to a liberation of geometrical research from the conventional straight-jacket of Euclidean orthodoxy. If Minkowski's geometry of convex bodies could prove so fertile for number theory, why should mathematicians not explore other new leads based on any potentially interesting system of axioms? Hilbert's approach thus sought to legitimize not only standardized Euclidean geometry, but also other geometries that satisfied some, but not all of his axioms. Seen in broader terms, Hilbert's *Grundlagen der Geometrie* represented a systematic approach toward understanding the precise scope of various geometries by deciding precisely when a given well-posed problem could be solved within them. Thus, approaching the turn of the century, the foundations of geometry had already assumed a prominent place within Hilbert's research program. Analysis, on the other hand, lay just over the horizon.

All of the above achievements reflected Hilbert's universality, but they also revealed that his work stood within clearly discernible mathematical traditions. Invariant theory, pursued originally by British mathematicians like Boole, Cayley, and Sylvester, had deep roots in the Königsberg school of Jacobi and Hesse. Algebraic number theory, as Hilbert's introductory remarks in his *Zahlbericht* made clear, had been a dominant theme within German mathematical circles from the time of Gauss. Even his contributions to the foundations of geometry evince how deeply Hilbert's axiomatic approach was rooted in an essentially algebraic vision. These fields should be contrasted with those parts of pure mathematics that lay well removed from Hilbert's primary research interests; indeed, doing so makes evident the importance of national schools and styles during this era. Classical analysis, for example, continued to be dominated by the French, whereas Italian mathematicians had assumed the leading role in algebraic and differential geometry. Like most of his German colleagues, Hilbert tended to follow developments in these areas from afar.

Munich, September 1899

It was with a sense of high adventure that Hilbert joined the 80-odd mathematicians who convened in Munich during mid-September 1899 to attend the annual meeting of the German Mathematicians Union (DMV).[5] Braving floods that left the city's transportation system virtually paralyzed, this group gathered in an upbeat atmosphere to celebrate the tenth anniversary of their organization. Max Noether, the DMV's presiding officer, took pleasure in congratulating the assembled throng, which included Georg Cantor. In opening the Munich meeting, Noether pointed with pride to its accomplishments over the last 10 years, not the least of which was the publication of Hilbert's *Zahlbericht*. In this brief period, the German Mathematicians Union had indeed come into its own, emerging as the core structure around which a fast-growing German mathematical community was taking form. Hilbert, still not yet 40, was about to move into its spotlight.

As already noted, it was at this Munich meeting that he first sketched his plan for a rigorous axiomatic approach to the real number continuum, one that entailed proving the consistency of his 18 axioms characterizing this mathemat-

[5]The information that follows is based on the report in *Jahresbericht der Deutschen Mathematiker-Vereinigung*, 8(1900): 3–5.

ical construct of such crucial importance. Hilbert delivered a second talk on this occasion that would prove even more important as a *leitmotiv* for many of his leading pupils. In "Über das Dirichlet'sche Prinzip" (Hilbert 1900b), he indicated how this classical tool in the calculus of variations could be revived and applied to special problems in geometry and physics. Riemann, who learned this technique while studying under Dirichlet in Berlin, had subsequently used the Dirichlet principle to establish his existence theorems for complex functions satisfying given boundary conditions. However, Weierstrass gave a counterexample undermining the validity of this argument, and afterward Schwarz, Carl Neumann, and Poincaré began developing other methods to replace it. Hilbert seems to have been the first mathematician who attempted to resuscitate the original, rather simple idea behind Dirichlet's principle. His methods later proved spectacularly successful, mushrooming into a major new field in analysis (Yandell 2002, 380–382). Indeed, this field and the theory of integral equations would become the dominant research areas for an army of talented students who came to Göttingen to do their doctoral studies under Hilbert during the period of his high fame, the crucial years from 1901 up until the outbreak of World War I.

Following the timeline of Hilbert's career, the conference that convened in Munich in September of 1899 emerges as a significant personal triumph. He was now riding the cusp of a wave that marked the arrival of a new era of national unity within the German mathematical community. Having witnessed these developments from their inception, he was more than ready to be catapulted forward as the leader of his country's next generation of mathematicians. Hilbert fully recognized the importance of reaping the kinds of rewards that only an organization like the DMV could bestow: official acknowledgement of the esteem he enjoyed among his fellow mathematicians. At the Munich meeting he was elected to succeed Noether as the new presiding officer, in which capacity he would chair the annual meeting held in Aachen the following year. His two lectures – presenting an axiom system for the real numbers and his plan for resuscitating the Dirichlet principle – clearly came at a propitious moment, adding luster to a rich scientific program that accented topics in analysis and mathematical physics. Beyond this feast of mathematical thought, the participants also formulated plans for advancing their interests on a number of educational and scientific fronts. Most of these involved the domestic scene, but with the Paris congress only a year away the membership took the opportunity to discuss this and other international events as well.

Hilbert returned from Munich in high spirits, writing to Hurwitz that this event was the best attended and most stimulating of all the DMV meetings to date (Hilbert to Hurwitz, 5–12 November 1899). Clearly pleased by the response to his two lectures, he also came away with vivid memories of another talk delivered by the inimitable Ludwig Boltzmann (1900). Compared with Hilbert's two presentations, Boltzmann's was far splashier, and the spellbound audience was afterward literally buzzing with excitement. Six months later, when Hilbert began work on the text of his now famous Paris lecture, he almost certainly did so with impressions of Boltzmann's performance still fresh in his mind. Indeed, the parallels between these two speeches are striking. Thematically, they may be regarded as belonging to the same genre: both offered sweeping accounts of past developments, but with an eye toward unsolved problems that would test the steel of the younger generation.

In Munich, Boltzmann surveyed major developments in theoretical physics over the course of the nineteenth century, ending his talk with a string of open questions, most of them connected in one way or another with the fate of the mechanical world picture. In closing, he beckoned the younger generation onward – "a Spartan war chorus calls out to its youth: be even braver than we!" (Boltzmann 1900, 95) – expressing his hopes that the new century would bring even greater surprises than the outgoing one had seen. For those who heard him, Boltzmann's performance was electrifying, but its effect proved as momentary as the nebulous images his poetic language evoked. His passionate message aroused real interest, but his ideas lacked specificity. Nor was Boltzmann willing to stick his neck out terribly far. At the very outset, he made it clear that he would not attempt to "lift the veil" (Boltzmann 1900, 73) covering the face of future events by making prognostications about the course of physical research. Hilbert would do just that one year later in Paris while borrowing Boltzmann's poetic imagery. As the mathematicians of Europe began making plans for the forthcoming International Congress in Paris they, too, cast their eyes toward the future.

Minkowski's Sage Advice

Back in Göttingen, the DMV's newly-elected president received an invitation from the ICM's organizing committee to deliver a plenary address on a topic of his choosing. Hilbert saw this as a splendid opportunity to speak as a leading representative of German mathematical culture. In keeping with that role, he toyed with the idea of presenting a kind of veiled challenge to the views set forth by the era's leading mathematician, Henri Poincaré, at the first ICM held three years earlier in Zürich.[6] There soon followed, in December 1899, a letter to Minkowski in which Hilbert disclosed some of his first thoughts about an appropriate

[6]The ICMs offered a large-scale stage for such jousting. At the 1908 ICM in Rome, Poincaré presented his own views on the future of mathematics, about which see Gray (2012).

theme. Since Poincaré had emphasized the role of physical conceptions in guiding fertile mathematical research, Hilbert had in mind to sing a hymn of praise to pure mathematics, perhaps with thoughts of how Jacobi had once rebuked his French contemporary, Joseph Fourier, for failing to recognize that the highest and true purpose of mathematics resided in nothing other than the quest for truth; this alone redounded honor on the human spirit. Thirty years later, in fact, Hilbert would recall this famous anecdote in his last major speech, delivered in Königsberg (Hilbert 1935, 378–387).

Minkowski did not respond to Hilbert's proposal immediately. He merely wrote back expressing these uplifting sentiments: "may the new century bring you much luck and fame and may the mathematics of the next century see itself robbed by you of even more of its deepest secrets than was already the case in the previous one" (Minkowski 1973, 30 December 1899). By early January, though, he had found time to reread Poincaré's speech, and so he wrote again, this time strongly counseling Hilbert to abandon his original plan. Minkowski found that Poincaré's assertions in no way compromised the integrity of pure mathematics and, furthermore, that the Frenchman's views had been so cautiously articulated that one could easily subscribe to them. Moreover, he recalled how Poincaré had not even been present in Zurich, so that his text had to be read by another party. Consequently, few would even remember what was then said, in contrast with the stirring impression left behind by Boltzmann's talk at the Munich meeting four months earlier.

After thus dowsing Hilbert's original idea with cold water, Minkowski threw out some fresh thoughts of his own about how his friend might best take advantage of the unusual opportunity that lay at hand. "Most alluring," he wrote, "would be the attempt to look into the future, in other words, a characterization of the problems to which the mathematicians should turn in the future. With this, you might conceivably have people talking about your speech even decades from now. Of course, prophecy is indeed a difficult thing" ((Minkowski 1973), 5 January 1900). Minkowski went on to reflect about the audience, assuring Hilbert that a lecture with real substance, like the one delivered by Hurwitz in Zurich, was far more effective than a more general presentation such as Poincaré's.[7] Grand visions swept before Minkowski's eyes, yet he assured Hilbert that success depended less on the theme of his address than the quality of his presentation. Still, he realized well enough that the "frame of the theme could have the effect that twice as many listeners turn out" as might otherwise.

[7]He even mentioned two earlier lectures that he thought might prove useful for Hilbert to read, one a speech delivered by Hermite in 1890, another by H. J. S. Smith entitled, "On the Present State and Prospects of some Branches of Pure Mathematics."

Hilbert vacillated, and Minkowski heard nothing more in reply. By the end of February he wrote to him in Göttingen, puzzled by his friend's silence: "why is it that we hear nothing from you? My last letter contained really nothing much more than the advice that if you deliver a nice lecture, then it will be very nice. But it wasn't easy to give good advice" (Minkowski (1973), 25 February 1900). A rendezvous took place in Switzerland during the early spring, but what they discussed during this brief sojourn remains a mystery. At any rate, Hilbert's plans remained uncertain, for in late March he wrote Hurwitz, "I must start preparing for a major talk in Paris, and I am hesitating about the theme.. .. The best would be a view into the future. What do you think about the likely direction in which mathematics will develop during the next century? It would be very interesting and instructive to hear your opinion about that" (Hilbert to Hurwitz, March 29 1900).

By June, Minkowski still knew nothing of Hilbert's plans, so when he received a program from the organizers of ICM II without any mention of a lecture by Hilbert he was deeply disappointed. He shot off a letter to Hilbert, lamenting that he had by now lost nearly all interest in the Paris congress: "participation from here [the German contingent] will be almost zero. Mostly one will get to see French school teachers and exotic mathematicians, Spaniards, Greeks, etc., for whom Paris in August still feels cool in comparison with their home countries" ((Minkowski 1973), 22 June 1900). Minkowski's infectious enthusiasm soon returned, however, when Hilbert wrote him that he had, in fact, been working all along on the lecture he would present in Paris. Apparently Hilbert's procrastination led the organizers to choose another representative from Germany as plenary speaker. They sent out an invitation to the Heidelberg historian of mathematics, Moritz Cantor, who got to deliver his lecture in a plenary session, whereas Hilbert was forced to present his in one of the sectional meetings.

By early summer, Hilbert's ideas for his talk had finally begun to crystallize: he would describe a set of problems intended as a challenge for the mathematicians of the coming twentieth century, but he still had yet to decide on the precise list. Thus he sounded out Hurwitz with regard to the state of uniformization theory in the wake of the most recent work by Poincaré and Picard. His former mentor replied that, in his view, France's two leading analysts had by no means spoken the last word on this subject. This response no doubt encouraged Hilbert to formulate his 22nd problem. Next he turned to Minkowski, tapping his expertise in mathematical physics to inquire whether he knew of appropriate physical theories with open foundational problems. Having spent some time during his days as a *Privatdozent* working with Heinrich Hertz in Bonn, Minkowski had considerably more experience with physics than Hilbert. Both, however, were familiar with Boltzmann's work on statistical mechanics and thermodynamics. This terrain, Minkowski hinted, was filled with "with many interesting mathematical questions also very useful for physics" ((Minkowski 1973), 10 July 1900) a remark that presumably had some influence when Hilbert described his rather amorphous sixth problem.

Only 1 week later, Minkowski found himself staring at page proofs of various portions of Hilbert's lecture, including drafts of what ultimately became the second and twenty-third problems. Minkowski's reaction to the text of the second problem reveals a keen sense of foresight into the kinds of controversies it would 1 day unleash.

> It is, indeed, highly original to pose as a problem for the future one which the mathematicians have believed already for a long time to possess completely, like the axioms of arithmetics. What will the large number of laymen in the audience say to that? Will their respect of us grow? And you will have a hard fight on your hands with the philosophers, too. ((Minkowski 1973), 17 July 1900)

Hilbert thought that with his axiomatization of the real numbers he had come very close to hitting bedrock, without apparently sensing that the logical foundations of his theory rested on terrain that could quickly turn into quicksand. The second problem merely reiterated in loftier language the challenge of proving the consistency of his $17 + 1$ axioms for arithmetic, which Minkowski knew both from their private discussions during the previous summer as well as from Hilbert's lecture "On the Number Concept" delivered at the DMV meeting in Munich. His reaction to this newest version thus suggests that he recognized clearly how Hilbert was suddenly playing for larger stakes. Four years later, at ICM III in Heidelberg, Hilbert would up the ante even higher. But while L. E. J. Brouwer quietly developed a radically different approach to the continuum, only the Jena philosopher Gottlob Frege openly criticized Hilbert's views and attempted to call his hand (Rowe 2000).

Hurwitz read these first proofs as well, and both he and Minkowski counseled Hilbert that major cuts would have to be made in the text, even though they had yet to receive all of it. But by the end of July, Minkowski had almost everything: the body of Hilbert's speech together with a description of 21 unsolved problems (it remains unclear which two problems he added afterward). By now, the presentation was clearly far too long, so Minkowski suggested that Hilbert read his general remarks in the plenary session, where he could then direct the audience to a second talk on the outstanding problems he had selected which would follow in one of the scientific sections. Aside from these reservations, however, Minkowski had no substantial criticisms to make. Indeed, he was nothing less than ecstatic by the time he finished reading this draft:

> I can only wish you luck on your speech; it will certainly be the event of the congress and its success will be very lasting. For I believe that this speech, which probably every mathematician without exception will read, will cause your powers of attraction on young mathematicians to grow still more, if that is even possible. . . . Now you have really wrapped up the mathematics for the twentieth century and in most quarters you will gladly be acknowledged as its general director. ((Minkowski 1973), 28 July 1900)

Minkowski's generous praise could only have lifted Hilbert's spirits and prepared him for the role he was destined to assume.

On Mathematical Knowledge

Decades later, few remembered Hilbert's larger messages from his Paris speech, even though he took great care in crafting it. Instead, his admirers (and a few critics) focused their attention on the list of 23 problems Hilbert set out for the consideration of his peers. Some of these were well known (like problem 1, Cantor's continuum hypothesis, or number 8, Riemann's conjecture concerning the zeroes of the zeta-function). Others were rather vague and programmatic (like the sixth problem, dealing with the axiomatization of certain branches of physics, or problem 23 calling for a fresh attack on the calculus of variations). In not a few cases, however, the problems attained new meanings, and some were soon shorn from the mathematical settings in which Hilbert had originally placed them (as, for example, the tenth problem on the decidability of Diophantine equations, which had to be sharpened before researchers could begin to tackle the question). Hilbert's problems thus took on lives of their own, many of them later to be regarded as markers for mathematical progress over the course of the twentieth century.

Seen, however, in its original context many long-since-forgotten aspects reappear. It should not be forgotten that Hilbert lived during an era in which philosophy – in particular, the theory of knowledge – played a central part in higher learning. Thus, it should come as no surprise that the main text of "Mathematical Problems" is strewn with allusions to general questions pertaining to the nature of mathematical knowledge. These kinds of general remarks were by no means uncommon in the ceremonial speeches of nineteenth-century mathematicians, particularly in Germany with its tradition of idealist philosophy so inclined to explain almost everything as a manifestation of *Geist*. Hilbert's speech had much in common with a genre of popular scientific lectures that employed quasi-religious language to espouse its message, in this case to sing the praises of mathematical knowledge and, by implication, those who pursue it. Indeed, it might even be said that Hilbert took this genre to a new level. For no one before him, and certainly none since, ever quite matched the fervor of his faith in the enduring vitality and value of mathematical knowledge.

Yet, however bold the words might have sounded, his fundamental message was in many respects a familiar one. Immanuel Kant had already proscribed a special role for mathematical knowledge within his epistemological system, typified by the assertion that the laws of Euclidean geometry transcend both those of logic as well as the principles of empirical inquiry. Thus, for Kant, the parallel postulate constituted a synthetic a priori proposition whose truth value could neither be deduced analytically nor verified as an empirical fact. Hilbert loved to cite the words of his famous fellow Königsberger, especially Kant's assertion that "one encounters in every special natural science just that degree of true science as one finds mathematics within it" (Hilbert 1935, 385). Kant had been deeply interested in exploring the limits of human reason and knowledge, a theme that took on significant new overtones afterward, particularly in the wake of debates over Darwin's evolutionary ideas.

Within Germany, one of the leading academic figures to address such epistemological issues head on was the Berlin physiologist Emil du Bois-Reymond, whose younger brother, Paul, happened to be a well-known mathematician. In one of his most widely read lectures, "Über die Grenzen des Naturerkennens" du Bois-Reymond (1886), delivered in Leipzig at the 1872 meeting of the Society of German Natural Scientists and Physicians, du Bois-Reymond took up a series of questions of deep human concern, some with clearly religious overtones. Mankind longed to know: what was the essence of force and matter? what were the origins of life and thought? was nature teleological? did human beings have free will? To each and every one of these, he asked another: can science provide the answer to these mysteries, if not today then some day in the future? Du Bois-Reymond's conclusion came in the form of a simple refrain – "*ignoramus et ignorabimus*" (we do not know, and we will not know) – a phrase that became a kind of watchword within the popular culture of Germany's educated elites.

Hilbert took no direct issue with Emil du Bois-Reymond when it came to the specific questions the latter had declared unanswerable, but he did want to make one point clear: in mathematics, he maintained, there is no *ignorabimus*. His optimistic assertion that every well-posed problem in mathematics was capable of being solved marks one of the high points in his Paris speech. After recounting how mathematicians had at last resolved two of the fundamental problems bequeathed upon them by the ancient Greeks—the status of the parallel postulate in Euclidean geometry and the possibility of squaring the circle—Hilbert claimed this only illustrated what was characteristic of mathematical problems in general, namely that each and every well-posed question in mathematics was capable of being answered with finality. For Hilbert, this belief offered not only a strong psychological support but even had the quality of a moral imperative: "the conviction in the solvability of every mathematical problem is a powerful incentive to the worker. We hear within us the perpetual call: there is the problem, seek its solution. You can find it by pure reason, for in mathematics there is no *ignorabimus*" (Hilbert 1935, 298).

This was in every way a truly remarkable pronouncement, one that only Hilbert could have made in such a forceful way. In fact, he even suggested in passing that "this conviction that every mathematician certainly shares" might itself be provable. This conjecture that came to be known as the decision problem, but in his Paris lecture Hilbert took this simply as an article of faith.[8] His arch-nemesis, Brouwer, would later declare this position epistemologically bankrupt since it was equivalent to claiming the unrestricted validity of the principle of the excluded middle (van Dalen 1999, 105–106). Thus, what Hilbert apparently regarded as a truism in 1900 (well before Brouwer stepped onto the scene), would later become a most problematic issue, one that came to preoccupy all his attention and even to haunt him throughout the final stages of his career. One mainly thinks of the first and second Hilbert problems when reflecting on this theme. Hilbert was especially vulnerable when it came to the second problem because he staked so much on its resolution, which eventually ran afoul of Gödel's undecidability theorems (Yandell 2002, 37–58). But the general fertility of Hilbert's vision has never been doubted, and this was clearly tied to his sense of where mathematics stood at the dawn of the new century.

As important as these pronouncements on the foundations of mathematics certainly were for Hilbert, they should be set against the views he expressed in his closing remarks, which address the architecture of mathematics taken as a whole (see also (Rowe 1994)). In broaching this subject, he first confronted the specter of specialization, the possibility that the field of mathematics might, like so many other sciences, dissolve into a number of narrow research disciplines that have little or nothing in common (Hilbert 1935, 329). Hilbert emphatically asserted that, in his opinion, this would not and should not happen. In his view, mathematics constituted a harmonious, "indivisible whole," an assertion that should be regarded as tantamount to the second article of Hilbert's mathematical faith. Surely, the brilliant successes he had already enjoyed during his still young career gave him compelling reasons to believe not only that the outstanding problems of mathematics can be resolved but that their solution would inevitably contribute to a simplification of complex theories and a recognition that *all* of mathematics can be derived from a unified set of fundamental principles. This bold claim was central to Hilbert's mathematical *Weltanschauung*, and certainly constituted one of the most daring visions he offered for the mathematics of the twentieth

[8] His private notes, however, make clear that he also saw this as a major unsolved meta-problem, about which see Thiele (2003).

Fig. 15.1 The Paris International Exposition in 1900.

century. In the heyday of Bourbaki, it appeared to many that this Hilbertian dream would soon come true.

Returning to the mundane events that Hilbert witnessed during his brief stay in Paris puts matters in a very different light (Fig. 15.1). He, in fact, expressed general disappointment with the ICM as a scientific event, a reaction that was widespread afterward. Charlotte Angas Scott reported in the *Bulletin of the American Mathematical Society* that "the arrangements excited a good deal of criticism" (Scott 1900, 75). Compared with the DMV meeting a year earlier in Munich, Hilbert found the Congress a rather lackluster affair. To Hurwitz he complained about the difficulties meeting other mathematicians in surroundings conducive to thoughtful discussion. Many organizational matters had to be improvised, with predictable results. Apparently the Parisian hosts had focused most of their attention on atmospherics, offering participants ample opportunity to gather for convivial chitchat. Thus, after the Tuesday sessions they retired for lunch at the Ecole Normale, where they drank a series of toasts to their chosen calling and mutual good fortune. Those who wished to take in the sights of the city and its Exposition found plenty of free time to do so. At the close of the Congress, Prince Roland Bonaparte threw a reception party, and many stayed on through Sunday August 12 to attend a noon banquet as well as a gala soiree at the opera as guests of the Minister of Public Education. Hilbert evinced little interest in all these peripheral activities, which he probably regarded as merely a senseless series of distractions. He had come to Paris with only one real purpose in mind – to talk mathematics – and he found rather little opportunity for serious exchanges with his colleagues.

As president-elect of the DMV, Hilbert vowed that Germany would do better in 1904 (ICM III was scheduled to be held in Baden-Baden, but eventually Heidelberg was chosen instead). Only one month after he had left Paris, he congregated with some of his German colleagues in Aachen, where he opened the annual meeting of the DMV as its newly elected president. In this capacity, he reported on the events of the Paris Congress, no doubt suppressing the critical commentary he conveyed to Hurwitz privately. The meeting in Aachen was only sparsely attended and its program fell far short of the one in Munich a year earlier. In his opening remarks, Hilbert warned the German mathematical community against the dangers of one-sidedness: "the mathematician tends to regard his views and methods as the most important and as the only appropriate ones... and the best preventative against this comes from scientific discourse with numerous other colleagues."[9] This, Hilbert understood to be the noblest purpose of the organization over which he had been elected to preside. His own efforts to realize that ideal within the confines of the Göttingen mathematical community would soon be marked by stunning new successes.

Despite his impressive achievements prior to 1900, Hilbert's glory years still lay ahead of him. His was a career that took many surprising turns, nearly all of which enhanced his fame, power, and influence. No single event contributed more to these than his Paris lecture with its famous list of unsolved problems. With the advantage of hindsight, this episode stands out as a virtual moment of metamorphosis in Hilbert's career, including the whole spectrum of

[9] *Jahresbericht der Deutschen Mathematiker-Vereinigung*, 9 (1901), S. 3.

mathematical interests that engaged him both before and afterward. Even within the close-knit circle of his friends and *Mitarbeiter*, none could well have imagined that the author of the *Zahlbericht* would after the turn of the century virtually turn his back on algebra and number theory in order to take up entirely new problems in geometry, integral equations, the calculus of variations, and mathematical physics (Courant and Hilbert 1968). At the same time, Hilbert would set forth a vision for a new axiomatic approach to mathematics and physics (Corry 2004), a veritable modern-day Leibnizian *mathesis universalis*. It was the stuff from which legends are born.

References

Alexandrov, Paul S. Hrsg. 1979. *Die Hilbertsche Probleme*, Ostwalds Klassiker der exakten Wissenschaften, 252, Leipzig: Teubner.

Biermann, Kurt-R. 1988. *Die Mathematik und ihre Dozenten an der Berliner Universität, 1810–1933*. Berlin: Akademie Verlag.

Blumenthal, Otto. 1935. Lebensgeschichte, in [Hilbert 1935, 388–429].

Boltzmann, Ludwig. 1900. Über die Entwicklung der Methoden der theoretischen Physik in neuerer Zeit. *Jahresbericht der Deutschen Mathematiker-Vereinigung* 8: 71–95.

Browder, Felix, ed. 1976. *Mathematical Developments arising from Hilbert's Problems*, Symposia in Pure Mathematics. Vol. 28. Providence: American Mathematical Society.

Corry, Leo. 2004. *David Hilbert and the Axiomatization of Physics (1898–1918): From Grundlagen der Geometrie to Grundlagen der Physik*. Dordrecht: Kluwer.

Courant, Richard, and David Hilbert. 1968. *Methoden der mathematischen Physik, I*. 3te Auflage ed. Heidelberg: Springer.

Dauben, Joseph W. 1979. *Georg Cantor. His Mathematics and Philosophy of the Infinite*. Cambridge: Harvard University Press.

du Bois-Reymond, Emil. 1886. Über die Grenzen des Naturerkennens (1872). In *Reden von Emil Du Bois-Reymond*, Bd. I ed. Leipzig: Veit.

Goldstein, Catherine and Norbert Schappacher. 2007. "Several Disciplines and a Book (1860–1901)," in [Goldstein/Schappacher/Schwermer 2007, 67–104].

Goldstein, Catherine, Norbert Schappacher, and Joachim Schwermer, eds. 2007. *The Shaping of Arithmetic after C. F. Gauss's Disquisitiones Arithmeticae*. Heidelberg: Springer.

Gray, Jeremy. 2000. *The Hilbert Challenge*. Oxford: Oxford University Press.

———. 2012. Poincaré replies to Hilbert on the Future of Mathematics, ca, 1908. *Mathematical Intelligencer* 34 (3): 15–29.

Hashagen, Ulf. 2000. Georg Cantor und die Gründung der Deutschen Mathematiker-Vereinigung, *Arbeitspapier, Münchner Zentrum für Wissenschafts- und Technikgeschichte*.

Hasse, Helmut. 1932. Zu Hilberts algebraisch-zahlentheoretischen Arbeiten, in [Hilbert 1932, 528–535].

Hilbert, David. 1899. *Grundlagen der Geometrie*, (Festschrift zur Einweihung des Göttinger Gauss-Weber Denkmals), Leipzig; revised 2nd ed. 1903, 3rd ed. 1909, 4th ed. 1913, 5th ed. 1922, 6th ed. 1923, 7th ed. 1930; Stuttgart, 8th ed. 1956, 9th ed. 1962, 10th ed. 1968, 11th ed. 1972, 12th ed. 1977.

———. 1900a. Über den Zahlbegriff. *Jahresbericht der Deutschen Mathematiker-Vereinigung* 8: 180–184.

———. 1900b. Über das Dirichlet'sche Princip. *Jahresbericht der Deutschen Mathematiker-Vereinigung* 8: 184–187.

———. 1901. Mathematische Probleme, *Archiv für Mathematik und Physik* 1(1901), 44–63 and 213–237; reprinted in [Hilbert 1935, 290–329]; Mathematical Problems: Lecture delivered before the International Congress of Mathematicians at Paris in 1900. Trans. Mary F. Winston. (1902), *Bulletin of the American Mathematical Society*, 8: 437–479.

———. 1932. *Gesammelte Abhandlungen*. Bd. 1 ed. Berlin: Springer.

———. 1935. *Gesammelte Abhandlungen*. Bd. 3 ed. Berlin: Springer.

———. 1998. *The Theory of Algebraic Number Fields*. Trans. Iain T. Adamson. Heidelberg: Springer.

———. 2004. *David Hilbert's Lectures on the Foundations of Geometry, 1891–1902*. Hallett, Michael, and Ulrich Majer, eds. Heidelberg: Springer.

Lemmermeyer, Franz and Norbert Schappacher. 1998. Introduction to [Hilbert 1998].

Minkowski, Hermann. 1973. In *Briefe an David Hilbert*, ed. Hg.L. Rüdenberg and H. Zassenhaus. New York: Springer.

Petri, B., and Schappacher, N.. 2007. On Arithmetization, in [Goldstein/Schappacher/Schwermer 2007, 343–374].

Reid, Constance. 1970. *Hilbert*. New York: Springer.

Rowe, David E. 1989. Klein, Hilbert, and the Göttingen Mathematical Tradition. In *Science in Germany: The Intersection of Institutional and Intellectual Issues*, ed. Kathryn M. Olesko, vol. 5, 189–213. Osiris.

———. 1994. The Philosophical Views of Klein and Hilbert. In *The Intersection of History and Mathematics*, ed. Sasaki Chikara, Sugiura Mitsuo, Joseph W. Dauben, Science Networks, vol. 15 (Proceedings of the 1990 Tokyo Symposium on the History of Mathematics), Basel: Birkäuser, 187–202.

———. 2004a. I Prolemi di Hilbert e la Matematica del XX Secolo. In *Storia della scienza*, ed. Sandro Petruccioli, vol. 8, 104–111. Rome: Istituto della Enciclopedia Italiana.

———. 2004b. Making Mathematics in an Oral Culture: Göttingen in the Era of Klein and Hilbert. *Science in Context* 17 (1/2): 85–129.

———. 2000. The Calm before the Storm: Hilbert's early Views on Foundations. In *Proof Theory: History and Philosophical Significance*, ed. Vincent Hendricks, 55–94. Dordrecht: Kluwer.

Sasaki, Chikara, Sugiura Mitsuo, Joseph W. Dauben, eds. 1994. *The Intersection of History and Mathematics*. Science Networks, vol. 15 (Proceedings of the 1990 Tokyo Symposium on the History of Mathematics), Basel: Birkhäuser.

Schappacher, Norbert. 2005. David Hilbert, Report on Algebraic Number Fields (*Zahlbericht*) (1897). In *Landmark Writings in Western Mathematics*, ed. Ivor Grattan-Guinness, 700–709. Amsterdam: Elsevier.

Scott, Charlotte Angas. 1900. Report on the International Congress of Mathematicians in Paris. *Bulletin of the American Mathematical Society* 7: 57–79.

Struik, Dirk J. 1987. *A Concise History of Mathematics*. 4th ed. New York: Dover.

Thiele, Rüdiger. 2003. Hilbert's Twenty-Fourth Problem. *The American Mathematical Monthly* 110: 1–24.

Tobies, Renate, and Klaus Volkert. 1998. *Mathematik auf den Versammlungen der Gesellschaft deutscher Naturforscher und Aerzte, 1843–1890*, (Schriftenreihe zur Geschichte der Versammlungen deutscher Naturforscher und Aerzte, Band 7), Stuttgart: Wissenschaftliche Verlagsgesellschaft.

Toepell, M.-M. 1986. *Über die Entstehung von David Hilberts ‚Grundlagen der Geometrie"*, Studien zur Wissenschafts-, Sozial-, und Bildungsgeschichte der Mathematik, Bd. 2, Göttingen.

van Dalen, Dirk. 1999. *Mystic, Geometer, and Intuitionist: The Life of L. E. J. Brouwer*. Vol. 1. Oxford: Oxford University Press.

Weyl, Hermann. 1932. Zu David Hilberts siebzigsten Geburtstag. *Die Naturwissenschaften* 20: 57–58.

———. 1944. David Hilbert and his Mathematical Work. *Bulletin of the American Mathematical Society* 50: 612–654.

Yandell, Ben H. 2002. *The Honors Class. Hilbert's Problems and their Solvers*. Natick: A K Peters.

Poincaré Week in Göttingen, 22–28 April 1909 16

(*Mathematical Intelligencer* 8(1)(1986): 75–77)

When Paul Wolfskehl died in 1906, his will established a prize for the first mathematician who could supply a proof of Fermat's Last Theorem, or give a counterexample refuting it. The interest from this prize money was later used to bring world-renowned mathematicians to Göttingen to deliver a series of lectures. Hilbert was apparently very pleased with this arrangement, and once jested that the only thing that kept him from proving Fermat's famous conjecture was the thought of killing the goose that laid these golden eggs.

The first guest of the Wolfskehl Foundation was Henri Poincaré, who had lectured at Göttingen once earlier in 1895, just after Hilbert's arrival. At that time he spoke on celestial mechanics, but for his Wolfskehl lectures he chose integral equations, Cantor's transfinite numbers, and relativity theory among his topics. Poincaré had actually been quite reluctant to accept this invitation due to a serious health problem that had plagued him since his collapse at the ICM in Rome the previous year. He alluded to these circumstances in a letter to Hilbert: "I am still recovering from an accident that befell me in Rome last year, and I am obliged to take some precautions. I can drink neither wine nor beer, only water. I can attend neither a banquet, nor a prolonged meal" (Gray 2013, 417). By November 1908 Hilbert and Poincaré had settled on the dates for his visit, which had to be moved back till late April when the summer semester began. As for the program, this remained unsettled until shortly before the honored guest arrived.[1]

There was, no doubt, a certain amount of tension surrounding Poincaré's presence in Göttingen, not only because of the traditional Franco-German rivalry but also because Poincaré had bested Klein in their race to launch the theory of automorphic functions during the early 1880s. More recently, in 1905, Poincaré was awarded the Bolyai Prize ahead of Hilbert, a decision reached by Klein and Gaston Darboux as the two most prominent members of the small commission assigned to judge which of the two had made the more significant contributions to mathematics as a whole. Moreover, this second trip to Göttingen coincided with a mournful period there, especially for Hilbert, as only four months earlier his dear colleague, Hermann Minkowski, had suffered a sudden and tragic death.

Poincaré offered six lectures to his Göttingen audience, covering a wide range of topics. The first five were delivered in German, a language he spoke tolerably well. Still, Hilbert made it a point to thank him for making the effort to do so. He began with three lectures on topics very close to Hilbert's own work and that of his many students at this time: Fredholmian analysis, followed by applications of integral equations to tidal motions as well as to Hertzian waves. Poincaré then moved to a Kleinian topic in complex analysis: Abelian integrals and the theory of Fuchsian functions. The fifth lecture on transfinite numbers brought him into the highly contested terrain of Cantor's *Mengenlehre*, the main topic to be discussed below. Two months earlier, Hilbert wrote him asking whether he would be willing to speak about a theme of "philosophical-logical coloration," a request to which Poincaré acceded, gladly or not (Fig. 16.1).

Hilbert also requested that his guest speak about a theme of his choosing either from mathematical physics or astronomy. In all likelihood, Hilbert had in mind the relativity principle, a topic that had been hotly debated ever since the publication of Einstein's "Zur Elektrodynamik bewegter Körper" in 1905.[2] Poincaré chose, in fact, to speak about this, but rather obliquely and in French for his final lecture on "La mécanique nouvelle" (Poincaré 1910). This talk, too, caused quite a stir, though more because of what Poincaré didn't say rather than what he actually said (Walter 2016). Poincaré

[1] For a detailed account of the planning along with transcriptions of the extant letters that Hilbert and Poincaré exchanged, see Walter (2016).

[2] Minkowski had been deeply interested in electron theory even before Einstein's paper appeared; see Pyenson (1979) and Walter (2008).

D.E. Rowe, *A Richer Picture of Mathematics*, https://doi.org/10.1007/978-3-319-67819-1_16

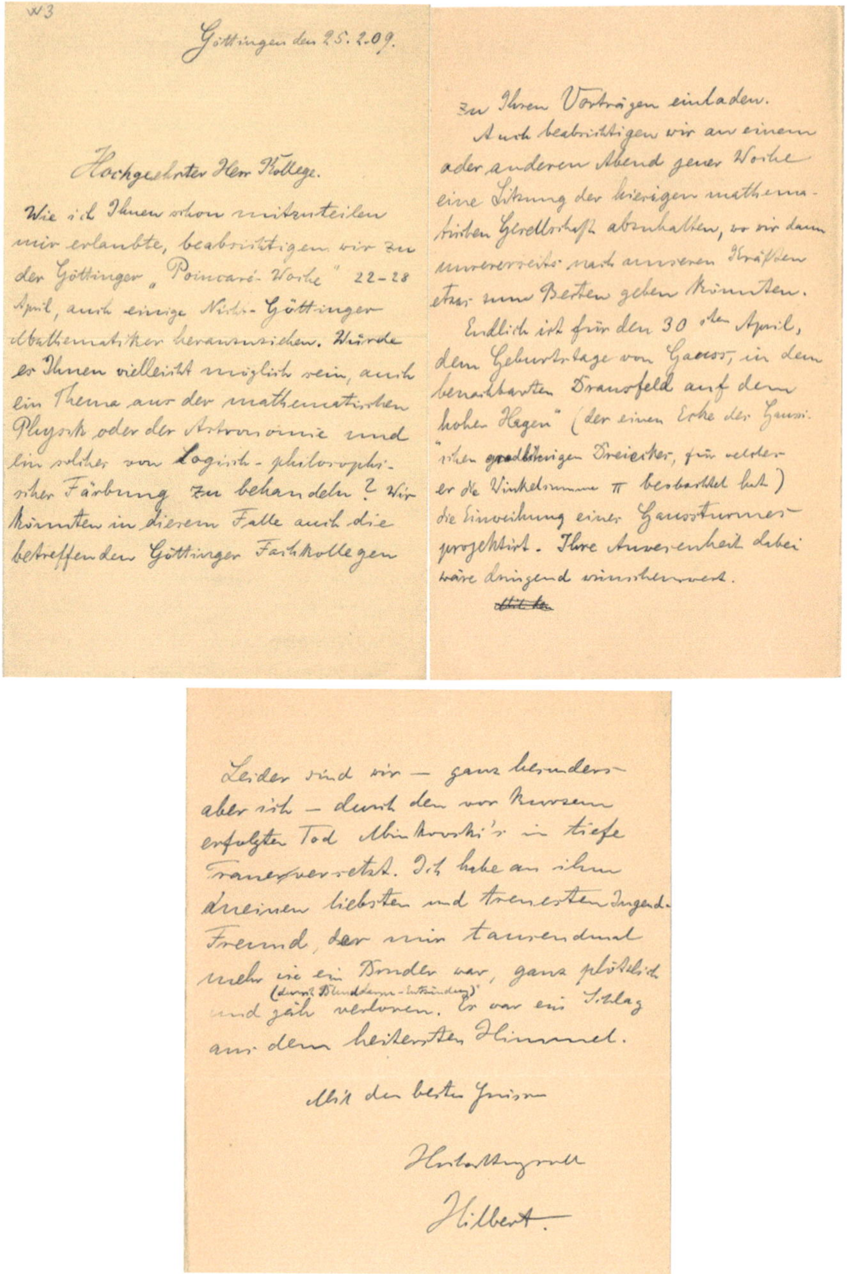

Göttingen den 25.2.09.

Hochgeehrter Herr Kollege.

Wie ich Ihnen schon mitzuteilen mir erlaubte, beabsichtigen wir zu der Göttinger „Poincaré-Woche" 22–28 April, auch einige Nicht-Göttinger Mathematiker heranzuziehen. Würde es Ihnen vielleicht möglich sein, auch ein Thema aus der mathematischen Physik oder der Astronomie und ein solches von logisch-philosophischer Färbung zu behandeln? Wir könnten in diesem Falle auch die betreffenden Göttinger Fachkollegen zu Ihren Vorträgen einladen.

Auch beabsichtigen wir an einem oder anderen Abend jener Woche eine Sitzung der hiesigen mathematischen Gesellschaft abzuhalten, so wir dann unsererseits nach unseren Kräften etwas zum Besten geben könnten.

Endlich ist für den 30sten April, dem Geburtstage von Gauss, in dem benachbarten Dransfeld auf dem „hohen Hagen" (der einen Ecke des Gauss'schen großdreiseitigen Dreiecks, für welches er die Winkelsumme π beobachtet hat) die Einweihung eines Gaussturmes projektirt. Ihre Anwesenheit dabei wäre dringend wünschenswert.

Leider sind wir — ganz besonders aber ich — durch den vor Kurzem erfolgten Tod Minkowski's in tiefe Trauer versetzt. Ich habe an ihm meinen liebsten und treuesten Jugendfreund, der mir tausendmal mehr wie ein Bruder war, ganz plötzlich (durch Blinddarm-Entzündung) und jäh verloren. Es war ein Schlag aus dem heitersten Himmel.

Mit den besten Grüssen

Hochachtungsvoll

Hilbert.

Fig. 16.1 Hilbert's letter to Poincaré, 25 February 1909 (Courtesy of Henri Poincaré Papers, Nancy).

had taken significant steps in the direction of Einstein's and Minkowski's ideas, but he remained all his life curiously silent about their work, even after others like Max Planck were proclaiming its revolutionary significance.[3] His lecture dealt with the possibility that mass increases with velocity, thereby making it impossible to accelerate bodies beyond the speed of light. Citing Laplace's argument that gravitation must necessarily propagate at a far greater velocity than light, he concluded that it would be premature to declare that classical physics was dead.

A glance at the program during Poincaré Week shows that he did not occupy the stage alone. In his letter to Poincaré, Hilbert had hinted that he and his colleagues planned to hold a special session of the Göttingen Mathematical Society in

[3]Some years after Poincaré's death in 1912, Felix Klein remembered Poincaré's silence and speculated that he might have merely been returning the favor, since Minkowski had ignored Poincaré's contributions in his famous Cologne lecture "Raum und Zeit."

order to highlight some topics of mutual interest. Evidently, he, Klein, Landau, and Zermelo felt they needed more than a single such session.

Program for Poincaré Week

22 April: Poincaré, On Fredholmian Equations
23 April: Poincaré, Applications of Integral Equations to Oceanic Tides
Hilbert, Dirichlet's Principle and General Mappings
Klein, Recent Work of Koebe and Hilb on Automorphic Functions
24 April: Poincaré, Applications of Integral Equations to Hertzian Waves
25 April: Celebration of Klein's 60th Birthday
26 April: Poincaré, On the Reduction of Abelian Integrals and the Theory of Fuchsian Functions
27 April: Poincaré, On Transfinite Numbers
Landau, On Analytic Number Theory (Poincaré's 1908 lecture in Rome)
Zermelo, A Rigorous Definition of $\aleph_1$ and the Assumptions and Methods used to prove the Well-Ordering Theorem
28 April: Poincaré, The New Mechanics

Hilbert delivered the first two speeches translated below on the first and third days of Poincaré's stay. Both were ceremonial occasions, and for them Hilbert found ways to pay indirect homage to, his revered friend, Minkowski. All three of the speeches that follow are translations from documents in the Hilbert *Nachlass* housed in the Niedersächsische Staats- und Universitätsbibliothek, Göttingen (Hilbert *Nachlass* 575, 579). As far as I know, the original handwritten texts are unpublished. It seems clear from the context and content of Hilbert's third speech that it was prepared some time after Poincaré's death in 1912, but I have been unable to determine where or when he presented these remarks. In any case, they constitute a striking tribute to Cantor's set theory.

Opening of Poincaré Week, April 22–29, 1909

Highly honored colleague,

In the name of the Wolfskehl Foundation, I greet you and thank you for accepting our invitation and for your willingness to make use of the German language, despite the discomfort involved, so that we and wider circles of students can more easily understand your lectures. Through the splendor of your name, you have inaugurated this first undertaking of the Wolfskehl Foundation.

You know, highly honored colleague, as do we all, how steady and close the mathematical interests of France and Germany have been and continue to be. Even when we recall only quickly the developments of the recent past, and out of the rich and many-voiced concert of mathematical science we take hold of the two fundamental tones of number theory and function theory, then we think perhaps of Jacobi, who had in Hermite the outstanding heir to his arithmetical ideas. And Hermite, who unfolded the flag of arithmetic in France, had our Minkowski, who brought it back to Germany again. Or if we only think of the names Cauchy, Riemann, Weierstrass, Poincaré, Klein, and Hadamard, these names build a chain whose links join one another in succession. The mathematical threads tying France and Germany are, like no two other nations, diverse and strong, so that from a mathematical perspective we may view Germany and France as a single land. Now you yourself have come to chosen our personal ties; we are most thankful for this and look forward to the promise of the richest and most wonderful fruits.

One easily recognizes these phrases as an expression of polite diplomacy. As a leading representative of German science, Hilbert was by all means a deeply committed internationalist, even during the most trying of times. In 1917, when Darboux passed away during the height of hostilities in the Great War, he took it upon himself to write a glowing obituary honoring the great French geometer, who had long been a corresponding member of the Göttingen Scientific Society. Quite obviously Klein would have been the obvious candidate to write this *Nachruf:* he, after all, knew both Darboux and his work far better than did Hilbert. But this, of course, only underscores the political significance of the latter's testimonial. Still, Hilbert had relatively few important ties with French mathematicians, quite unlike his close protégé Otto Blumenthal, who travelled to Paris regularly. His speech thus reflected more a cherished ideal than something substantial and real. No doubt many in the audience would have smiled at Hilbert's pronouncement that "Germany and France [form] a single land" from a mathematical perspective, knowing his propensity for exaggeration.

One finds some of these diplomatic niceties again in the speech Hilbert delivered two days later, when honoring Klein on his 60th birthday. Still, this was clearly an altogether different sort of occasion, one that called for serious reflection given the many experiences that Hilbert and Klein had shared. This speech is particularly interesting as an indication of the high regard the younger man showed for Klein's highly intuitive style of mathematics, which was so different from his own. Of course no one in 1909 would have questioned Klein's importance as an educator, and Hilbert's references to drawing, models, etc., have mainly to do with Klein's educational innovations. Yet he also invokes the

Fig. 16.2 Poincaré and Mittag-Leffler with Edmund Landau and Carl Runge at the home of David and Käthe Hilbert on the occasion of Klein's sixtieth birthday (Oberwolfach Photo Collection).

honored name of Riemann, and here he is praising Klein's contributions to mathematical knowledge.

An even more striking aspect of this speech is the way Hilbert confronts the sensitive topic of Klein's competition with Poincaré in their work on automorphic functions when speaking in the presence of both men (Gray 2000). His remarks surely inflate Klein's role, although perhaps he felt the occasion warranted such exaggeration. Only a short time before, Koebe and Brouwer had shown how the proof Klein had sketched 25 years earlier for his uniformization theorems could be made solid. Finally, Hilbert sings the praises of Klein's pet project, *Die Encyklopädie der mathematischen Wissenschaften,* a gigantic undertaking that sought to assimilate all of the leading developments in mathematics and its areas of immediate application. It would seem rather ironic that, in retrospect, this huge enterprise ended up having less influence on the course of mathematics than a single speech that Hilbert himself gave nine years earlier in Paris, when he posed 23 unsolved problems for mathematicians to contemplate in the future (Alexandrov 1979).

For Klein on His 60th Birthday, April 25, 1909

My dear and honored guests,

I have first, together in the name of my wife, to greet in my house our highly valued colleague Poincaré and to thank him for accepting our invitation (Fig. 16.2). Yet there is another reason for me and all of us to thank him, and for which he himself is perhaps unaware, although it is only because of his coming here that this has come about. Namely, that we have the joy of seeing our colleague Klein here on his 60th birthday. For without the Poincaré Festivities Klein would be hiding out today in some remote place far from Göttingen. It is indeed far better for us at least, his closest colleagues and friends, that we can congratulate him today and this for good reason. For what a delight it is to be a mathematician today, when mathematics is sprouting up everywhere and the sprouts blossoming; when it is being brought to bear on applications to the natural sciences as well as in the philosophical direction; and it is on the point of reconquering its former central position. But what a delight it especially must be, to be the mathematician Felix Klein on his 60th birthday.

As signposts of your scientific success, I would like to take just three points as typical examples. First, right from the beginning you stressed the general role of geometric intuition, placed it in the foreground, and cultivated it through drawing and models, and by emphasizing the physical, kinematical, and mechanical aspects of mathematical thinking. Riemann was the name that stood on your flag and under whose sign you have been victorious right down the line – victorious against opponents because of the correctness of your ideas, which brought you support from altogether unexpected quarters, from Minkowski, for example, who continually utilized geometric intuition as an arithmetical method. If I should select, secondly, a special mathematical area, we only need to hear the names Poincaré and Klein together and what mathematician would not be reminded of the automorphic functions, whose main theory was founded by Poincaré, but whose rich design is due to you. You foresaw with a lively presentiment precisely those deepest aspects of the theory and provided the ideas for their proof. Now, exactly at the present time, these ideas have finally reached completion. Thirdly, when all our names are no longer heard of, or perhaps one or another have some historical interest, distant generations will remain grateful to you for the magnificent work of the *Encyklopädie,* whose realization required exactly a man like you with so much selflessness and self-sacrifice. And here I should say that, much more than all your successes, you must be especially satisfied in knowing that all these accomplishments were only due to your time-sparing dedication, your energy, your powers of application, your character, and not least, your brilliant gifts. You were never interested in personal advantage, nor were your actions directed in the interests of another party, but rather continuously in pursuit of the cause. And thus everyone today, including your opponents – for a man like you always has these – have accorded you full justice; and within the circle of your colleagues you have earned the highest recognition and many grateful feelings.

Yet your life's work is still not complete. Your little ship is still moving ahead with youthful vigor at a full clip. Minkowski taught us that the concept of simultaneity.

(Gleichzeitigkeit) is relative. This applies even more so to the concept of age *(Gleichaltrigkeit)*. Age, in the only sense this term matters, is not a simple function of time alone, but rather depends on many imponderable parameters – a man of 70 years can have the same vitality, plans, and living power as a youngster. That this will 1 day be the case with you, to that speaks all probability. And with this wish I close, namely that this probability shall become a reality.

______.

The following day was hardly marked by nostalgia and sweet harmony. On the contrary, it occasioned a clash between Poincaré and Zermelo over some controversial aspects of Cantorian set theory. By the time Poincaré's lecture on transfinite numbers took place, the recently discovered antinomies in logic and set theory had already created a great stir. During these years, Zermelo was working closely with Hilbert in Göttingen, and in Zermelo (1904) he took a bold new step forward by showing how any arbitrary set could be given a well-ordering (Peckhaus 1990). His proof, however, invoked a principle – the axiom of choice – that sparked a series of controversies, especially among French analysts (Moore 1982). Other paradoxes involving large infinite sets remained unresolved, a circumstance that led Zermelo in 1908 to present the first axiom system for set theory (Zermelo 1908b). This was later refined into the now standard Zermelo-Fraenkel system, which successfully evaded such difficulties as Russell's paradox.

Poincaré's skeptical views regarding Cantor's theory were very well known before his arrival in Göttingen. The year before, he touched on "the disease afflicting set theory" in a lecture – read in his absence by Darboux – for the International Congress in Rome (Gray 2012). As Jeremy Gray has emphasized, his statement that "we may look forward with the joy of a doctor called to a beautiful pathological case" has often been misquoted or misconstrued to say that mathematicians would later regard set theory as a disease that had to be overcome (Gray 2008, 262). Gray traced this misinterpretation back to 1924, when it appears in a text written by Otto Hölder. As can be seen below, Hilbert made the very same assertion in an undated text that presumably predated Hölder's. Hilbert even claimed that Poincaré "showed complete incomprehension concerning Cantor's scheme of ideas" for which he "invented the invidious term 'Cantorism'."

Regarding the antinomies of set theory, Poincaré was particularly influenced by the ideas first published by Jules Richard in Richard (1905). In Göttingen, he began by asserting that an object is conceivable only if it can be defined with a finite number of words, thus an expressible law. He then noted why this position carries the implication that the collection of all such objects is countable, as Richard had proved. Not just any such formulation in words will do, however, as Richard's paradox made clear. Poincaré cited this simple example, due to Russell: Let S be the set of natural numbers definable by no more than 100 words. Since S must be finite, let m be the smallest number in $N - S$: but then m must be in S! Following Russell, Poincaré banned such self-referential or *impredicative definitions* from set theory. His Göttingen lecture thus began be reiterating these familiar views:

Every law of correspondence assumes a definite classification. I call a correspondence predicative when the classification on which it rests is predicative. However, I call a classification predicative when it is not altered by the introduction of new elements. With Richard's classification this, however, is not the case. Rather, the introduction of the law of correspondence alters the division into sentences which have a meaning and those which have none. What is meant here by the word "predicative" is best illustrated by an example. When I arrange a set of objects into various boxes, then two things can happen. Either the objects already arranged are finally in place. Or, when I arrange a new object, the existing ones, or a least a part of them, must be taken out and rearranged. In the first case I call the classification predicative, and in the second non-predicative (Poincaré 1910, 47).

Richard's proof that the number of mathematically definable objects is countable seemed to fly in the face of Cantor's diagonal argument proving that the real number continuum is uncountable. Poincaré showed, however, why this was not a true contradiction. Richard's procedure was always capable of extension, so that "wherever I break off the process a corresponding law arises, while *Cantor* proves that the process can always be continued arbitrarily far. There is thus no contradiction between these conclusions." (Poincaré 1910, 46–47).

Poincaré next noted how Zermelo had argued that impredicative definitions were necessary in order to prove several key theorems in mathematics, including the fundamental theorem of algebra. Faced with this contention, Poincaré was intent on refuting Zermelo's argument. After sketching the latter's proof, he showed how it could be easily modified in order to avoid any reference to an impredicative definition. Such a modification, he went on to claim, must always be possible; otherwise the proof would not be strictly rigorous. With this in mind, he briefly commented on the status of various key results in set theory, beginning with the classical proof of Bernstein's Theorem, which Poincaré formulated as follows: If for three sets A, B, C with $A \subseteq B$, $B \subseteq C$, $A \cong C$, then $A \cong B$. In other words, the existence of a bijective correspondence between A and C implies the existence of a bijection between A and B. The proof for this, he noted, can be rigorously established under the assumption that the first correspondence is predicative, in which case there will always be a predicative bijective correspondence between A and B.

Poincaré's final examples swirled around Zermelo's proof for the well-ordering theorem and Cantor's continuum problem. Hilbert had placed these together at the top of his list of outstanding unsolved problems in Hilbert (1900), while hinting strongly that he thought they were both true. If one could prove that every subset of the continuum were either countable or else had the cardinality of the continuum itself, then this would establish that $2^{\aleph_0} = \aleph_1$, i.e. "that the continuum has the next cardinal number beyond that of the countable sets." Cantor had shown how to define the sequence of transfinite alephs, but the possibility of assigning a definite aleph to the cardinality of a set depended on the possibility of giving every set a well-ordering. In 1900 Hilbert called attention to the fact that Cantor was confident the real numbers could be well-ordered, but added: "It appears to me most desirable to obtain a direct proof of this remarkable statement of Cantor's, perhaps by giving an actual arrangement of numbers such that in every partial system a first number can be pointed out." Poincaré surely knew this pronouncement, and he also knew that Zermelo's proof by means of the axiom of choice did nothing of the kind.

In Göttingen, he cast doubt on whether the second transfinite cardinal number $\aleph_1$ actually even exists from the standpoint of his foundational position. Since it arises by taking the collection of all ordinal numbers with the cardinality $\aleph_0$, its cardinality was clearly greater, but was it really well defined? Poincaré doubted whether one could speak about such a next greater cardinal without contradiction, to which he added: "in any case, there is no actual infinity." This cryptic remark led him to some equally mysterious statements that related directly to Zermelo's work:

What should we think of the famous continuum problem? Can one well-order the points of space? What do we mean by that? There are two possible cases: either one claims that the law of well-ordering can be formulated in finitely many words, then the assertion is unproved; Mr. Zermelo makes no claim to have proved such an assertion. Or else we admit the possibility that this law cannot be finitely expressed. But then I can no longer make sense of this statement; for me, these are only empty words. Herein lies the difficulty, and that is probably also the cause of the dispute over the almost ingenious theorem of Zermelo. This dispute is very strange: one side rejects the axiom of choice, but regards the proof [of well-ordering] as correct; the other side accepts the axiom, but not the proof. But I could talk about this for many hours without solving the question (Poincaré 1910, 49).

One can easily imagine Zermelo's mood when he stepped to the podium later that day to talk about these very issues. Poincaré's remarks were sufficiently vague that one can only guess whether he had actually read (Zermelo 1908a), which gave a new proof for the well-ordering theorem that evaded some of the criticisms brought against his original argument in Zermelo (1904). Kanamori describes this proof as inspired by methods first introduced by Dedekind, so that "instead of initial segments of the desired well-ordering, Zermelo switched to final segments and proceeded to define the maximal reverse inclusion chain by taking an intersection in a larger setting." This intersection approach evaded the problem of "classes deemed too large, such as the class of all ordinal numbers.. .. As set theory would develop, however, the original [1904] approach would come to be regarded as unproblematic and more direct, leading to incisive proofs of related results" (Kanamori 1997, 296).

Unfortunately, there seem to be no extant sources that would throw any light on how Zermelo dealt with Poincaré's criticisms. What has survived, though, is a later rather brief statement by Hilbert that shows how little patience he had for Poincaré's general standpoint, which he dismissed out of hand.

Hilbert on Poincaré's Conventionalism and Cantor's Theory of Transfinite Numbers

Poincaré was certainly, not merely among his contemporaries but during one of the most far-reaching and significant developmental periods for our science, the most versatile, imaginative, and creative mathematician. With regard to fundamental questions and philosophical foundations, however, he was not so fortunate. In his investigations of the role played by geometry in the pursuit of knowledge about nature, Poincaré adopted a position that he himself described as "conventionalism." When a year ago, on the occasion of Sommerfeld's stay, I had the honor of lecturing before you, I believed I had convinced you that this "conventionalism" of Poincaré's was in no sense justifiable. And yet the great name of Poincaré has brought it to pass that notable scholars have come to agree with this viewpoint, while its opponents only appear timidly and as such are hardly recognizable.

But still much more fateful are Poincaré's conceptions and viewpoints on the foundations of pure mathematics. Georg Cantor is the creator of set theory (Fig. 16.3) and this has long since become a generally recognized instrument for the mathematician. However, when one simply speaks

Fig. 16.3 Georg Cantor (1845–1918).

of the diverse applications of set theory, one usually means the so-called point set theory, that is, that subfield of the Cantorian theory that takes in immediate problems from analysis, function theory, and geometry, and that essentially aims and attains nothing other than the refinement and extension of objects that were already handled by Cauchy and Weierstrass. No, I see the truly singular and original aspect, the real essence of the Cantorian theory in the theory of transfinite numbers. Now Poincaré, who already misunderstood the significance of Dedekind's efforts to develop the principle of strong induction, showed complete incomprehension concerning Cantor's scheme of ideas. Poincaré invented the invidious term "Cantorism" and went so far as to prophesy that in the future it would be considered as nothing but an interesting pathological case. For me, on the contrary, the Cantorian theory of transfinite numbers appears as a wonderful and admirable flower and one of the greatest achievements of the mathematical spirit.

Turning back to the immediate aftermath of Poincaré's visit, the members of the Göttingen Mathematical Society convened again on 4 May 1909 to hear Klein's report on the events of Poincaré Week. The following month, Max Born presented two lectures on his theory of rigid-body motions in special relativity, a problem inspired by Minkowski's space-time geometry. Then, at the meeting on 13 July Paul Koebe and Richard Courant elaborated on the applications of Dirichlet's principle that were discussed by Hilbert in his lecture for Poincaré Week. Finally, just four days before Christmas, Hilbert spoke about meta-mathematics as well as Poincaré's lecture on set theory. The following year, the Dutch physicist H. A. Lorentz was chosen to deliver the second set of Wolfskehl lectures. Shortly before his arrival, Hermann Weyl wrote to his Dutch friend Pieter Mulder about how everyone was looking forward to this visit with great anticipation and much higher expectations after the disappointment over Poincaré's lacklustre performance.[4]

After Poincaré's sudden and altogether unexpected death in 1912, Gösta Mittag-Leffler went to great lengths in an effort to memorialize him. He thus asked Poincaré's nephew, the mathematician Pierre Boutroux, to write about his personal impressions of his uncle. Boutroux responded by saying that he had little noteworthy to tell him, as Poincaré's daily life was largely passed in solitude, often deep in thought or quietly composing a memoir in the shade of a tree in his garden. What struck him most, though, was the sharp contrast between his uncle's preferred way of doing mathematics and the altogether different atmosphere he encountered when he spent several weeks in Göttingen. There it seemed everyone was constantly talking about mathematics, as if the whole town knew what problems were then current, who was working on what, and even where they had gotten stuck. Poincaré had no such needs, and so "he kept his thoughts jealously for himself. Unlike some scholars, he did not believe that oral communication or the verbal exchange of ideas could promote discovery" (Boutroux 1914, 198).

One should not imagine that Boutroux's picture of the contemplative Henri Poincaré captures him fully. Quite the opposite is the case, as Scott Walter suggested when he sent me these remarks: "A big wheel in the French scientific machine, Poincaré was constantly attending meetings of the Académie des sciences, the Bureau des longitudes, the Bulletin astronomique, the Sorbonne, etc., writing reports and lecturing, serving on thesis juries, when he wasn't going to the opera or the symphony, or attending Gustave Le Bon's luncheons and Marie de Roumanie's soirées. It is true that he didn't actively seek inspiration from his Parisian colleagues, but he found enough in his daily regime at the pinnacle of French science." Still, Bourtroux was an eye witness, and the sharp contrast he drew between Poincaré's cool intellectual independence and the hothouse atmosphere both Frenchmen encountered in Göttingen tells us something important about what had changed since the days of Gauss

[4] Weyl to Mulder, 29 July 1910, Brouwer Archive.

and Riemann. Hilbert's idealism notwithstanding, Paris and Göttingen were, after all, two quite different worlds.

References

Alexandrov, P. S., Hg. 1979. *Die Hilbertschen Probleme*, Ostwalds Klassiker der exakten Wissenschaften, vol. 252, Leipzig: Akademische Verlagsgesellschaft Geest & Portig.

Boutroux, Pierre. 1914. *Lettre de M Pierre Boutroux à M Mittag-Leffler. Acta Mathematica* 38: 197–201.

Gray, Jeremy. 2000. *Linear Differential Equations and Group Theory from Riemann to Poincaré*. 2nd ed. Boston: Birkhäuser.

———. 2008. *Plato's Ghost: the Modernist Transformation of Mathematics*. Princeton: Princeton University Press.

———. 2012. Poincaré replies to Hilbert: On the Future of Mathematics ca. 1908. *Mathematical Intelligencer* 33: 15–29.

———. 2013. *Henri Poincaré: A Scientific Biography*. Princeton: Princeton University Press.

Hilbert, David. 1900. Mathematische Probleme. Vortrag, gehalten auf dem internationalen Mathematiker-Kongress zu Paris 1900, *Nachrichten von der Königlichen Gesellschaft derWissenschaften zu Göttingen. Mathematisch-physikalische Klasse aus dem Jahre 1900*, 253–297.

Kanamori, Akihiro. 1997. The Mathematical Import of Zermelo's Well-Ordering Theorem. *Bulletin of Symbolic Logic* 3 (3): 281–311.

Moore, Gregory H. 1982. *Zermelo's Axiom of Choice: its Origins, Development, and Influence*. New York: Springer.

Peckhaus, Volker. 1990. *Hilbertprogramm und Kritische Philosophie*, Studien zur Wissenschafts-,Sozial- und Bildungsgeschichte der Mathematik, vol. 7, Göttingen: Vandenhoeck & Ruprecht.

Poincaré, Henri. 1910. Über transfinite Zahlen. In *Sechs Vorträge über ausgewählte Gegenstände aus der reinen Mathematik und mathematischen Physik*. Leipzig: B. G. Teubner.

Pyenson, Lewis. 1979. Physics in the Shadow of Mathematics: the Göttingen Electron-theory Seminar of 1905. *Archive for History of Exact Sciences* 21 (1): 55–89.

Richard, Jules. 1905. *Les principes des mathématiques et le problème des ensembles*, in *Revue générale des sciences pures et appliquées 16*, 541–543; English translation also in (van Heijenoort 1967, 142–144).

van Heijenoort, Jan, ed. 1967. *From Frege to Gödel: A Source Book in Mathematical Logic*. Cambridge, MA: Harvard University Press.

Walter, Scott A. 2008. Hermann Minkowski's Approach to Physics. *Mathematische Semesterberichte* 55 (2): 213–235.

———. 2016. Poincaré-Week in Göttingen, in Light of the Hilbert-Poincaré Correspondence of 1908–1909, online version: http://scottwalter.free.fr/papers/2016-borgato-walter.html.

Zermelo, Ernst. 1904. Beweis, daß jede Menge wohlgeordnet werden kann. Aus einem an Herrn Hilbert gerichteten Briefe. *Mathematische Annalen*, 59: 514–516.

———. 1908a. Neuer Beweis für die Möglichkeit einer Wohlordnung. *Mathematische Annalen*, 65:107–128. English translation in (van Heijenoort 1967, 183–198).

———. 1908b. Untersuchungen über die Grundlagen der Mengenlehre, I. *Mathematische Annalen*, 65:261–281. English translation also in (van Heijenoort 1967, 200–215).

Part IV

Mathematics and the Relativity Revolution

17 Introduction to Part IV

When looking at the early development of relativity theory, one finds an astonishing number of contributions by mathematicians, some of which deeply influenced the work of leading theoretical physicists. Within the context of special relativity, Hermann Minkowski's writings come immediately to mind (Walter 2008). Klein and Hilbert followed Minkowski's ideas from their infancy, and both pursued some of their consequences after the latter's premature death in January 1909. Two other figures with close ties to Göttingen, Max Born and Arnold Sommerfeld, were both instrumental in elaborating Minkowski's 4-dimensional approach for physicists (Walter 2007). Born had been Minkowski's assistant for a brief time, and so he was given the task of publishing his mentor's final uncompleted work on electrodynamics (Staley 2008). Sommerfeld was since 1906 the head of a leading school for theoretical physics in Munich, where he afterward played a key role in promoting special relativity theory in Germany, including mathematical aspects of the theory (Eckert 2013).

The first two essays below are concerned with Minkowski's famous lecture "Space and Time" (Minkowski 1909) and subsequent, less heralded contributions of Max von Laue to special relativity (Janssen and Mecklenburg 2006; Janssen 2009). This work owes much to Einstein's famous paper, "Zur Elektrodynamik bewegter Körper" (Einstein 1905), but also to the larger background of developments in electrodynamics, as described in Darrigol (2000). Throughout Part III, I emphasized the importance of Minkowski's support as a major intellectual and psychological factor behind Hilbert's grandiose success story. In Chap. 18, I take up this theme once again, stressing the highly symbiotic nature of their relationship as two devotees of pure mathematics, a calling that had a special significance in the German tradition. Together Hilbert and Minkowski pursued their purist vision of mathematical knowledge, at first in the fields of algebra and number theory, but later by moving into the heart of the new physics (Minkowski 1973). For them, these were not distinct disciplines: this was all just part of the larger mathematical realm.

Two months before Minkowski delivered his lecture in Cologne, he contacted Sommerfeld to ask if he could visit Göttingen in early August to take part in an "electron debate." The semester would be ending by then, but Minkowski still thought various colleagues might still be able to join in, among them: Klein, Hilbert, Schwarzschild, Woldemar Voigt, Ludwig Prandtl, Carl Runge, and Emil Wiechert. This plan proved unfeasible, however, so instead Minkowski spoke instead about his own latest thoughts on electron theory as a warm-up for his lecture in Cologne. Sommerfeld would later edit the text of an earlier lecture on "Das Relativitätsprinzip" that Minkowski delivered before the Göttingen Mathematical Society in November 1907, published as Minkowski (1915). He and Max Born were among the 75 auditors who heard Minkowski's lecture on 21 September in Cologne. Another was the Canadian-American mathematician, Roland G. D. Richardson, who was taking a sabbatical from Brown University (on his later career at Brown, see Chap. 36). Richardson later reported on Minkowski's "Space and Time" for the *Bulletin of the American Mathematical Society*:

> Professor Minkowski discussed the complete revolution that has taken place in our conception of time and space, owing to the exact mathematical deductions from the latest investigations of physics. Lagrange has called physics a four-dimensional geometry because three dimensions of space, and one of time are introduced, and this definition seems to-day to be applicable in a deeper, unhoped-for sense. Minkowski shows that the remarkable hypothesis of H. A. Lorentz concerning the contraction of the electrons and the alleged contradiction between the Newtonian mechanics and the modern theory of electricity can be completely accounted for if we assume that we live in a four-dimensional world of which the fourth dimension (time) may be neglected with a certain amount of freedom. The axiom that the velocity of matter cannot exceed that of light in ether plays here an important role. This new comprehension of the world as a sort of union of time and space makes possible great strides in the theory of electricity and magnetism, and requires finally a revision of all physical theories. That these revisions are capable of being carried out is due alone to the many-sided advances made by pure mathematics in the past century. (Richardson 1908)

D.E. Rowe, *A Richer Picture of Mathematics*, https://doi.org/10.1007/978-3-319-67819-1_17

Richardson's final sentence reflects a common opinion, but one that is easily refuted simply by citing the very first words from Minkowski's speech: "The views of space and time which I wish to lay before you have sprung from the soil of experimental physics, and therein lies their strength. They are radical. Henceforth, space by itself, and time by itself, are doomed to fade away into mere shadows, and only a kind of union of the two will preserve an independent reality." (Minkowski 1909, 104).

As for Einstein's general theory of relativity theory, this led to an even greater flurry of activity within the mathematical community. Some of this work is recounted in Chaps. 21 and 22, which move beyond Göttingen to other important centers, including Leiden, Delft, Zürich, Cambridge, Rome, and Princeton. Although long recognized, the central role of the Göttingen scientific community in promoting relativity theory has never been studied in a systematic fashion. A full picture would require taking not just Minkowski, Hilbert, Klein, Sommerfeld, and Born into account, but several additional figures as well, including Emil Wiechert, Karl Schwarzschild (Schemmel 2007), Hermann Weyl (Scholz 2001b), and Erwin Freundlich (Hentschel 1997). This intense activity evoked, in turn, a broad spectrum of responses among physicists, not all of whom reacted enthusiastically. Most theoreticians, however, showed little concern over the intrusion of mathematicians into their discipline. Einstein, in particular, openly welcomed this kind of cooperation; the only noteworthy exception came when he felt that Hilbert had overstepped the bounds of propriety. That came about when the two became entangled in November 1915, a story I briefly sketch below.

Einstein's quest for a generalized theory of relativity began in earnest in 1912 when he collaborated with the Swiss geometer Marcel Grossmann, a personal friend (Sauer 2015). Together they sketched an early version that came to be called the *Entwurf* theory (Einstein and Grossmann 1913).[1] Afterward Einstein rarely worked on field physics without the assistance of a mathematical expert. Many of the various interactions between physicists and mathematicians that took place afterward were, like Einstein's theory itself, complex and mutually reciprocal. As a case in point, when Max von Laue wrote the first detailed textbook on general relativity (Laue 1921) – Weyl's monograph *Raum-Zeit-Materie* (Weyl 1918) can hardly be considered a *Lehrbuch* – he relied heavily on the works of mathematicians, including David Hilbert's unpublished lectures on the foundations of physics from 1916 to 1917 (see "The Mathematicians' Happy Hunting Ground," Chap. 21). As Laue's text and Wolfgang Pauli's famous encyclopedia article (Pauli 1921/1981) from the same year illustrate, physicists who wanted to learn general relativity did not simply go shopping along a one-way street whose store windows were filled with ready-made mathematical methods.[2] In fact, many of the standard tools of the trade had to be fashioned on the spot using whatever was then available. What ensued was, not surprisingly, an intensely interactive and interdisciplinary search to develop new methods, a process that gained momentum after 1915 with the work of Tullio Levi-Civita, Weyl (Scholz 2001a), and several others who decisively reshaped modern differential geometry.

During the summer of 1915, Einstein spent a week in Göttingen to deliver his six Wolfskehl lectures on general relativity. It was on this occasion that he met Hilbert for the first time (Rowe 2001). Although little documentary evidence has survived pertaining directly to these lectures, it is safe to conclude that Einstein's visit gave both him and Hilbert fresh impetus to explore this new approach to gravitation. Einstein was warmly received in Göttingen, not only by Hilbert but also by the larger community of mathematicians and physicists he met there. Einstein particularly appreciated Hilbert's liberal attitude and strong commitment to international scientific ideals during wartime. This left a deep impression on Einstein, who afterward wrote his Zürich friend, Heinrich Zangger:

> The longer this dreadful state of war continues, the more grimly people hang on to insensible hatred founded on nothing. When one is young, one admires vibrant emotion and disdains cold calculation. But now I think that the blunders originating from blind feeling cause much more bitter unhappiness in the world that the most heartless of calculating minds.
>
> But in these times you appreciate doubly the few people who rise well above the situation and do not let themselves get carried away in the turbid current of the times. One such person is Hilbert, the Göttingen mathematician. . . . I met him there and became quite fond of him. I held six two-hour lectures on gravitation theory, which is now clarified very much already, and had the pleasurable experience of convincing the mathematicians there thoroughly. Berlin is no match for Göttingen, as far as the liveliness of academic interest is concerned, in this field at least. (Einstein-CPAE 8A 1998, 145)

Aside from European affairs, Einstein and Hilbert surely talked about local matters, including Hilbert's running battles in the philosophical faculty where he was the *bête noire* of a large bloc of conservative humanists (Rowe 1986, 445–448). Quite probably Einstein also learned that a short time earlier Hilbert succeeded in blocking the appointment of the rabid nationalist, Johannes Stark, to the chair in experimental physics, though not for scientific, but rather for political reasons. But whatever he heard, Einstein found in Hilbert a true "comrade of convictions" ("*Gesinnungsgenosse*").

[1] The literature on how Einstein found his field equations is by now vast; see the essays in Renn (2007), in particular (Janssen and Renn 2007). For an overview, see Janssen and Renn (2015) and the references cited therein.

[2] The same point can be made for special relativity, as nicely illustrated in Walter (1999).

For Einstein this meant someone who believed not only that science knew no national boundaries but also that the honor of belonging to the international community of scientists deserved precedence over any sense of patriotic duty. Hilbert, unlike the vast majority of German professors, never equivocated when it came to that principle (Rowe 2004). A week later, while vacationing in Rügen, Einstein wrote Sommerfeld, "in Göttingen I had the great pleasure of seeing that everything was understood down to the details. I am quite enchanted with Hilbert. That's an important man for you!" (Einstein-CPAE 8A 1998, 147).

While little is known about the substance of Einstein's lectures in Göttingen, Leo Corry did discover notes taken from his first talk in Hilbert's papers (Einstein-CPAE 6 1996, 586–590). These indicate that Einstein began with some general remarks about the significance of Michelson's experimental result, which suggested that the laws of physics were unaffected by the earth's motion relative to the sun. This finding implied that the speed of light had to be the same in all directions, and furthermore that the description of physical phenomena must be equivalent in all inertial frames. This spelled the end of ether physics and inaugurated the special theory of relativity. From here, Einstein went on to discuss rotational motions and the centrifugal forces that arise from these. Not surprisingly, he paid tribute to Ernst Mach's earlier treatment of this theme, in particular his critique of Newton's argument that such forces resulted from the properties of absolute space rather than the relative motions of physical substances within space.

Einstein asserted further that, based on his equivalence principle, centrifugal forces were special types of gravitational fields, but he also noted that one could distinguish between those fields that resulted merely from coordinate transformations and those that arose due to the presence of matter. The former were characterized by the vanishing of the Riemann-Christoffel tensor (Christoffel 1869), a condition that, in general, was only satisfied in local terrestrial frames. Einstein also alluded to the bending of light rays in the neighborhood of large masses before proceeding on to the mathematical features of his theory. For these he began with the flat space-time metric introduced by Minkowski for special relativity. In passing to a new coordinate system by means of an arbitrary transformation, this metric passes over to a quadratic differential form whose coefficients $g_{\mu\nu}$ form the 10 components of the gravitational potential.

At the time of his visit, Einstein still showed a certain reluctance to put much stock in what he called "formal methods." This attitude gradually weakened, however, as he began to delve more deeply into the complexities of Ricci's tensor calculus. Here, Grossmann clearly played a key role in reorienting Einstein's outlook. A striking illustration of this can be found in Einstein's unpublished article on special relativity, which he began in Prague during 1912 (see the discussion of this in Chap. 19). Arriving in Zurich, the brown ink of his Prague manuscript barely dry, so to speak, he began anew, only now with black ink replacing the drab brownish tint. Moreover, matching this contrast in colors came an even more striking shift in content as, almost prophetically, Einstein began producing pages describing the formal properties of tensors. Three years later, he was far better versed in this trade than were most mathematicians, very few of whom had more than an inkling of the new discipline Ricci had dubbed the absolute differential calculus.

By the time he arrived in Göttingen, Einstein was well prepared to inform the mathematicians there about this esoteric branch of their subject and its indispensable importance for general relativity. Indeed, his fundamental paper on "The Formal Foundation of the General Theory of Relativity" (Einstein 1914) had the appearance of being just pure mathematical physics, even if its driving ideas came from physical and epistemological considerations. Since the *Entwurf* theory lacked any direct empirical support, Einstein employed new mathematical arguments to bolster its plausibility. This brought him at least one step closer to Hilbert, who had been strongly pushing his axiomatic approach to physical theories and problems (Corry 2004). Sommerfeld acted as an intermediary between Einstein and Hilbert, particularly during the fateful month of November 1915. The first direct contact came on November 7th when Einstein sent page proofs of Einstein (1915a), the first of his four November notes, to Hilbert, informing him that he had changed the gravitational equations "after having noticed about four weeks ago that my method of proof was fallacious" (Einstein-CPAE 8A 1998, 191). Five days later, Einstein wrote Hilbert again, this time to explain his second note (Einstein 1915b): "if my present modification. .. is legitimate, then gravitation must play a fundamental role in the composition of matter" (Einstein-CPAE 8A 1998, 194). This new idea turned out to be very short-lived, however (Janssen and Renn 2015).

Einstein was fully aware that Hilbert hoped to apply the principle of general relativity to Mie's theory of matter, so he realized that this new feature of his (provisional) equations would be of great interest to him. Only one day later, Hilbert wrote back, "actually, I first wanted to think of a very concrete application for physics, namely, reliable relations between the physical constants, before obliging with my axiomatic solution to your great problem. But since you are so interested, I would like to lay out my th[eory] in very complete detail on the coming Tuesday." He even invited Einstein to Göttingen to hear his lecture, but the physicist probably had real misgivings when he read this description of what Hilbert claimed he had achieved: "I find [the theory] ideally handsome mathematically and absolutely compelling according to axiomatic method, even to the extent that not quite transparent calculations do not occur at all and therefore rely on its factuality. As a result of a general mathematical law, the (generalized Maxwellian) electrody[namical] eq[uation]s appear as a math[ematical] consequence of the gravitational eq[uation]s, such that gravitation and electrodynamics are actually nothing different at all" (Einstein-CPAE 8A 1998, 195).

Einstein declined the invitation, but he sent this reply: "Your analysis interests me tremendously, especially since I have often racked my brains to construct a bridge between gravitation and electrodynamics. The hints you give in your postcards raise the greatest expectations." (Einstein-CPAE 8A 1998, 199). Hilbert apparently then sent Einstein a copy of the manuscript he would present two days later to the Göttingen Scientific Society. After receiving this, Einstein wrote:

> The system you furnish agrees – as far as I can see – exactly with what I found in the last few weeks and have presented to the Academy. The difficulty was not in finding generally covariant equations . . . for this is easily achieved with the aid of Riemann's tensor. Rather it was hard to recognize that these equations are a generalization, that is, a simple natural generalization of Newton's law. It has just been in the last few weeks that I succeeded in this . . . whereas three years ago with my friend Grossmann I had already taken into consideration the only possible generally covariant equations, which have now been shown to be the correct ones. We had distanced ourselves only with heavy hearts from it because it seemed to me that the physical treatment was incompatible with Newton's law. (Einstein-CPAE 8A 1998, 201)

Hilbert's foremost objective was to link Einstein's theory of gravitation with Mie's theory of matter by exploiting invariant theory and variational methods. Nevertheless, Einstein, who was intent on finding generally covariant field equations, totally ignored this aspect of Hilbert's paper. Moreover, the tone of his remarks strongly suggests that he was worried about protecting his priority claim with respect to generally covariant field equations. Had Einstein gone to Göttingen, he would have learned firsthand what Hilbert had accomplished and what he still hoped to achieve. Instead, reading Hilbert's postcards and manuscript, he became annoyed with his rhetoric and somewhat worried about the motives behind the Göttingen mathematician's latest breakthrough. Einstein had been forewarned by the physicist Max Abraham, who was famous in Göttingen for his backstabbing behavior, about a certain tendency that some had to borrow the ideas of others. In Göttingen, this was known humorously as *nostrifizieren*, a process whereby good ideas had a way of spreading once they got into the local atmosphere (Reid 1976, 121).

In this letter from November 18, Einstein went on to make the following stunning announcement: "the important thing is that the difficulties have now been overcome. Today I am presenting to the Academy a paper [(Einstein 1915c)] in which I derive quantitatively out of general relativity, without any guiding hypothesis, the perihelion motion of Mercury discovered by LeVerrier. No gravitation theory has achieved this until now" (Einstein-CPAE 8A 1998, 201–202). Hilbert was more than a little impressed and wrote back the next day to congratulate Einstein, adding that, "if I could calculate as rapidly as you, in my equations the electron would have to capitulate accordingly, and at the same time the hydrogen atom would have to produce its note of apology for why it does not radiate" (Einstein-CPAE 8A 1998, 202).

Einstein was working under great duress at this time, no doubt aggravated by Hilbert's sudden appearance in the arena. Still, he must have been in an agitated state of mind if he imagined that Hilbert might stake a priority claim with regard to the gravitational field equations. His letter certainly suggests that Hilbert's manuscript contained Einstein's field equations in an *explicit form*, but this is surely not the case. Hilbert later added the explicit field equations to the page proofs for his first note, a discovery made by Leo Corry in the mid-1990s (see Corry et al. 1997). In fact, a careful reading of Hilbert's paper makes clear that his agenda had very little to do with the problem Einstein had struggled with for so long. Thus, the whole idea that Einstein and Hilbert competed with each other over the correct generally covariant field equations has little to do with historical reality.

Since Hilbert's goal was to establish a link between Einstein's theory of gravitation and Mie's theory of matter, he only needed to derive Euler-Lagrange equations from his variational formalism. Nothing in his paper indicates he had any interest in finding a more explicit form for these equations, which would only have been necessary if had been investigating gravitational phenomena. But that was not his aim, which explains why the page proofs of his original paper contain nothing corresponding to the equations Einstein published in his final note from 25 November (Einstein 1915d). Hilbert had submitted his manuscript (Hilbert 1915) five days earlier, and the submission date of 20 November contributed mightily toward confusing later commentators who have usually interpreted the competition between Einstein and Hilbert as a "race for the field equations." Einstein clearly found and published the "Einstein equations" before Hilbert, though the latter was the first to indicate how they could be derived by variational methods (Sauer 1999). Einstein's last November note was printed almost immediately, whereas Hilbert's paper would only appear in March 1916. He received proofs that bore the date 6 December, after which Hilbert made quite extensive revisions (Sauer 2005). Among these changes, he added his own version of the field equations, while noting their compatibility with those already published by Einstein. This brief statement, however, was all he had to say about the explicit form of Einstein's equations, which play no role in the remainder of his text.

From a methodological standpoint, the approaches to gravitation taken by Hilbert and Einstein were glaringly different and these differences were reflected in their respective derivations of gravitational field equations (Renn and Stachel 2007). In Hilbert's case, these equations followed from a variational principle, whereas Einstein produced his equations through a combination of complex physical arguments and heuristic reasoning. Hilbert's approach was simpler and more elegant, but its physical underpinnings were far less transparent (Brading and Ryckman 2008). As an expert in invariant theory, Hilbert recognized the advantages of constructing a generally covariant theory. In particular, he knew that this would dramatically reduce the candidates for an invariant Lagrangian from which he could derive the field equations. Einstein no doubt appreciated such advantages, too, but his own reasons for adopting general covariance were mainly rooted in physical arguments and epistemological concerns.

Shortly after he submitted the last of his four November notes (Einstein 1915d), Einstein wrote apologetically to Sommerfeld. "You must not be cross with me that I am answering your kind and interesting letter only today. But in the last month, I had one of the most stimulating and exhausting times of my life, and indeed one of the most successful as well. I could not think of writing, for I realized that my existing gravitational field equations were entirely untenable!" (Einstein-CPAE 8A 1998, 206). After explaining how he reached this conclusion and subsequently found the new generally covariant equations, Einstein added that he "had considered these equations with Grossmann already three years ago, with the exception of the second term on the right-hand side, but at that time had come to the erroneous conclusion that it did not fulfill Newton's approximation" (Einstein-CPAE 8A 1998, 207). Einstein made no mention of Hilbert's concurrent activities in this letter, but in a letter to Heinrich Zangger, he expressed bitterness over Hilbert's behavior: "The theory is beautiful beyond comparison. However, only *one* colleague has really understood it, and he is seeking to nostrify it. .. in a clever way. In my personal experience I have hardly come to know the wretchedness of mankind better than as a result of this theory and everything connected to it" (Einstein-CPAE 8A 1998, 205). Four days later, he wrote his friend Michele Besso: "the problem has now finally been solved (general covariance). My colleagues are acting hideously in this affair." (Einstein-CPAE 8A 1998, 210).

It remains unclear whether Einstein conveyed any such feelings to Hilbert, but 1 month later, he found an opportunity to put their relationship back on a cordial basis. After his election as a corresponding member of the Göttingen Scientific Society, he sent him a note of thanks to which he added:

> On this occasion I feel compelled to say something else to you that is of much more importance to me. There has been a certain ill-feeling between us, the cause of which I do not wish to analyze. I have struggled against the feeling of bitterness attached to it, and this with complete success. I think of you again with unmixed congeniality and ask that you try to do the same with me. Objectively it is a shame when two real fellows who have managed to extricate themselves somewhat from this shabby world do not give one another pleasure. (Einstein-CPAE 8A 1998, 222)

Hilbert evidently responded in kind, as he and Einstein got along very well after this tense episode. Of course, the restoration of their friendship did nothing to change Einstein's generally negative view of Hilbert's program for axiomatizing physics, including what he considered Hilbert's naive optimism when it came to Mie's theory. In a letter to Hermann Weyl, he wrote: "Hilbert's assumption about matter appears childish to me, in the sense of a child who is innocent of the tricks of the outside world" (Einstein-CPAE 8A 1998, 366). Still, he must have been annoyed that Hilbert had found what looked like a royal road to the gravitational field equations.

Einstein's longstanding preoccupation with this problem helps to explain why he should have been so concerned about protecting his priority rights (Janssen 2014). By the same token, not a shred of documentary evidence exists indicating that this was ever an issue of direct contention between him and Hilbert. This only came up a few years later, and then very quietly. In 1918 in the first edition of his *Raum – Zeit – Materie*, Hermann Weyl noted that Hilbert had found the gravitational field equations more or less simultaneously with Hilbert, albeit within the special context of Mie's theory (Weyl 1918, 230), an attribution he later dropped in the fourth edition (Weyl 1922, 322, note 8). Weyl's rather difficult relationship with Hilbert, his former mentor, is discussed in Chap. 27.

Three years later, at the insistence of Felix Klein, Wolfgang Pauli added a reference to Hilbert's independent route to the Einstein equations in his highly authoritative encyclopedia article on relativity theory (Pauli 1981, 145). Klein, who imparted a considerable amount of advice to Pauli during the preparation this article, also counseled the young physicist to read what he had written about the "priority" issue in the first volume of his *Gesammelte Mathematische Abhandlungen*. There Klein voiced the firm opinion that Hilbert and Einstein had been guided by such fundamentally different conceptual viewpoints that there could be no question of adjudicating their respective priority claims to the derivation of the gravitational field equations (Klein 1921, 566). Although he had the highest respect for Einstein, Klein wanted mostly to showcase the many contributions of Göttingen mathematicians to general relativity, including his own. Clearly, Hilbert's first note was among the more spectacular of these, even though Klein had no sympathy for Hilbert's "fanatical belief in variational principles" any more than he took seriously his attitude that "the essence of nature could be explained by mere mathematical reflection" (Pauli

1979, 31). In an earlier letter to Pauli, Klein complained that the "physicists mostly pass over Hilbert's contributions in stony silence" (Pauli 1979, 27).

Klein and Hilbert, in fact, differed sharply when it came to foundational issues, and these differences are reflected in their respective contributions to the theory of relativity. Hermann Weyl was perhaps uniquely qualified to judge the respective philosophical outlooks of both men, given his deep roots in Göttingen mathematics as well as his subtle grasp of philosophical issues.[3] As Weyl emphasized, Klein's work sparkles with an intuitive, aesthetic feel for mathematical ideas. He was the geometer *par excellence*, a grand synthesizer, forever seeking out the visual elements that bring a theory to life, leaping over the conventional boundaries that separate algebra, number theory, and analysis from geometry proper. Klein's style, based on a genetic approach to mathematical ideas, was highly eclectic, and the breadth of his knowledge truly astounding, as may be judged by the mammoth *Encyklopädie der mathematischen Wissenschaften*, a project he managed for more than 20 years.

But whereas Weyl found Hilbert's views on the foundations of mathematics rigid and one-sided, he had even less sympathy for Klein's opinions, which he regarded as nothing more than a spin-off from a kind of naive positivism that Weyl associated with the ideas of Ernst Mach. Unlike Hilbert and most everyone else, Klein held a fundamentally dualistic view of mathematical knowledge that made a sharp distinction between pure and applied mathematics, or what he called *Präsizions- und Approximationsmathematik*; the former reflects "refined", the later "naïve intuition." For Klein, refined intuition was just a word for abstract reasoning without recourse to measurement or other empirical supports, and he freely admitted that this was not really a form of intuitive knowledge. Weyl could not abide Klein's attitude toward modern axiomatics. He once characterized his views on this subject by recalling a conversation in which Klein had belittled the work of such people, while complaining about how they failed to appreciate properly his own work. After he had shown how to jump over a certain ditch, these axiomatists would come along and show that one can still clear it with a table tied to one leg (Weyl 1985, 16). Hilbert, of course, stood at the very forefront of the movement that spawned modern axiomatic researches (Corry 2004).

The early history of general relativity contains many surprising twists and innumerous false turns. Unfortunately, the fabulous story of Einstein's search for gravitational field equations has received so much attention in the annals of physics that much of the rest of this rich story has been cast into the shadows. While recent scholarship has called attention to Hilbert's original research agenda, this was only one facet of the Göttingen response to Einstein's theory. Other aspects are discussed in Chaps. 21, 22 and 24. One of these concerns the purely mathematical properties underlying the derivation of the field equations. This refers specifically to certain folklore results in the theory of differential invariants that I describe in Chap. 21, "The Mathematicians' Happy Hunting Ground." As emphasized there, both Einstein and Hilbert made use of *unproven* results in their derivations. Hilbert's argument, based on a variational principle, depended on a uniqueness property of the Riemannian scalar R. Einstein claimed that $g_{\mu\nu}$, $Rg_{\mu\nu}$, $R_{\mu\nu}$ are the only second-order tensors that can be built from the $g_{\mu\nu}$ and their first and second derivatives, and which are linear in the latter.

Felix Klein may well have been the first to notice that these invariant-theoretic results could not be found anywhere in the mathematical literature. Recognizing this, he challenged his assistant, Hermann Vermeil, to come up with a direct proof for the uniqueness of the Riemannian curvature scalar R, i.e. to show that R was the sole invariant containing derivatives of the $g_{\mu\nu}$ only to the second order, and these linearly (Vermeil 1917, 335–337). Vermeil's proof was little more than a routine exercise; nevertheless, it shored up the foundations of Hilbert's derivation of the field equations by proving that R was the only possible invariant Lagrangian of the prescribed type. Four years later, in 1921, Weyl gave a somewhat different proof of this theorem for Riemannian spaces, noting that his argument could easily be adapted to spaces with a Lorentzian metric. Weyl further added that the same reasoning could be carried over directly to prove the folklore result cited by Einstein and Hilbert (Weyl 1922, Appendix II, 315–317).

More pressing for Einstein's theory, however, were the problems that arose in trying to solve his gravitational field equations even in the simplest cases. The astronomer Karl Schwarzschild, who taught in Göttingen from 1901 to 1909, had been quite skeptical of Einstein's new theory right up until the appearance of Einstein (1915c), the paper in which he derived the long puzzling deviation of Mercury's perihelion from what was predicted on the basis of Newtonian gravitational theory (Earman and Janssen 1993). Although he was then stationed on the eastern front, Schwarzschild found the time to work on this problem, which Einstein had solved by approximation methods. One month later, Schwarzschild had produced an exact solution, which he communicated to

[3]His contacts with Klein intensified during the years leading up to the outbreak of World War I when Weyl was teaching as a *Privatdozent* in Göttingen. During the winter semester of 1911–1912 he lectured on Riemann surfaces, one of Klein's favorite subjects, and this course led him to publish *Die Idee der Riemannschen Fläche,* the book that grounded this theory by introducing the modern notion of a (two-dimensional) complex manifold. Nevertheless, Weyl stressed his indebtedness to Klein's booklet of 1882, *Über Riemanns Theorie der algebraischen Funktionen und ihrer Integrale*. According to Weyl, it was Klein's more general conception of a Riemann surface—in which he replaced the complex plane by an arbitrary closed surface—that gave the modern theory its decisive vitality and power (Weyl 1913, iv–v).

Einstein on 22 December 1915 (Eisenstaedt 1989). He was clearly elated to have found this result: "[i]t is a wonderful thing that the explanation for the Mercury anomaly emerges so convincingly from such an abstract idea. As you see, the war is kindly disposed toward me, allowing me, despite fierce gunfire at a decidedly terrestrial distance, to take this walk into this your land of ideas." (Einstein-CPAE 8A 1998, 225).

In reviewing these technical problems, one naturally encounters the name of Riemann, whose famous habilitation lecture of 1854 has often been seen as anticipating Einstein's theory. The latter, after all, interprets gravitation in terms of curvature effects in a space-time manifold of four dimensions, whereas Riemann showed that the notion of Gaussian curvature could be extended to manifolds of arbitrary dimension. A quite readable account of Riemann's concept of curvature can be found in Klein (1927, 164–188), a text based on Klein's wartime lectures. Interested readers can also turn to a recent study by the historian of physics Olivier Darrigol, who undertook a careful investigation of Riemann's texts in Darrigol (2015). These include the technical arguments found in Riemann's "Commentatio," another unpublished manuscript that was only published in 1876. In the wake of Einstein's exploitation of the Ricci calculus, Levi-Civita studied this text after he came up with a new geometric interpretation of curvature based on infinitesimal parallel displacements. As Darrigol describes, Levi-Civita at first assumed this was already known to Riemann, so he was surprised when he was unable to find this concept in his "Commentatio" (see further (Cogliati 2014)).

In Chap. 20, I describe the longstanding tradition that implicitly linked Euclidean geometry with physical space. This viewpoint was central to Newtonian mechanics and it was later defended by both Kant and Poincaré, though on quite different grounds. Riemann was one of the first to break away from this tradition. Still, Einstein owed very little to Riemann and much to Ernst Mach, whose critique of Newtonian absolute space served as a touchstone principle for relativity. Already in 1907, Einstein formulated one of the key ideas for a generalized theory of relativity, his equivalence principle. This associated gravitational effects with non-inertial reference frames and vice-versa (Janssen 2012). By 1912, Einstein saw that such a theory would require mathematical tools that could describe physical processes in arbitrary moving frames. His friend Marcel Grossmann then acquainted him with the Ricci calculus, which later came to be called tensor analysis. Grossmann had studied at the ETH alongside Einstein, and he later succeeded his mentor, Wilhelm Fiedler, as professor of geometry there (Sauer 2015). Like Fiedler, he was an expert in classical synthetic and projective geometry, but hardly an authority on differential geometry, much less the rather esoteric Ricci calculus. That theory was barely known outside Italy, but Grossmann was aware of a paper that Ricci and Levi-Civita had published in *Mathematische Annalen* back in 1901. He read it, taught Einstein the basic concepts, and by 1913 they were able to produce a first sketch for a new theory of gravitation.

Chapter 21 also touches on the strong resistance Einstein's ideas later encountered, especially after Germany's defeat in the Great War. This topic involves complex matters that have recently received a good deal of attention from historians. Klaus Hentschel incorporated the anti-relativists into his vast survey of philosophical reactions to relativity in Hentschel (1990). Hubert Goenner focused on the literature that accompanied the anti-relativity campaign in Germany during the early 1920s (Goenner 1993); see also van Dongen (2007). In several of my own studies (Rowe 2002, 2006, 2012) I probed into this theme more deeply, whereas Milena Wazeck explored two different networks of anti-relativists in Wazeck (2009/2014): one centered in German-speaking countries, the other in the United States. She also focused attention on their two respective ringleaders, the Berlin physicist Ernst Gehrcke, and Arvid Reuterdahl, a polymath engineer who taught at the College of St. Thomas in St. Paul, Minnesota. In "The Mathematicians' Happy Hunting Ground" (Chap. 21), I recall Reuterdahl's less than happy exploits in the pages of Henry Ford's notorious newspaper, the *Dearborn Independent.*

Chapter 22 discusses one of the curious episodes in the subsequent story, namely the role played by certain differential identities found long before by Luigi Bianchi, a leading authority on differential geometry in Italy. Einstein's biographer, Abraham Pais, highlighted their importance in Pais (1982); though he mainly wanted to show that general ignorance of the Bianchi identities led Einstein and others seriously astray. In effect, this purely retrospective analysis reduces the early history of GR to a comedy of errors. Rather than engaging in counter-factual history, my account attempts to describe what actually happened. Pais was surely right when he put his finger on this problem, but he was wrong to treat the issues then at stake so perfunctorily. Lack of knowledge of the Bianchi identities seems to have contributed to widespread confusion over the status of energy conservation in general relativity, one of the thorniest issues debated during the last years of the Great War. Here, as elsewhere, sorting out the physical content from purely mathematical elements in Einstein's theory proved exceedingly difficult. These problems only added fuel to the fiery arguments of those who claimed that GR was nothing but pure mathematical speculation.

Einstein's treatment of gravitational energy caused considerable consternation among experts, but Hilbert's approach via his so-called energy vector was so complicated that no ordinary mortal could understand it. In fact, Klein indicated as much to Pauli. Einstein had great difficulty grasping the arguments in Hilbert's first note. He struggled with these while preparing a lecture for the Berlin physics colloquium, and then had to write Hilbert twice asking him to explain certain steps in his reasoning. He thanked him for his help, but confided his true feelings to his friend Paul Ehrenfest. "Hilbert's description doesn't appeal to me. It is unnecessarily specialized regarding 'matter,' is unnecessarily complicated, and not straightforward (=Gauss-like) in setup (feigning the super-human through concealment of the methods)" (Einstein-CPAE 8A 1998, 288). Still, Hilbert could not feign that he understood the connection between his approach to energy conservation and Einstein's. About this, he intimated to Einstein that "[m]y energy law is probably related to yours; I have already given this question to Frl. Noether." She apparently made some progress on this problem at the time, as Hilbert later acknowledged. "Emmy Noether, whose help I called upon more than a year ago to clarify these types of analytical questions pertaining to my energy theorem, found at that time that the energy components I had set forth – as well as those of Einstein – could be formally transposed by means of the Lagrangian differential equations. .. into expressions whose divergence vanished *identically*" (Klein 1921, 560–561) (Fig. 17.1).

Emmy Noether, who later became famous in Göttingen as the mother of modern algebra, was the daughter of the Erlangen mathematician Max Noether, a close friend of Klein's. She had briefly studied in Göttingen, but then returned to Erlangen, where she took her doctorate under Paul Gordan, who died in 1912. Gordan's successor, Ernst Fischer, exerted a major influence in turning Noether's head toward modern trends in algebra. Encouraged by Klein and Hilbert, she returned to Göttingen just a few months prior to Einstein's visit, hoping to become the first woman to habilitate at a Prussian university. This plan quickly floundered, but not before Hilbert caused a minor scandal in the philosophical faculty when he attacked his colleagues in the humanities who blocked her nomination (Tollmien 1991). This episode would enter the annals of Göttingen lore in the form of a typical Hilbertian wisecrack: he supposedly reminded the dissenters that they were speaking of membership in a scientific institution and not a public bathhouse. After the collapse of the *Kaiserreich*, Einstein offered to intercede on her behalf, though this proved unnecessary. With the new Weimar Ministry of Culture paving the way, she joined the faculty in 1919; still, she never did obtain a professorship in Germany.

Fig. 17.1 Emmy Noether (1882–1935).

During the war years, Noether assisted Hilbert as well as Klein, both of whom relied heavily on her expertise in the field of differential invariants. She also taught advanced courses that were offered under Hilbert's name. One of her auditors at this time was a Swiss doctoral student named Rudolf Humm, a somewhat shadowy figure who spent the war years studying physics in Munich, Göttingen, and Berlin before returning to Zürich. He later made a name for himself there as a writer, though probably few in Göttingen read his works. Instead, he was remembered anonymously in various versions of an oft-told Hilbert anecdote. When the illustrious mathematician once inquired about what had become of a certain Herr X, he was at first surprised to learn he was now earning his living as a writer (or as a poet). Upon reflecting about this for a moment, Hilbert replied, "Naja, that's a very good thing . . . he didn't have enough imagination to do mathematics."

Perhaps he was right, but Humm's decision to leave Göttingen in 1918 had little to do with his aspirations to become a writer. Rather his departure resulted from Hilbert's callous behavior toward him, as reflected in various anecdotes recorded in Humm's diaries. Humm was very lonely in Göttingen, and he resented how Hilbert appropriated his ideas. In Berlin, his interactions with Einstein were altogether different. Another Swiss writer, Carl Seelig, cited one such encounter in his biography of Einstein. Having studied under Hilbert, Humm was anxious to know how Einstein thought about his mentor's approach to physics. Their conversation from May 1917 proceeded, according to Humm's notes, as follows (Fig. 18.2):

> He [Einstein] is a bad calculator, he said; he rather works conceptually. He does not seem to believe that what we are doing in Göttingen is correct. He himself has never thought so formalistically. His imagination is firmly tied to reality. He is very careful, and entirely a physicist. He does not rush immediately to generalize as we do in Göttingen. He explains this [attitude] by saying that he had to rid himself of his prejudices very deliberately. That's why he did not grasp straight away how general covariance could exist. Rather, he had to come to this view step by step, which subsequently seemed to be very plausible indeed. But before that he had a real aversion to it

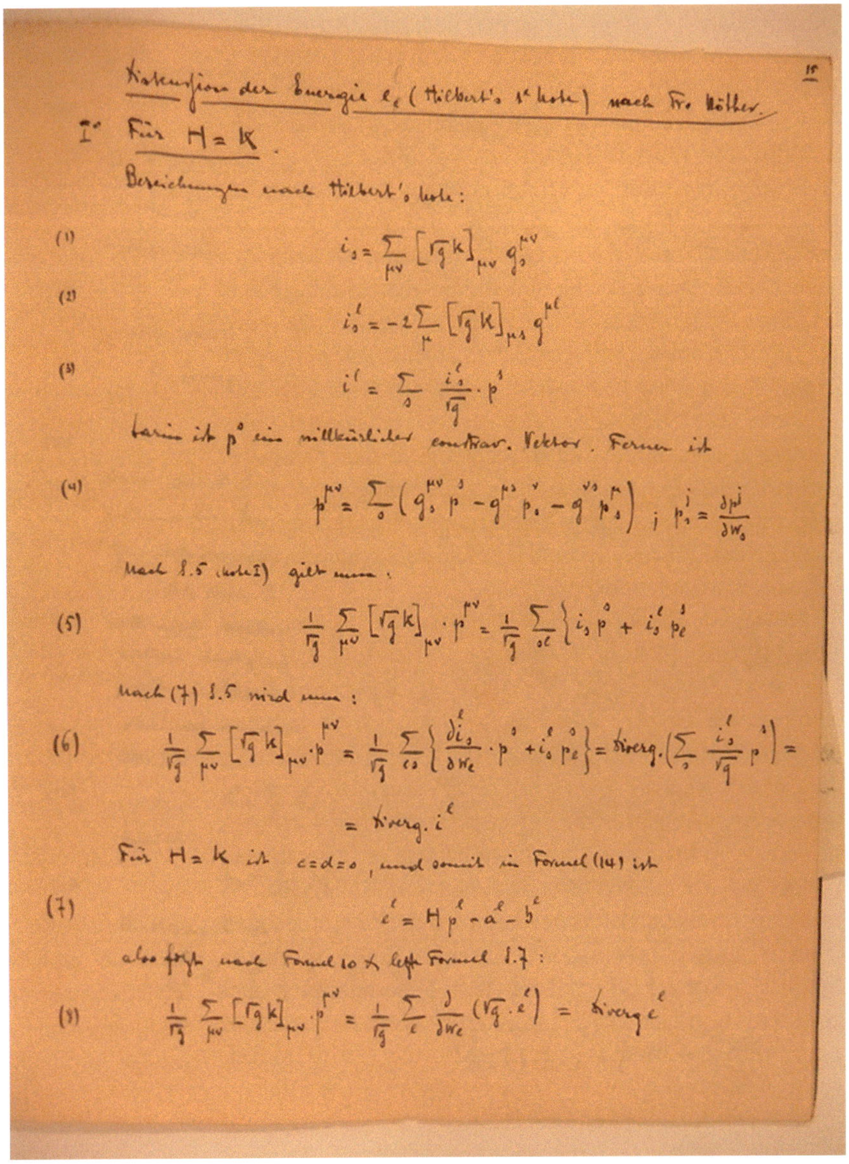
Diskussion der Energie e_ℓ (Hilbert's 1te Note) nach Fr. Nöther. 15

I° Für H = K.

Bezeichnungen nach Hilbert's Note:

(1) $$i_s = \sum_{\mu\nu} [\sqrt{g}K]_{\mu\nu}\, g_s^{\mu\nu}$$

(2) $$i_s^l = -2\sum_{\mu} [\sqrt{g}K]_{\mu s}\, g^{\mu l}$$

(3) $$i^l = \sum_s \frac{i_s^l}{\sqrt{g}} \cdot p^s$$

darin ist p^s ein willkürlicher contrav. Vektor. Ferner ist

(4) $$p^{\mu\nu} = \sum_s \left(g_s^{\mu\nu} p^s - g^{\mu s} p_s^{\nu} - g^{\nu s} p_s^{\mu}\right) ; \quad p_s^j = \frac{\partial p^j}{\partial w_s}$$

nach 1.5 (Note I) gilt nun:

(5) $$\frac{1}{\sqrt{g}} \sum_{\mu\nu} [\sqrt{g}K]_{\mu\nu} \cdot p^{\mu\nu} = \frac{1}{\sqrt{g}} \sum_{sl} \left\{ i_s p^s + i_s^l p_l^s \right\}$$

nach (7) 1.5 wird nun:

(6) $$\frac{1}{\sqrt{g}} \sum_{\mu\nu} [\sqrt{g}K]_{\mu\nu} \cdot p^{\mu\nu} = \frac{1}{\sqrt{g}} \sum_{ls} \left\{ \frac{\partial i_s^l}{\partial w_l} \cdot p^s + i_s^l p_l^s \right\} = \text{Diverg.}\left(\sum_s \frac{i_s^l}{\sqrt{g}} p^s\right) = \text{Diverg.}\, i^l$$

Für H = K ist c = d = 0, und somit in Formel (14) ist

(7) $$e^l = H p^l - a^l - b^l$$

also folgt nach Formel 10 & letzt. Formel 1.7:

(8) $$\frac{1}{\sqrt{g}} \sum_{\mu\nu} [\sqrt{g}K]_{\mu\nu} \cdot p^{\mu\nu} = \frac{1}{\sqrt{g}} \sum_l \frac{\partial}{\partial w_l} (\sqrt{g} \cdot e^l) = \text{Diverg.}\, e^l$$

Fig. 17.2 Notes taken by Rudolf Humm from a lecture by Emmy Noether, "Diskussion der Energie e_ℓ (Hilberts 1. Note) nach Fr. Nöther", Humm Nachlass, Zentralbibliothek Zürich.

because the quantities employed there – the curvature tensors – had seemed to him very unclear.

> I suggested to him that he had taken recourse to the quantum theory in order to modify the theory of gravitation, while Hilbert on the other hand wanted to derive the quantum theory from the theory of gravitation. At this he made a roguish grin. This would not do, he said, even though the theory of gravitation was more general. But [in his view] the ideas of relativity could not yield anything more than gravitation (Seelig 1956, 155).

Around this time, Humm began acquainting himself with relativistic cosmology. After returning to Göttingen in the fall, he spoke in Hilbert's seminar about the divergent views of Einstein and de Sitter with regard to conditions at infinity (Hilbert Nachlass 702), a central issue for the discussion in Chap. 24. Earlier Humm worked on issues involving energy conversation in GR, a topic he studied under Emmy Noether in 1916 (Fig. 17.2). He took careful notes on her lectures dealing with Hilbert's invariant energy vector, the first page of which is reproduced below. Humm's interest in this topic led to a paper on the equations of motion of matter in GR published in a 1918 issue of *Annalen der Physik*. Presumably, he had hoped to submit this as his doctoral dissertation, but Hilbert felt he was insufficiently prepared and so refused to approve his application to take the final oral exams.

During the years 1917–1918, Emmy Noether was working closely with Felix Klein, who was then nearly seventy and ailing. His youthful vigor seemed to return, however, after reading papers by Einstein and the Leiden astronomer Willem de Sitter on relativistic cosmology. Klein began corresponding with both authors, and then with Hermann Weyl, leading to a fascinating four-cornered debate discussed by Michel Janssen in his commentary for volume 8 of the Einstein edition (Einstein-CPAE 8A 1998, 351–356). Spurred on by Einstein, Klein struggled with several of general relativity's most demanding problems (Rowe 1999). One of these involved Einstein's controversial handling of gravitational energy; another problem, related to this, was how to understand the various forms of energy conservation introduced by Einstein, Hilbert, and Lorentz.

Finally, there was the central question that Klein posed to Hilbert. How can one formally distinguish traditional conservation laws – like those of classical mechanics or special relativity – from the sorts of laws that arise in general relativity as consequences of the field equations for gravitation? The answer, worked out in stunning generality by Emmy Noether, can be found by reading the closing remarks in Noether (1918, 255–257). This paper is remembered today because it contains two deep theorems in the calculus of variations that now bear Noether's name. Only afterward did it become clear that energy-momentum conservation in GR could be derived directly from the Bianchi identities without any appeal to variational principles.

Probably it was for this reason that Noether's paper made little lasting impression on relativists. Decades later, however, it was rediscovered by particle physicists, who came to regard the Noether theorems as standard tools for their research (Kosmann-Schwarzbach 2004). Few were aware, of course, that Noether's paper had a very specific motivation, namely to characterize the different kinds of laws that can be derived from an invariant action integral as these arise in typical physical theories. This stemmed from Klein's interest in energy-momentum conservation as derived in general relativity through a variational principle (Rowe 1999). His familiarity with Lie's work on systems of partial differential equations invariant under Lie groups served as a general guide for Noether's work. Her paper Noether (1918), in fact, broke new ground on this front by applying Lie's theory to the Euler-Lagrange equations derived from a variational principle in which the Lagrangian was a function of several variables that included field quantities. Eugene Wigner was apparently the first to recognize the importance of Noether's theorems for quantum electrodynamics.

Energy conservation in general relativity was a matter hotly debated by experts, though clearly this held no real interest for others. By contrast, the significance of Einstein's theory for cosmology drew much attention from a larger public eager to learn what the fuss was all about. Moreover, Einstein's guiding ideas regarding the relativity of inertia carried strong implications for his Machian conception of space. He explained these in an interesting letter, written to Karl Schwarzschild on 9 January 1916:

> The statement that "the fixed-star system" is rotation-free is meant in a relative sense. . . . On a small scale the individual masses produce gravitational fields that . . . reflect the character of a quite irregular small-scale distribution of matter. If I regard larger regions, as those available to us in astronomy, the Galilean reference system provides [an adequate approximation]. But if I consider even larger regions, a continuation of the Galilean system providing the description of the universe in the same dimensions as on a smaller scale probably does not exist, that is, where throughout a mass-point sufficiently removed from other masses moves uniformly in a straight line.
>
> Ultimately, according to my theory, inertia is simply an interaction between masses, not an effect in which "space" of itself were involved, separate from the observed mass. The essence of my theory is precisely that no independent properties are attributed to space on its own. It can be put jokingly this way. If I allow all things to vanish from the world, then following Newton, the Galilean inertial space remains; following my interpretation, however, nothing remains. (Einstein-CPAE 8A 1998, 240–241)

Schwarzschild, though he still did not realize it, had only a few more months to live. Writing from the Russian front on 6 February, he expressed fundamental agreement with Einstein's position:

> With regard to the inertial system, we are in agreement. You say that beyond the Milky Way conditions could exist under which the Galilean system is no longer the simplest. I only contend that within the Milky Way such conditions do not exist. As far as very large spaces are concerned, your theory takes an entirely similar position to Riemann's geometry, and you are certainly not unaware that elliptic geometry is derivable from your theory, if one has the entire universe under uniform pressure (energy tensor $-p$, $-p$, $-p$, 0). I cannot deny that you have put the freedom extending beyond it to most fortunate use. (Einstein-CPAE 8A 1998, 258–259)

Einstein was certainly not yet aware of the possibility of elliptic geometry at this time, but it seems quite likely that he remembered Schwarzschild's indications one year later when he published (Einstein 1917), his famous paper on relativistic cosmology. Schwarzschild went on to describe another special case he had recently solved, the gravitational field of an incompressible liquid sphere, where the energy tensor is $(-p, -p, -p, +\rho_0)$ (Schwarzschild 1916b). He was puzzled in this case that there was a lower bound on the radius for a given mass, and also struck by the bifurcated nature of the line element, which showed that the interior solution was a portion of a spherical geometry.

Felix Klein would pick up some of these threads one year later in his correspondence with Einstein. Klein had longstanding interests in the geometry of space, the topic sketched in Chap. 20, but discussed in detail in the seventh, which deals with relativistic cosmology in the period 1917 to 1924. This began with Einstein's famous paper (Einstein 1917), in which he took the position that the global structure of space was both homogeneous and static. Initially he tried to develop a cosmology based on a conventional Euclidean model, an *Ansatz* he explored with the help of Hilbert's former student, Jakob Grommer. They soon realized, however, that this was a dead end: one could never obtain static solutions to the field equations that took on reasonable boundary conditions at spatial infinity. By this time, Einstein was of course very familiar with Riemann's famous *Habilitationsvortrag* (Riemann 1868), a text that ends with speculations about a spherical space of small constant curvature. If ever he felt inspired by Riemann's writings, then it was this idea that clicked – a concept of physical space that was finite but unbounded. This conception obviously required no spatial boundary conditions, but Einstein did need to modify his field equations by adding a new cosmological term in order to obtain static solutions.

Leiden's Willem de Sitter was one of the few astronomers who took a keen interest in Einstein's gravitational theory right from the beginning. A self-styled skeptic, he drew the line when Einstein spoke about what he called Mach's principle, the notion that inertial properties of massive bodies had something to do with distant masses (de Sitter 1917a). They debated the legitimacy of such a doctrine for cosmological speculations, only to reach agreement that they disagreed with each other. Soon after Einstein unveiled his solution to the new cosmological equations, de Sitter presented him with a second solution with several mysterious properties (de Sitter 1917b). For Einstein, this so-called de Sitter universe was simply anathema, since it blatantly contradicted Mach's principle, as de Sitter pointedly emphasized. He had found a vacuum solution to the new cosmological equations, a universe devoid of matter. A new question thus now arose: which solution was the more physically feasible?

During the initial stages of this debate, Weyl took Einstein's side, whereas Klein remained neutral. Weyl even thought he could prove that the de Sitter solution contained "hidden masses" along the horizon of a given coordinate system, which he interpreted as singularities of the space-time manifold (Goenner 2001). Klein disputed this and showed that de Sitter space-time was free of singularities. His approach was based on projective geometry, a by now neglected field of research that Klein had mastered in his youth. His missionary zeal now reawakened, he sought to show the other combatants that cosmology could profit from the fruits of projective thinking. Thus, after reading Einstein's paper that introduced a cosmology based on Riemannian spherical space, he wrote him in March 1917 in order to point out that a second possible model, elliptic space, was also a viable candidate for a space of constant positive curvature. Klein had already discussed these issues many decades earlier in correspondence with Eugenio Beltrami.

Einstein had not been aware of elliptic geometry, but the astronomer Erwin Freundlich was quick to inform him. Freundlich had studied in Göttingen, so he very likely knew that Karl Schwarzschild had published a paper estimating the size of the universe, assuming it had a non-zero constant curvature (Schwarzschild 1900). Schwarzschild's analysis was probably the first in which accurate measurements of stellar parallax entered into the picture. To be sure, Lobachevsky had discussed stellar parallax in connection with hyperbolic geometry, but this was before Bessel's observations became available. By 1917, the distinction between spherical and elliptic geometry was probably common knowledge within the astronomical community, thanks to a paper by Simon Newcomb published in Crelle's Journal (Newcomb 1877). Newcomb's paper apparently became fairly well known among astronomers, few of whom knew of Klein's earlier contributions to this topic, which managed to escape Newcomb's attention as well. De Sitter referred exclusively to Newcomb's paper when he published his first deliberations on a mass-free model of a global space-time satisfying Einstein's cosmological equations with elliptic space-like cross-sections. Klein's involvement in the discussions between Einstein, de Sitter, and Weyl with regard to the intrinsic properties of de Sitter's model should thus be seen against the backdrop of his longstanding acquaintance with the projective theory of manifolds of constant curvature. His role in the early history of relativistic cosmology, described in Chap. 24, reveals some of the surprising twists and turns in this now largely forgotten story.

Chapter 23, written in appreciation of Martin Gardner, traces the historical background behind a few of the many gems in recreational mathematics that Gardner wrote about over the course of his unusual career. At first glance, this piece might appear out of place, but I invite the reader to think about mathematical paradoxes as special types of thought experiments. Seen in this way, it may not be so surprising to encounter several doodlings of classic examples in one of Einstein's notebooks. It turns out that quite a few of these can be found in a popular book on recreational mathematics, entitled *Mathematische Mußestunden* (*Mathematical Pastimes*, 1898), that was written by Hermann Schubert, remembered today as the inventor of the Schubert calculus in enumerative geometry. He was also the man who first discovered a talented youngster named Adolf Hurwitz (as described in Chap. 14). Moreover, these threads also connect with Hurwitz's protégé, David Hilbert, who during his Königsberg days had already begun recasting Euclid's theory of plane rectilinear figures. Hilbert, too, was familiar with a paradox involving the areas of dissected figures, a puzzle first presented by Victor Schlegel and then later popularized by Martin Gardner. Indeed, Hilbert recognized its profounder implications for a theory of area based on dissections but that also avoided invoking Euclid's assumption that "the whole is greater than the part."

Gardner's interest in thought experiments extended to physics as well, and so this essay ends with a picture taken from his book *The Relativity Explosion* (1976). There he discusses Dennis Sciama's cosmological speculations concerning one of Einstein's original guiding ideas, dubbed Mach's Principle back in 1918. Einstein and de Sitter debated the merits of this appeal to "distant masses" – meaning the stars in our galaxy viewed as a static universe – the controversy described in Chap. 24. Sciama hoped to rehabilitate Mach's Principle within the modern setting of an expanding universe in which the distant matter was almost entirely extragalactic. According to his estimates, the stars in our galaxy account for only one ten-millionth of the inertial effects measured in the vicinity of our planet. In Gardner's playful mind, these ideas brought back memories of a physical puzzle that required moving a glass game board so that four steel balls would simultaneously occupy its corners. He found this task to be virtually impossible until he hit on the idea of just rotating the board and letting centrifugal force do the rest. Following Sciama's lead, he now pictured this solution as resulting from the extended hands of distant galaxies that carefully guided the balls to their required destinations.

Chapter 25, the final selection for "Part IV: The Relativity Revolution," presents an excerpt from a classic paper by one of Dennis Sciama's numerous protégés, Sir Roger Penrose. "Gravitational Collapse: the Role of General Relativity" was originally published in 1969; it was reprinted with the author's permission as part of a special 30th anniversary issue of *The Mathematical Intelligencer.* Flipping over to Chap. 19, one finds two diagrams that show the trajectories of objects in the gravitational field surrounding a massive body. These drawings picture Hilbert's interpretation of the Schwarzschild solution (Schwarzschild 1916a) and its reproduction in the textbook on GR published by Max von Laue in 1921, both showing the impenetrable spherical shell that supposedly formed at the length of the Schwarzschild radius. This longstanding mathematical interpretation lent support for A. S. Eddington's influential view that giant stars, once they burn out, can never shrink beyond a certain fixed size. In other words, the formation of a black hole is a physical impossibility. Thanks to Penrose's work on gravitational collapse and its influence on subsequent research, which surely require no further introduction here, we now know better. His ideas helped inaugurate what has come to be known as the "Renaissance of General Relativity" of the 1960s, an era when the Penrose-Hawking singularity theorems gave astronomers new reason to take Einstein's theory more seriously. For those who have not yet explored this rich territory, a good place to begin is Kip Thorne's *Black Holes & Time Warps: Einstein's Outrageous Legacy* (Thorne 1994).

References

Brading, Katherine, and Thomas Ryckman. 2008. Hilbert's "Foundations of Physics": Gravitation and Electromagnetism within the Axiomatic Method. *Studies in History and Philosophy of Modern Physics* 29: 102–153.

Christoffel, Erwin B. 1869. Ueber die Transformation der homogenen Differentialausdrücke zeiten Grades. *Journal für die reine und angewandte Mathematik* 70: 46–70.

Cogliati, Alberto. 2014. Riemann's Commentatio Mathematica, a Reassessment. *Revue d'histoire des mathématiques* 20(1), 73–94.

Corry, Leo. 2004. *David Hilbert and the Axiomatization of Physics (1898–1918): From Grundlagen der Geometrie to Grundlagen der Physik*. Dordrecht: Kluwer.

Corry, Leo, Jürgen Renn, and John Stachel. 1997. Belated Decision in the Hilbert–Einstein Priority Dispute. *Science* 278: 1270–1273.

Darrigol, Olivier. 2000. *Electrodynamics from Ampère to Einstein*. Oxford: Oxford University Press.

———. 2015. The Mystery of Riemann's Curvature. *Historia Mathematica* 42: 47–83.

Sitter, Willem de. 1917a. On the Relativity of Inertia. Remarks Concerning Einstein's Latest Hypothesis, *Koninklijke Akademie van Wetenschappen te Amsterdam, Proceedings*, 19 (1916–17): 1217–1225.

———. 1917b. On the Curvature of Space, *Koninklijke Akademie van Wetenschappen te Amsterdam, Proceedings* 20 (1917–18): 229–243.

Earman, John, and Michel Janssen. 1993. Einstein's Explanation of the Motion of Mercury's Perihelion. In *The Attraction of Gravitation: New Studies in the History of General Relativity*, ed. John Earman et al., 129–172. Boston: Birkhäuser.

Eckert, Michael. 2013. *Arnold Sommerfeld:Science, Life and Turbulent Times 1868–1951*. New York: Springer.

Einstein, Albert. 1905. Zur Elektrodynamik bewegter Körper. *Annalen der Physik* 17: 891–921.

———. 1914. Die formale Grundlage der allgemeinen Relativitätstheorie, *Königlich Preußische Akademie der Wissenschaften, Sitzungsberichte*: 1030–1085; Reprinted in (Einstein-CPAE 6 1996, 72–130).

———. 1915a, Zur allgemeinen Relativitätstheorie, *Königlich Preußische Akademie der Wissenschaften, Sitzungsberichte*, 778–786; Reprinted in (Einstein-CPAE 6 1996, 214–224).

———. 1915b. Zur allgemeinen Relativitätstheorie. (Nachtrag), *Königlich Preußische Akademie der Wissenschaften, Sitzungsberichte* (1915), 799–801; Reprinted in (Einstein-CPAE 6 1996, 225–229).

———. 1915c. Erklärung der Perihelbewegung des Merkur aus der allgemeinen Relativitätstheorie, *Königlich Preußische Akademie der Wissenschaften, Sitzungsberichte* (1915), 831–839; Reprinted in (Einstein-CPAE 6 1996, 233–243).

———. 1915d. Die Feldgleichungen der Gravitation, *Königlich Preußische Akademie der Wissenschaften, Sitzungsberichte* (1915), 844–847; Reprinted in (Einstein-CPAE 6 1996, 244–249).

———. 1917. Kosmologische Betrachtungen zur allgemeinen Relativitätstheorie, *Königlich Preußische Akademie der Wissenschaften* (Berlin). *Sitzungsberichte* (1917), 142–152; reprinted in (Einstein-CPAE 6 1996, 540–552).

Einstein, Albert and Marcel Grossmann. 1913. *Entwurf einer verallgemeinerten Relativitätstheorie und einer Theorie der Gravitation* (Leipzig: Teubner, 1913); reprinted in (Einstein-CPAE 4 1995, 302–343).

Einstein-CPAE 6. 1996. In *The Collected Papers of Albert Einstein, Vol. 6: The Berlin Years: Writings, 1914–1917*, ed. A.J. Kox et al. Princeton: Princeton University Press.

Einstein-CPAE 8A. 1998. In *The Collected Papers of Albert Einstein: the Berlin Years: Correspondence, 1914–1917*, ed. Robert Schulmann et al. Princeton: Princeton University Press.

Eisenstaedt, Jean. 1989. The Early Interpretation of the Schwarzschild Solution. In *Einstein and the History of General Relativity*, Einstein Studies, ed. D. Howard and J. Stachel, vol. 1, 213–233. Basel: Birkhäuser.

Goenner, Hubert. 1993. The Reaction to Relativity Theory I: The Anti-Einstein Campaign in Germany in Germany. *Science in Context* 6: 107–133.

———. 2001. Weyl's Contributions to Cosmology, in (Scholz 2001b, 105–137).

Hentschel, Klaus. 1990. *Interpretationen und Fehlinterpretationen der speziellen und der allgemeinen Relativitätstheorie durch Zeitgenossen Albert Einsteins*. Basel: Birkhäuser.

———. 1997. *The Einstein Tower: An Intertexture of Dynamic Construction, Relativity Theory, and Astronomy*. Palo Alto: Stanford University Press.

Hilbert, David. 1915. Die Grundlagen der Physik. (Erste Mitteilung), *Königliche Gesellschaft der Wissenschaften zu Göttingen. Mathematisch-physikalische Klasse. Nachrichten*, 395–407.

Janssen, Michel. 2009. Drawing the Line between Kinematics and Dynamics in Special Relativity. *Studies in History and Philosophy of Modern Physics* 40: 26–52.

———. 2012. The Twins and the Bucket: How Einstein Made Gravity rather than Motion Relative in General Relativity. *Studies in History and Philosophy of Modern Physics* 43: 159–175.

———. 2014. No Success Like Failure ...': Einstein's Quest for General Relativity, 1907– 1920. In *The Cambridge Companion to Einstein*, ed. Michel Janssen and Christoph Lehner, 167–227. Cambridge: Cambridge University Press.

Janssen, Michel, and Mathew Mecklenburg. 2006. From Classical to Relativistic Mechanics: Electromagnetic Models of the Electron. In *Interactions: Mathematics, Physics and Philosophy, 1860–1930*, ed. Jesper Lützen et al., 65–134. New York: Springer.

Janssen, Michel and Jürgen Renn. 2007. Untying the knot: How Einstein found his way back to field equations discarded in the Zurich notebook, in (Renn 2007, vol. 2, 839–925).

Janssen, Michel, and Jürgen Renn. 2015. Arch and scaffold: How Einstein found his field equations. *Physics Today, November*: 30–36.

Klein, Felix. 1921. *Gesammelte Mathematische Abhandlungen*. Bd. 1 ed. Berlin: Springer.

———. 1927. *Vorlesungen über die Entwicklung der Mathematik im 19. Jahrhundert*, hrsg. von R. Courant u. St. Cohn-Vossen, Bd. 2, Berlin: Julius Springer.

Kosmann-Schwarzbach, Yvette. 2004. *Les Théorèmes de Noether. Invariance et lois de conservation au XXe siècle*. Paris: Les Éditions de L'école Polytechnique.

Minkowski, Hermann. 1909. Raum und Zeit. *Physikalische Zeitschrift* 10: 104–111.

———. 1915. Das Relativitätsprinzip. *Annalen der Physik* 352 (15): 927–938.

———. 1973. In *Hermann Minkowski, Briefe an David Hilbert*, ed. von Hans Zassenhaus and L. Rüdenberg. New York: Springer.

Newcomb, Simon. 1877. *Elementary theorems relating to the Geometry of a space of three dimensions and of uniform positive curvature in the fourth dimension. Journal für die reine und angewandte Mathematik* 83: 293–299.

Noether, Emmy. 1918. Invariante Variationsprobleme, *Königliche Gesellschaft der Wissenschaften zu Göttingen. Mathematisch-physikalische Klasse. Nachrichten*, 235–257.

Pais, Abraham. 1982. *Subtle is the Lord...' The Science and the Life of Albert Einstein*. Oxford: Clarendon Press.

Pauli, Wolfgang. 1921/1981. Relativitätstheorie, in *Encyklopädie der mathematischen Wissenschaften, mit Einschluss ihrer Anwendungen*. Arnold Sommerfeld, ed. vol. 5, *Physik*, part 2 (Leipzig: Teubner, 1921), 539–775. W. Pauli, *Theory of Relativity*. Trans. G. Field. (New York: Dover, 1981).

Pauli. 1979. In *Wolfgang Pauli. Wissenschaftlicher Briefwechsel*, ed. A. Hermann, K.V. Meyenn, and V. Weisskopf, Bd. 1 ed. New York, Springer.

Reid, Constance. 1976. *Courant in Göttingen and New York: the Story of an Improbable Mathematician*. New York: Springer.

Renn, Jürgen, ed. 2007. *The Genesis of General Relativity*. Vol. 4. Dordrecht: Springer.

Renn, Jürgen and John Stachel. 2007. Hilbert's foundation of physics: From a theory of everything to a constituent of general relativity, in (Renn 2007, vol. 4, 858–973).

Richardson, R.G.D. 1908. The Cologne Meeting of the Deutsche Mathematiker-Vereinigung. *Bulletin of the American Mathematical Society* 15: 117–119.

Riemann, Bernhard. 1868. Ueber die Hypothesen, welche der Geometrie zu Grunde liegen. *Abhandlungen der Königlichen Gesellschaft der Wissenschaften zu Göttingen* 13: 133–150.

Rowe, David E. 1986. "Jewish Mathematics" at Göttingen in the Era of Felix Klein. *Isis* 77: 422–449.

———. 1999. The Göttingen Response to General Relativity and Emmy Noether's Theorems, in (Gray 1999, 189–233).

———. 2001. Einstein meets Hilbert: At the Crossroads of Physics and Mathematics. *Physics in Perspective* 3: 379–424.

———. 2002. Einstein's Encounters with German Anti-Relativists, in (Einstein-CPAE 7 2002, 101–113).

———. 2004. Making Mathematics in an Oral Culture: Göttingen in the Era of Klein and Hilbert. *Science in Context* 17 (1/2): 85–129.

———. 2006. Einstein's Allies and Enemies: Debating Relativity in Germany, 1916–1920, in (Hendricks 2006, 231–280).

———. 2012. Einstein and Relativity. What Price Fame? *Science in Context* 25 (2): 197–246.

Sauer, Tilman. 1999. The Relativity of Discovery: Hilbert's First Note on the Foundations of Physics. *Archive for History of Exact Sciences* 53: 529–575.

———. 2005. Einstein equations and Hilbert action: What is missing on page 8 of the proofs for Hilbert's first communication on the foundations of physics? *Archive for History of Exact Sciences* 59: 577–590.

———. 2015. Marcel Grossmann and his contribution to the general theory of relativity, in *Proceedings of the 13th Marcel Grossmann meeting on Recent Developments in Theoretical and Experimental General Relativity, Astrophysics and Relativistic Field Theories, July 2015*, Robert T. Jantzen, et al., eds., Singapore: World Scientific, 2015, 456–503.

Schemmel, Matthias. 2007. The *Continuity between Classical and Relativistic Cosmology in the Work of Karl Schwarzschild*, in (Renn 2007, vol. 3, 155–182).

Scholz, Erhard. 2001a. Weyls Infinitesimalgeometrie, 1917–1925, in (Scholz 2001b, 48–104).

———. 2001b. *Hermann Weyl's Raum-Zeit-Materie and a General Introduction to his Scientific Work*. Basel: Birkhäuser.

Schwarzschild, Karl. 1900. *Über das zulässige Krümmungsmaß des Raumes, Vierteljahresschrift der astronomischen Gesellschaft*, 35. Jahrgang.

———. 1916a. Über das Gravitationsfeld eines Massenpunktes nach der Einstein'schen Theorie, *Sitzungsberichte der Königlich-Preussischen Akademie der Wissenschaften*, 189–196.

———. 1916b. Über das Gravitationsfeld einer Kugel aus inkompressibler Flüssigkeit. *Sitzungsberichte der Königlich-Preussischen Akademie der Wissenschaften*: 424–434.

Seelig, Carl. 1956. *Albert Einstein: A Documentary Biography*. Staples Press.

Staley, Richard. 2008. *Einstein's Generation: The Origins of the Relativity Revolution*. Chicago: University of Chicago Press.

Thorne, Kip. 1994. *Black Holes & Time Warps: Einstein's Outrageous Legacy*. New York: Norton.

Tollmien, Cordula. 1991. Die Habilitation von Emmy Noether an der Universität Göttingen. *NTM Zeitschrift für Geschichte der Naturwissenschaften, Technik und Medizin* 28: 13–32.

van Dongen, Jeroen. 2007. Reactionaries and Einstein's fame: 'German Scientists for the Preservation of Pure Science,' relativity and the Bad Nauheim conference. *Physics in Perspective* 9: 212–230.

Vermeil, Hermann. 1917. Notiz über das mittlere Krümmungsmass einer *n*-fach ausgedehnten Riemann'schen Mannigfaltigkeit, *Königliche Gesellschaft der Wissenschaften zu Göttingen. Mathematisch-physikalische Klasse. Nachrichten*: 334–344.

von Laue, Max. 1921. *Die Relativitätstheorie, zweiter Band: Die allgemeine Relativitätstheorie und Einsteins Lehre von der Schwerkraft*. Braunschweig: Vieweg.

Walter, Scott. 1999. The Non-Euclidean Style of Minkowskian Geometry, in (Gray 1999, 91–126).

———. 2007. Breaking in the 4- Vectors: the Four-Dimensional Movement in Gravitation, 1905–1910, in (Renn 2007, vol. 3, 193–252).

———. 2008. Hermann Minkowski's Approach to Physics. *Mathematische Semesterberichte 55*: 213–235.

Wazeck, Milena. 2009/2014. *Einsteins Gegner: Die öffentliche Kontroverse um die Relativitätstheorie in den 1920er Jahren*, Berlin: Campus 2009; *Einstein's Opponents. The Public Controversy about the Theory of Relativity in the 1920s*, Cambridge: Cambridge University Press.

Weyl, Hermann. 1913. *Die Idee der Riemannschen Fläche*. Leipzig: Teubner.

———. 1918. *Raum - Zeit - Materie. Vorlesungen über allgemeine Relativitätstheorie*. 1st ed. Berlin: Springer.

———. 1921/1922. *Raum–Zeit–Materie. Vorlesungen über allgemeine Relativitätstheorie*, 4th rev. ed. (Berlin: Springer, 1921); H. Weyl, *Space—Time—Matter*. Trans. Henry L. Brose. London: Methuen, 1922.

———. 1985. Axiomatic versus Constructive Procedures in Mathematics, edited by Tito Tonietti. *Mathematical Intelligencer* 7 (4): 10–17.

18 Hermann Minkowski's Cologne Lecture, "Raum und Zeit"

(*Mathematical Intelligencer* 31(2)(2009): 27–39)

A century ago, David Hilbert stepped to the podium at a special meeting of the Göttingen Academy of Sciences to recall the achievements of his close friend Hermann Minkowski. Just one month earlier, on 12 January 1909, the 43-year-old Minkowski had died unexpectedly after suffering a ruptured appendix, leaving those close to him in a state of shock. None was more deeply affected than Hilbert, whose memorial lecture (Hilbert 1910) reflects the deep sense of personal loss he felt at that time:

> Our science, which we loved above everything, had brought us together. It appeared to us as a flowering garden. In this garden there are beaten paths where one may look around at leisure and enjoy one self without effort, especially at the side of a congenial companion. But we also liked to seek out hidden trails and discovered many a novel view, beautiful to behold, so we thought, and when we pointed them out to one another our joy was perfect (Hilbert 1910, 363).

Hilbert loved to think about mathematics while tending the spacious garden behind his house, a sanctuary that only a few were allowed to enter. Minkowski, who lived just a stone's throw away, was a regular visitor. So far as we know, he was not, like Hilbert, a passionate gardener. But both loved nothing more than to rendezvous in order to talk about mathematics (Fig. 18.1).

Minkowski's tragic death came at a time when he and his friends were still celebrating one of the high points of his career, his lecture on "Raum und Zeit" (Minkowski 1909) delivered in Cologne on 21 September, 1908. Speaking on that occasion before a mixed audience of seventy-one mathematicians, physicists, and philosophers (Wangerin 1909, 4), Minkowski wanted to make clear right from the outset that he had something important to say. So he began with these stirring words:

> The views of space and time which I wish to lay before you have sprung from the soil of experimental physics, and therein lies their strength. Their tendency is radical. From this moment onward, space by itself and time by itself will totally fade into shadows and only a kind of union of both will preserve independence (Minkowski 1909, 431) (Fig. 18.2).

The physics Minkowski had in mind was itself something radically new. It sprang from Lorentz's electron theory, Einstein's ether-free electrodynamics based on the principle of relativity, and Planck's further extension of Einstein's findings regarding the inertia of energy. Minkowski not only found this line of physical research promising, he sought to convince his audience that these developments demanded a totally new conception of space and time, one that departed fundamentally from the premises of classical Newtonian physics. His lofty rhetoric in "Raum und Zeit" no doubt made an immediate impression, even if few in his audience were in a position to appreciate the deeper points that propelled his argument forward. As it tuned out, two highly knowledgeable individuals were present on this occasion: Max Born and Arnold Sommerfeld.[1] Both went on to pursue Minkowski's ideas immediately after his death, thereby bringing space-time physics into the relativistic arena. Hilbert, too, was hardly a disinterested party; he, too, was determined to help ensure that his friend's contributions to relativity theory would not be forgotten.[2] In his memorial lecture, he recalled watching Minkowski put the last touches on the final page proofs as he lay on his deathbed. He further related how Minkowski took solace in the thought that his passing might induce others to study his contributions to the new physics more carefully (Fig. 18.3).

During their lifetimes, Hilbert and Minkowski were likened with Castor and Pollux, the inseparable twins (Born 1978, 80). Their careers first became intertwined in the 1880s as fellow students at the university in Königsberg, the *Albertina,* where Jacobi had once taught. From this remote outpost in East Prussia, they made their way into the German mathematical community that was then taking form. They were among the small group of mathematicians who

[1] Born recalled attending Minkowski's lecture in his autobiography (Born 1978, 131); Sommerfeld was listed as one of the discussants after the lecture in Wangerin (1909, 9).

[2] Hilbert took charge of publishing Minkowski's collected works (Minkowski 1911) soon after the latter's death.

D.E. Rowe, *A Richer Picture of Mathematics*, https://doi.org/10.1007/978-3-319-67819-1_18

Fig. 18.1 This photo, dating from around 1907, shows Minkowski on an excursion with the Hilberts: Käthe, David, and their son Franz. Flanking the group are two of Hilbert's many gifted doctoral students: Alfréd Haar, to the left, and Ernst Hellinger. The latter took his degree in 1907 and thereafter became Hilbert's first *Assistent*. The young woman in the middle of the back was probably the Hilberts' housekeeper (Reid 1976).

gathered in 1890 at a meeting of the Society of German Natural Scientists and Physicians in order to launch a new national organization, the Deutsche Mathematiker-Vereinigung (DMV). The DMV remained, in fact, for some time under the umbrella of this larger scientific society, which helps to explain why several physicists were in the audience in Cologne when Minkowski delivered his famous lecture on "Space and Time." Göttingen's Felix Klein also attended that very first meeting of the DMV, and he came away from it firmly convinced that Hilbert was the "rising man" among the younger generation of German mathematicians (Rowe 1989, 196). Klein was not merely a talent scout for his university. Already well past his prime as a researcher, he saw himself as uniquely qualified to judge prospective appointments throughout the entire German system of higher education. So soon after the Bremen meeting he imparted his opinion to Friedrich Althoff, the key figure in the Prussian Ministry of Culture who would later play a decisive role in promoting the careers of both Hilbert and Minkowski. After 1902, as colleagues in Göttingen, both would put their unmistakable imprint on the mathematics of the new century.

One of those most deeply affected by this Göttingen atmosphere was Hermann Weyl. Just one month after the German army surrendered to the Soviets in Stalingrad in early February, 1943, Weyl wrote a letter to Minkowski's widow recalling the mathematical world he had known in his youth:

> Last Sunday while attending a mathematical meeting in New York I learned that Hilbert has died. There was a short report about this, dated Bern, 19 February, which appeared in the New York Times on 20 February, but I missed this and perhaps you did, too. The news brought back powerfully again the whole Göttingen past . . . The two friends' activities, their influence on the younger generation complemented one another in the most harmonious way, Hilbert perhaps the more blazing and buoyant (Lichtere und Leichtere), your husband the warmer . . . and certainly the kinder . . .. Nowhere today is there anywhere anything even remotely comparable . . ..[3]

Soon thereafter, Weyl paid tribute to Hilbert in two obituary articles, the longer of which appeared in the *Bulletin of the American Mathematical Society* (Weyl 1944). It began on the same note, though without mentioning Minkowski's role in Hilbert's success story. "In retrospect," he reflected, "it seems to us that the era of mathematics upon which [Hilbert] impressed the seal of his spirit and which is now sinking below the horizon achieved a more perfect balance than

[3]Hermann Weyl to Auguste Minkowski, March 1943; transcription courtesy of Gunther Rüdenberg.

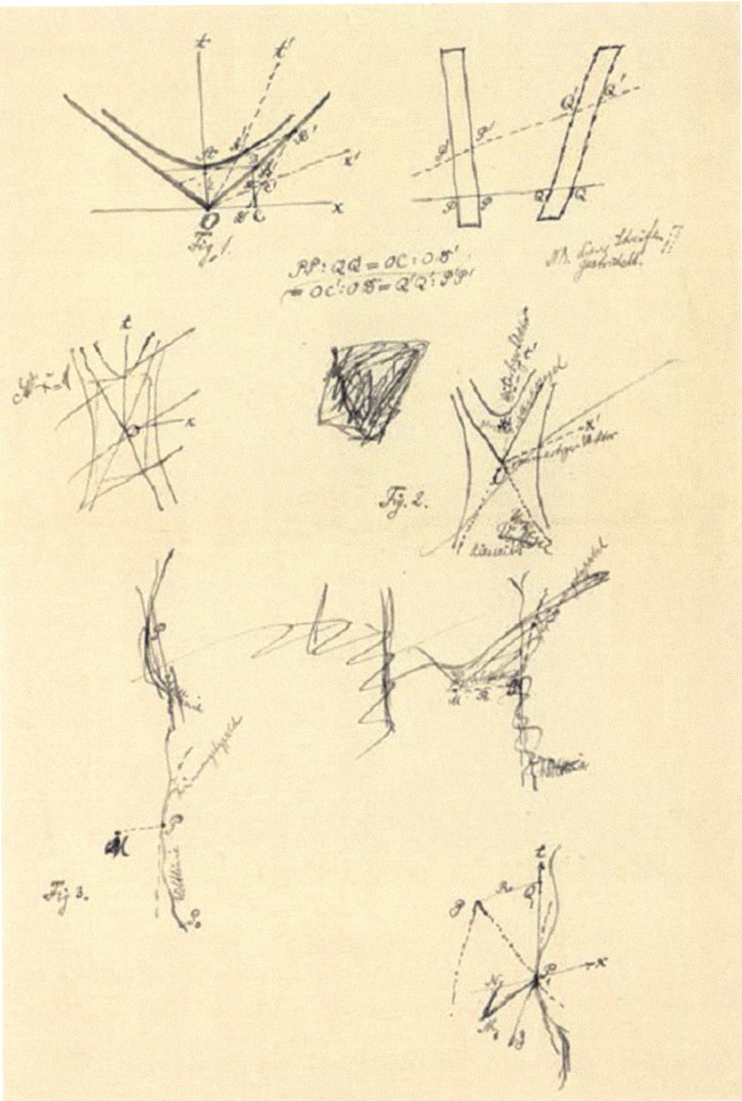

Fig. 18.2 Minkowski made these five sketches for the illustrations that appear in the published version of his "Raum und Zeit" (Mathematisches Archiv 60.2, SUB Göttingen).

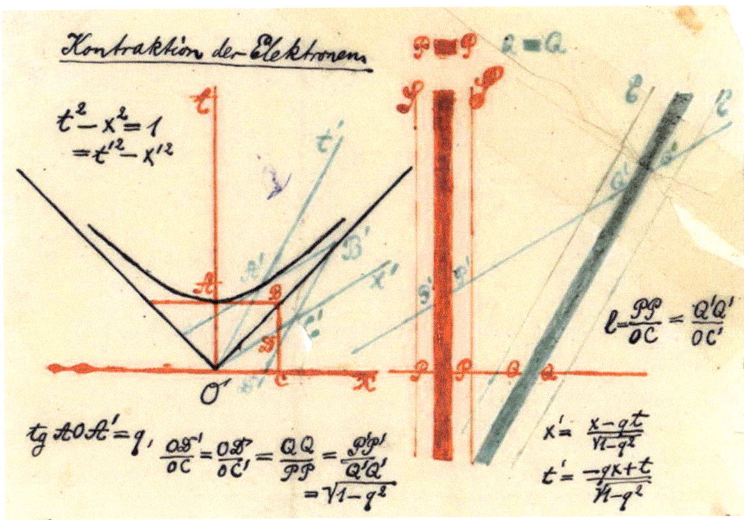

Fig. 18.3 Minkowski probably prepared this drawing of a spacetime diagram to accompany his Cologne lecture delivered on 21 September, 1908. It closely resembles the first diagram in the published version of "Raum und Zeit," though he used it here to explain the Lorentz contraction of an electron (Mathematisches Archiv 60.2, SUB Göttingen).

Fig. 18.4 This portrait of Minkowski, reproduced in his *Gesammelte Abhandlungen*, was made at the time he was teaching in Zürich, 1896–1902 (Courtesy of Niedersächsische Staats- und Universitätsbibliothek Göttingen).

prevailed before and after. . . . No mathematician of equal stature has risen from our generation" (Weyl 1944, 130).

In recalling these events and circumstances, one cannot help but notice a certain imbalance in the literature documenting the lives of these two stellar figures. Whereas accounts of Hilbert's singular career abound (for example Blumenthal (1935) and Reid (1970)), relatively little has been written about Minkowski's life, personality, or mathematical style (Fig. 18.4). Yet despite some striking affinities, one can hardly overlook the contrasts between them. As a researcher, Hilbert pursued clearly posted programs undertaken during periods of intense activity. Most of his work focused on building a theory from the ground up: starting with invariant theory and algebraic number fields, he moved on to Euclidean and non-Euclidean geometry, then variational methods, integral equations, and foundations of physics, and finally proof theory and the axiomatization of arithmetic. He was a master builder, a systematizer who only seldom broke entirely new ground. Minkowski no doubt appreciated that approach, yet his own style was anything but systematic. More original and highly inventive, he had the temperament of an artist, one who sought to commune with his material rather than impose a preconceived design on it. One can sense these differences in the letters he wrote to Hilbert, especially those that pertain to their nearly abortive joint venture from the 1890s, the *Zahlbericht*.

Yet despite this contrast in their styles, Hilbert had a deep understanding of Minkowski's work; indeed, his memorial lecture (Hilbert 1910) remains even today the best single overview of the latter's mathematical achievements. As a human portrait, on the other hand, it appears to tell us more about Hilbert's personality than about Minkowski's. Insight into the latter's persona, however, can be gained from his letters to Hilbert (Minkowski 1973), published by his daughter Lily Rüdenberg, who also added a short family biography. These letters shed considerable light on Minkowski's relationship with Hilbert, though unfortunately only this half of their correspondence has survived. Hilbert's letters – a few of which were cited in Otto Blumenthal's biographical essay on Hilbert (Blumenthal 1935) – disappeared sometime after Hilbert's death. Another useful resource is Max Born's autobiography (Born 1978), which abounds with lively reminiscences of Göttingen in the era of Klein, Hilbert, and Minkowski, though Born is not always reliable regarding historical details.

Drawing on these and other sources, I begin this essay with some key episodes in the partnership that developed between Hilbert and Minkowski, a familiar enough theme, though one that nevertheless deserves closer attention.[4] In particular, I would like to focus on the symbiotic nature of their spiritual relationship as devotees of pure mathematics, a calling of special significance within the context of German culture. Both were intent on promoting their purist vision of mathematical knowledge, at first within the realm of algebra and number theory, but eventually permeating into the new physics. Minkowski's ideas were to have an especially deep and lasting impact on relativity theory, though it would be mistaken to regard this final chapter of his life in isolation from the rest. Indeed, the episodes related here are meant to suggest that his Cologne lecture on "Space and Time," the culminating triumph of Minkowski's career, was shaped by earlier experiences he and Hilbert shared in public mathematical arenas.

[4]For a thoughtful assessment of how Hilbert's views on axiomatics related to Minkowski's approach to the foundations of physics, see Corry (1997, 2004).

Minkowski's Partnership with Hilbert

Minkowski and Hilbert grew up in Königsberg where they developed a deep aesthetic appreciation of mathematical ideas. Both became number theorists, which meant that they tended to identify good mathematics with beautiful mathematical ideas. In his memorial lecture, Hilbert described his friend's tastes in these words: "Minkowski was deeply captivated by the significance of number theory, as reflected in the works and enthusiastic expressions of its heroes: Fermat, Euler, Lagrange, Legendre, Gauss, Hermite, Dirichlet, Kummer, and Jacobi. At all times he felt its charms in the liveliest way, for the virtues of number theory – the simplicity of its foundations, the precision of its concepts, and the purity of its truths – accorded completely with his innermost sensibilities" (Hilbert 1910, 351).

Hilbert shared Minkowski's sense of belonging to an exalted elite: they were not just pure mathematicians but enthusiasts who cultivated that lovely corner of their garden in which the delicate flowers of number theory grew. Only a small circle of mathematicians belonged to this group because, in the opinion of Hilbert and Minkowski, most of their colleagues lacked the requisite qualifications. Yet even if the noble creations of Gauss and other greats were "too sublime" to be appreciated by most mathematicians, this did not prevent them from reaching down to "the masses." Indeed, Minkowski's lecture course from 1903 to 1904, later published in book form under the title *Diophantische Approximationen*, had this very objective: to open the ears of its Zuhörer to this "gewaltige Musik," as he called it (Hilbert 1910, 351–352).

Yet if number theory stood at the center of Minkowski's research interests, this hardly kept him from exploring other intellectual terrain. Thus, he took up a serious interest in physics during the early 1890s while teaching as a Privatdozent in Bonn.[5] There he even went so far as to ask his brilliant colleague, Heinrich Hertz, for permission to take part in the laboratory exercises offered for beginning students. At home, he studied the works of leading theorists, Kelvin, Helmholtz, and co., motivated in part by the circumstance that he had so little intellectual intercourse with his colleagues in mathematics. After donning a lab frock for a while, he quickly sensed the vast gulf that separated the world of the pure mathematician from that of the natural scientist. Shortly before Christmas, he sent Hilbert a humorous account of his activities, in which he described himself as a constructor of instruments, toiling away in his blue lab uniform, a "Praktikus of the worst sort imaginable" (Minkowski to Hilbert, 22 December 1890, Minkowski 1973, 39). He let Hilbert know that he would not be returning to Königsberg for the holidays, but consoled his friend with the thought that he and Adolf Hurwitz would have found him unfit for proper mathematical discourse. For he had become "thoroughly physically contaminated" to such a degree that he would have "perhaps even had to undergo a 10-day quarantine" before they would have "found him sufficiently pure and unapplied mathematically to be allowed to take part in their walking excursions" (*ibid.*).

By 1891 Minkowski's letters began to take on a more intimate tone as he and Hilbert became "Dutzfreunde" and close allies. From now on they would share all kinds of news, both mathematical as well as purely personal, including plenty of innocent gossip. Many of the allusions in these letters are obscure or even totally incomprehensible today, but their witty light-heartedness nevertheless makes for delightful reading. Minkowski undertook plenty of strolls through favorite paths in their impressive mathematical garden, but most of the time he reported about people and events of mutual interest. Thus he writes often about their former teachers in Königsberg, Ferdinand Lindemann and Adolf Hurwitz, especially the latter after Minkowski became his colleague in Zurich from 1896 to 1902.

Hurwitz's mathematical tastes had much in common with Minkowski's own; they also occasionally shared painful experiences as German Jews who had managed to rise to prominence in a highly competitive academic environment. Minkowski's brother, Oscar, was a distinguished medical researcher who discovered the role of pancreatic dysfunction in diabetics (Minkowski 1973, 12). His work, however, came under attack and he was already reached age 50 by the time he became a professor. Hurwitz was Klein's most gifted pupil, yet even this was not enough for him to become an *Ordinarius* in Prussia (see Chap. 15). Still, such milder forms of anti-Semitism were so pervasive at this time that both Hurwitz and Minkowski seem to have taken such slights for granted; in fact, this theme rarely even surfaces in their correspondence. What they wrote to Hilbert and to each other reflected instead their deep love for the mathematical quest and a strong personal ambition to crack a problem or create a beautiful theory. Hurwitz and Minkowski were artists who lacked Hilbert's ferocious career ambition or his striking ability to reform a subject and then conquer it.

Hurwitz had a genuinely serene personality that won him many friends, but he also suffered from serious health problems that sometimes left him incapacitated. Minkowski enjoyed his company thoroughly, but he was nevertheless eager to leave Zurich, as he took little pleasure teaching there. After only one semester he complained to Hilbert that the students at the ETH expected their professors to

[5] Here, the example of Peter Gustav Lejeune-Dirichlet readily comes to mind. As the successor of C. F. Gauss in Göttingen, Dirichlet managed to combine his research interests in number theory with impressive contributions to mathematical physics. Minkowski documented his close affinity with Dirichlet in Minkowski (1905), a glowing tribute delivered before the Göttingen Academy to celebrate the centenary of his birth.

spoon-feed them (Minkowski to Hilbert, 31 January 1897, Minkowski 1973, 94), an attitude he had seldom encountered at the two Prussian universities, Bonn and Königsberg, where he had taught earlier. Minkowski tried to accommodate his lecture style to the students' expectations, but he soon gave up. Unlike Hurwitz, he never showed much talent as a lecturer, and even in Göttingen his courses were not a great success. His position at the ETH, by the way, remained vacant for many years after his departure in 1902, but it was eventually filled by a former student whom Minkowski had found particularly lazy; his name was Albert Einstein (Fölsing 1993, 281). By this time, Minkowski and Hilbert were together again in Göttingen, where they found themselves deeply immersed in the new physics of the electron. But back in the early 1890s their friendship had only begun to intensify on the way to becoming collaborators.

Minkowski's early letters to Hilbert reveal that he followed the latter's work on invariant theory closely and with enthusiasm. He was, of course, well aware of the controversy that erupted when Paul Gordan took issue with Hilbert's non-constructive methods (see Chap. 14). So he took delight when he saw how his friend had managed to find a new approach that was less objectionable on these grounds. "For a long while," Minkowski wrote, "it has been clear to me that it would only be a question of time before you settled the old invariant questions so that hardly more than the dotting of the "i" remained. But that it all went so easily and simply makes me truly happy, and I congratulate you on this" (Minkowski to Hilbert, 9 February 1892, Minkowski 1973, 45). And so he went on: the smoke from Hilbert's earlier paper might still have been bothering Gordan's old eyes, but now that Hilbert had invented smokeless gunpowder it was high time to decimate the castles of the robber knights like Gordan, who went around capturing the individual invariants and stowing them in their dungeons, before the danger mounted that no new life would ever emerge from these quarters again. Minkowski went on teasing his friend that if he were not "so sehr radikal" he would consider doing his fellow mathematicians the favor of setting forth his new results in the form of a monograph, a book that would enable them to build on the newly reformed theory. He had in mind something like the edition of Paul Gordan's lectures on invariant theory prepared by Georg Kerschensteiner, but jested that if Hilbert waited for his Kerschensteiner to come along he might find that his book contained a lot of cherry stones that could spoil the reader's appetite.

Needless to say, Hilbert never entertained the idea of writing such a book, nor did a young Kerschensteiner emerge to write one for him. Six months later he completed his research program in invariant theory by submitting a definitive 70-page paper to the *Mathematische Annalen* (Hilbert 1893). With it, he wrote Minkowski, he would leave the field and turn instead to the theory of algebraic number fields. Indeed, the methods he used in this paper were largely taken from algebraic number theory together with powerful new results in algebraic geometry.

The following year Hilbert offered some general reflections on invariant theory in the form of a short survey article (Hilbert 1896). This had been solicited by Felix Klein, who packed it in his luggage along with the several other papers he later presented at the Mathematical Congress held in conjunction with the Chicago World's Fair. Hilbert's text makes for interesting reading, even for historians, because in it he describes three phases in research on classical invariants: a naïve period (Cayley and Sylvester), followed by the development of sophisticated formal methods (Clebsch and Gordan), and finally a critical phase, dominated of course by his own work. No one who read this could have had any doubt that Hilbert was a mathematician whose immense talent was only matched by his formidable *chutzpah*.

While Klein was busy in Chicago showcasing his own research and that of his German colleagues, Hilbert and Minkowski were attending the DMV meeting in Munich. From its founding, the society had signaled that a major goal would be the preparation by its membership of official reports on recent and not so recent research in various fields. Thus, Alexander Brill and Max Noether, who were charged with the task of writing a compendium on algebraic functions, would soon unveil a monumental study on this vast field of research (Brill and Noether 1893). Clearly, they had splendid credentials for taking on such a task. Brill and Noether had been collaborators for many years, whereas Hilbert and Minkowski were still very young and had never collaborated at all. Nevertheless, they agreed to prepare a report on recent developments in number theory, even though Hilbert had yet to publish a single paper in this field. Minkowski, on the other hand, had already as a teenager made a name for himself as a number theorist when the Paris Academy awarded him first prize for his essay on the representations of a number as the sum of five squares. Hilbert evidently had more to gain from this challenging venture than did his friend.

Minkowski's letters over the next several years contain numerous allusions to his efforts to make headway on this project, though their sometimes exasperated tone also made clear enough that he viewed this as a largely thankless task. His main interest was to finish his monograph on the geometry of numbers, a topic he pursued with the encouragement of the leading French number theorist, Charles Hermite. Still, he did not neglect to do his homework for the DMV report. In fact, he managed to produce 75 abstracts for the *Jahrbuch über die Fortschritte der Mathematik* on papers from the mid 1890s, thereby attesting to his willingness to delve into the literature. His report, so he wrote Hilbert, would highlight results from his forthcoming book as well as his earlier papers. "Whether beyond that I bring people closer to comprehending the results of several till now hardly appreciated works, which after all is the purpose of such reports, this is for me a pleasant enough task, even though not the type that I value the most" (Minkowski to Hilbert, 10 February 1896, Minkowski 1973, 78).

By early 1896 Minkowski's book was still not finished, whereas Hilbert's portion of the report, dealing with the modern theory of algebraic number fields, was nearing completion. Not surprisingly, Hilbert grew impatient and suggested to Minkowski that he might want to consider publishing his part of the report later, a plan his friend greeted with open arms. As it turned out, Hilbert's report did not appear until the following year whereas Minkowski managed to finish a truncated version of his *Geometrie der Zahlen* in 1896 (Minkowski 1896). His report for the DMV, on the other hand, never saw the light of day at all.

The following autumn Minkowski joined Hurwitz in Zurich, and both read the galley proofs of Hilbert's *Zahlbericht* with great care and considerable enthusiasm. Minkowski was shocked to see that Hilbert had neglected to thank his wife for all her work in copying the entire manuscript and preparing the index (Minkowski to Hilbert, 17 March 1897, Minkowski 1973, 98). After receiving this thorough scolding, Hilbert added a note of thanks to her in the preface along with these words: "My friend Hermann Minkowski has read the proofs of this report with great care; he also read most of the manuscript. His suggestions have led to many significant improvements, both in content and presentation. For all this help I offer him my most hearty thanks" (Hilbert 1897, 64).

When he received the published text a month later, Minkowski wrote to congratulate Hilbert, offering the prognosis that it would not be long before he too would be counted among the great classical authors in the field of number theory, the discipline both men revered so much (Minkowski to Hilbert, 14 May 1897, Minkowski 1973, 100). But in this same letter, Minkowski also cautioned that when such recognition came it might be confined to a select group of connoisseurs. In this connection, he related his impressions of Gösta Mittag-Leffler, the politically astute pupil of Weierstrass who had recently visited Zurich in connection with the forthcoming International Congress. Minkowski found him dull and unimpressive, the more so because he evinced no real appreciation for Minkowski's new book; nor did he imagine that Hilbert's report would fare any better in the eyes of this worldly-wise Swede. But as a consolation he offered Hilbert the thought that "the facility for mental abstraction indeed causes many people headaches" (*ibid.*).

Plans for Hilbert's Paris Lecture

Minkowski's role as Hilbert's silent collaborator during the production of the *Zahlbericht* ought not to be exaggerated, but he certainly did offer his friend a great deal of intellectual and moral support. The same pattern can be observed a few years later when Hilbert called on his advice again during a time when both were making plans to attend the Second International Congress of Mathematician in Paris. It was on this occasion, of course, that Hilbert delivered his famous lecture on "Mathematical Problems" (Hilbert 1901), a performance that soon brought him lasting fame. Had it not been for Minkowski's wise counsel, however, he might well have chosen a more controversial, but less ambitious topic for this lecture. Once again, Minkowski's letters shed considerable light on the motivations and behaviors of both friends.

As the newly-elected president of the DMV, Hilbert received an invitation in December 1899 to deliver a plenary address in Paris on a theme of his choosing (see Chap. 16). This was a splendid opportunity, or at least so Hilbert thought, to throw down the gauntlet to the dominant mathematician of the era, Henri Poincaré. At the previous ICM, held in Zurich in 1897, Poincaré had been designated as one of the plenary speakers, and for this occasion he took as his theme the role of physical conceptions in guiding fertile mathematical research. With that speech still in mind, Hilbert thought he should counter by singing a hymn of praise to pure mathematics. No doubt he also recalled how Jacobi had once rebuked his French contemporary, Joseph Fourier, for failing to recognize that the highest and only true purpose of mathematics resided in nothing other than the quest for truth; this alone redounded honor on the human spirit (Klein 1926, 114).

But Minkowski strongly counseled his friend to abandon this plan (Minkowski to Hilbert, 5 January 1900, Minkowski 1973, 119). After rereading Poincaré's text, he found that its assertions in no way compromised the integrity of pure mathematics. Furthermore, he felt that the Frenchman's views were formulated so cautiously that Hilbert and he could easily subscribe to them. Moreover, he recalled how Poincaré had not even been present in Zurich, so his text was actually read by another party. Consequently, Minkowski was sure that there was little point in drawing attention to Poincaré's opinions, since few would remember what he had written for that occasion. He then contrasted this rather dull speech with the stirring lecture delivered by Ludwig Boltzmann at the annual meeting of German scientists held in Munich just 4 months earlier (Boltzmann 1900).

Hilbert clearly took this all to heart, even if he never said so. Indeed, he had returned from that Munich meeting in high spirits, writing to Hurwitz that this event was the best attended and most stimulating of all the DMV meetings held thus far.[6] Furthermore, he was clearly pleased by the response to his two talks (one on the axioms of arithmetic, the other on Dirichlet's principle). Still, these could hardly be compared with Boltzmann's dazzling performance (Boltzmann 1900), which left the spellbound audience literally buzzing with excitement afterward. Clearly it left a strong impression on both Hilbert and Minkowski as well, not only theatrically but also thematically; in fact the parallels between Boltzmann's speech and Hilbert's far more famous Paris lecture are simply impossible to overlook (although it seems nearly everyone since has done just that). Both texts offered sweeping accounts of past developments, undertaken with an eye toward unsolved problems that would test the steel of the younger generation. Boltzmann had beckoned the younger generation on – "a Spartan war chorus calls out to its youth: be even braver than we!" – but without attempting to "lift the veil" (Boltzmann 1900, 73) by making prognostications about the course of future research as Hilbert would do one year later in Paris (see Hilbert 1901).

This brings us back to Minkowski's letter in which he wrote, "most alluring would be the attempt to look into the future, in other words, a characterization of the problems to which the mathematicians should turn in the future. With this, you might conceivably have people talking about your speech even decades from now. Of course, prophecy is indeed a difficult thing" (Minkowski to Hilbert, 5 January 1900, Minkowski 1973, 120). Minkowski was sure that a lecture with real substance, like the one delivered by Hurwitz at the ICM in Zurich, would be far more effective than a merely general presentation, such as the one given by Poincaré. He even mentioned two earlier lectures that he thought might prove useful for Hilbert to read, one a speech delivered by Hermite in 1890, another by H.J.S. Smith entitled, "On the Present State and Prospects of some Branches of Pure Mathematics."

Hilbert later turned to Minkowski for advice about physical theories with open foundational problems (Walter 1999). His friend pointed to Boltzmann's work on statistical mechanics and thermodynamics, noting that this terrain offered "many interesting mathematical questions also very useful for physics," (Minkowski to Hilbert, 10 July 1900, Minkowski 1973, 128). This tip clearly influenced Hilbert's formulation of his sixth Paris problem, which addresses the axiomatization of physics. One week later, Minkowski was busy reading page proofs of various portions of Hilbert's lecture, including drafts of what ultimately became the second and twenty-third problems. He and Hurwitz continued to make various suggestions, including cuts that would be necessary to stay within the time limits allowed. By the time he finished reading the final draft, Minkowski was nothing less than ecstatic:

> I can only wish you luck on your speech; it will certainly be the event of the congress and its success will be very lasting. For I believe that this speech, which probably every mathematician without exception will read, will cause your powers of attraction on young mathematicians to grow still more, if that is even possible. . .. Now you have really wrapped up the mathematics for the twentieth century and in most quarters you will gladly be acknowledged as its general director (Minkowski to Hilbert, 28 July 1900, Minkowski 1973, 129–130).

These words could hardly have been more prophetic.

Boltzmann and the Energetics Debates

Soon after Minkowski's appointment to a new chair in Göttingen in 1902, he and Hilbert began pursuing various topics in the foundations of physics. In 1905 – the very year Einstein published his famous paper, "On the electrodynamics of moving bodies" (Einstein 1905) marking the birth of special relativity – they conducted a joint seminar on electron theory, a field that was already buzzing with activity in Göttingen. Two years later, Minkowski unveiled his own 4-dimensional approach to electrodynamics in Minkowski (1908), a paper widely regarded by contemporaries as unreadable (Walter 2008, 214, 229). Einstein and his young collaborator, Jakob Laub, rewrote some of its key results in the language of 3-vectors, while disputing some of Minkowski's physical claims.

In all likelihood, Minkowski took little or no notice of this critique. Einstein, after all, was still just an obscure patent clerk in Bern. A far more significant rival was Poincaré, who had already utilized a 4-dimensional formalism in connection with the Lorentz transformations. Poincaré was, in fact, the first to note that these form a group, thereby calling attention to the importance of group invariants for electrodynamics. On the other hand, Poincaré's conventionalist views led him to attach far less importance to space-time physics, a viewpoint that Minkowski meant to challenge as forcefully as possible.[7]

It is not hard to imagine that Hilbert's Paris address would have been in the back of Minkowski's mind when he began thinking about his Cologne lecture, "Space and Time." He knew that such opportunities were rare and that the time was ripe for promoting his recent work on electrodynamics. Undoubtedly disappointed by the cool reception his work

[6]Hilbert also noted that Lindemann failed to appear at the Munich meeting, even though it was held in the city where he taught. Hilbert to Hurwitz, 5–12 November 1899, Mathematisches Archiv 76, 975, SUB Göttingen.

[7]Scott Walter has pointed out that Minkowski paid very close attention to Poincaré's publications (Walter 2008, 223).

had received, he clearly saw this as a chance to make a real splash. Hilbert's speech in Paris had accomplished just that and, for Minkowski, its lofty rhetoric was surely inspirational. Yet, just as clearly, the scope and purpose of Hilbert's speech had to be measured on an entirely different scale. So this could hardly serve as a model for Cologne. Minkowski's topic was the foundations of physics, and on that score he surely remembered the brilliant public performances of Ludwig Boltzmann, the Austrian physicist who shortly before this time had committed suicide while vacationing with his family.

Boltzmann was in many ways a singular figure. During his lifetime, no physicist in Germany cultivated contacts with mathematicians with the same intensity as did he. A decade prior to his death, Minkowski and Hilbert were both present at a long-remembered conference in Kiel during which Boltzmann caused a great stir over the energetics program promoted by the chemist Wilhelm Ostwald, among others. A glimpse of the atmosphere in Kiel can be found in one of Minkowski's letters to Hilbert (Minkowski to Hilbert, 24 September 1895, Minkowski 1973, 70–71). The two friends had met in Göttingen shortly before the conference convened, which gave them time to prepare a preliminary report on their number theory project. This was presented by Hilbert during the first days of the meeting, after which he apparently left the conference. Minkowski thus described the final Friday, when he bumped into Klein and had lunch with him at their hotel, surrounded by three physicists from the anti-energeticist camp: Boltzmann, Walther Nernst, and Arthur Oettingen. All five continued their passionate discussion. Minkowski even suggested to Hilbert that they should consider ending their *Zahlbericht* with some general theses regarding the essence and significance of number theory, just as the physicists had done in closing their debates on energetics.

Another eye witness at the Kiel conference was young Arnold Sommerfeld, who was then Klein's assistant in Göttingen. Sommerfeld left this brief, but memorable account of the climactic confrontation in Kiel:

> [Georg] Helm from Dresden gave the report on energetics, behind him stood Wilhelm Ostwald and behind both stood the Naturphilosophie of Ernst Mach, who was not present. The opponent was Boltzmann, seconded by Felix Klein. The conflict between Boltzmann and Ostwald resembled, both outwardly and inwardly, a fight between a bull and a subtle fencer. But on this occasion, despite all his swordsmanship, the toreador was defeated by the bull. Boltzmann's arguments won out. At that time all of us younger mathematicians stood on the side of Boltzmann (Deltete 1999, 56).

This version is about all that remains of this debate in the collective memory of physicists, though one can easily reconstruct many details that reveal a far more complicated and interesting story (see Deltete (1999) for such an account). Indeed, the debates that took place during that particular meeting echoed on for many years afterward, particularly in Göttingen circles. Moreover, one of the critical issues then under discussion – namely, the link between conservation of energy and the laws of classical Newtonian mechanics – was of direct and immediate relevance for Minkowski when he composed his 1908 lecture.

At the Kiel conference, as Minkowski well knew, Boltzmann had contested Georg Helm's claim that the laws of motion in classical mechanics could be deduced from the energy conservation law. In fact, what Minkowski later wrote in "Raum und Zeit" flies in the face of Boltzmann's earlier claim: "At the limiting transition. .. to $c = \infty$, this fact [that the laws of motion follow from energy conservation] retains its importance for the axiomatic structure of Newtonian mechanics as well, and has already been appreciated in this sense by I. R. Schütz." In a footnote, Minkowski cited Schütz's paper on "The Principle of the absolute conservation of energy," published in the *Göttinger Nachrichten* (Schütz 1897).

Oddly enough, this passage from Minkowski's lecture has been largely overlooked in recent studies and commentaries. Historians have likewise ignored this paper by Ignatz Schütz, himself a forgotten figure. Schütz was a young Dozent in Göttingen at this time, a protégé of the physicist Woldemar Voigt. He spoke at a number of meetings of the Göttingen Mathematical Society and was also a regular speaker at the annual *Naturforscher* conferences. So Minkowski was not the only one who knew and appreciated Schütz's work; his paper on conservation of energy was later highlighted by Max von Laue in a survey article on the energy concept (Laue 1955, 374). Moreover, Felix Klein had a longstanding interest in these matters. Indeed, shortly after the Kiel conference he reported on the energetics debates at a meeting of the Göttingen Mathematical Society. His protocol notes indicate that on this occasion he explained the alleged mistake Boltzmann had found in Georg Helm's argument deducing the laws of motion in Newtonian mechanics from the energy principle. Since Ignatz Schütz also attended the Kiel conference, it seems safe to assume that his work was part of a larger Göttingen effort aimed at resolving this particular point of contention.[8] In effect, Schütz's paper came to be regarded as the last word on this central issue in the debates. More than likely, Minkowski followed the ensuing literature stemming from the Kiel debates quite closely, so he probably read

[8]Klein's notes are in the Mathematisches Archiv, Niedersächsische Staats- und Universitätsbibliothek, Göttingen. In the second volume of his lectures on nineteenth-century mathematics, Klein noted how Schütz had managed to clarify the relationships between the 10 first integrals of classical mechanics (Klein 1927, 58). Klein went further, enlisting the support of Friedrich Engel, who published two papers on applications of Lie's theory of infinitesimally generated groups to classical mechanics.

Schütz's paper soon after its publication in 1897. Clearly, it was very much on his mind when he was preparing his Cologne lecture.

Minkowskian Physics

Minkowski was guided by a unified view of physics based on the universal validity of Einstein's principle of relativity. In "Raum und Zeit" he made no pronouncements about the ultimate inertial properties of matter. Moreover, his whole approach led to a deeper understanding of how in special relativity the laws of motion are wedded to energy conservation and vice-versa. In his memorial lecture, Hilbert describes how Minkowski was guided by his world postulate, according to which the laws of physics are based on Lorentz covariant concepts. Since the velocity components together with the density transform just like the space-time coordinates, these four entities form a 4-vector. Likewise, the electric and magnetic 3-vectors, taken together form an anti-symmetric tensor; they therefore transform just like the six Plücker coordinates familiar from line geometry. Thus these two pairs of physical concepts – velocity and density, electricity and magnetism – cannot be separated, but rather each pair depends on the choice of a spacetime coordinate system. These insights naturally emerge as consequences of Minkowski's world postulate, in fact so naturally that it is perhaps difficult to realize how fundamentally new this way of thinking was in 1908.

To appreciate the physical implications Minkowski drew for spacetime physics, one should turn to the fourth section of "Raum und Zeit." There he proposed to reform the concepts of mechanics to ensure their compatibility with the principle of relativity pertaining to the Lorentz group. For this purpose, he took the invariant motive force vector from electrodynamics and set this equal to the mechanical force vector, thereby obtaining four relativistic laws of motion. The first three equations yield a generalization of the Newtonian laws associated with the conservation of center of mass, whereas the fourth equation expresses the relativistic version of conservation of energy. Since these two force vectors are both perpendicular to the velocity vector, Minkowski immediately deduced that the fourth equation, expressing energy conservation, is a consequence of the other three.

But he also noted – and this was the decisive point that emerged from our story about the energetics debates in Kiel – that since the time axis can be taken freely in the direction of any time-like vector, one can also deduce the laws of motion from the energy principle alone. Hilbert underscored the importance of this result in his memorial tribute when he wrote that "Minkowski's investigation led moreover to the principally interesting fact that by virtue of the world postulate the complete laws of motion can be deduced from the conservation of energy theorem alone" (Hilbert 1910, 359).[9] Hilbert also attached great importance to the manner in which Minkowski derived his phenomenological equations for the motion of bodies in electromagnetic fields. This derivation required only three axioms: 1. that moving bodies can never attain the speed of light; 2. that Maxwell's equations are valid for bodies in a state of rest; and 3. that when a single point of a body is motionless, then this point may be treated as if the entire body were at rest. Since the validity of Maxwell's equations for matter at rest was considered beyond question, Minkowski's derivation was based on a truly minimal set of assumptions that gave an impressive demonstration of the power of the principle of relativity. Hilbert was just one of several contemporaries who appreciated this. What he seems not to have noticed, however, is that Göttingen's Max Abraham had come up with a different set of equations that was nevertheless compatible with Minkowski's axioms (see Walter (2007); Fig. 2).

Minkowski's systematic reform of electrodynamics and mechanics involved important revisions to both fields. His fundamental equations for the electrodynamics of moving bodies were derived directly from the relativity principle and differed from the earlier results of Lorentz and Emil Cohn, neither of which was Lorentz-invariant. Later investigations showed that Lorentz's derivation based on his electron theory was erroneous; afterward Minkowski's equations gained general acceptance. In mechanics, the same was true of the Minkowski force, which was taken up by Lorentz in his Leiden lectures of 1910–1912 on special relativity, published in German translation in (Lorentz 1929). Max von Laue's influential textbook on special relativity was even more decisively influenced by Minkowski's new approach, which led Laue to formulate a new world tensor for expressing the totality of energy and matter in a physical system (see the essay that follows). It did not take Einstein long to absorb these new developments, as can be seen from his unfinished draft of a review article on special relativity, written around 1912 (Einstein 1996).

Minkowski's research program for mechanics was based on a strong analogy between mechanical concepts and those he had developed in his 4-dimensional approach to electrodynamics. Some historians, however, have regarded him as an exponent of the electromagnetic worldview, according to which the inertial properties of matter stem entirely from electromagnetic forces. This view, as Leo Corry has

[9]One could also cite Wolfgang Pauli, who in his report on relativity theory for the German encyclopedia (Pauli 1921) noted that similar consequences can be drawn for Minkowskian electrodynamics. Felix Klein also alluded to this in his commentary on Hilbert's first note on the foundations of physics. Klein thereby launched the investigations into the status of energy conservation theorems in special and general relativity, the topic that led to Emmy Noether's fundamental theorems on invariant variational problems (see Rowe 1999).

emphasized, is quite mistaken (Corry 2004, 191). Very possibly Minkowski's own rhetoric contributed to this faulty impression. In the closing paragraph of "Raum und Zeit" he made this pronouncement: "The unexceptional validity of the world postulate is, in my opinion, the true nucleus of an electromagnetic image of the world, which was discovered by Lorentz and further revealed by Einstein, and which now lies open in the full light of day" (Minkowski 1909, 444).

In later years, Einstein distinguished between two main types of physical theories: those that are constructive, like Lorentz's electron theory, and those of principle, thermodynamics being a prime example.[10] The relativity principle, which later evolved into the special and general theories of relativity, was of the second type, which helps explain why Minkowski was so attracted to it. Still, one should not imagine that he and Hilbert were altogether out of sync with mainstream developments in theoretical physics. In his magisterial study, *Electrodynamics from Ampere to Einstein* (Darrigol 2000), Olivier Darrigol highlights the importance of a new trend in late-nineteenth-century theorizing, one in which general physical principles began to displace the more concrete styles that had been prevalent theretofore. Among the strongest representatives of this new approach were such luminaries as Helmholtz, Hertz, Poincaré, and Planck. Hilbert, on the other hand, swung to the far extreme by attaching great faith in axiomatic analysis and general mathematical methods, in particular variational principles. Minkowski's success with his world postulate no doubt reinforced these convictions. Both he and Minkowski were extreme optimists in this regard, and their views with regard to the efficacy of mathematics in the pursuit of knowledge of the natural world differed sharply from the stance taken by Poincaré.

Thus, while advancing the new physics, Minkowski parted company with Poincaré regarding the importance of a 4-dimensional formalism. For Poincaré, who was always sceptical of arguments that purported to establish the truth of a physical theory, the relativity principle was nothing more than a convenient hypothesis that helped the natural philosopher organize important knowledge. By contrast, Minkowski saw it as nothing less than a revealed truth; hence, he argued that the mathematical physicist should strive to develop a firmly grounded theoretical structure as a framework for guiding experimental work. His final rhetorical flourish in "Space and Time" captures this vividly: "In the development of its mathematical consequences there will be ample suggestions for experimental verifications of the world postulate, which will suffice to conciliate even those for whom the abandonment of older established views is unsympathetic or painful by the idea of a pre-established harmony between pure mathematics and physics" (Minkowski 1909, 444).

On Canonizing a Classic

Max Born once commented that relativity theory was not the creation of one man, but rather of several, most notably Lorentz, Poincaré, Einstein, and Minkowski (Born 1959, 501). That it came to be associated almost exclusively with Einstein's name can mainly be attributed to the events that propelled him to world fame, beginning in November 1919. Born was in an unusually advantageous position when it came to judging the merits of Einstein's and Minkowski's respective contributions. He began his academic career in Göttingen and was serving as Minkowski's assistant at the time of his death. Later he befriended Einstein when both were teaching in Berlin, and in Frankfurt he went on to write one of the most successful of the many popular books on relativity (Born 1920). Though he made very ample use of Minkowski's spacetime diagrams throughout his text, Born entitled it *Die Relativitätstheorie Einsteins.* Still, Minkowski's Cologne lecture continued to be read and cited, in large part due to the promotional efforts of Göttingen allies and friends. Thanks to them, "Raum und Zeit" was well on its way to becoming a classic. To see how this happened, let us turn back to that work and its reception.

In 1910 Hilbert's ever-faithful pupil, Otto Blumenthal, began a new series with the Leipzig publishing house of B. G. Teubner entitled *Fortschritte der mathematischen Wissenschaften in Monographien.* This venture turned out to be a total flop, but initially it served its main purpose, namely to promote Minkowski's contributions to physics. Thus, the first volume (Blumenthal 1910) contained a reprint of his 1907 paper, along with a related study on the properties of the electron prepared by his former assistant, Max Born, on the basis of Minkowski's extant texts. Earlier, Teubner had also published "Raum und Zeit" as a separate brochure; this contained a brief introduction by the president of the DMV, August Gutzmer.

By 1913, however, this brochure was already out of print, and when Arnold Sommerfeld learned of this he suddenly came up with a brilliant idea. Rather than requesting a new printing, he approached Blumenthal with a proposal for a second volume in his nearly still-born series. Sommerfeld's idea was to publish a compendium of papers on relativity that featured Minkowskis's "Raum und Zeit" as its centrepiece. It did not take long for Blumenthal to see the light. After gaining the approval of Lorentz and Einstein, he was able to convince Teubner that this anthology (Blumenthal 1913), bearing the simple title *Das Relativitätsprinzip*, represented a promising commercial venture. Just how promising, no one could have imagined. After going through numerous

[10]He emphasized this distinction, for example, in his widely read article for the *Times* (London), Einstein (1919b).

printings and some major changes, it emerged in the early 1920s in its present-day form, a staple item in bookstores worldwide.

To a considerable degree, *Das Relativitätsprinzip* has since served to canonize the classics of relativity theory, especially Einstein's works, though this was clearly not its original intent. The 1913 edition was, in fact, primarily conceived as another tribute to Minkowski, as evidenced by the inclusion of a photographic portrait of the deceased mathematician. Inside the reader was exposed to the relevant prehistory: the two classic papers by Lorentz from 1894 and 1904, followed by Einstein's famous 1905 article "On the Electrodynamics of Moving Bodies" as well as his brief note on the relationship between energy and inertial mass. These texts set the scene for Minkowski's "Raum und Zeit," which Blumenthal described in the foreword as the speech that set the great wave of popular interest in relativity theory in motion.

A reader in 1913 would presumably notice the difference between Minkowski's soaring rhetoric and the lean and pithy style that marked Einstein's presentations. Not that Minkowski took great pains to spell out the details, but for this the reader could turn to Sommerfeld's notes for help, with the reassurance that these were merely added to remove "small, formal mathematical difficulties that could stand in the way of penetrating into the great thoughts of Minkowski." Sommerfeld rounded out this slim volume with Lorentz's Wolfskehl lecture, "Das Relativitätsprinzip und seine Anwendungen auf einige besondere physikalische Erscheinungen," delivered in Göttingen one year after Minkowski's death. It was a fitting choice, as on that occasion the gracious Dutch physicist paid tribute to the distinguished mathematician's contributions to his own field.

Given the agenda Blumenfeld and Sommerfeld had in mind, this choice of texts hardly seems surprising, though in retrospect its omissions are striking. The most glaring of these, of course, was the absence of any of Poincaré's writings. Could this have had something to do with the longstanding rivalry between France's leading mathematician and Göttingen's three leading mathematicians: Klein, Hilbert, and Minkowski? Less surprising, but nevertheless noteworthy was the lack of any mention of Einstein's generalized theory of relativity. By 1913, Einstein had been struggling for some six years to go beyond special relativity in order to link gravitation with the effects that arise in noninertial frames. Yet the only relativistic treatment of gravitation to be found in the first edition of *Das Relativitätsprinzip* was the one given by Minkowski in the closing section of "Raum und Zeit" (Walter 2007). There he briefly described how one can formulate a simple 4-dimensional force law between point masses that corresponds to Newton's law for low velocities. Minkowski then concluded that this law of attraction "when combined with the new mechanics is no less well adapted to explain astronomical observations than the Newtonian law of attraction combined with Newtonian mechanics" (Minkowski 1909, 440).

Einstein had long been promoting a very different approach to gravitation. In 1911 he published his calculations for the bending of light in the vicinity of the sun as well as for gravitational red shift. By 1915 he wrote an introductory article for the series *Die Kultur der Gegenwart* in which he expressed the opinion that the fundamental principle expressing the constancy of the speed of light had to be dropped since it appeared to be valid only in reference systems in which the gravitational potential was constant. A year earlier he published a 55-page article (Einstein 1914) on "The Formal Foundations of General Relativity," a highly mathematical presentation that took great pains to show that the Einstein-Grossmann field equations were the only feasible candidates for a sound relativistic theory of gravitation.

Interestingly enough, even in 1915 Einstein was still not satisfied with what he had written about general relativity. Only about a week after he delivered his Wolfskehl lectures in Göttingen, he was contacted by Arnold Sommerfeld, who was then laying plans for an expanded second edition of *Das Relativitätsprinzip* for which he wanted to include texts by Einstein dealing with his general theory of relativity (Sommerfeld to Einstein, 15 July 1915, *CPAE* 8A 1998, 147). But Einstein responded without enthusiasm, though he did note that for this purpose his paper (Einstein 1911) and the longer one (Einstein 1914) might have been the most suitable among his recent papers. Still, he preferred that the collection be reprinted without change since he was planning to write an introductory booklet that led to the general theory anyway. This letter appears to be the earliest evidence we have of Einstein's own plan to produce a semi-popular textbook, an idea that evidently antedated the breakthrough he made in November 1915. Two years later Vieweg published (Einstein 1917b), an instant bestseller that soon went through numerous editions and translations.

Sommerfeld waited until 1920 before introducing a new edition of *Das Relativitätsprinzip*, one that finally included writings by Einstein on general relativity. Since Lorentz's 1910 lecture was by now thoroughly antiquated, he replaced this with Einstein (1911), the paper in which Einstein for the first time drew optical consequences from his equivalence principle. The new volume also contained Einstein's canonical paper on the mature theory of general relativity (Einstein 1916a), his brief sequel on Hamilton's principle from the same year (Einstein 1916b), the essay on relativistic cosmology (Einstein 1917a), and a speculative paper (Einstein 1919a), in which he modified the field equations in order to link gravity with an electromagnetic theory of matter. Clearly, this new anthology greatly enhanced Einstein's reputation as the central, if not sole creator of the theory of relativity. This picture remained unaltered with the definitive

1922 edition, which added Hermann Weyl's paper (Weyl 1918) on "Gravity and Electricity" to this compendium of classics. One year later, this text appeared in English translation, published by Methuen; this edition later served as the basis for the popular Dover paperback edition (Lorentz et al. 1952), which can still be found in bookstores today.

An ironic twist took place when this edition came out: for some reason the introductory passage in Einstein's fundamental paper, Einstein (1916a), was omitted from the text, though it was faithfully reproduced in the German original. Therein he acknowledged his debt not only to Minkowski but also to a whole series of other mathematicians, including his friend Marcel Grossmann. "The generalization of the theory of relativity," he wrote, "was facilitated considerably by Minkowski, the mathematician, who was the first to recognize clearly the formal equivalence of the space and time coordinates, and who utilized this in the construction of the theory" (Einstein 1916a, 283). Einstein noted that the "mathematical tools necessary for general relativity were readily available in the absolute differential calculus," a theory developed by Ricci and Levi-Civita and which was based on earlier research on non-Euclidean manifolds undertaken by Gauss, Riemann, and Christoffel. He also pointed out that the Italian inventors of the absolute differential calculus had already applied it to a number of problems in theoretical physics. As for Grossmann, Einstein wrote that his "help not only saved me the effort of studying the pertinent mathematical literature" but "also aided me in my search for the field equations of gravitation" (*ibid.*). The omission of all this from the English language edition, (Einstein et al. 1952), surely left a faulty impression, as the remainder of the text, that is the portion that did get printed, contains only slight hints of the contributions of others. What such readers thus missed was that Einstein clearly did make a sincere effort to discharge his intellectual debts.

Much else slipped away with the passage of time, as key witnesses and their memories passed from the scene. Hilbert continued to pursue his dream of axiomatizing parts of physics until well into the 1920s, but his contributions to Einstein's general theory of relativity were viewed by his contemporaries as being only of marginal importance. With the advent of the Nazi regime in 1933, Göttingen became just another provincial German university, though the allure of its glorious past was by no means forgotten. In his obituary article for his former mentor, Hermann Weyl briefly alluded to Hilbert's friendship with Minkowski during their Königsberg days, and then went on to recall wistfully the atmosphere he remembered in Göttingen after Minkowski's arrival:

The two friends became the heroes of the great and brilliant period, which our science experienced during the following decade in Göttingen, unforgettable to those who lived through it. Klein, for whom mathematical research had ceased to be the central interest, ruled over it as a distant but benevolent god in the clouds. Too soon was this happy constellation dissolved by Minkowski's sudden death in 1909 (Weyl 1944, 131–132).

Acknowledgments This paper is based on a lecture delivered at the conference "Space and Time 100 Years after Minkowski," held from 7–12 September 2008 at the Physics Center in Bad Honnef, Germany. My thanks go to the conference organizers, Claus Kiefer and Klaus Volkert, as well as the other participants, especially Engelbert Schücking and Scott Walter, who offered several insightful remarks.

References

Blumenthal, Otto, ed. 1910. *Zwei Abhandlungen über die Grundgleichungen der Elektrodynamik. Mit einem Einführungswort von Otto Blumenthal*, Fortschritte der mathematischen Wissenschaften in Monographien. Vol. Heft 1. Leipzig: Teubner.

———, ed. 1913. *Das Relativitätsprinzip. Eine Sammlung von Abhandlungen. Mit Anmerkungen von A. Sommerfeld und Vorwort von O. Blumenthal*, Fortschritte der mathematischen Wissenschaften in Monographien. Vol. Heft 2. Leipzig: Teubner.

———. 1935. Lebensgeschichte. In *David Hilbert, Gesammelte Abhandlungen*, Bd. 3 ed., 388–429. Berlin: Springer.

Boltzmann, Ludwig. 1900. Über die Entwicklung der Methoden der theoretischen Physik in neuerer Zeit. *Jahresbericht der Deutschen Mathematiker-Vereinigung* 8: 71–95.

Born, Max. 1920. *Die Relativitätstheorie Einsteins und ihre physikalischen Grundlagen*. Berlin: Springer.

———. 1959. Erinnerungen an Hermann Minkowski zur 50. Wiederkehr seines Todestages. *Die Naturwissenschaften* 17: 501–505.

———. 1978. *My Life. Recollections of a Nobel Laureate*. New York: Charles Scribner's Sons.

Brill, Alexander, and Max Noether. 1893. Die Entwicklung der Theorie der algebraischen Functionen in älterer und neuerer Zeit. *Jahresbericht der Deutschen Mathematiker-Vereinigung* 3: 197–566.

Corry, Leo. 1997. Hermann Minkowski and the Postulate of Relativity. *Archive for History of Exact Sciences* 51: 273–314.

———. 2004. *David Hilbert and the Axiomatization of Physics (1898–1918): From Grundlagen der Geometrie to Grundlagen der Physik*. Dordrecht: Kluwer.

CPAE 6. 1996. *Collected Papers of Albert Einstein*, Vol. 6: *The Berlin Years: Writings, 1914–1917*, A. J. Kox, *et al.*, eds., Princeton: Princeton University Press.

CPAE 7. 2002. *Collected Papers of Albert Einstein*, Vol. 7: *The Berlin Years: Writings, 1918–1921*, Michel Janssen, *et al.*, eds., Princeton: Princeton University Press.

CPAE 8A. 1998. *Collected Papers of Albert Einstein*, Vol. 8A: *The Berlin Years: Correspondence, 1914–1917*, Robert Schulmann, *et al.*, eds., Princeton: Princeton University Press.

Darrigol, Olivier. 2000. *Electrodynamics from Ampere to Einstein*. Oxford: Oxford University Press.

Deltete, Robert. 1999. Helm and Boltzmann: Energetics at the Lübeck Naturforscherversammlung. *Synthese* 119: 45–68.

Einstein, Albert. 1905. Zur Elektrodynamik bewegter Körper. *Annalen der Physik* 17: 891–921.

———. 1911. Über den Einfluß der Schwerkraft auf die Ausbreitung des Lichtes. *Annaeln der Physik* 35: 898–908.

———. 1914 Die formale Grundlage der allgemeinen Relativitätstheorie, *Königlich Preußische Akademie der Wissenschaften* (Berlin). *Sitzungsberichte*, 1030–1085; reprinted in [*CPAE*, 6, 72–130].

———. 1916a. Die Grundlage der allgemeinen Relativitätstheorie, *Annalen der Physik* 49, 769–822; reprinted in [*CPAE*, 6,.283–339].

———. 1916b. Hamiltonsches Prinzip und allgemeine Relativitätstheorie, *Königlich Preußische Akademie der Wissenschaften (*Berlin*). Sitzungsberichte* (1916), 1111–1116; reprinted in [*CPAE*, 6, 409–416].

———. 1917a. Kosmologische Betrachtungen zur allgemeinen Relativitätstheorie, *Königlich Preußische Akademie der Wissenschaften (Berlin). Sitzungsberichte*, 142–152; reprinted in [*CPAE*, 6,.540–552].

———. 1917b. *Über die spezielle und die allgemeine Relativitätstheorie. (Gemeinverständlich.)* Braunschweig: Vieweg; reprinted in [CPAE 6, 420–539].

———. 1919a. Spielen Gravitationsfelder im Aufbau der materiellen Elementarteilchen eine wesentliche Rolle?, *Königlich Preußische Akademie der Wissenschaften (Berlin). Sitzungsberichte*, 349–356.

———. 1919b. Time, Space, and Gravitation, *The Times* (London), 28 November 1919; reprinted in [*CPAE*, 7, 212–215].

———. 1996. *Einstein's 1912 Manuscript on the Special Theory of Relativity. A Facsimile.* English translation by Anna Beck. George Braziller.

Fölsing, Albrecht. 1993. *Albert Einstein. Eine Biographie*. Frankfurt am Main: Suhrkamp.

Goenner, Hubert, et al., eds. 1999. *The Expanding World of General Relativity*, Einstein Studies. Vol. 7. Boston: Birkhäuser.

Hilbert, David. 1893. Über die vollen Invariantensysteme. *Mathematische Annalen* 42: 313–373; Reprinted in [Hilbert 1933, 287–344].

———. 1896. Über die Theorie der algebraischen Invarianten, *Mathematical Papers Read at the International Mathematical Congress Chicago 1893*, New York: Macmillan, 116–124; reprinted in [Hilbert 1933, 376–383].

———. 1897. Die Theorie der algebraischen Zahlkörper. *Jahresbericht der Deutschen Mathematiker-Vereinigung* 4: 175–546; Reprinted in [Hilbert 1932, 63–363].

———. 1901. Mathematische Probleme, *Archiv für Mathematik und Physik*, 44–63 and 213–237; Reprinted in [Hilbert 1935,. 290–329].

———. 1910. Hermann Minkowski, *Mathematische Annalen* 68, 445-471; Reprinted in [Hilbert 1935, 339–364].

———. 1932, *Gesammelte Abhandlungen*, 1, Berlin: Springer.

———. 1933, *Gesammelte Abhandlungen*, 2, Berlin: Springer.

———. 1935, *Gesammelte Abhandlungen*, 3, Berlin: Springer.

Klein, Felix. 1926/1927. *Vorlesungen über die Entwicklung der Mathematik im 19. Jahrhundert*. 2 Bde. ed. Berlin: Springer.

Laue, Max von. 1955. *Trägheit und Energie, Philosophen des 20. Jahrhunderts, Albert Einstein*, 364–388. Stuttgart: Kohlhammer.

Lorentz, H. A. 1929. *Die Relativitätstheorie für gleichförmige Translationen (1910–1912)*, bearbeitet von A. D. Fokker, Leipzig: Akademische Verlagsgesellschaft.

Lorentz, H.A., et al. 1952. *The Principle of Relativity: A Collection of Original Memoirs on the Special and General Theory of Relativity*. New York: Dover.

Minkowski, Hermann. 1896. *Geometrie der Zahlen*. Leipzig: Teubner.

———. 1905. Peter Gustav Lejeune Dirichlet und seine Bedeutung für die heutige Mathematik, *Jahresbericht der Deutschen Mathematiker-Vereinigung*, 14, 149–163; Reprinted in [Minkowski 1911, vol. 2, 447–461].

———. 1908. Die Grundgleichungen für die elektromagnetischen Vorgänge in bewegten Körpern, *Nachrichten von der Gesellschaft der Wissenschaften zu Göttingen, Mathematisch-Physikalische Klasse*, 53–111; reprinted in [Minkowski 1911, vol. 2, 352–404].

———. 1909. Raum und Zeit, *Physikalische Zeitschrift* 10, 104–111; reprinted in [Minkowski 1911, vol. 2, 431–444].

———. 1911. In *Gesammelte Abhandlungen*, ed. D. Hilbert, vol. 2. Leipzig: Teubner.

———. 1973. In *Briefe an David Hilbert*, ed. Lily Rüdenberg and Hans Zassenhaus. New York: Springer.

Pauli, Wolfgang. 1921. Relativitätstheorie, *Encyklopädie der mathematischen Wissenschaften*, VI.2, 539–775.

Reid, Constance. 1970. *Hilbert*. New York: Springer.

———. 1976. *Courant in Göttingen and New York: the Story of an Improbable Mathematician*. New York: Springer.

Renn, Jürgen, ed. 2007. *The Genesis of General Relativity*. Vol. 4. Dordrecht: Springer.

Rowe, David. 1989. Klein, Hilbert, and the Göttingen Mathematical Tradition. *Osiris* 5: 186–213.

———. 1999. The Göttingen Response to General Relativity and Emmy Noether's Theorems. In *The Symbolic Universe. Geometry and Physics, 1890–1930*, ed. Jeremy Gray, 189–234. Oxford: Oxford University Press.

Schütz, I. R. 1897. Prinzip der absoluten Erhaltung der Energie, *Nachrichten von der Gesellschaft der Wissenschaften zu Göttingen, Mathematisch-Physikalische Klasse*, 110-123.

Walter, Scott. 1999. Minkowski, Mathematicians, and the Mathematical Theory of Relativity, in [Goenner et al. 1999, 45–86].

———. 2007. Breaking in the 4-Vectors: the Four-Dimensional Movement in Gravitation, 1905–1910, in [Renn 2007, vol. 3, 193–252].

———. 2008. Hermann Minkowski's Approach to Physics. *Mathematische Semesterberichte* 55: 213–235.

Wangerin, Alfred, ed. 1909. *Verhandlungen der Gesellschaft Deutscher Naturforscher und Ärtzte. 80. Versammlung zu Cöln, 20.—26. September 1908*. Leipzig: Vogel.

Weyl, Hermann. 1918. Gravitation und Elektrizität, *Königlich Preußische Akademie der Wissenschaften (Berlin). Sitzungsberichte* 465–478: 478–480.

———. 1944. David Hilbert and his Mathematical Work. *Bulletin of the American Mathematical Society* 50: 612–654; Reprinted in [Weyl 1968, vol. 4, 130–172].

———. 1968. In *Gesammelte Abhandlungen*, ed. K. Chandrasekharan, vol. 4. Berlin: Springer.

19 Max von Laue's Role in the Relativity Revolution

(*Mathematical Intelligencer* 30(3)(2008): 54–60)

Whereas countless studies have been devoted to Einstein's work on relativity, the contributions of several other major protagonists have received comparatively little attention. Within the immediate German context, no single figure played a more important role in developing the consequences of the special theory of relativity (SR) than Max von Laue (1879–1960). Although remembered today mainly for his discovery of x-ray diffraction in 1912 – an achievement for which he was awarded the Nobel Prize – Laue's accomplishments in promoting the theory of relativity were of crucial importance.[1] They began early, well before most physicists even knew anything about a mysterious Swiss theoretician named Einstein (Fig. 19.1).

As a student of Max Planck in Berlin, Laue was one of the first to appreciate the novelty and significance of Einstein's fundamental paper "On the Electrodynamics of Moving Bodies" (Einstein 1905). Following Planck's advice he decided to visit Bern during the summer of 1907 to make the acquaintance of this barely known author. When he learned that Einstein was not a member of the faculty at the University of Bern but rather a mere patent clerk, he was more than a little astonished. Laue made his way to the Post and Telegraph building, where Einstein and his colleagues fingered strange gadgets and examined patent applications all day. Waiting for Einstein in the reception area, he was told to go down the corridor where Einstein would be coming from the opposite direction to meet him. Laue then got a second surprise, as he later recalled: "I did as told, but the young man who came toward me made so unexpected an impression on me that I did not believe he could be the father of the relativity theory, so I let him pass" (Seelig 1960, 130). After Einstein returned from the reception room, Laue went back to introduce himself. Still, the really big surprise came when he heard what this young patent clerk had to say about the state of modern physical research. "During the first two hours of our conversation," Laue recalled, "he overthrew everything in mechanics and electrodynamics" (Fölsing 1993, 240). Thus began a lifelong friendship.

Laue soon became convinced that Einstein's principle of relativity provided theoretical physics with a new foundation that promised to unite the two great theories of the day. Maxwellian electrodynamics and Newtonian mechanics had reached an impasse it seemed, and yet with one stroke Einstein had established a new foundation for both by creating a new relativistic kinematics that he claimed was valid for all physical phenomena in all inertial frames. One of the earliest and most important triumphs for Einstein's special theory came in 1907 when Laue showed how the Fresnel drag coefficient could easily be derived as a kinematic effect using Einstein's formula for the addition of velocities for parallel moving frames. This derivation was so natural, in fact, that Laue was taken aback when he realized Einstein had overlooked such a fundamental result. But after conducting a thorough search of the literature, beginning with Einstein's 1905 paper, he convinced himself that his derivation was indeed new, and so he wrote it up for publication in the *Annalen der Physik* (Laue 1907).

Four years later came Laue's single most important contribution to relativity, his monograph *Das Relativitätsprinzip* (Laue 1911). This volume went through numerous revised editions, the last of which appeared in 1955. Following Minkowski's lead, Laue herein developed a 4-dimensional Lorentz-invariant electrodynamics, which he then used as a foundation for his relativistic dynamics based on a general world tensor. Laue's approach provided Einstein with a whole new theoretical Ansatz, one that seems to have had little in common with Einstein's original viewpoint, which was linked with Lorentz's theory of the electron. Furthermore, Laue's formulation provided Einstein with some of the essential physical and mathematical concepts he would thereafter employ in his search for gravitational field equations.

After assimilating the essential insights Laue brought forth in the second edition of *Das Relativitätsprinzip* (Laue

[1]For a recent account of his career with special attention to Laue's importance for rebuilding German science after the Second World War, see (Zeitz 2006).

D.E. Rowe, *A Richer Picture of Mathematics*, https://doi.org/10.1007/978-3-319-67819-1_19

Fig. 19.1 Max von Laue.

1913), Einstein was off again on his own track. Up until his departure for Berlin in the spring of 1914, he worked closely with his Swiss friends Marcel Grossmann and Michele Besso on this new theory of gravitation, which later became known as the general theory of relativity. Eight years later, Max von Laue published a companion volume on general relativity (Laue 1921), noteworthy for being the first advanced textbook on this subject written by a theoretical physicist. Though its impact was hardly comparable to his text on special relativity, it nevertheless offers important clues to Laue's understanding of Einsteinian gravitation and also for appreciating the climate of reception in Germany during these tumultuous years. As such, it too represents a document of central importance for understanding the relativity revolution. I will briefly discuss both of these volumes below, trying to indicate their significance as guides to important developments in which Laue and others built upon Einstein's ideas. But let me begin with a few words about how the not yet famous Albert Einstein viewed young Max Laue.

Before returning to Zurich's ETH after a year in Prague, Einstein was asked to comment on a number of candidates for a professorship in theoretical physics at the university. Writing to Alfred Kleiner, he characterized Max Born as a "good calculator" who had not yet "demonstrated much acumen for physical matters." Born's work on relativity followed in the wake of his former mentor, Hermann Minkowski, whose formalism Einstein initially found both unwieldy and unnecessary. Einstein was far more enthusiastic when it came to Laue, calling him simply "the most important of the younger German theoreticians." He also praised Laue's book on relativity as "a real masterpiece, much of it being his own intellectual property" (Einstein to Kleiner, 3 April 1912, *CPAE 4*. 1995, 445). Needless to say, Max Laue got the job, and his stay in Zurich virtually coincided with Einstein's own. By the spring of 1914, when Einstein joined the Prussian Academy, Laue was already on his way to the newly opened University of Frankfurt. After the war, he would rejoin Einstein in Berlin, where both men were on close terms with Max Planck.

Their brief time together in Zurich, however, was of crucial importance for Einstein's next bold steps forward. This nexus of events has somehow escaped notice in the historical literature. In Zurich Laue continued work on the revised second edition of his textbook on special relativity (Laue 1913). We can be sure that he kept Einstein fully abreast of the novelties it contained, though no documentary evidence of their conversations from this period has survived. Still, some striking clues can be found in a long-forgotten manuscript on special relativity that first surfaced in 1995 with the publication of the fourth volume of *The Collected Papers of Albert Einstein*, (*CPAE 4*. 1995, 3–108). Einstein wrote this text at the request of the Leipzig physicist, Erich Marx, who hoped to have a contribution on relativity theory from Einstein for his *Handbuch der Radiologie*. While the precise circumstances remain obscure, it appears that Einstein worked on this article off and on in Prague and Zurich from 1912 to 1914, producing a 72-page text that I will refer to as his Marx manuscript.

Einstein's Obsolete Account of SR

Publication plans for this manuscript were interrupted by the war, during which time the full-blown general theory of relativity emerged. Afterward Einstein decided to withdraw permission to publish his by now dated text, considering it to be scientifically obsolete. So after much pleading and various aborted plans Erich Marx finally gave up. After he died in 1956, his relatives eventually salvaged the long-forgotten manuscript, which represents Einstein's most detailed presentation of the special theory of relativity. In 1987, the family put the manuscript up for auction, fetching a tidy $1.2 million, twice as much as Kafka's letters to his fiancée, which were auctioned that same year. Eight years later, the anonymous owner put the manuscript up for auction again. This time the bidding reached $3.3 million, but the experts at Sotheby's thought an original 72-page manuscript by Einstein should have brought at least $4 million if not $6. So the owners refused the bid, and soon thereafter the press reported that Einstein's lost relativity manuscript was purchased for an undisclosed amount by the Jacob E. Safra Foundation. This institution then donated it to the Israel Museum in Jerusalem, which published a very handsome facsimile edition in 1996 (Einstein 1996).

This by now not so unusual convergence of intellectual and commercial interests was accompanied by a certain amount of media hype of the kind by now long associated with Einstein's famous name. On the eve of the Sotheby's auction, a reporter for the *New York Times* tried to dramatize the event by predicting that when the hammer came down Einstein's manuscript was likely to go for more than a recently auctioned version of Monet's "Water Lilies." This presumably more accessible work of art no doubt lacked those qualities singled out by the *Times* reporter, who noted that:

> Its value lies as much in its form as in its substance. In perhaps the manuscript's most striking example of Einstein's scientific gymnastics, he takes the equation $EL = mc^2$ and crosses out the "*L*," thus rendering the historic special theory of relativity – energy equals mass times the square of the speed of light – right before the reader's eyes" ("Einstein Manuscript up for Auction shows Science can be Art," *New York Times,* 15 March, 1996).

Apparently this journalist thought that *EL* was an abbreviation for electricity, so on this interpretation Einstein's brilliant insight was to have recognized that he should have written only an *E* instead. However inane this explanation may be, the Israel Museum chose this altered equation as the logo for its 1996 facsimile edition.

Since this time, Einstein scholars have made very careful study of two other unpublished manuscripts of signal importance for the crucial period 1912 to 1914. Like the Marx manuscript, both were published for the first time in volume 4 of the Einstein edition. Einstein's Zurich Notebook has subsequently received exhaustive analysis, thanks to the efforts of a research group at the Max Planck Institut für Wissenschaftsgeschichte in Berlin led by Jürgen Renn (see (Renn 2007, vols. 1, 2)). Likewise, Michel Janssen undertook a careful study of the Einstein-Besso manuscript, which contains a vain attempt to derive the perihelion of Mercury from the Einstein-Grossmann field equations (*CPAE 4*. 1995, 344–474). Both of these documents constitute working notes that were obviously never intended for publication. For historians, however, they represent significant markers along the difficult road that Einstein traveled before his breakthrough in the fall of 1915, when he cast aside that first set of field equations in favor of the generally covariant equations that now bear his name.

The Marx manuscript has an entirely different character. Moreover, unlike these other two documents, it has been virtually ignored by the very scholars who have taken such care to reconstruct Einstein's tortuous path from special to general relativity. It is not hard to understand why. Intended as an expository article on special relativity, it contains virtually nothing original, though it does offer a clear picture of how Einstein saw the theory at this time. My contention, however, is that the Marx manuscript really *does* constitute an important historical document, despite the fact that it contains no new groundbreaking results. Rather, what it offers us is a picture of Einstein catching up with what his contemporaries had been doing during the period from 1908 to 1912.

One should keep in mind that between 1908 and 1911 Einstein published almost nothing on relativity. In 1907, however, Planck made an important breakthrough by associating a momentum density with *any* energy flow (elastic, heat, chemical, gravitational). Soon afterward, Minkowski's work inspired the development of a new framework for integrating relativistic physics. Then, in 1910 Sommerfeld published a vector analysis for SR based on the Lorentz group, and one year later Laue brought out the first edition of his textbook (Laue 1911) containing major theoretical results based on the work of Minkowski and Sommerfeld.

In 1912, when Einstein started writing the Marx manuscript, he was also preparing the ground for a fresh new attack on the problem of gravitation.[2] For this purpose, he began exploring the possibilities of using the Ricci calculus for creating a generalized theory of relativity. The timing for all this would appear quite crucial. As it turns out, the Marx manuscript contains a textual emendation that helps pinpoint the link between Einstein's consolidation of Minkowskian relativity and the new mathematical formalisms he was learning in order to generalize this theory to cover arbitrary frames of reference.

Soon after arriving in Zurich on 25 July 1912, Einstein obtained new paper and ink. This circumstance makes it possible to distinguish that portion of the text written earlier in Prague from the pages he composed in Zurich. According to Einstein's own testimony, his return marked the beginning of the last phase in his struggle to incorporate gravitation into a generalized theory of relativity (Einstein 1956, 15–16). Soon thereafter his friend Marcel Grossmann introduced him to the general methods of what came to be called the tensor calculus, the crucial tool required in order to deal with non-inertial frames. At this very same time, Einstein crossed out his definition of four-vectors, written in Prague, and began anew using a darker ink and heavier paper on which he began

[2]Despite many years of intense efforts to reconstruct Einstein's intellectual journey in detail, leading experts still disagree about some of the key problems he had to overcome along the way. See the commentary and essays by Michel Janssen, John Norton, Jürgen Renn, Tilman Sauer, and John Stachel in (Renn 2007, vol. 2).

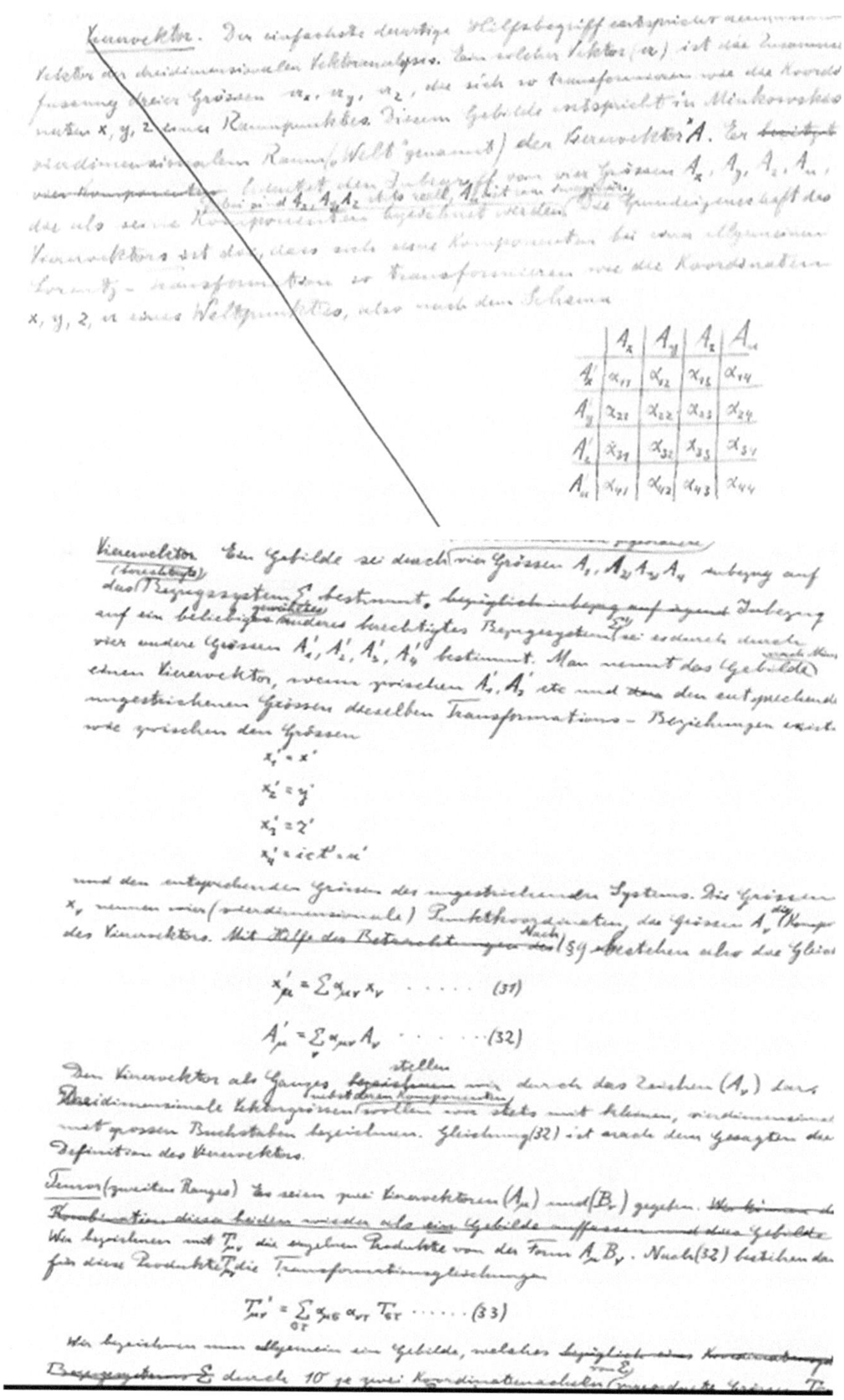

Fig. 19.2 Einstein's revision of his Marx manuscript introducing the language of tensors.

writing a rather lengthy exposition of the relevant tensorial concepts and operations for special relativity (Einstein 1996, 135, 137) (Fig. 19.2):

New Math for Physicists

Historians of physics have long recognized that with the inception of relativity theory German theoretical physics underwent a profound transformation marked by a strong infusion of mathematical techniques and ideas. Jungnickel and McCormmach, in their magisterial *Intellectual Mastery of Nature* (Jungnickel and McCormach 1986), describe the earlier process of discipline formation, showing how theoretical physics emerged around 1890 just as the German mathematicians were setting down disciplinary boundaries by founding their national society, the Deutsche Mathematiker-Vereinigung. Soon thereafter, two young experts in number theory, Minkowski and Hilbert, agreed to produce a report on developments in this field for the DMV. Later, however, they grew restless and began sniffing around for open problems within the terrain earlier occupied jointly by mathematicians and physicists.

This mathematicians' quest soon turned into something like an imperialist campaign, especially for Hilbert. After Minkowski's sudden death in 1909, he began regularly inviting leading theoreticians to deliver special Wolfskehl lectures in Göttingen. Beginning with Poincaré and Lorentz, but culminating in 1922 with the famous Bohr Festspiel, practically every famous figure took this opportunity to address some of the most pressing recent developments. In Einstein's case, his six lectures on the new theory of gravitation mark the beginning of the dramatic turning point that reached its crest in November 1915 and led Einstein to his first triumphs in the general theory of relativity.[3]

Something truly remarkable took place in this Göttingen setting. Hilbert, who dominated these proceedings, was even among mathematicians the purest of the pure. And of course he knew full well that many viewed his forays into physics with suspicion, perhaps even disdain. He took most such criticism in stride, letting his critics know that "physics had become too difficult for the physicists." That oft repeated quip conveyed an essential part of Hilbert's personality. Whenever repeated, it tended to conjure up – among those who knew the source – an instant image of the man whose easy mixture of self-confidence and disrespect for established norms of thought and behavior were legend. Hilbert, that "Pied Piper of Göttingen," helped instill a collective hubris within the community of young talent that flocked there. Most who belonged to the small elect with close ties to the master and his wife Käthe identified with this mentality and shared a sense of superiority over outsiders.

These atmospherics should certainly be kept in mind when thinking about the careers of figures like Max Born, Richard Courant, and Hermann Weyl, despite their divergent trajectories. All three were very familiar with another saying that Hilbert was fond of repeating: "Das Wissen kennt keine Fächer" ("Knowledge knows no disciplinary boundaries"). Relativity theory was, of course, a prime example of what he had in mind, though the interplay between mathematical and physical conceptions was and is by no means easy to trace. Only recently, in fact, have any really detailed studies shed much light on how this cross-fertilization between disciplines took place.[4]

Scott Walter has carefully studied how physicists reacted to these new mathematical methods, and his findings indicate that this was anything but a royal road (see (Walter 2007)). Those familiar with the even messier historical process that led to 3-dimensional vector analysis should not be surprised. Still, the rhetoric of pre-established harmony or Wigner's image of abstract mathematics as "unreasonably effective" for modern physics requires many more focused studies like Walter's "Breaking in the 4-vectors" (Walter 2007), which shows quite clearly how fragile Minkowski's project really was. Had Arnold Sommerfeld not come along to reconcile his 4-dimensional formalism with the by now standardized operations of the 3-D vector calculus, the history of relativity might have taken a quite different turn. Instead, Laue and then Einstein could embrace this new approach without which, as Einstein later said, relativity would have remained in diapers (Einstein 1917, 39; *CPAE 6*. 1996, 463).

Laue's Influence on Einstein

But it was Laue who first saw the fertility of Minkowski's *physical* conceptions – and not just his mathematical tools – for realizing Einstein's fundamental program from 1905, namely to show that *all* the laws of physics can be expressed in Lorentz-covariant form. Laue's approach became familiar to Einstein through the 1911 edition of *Das Relativitätsprinzip*, although the discussion there is still rather sketchy. However, in the second edition (Laue 1913) – which appeared when Laue and Einstein both taught in Zurich – we find a full-blown treatment that shows how Laue's macroscopic *Ansatz* leads to sharp microscopic conclusions with regard to the mechanical and electromagnetic properties of the electron. Working next door, Laue and Einstein were obviously in contact with one another during this time. We also know the significance Einstein attached to the line of ideas developed in the closing section of Laue's text on relativistic dynamics (Laue 1913, 174–253). In the Marx manuscript, Einstein gave presented similar ideas in a much abbreviated form. Like Laue, he began with a discussion of the general form of the momentum-energy law in electrodynamics as introduced by Minkowski using the symmetric tensor $T_{\mu\nu}$:

$$T_{\mu\nu} = \begin{pmatrix} p_{xx} & p_{xy} & p_{xz} & icg_x \\ p_{yx} & p_{yy} & p_{yz} & icg_y \\ p_{zx} & p_{zy} & p_{zz} & icg_z \\ \frac{i}{c}s_x & \frac{i}{c}s_y & \frac{i}{c}s_z & -\eta \end{pmatrix}$$

Its spatial components are the Maxwell stresses, whereas the symmetric space-time components link two fundamental physical entities: the 3-vector g representing momentum density and the 3-vector s which represents the energy flow. The pure time scalar denotes the negative energy density.[5] The notation I have used here is identical to that found on page 62 of Einstein's original manuscript (Einstein 1996,

[3]This interpretation of these events is advanced in (Rowe 2001, 2004).

[4]For a detailed examination of Hilbert's interests in physics see (Corry 2004).

[5]For a discussion of Laue's ideas and their historical importance, see (Norton 1992).

167), and this is nearly exactly the same as the notation found on page 182 of the 1913 edition of Laue's text. Moreover, no corresponding array appears in the 1911 edition, which suggests that Einstein either took this formulation from the second edition or, perhaps more likely, was made aware of this line of development by Laue himself.[6]The comments Einstein provides in the manuscript are, here as elsewhere, very terse. Nevertheless, they make abundantly clear that he attached great significance to this new 4-D formulation due to Minkowski. He thus notes that the symmetry conditions lead directly to the equation

$$\vec{g} = \frac{1}{c^2}\vec{s},$$

which is "closely related to the circumstance that, according to the theory of relativity, an inertial mass must be ascribed to energy. For this entails that the energy flow is always associated with a momentum" (Einstein 1996, 168). Planck had been the first to point this out, but here this fundamental physical relationship becomes naturally embedded in the mathematical formalism.

From here, Einstein goes on to take the divergence of $T_{\mu\nu}$, obtaining thereby a four-vector that represents the force and energy that the electromagnetic field delivers to charged bodies. He then passes over to the general dynamical situation, as elucidated by Laue in the 1913 edition of his textbook. Here is what Einstein wrote about the significance of this framework for special relativity:

> The general validity of the conservation laws and of the law of the inertia of energy . . . suggests that the relations [deduced for electrodynamics] are to be ascribed a *general* significance, even though they were obtained in a very special case. We owe this generalization, which is the most important new advance in the theory of relativity, to the investigations of Minkowski, Abraham, Planck, and Laue. To every kind of material process we might study, we have to assign a symmetric tensor $T_{\mu\nu}$, the components of which have the physical meaning indicated [by the schema given above] [Einstein 1996, 168].

Had Einstein actually published this statement at the time he wrote it, there can be little doubt that it would have received prominent attention in the historical literature.[7] Instead, the rather misleading impression has arisen that the geometry of a curved spacetime was Einstein's nearly sole preoccupation in his search for a generalized theory of relativity. Clearly this was a central concern, but one should not forget the right-hand-side of the gravitational field equations. For without a general stress-energy tensor Einstein could never have begun to envision the possibility of generally covariant field equations. This alone should make clear that the link Einstein was able to make to the special relativistic framework for dynamics – as set forth in (Laue 1913) – was of the utmost importance for his quest to establish a general theory of relativity.

Laue's Slow Acceptance of General Relativity

As a leading expert in optics, Max von Laue had been one of the first to accept special relativity and to pursue its consequences. But like many other theoretical physicists, including Max Planck, he found it difficult to accept the premises of Einstein's general theory of relativity. Initially, Laue rejected Einstein's equivalence principle out of hand after giving due consideration to the empirical implications Einstein drew from it in 1911. On 27 December, 1911, Laue wrote to Einstein:

> I have now carefully studied your paper on gravitation and have also lectured about it in our colloquium [Arnold Sommerfeld's colloquium in Munich]. I do not believe in this theory because I cannot concede the full equivalence of your systems *K* and *K′* . After all, a body causing the gravitational field must be present for the gravitational field in system *K*, but not for the accelerated system *K'* [*CPAE* 5. 1993, 384].

When Laue changed his mind is not very clear, but he showed mounting interest after the initial triumphs of November 1915, when Einstein was able to account for the 43 s of missing arc in Mercury's perihelion, and especially the spectacular British confirmation of his prediction for light deflection in the vicinity of the sun's gravitational field. It was only after this latter event in November 1919 that Laue took up GRT in earnest. Although he had long been the most outspoken defender of Einstein's theory of relativity among theoretical physicists in Germany, he always did so in a dignified way, ignoring the polemical language of the anti-relativists.[8] His 1921 textbook on general relativity provides just one of many such examples.

Seldom has an author taken such pains to describe the audience for whom he has written, beginning with a few references to the popular literature on relativity. Besides Einstein's booklet (Einstein 1917) Laue called attention to a book by Paul Kirchberger with the alluring title, *Was kann man ohne Mathematik von der Relativitätstheorie verstehen?* In fact, he even wrote a preface to this volume. Leaping to the other end of the spectrum, Laue called attention to Hermann Weyl's classic monograph, *Raum-Zeit-Materie*, a work that even mathematicians could hardly read with ease.

[6]As Scott Walter pointed out to me, however, another likely possibility is that Einstein learned about these developments already in 1911 by reading Laue's "Zur Dynamik der Relativitätstheorie," *Annalen der Physik* 35: 524–542; the matrix given above appears there on p. 529.

[7]Scott Walter pointed out to me that this passage is also cited in (Janssen and Mecklenburg 2006).

[8]On Laue's role in these debates, see (Beyerchen 1977), (Hentschel 1990), and (Rowe 2006).

Finally, for those seeking a sophisticated philosophical analysis, Laue recommended Ernst Cassirer's *Zur Einsteinschen Relativitätstheorie*.. He then went on to write: "But up till now there has been no book written by a physicist that is both rigorously scientific and fairly complete; and we contend that only a physicist can truly comprehend and try to remove those difficulties that have left the majority of his colleagues in the dark regarding the general theory of relativity" (Laue 1921, v).

Laue also noted that this was not a book for everyone, and most certainly not for Einstein's outspoken opponents, led by Philipp Lenard. He referred to these anti-relativists as a group of "in part very important men" who rejected relativity for reasons not unlike Goethe's attitude toward Newtonian optics. Still, Laue was convinced that most German physicists had taken no clear position largely because they lacked familiarity with non-Euclidean geometry and the tensor calculus. He offered them a whole chapter on Gaussian curvature and projective geometry, mathematical topics that went well beyond the physicists immediate needs. His justification for this was simple: a theoretician who knows only what he absolutely needs is a physicist who knows too little.

Laue clearly had learned a great deal, and he ended his preface by thanking a whole series of mathematicians – Ludwig Bieberbach, Friedrich Schur, Georg Hamel, and Emil Hilb – for the personal assistance they gave him (Laue 1921, vii). He particularly acknowledged the debt he owed to David Hilbert, who lent him a copy of his lectures on general relativity from 1916–17. Laue's book contains numerous references to Hilbert's brief foray into this field, far more than can be found in other contemporary sources likes Weyl's *Space-Time-Matter* or Pauli's article for the *Encyklopädie der mathemaischen Wissenschaften.* A striking example is his discussion of the Schwarzschild metric, in which he describes the trajectories of test particles replicating the figure found in Hilbert's lecture notes.[9]

During the ensuing controversies that swirled around Einstein and his theory, Laue defended both the man and his ideas against the attacks of anti-relativists. But he also criticized pro-relativists, in particular Max Born, for promoting the new "Einstein cult" (Born 1969, 67). Laue was present at the sensationalized meeting held at the Berlin Philharmonic Hall in August 1920 when a small group of anti-relativists launched a politically motivated attack that nearly caused Einstein to leave Germany. It was Laue who alerted Arnold Sommerfeld, the newly elected president of the German Physical Society, that a major scandal was brewing that threatened to rip their fragile community apart. Laue was shocked by the crudity of this pseudo-scientific gathering, at which the opening speaker called the theory of relativity "scientific Dadaism" (Rowe 2006). Little did he know that Einstein's now famous visage had been on prominent public display in Berlin as part of a collage featured at the recently held exhibition of Dada art (see Fig. 19.3).

When it became too dangerous for Einstein to appear at the 1922 centenary Naturforscher meeting in Leipzig, he asked Laue to speak in his place. Lenard and his followers responded by circulating a flyer protesting the celebration of a theory anti-relativists viewed as non-scientific. Their protest had clear political overtones, coming only a few months after the assassination of foreign minister Walther Rathenau, who like Einstein was targeted as an enemy of the German people. The rabid nationalists eventually felt vindicated when the Nazis swiftly destroyed the last vestiges of democracy in the Weimar Republic, vilifying all those who dared to oppose them. The proponents of Deutsche Physik, led by Lenard and Johannes Stark, now stood poised to assume scientific power. They were determined to see that the Berlin clique – Planck, Laue, Haber, and co., who had embraced Einstein and his despised theory of relativity – would henceforth play a minor role in German academic affairs.

In September 1933 – exactly 11 years after his relativity lecture in Leipzig – Laue opened the annual meeting of the German Physical Society in Würzburg by recalling the events surrounding Galileo's trial, which had taken place 300 years earlier.[10] He reminded his audience of the legendary words of defiance – "And yet, it moves!" ("Eppur si muove!") – the words supposedly uttered by Galileo after his recantation. A lovely myth, as Laue described, since it is both "historically unverifiable and intrinsically implausible—and yet it is ineradicable in common lore." Its power, at least for those who trusted and believed in scientific truth, was plain enough, and of course the Church utterly failed in its effort to stamp out Copernicanism.

Had he ended there, probably no one would have taken much notice. But Laue then alluded to the unfavourable scientific climate in Prussia under Friedrich Wilhelm I (1688–1740), the "Soldier-King" who laid the groundwork for the Prussian military tradition. Friedrich Wilhelm ran a clean ship of state that had no place for freethinkers like Christian Wolff, then Germany's leading natural philosopher. The king not only forced Wolff out of his professorship in Halle, he gave him just 24 h to leave Prussia altogether (Wolff took up a post in Marburg). After Friedrich II ascended to the throne, he granted Wolff amnesty and allowed him to return to his chair in Halle. Every educated German was surely familiar

[9]For a comparison of Laue's diagram of the trajectories with the one in Hilbert's lecture notes, see Chap. 21.

[10]Laue's remarks can be found in English translation in (Hentschel and Hentschel 1996, 67–71).

Fig. 19.3 Hannah Höch's collage, *Cut with the kitchen knife through the beer-belly of the Weimar Republic*, was part of the Dada exhibition on display in Berlin in 1920. Left, the artist with Raoul Hausmann (Photo by Robert Sennecke, Hannah-Höch-Archiv, Berlin).

with this story and its simple moral: "Yet in the face of all the repression," Laue concluded, "the supporters of science could stand steadfast in the triumphant certainty expressed in the modest phrase: And yet, it moves!" (Hentschel & Hentschel 1996, 71).

Lenard and Stark were incensed by this open provocation, but Laue had chosen his words carefully and had left the anti-relativists little room to attack him as politically unreliable. His enemies tried to denounce him in memoranda sent to political authorities, but these efforts proved futile: the so-called Einstein clique was not so vulnerable after all.

Acknowledgments An earlier version of this paper was presented in November 2007 at the annual meeting of the History of Science Society, held in Arlington, Virginia. This was part of a session organized by Scott Walter, "Beyond Einstein: Contextualizing the Theory of Relativity." My thanks go to him not only for extending an invitation to speak on that occasion but also for his valuable comments and critique of that earlier paper.

References

Beyerchen, Alan. 1977. *Scientists under Hitler: Politics and the Physics Community in the Third Reich*. New Haven: Yale University Press.

Born, Max, ed. 1969. *Albert Einstein/ Max Born. Briefwechsel, 1916–1955*. Munich: Nymphenburger.

Corry, Leo. 2004. *David Hilbert and the Axiomatization of Physics (1898–1918): From Grundlagen der Geometrie to Grundlagen der Physik*. Dordrecht: Kluwer.

CPAE 4. 1995. *Collected Papers of Albert Einstein*, vol. 4: *The Swiss Years: Writings, 1912–1914*, Martin J. Klein, *et al.*, eds., Princeton: Princeton University Press.

CPAE 5. 1993. *Collected Papers of Albert Einstein*, Vol. 5: *The Swiss Years: Correspondence, 1902–1914*, Martin J. Klein, *et al.*, eds., Princeton: Princeton University Press.

CPAE 6. 1996. *Collected Papers of Albert Einstein*, Vol. 6: *The Berlin Years: Writings, 1914–1917*, A. J. Kox, *et al.*, eds., Princeton: Princeton University Press.

Einstein, Albert. 1905. Zur Elekrtrodynamik bewegter Körper. *Annalen der Physik* 17: 891–921.

———. 1917. *Über die spezielle und die allgemeine Relativitätstheorie*. (Gemeinverständlich.) Braunschweig: Vieweg; reprinted in CPAE 6. 1996, 420–539.

———. 1956. Autobiographische Skizze. In *Helle Zeit – Dunkle Zeit. In Memoriam Albert Einstein*, ed. Carl Seelig, 9–17. Zürich: Europa Verlag.

———. 1996. *Einstein's 1912 Manuscript on the Special Theory of Relativity. A Facsimile.* English translation by Anna Beck. George Braziller.

Fölsing, Albrecht. 1993. *Albert Einstein. Eine Biographie*. Frankfurt am Main: Suhrkamp.

Hentschel, Klaus. 1990. *Interpretationen und Fehlinterpretationen der speziellen und der allgemeinen Relativitätstheorie durch Zeitgenossen Albert Einsteins*. Basel: Birkhäuser.

Hentschel, Klaus, and Ann Hentschel, eds. 1996. *Physics and National Socialism. An Anthology of Primary Sources*. Basel: Birkhäuser.

Janssen, Michel, and Matthew Mecklenburg. 2006. From Classical to Relativistic Mechanics: Electromagnetic Models of the Electron. In *Interactions: Mathematics, Physics and Philosophy, 1860–1930*, ed. Jesper Lützen et al., 65–134. New York: Springer.

Jungnickel, Christa and McCormach, Russell. 1986. *Intellectual Mastery of Nature: Theoretical Physics from Ohm to Einstein*, vol. 2 (*The Now Mighty Theoretical Physics, 1870–1925*), Chicago: University of Chicago Press.

Laue, Max. 1907. Die Mitführung des Lichtes durch bewegte Körper nach dem Relativitätsprinzip. *Annalen der Physik* 23: 989–990.

———. 1911. *Das Relativitätsprinzip*. Braunschweig: Vieweg.

———. 1913. *Das Relativitätsprinzip*. 2te vermehrte Auflage ed. Braunschweig: Vieweg.

———. 1921. *Das Relativitätsprinzip. Zweiter Band: Die allgemeine Relativitätstheorie und Einsteins Lehre von der Schwerkraft*. Braunschweig: Vieweg.

Norton, John. 1992. Einstein, Nordström and the early Demise of Lorentz-covariant, Scalar Theories of Gravitation, *Archive for History of Exact Sciences*, 45: 17–94; reprinted in [Renn 2007, 413-488].

Renn, Jürgen, ed. 2007. *The Genesis of General Relativity*. Vol. 4. Dordrecht: Springer.

Rowe, David. 2001. Einstein meets Hilbert: At the Crossroads of Physics and Mathematics. *Physics in Perspective* 3: 379–424.

———. 2004. Making Mathematics in an Oral Culture: Göttingen in the Era of Klein and Hilbert. *Science in Context* 17 (1/2): 85–129.

———. 2006. Einstein's Allies and Enemies: Debating Relativity in Germany, 1916–1920. In *Interactions: Mathematics, Physics and Philosophy, 1860–1930*, ed. Jesper Lützen et al., 231–280. New York: Springer.

Seelig, Carl. 1960. *Albert Einstein. Leben und Werk eines Genies unserer Zeit*. Zürich: Europa Verlag.

Walter, Scott. 2007. Breaking in the 4-Vectors: the Four-Dimensional Movement in Gravitation, 1905–1910, in [Renn 2007, vol. 3, 193–252].

Zeitz, Katharina. 2006. *Max von Laue (1879–1960).Seine Bedeutung für den Wiederaufbau der deutschen Wissenschaft nach dem Zweiten Weltkrieg*. Wiesbaden: Franz Steiner.

Euclidean Geometry and Physical Space

20

(*Mathematical Intelligencer* 28(2)(2006): 51–59)

It takes a good deal of historical imagination to picture the kinds of debates that accompanied the slow process, which ultimately led to the acceptance of non-Euclidean geometries little more than a century ago. The difficulty stems mainly from our tendency to think of geometry as a branch of pure mathematics rather than as a science with deep empirical roots, the oldest natural science so to speak. For many of us, there is a natural tendency to think of geometry in idealized, Platonic terms. So to gain a sense of how late nineteenth-century authorities debated over the true geometry of physical space, it may help to remember the etymological roots of geometry: "geo" plus "metria" literally meant to measure the earth, of course. In fact, Herodotus reported that this was originally an Egyptian science; each spring the Egyptians had to re-measure the land after the Nile River flooded its banks altering the property lines. Among those engaged in this land surveying were the legendary Egyptian rope-stretchers, the "harpedonaptai" who were occasionally depicted in artwork relating to Egyptian ceremonials.

We are apt to smile when reading Herodotus's remarks, dismissing these as just another example of the Greek tendency to think of ancient Egypt as the fount of all wisdom. Herodotus was famous for repeating such lore, and here he was apparently confusing geometry with the science of geodesy, when the latter has little to do with the former; at least not anymore. We don't think of circles, triangles, or the five Platonic solids as real figures. They are far too perfect, the products of the mind's eye. Of course, there's still plenty of room for disagreement. A formalist will stress that geometrical figures are mere conventions or, at best, images we attach to fictive objects that have no purpose other than to illustrate a system of ideas ultimately grounded in undefined terms and arbitrary axioms. A modern-day Platonist would vehemently object to this characterization, which puts all too much emphasis on purely arbitrary constructions rather than conceiving of geometrical figures as idealized instantiations of perfect forms.

For those of you who might like to see what such a debate looks like around 1900, let me recommend that you read the correspondence between the philosopher Gottlob Frege and the mathematician David Hilbert (Gabriel et al. 1980). Their dispute began when Frege wrote Hilbert after reading the opening pages of Hilbert's *Foundations of Geometry* (Hilbert 1899), the book that helped make the modern axiomatic approach so fashionable. Although one of the founding fathers of modern logic, Frege simply could not accept Hilbert's notion that the fundamental concepts of geometry had no intrinsic meaning. Points, lines, and planes were not simply empty words for Frege; they were in some deep sense real and geometry was that body of knowledge, which uncovered the properties of real figures composed of them. Hilbert's claim that, at bottom, all that mattered was the consistency of a certain set of axioms was anathema in Frege's eyes. Most mathematicians, to the extent that they grasped what was at stake, sided with Hilbert in this debate. Over the course of the next three decades the status of the continuum nevertheless played a major role in the larger foundations debates between formalists and intuitionists. Those rather esoteric discussions, however, left traditional realist assumptions about the nature of geometrical knowledge behind. For despite their strong differences, the proponents of formalism and intuitionism were both guided by their respective visions of pure mathematics, independent of its relevance to other disciplines, like astronomy and physics.

So what I'd like to emphasize here is that a fundamental shift took place around 1900 regarding the status of geometrical knowledge. This reorientation was certainly profound, but it seems to have been rather quickly forgotten in the wake of other even more dramatic developments. Soon afterward Einstein's general theory of relativity would lead to a flurry of new discussions about the interplay between space, time, and matter. Leading mathematicians like Hilbert and Hermann Weyl became strong proponents of Einstein's ideas even as they sharply disagreed about epistemological issues relating to the mathematical continuum, a concept of central importance for the geometer (Corry 2004). Looking back to the 1920s, it would seem that the opposing views of formalists and intuitionists actually reflect distinctly modern

D.E. Rowe, *A Richer Picture of Mathematics*, https://doi.org/10.1007/978-3-319-67819-1_20

attitudes about the nature of geometrical knowledge that would have been scarcely thinkable prior to 1900. Up until then, geometry was always conceived as somehow wedded to a physical world that displayed discernible geometrical features. Take the developments that led to the birth of modern science in the seventeenth century. Anyone who studies the works of Copernicus, Kepler, Galileo, or Newton cannot help but notice the deep impact of geometry on their conceptions of the natural world. But the same can be said of Gauss, whose career ought to make us rethink what Herodotus wrote about the Egyptian roots of the science of geometry.

Gauss, after all, was not only a mathematician and astronomer, he was also a professional surveyor who at least occasionally waded through the marshy hinterlands of Hanover taking sightings in order to construct a net of triangles that would span this largely uncharted region. This work helped inspire a profound contribution to pure geometry: Gauss's study of the intrinsic geometry of surfaces. As I'll describe a bit later, this theory helped launch a theory of measurement in geometry that helped open the way to probing the geometry of space itself. Only about a half century earlier, two leading French mathematicians, Clairaut and Maupertuis, had studied the shape of the earth's surface, showing that it formed an oblate spheroid. As one moved northward, they discovered, the curvature of the earth flattened, just as Newtonian theory predicted. Maupertuis' celebrated expedition to Lapland brought him fame and the nickname of "the earth flattener" (Terrall 2002) (Fig. 20.1). It also provided the French Academy with stunning proof that Descartes' theory of gravity could not be right, thereby overcoming the last major bastion to Newtonianism in France. Thus precise measurements of the earth's curvature had already exerted a deep impact on modern science.

Fig. 20.1 Maupertuis, the "earth flattener," from a portrait by Robert Levrac-Tournières made in 1740.

In the 1820s, Gauss took the measurement of the earth as his point of departure for an abstract theory of surfaces that marks the birth of modern differential geometry. Gauss asked whether and how a scientist could determine the curvature of an arbitrary surface by means of measurements made only along its surface, thus without knowing anything at all about the way in which the surface might be embedded in space. In this pioneering work on the intrinsic geometry of surfaces he introduced a concept we today call Gaussian curvature. In one sense, this notion was a refinement of the classical notion of curvature introduced by Euler in the eighteenth century. In Euler's theory, there are two principal curvatures associated with each point of a surface. He obtained these by taking the surface normal at each point as the axis for a pencil of planes. Each of these planes cuts out a curve with a certain planar curvature at the given point. Euler then proved that fot two particular planes this planar curvature will be maximum or minimum, and furthermore that these two planes will always be perpendicular to one another. They thus determine the two principal directions with plane curvatures: κ_1, κ_2. These, however, are not intrinsic invariants of the surface, since they depend on knowing how the surface sits in the surrounding space. Remarkably, however, the product $\kappa_1 \cdot \kappa_2 = \kappa$ turns out to be an intrinsic invariant, as Gauss was able to prove in his

Theorema Egregium: If two surfaces are isometric then they have the same Gaussian curvature $\kappa_1 \cdot \kappa_2 = \kappa$ at corresponding points.

Cartographers had, of course, long ago realized that it was impossible to find an isometric mapping that projected a sphere – a surface of constant positive curvature – onto a plane. Thus something had to be sacrificed in representing the surface of the earth on a flat map, either by distorting the distances between points or the angles between curves. On a sphere the surface normal at each point passes through the sphere's center and each plane section determines a great circle. So $\kappa_1 = \kappa_2 = \frac{1}{r}$, and the Gaussian curvature is therefore just $\kappa = \kappa_1 \cdot \kappa_2 = \frac{1}{r^2}$. Since this scalar is an intrinsic

Fig. 20.2 Rope-stretchers carrying out a crop survey under the supervision of a scribe, dated ca. 1400 B.C (Photo courtesy of Jacques Livet (OsirisNet 2001)).

invariant, two spheres admit the same geometry if and only if their radii are equal.

To talk about curvature as an intrinsic property of a surface requires a careful reconsideration of concepts like the measure of distances between points and angles between curves. So let's briefly review the role of measurement in geometry historically. Consider the Pythagorean Theorem, a ubiquitous result familiar to many cultures and already found by Babylonian mathematicians more than a millennium before the time of Pythagoras. It tells us something fundamental about planar measurements in right triangles: the square on the hypotenuse equals the sum of the squares on the two other sides. Some have conjectured that the Egyptian harpedonaptai, whom Democritus once praised, used the converse of the Pythagorean Theorem to lay out right angles at the corners of temples and pyramids. The claimants suggest that these professional surveyors used a rope with 12 knots of equal length. By pulling the rope taut, they could form a 3-4-5 right triangle. It's a nice idea, but nothing more. Archaeologists can still measure the angles of Egyptian buildings, of course, but our access to the mathematical knowledge that lay behind the architectural splendours of ancient Egypt is highly limited. Papyri can easily disintegrate with time, and only two have been found that provide much insight into the mathematical methods of the time: the Rhind and Moscow papyri, which were presumably used as training manuals for Egyptian scribes. Neither contains anything close to the Pythagorean Theorem, but with such scanty evidence available it would surely be unwarranted to claim that the Egyptians were unfamiliar with it (Fig. 20.2).

Historians of Mesopotamian mathematics have been luckier; they have had plenty of source material available ever since it became possible to decipher the cuneiform tablets archaeologists began turning up a little more than a century ago. These tablets were prepared by using a wedge-shaped instrument to gouge out marks on a clay tablet, after which it was baked in an oven. A number of them contain sophisticated mathematics, written in the familiar base 60 number system that Greek astronomers like Ptolemy eventually adopted for angle measurements. Some of the Babylonian mathematical texts reveal not just a passing familiarity with the Pythagorean Theorem but even a masterful use of it for numerical computations, like approximating the value of $\sqrt{2}$ or calculating Pythagorean triples.

Still, whether justified or not, we tend to credit the Greeks with being the first to give a proof of this ancient theorem. They were the first culture to study mathematics as a pure science based on first principles and rigorous reasoning, and presumably the Pythagoreans knew how to prove the theorem named after the founder of their sect by the late sixth century. This would place this discovery at least two centuries before Euclid compiled the 13 books of his Elements (Heath 1956) around 300 B.C. But since nearly all information about early Greek mathematical texts is lost we can only speculate about the context of discovery; we know nothing about the original proof itself. What we can observe is that right triangles play a central role in Euclid's *Elements*. Right angles are even accorded a special place in the fourth of his five postulates, which asserts that all such angles are equal. So the geometer is entitled to transport them from place to place. Moreover, the Pythagorean Theorem and its converse appear as I. 47 and I. 48 (Heath 1956, I: 349–369), the two culminating propositions in Book I of the *Elements*. Thereafter Euclid makes use of it in several of the most important propositions of Books II and III.

Today we are accustomed to thinking of the Pythagorean Theorem as the centerpiece of a special type of geometry, the classical Euclidean model. We talk about various models for all kinds of geometries and spaces today, but probably few people sense that this is a distinctly modern way of thinking. Classically, geometry was always about figures in space, whereas space itself was never the object of study. One didn't go about thinking of different kinds of spaces, or even space in the plural. True, in the nineteenth century mathematicians studied different kinds of properties of figures in space. They differentiated between the metrical and projective properties of curves and surfaces, and they used calculus to study their differential properties. By mid-century they had even begun to leap by analogy into higher dimensions, and above all they liked to use complex numbers in connection with a mathematical realm of four dimensions. But to the extent they identified themselves as geometers, mathematicians drew their inspiration from phenomena they could somehow visualize or imagine in ordinary Euclidean space.

In the German tradition one spoke of *anschauliche Geometrie*, a term that doesn't really translate well into English. The popular textbook with this title by Hilbert and Cohn-Vossen was published in the United States as *Mathematics and the Imagination* (Hilbert and Cohn-Vossen 1965). It would be exaggerated to call this a "picture-book" approach to geometry, but without the visuals the book would certainly lose all its charm. Throughout much of Western history, the discipline of geometry was treated as closely linked with logic. But many experts appreciated that the *source* of geometrical knowledge owed nothing to logical rigor. The Hilbert and Cohn-Vossen text makes this abundantly clear; it seldom even presents a chain of arguments. The authors mainly just describe and explain what the reader is supposed to *see*. This kind of seeing, though, requires imagination, and to be imagined a figure or configuration must have some relation to objects in real space. Which takes us back to Euclid again.

Euclidean geometry, in its original garb, was long regarded not as *one model* among many, but rather as *the model* for a rigorous scientific system based on deductive argument. The English universities, above all Oxford and Cambridge, cultivated this tradition for centuries. Even in the 1880s, Charles Dodgson, better known today as Lewis Carroll, was a valiant spokesman for this conservative approach to teaching geometry (Fig. 20.3). Like so many Victorian intellectuals, Dodgson primarily valued geometry as a discipline for training the mind to think logically. It was the tightly constructed arguments in the first two books of Euclid's *Elements* that left an indelible impression on the logician Lewis Carroll.

In *Euclid and his Modern Rivals* (Carroll 1885), one of the oddest dramatic works ever written (and surely never performed), Carroll brought Euclid's ghost back to life to face his challengers. This play without a plot quickly turns into a bizarre mathematical dialogue in which Euclid defends his text and leaves it to the judges in Hades to decide whether any of the thirteen other modernized treatments of plane geometry deserved to take its place. The showdown that ensues has all parties citing the *Elements*, chapter and verse, scurrying to discover which textbook most effectively honed the minds of young Englishmen. Just to be sure that you have the right picture here: the authors whose books come under discussion are all respectable English gentlemen, among the leading mathematicians of their day. So these rival texts obviously did not breath a word about the new fangled non-Euclidean geometries that had since found their way across the channel from the Continent. Such monstrosities clearly had no place in the college curriculum. Euclid's rivals merely sought to upgrade the very same body of knowledge one found in the *Elements*. Needless to say, the author of *Alice in Wonderland* gave Euclid's ghost full satisfaction, routing the thirteen rivals with scholarly acumen and witty jibes. Nor should we be surprised that the meatiest arguments on both sides were reserved for Euclid's controversial fifth postulate concerning parallel lines.

Historically, mathematicians had long focused attention on this parallel postulate as the crux of what made Euclid's presentation of geometry Euclidean. Even in antiquity, many were dissatisfied with this very feature of the *Elements*. Some of Euclid's modern rivals preferred Playfair's more elegant formulation, but Lewis Carroll clearly disagreed with them. His arguments seem to me both sound and convincing. Playfair's version of the Parallel Postulate is in the spirit of an existence theorem in modern mathematics; it lacks the constructive character that makes Euclid's rendition so useful. In Greek geometry the emphasis was always on constructability, though Euclid also proves theorems, which are systematically distinguished from propositions which call for a construction. Euclid's Fifth Postulate enables the geometer to know in advance not merely that two lines, when extended, will intersect, but also *where the point of intersection will occur*, namely on the side where the angle sum is smaller than two right angles (Fig. 20.4). Euclid makes crucial use of this property, for example, in Prop. I.44 while executing a parabolic application of area for a given triangle. Carroll clearly understood the distinction between theorems and problems in geometry. He even has Euclid's ghost address the proposal by the British Committee for the Improvement of Geometrical Teaching that called for presenting theorems and problems separately.

A more serious objection, however, to Euclid's fifth postulate came from those mathematicians who maintained that it was not a postulate at all, but rather a theorem of plane geometry. If you know anything about the history of non-Euclidean geometry, then you will already know that there was a centuries-long effort to prove the parallel postulate, an effort that only turned up more and more connections between it and other familiar theorems of Euclidean geometry. Thus, over the course of time, mathematicians came

Fig. 20.3 Charles Dodgson, alias Lewis Carroll.

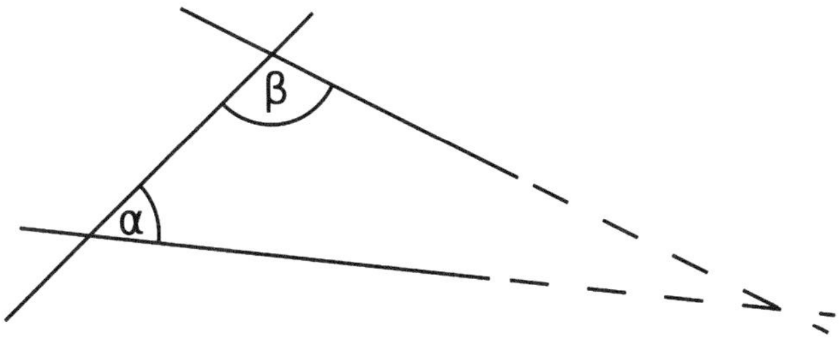

Fig. 20.4 In Euclid's formulation of the 5th Postulate two line segments are cut by a third forming interior angles α and β on one side of the transversal whose sum is less than two right angles. Under these conditions, extending the segments on this same side of the transversal sufficiently far (which can be done by virtue of the 2nd Postulate) they will eventually intersect one another.

Fig. 20.5 Leonhard Euler.

to realize that the Pythagorean Theorem is mathematically equivalent to the parallel postulate, and both are equivalent to the proposition that the sum of the angles in a triangle equals 180°. For us, it is a simple matter to replace this last result with another, for example that the angles in a triangle always have a sum exceeding two right angles. This being so, it is worth asking: why weren't the ancient Greeks willing to entertain this possibility?

Clearly Greek astronomers were familiar with this case, too, since we can still read ancient works on spherics, or what came to be known as spherical geometry. The shortest distance between two points on the surface of a sphere forms the arc of a great circle, a plane section of the sphere where the plane passes through the sphere's center. Taking such geodesics as the counterparts to straight lines in the plane, it is easy to see that the sum of the angles in a spherical triangle exceeds 180°. In fact, this angle sum varies: the smaller the triangle, the smaller will be the sum of its angles, a result that was elegantly proved by Leonhard Euler in the eighteenth century (Fig. 20.5).

But this leads us to ask: if the Greeks already knew several fundamental results of spherical geometry, why did it take so long before mathematicians were willing to accept the validity of non-Euclidean geometries? Wasn't the geometry on the surface of a sphere already a counterexample that refuted the absolute validity of Euclid's theory of parallels? Apparently not; in fact, the answer to this puzzle deserves serious scrutiny as it will tell us a good deal about the status of geometrical investigations not only in antiquity but for many centuries afterward. If we widen the scope of our inquiry into the ancient sciences, then we'll see that spherics was for many centuries studied as a foundation for astronomy and astrology. In fact, the *Sphaerica* of Autolycus, the oldest extant mathematical text available in its entirety, was inspired by a geometrical analysis of the risings and settings of the sun and stars throughout the year. A more poetic account of the geometry of the heavens can be found near the beginning of Plato's *Timaeus*, where he describes how the cosmos was constructed by the Demiurge using two great circles: the celestial equator and the ecliptic. In the ancient world, the motions of heavenly bodies were circular, unlike the natural motions of terrestrial objects, which rise or fall as they seek their natural place in the world. Physical objects, as we know them here on earth, move about in the space surrounding us.

Plane and solid geometry studied the properties of figures in this terrestrial realm. Whether or not the roots of this science were Egyptian, they surely drew on centuries of human experience with physical objects in space. To fill up parts of space, builders used bricks of a uniform size and shape, rectangular solids, not round ones or solids with curved surfaces. Straightness and flatness were the primary spatial qualities one imagined in everyday life. A solid, such a sphere, was obviously in some basic sense a more complex object than a solid figure bounded by plane figures. Both are treated in the last three books of Euclid's Elements, but it is clear that the latter figures, particularly the five Platonic solids, were regarded as fundamental. Some of these regular solids occur in nature in crystalline forms, and Plato identified four of them as the shapes of the four primary elements: earth, water, air, and fire. For the Greeks, the sphere had a deep cosmic significance: the planets and fixed stars were conceived as carried about on giant invisible spheres. The natural rotational motion of spheres could be simulated here on earth, of course, but the truly natural motions of terrestrial objects were rectilinear: straight up and down. Of course the earth itself was not seen as flat, but it was mainly on a cosmic scale that the sphere came forcefully into play in Greek science.

These distinctions were described in detail by Aristotle, who drew heavily on earlier authors. By the period of Euclid, who lived shortly after Aristotle around 300 B.C., these ideas had become firmly established categories of understanding for Greek philosophers and scientists. Astronomy and physics, which was confined to the sublunar sphere, had very little to do with one another. Whereas spherics belonged to the former, geometry had close ties to ancient mechanics, which was not a branch of natural science at all, but rather was synonymous with ancient technology. Mechanicians constructed machines just as geometers constructed figures and diagrams. Among the Greeks, Archimedes was a virtuoso in both disciplines. Indeed, his work successfully bridged the gap between mechanics and geometry, as he relied heavily on the law of the lever to "weigh" geometrical figures before proving theorems about their areas and volumes. Thus, in the ancient world, geometry was closely linked to technology, machines, and the science of mechanics. These links were, if anything, made even stronger when the works of Archimedes were taken up by the mathematicians of the Renaissance.

As everyone familiar with the history of science knows, Isaac Newton was celebrated for demonstrating that the whole universe operates like a giant machine. Newton effectively created the discipline of celestial mechanics when he published his *Principia* (Newton 1687). Since he lived another 40 years, Newton had ample time to comment on his legendary success, and his assessments have been repeated countless times (e.g. in Newton 1726). One is tempted to discount his famous refrain that he "stood on the shoulders of giants" – on most occasions Newton wasn't prone to such modesty – but perhaps we can make sense of this pronouncement by asking just whose shoulders he might have had in mind. To gain a better idea of what he thought he had accomplished, one might start by reading his preface to the *Principia*, one of many places where Newton writes about the relationship between geometry and mechanics: "The description of right lines and circles, upon which geometry is founded, belongs to mechanics. Geometry does not teach us to draw these lines, but requires them to be drawn." Newton went on to mention briefly the conception of mechanics set forth by Pappus of Alexandria around 300 A.D. In his *Collection* Pappus described five types of machines designed to save work, a concept that Newton's system of mechanics would help quantify. This is the tradition of terrestrial mechanics represented by Archimedes and later by Leonardo da Vinci and Galileo. None of these Renaissance figures ever dreamed, of course, that the heavens themselves might be understood as a giant machine. That was Newton's grand vision, made famous by the image of a Deity that designed the world like an intricate clock (Fig. 20.6).

This story has been told many times, for example by the eminent Dutch historian of science E. J. Dijksterhuis in *The Mechanization of the World Picture*. The Copernican heliocentric system simplified Ptolemaic astronomy, but it raised new physical problems that looked insurmountable to many. How could Copernicans ever account for the earth's daily rotation and yearly movement around the sun when we find no direct evidence that the earth really moves? Galileo struggled to answer this in his famous *Dialogue* dealing with the "two chief world systems, Ptolemaic and Copernican" (he ignored the systems of Tycho Brahe and Kepler, the leading astronomers of their day). Although he introduced the principle of relativity, Galileo's arguments reveal that he did not posit a principle of inertia like that formulated by Newton in the *Principia* as the first of his three laws linking mechanical forces and the motion of bodies. In fact, Galileo developed a theory of the tides that explained their daily ebb and flow as a result of the earth's rotation, rejecting Kepler's lunar theory as an occult explanation.

Fig. 20.6 Isaac Newton.

Newton was ahead of his time in so many respects that it is easy to overlook how steeped he was in the traditions of the past. Unlike Descartes, he revered Euclid and the ancients, and this surely accounts for the arcane geometric style of the *Principia.* Not that he was merely paying homage to the ancients or trying to accommodate modern readers, as some historians of physics have suggested. Since most physicists learn celestial mechanics based on the analytic methods first developed by Euler, D'Alembert, and others, they have a hard time squaring what they find in the *Principia* with their image of Newton as co-founder of the calculus. Thus some have occasionally argued that Newton must have originally derived the results in his *Principia* by using the calculus. He then supposedly chose to couch the whole thing in the language of traditional geometry in order not to overwhelm readers with mathematical terminology and techniques that were new and unfamiliar. There seems not to be a scintilla of evidence to support this claim, and I don't know of any leading Newton scholar who thinks that Newton just dressed up the *Principia* in geometry to make it easier to swallow (if he had done so, we would have to conclude that he failed pretty miserably, since his contemporaries found it a very tough read, too). What we do find in Newton's published and unpublished writings are numerous explicit statements and arguments expressing why he preferred to use geometry as the natural mathematical language for treating problems in mechanics. And by geometry, he meant traditional synthetic geometry in the tradition of Euclid and the Greeks.

Newton believed that space had an absolute reality that endowed it with physical properties and geometrical structure. He accounted for inertial effects like the centrifugal forces that accompany rotations as due to the effort required to move against the grain of space, so to speak. His first law, describing the nature of force-free motion, tells us what it means to move with the grain of absolute space, namely in a straight line with uniform speed. The principle of relativity, in its original form, is then a simple consequence of Newton's first two laws. Since the laws of mechanics all deal with forces acting on bodies, and since these forces are directly linked to accelerations by Newton's second law, no physical experiment can distinguish between two inertial frames of reference. Both move with uniform velocity with respect to absolute space and hence with respect to each other. So no accelerations arise, unlike Newton's famous example of the rotating water bucket.

Leibniz and others objected vociferously to Newton's quasi-theological doctrines regarding absolute space and time. But toward the end of the eighteenth century Immanuel Kant gave them a central place in his epistemology. Kant's *Critique of Pure Reason* was widely regarded as a *tour de force* that tamed the excesses of Continental rationalism and metaphysics while overcoming the scepticism of Humean empiricism. Newtonian space and time provided Kant with the keys that led to a new synthesis. He argued that our knowledge of space and time had an utterly different character than all other forms of knowing: it was neither analytic nor a posteriori – meaning that we cannot know the properties of space and time by means of deductive reasoning nor by appealing to sense experience. Nevertheless, we can formulate true synthetic a priori propositions about them because they provide the foundations for all other forms of knowledge. Thus, according to Kant, space and time are the necessary preconditions for knowing; they supply the transcendent categories that give mankind the ability to know.

Gauss and the Advent of Non-Euclidean Geometry

In different ways Euclid, Newton, and Kant were still massive authorities during the nineteenth century, and each reinforced the established view that space carried a geometrical structure that was Euclidean. That position seemed invulnerable throughout most of the century, in part because no other mathematical alternative seemed conceivable. It was not until the 1860s that mathematicians began to take the possibility of a non-Euclidean geometry seriously, this despite the fact that Carl Friedrich Gauss had entertained this idea throughout much of his career. In 1817 Gauss wrote to a colleague: "I am coming more and more to the conviction that the necessity of our geometry cannot be demonstrated. . .. Geometry should be ranked not with arithmetic, which is purely aprioristic, but with mechanics." And in 1824 he went further:

> The assumption that the sum of the three angles of a triangle is less than 180° leads to a curious geometry, quite different from ours [i.e. Euclidean geometry] but thoroughly consistent, which I have developed to my entire satisfaction, so that I can solve every problem in it excepting the determination of a constant, which cannot be fixed a priori. . . . the three angles of a triangle become as small as one wishes, if only the sides are taken large enough, yet the area of the triangle can never exceed, or even attain a certain limit, regardless of how great the sides are.

Soon after his death in 1855, Gauss's friend Sartorius von Waltershausen wrote that he had actually attempted to test the hypothesis that space might be curved by measuring the angles in a large triangle that he used in his survey of Hanover. The triangle had vertices located on the mountain peaks of Hohenhagen, Inselberg, and Brocken, which served as a reference system for the system of smaller triangles. Several writers have argued that this famous story is just a myth, but even if true, Gauss apparently concluded that the deviation of the sum of the angle measurements from 180° was smaller than the margin of error. So this test would have merely confirmed that the geometry of space is either flat or else its curvature was too small to be detected. Nearly unapproachable during his lifetime, Gauss passed from the scene without so much as once publicly addressing the dogma that the geometry of space had to be Euclidean. Let me conclude with some last comments on the history of non-Euclidean geometry.

If one is tempted to speak of revolutions in the history of mathematics (a debatable point), then one might well regard the advent of non-Euclidean geometry in the nineteenth century as a striking example (Bonola 1955, Rosenfeld 1988). Mathematicians tinkered for centuries trying to find a completely elementary proof that Euclid's fifth postulate was true. Most were convinced that it wasn't a postulate at all, but rather a theorem. By the 1820s and 1830s, the Russian mathematician Nicholas Lobachevsky and a young Hungarian named Janos Bolyai finally solved this perplexing problem. They showed that one could develop an exotic system of geometry in which the fifth postulate was false. In what came to be called hyperbolic geometry the parallel postulate fails because instead of having only one line in the plane that passes through a given point without meeting a given line there are infinitely many.

Yet oddly enough, this revolutionary discovery caused barely a stir within the mathematical world. In fact, it was not until the 1860s that mathematicians began to take the new theory seriously. Quite possibly this had something to do with the obscurity of the findings involved, but more likely this general neglect resulted from a lack of intellectual courage on the part of the leading mathematical minds of Europe. It would take over three decades before the publications of Lobachevsky and Bolyai gained belated recognition.

Fig. 20.7 Bernhard Riemann.

The history of non-Euclidean geometry surely would have unfolded quite differently had Gauss made his views on the theory of parallels publicly known (Kaufmann-Bühler 1981). Wolfgang Bolyai, the father of Janos, was an old friend of Gauss. In 1831 the elder Bolyai sent Gauss his son's publication the moment it was printed, only to receive this reply: "To praise it would amount to praising myself. For the entire content of the work . . . coincides almost exactly with my own meditations which have occupied my mind for the past thirty or thirty-five years." Not surprisingly, Janos Bolyai was deeply embittered by this reaction, but Gauss refused to make any public pronouncements about non-Euclidean geometry despite the fact that he praised the works of Lobachevsky and Bolyai in private correspondence. Bolyai only learned about Lobachevsky's work in 1848. By 1860 all three men were dead while their contributions still remained virtually unknown.

Shortly before his death, Gauss had one last chance to make his views about the geometry of space known. In December 1853, Bernhard Riemann submitted his post-doctoral thesis to the Göttingen philosophical faculty along with three proposed topics for the final lecture required of all new members (Fig. 20.7). The elderly Gauss presided on this occasion and requested that Riemann speak about the third topic on the list: "The Hypotheses that lie at the Foundations of Geometry." Riemann was undoubtedly surprised by that decision and none too pleased about this either. He wrote his brother, complaining that this was the only topic he had not properly prepared at the time he submitted his thesis. The lecture took place the following June, and according to Richard Dedekind's later account it made a deep impression on Gauss, as it "surpassed all his expectations. In the greatest astonishment, on the way back from the faculty meeting he spoke to Wilhelm Weber about the depth of the ideas presented by Riemann, expressing the greatest appreciation and an excitement rare for him."

This was the famous lecture in which Riemann explained how the notion of Gaussian curvature could be extended beyond surfaces to manifolds with an arbitrary number of dimensions (Riemann 1854). In particular, this meant that one could study the intrinsic geometry of three-dimensional spaces. Riemann emphasized further that our intuition made it hard to conceive of physical space as bounded, whereas a space of infinite extent posed real difficulties for cosmology. This suggested the possibility that our cosmos might have the structure of an unbounded 3-dimensional manifold of constant positive curvature. Riemann said all this and much more in 1854, but he took these thoughts with him to his grave. Neither Gauss nor anyone else urged him to publish his manuscript or pursue its consequences further. After Riemann's death in 1866, Dedekind was appointed as co-editor of his collected works (Riemann 1892), and it was he who stumbled upon the manuscript among his deceased friend's papers. Its publication in 1868 sparked immediate interest not only in Germany, but in Italy and Great Britain as well. Still, it would take several more years before non-Euclidean geometry found widespread acceptance among mathematicians, many of whom remained convinced that Euclid's geometry of space would always reign supreme.

Acknowledgments This essay was written in preparation for giving the N. A. Court lecture at the March 2006 meeting of the Oklahoma-Arkansas Section of the Mathematical Association of America. That pleasant event took place at the University of Arkansas in Fayetteville. My thanks go to Paul Goodey for extending the invitation.

References

Bonola, Roberto. 1955. *Non-Euclidean Geometry: A Critical and Historical Study of its Development*. New York: Dover.

Carroll, Lewis. 1885. *Euclid and his Modern Rivals*. London: Macmillan; (Reprinted New York: Dover, 1973).

Corry, Leo. 2004. *David Hilbert and the Axiomatization of Physics (1898–1918)*. Dordrecht: Kluwer.

Gabriel, G., et al., eds. 1980. *Gottlob Frege – Philosophical and Mathematical Correspondence*. Chicago: University of Chicago Press.

Gauss, C. F. 1828. *Disquisitiones generales circa superficies curvas.* Gottingen: Dieterich, 1900. Gauss Werke Bd, VIN, Gottingen: Konigliche Gesellschaft der Wissenschaften.

Heath, Thomas L., ed. 1956. *The Thirteen Books of Euclid's Elements.* 3 vols., New York: Dover.

Hilbert, David. 1899. Die Grundagen der Geometrie. In *Festschrift zur Feier der Enfhuiiung des Gauss-Weber Denkmais*, 3–92. Lepzig: B,G, Teubner.

Hilbert, D., and S. Cohn-Vossen. 1965. *Geometry and the Imagination.* Chelsea: NewYork.

Kaufmann-Bühler, Walter. 1981. *Gauss: A Biographical Study*. New York: Springer.

Newton, Isaac. 1687. *Philosophiae Naturalis Principia Mathematica.* London: Joseph Streater.

———. 1726. In *Philosophiae Naturalis Principia Mathematica*, ed. I. Alexandre Koyre, Bernard Cohen, and Anne Whitman, 3rd ed. Cambridge, Mass: Harvard University Press.

Riemann, Bernhard. 1854. Über die Hypothesen, weiche der Geometrie zugrunde liegen. *Abhandiungen der Kgl. Gesellschaft der Wissenschaften zu Gottingen* 13 (1867): 133–152.

———, 1892. *Bernard Riemann's Gesammelte Mathematische Werke und Wissenschaftlicher Nachlass*, hrsg. unter Mitwirkung von R. Dedekind und H. Weber, 2. Aufl. Leipzig: Teubner.

Rosenfeld, B. A. 1988. *A History of Non-Euclidean Geometry: Evolution of the Concept of a Geometric Space*. Trans. Abe Shenitzer. New York: Springer-Veriag.

Terrall, Mary. 2002. *The Man who flattened the Earth: Maupertuis and the Sciences in the Enlightenment.* Chicago: University of Chicago Press.

21 The Mathematicians' Happy Hunting Ground: Einstein's General Theory of Relativity

(*Mathematical Intelligencer* 26(2)(2004): 58–66)

There is hardly any doubt that for physics special relativity theory is of much greater consequence than the general theory. The reverse situation prevails with respect to mathematics: there special relativity theory had comparatively little, general relativity theory very considerable, influence, above all upon the development of a general scheme for differential geometry.

– Hermann Weyl, *Relativity as a Stimulus to Mathematical Research*, (Weyl 1949, 536–537).

No one was more familiar with the impact of Einstein's general theory of relativity on mathematics and mathematicians than Hermann Weyl, who threw himself headlong into this new field shortly after the appearance of Einstein's classic paper (Einstein 1916). In the summer semester of 1917 Weyl taught a 3 h course at the ETH in Zurich on "Raum, Zeit, Materie." The idea to publish a book based on these lectures came from Einstein's close friend, Michele Besso (CPAE 8B 1998b, 663). One year later, the first edition of Weyl's classic *Raum-Zeit-Materie* (Weyl 1918) was already in print. Einstein praised it to the skies:

> I am reading with genuine delight the page proofs of your book, which I am receiving piece by piece. It is like a symphonic masterpiece. Every word has its relation to the whole, and the design of the work is grand. What a magnificent method the infinitesimal parallel displacement of vectors is for deriving the Riemann tensor! How naturally it all comes out. And now you have even given birth to the child I absolutely could not muster: the construction of the Maxwell equations out of the $g_{\mu\nu}$'s! (CPAE 8B 1998b, 669–670).

Weyl, for his part, always held Einstein's theory of general relativity in the highest esteem, and *Raum-Zeit-Materie* did much to promote Einstein's fame as the "New Copernicus." Thirty years later, when Weyl reassessed its impact in "Relativity as a Stimulus to Mathematical Research" (Weyl 1949), he spoke far more soberly, though still with decided enthusiasm. That lecture and the passage from it cited above came on an auspicious occasion.

On March 19, 1949 over 300 scientists – including such eminent figures as J. R. Oppenheimer, Eugene Wigner, I. I. Rabi, and H. P. Robertson – gathered in Princeton to celebrate Albert Einstein's 70th birthday, which fell five days earlier. The celebrant's young assistant, John Kemeny, later recalled the excitement in Princeton during the days leading up to this stellar event, held as a tribute to Einstein's scientific achievements. "People fought over tickets like mad," Kemeny remembered. "I had nothing to do with the tickets, but people somehow thought that being Einstein's assistant I had some pull, and more big shots came to me begging for an extra ticket. They were absolutely dying to get in, and Einstein just had no sense at all about what absolute reverence there was for him" (Sayen 1985, 227).

Einstein in Berlin and Göttingen

For Einstein, Weyl's address must have brought back some fond memories of the enthusiastic response his general theory of relativity received from leading mathematicians, especially after November 1915 when he succeeded in finding an elegant set of generally covariant gravitational field equations:

$$R_{\mu\nu} = -\kappa\left(T_{\mu\nu} - \frac{1}{2}g_{\mu\nu}T\right) \tag{21.1}$$

Most physicists, by contrast, found Einstein's whole approach to gravitation a daring speculation, at best. Max von Laue, who later became one of the strongest advocates of GR in Germany, was originally dismayed even by Einstein's original theory based on the equivalence principle alone. Laue related his misgivings to Einstein in a letter dated 27 December 1911: "I have now carefully studied your paper on gravitation and have also lectured about it in our colloquium [Arnold Sommerfeld's colloquium in Munich]. I do not believe in this theory because I cannot concede the full equivalence of your systems K and K'. After all, a body causing the gravitational field must be present for the gravitational field in system K, but not for the accelerated system K'" (CPAE 5 1993, 384).

Laue's criticism came before Einstein had wandered into the thickets of the Ricci calculus, an adventure made possible through his collaboration with the Zurich mathematician, Marcel Grossmann (Pais 1982, 208–227). After Einstein left the ETH in Zurich to become a member of the Prussian Academy in Berlin, he found virtually no sympathy in his new surroundings for the Einstein-Grossmann approach, often called the *Entwurf* theory. Indeed, the only encouragement he received came from an Italian mathe-

D.E. Rowe, *A Richer Picture of Mathematics*, https://doi.org/10.1007/978-3-319-67819-1_21

matician who happened to be the world's leading authority on Ricci's absolute differential calculus, Tullio Levi-Civita. As Einstein mentioned to his friend Heinrich Zangger in April 1915, corresponding with Levi-Civita gave him great pleasure:

> The theory of gravitation will not find its way into my colleagues' heads for a long time yet, no doubt. Only *one*, Levi-Civita in Padua, has probably grasped the main point completely, because he is familiar with the mathematics used. But he is seeking to tamper with one of the most important proofs in an incessant exchange of correspondence. Corresponding with him is unusually interesting; it is currently my favorite pastime (CPAE 8A 1998a, 177–118).

Meanwhile Berlin's two leading theoreticians, Laue and Max Planck, remained skeptical. Einstein's fortunes in Germany began to change, however, when he visited Göttingen to deliver six lectures on general relativity during the summer of 1915. Afterward, Arnold Sommerfeld and David Hilbert both began to take a keen interest in Einstein's daring approach to gravitation and inertia. The three of them corresponded fairly regularly throughout the autumn, by which time Hilbert had developed a strategy for combining Einstein's gravitational theory and Mie's electromagnetic theory of matter within the framework of a generally covariant mathematical formalism (Rowe 2001).

Einstein was less than enthused but saw that he had to abandon the restricted covariance of the Einstein-Grossmann theory in order to make progress. This led to his four famous notes on general relativity from November 1915, the last of which contained the equations (21.1). Writing to Sommerfeld three days later, he described his dramatic struggle: "you must not be cross with me that I am answering your kind and interesting letter only today. But in the last month I had one of the most stimulating and exhausting times of my life, and indeed one of the most successful as well" (CPAE 8A 1998a, 206). His unhappiness with Hilbert at this time can be seen most clearly from another letter to Zangger in which Einstein wrote:

> The theory is beautiful beyond comparison. However, only *one* colleague has really understood it, and he is seeking to 'partake' [nostrifizieren][1] in it . . . in a clever way. In my personal experience I have hardly come to know the wretchedness of mankind better than as a result of this theory and everything connected to it. But it does not bother me (CPAE 8A 1998a, 205).

It bothered him enough, however, that he wrote to Hilbert suggesting that they forget about this momentary disturbance to their friendship (CPAE 8A 1998a, 222). Afterward they remained on excellent terms, in large part because of their mutual dedication to the ideal of an international scientific community free from political pressures.

The Schwarzschild Solution

The Einstein equations (21.1) are notoriously difficult to solve, a circumstance that led Einstein to devise methods of approximation for handling the special physical problems of greatest interest. He applied these methods in deriving first results for the three classic tests of GRT: the perihelion motion of Mercury, gravitational redshift, and the bending of light rays in the sun's gravitational field. All of these findings faced challenges from the scientific community during the decade following Einstein's breakthrough in 1915, but leading proponents of GRT found a series of improvements that helped strengthen Einstein's case.

The first important result was conveyed to Einstein in December 1915 by the astronomer Karl Schwarzschild, who was then stationed at the Russian front (Figs. 21.3, 21.4). Einstein replied in late December and then in a longer letter dated 9 January 1916, which began: "I examined your paper with great interest. I would not have expected that the exact solution to the problem could be formulated so simply. The mathematical treatment of the subject appeals to me greatly." (CPAE 8A 1998a, 239).

Schwarzschild gave the first exact solution of Einstein's equation (21.1) in the vacuum case, where the right hand side is zero, by showing how to calculate the exterior gravitational field of a static massive body that was spherically symmetric, like the sun, but treated as a mass point (Schwarzschild 1916). The resulting metric thus gave a precise means of calculating the tiny deviations from Newton's theory that were of such crucial importance for the success of general relativity. An even more useful derivation of this result was obtained by Johannes Droste, a student of H. A. Lorentz, in May 1916, the very month in which Schwarzschild died after contracting a terminal skin disease in Russia. Jean Eisenstaedt has pointed out in (Eisenstaedt 1989, 216) that the famous Schwarzschild solution found in all standard textbooks on general relativity

$$\begin{aligned} ds^2 = &-\left(1-\frac{2M}{r}\right)dt^2 + \left(1-\frac{2M}{r}\right)^{-1}dr^2 \\ &+ r^2\left(d\theta^2 + \sin^2\theta\, d\phi^2\right) \end{aligned} \tag{21.2}$$

was actually first presented in modern form in Droste's dissertation (Droste 1916). In Droste coordinates, one sees immediately that the metric breaks down at $r = 0$ and $r = 2M$. Converting from geometrized units, the condition $r = 2M \Rightarrow r = \frac{2GM}{c^2} \approx 3\left(\frac{M}{M_S}\right)$ *km.*, where M_S is the solar

[1]Einstein identified the term *nostrifizieren* with the name Max Abraham, who was famous in Göttingen for his back-stabbing behavior. But, if Einstein learned of this expression from Abraham, it was, in fact, a familiar one in Göttingen mathematical circles, where visitors were often warned that good ideas had a way of getting "nostrified" once they got into the local atmosphere. See the introduction for more about this, and also (Reid 1976).

mass. Clearly, for any conventional astronomical object, the Schwarzschild radius r lies well inside the body, so that the vacuum equations no longer apply. Indeed, for several decades all leading experts agreed that the Schwarzschild radius had no physical relevance whatsoever (see (Eisenstaedt 1989)). The kind of catastrophic gravitational collapse that leads to black holes was simply unthinkable back in those days.

Arthur Stanley Eddington was emphatic about this point in *Space, Time and Gravitation*, where he described what would happen if an observer were to approach the surface $r = 2M$ with a measuring rod. "We can go on shifting the measuring rod through its own length time after time," he wrote, "but dr is zero; that is to say we do not reduce r. There is a magic circle which no measurement can bring us inside" (Eddington 1920, 98). Hermann Weyl, who derived the Schwarzschild solution in Droste coordinates in *Raum-Zeit-Materie*, noted that the gravitational radius of the earth's mass was a mere 5 millimeters. David Hilbert claimed that the Schwarzschild solution contained two singularities, the obvious one at $r = 0$ and the more mysterious one located on the two-sphere with radius $r = 2M$. Following the lead of Ludwig Flamm, Hilbert calculated the trajectories of non-radial light rays that pass near this imaginary sphere, concluding that those which approach it can never penetrate its surface since their velocity will approach zero before arrival (see Fig. 21.1).

Hilbert presented these findings in an unpublished lecture course offered in Göttingen (Hilbert 1916–17), but Max von Laue borrowed a copy of the course notes when he was writing (Laue 1921), the first textbook on general relativity in the German language. Laue repeated Hilbert's pronouncement regarding the singular nature of the two-sphere with $r = 2M$ and even replicated Hilbert's diagram showing how light rays fail to penetrate the sphere (Fig. 21.2). Like the Göttingen mathematician, he was convinced of the essential nature of this singular region, which "cannot be eliminated by using any other coordinates; it is essential to the *nature of the thing*" (Laue 1921, 215). Physically, this seemed patently obvious, since "every mass m. .. has a radius greater than $\frac{2Gm}{c^2}$. . . in fact we do not know as yet any counter-example even in the nucleus of atoms." (For a modern treatment of the trajectories associated with the Schwarzschild solution, see (Wald 1984, 136–148).)

Johannes Droste had actually presented solutions to the Einstein vacuum equations for a mass point in three different coordinate systems, and he was well aware that the r coordinate had no direct physical significance (Eisenstaedt 1989, 216–217). Einstein, too, was completely clear about this circumstance. His friend, Paul Painlevé, had been troubled by the plethora of different representations of the Schwarzschild solution, and passed his misgivings on to Einstein. Noting that one could obtain an infinite number of different formulas for the metric, he suggested this "gave a clear indication of the hazardous character of such predictions," concluding that "it is pure imagination to claim that such consequences can be derived from the ds^2" (Eisenstaedt 1989, 222). Einstein response illuminated the mathematical and physical issues that bothered Painlevé and others:

> When in the ds^2 of the static solution with central symmetry you introduce any function of r instead of r, you do not obtain a *new* solution because the quantity r in itself has no physical meaning . . . only conclusions reached after the elimination of coordinates may pretend to an objective significance. Furthermore, the metrical interpretation of the quantity $ds$is not "pure imagination" but rather the inner-most core of the theory itself. (Quoted in [Eisenstaedt 1989, 223].)

Einstein's original treatment of the Schwarzschild problem was based on a specialized class of coordinate systems. He also postulated that the metric tensor be spherically symmetric, independent of time, stationary, and asymptotically flat at spatial infinity. These conditions hold, of course, for the Schwarzschild-Droste metric (21.2) and Hilbert showed that the last condition was not required in order to deduce this solution. Then, in 1923, G. D. Birkhoff proved a far stronger result:

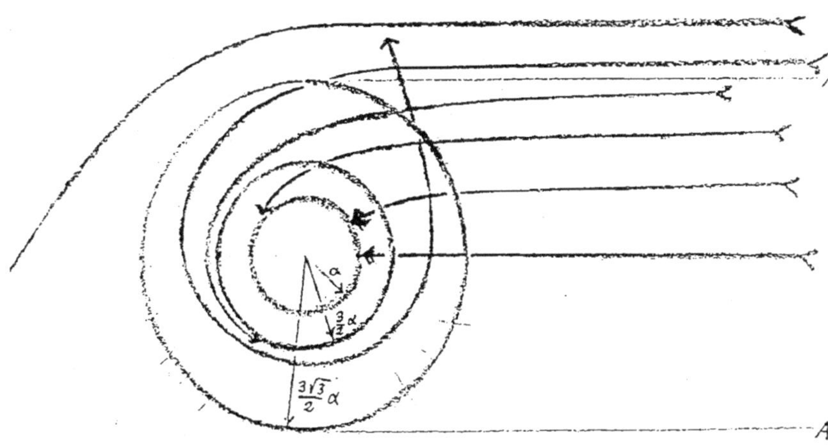

Fig. 21.1 Hilbert's visualization in (Hilbert 1916–17) of the trajectories of light rays in the neighborhood of a strong gravitational field. Note how the rays fail to penetrate the "magic circle" (Eddington's term) determined by the Schwarzschild radius.

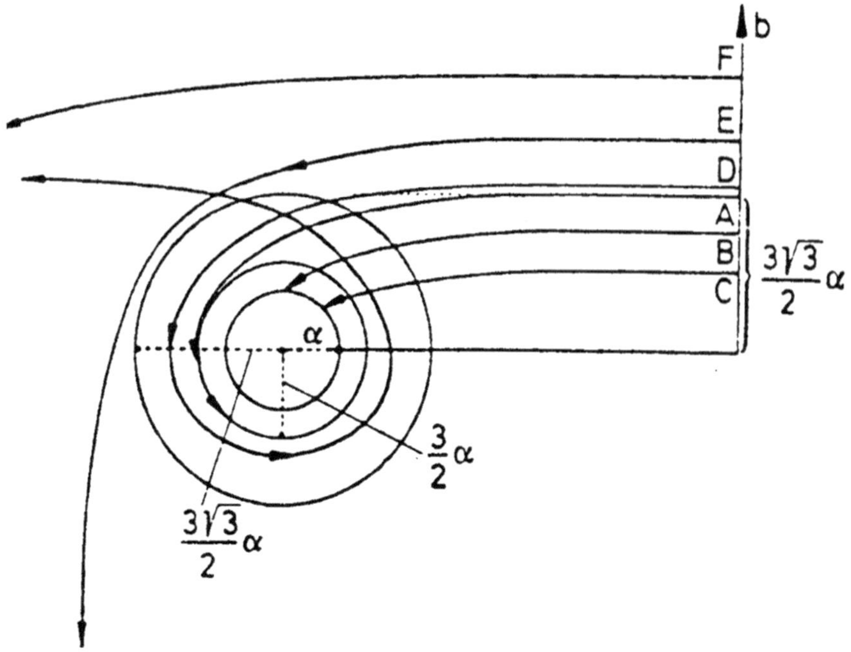

Fig. 21.2 Max von Laue's diagram of the same phenomena in (Laue 1921) was clearly taken from (Hilbert 1916–17).

Birkhoff's Theorem: Any spherically symmetric solution of Einstein's vacuum field equations is necessarily static and has a Schwarzschild metric.

In *Relativity and Modern Physics* (Birkhoff 1923), the Harvard mathematician showed that the assumption of spherical symmetry alone was sufficient to ensure the existence of a coordinate system in which the solution (21.2) holds for the vacuum case. The Schwarzschild solution thus bears a strong analogy with the Coloumb field associated with a static charged particle in electrodynamics: in both cases the fields are unique, showing that a monopole cannot emit radiation. (For a modern treatment of Birkhoff's Theorem, see (Hawking and Ellis 1973, Appendix B)).

Relativity and Differential Geometry

Einstein had met Birkhoff on his very first visit trip to the United States in 1921, a whirlwind tour that ended with a stop in Princeton. There, in May, Einstein delivered five Stafford Little Lectures on the theory of relativity. The first two of these were of an introductory nature, whereas the latter three entered into the technicalities of the theory. Later that year, Einstein worked on a revision of his last three talks and these were published in booklet form under the title *The Meaning of Relativity* (Einstein 1922) one of his best known works. Transcripts of his first two talks, on the other hand, were only recently published in volume 7 of the *Collected Papers of Albert Einstein* (CPAE 7 2002, App. C).

Fig. 21.3 Karl Schwarzschild in academic attire (Schwarzschild 1992).

One of those who followed Einstein's lectures closely was the differential geometer Luther Pfahler Eisenhart, Princeton's senior mathematician. In fact, Eisenhart instigated the

Fig. 21.4 Schwarzschild (left) with fellow officers during the Great War (Schwarzschild. Nachlass, SUB Göttingen).

Fig. 21.5 Einstein showing off a favorite tensor equation (Courtesy of the Albert Einstein Archives).

invitation that led Einstein to visit Princeton that spring, although he originally hoped to coax him into coming as a guest lecturer during the winter semester of 1920–21 (Eisenhart to Einstein, 20 October 1920, (CPAE 7 2002, 231)). Einstein declined this offer, but in February 1920 Kurt Blumenfeld persuaded the now famous physicist to undertake a trip to the United States to support the Zionist movement and raise funds for Hebrew University (CPAE 7 2002, 231: note 48), a circumstance that led Einstein to reconsider Eisenhart's invitation (Fig. 21.5).

Einstein surely recalled Eisenhart's avid interest in general relativity when he heard Hermann Weyl's lecture on "Relativity as a Stimulus to Mathematical Research." Weyl noted that a major part of this stimulation began with investigations of affinely connected manifolds, which generalized Riemannian manifolds but still admitted infinitesimal parallel transport of vectors. Since covariant differentiation and the whole apparatus of tensor analysis carried over to this more general setting, this meant it was not necessary to assume the existence of a metric tensor. One area of research focused on the conformal structure of Riemannian spaces, which led to the consideration of systems of geodesics, but another vast terrain was opened by the study of general path spaces, a specialty of the Princeton geometers. Weyl described this research activity in the following words:

> ... here is clearly rich food for mathematical research and ample opportunity for generalizations. Thus schools of differential geometers sprang up in the wake of general relativity. Here in Princeton Eisenhart and Veblen took the lead, Schouten in Holland. In France, E. Cartan's fertile geometric imagination disclosed many new aspects of the subject. Some of their outstanding pupils are Tracy Thomas and J. M. Thomas in Princeton, van Dantzig in Holland and Shiing Shen Chern of the Paris school. A lone wolf in Zurich, Hermann Weyl, also busied himself in this field; unfortunately he was all too prone to mix up his mathematics with physical and philosophical speculations [Weyl 1949, 538].

The work of Élie Cartan and Weyl eventually opened up the whole vast theory of fiber spaces of differential manifolds. Cartan's investigations dealt with geometries possessing transitive transformation groups that act on their tangent spaces, an approach that drew much inspiration from Lie theory and Klein's Erlangen Program (see (Cartan 1974)). Weyl developed an axiomatic treatment of maps of tangent spaces by utilizing parallelism along smooth curves. The latter idea was first developed by Levi-Civita in 1917 for Riemannian spaces, but soon generalized by Weyl to manifolds with an affine connection. Using these constructs, Weyl sought to develop a purely infinitesimal geometry that could serve as a basis for a unified field theory for gravity and electromagnetism (Scholz 2001).

Eisenhart took his inspiration from the generalization to general affine spaces of geodesics in Riemannian geometry, namely the paths satisfying the differential equations

$$\frac{d^2x^i}{dt^2} + \Gamma^i_{jk}\frac{dx^j}{dt}\frac{dx^k}{dt} = 0 \tag{21.3}$$

In Riemannian geometry one uses the metric tensor to define the Christoffel symbols Γ^i_{jk}, from which one then proves that the solutions of (21.3) are geodesics, the shortest paths joining any two points within an appropriate neighborhood of the manifold. In an affine geometry, the Γ^i_{jk} are defined independent of a metric, and the solutions of (21.3) determine a congruence of paths with special properties. The geometry of paths in affine spaces can also be regarded as a generalization of methods introduced by Riemann in his famous *Habilitationsschrift* of 1854, in particular his systems of normal coordinates. Eisenhart worked directly with objects of the manifold, in contrast to the methods of Cartan and Weyl, which dealt with the associated tangent spaces. Unfortunately, the former approach leads to complications since the intrinsic geometric objects associated with paths do not satisfy the transformation law for generalized Christoffel symbols, a central feature in the affine theories of Weyl and Cartan. Eisenhart summarized his contributions to the geometry of paths as well as those of his students in (Eisenhart 1927), whereas the related theory of differential invariants was sketched by Oswald Veblen in (Veblen 1927, Chap. 6).

Einstein's Enemies: The Anti-relativists

Not surprisingly, these mathematical developments were largely ignored by physicists, few of whom had sufficient knowledge of differential geometry to make any sense of them. Opponents of relativity, on the other hand, had long claimed that Einstein's theory was of purely mathematical interest, pointing to Minkowski's four-space formalism to make their case. Aside from Weyl, most prominent mathematicians chose to avoid engaging in polemics with vituperative anti-relativists. Still, just two years after Einstein's visit to Princeton in 1921, the soft-spoken Eisenhart became embroiled in controversies surrounding Einstein's theory, which by then had spilled over from Germany to the United States.

In an article in *Science* (December 21, 1923) the Princeton geometer came to Einstein's defense after Philipp Lenard republished portions of an older theory of gravitation presented by Georg Soldner. Lenard claimed that Soldner had derived precisely the same result for the deflection of light in the vicinity of the sun as that found by Einstein in 1911 (Einstein later revised this figure to 1.75″ of arc in 1915, twice the original value, due to the curvature of space caused by the sun's gravitational field). Eisenhart's article appeared in the wake of a similar defense of Einstein's priority published by David Hilbert and Max Born in the widely read *Frankfurter Zeitung* (see the appendix below).[2] Hermann Weyl also countered the criticisms of anti-relativists, but this only hardened the views of experimental physicists like Lenard and co., who regarded general relativity as a theory devoid of physical content. (For details, see the editorial note "Einstein's Encounters with German Anti-Relativists" in (CPAE 7 2002, 101–113)).

These were episodes Einstein surely had no reason to think about in 1949, now that Lenard was dead and with him the whole Aryan physics movement, too. During the early 1920s, in fact, Einstein had mainly been content to let others battle the anti-relativists. The one time, in August 1920, that he did take pen in hand to attack his opponents in the pages of the *Berliner Tageblatt* (CPAE 7 2002, 344–349), his words caused such a stir that he seriously contemplated leaving Berlin for quieter quarters. The streets of the Prussian capital were, indeed, amuck with chaos, but even in the quiet backwoods of Minnesota an American anti-relativist was keeping a close eye on the latest events (Fig. 21.6).

Dr. Arvid Reuterdahl, who came to the United States from Sweden as a young lad, was President of Ramsay Institute of Technology in St. Paul. He was also a regular contributor to Henry Ford's newspaper, *The Dearborn Independent*, remembered today for its blatantly anti-Semitic editorials. Reuterdahl wrote a regular column called "International

[2] Thanks go to Klaus Hentschel for sending me a copy of this newspaper article.

12 THE DEARBORN INDEPENDENT March 25, 1922

Reuterdahl vs. Einstein: Nailing a Fallacy

Former Insists Latter's Theory Wrong and Questions Its Originality

By E. LEE HEIDENREICH

AS THE Einstein enthusiasm is waning, we may well ask what thoughts and influences have caused this slowing down of the meteorical rise and advance of Einstein in giving the study of relativity its tremendous and unexpectedly sudden impetus.

The History of Relativity shows us that in 1889 Rudolf Mewes, in an article entitled, "Das Wesen der Materie und des Naturerkennens," gave a brief outline of his theory of relativity. Mewes held that the essence of matter consists in its action as manifest in continuous transformation in accordance with causality. He found in his basic concept a source of the relativity of space, time and matter. For him all reality depends upon the combination of space and time.

From 1892 to 1894 Mewes made a careful analysis of the derivation of *Weber's Law* from the *Doppler Principle*. This resulted in the evaluation by Mewes of the transformation formula which later made Lorentz famous through its use by Einstein. Mewes clearly antedates Larmor, Lorentz and Einstein in this development. In addition, Mewes grounds his system in sound philosophy.

Arvid Reuterdahl presented a brief outline of his Space-Time Potential and Theory of Interdependence at the inaugural meeting of the American Electrochemical Society, held in Philadelphia, April 5, 1902, under the title of "The Atom of Electrochemistry." In this lecture Reuterdahl showed that the physical universe is ultimately reducible to centers of activity which undergo compensating changes and displacements in conformity with the requirement of the whole cosmos regarded as a unitary, interacting and interdependent system of multiplicity, which is its own continuum.

What the Records Show

J. HENRI ZIEGLER, in September, 1902, laid the foundation of a cosmic theory in a lecture entitled "Die Universelle Weltformel und ihre Bedeutung fur die wahre Erkenntniss aller Dinge." Ziegler's cosmology is based upon the fundamental conception that the world is a unitary structure generated from the universal trinity of space, time and force, with the inclusion of an Absolute Principle, the constancy of the velocity of light.

In "Einstein and the New Science," Arvid Reuterdahl gives us the history of relativity, and gives a most concise chronological account of the authors and scientists preceding Einstein in the study of the theory of relativity—but also at the same time gives a serious blow to both the originality and the validity of "Einstein's Theory of Relativity." In this little book of 26 pages of concentrated facts, we find that Reuterdahl commends, lauds, encourages and acknowledges with the greatest freedom the works of the many scientific contributors, and in an article in THE DEARBORN INDEPENDENT of April 30, 1921, entitled, "Kinertia vs. Einstein," gloriously defends the great work of "Kinertia," although we know that Reuterdahl and "Kinertia" do not entirely agree in their ideas. But it shows Reuterdahl's unselfishness and fairness in contraposition to Einstein's avoidances of acknowledgments to the most important of his predecessors.

Reuterdahl has, in the past, been wilfully ignored. Attempts have been made to intimidate him and his work has been practically suppressed by the machinations of his jealous enemies. Undaunted by these obstacles, Reuterdahl's faith in the ultimate victory of truth remains unshaken. Now the time has come for someone who knows the facts to present to the world the truth concerning Reuterdahl—the man who dared to challenge Einstein, the alleged superman of science.

Back in 1913, Reuterdahl discussed with me in Kansas City his "Space-Time Potential, a conception of Gravitation and Electricity," and had his manuscript typewritten in my office. I spoke about the subject on my lecture tours covering other matters, and with the result, that in December, 1914, Reuterdahl was invited to lecture at the Kansas State College of Agriculture and the University of Kansas, which lectures were given early in 1915.

ARVID REUTERDAHL, dean of the department of engineering and architecture at the College of St. Thomas, St. Paul, Minnesota, since 1918, as far back as 1896, while a student at Brown University, Providence, Rhode Island, conceived the basic outline of his work, "Scientific Theism Versus Materialism, the Space-Time Potential." He has lectured extensively on engineering, scientific, mathematical and philosophical subjects, and has published a number of books on scientific subjects.

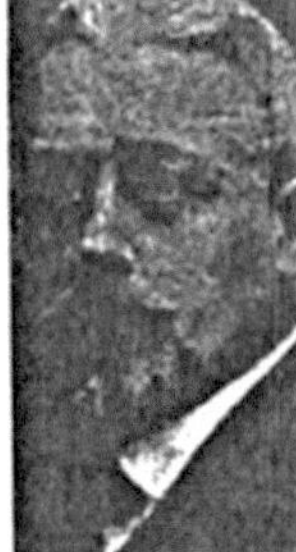
ARVID REUTERDAHL

When the trumpets of Einstein began to sound the blasts of glory for Relativity and its discoverer, the American scientists were swept off their feet because they were unfamiliar with these strange doctrines, but Dr. Reuterdahl had been thinking in "Space-Time" terms since 1896; in fact, he was the first to employ the hyphen between the two words, space and time, and his Space-Time Potential antedates Einstein's Space-Time Continuum. Professor Reuterdahl was the first in the United States to denounce Einsteinism as a fallacy.

Dean Reuterdahl was born February 15, 1876, at Karlstad, Sweden, the son of Jonas and Christina Reuterdahl. In 1882 the family came to the United States, eventually locating in Providence. The son's early education consisted in private instruction given by his father, a former officer in the Swedish army, and an excellent mathematician and philosopher.

between "space" and "time" is a mere *coincidence*. In a statement given to the New York *World* (June 12, 1921), he again disclaims all knowledge of Reuterdahl. This disclaimer is irreconcilable with his own previous statement to the *Aftenposten*. The question naturally arises: "Why was Einstein so anxious to efface the name of Reuterdahl in connection with the Theory of Relativity? Does the five years' sojourn in Europe of Reuterdahl's manuscript answer this question?

to an infinitesimal portion of the great frame are the same as hold for a part conceived as great as the imagination may permit in conformity with the frame.

The Einsteinean *world reference frame* is Gaussian and four-dimensional in type, the co-ordinate system being an attempt at the *amalgamation* of space and time. Any reference section arbitrarily cut out of the frame is a veritable co-ordinate fabric. Such section has been called a "*Reference-Mollusk*" by Einstein. We may conceive that a large reference section or "Mollusk" shrivels or contracts into a small section or vice versa.

During such diminutions or augmentations the laws of the relational factors remain the same, and the resulting section is either a reduced or increased replica of the original. The changes are similar to the effect produced by looking at an original, first with a reducing glass, and then with a magnifying glass. All the parts of the co-ordinate fabric will remain in the same relative juxtaposition irrespective of the type of the glass through which it is observed.

In such a system force regarded as an independent entity does not exist. Action at a distance no longer becomes a problem because the world frame is a unitary interacting system.

Einstein holds that a further consequence of his worldsystem is the disappearance of the mass *M* (which pertains to a particular point) from the law of motion. He reduces the motion of a particle under the so-called influence of gravitation to a mere "Following" of a geodetic line in his four-dimensional world.

Nowhere has Einstein stated his own case as clearly as it has been presented in the above analysis, which is due to Reuterdahl. It is even doubtful if Einstein has grasped the significance of his own vague and stumbling attempts at formulating this great world synthesis. His pitiable inability to express the essentials of the system in a clear and intelligent manner can readily be accounted for on the supposition which grows upon us—that Einstein has taken his entire thought of this synthesis from the work of Reuterdahl.

Not grasping clearly the intent of Reuterdahl's Space-Time Potential, Einstein camouflaged it in a distorted manner into his own Space-Time Continuum—by transferring the conception from Euclidean to non-Euclidean. Reuterdahl's Space-Time Potential is a synthesis, not only of the same intent as Einstein's, but of a vastly more comprehensive order. Reuterdahl's complete cosmology proceeds direct from physical facts, whereas Einstein presents us with a skeleton, incapable of harmonious functioning, because it is constructed from the fibers of inconsistent mathematical speculation, which predict coming events with the hope that they will conform—and if they do not, they are quietly thrown out of court. Reuterdahl has distinctly stated, that force does not exist in the physical universe as an independent and distinct entity, and action-at-a-distance offers no obstacles, because the cosmos is an interacting unitary system. While Einstein numbers his various co-ordinate curves which constitute the fabric of his reference-mollusks, Reuterdahl also numbers his potential loci and shows carefully, that relative conditions existing between any two elements of a section, also pertain without change between the similarly numbered elements in any other section, irrespective of its size, and whether located in the realm of microcosm or that of macrocosm. His Space-Time Potential is a chart or diagram defining the motion of every activity center in the entire interacting system.

We realize the similarity of the two systems, the one complete and the other an incomplete instantaneous static view of the dynamics in the former.

Reuterdahl's priority is undisputed, the relative dates being nailed down chronologically. If the Space-Time Continuum is not a camouflaged edition of Space-Time Potential, it certainly looks like *another coincidence*.

Einstein's version suffers sorely when compared with the earlier synthesis of Reuterdahl. Einstein abandons the classical mechanics, while Reuterdahl adheres strictly to its experimentally demonstrated principles. Einstein resorts to non-Euclidean geometry for the fibers of his Space-Time Continuum, whereas Reuterdahl contents himself with observed physical data.

Fig. 21.6 Arvid Reuterdahl, author of a booklet *Einstein and the New Science*, used Henry Ford's *Dearborn Independent* to polemicize against relativity theory.

Science Briefs," in which he regularly inveighed against the Einsteinians as corrupters of scientific truth. Even Eisenhart was taken to task for his defense of Einstein's theory in the face of Soldner's ancient derivation of light deflection *sans* a principle of relativity. A column entitled "The Einstein Film and the Debacle of Relativity" caught the eye of G. D. Birkhoff, who sent a copy on to Eisenhart asking: "Did you see this article!" (Einstein Archive 1973–6).

In it, Reuterdahl unleashes a flurry of invective against the Einsteinians, singling out Eisenhart for supposedly claiming that *both* of Einstein's derivations – the 1915 value based on GR and the 1911 derived from the equivalence principle alone – yielded the correct value for the deflection of a light ray in the neighborhood of the sun. This, he noted, contradicted ordinary common sense, since GR predicted a value twice as great as the one based on SR. Yet "[b]oth are right according to the Einsteinians:–*two equals one.* In the presence of the mighty "invariants" and "covariants" of modern relativity such mere trivial inconsistencies must be ignored as negligible "Einsteinian effects" (Reuterdahl 1924). Continuing in this vein, Reuterdahl claims that Eisenhart gave an hysterical argument to defend Einstein from the specter of Soldner by asserting that the velocity of light decreases near the sun, thereby contradicting the fundamental tenet of (special) relativity which asserts the constancy of the speed of light.

Clearly, Reuterdahl's voice carried more than a little hysteria of its own, not to mention an apparently abysmal understanding of the fundamental ideas of Einstein's theory of gravitation. It was all just some mathematical hocus-pocus to him. He called the relativity film "balderdash" and "an insult to common sense"; Max von Laue criticized it, too, but of course for different reasons. According to Reuterdahl, relativists ask us "to admit that we live in a four-dimensional space without knowing it. We admit that we are not aware of this four-dimensional space, but we refuse to accept [it] as physically real .... Mathematical expressions, supposed to portray this four-dimensional space, depict nothing but conceptual speculations having no counterpart in a real physical world."

This dogmatic anti-mathematical stance was the common currency of nearly all anti-relativists from this time. They came out in droves during the early 1920s and cited each other's publications approvingly. In fact, Reuterdahl corresponded with a number of leading figures within the anti-relativists' network, including Ernst Gehrcke, the most tenacious of the German anti-relativists. Thus it comes as no surprise that Reuterdahl praises Gehrcke for having "exposed the Einsteinian use of Gerber's earlier formula of 1898," an allusion to Gehrcke's contention that Einstein had plagiarized an earlier paper written by a Gymnasium instructor named Paul Gerber.

The anti-relativists attached great significance to the fact that Gerber had come up with a formula that Einstein later derived in accounting for the perihelion motion of Mercury. Immediately after the appearance of (Einstein 1916) in *Annalen der Physik*, Gehrcke published a "rebuttal" in the same journal, where he charged that Einstein had surely taken the formula from Gerber without citing him (Gehrcke 1916). Both Gehrcke and Lenard tried to make this plagiarism charge stick, but Max von Laue easily showed that Gerber's whole approach to gravitation was faulty, whatever the status of his formula (Laue 1917). Nevertheless, Reuterdahl was still trumpeting this news seven years later. Few readers of *The Dearborn Independent* could have suspected that Gehrcke had turned up nothing more than a curiosity in the physics literature. In case some might have wondered about Gehrcke's scientific credentials, Reuterdahl led them to believe he was one of the giants of German physics by claiming that Gehrcke had succeeded Hermann von Helmholtz as Director of the Physikalisch-Technische Reichsanstalt in Charlottenburg. In fact, he was only one of several scientists working on the PTR's staff, not its director.

Reuterdahl and Gehrcke shared a zealous passion not only to refute the basic tenets of Einstein's theory but to reveal that its creator and his followers had perpetuated a massive fraud on the scientific world as well as the general public. Thus, Reuterdahl was proud to claim that he "was the first in the United States" to call attention to Georg Soldner's ancient result on light deflection from 1801. Indeed, he was convinced – thanks to the revelations of the works by Gerber and Soldner – that the Einsteinian relativists were now more vulnerable than ever, and once "forced into the open, the complete debacle of Einsteinianism is inevitable in the near future."

This was a strange vision coming from an American writing in 1924, but hardly an isolated opinion. In Germany, it became part of the dogma of the German physics movement whose hero was Philipp Lenard. By the time Einstein joined the faculty at Princeton's new Institute for Advanced Study in 1933, only the political hostilities remained as a residue of the anti-relativity movement from the early 1920s. Once the Nazis gained power, the original aim of refuting relativity on scientific grounds gave way to racial arguments. Lenard and Johannes Stark no longer claimed that relativity theory and quantum mechanics were wrong scientifically. They protested instead that these theories represented an alien spirit and hence posed a deadly threat to "true" German physics, which had no room for empty mathematical abstractions.

Appendix: Text of an Article by Hilbert and Born in the *Frankfurter Zeitung*

In their rebuttal, Hilbert and Born defended the originality of Einstein's prediction for the deflection of light in the sun's gravitational field. They emphasize that Einstein's original prediction from 1911, which agreed with the value earlier obtained by Georg Soldner, was incorrect, and that Einstein's revised figure from 1915 represented the true relativistic result. Strangely, Hilbert and Born state that the British confirmed Einstein's result during the solar eclipse of 1920; the eclipse took place in May 1919.

Soldner und Einstein

Von Prof. Dr. Hilbert und Prof. Dr. Born – Göttingen

Der Aufsatz von L. Baumgardt veranlaßt uns zu folgender sachlicher Berichtigung:

1. Die von Soldner 1801 abgeleitete Formel für die Lichtablenkung stimmt mit der überein, die Einstein im Jahre 1911 als Resultat einer vorläufigen Überlegung über den Einfluß der Schwere auf das Licht veröffentlicht hat. Diese Arbeit von 1911 enthält aber noch nicht diejenigen Grundgesetze der Physik, die man als „allgemeine Relativitätstheorie" bezeichnet.
2. Die richtigen Grundgleichungen dieser allgemeinen Relativitätstheorie wurden erst Ende 1915 von Einstein gefunden. Dabei ergab sich, daß der von Einstein vorher 1911 vermutete Wert, der mit der Soldnerschen Angabe übereinstimmt, vom Standpunkt der Relativitätstheorie falsch ist; vielmehr stellte sich die Lichtablenkung doppelt so groß heraus. Daß Einstein bei seinem Suchen nach den richtigen Gesetzen der Schwere zunächst durch vorläufige Abschätzungen zu dem falschen Wert gelangt war, ist leicht begreiflich; das Genie sieht lange, ehe ein Gedanke begrifflich bis in alle Einzelheiten geklärt und formuliert ist, die wichtigsten Zusammenhänge voraus und schätzt die experimentell kontrollierbaren Wirkungen ihrer Größe nach zunächst mit rohen Mitteln ab.
3. Der endgültige Wert der Lichtablenkung von Einstein folgt völlig eindeutig aus der allgemeinen Relativitätstheorie. Es ist ein Irrtum zu glauben, daß Soldner eine Folge der Relativitätstheorie richtig vorausgesehen habe; vielmehr ist es umgekehrt: wenn die von Soldner auf Grund der Newtonschen Theorie berechnete Größe der Lichtablenkung (die mit Einsteins vorläufiger Schätzung von 1911 zufällig übereinstimmt) durch die Beobachtungen als richtig erwiesen würde, so wäre damit eine vollgültige Widerlegung der Einsteinschen Relativitätstheorie erzielt.

Die englischen Astronomen, die bei der Sonnenfinsternis des Jahres 1920 die Messungen der Lichtablenkung ausgeführt haben, sind der Meinung, daß nicht der auf der Newtonschen Attraktionstheorie fußende Wert von Soldner, sondern der von der Einsteinschen Relativitätstheorie vorausgesagte Wert tatsächlich gilt. Neue Untersuchungen werden bei der Sonnenfinsternis des Jahres 1922 stattfinden.

References

Birkhoff, George D. 1923. *Relativity and Modern Physics, Cambridge*. Mass.: Harvard University Press.

Cartan, Élie. 1974. *Notice sur les travaux scientifiques*. Paris: Gauthier-Villars.

CPAE 5. 1993. *Collected Papers of Albert Einstein*, Vol. 5: *The Swiss Years: Correspondence, 1902–1914*, Martin J. Klein, et al., eds., Princeton: Princeton University Press.

CPAE 7. 2002. *Collected Papers of Albert Einstein*, Vol. 7: *The Berlin Years: Writings, 1918–1921*, Michel Janssen, *et al.*, eds., Princeton: Princeton University Press.

CPAE 8A. 1998a. *Collected Papers of Albert Einstein*, Vol. 8A: *The Berlin Years: Correspondence, 1914–1917*, Robert Schulmann, *et al.*, eds., Princeton: Princeton University Press.

CPAE 8B. 1998b. *Collected Papers of Albert Einstein*, Vol. 8B: *The Berlin Years: Correspondence, 1918*, Robert Schulmann, *et al.*, eds., Princeton: Princeton University Press.

Droste, Johannes. 1916. The Field of a Single Centre in Einstein's Theory of Gravitation and the Motion of a Particle in that Field, *Proceedings of the Section of Sciences, Koninklijke Akademie van Wetenschappen te Amsterdam*, 19: 197–215.

Eddington, A.S. 1920. *Space, Time and Gravitation*. Cambridge: Cambridge University Press.

Einstein, Albert. 1916. Grundlagen der allgemeinen Relativitätstheorie. *Annalen der Physik* 49: 769–822.

———. 1922. *Vier Vorlesungen über Relativitätstheorie,* Braunschweig: Vieweg. (Reprinted in translation as *The Meaning of Relativity. Princeton*: Princeton University Press, 1923).

Eisenhart, Luther P. 1927. *Non-Riemannian Geometry*. Princeton: Princeton University Press.

Jean Eisenstaedt, The Early Interpretation of the Schwarzschild Solution, in *Einstein and the History of General Relativity*, Einstein Studies, vol. 1, D. Howard and J. Stachel, eds., Basel: Birkhäuser, 1989, pp. 213-233.

Gehrcke, Ernst. 1916. Zur Kritik und Geschichte der neueren Gravitationstheorien. *Annalen der Physik* 51: 119–124.

Hawking, Stephen W., and George F.R. Ellis. 1973. *The Large Scale Structure of the Universe*. Cambridge: Cambridge University Press.

Hilbert, David. 1916–1917. *Die Grundlagen der Physik II, Vorlesung, Wintersemester 1916–17, ausgearbeitet von R.* Bär: Mathematisches Institut der Universität Göttingen.

von Laue, Max. 1917. Die Fortpflanzungsgeschwindigkeit der Gravitation. Bemerkungen zur gleichnamigen Abhandlung von P. Gerber. *Annalen der Physik* 52: 214–216.

———. 1921. *Die Relativitätstheorie, zweiter Band: Die allgemeine Relativitätstheorie und Einsteins Lehre von der Schwerkraft*. Braunschweig: Vieweg.

Pais, Abraham. 1982. *'Subtle is the Lord... ' The Science and the Life of Albert Einstein*. Oxford: Clarendon Press.

Reid, Constance. 1976. *Courant in Göttingen and New York: the Story of an Improbable Mathematician*. New York: Springer.

Reuterdahl, Arvid. 1924. The Einstein Film and the Dabacle of Relativity. *The Dearborn Independent* 22: 15.

Rowe, David. 2001. Einstein meets Hilbert: At the Crossroads of Physics and Mathematics. *Physics in Perspective 3*: 379–424.

Sayen, Jamie. 1985. *Einstein in America. The Scientist's Conscience in the Age of Hitler and Hiroshima*. New York: Crown Publishers.

Scholz, Erhard. 2001. Weyls Infinitesimalgeometrie, 1917–1925, in *Hermann Weyl's Raum-Zeit-Materie and a General Introduction to his Scientific Work*, ed. E. Scholz (DMV Seminar, 30.) Basel/Boston: Birkhäuser Verlag, 2001. 48–104.

Schwarzschild, Karl. 1916. Über das Gravitationsfeld eines Massenpunktes nach der Einstein'schen Theorie. *Sitzungsberichte der Königlich-Preussischen Akademie der Wissenschaften*: 189–196.

———. 1992. *Gesammelte Werke*. Vol. 1. Heidelberg: Springer-Verlag.

Veblen, Oswald. 1927. *Invariants of Quadratic Differential Forms*. Cambridge: Cambridge University Press.

Wald, Robert M. 1984. *General Relativity*. Chicago: University of Chicago Press.

Weyl, Hermann. 1918. *Raum–Zeit–Materie. Vorlesungen über allgemeine Relativitätstheorie*. Berlin: Springer.

———. 1949. Relativity Theory as a Stimulus in Mathematical Research. *Proceedings of the American Philosophical Society* 93: 535–541.

22 Einstein's Gravitational Field Equations and the Bianchi Identities

(*Mathematical Intelligencer* 26(2)(2004): 58–66)

Introduction

In his highly acclaimed biography of Einstein, Abraham Pais gave a fairly detailed analysis of the many difficulties his hero had to overcome in November 1915 before he finally arrived at generally covariant equations for gravitation (Pais 1982, 250–261). This story includes the famous competition between Hilbert and Einstein, an episode that has recently been revisited by several historians in the wake of newly discovered documentary evidence, first presented in Corry et al. (1997).

In his earlier account, Pais emphasized that "Einstein did not know the [contracted] Bianchi identities

$$\left(R^{\mu\nu} - \frac{1}{2}g^{\mu\nu}R\right)_{;\nu} = 0 \qquad (22.1)$$

when he wrote his work with Grossmann." (The symbol ';' denotes covariant differentiation, which here is used as the generalized divergence operator.) In 1913 Einstein and the mathematician Marcel Grossmann presented their *Entwurf* for a new general theory of relativity. Guided by hopes for a generally covariant theory, they nevertheless resolved to use a set of differential equations for the gravitational field that were covariant only with respect to a more restricted group of transformations. However, when Einstein abandoned this *Entwurf* theory in the fall of 1915, he once again took up the quest for generally covariant field equations. By late November he found, though in slightly different form, the famous equations:

$$G^{\mu\nu} = -\kappa T^{\mu\nu}, \qquad \mu, \nu = 1, \ldots, 4 \qquad (22.2)$$

where

$$G^{\mu\nu} \equiv R^{\mu\nu} - \frac{1}{2}g^{\mu\nu}R \qquad (22.3)$$

is the Einstein tensor. (Here $T^{\mu\nu}$ is the energy-momentum tensor and $g^{\mu\nu}$ the metric tensor that determines the properties of the space-time geometry. The contravariant Ricci tensor $R^{\mu\nu}$ is obtained by contracting the Riemann-Christoffel tensor; contracting again yields the curvature scalar $R^{\mu\nu}g_{\mu\nu} = R$. The symmetry of $g^{\mu\nu}, R^{\mu\nu}$, and $T^{\mu\nu}$ means that (22.2) yields only 10 equations rather than 16.)

Applying the covariant divergence operator to both sides of the Einstein equations (22.2) yields, according to (22.1),

$$G^{\mu\nu}_{;\nu} = T^{\mu\nu}_{;\nu} = 0. \qquad (22.4)$$

This tells us that actually only $10 - 4 = 6$ of the field equations (22.2) are independent, as should be the case for generally covariant equations. Ten equations for the 10 components of the metric tensor $g^{\mu\nu}$ would clearly over-determine the latter, for general covariance requires that to any single solution $g^{\mu\nu}(x_i)$ of (22.2) corresponds a 4-parameter family of solutions obtained simply as the $g^{\mu\nu}(x'_i)$ induced by arbitrary coordinate transformations. Choosing a specific coordinate system thus singles out a unique solution among this family.

Yet for a long time Einstein resisted drawing this seemingly obvious conclusion. Instead he concocted a thought experiment—his infamous hole argument—that purported to show how generally covariant field equations will lead to multiple solutions within one and the same coordinate system (see Norton 1989 and Stachel 1989). His initial efforts therefore aimed to circumvent this paradox of his own making. For, on the one hand, physics demanded that generally

D.E. Rowe, *A Richer Picture of Mathematics*, https://doi.org/10.1007/978-3-319-67819-1_22

covariant gravitational equations *must* exist, whereas logic (mixed with a little physics) told him that no such equations can be found (see his remarks in Einstein 1914, 574). Luckily, Einstein had the ability to suppress unpleasant conceptual problems with relative ease. And so in November 1915 he plunged ahead in search of generally covariant equations, unfazed by his own arguments against their existence! Once he had them, he quickly found a way to climb out of the hole he had created (as explained in Norton 1989 and Stachel 1989).

By 1916 Einstein was also quite aware that his field equations led directly to the conservation laws for matter $T^{\mu\nu}_{;\nu} = 0$. Nevertheless, he was rather vague about the nature of this connection in Einstein (1916a), his first summary account of the new theory. There he wrote: "the field equations of gravitation contain four conditions which govern the course of material phenomena. They give the equations of material phenomena completely, if the latter is capable of being characterized by four differential equations independent of one another" (*ibid.*, p. 325). He then cited Hilbert's note (Hilbert 1915) for further details, suggesting that he was not yet ready to make a final pronouncement on these issues.

Not until early 1916 did Einstein come to realize that energy conservation can be deduced from the field equations and not the other way around. Pais remarked about this in connection with the tumultuous events of November 1915:

> Einstein *still* did not know [the contracted Bianchi identities] on November 25 and therefore did not realize that the energy-momentum conservation laws
>
> $$T^{\mu\nu}_{;\nu} = 0 \tag{22.5}$$
>
> follow automatically from [(22.1)] and [(22.2)]. *Instead, he used these conservation laws* [(22.5)] *as a constraint on the theory!* (Pais 1982, 256).

Such 20–20 hindsight was no doubt helpful when it came to identifying this particular source of Einstein's difficulties, but it hardly explains what happened in November 1915. Nor does it shed much light on subsequent developments. Paging ahead, we see that Pais returns to the Bianchi identities in his discussion of energy-momentum conservation, which in 1918 was one of the most hotly debated topics in GRT (see Cattani and De Maria 1993). There he points out that, "from a modern point of view, the identities [(22.1)] and [(22.5)] are special cases of a celebrated theorem of Emmy Noether, who herself participated in the Göttingen debates on the energy-momentum law" (Pais 1982, 276). But back in November 1915 "neither Hilbert nor Einstein was aware of this royal road to the conservation laws" (Pais 1982, 274). Pais might have added that even in 1918 no one in Göttingen seems to have realized the connection between Noether's results on identities derived from variational principles and the classical Bianchi identities.

A little bit of contextualization can go a long way here. During the period 1916–1918 only a few individuals were in a position to see the connection between Einstein's equations and the Bianchi identities, even though the latter were quite familiar to those immersed in Italian differential geometry. Among these experts, only Levi-Civita seems to have seen the relevance of the Bianchi identities immediately (Fig. 22.1). But, as we shall see below, by 1918 a handful of others began to rediscover what the Italians had already largely forgotten. Emmy Noether, however, was not among them. Her work on GRT was mainly rooted in Sophus Lie's theory of differential equations, as applied to variational problems, an area Lie left untouched. Most importantly, Noether's efforts came as a response to a set of problems first raised by Hilbert, who tried to synthesize Einstein's theory of gravitation with Gustav Mie's theory of matter (see Rowe 1999).

Hilbert's approach to energy conservation in 1915–1916 used a generally invariant variational principle, which he claimed (without proof) led to four differential identities linking the Lagrangian derivatives (Hilbert 1915). No one understood this argument at the time, and only recently have historians managed to disentangle its many threads (see Sauer 1999 and Renn and Stachel 1999). Several other related issues remained murky, as well. The relationship between Einstein's theory and Hilbert's adaptation of Mie's matter theory, for example, was by no means clear. Nor was it easy to discern whether one could formulate conservation laws in general relativity that were fully analogous to those of classical physics. Aided by Emmy Noether, in 1918 Felix Klein eventually managed to find a simpler way to construct Hilbert's invariant energy vector in Klein (1918a) and Klein (1918b). He also urged Noether to explore Hilbert's assertion regarding the four identities that he saw as the key to energy conservation in GRT. In July 1918 she generalized and proved this result as one of *two* fundamental theorems on invariant variational problems (Noether 1918). Although famous today, Noether's theorems evoked very little interest at the time they were published. Moreover, unlike Pais, no contemporary writer linked Noether's results with the Bianchi identities so far as I have been able to discern. Here I should say a few words about how these identities arise in Riemannian geometry.

If $R^{\sigma}_{\mu\nu\kappa}$ is the Riemann-Christoffel tensor, then lowering the index σ yields the purely covariant curvature tensor

$$R_{\lambda\mu\nu\kappa} \equiv g_{\lambda\sigma} R^{\sigma}_{\mu\nu\kappa}.$$

The latter satisfies the following three properties:

$$R_{\lambda\mu\nu\kappa} = R_{\nu\kappa\lambda\mu} \tag{22.6}$$

$$R_{\lambda\mu\nu\kappa} = -R_{\mu\lambda\nu\kappa} = -R_{\lambda\mu\kappa\nu} = +R_{\mu\lambda\kappa\nu} \tag{22.7}$$

$$R_{\lambda\mu\nu\kappa} + R_{\lambda\kappa\mu\nu} + R_{\lambda\nu\kappa\mu} = 0 \tag{22.8}$$

Fig. 22.1 Tullio Levi-Civita was the leading expert on the absolute differential calculus in Italy. Together with his teacher Gregorio Ricci, he coauthored an oft-cited paper on the Ricci calculus published in Mathematische Annalen in 1901. In 1915, Einstein confided to a friend that Levi-Civita was probably the only one who grasped his gravitational theory completely: "because he is familiar with the mathematics used. But he is seeking to tamper with one of the most important proofs in an incessant exchange of correspondence. Corresponding with him is unusually interesting; it is currently my favorite pastime" (Einstein 1998, 117–118). Their correspondence broke off, however, about one month later when Italy entered the war against the Axis powers.

These algebraic conditions imply that $R_{\lambda\mu\nu\kappa}$ has only $C_n = \frac{1}{12}n^2(n^2-1)$ independent components. For the space-time formalism of GRT, where $n = 4$, the Riemann-Christoffel tensor thus depends on $C_4 = 20$ parameters. Using its covariant form, the classical Bianchi identities read:

$$R_{\lambda\mu\nu\kappa;\eta} + R_{\lambda\mu\eta\nu;\kappa} + R_{\lambda\mu\kappa\eta;\nu} = 0, \qquad (22.9)$$

where the last three indices ν, κ, η are permuted cyclically. The connection between these identities and Einstein's equations follows immediately from two basic results of the tensor calculus: (1) raising and lowering indices commutes with covariant differentiation, and (2) Ricci's lemma, which asserts that the covariant derivative of the fundamental tensor $g^{\mu\nu}$ vanishes, $g^{\mu\nu}_{;\kappa} = 0$. Thus, by multiplying (22.9) by $g^{\lambda\nu}$ and contracting, we obtain in view of (22.6) and (22.7):

$$R_{\mu\kappa;\eta} - R_{\mu\eta;\kappa} + R^{\nu}_{\mu\kappa\eta;\nu} = 0. \qquad (22.10)$$

Multiplying by $g^{\mu\kappa}$ and contracting again yields

$$R_{;\eta} - R^{\mu}_{\eta;\mu} - R^{\nu}_{\eta;\nu} = 0,$$

$$(R^{\nu}_{\eta} - \frac{1}{2}\delta^{\nu}_{\eta}R)_{;\nu} = 0,$$

which are the contracted Bianchi identities (22.1):

$$(R^{\mu\nu} - \frac{1}{2}g^{\mu\nu}R)_{;\nu} = G^{\mu\nu}_{;\nu} = 0.$$

These conditions simply assert that the divergence of the Einstein tensor vanishes, a relation already derived by Weyl in 1917 using variational methods. Yet neither he nor his Göttingen mentors, Hilbert and Klein, recognized that these identities could be obtained directly as above using elementary tensor calculus. As I will indicate below, disentangling these differential-theoretic threads from variational principles took considerable time and effort. Even as late as 1922 the Bianchi identities and their significance for GRT were still being "rediscovered" anew.

Clearly, Pais's retrospective account skirts the real historical difficulties, telling us more about what didn't happen between November 1915 and July 1918 than about what actually did. In the meantime, however, a number of new studies have cast fresh light on the early history of GRT (see especially the articles in *Einstein Studies*, vols. 1, 3, 5, 7). Moreover, the crucial period 1914–1918 has become more readily accessible through the publication of volumes 6 and 8 of the *Collected Papers of Albert Einstein* (Einstein 1996 and Einstein 1998). Michel Janssen's commentaries and annotations in Einstein (1998) are particularly helpful when it comes to contextualizing the topics I address below. This new source material has helped sustain a flurry of recent research on the early history of general relativity, some of which has managed to fill important gaps left open by earlier researchers. Still, no one since Abraham Pais has addressed the issues surrounding the interplay (or lack thereof) between the Bianchi identities and GRT, especially Einstein's equations, energy-momentum conservation, and Noether's theorems. So without any pretense of doing justice to this rich topic, let me take it up once again here.

Einstein's Field Equations

Roughly a half year after Einstein delivered six lectures on general relativity in Göttingen, Hilbert entered the field with his famous note Hilbert (1915) on the foundations of physics, dated 20 November 1915. Until quite recently, historians had paid little attention to the substance of this paper, which makes horribly difficult reading. Its fame stems from one brief passage in which Hilbert asserted that his gravitational equations, derived from an invariant Lagrangian, were identical to Einstein's equations (22.2), submitted to the Berlin Academy five days later. It was long believed that Hilbert's "derivation" was more elegant, and to many it appeared he and Einstein had found the same equations virtually simultaneously. As they had corresponded with one another during November 1915, it was natural to speculate about who might have influenced whom. Curiously, this interest centered exclusively on a *post hoc* adjudication of priority claims, as historians pondered over who should get credit for finding and/or deriving the Einstein equations: Einstein, Hilbert, or both? Probably Jagdish Mehra went furthest in pushing the case for Hilbert in Mehra (1973), but until recently the balance of opinion was aptly summarized by Pais, who wrote: "Einstein was the sole creator of the physical theory of general relativity and ... both he and Hilbert should be credited for the discovery of the fundamental equation" (Pais 1982, 260).

Today we know better: when Hilbert submitted his text on 20 November 1915, it did not contain the equations (22.2) (see Rowe 2001). In fact, Hilbert's original text contained no explicit form for his 10 gravitational field equations, which he derived from a variational principle by varying the components of the metric tensor $g^{\mu\nu}$. Only later, some time after 6 December, did he add the key passage containing a form of (22.2) into the page proofs. Presumably he did so without any wish to stake a priority claim, for he cited Einstein's paper of 25 November. Moreover, the explicit field equations play no role whatsoever in the rest of Hilbert's paper. It therefore seems likely that he added this paragraph merely in order to make a connection with Einstein's results, which were by no means clear or easily accessible at this time (only in his fourth and final November note did Einstein present generally covariant field equations with the trace term). So the equations (22.2) are rightly called "Einstein's equations" and not the "Einstein-Hilbert equations." This "belated priority issue" was definitely put to rest in Corry et al. (1997). Unfortunately, some of the authors' other more speculative claims have now been spun into a highly romanticized popular account in *God's Equation* (Aczel 1999).

To the question who first derived the Einstein equations, the answer is less clear. If by a derivation we mean an argument showing that the equations (22.2) uniquely satisfy a certain number of natural properties, then for Einstein and Hilbert one can only reach the conclusion "none of the above" (see Rowe 2001, 416–418). Hilbert, in particular, failed to show how the Einstein equations could be obtained from those of his own theory, citing a bit of folklore about second rank tensors that he probably got from Einstein. One finds scattered hints in Einstein's published and unpublished papers indicating that the only possible second-rank tensors obtainable from the metric tensor and its first and second derivatives and linear in the latter must be of the form:

$$a\,R^{\mu\nu} + b\,g^{\mu\nu}R + c\,g^{\mu\nu} = 0. \tag{22.11}$$

Einstein knew very well that this mathematical result was important for narrowing down the candidates for generally covariant field equations. And without these mathematical underpinnings, his claim (Einstein 1996, 318–319) that the equations (22.2) represent the most natural generalization of Newton's theory would have been seriously weakened. Still, he simply took this for granted, possibly because he relied on Grossmann's (limited) expertise in the theory of differential invariants.

By 1917 Felix Klein asked his assistant Hermann Vermeil to give a direct proof of this fundamental result to which both Einstein and Hilbert had appealed. By employing so-called normal coordinates, as first introduced by Riemann, Vermeil was able to prove that the Riemannian curvature scalar R was the only absolute invariant that satisfied the above conditions (see Vermeil 1917). In 1921 Max von Laue completed Vermeil's argument in von Laue (1921, 99–104), and Hermann Weyl gave an even more direct proof of Vermeil's result in Weyl (1922, Appendix II, 315–317). Wolfgang Pauli also referred to Vermeil's work in his definitive report (Pauli 1921), but one otherwise finds very few references to such formal issues in the vast literature on GRT.[1]

So who first proved Einstein's equations? If this were a game show question, one might be tempted to answer: Hermann Vermeil. But a more serious response would begin by reformulating the question: when and why did physicists and mathematicians become interested in foundational issues like proving Einstein's equations? To answer this it is again helpful to look carefully at local contexts. Among the more important centers for research on GRT were Leiden, Rome, Cambridge, and Vienna. In the case of Einstein's equations, this was largely a mopping-up operation, part of a communal effort orchestrated by Felix Klein in Göttingen. Klein's initial interest in general relativity focused on the geometrical underpinnings of the theory, including the various "degrees of curvature" in space-times (Eddington's terminology in

[1] An exception is the work of David Lovelock, who proved that the only divergence-free, contravariant second-rank tensor densities in dimension four are of the form $a\sqrt{g}G^{\mu\nu} + b\sqrt{g}g^{\mu\nu}$ in Lovelock (1972).

Eddington 1920, 91). By early 1918, however, Klein became even more puzzled by the various results on energy conservation in GRT that had been obtained by Einstein, Hilbert, Lorentz, Weyl, and Emmy Noether. He was not alone in this.

General Relativity in Göttingen

As Einstein himself conceded, energy-momentum conservation was the one facet of his theory that caused virtually all the experts to shake their heads. Back in May 1916, he had struggled to understand Hilbert's approach to this problem, the topic of a lecture he was preparing for the Berlin physics colloquium. Twice he wrote Hilbert asking him to explain various steps in his complicated chain of reasoning (24 and 30 May, 1916, Einstein 1998, 289–290, 293–294). Einstein expressed gratitude for Hilbert's illuminating replies, but to his friend Paul Ehrenfest he remarked: "Hilbert's description doesn't appeal to me. It is unnecessarily specialized regarding 'matter,' is unnecessarily complicated, and not straightforward (=Gauss-like) in set-up (feigning the super-human through concealment of the methods)" (24 May, 1916, Einstein 1998, 288). But Hilbert couldn't feign that he understood the connection between his approach to energy conservation and Einstein's. About this, he intimated to Einstein that "[m]y energy law is probably related to yours; I have already given this question to Frl. Noether" (27 May 1916, Einstein 1998, 291). She apparently made some progress on this problem at the time, as Hilbert later acknowledged: "Emmy Noether, whose help I called upon more than a year ago to clarify these types of analytical questions pertaining to my energy theorem, found at that time that the energy components I had set forth—as well as those of Einstein—could be formally transposed by means of the Lagrangian differential equations . . . into expressions whose divergence vanished *identically*. . . " (Klein 1918a, 560–561).

By late 1917 Klein reengaged Noether in a new round of efforts to crack the problem of energy conservation (see Rowe 1999, 213–228). Klein's discomfort with energy conservation in GRT had to do with his knowledge of classical mechanics in the tradition of Jacobi and Hamilton. There conservations laws help to describe the equations of motion of physical systems which would otherwise be too hopelessly complicated to handle as an n-body problem. In GRT, by contrast, the conservation laws for matter (22.5) could be derived directly from the field equations (22.2) without any recourse to other physical principles. Hilbert's work pointed in this direction, but his "purely axiomatic" presentation only obscured what was already a difficult problem. Klein later described Hilbert's presentation as "completely disordered (evidently a product of great exertion and excitement)" (Lecture notes, 10 December 1920, Klein Nachlass XXII C, p. 18).

In early 1918 Klein succeeded in giving a simplified derivation of Hilbert's invariant energy equation, which involves a very complicated entity e^{ν} known as Hilbert's energy vector satisfying $\mathrm{Div}(e^{\nu}) = 0$. Klein emphasized that this relation should be understood as an identity rather than as an analogue to energy conservation in classical mechanics. He noted that in mechanics the differential equation

$$\frac{d(T+U)}{dt} = 0 \tag{22.12}$$

cannot be derived without invoking specific physical properties, whereas in GRT the equation $\mathrm{Div}(e^{\nu}) = 0$ follows from variational methods, the principle of general covariance, and Hilbert's 14 field equations for gravity and matter.

Klein's article was written in the form of a letter to Hilbert. After discussing the main mathematical points, Klein remarked that "Frl. Noether continually advises me in my work and that actually it is only through her that I have delved into these matters" Klein (1918a, 559). Hilbert expressed total agreement not only with Klein's derivation but with his interpretation of it as well. He even claimed one could prove a theorem that ruled out conservation laws in GRT analogous to those that hold for physical theories based on an orthogonal group of coordinate transformations. Klein replied that he would be very interested "to see the mathematical proof carried out that you alluded to in your answer." He then turned to Emmy Noether, who resolved the issue six months later in her fundamental paper Noether (1918).

In the meantime, Einstein had taken notice of this little published exchange, and in March 1918 he wrote Klein: "With great pleasure I read your extraordinarily penetrating and elegant discussion on Hilbert's first note. Nevertheless, I regard what you remark about my formulation of the conservation laws as incorrect" (13 March, 1918, Einstein 1998, 673). Einstein objected to Klein's claim that his approach to energy-momentum conservation could be derived from the same formal relationships that Klein had applied to Hilbert's theory. Instead, Einstein insisted that "exactly analogous relationships hold [in GRT] as in the non-relativistic theories." After explaining the physical import of his own formalisms, he added: "I hope that this anything but complete explanation enables you to grasp what I mean. Most of all, I hope you will alter your opinion that I had obtained for the energy theorem an identity, that is an equation that places no conditions on the quantities that appear in it" (*ibid.*, p. 674).

Eight days later, Klein replied with a ten-step argument aimed at demolishing Einstein's objections. His main point was that Einstein's approach to energy-momentum conservation expressed nothing beyond the information deducible from the variational apparatus and the field equations that can be derived from it. Einstein countered by asserting that his version of energy-momentum conservation was not a trivial consequence of the field equations. Furthermore, if one has a physical system where the energy tensors for matter and the gravitational field, T^{ν}_{σ} and t^{ν}_{σ}, vanish on the boundary, then from the differential form of Einstein's conservation laws

$$\sum_{\nu} \frac{\partial(T^{\nu}_{\sigma} + t^{\nu}_{\sigma})}{\partial x_{\nu}} = 0 \tag{22.13}$$

one could derive an integral form that was physically meaningful:

$$\frac{d}{dx^4}\{\int (T^4_{\sigma} + t^4_{\sigma})dV\} = 0, \quad \text{for} \quad \sigma = 1, 2, 3, 4. \tag{22.14}$$

Einstein stressed to Klein that the constancy of these four integrals with respect to time could be regarded as analogous to the conservation of energy and momentum in classical mechanics.

Klein eventually came to appreciate Einstein's views, though only after giving up on an alternative approach suggested by his colleague Carl Runge. Several experts, including Lorentz and Levi-Civita, objected to Einstein's use of the pseudo-tensor t^{ν}_{σ} to represent gravitational energy. Klein and Runge briefly explored the possibility of dispensing with this t^{ν}_{σ}, but Noether threw cold water on Runge's proposal for doing so (Rowe 1999, 217–218). By July 1918, Klein wrote Einstein that he and Runge had withdrawn their publication plans, and that he was now investigating Einstein's formulation of energy conservation based on $T^{\nu}_{\sigma}+t^{\nu}_{\sigma}$. To this, Einstein replied: "It is very good that you want to clarify the formal significance of the t^{ν}_{σ}. For I must admit that the derivation of the energy theorem for field and matter together appears unsatisfying from the mathematical standpoint, so that one cannot characterize the t^{ν}_{σ} formally" (Einstein 1998, 834).

Einstein and Klein quickly got over their initial differences regarding the status of Einstein's (22.13), and afterward Klein dealt with this topic and the various approaches to energy conservation adopted by Einstein, Hilbert, and Lorentz in Klein (1918b). Einstein responded with enthusiasm: "I have already studied your paper most thoroughly and with true amazement. You have clarified this difficult matter fully. Everything is wonderfully transparent" (A. Einstein to F. Klein, 22 October, 1918, Einstein 1998, 917). The contrast between this response and Einstein's reaction to Hilbert's work on GRT (noted above) could hardly have been starker. Perhaps the supreme irony in this whole story lies here. For Klein (1918b) is nothing less than a carefully crafted *axiomatic* argument, set forth by a strong critic of modern axiomatics largely in order to rectify the flaws in Hilbert's attempt to wed GRT to Mie's theory of matter via the axiomatic method.

In this paper Klein developed ideas that were closely linked with Noether (1918), though he mentions this parallel work only in the concluding paragraph. Emmy Noether, on the other hand, gave several explicit references to Klein (1918b) that make these interconnections very clear. In his private lecture notes, Klein later wrote that it was only "through the collaboration of Frl. Noether and me" that (Hilbert 1915) "was completely decoded" (Lecture notes, 10 December 1920, Klein Nachlass XXII C, p. 18). Today, this jointly undertaken work would normally appear under the names of both authors, but back in 1918 Emmy Noether wasn't even allowed to habilitate in Göttingen, despite the backing of both Hilbert and Klein.

On Rediscovering the Bianchi Identities

All of these events, it must not be forgotten, took place against the backdrop of the Great War that nearly brought European civilization to its knees. Einstein's revolutionary theory of gravitation interested almost no one prior to 1916, and before November 1919 only a handful of experts had written about it. But afterward, Einstein and relativity emerged as two watchwords for modernity. On 6 November, just before the Versailles Treaty was to take effect, the British scientific world announced that Einstein's prediction regarding the bending of light in the sun's gravitational field had been confirmed. Thereafter, the creator of general relativity was no longer merely a famous physicist: he emerged as one of the era's leading cultural icons. But let's now wind back the reel and look again at GRT during the Great War.

Once Einstein's mature theory came out in 1916 – alongside Hilbert's paper and the pioneering work of Karl Schwarzschild containing the first exact solutions of the Einstein equations – many mathematicians and physicists began to take up GRT and the Ricci calculus. Doing so in wartime, however, presented real difficulties. Communication between leading protagonists in Italy and Germany proved next to impossible, as the lapse in Einstein's correspondence with Tullio Levi-Civita demonstrated. Through his friend Adolf Hurwitz, whom he visited in Zurich in August 1917, Einstein managed to get his hands on Levi-Civita's paper Levi-Civita (1917b), which briefly reignited their earlier correspondence. Like many others, Levi-Civita found Einstein's formulation of energy conservation unacceptable due to his use of the pseudotensor t^{ν}_{σ} for gravitational energy (Cattani and De Maria 1993). What Einstein (and presumably everyone else in Germany) overlooked was that in this paper Levi-Civita employed the classical Bianchi identities. In Levi-Civita (1917a) he introduced an even more fundamental concept: parallel displacement of vectors in Riemannian spaces, a notion quickly taken up by Hermann Weyl and Gerhard Hessenberg.

These fast-breaking mathematical developments raised staggering difficulties, and not just for physicists like Einstein. None of the mathematicians in Göttingen was a *bona fide* expert in differential geometry, which helps explain why no one in the Göttingen crowd recognized the central importance of the Bianchi identities. Had Klein suspected that the Einstein tensor satisfied four simple differential identities (corresponding to the four parameters in a generally covariant system of equations), he might have turned to his old friend Aurel Voss for advice. Had he asked him, Voss likely would have remembered that he had published a version of the contracted Bianchi identities back in Voss (1880)! Thus, the Göttingen mathematicians clearly could have found references to the Bianchi identities, in either their general or contracted form, in the mathematical literature. They just didn't know where to look. As Pais pointed out, the name Bianchi does not appear in any of the five editions of Weyl's *Raum, Zeit, Materie*, nor did Wolfgang Pauli refer to it in his Encyclopädie article Pauli (1921).

With regard to the Bianchi identities in their full form (22.9), we have it on the authority of Levi-Civita that these were known to his teacher, Gregorio Ricci (Levi-Civita 1926, 182). Ricci passed this information on to Ernesto Padova, who published the identities without proof in Padova (1889). They were thereafter forgotten, even by Ricci, and then rediscovered by Luigi Bianchi, who published them in Bianchi (1902). Both Ricci and Bianchi had earlier studied under Felix Klein, who solicited the now famous paper Ricci and Levi-Civita (1901) for *Mathematische Annalen*; but this classic apparently made little immediate impact. Indeed, before the work of Einstein and Grossmann, Ricci's absolute differential calculus was barely known outside Italy (Reich 1992).

By 1918 a number of investigators outside Italy had begun to stumble upon various forms of the full or contracted Bianchi identities. Two of them, Rudolf Förster and Friedrich Kottler, even passed their findings on to Einstein in letters (see Einstein 1998, 646, 7049). Förster, who published under the pseudonym Rudolf Bach, took his doctorate under Hilbert in Göttingen in 1908 and later worked as a technical assistant for the Krupp works in Essen. In explaining to Einstein that the identities (22.1) follow directly from (22.9), he noted that the latter "relations appear to be still completely unknown" (*ibid.*). Förster contemplated publishing these results, but only did so in Bach (1921), which dealt with Weyl's generalization of Riemannian geometry. Even at this late date he presented these identities as "new" (*ibid.*, p. 114).

During the war years, the Dutch astronomer Willem De Sitter introduced the British scientific community to Einstein's mature theory. This helped spark Arthur Stanley Eddington's interest in GRT and the publication of Eddington (1918), which contains the contracted Bianchi identities. Two years later, in Eddington (1920), he expressed doubt that anyone had ever verified these identities by straightforward calculation, and so he went ahead and carried this out himself for the theoretical supplement in the French edition of Eddington (1920).

In 1922 Eddington's calculations were simplified by G. B. Jeffery, and almost immediately afterward the English physicist A. E. Harward reproved Bianchi's identities (22.9) and used them to derive the conservation of energy-momentum in Einstein's theory in Harward (1922). He also conjectured that he was probably not the first to have discovered (22.9). In Schouten and Struik (1924), an open letter to the *Philosophical Magazine*, dated 28 April 1923, J. A. Schouten and Dirk Struik (Fig. 22.2) confirmed Harward's conjecture, noting that (22.9) "is known, especially in Germany and Italy, as 'Bianchi's Identity.'" More importantly, they emphasized that similar identities hold in affine spaces (those that do not admit a Riemannian line element $ds^2 = g_{\mu\nu}dx^\mu dx^\nu$). Regarding these, they referred explicitly to Bach (1921) for Weyl's gauge spaces as well as a 1923 paper by Schouten for non-Riemannian spaces with a symmetric connection $\Gamma^\lambda_{\mu\nu} = \Gamma^\lambda_{\nu\mu}$ (Fig. 22.3). These results and many more appeared soon afterward in Schouten's 1924 textbook *Der Ricci-Kalkül*. By this time, of course, the dust had largely cleared, as a number of leading experts—including Schouten and Struik, Veblen, Weitzenböck, and Berwald—had by now shown the importance of Bianchi-like identities in non-Riemannian geometries.

Fig. 22.2 Dirk Jan Struik, ca. 1920, when he was working as an assistant to Jan Arnoldus Schouten in Delft. Earlier Struik had studied in Leiden with Paul Ehrenfest, a close personal friend of Einstein's who therefore realized the importance of Ricci's calculus for general relativity at an early stage. Ehrenfest arranged for Struik to meet with Schouten, Holland's leading differential geometer. Thus began a collaboration that led to several books and articles during the 1920s and 1930s.

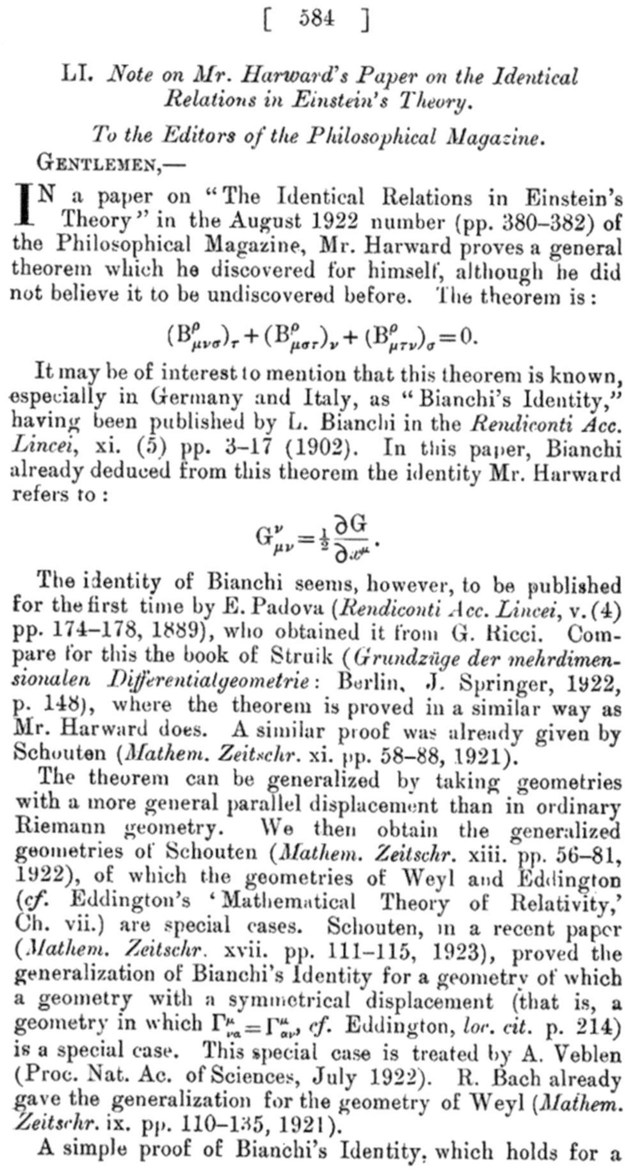

[584]

LI. *Note on Mr. Harward's Paper on the Identical Relations in Einstein's Theory.*

To the Editors of the Philosophical Magazine.

GENTLEMEN,—

IN a paper on "The Identical Relations in Einstein's Theory" in the August 1922 number (pp. 380–382) of the Philosophical Magazine, Mr. Harward proves a general theorem which he discovered for himself, although he did not believe it to be undiscovered before. The theorem is:

$$(\mathrm{B}^{\rho}_{\mu\nu\sigma})_{\tau}+(\mathrm{B}^{\rho}_{\mu\sigma\tau})_{\nu}+(\mathrm{B}^{\rho}_{\mu\tau\nu})_{\sigma}=0.$$

It may be of interest to mention that this theorem is known, especially in Germany and Italy, as "Bianchi's Identity," having been published by L. Bianchi in the *Rendiconti Acc. Lincei*, xi. (5) pp. 3–17 (1902). In this paper, Bianchi already deduced from this theorem the identity Mr. Harward refers to:

$$\mathrm{G}^{\nu}_{\mu\nu}=\tfrac{1}{2}\frac{\partial \mathrm{G}}{\partial x^{\mu}}.$$

The identity of Bianchi seems, however, to be published for the first time by E. Padova (*Rendiconti Acc. Lincei*, v. (4) pp. 174–178, 1889), who obtained it from G. Ricci. Compare for this the book of Struik (*Grundzüge der mehrdimensionalen Differentialgeometrie*: Berlin, J. Springer, 1922, p. 148), where the theorem is proved in a similar way as Mr. Harward does. A similar proof was already given by Schouten (*Mathem. Zeitschr.* xi. pp. 58–88, 1921).

The theorem can be generalized by taking geometries with a more general parallel displacement than in ordinary Riemann geometry. We then obtain the generalized geometries of Schouten (*Mathem. Zeitschr.* xiii. pp. 56–81, 1922), of which the geometries of Weyl and Eddington (*cf.* Eddington's 'Mathematical Theory of Relativity,' Ch. vii.) are special cases. Schouten, in a recent paper (*Mathem. Zeitschr.* xvii. pp. 111–115, 1923), proved the generalization of Bianchi's Identity for a geometry of which a geometry with a symmetrical displacement (that is, a geometry in which $\Gamma^{\mu}_{\nu\alpha}=\Gamma^{\mu}_{\alpha\nu}$, *cf.* Eddington, *loc. cit.* p. 214) is a special case. This special case is treated by A. Veblen (Proc. Nat. Ac. of Sciences, July 1922). R. Bach already gave the generalization for the geometry of Weyl (*Mathem. Zeitschr.* ix. pp. 110–135, 1921).

A simple proof of Bianchi's Identity, which holds for a

Fig. 22.3 First page from an open letter Schouten and Struik wrote for the Philosophical Magazine on 28 April, 1923. Their account clarified several historical issues involving the Bianchi identities. It also contained a simple proof of these identities for spaces with a symmetrical connection as suggested by the Prague mathematician, Ludwig Berwald. Struik presumably knew Berwald through his wife, Ruth, who studied mathematics in Prague during happier days. In 1941, Berwald and his wife were transported to the ghetto in Lodz, where they died from malnourishment.

What should be made of all this groping in the dark? No doubt a certain degree of confusion arose due to the importance Hilbert and others attached to variational principles in mathematical physics. Not surprisingly, within Göttingen circles there was considerable expertise in the use of sophisticated variational methods. Emmy Noether coupled these with invariant theory to obtain her impressive results. But she and her mentors had relatively little familiarity with Italian differential geometry. Ironically, this widespread lack of fundamental knowledge of tensor analysis had at least one important payoff. It gave the aged Felix Klein an inducement to explore the mathematical foundations of general relativity theory. He did so by drawing on ideas familiar from his youth, most importantly Sophus Lie's work on the connection between continuous groups and systems of differential equations. Moreover, his efforts helped clarify one of the most baffling and controversial aspects of Einstein's theory: energy-momentum conservation. Even Hilbert's muddled derivation of an invariant energy vector found its proper place in the scheme set forth in Klein (1918b). Through Klein, Emmy Noether became deeply immersed in these complicated problems, and she succeeded in extracting from them two fundamental theorems in the calculus of variations that would later provide field physicists with an important tool for the derivation of conservations laws.

Mathematicians often prefer to figure out something on their own rather than read someone else's work, so we need not be surprised that the classical Bianchi identities escaped the notice of such eminent mathematicians as Hilbert, Klein, Weyl, Noether, and of course Einstein himself. Had they known them, the early history of the general theory of relativity might have looked less confused.

Acknowledgements The author is grateful to Michel Janssen and Tilman Sauer for their perceptive remarks on an earlier version of this column. Chandler Davis deserves a note of thanks, too, for posing questions that helped clarify some obscure points. Remaining errors and misjudgments are, of course, my own, and may even reflect a failure to heed wise counsel.

References

Aczel, Amir D. 1999. *God's equation. Einstein, relativity and the expanding universe*. New York: Delta.

Bach, Rudolf. 1921. Zur Weylschen Relativitätstheorie und die Weylschen Erweiterung des Krümmungstensorbegriffs. *Mathematische Zeitschrift* 9: 110–135.

Bianchi, Luigi. 1902. Sui simboli a quattro indici e sulla curvature di Riemann. *Rendiconti della Reale Accademia dei Lincei (V)* 11(11): 3–7.

Cattani, Carlo, and Michelangelo De Maria. 1993. Conservation laws and gravitational waves in general relativity (1915–1918). In [Earman, Janssen, Norton 1993], 63–87.

Corry, Leo, Jürgen Renn, and John Stachel. 1997. Belated decision in the Hilbert-Einstein priority dispute. *Science* 278: 1270–1273.

Earman, John, Michel Janssen, and John D. Norton, eds. 1993. *The attraction of gravitation: New studies in the history of general relativity*, Einstein studies, vol. 5. Boston: Birkhäuser.

Eddington, A. S. 1918. *Report on the relativity theory of gravitation*. London: Fleetway Press.

Eddington, A. S. 1920. *Space, time and gravitation*. Cambridge: Cambridge University Press.

Einstein, Albert. 1914. Prinzipielles zur verallgemeinerten Relativitätstheorie und Gravitationstheorie. *Physikalische Zeitschrift* 15: 176–180; reprinted in (Einstein 1996, 572–578).

Einstein, Albert. 1916a. Grundlagen der allgemeinen Relativitätstheorie. *Annalen der Physik* 49: 283–339. Reprinted in (Einstein 1996, 283–348).

Einstein, Albert. 1996. *The Collected Papers of Albert Einstein*, The Berlin years: Writings, 1914–1917, vol. 6, ed. A. J. Kox, Martin J. Klein, and Robert Schulmann. Princeton: Princeton University Press.

Einstein, Albert. 1998. *The Collected Papers of Albert Einstein*, The Berlin years: Correspondence, 1914–1918, vol. 8, ed. Robert Schulmann, A. J. Kox, Michel Janssen, Jósef Illy. Princeton: Princeton University Press.

Harward, A. E. 1922. The identical relations in Einstein's theory. *Philosophical Magazine* 44: 380–382.

Hilbert, David. 1915. Die Grundlagen der Physik (Erste Mitteilung), *Nachrichten der königlichen Gesellschaft der Wissenschaften zu Göttingen, Mathematisch-physikalische Klasse,* 395–407.

Howard, Don, and John Stachel, eds. 1989. *Einstein and the history of general relativity*, Einstein Studies, vol. 1. Boston: Birkhäuser.

Klein, Felix. 1918a. Zu Hilberts erster Note über die Grundlagen der Physik, *Nachrichten der Königlichen Gesellschaft der Wissenschaften zu Göttingen, Mathematisch-Physikalische Klasse,* vol. 1, 553–565. Reprinted in [Klein 1921–1923], 553–567.

Klein, Felix. 1918b. Über die Differentialgesetze für die Erhaltung von Impuls und Energie in der Einsteinschen Gravitationstheorie, *Nachrichten der Königlichen Gesellschaft der Wissenschaften zu Göttingen, Mathematisch-Physikalische Klasse,* vol. 1, 568–584. Reprinted in [Klein 1921–1923], 568–585.

Klein, Felix. 1921. *Gesammelte Mathematische Abhandlungen*, Bd. 1. Berlin: Julius Springer.

Levi-Civita, Tullio. 1917a. Nozione di parallelismo in una varietà qualunque e conseguente specificazione geometrica della curvatura Riemanniana, *Circolo Matematico di Palermo. Rendiconti* 42: 173–205.

Levi-Civita, Tullio. 1917b. Sulla espressione analitica spettante al tensore gravitazionale nella teoria di Einstein, *Rendiconti della Reale Academia dei Lincei. Atti* 26: 381–391.

Levi-Civita, Tullio. 1926. The absolute differential calculus, trans. Marjorie Long. London: Blackie & Son.

Lovelock, David. 1972. The four-dimensionality of space and the Einstein tensor. *Journal of Mathematical Physics* 13: 874–876.

Mehra, Jagdish. 1973. Einstein, Hilbert, and the theory of gravitation. In *The physicist's conception of nature*, ed. Jagdish Mehra, 92–178. Dordrecht: Reidel.

Noether, Emmy, 1918. Invariante Variationsprobleme, Nachrichten der Königlichen Gesellschaft der Wissenschaften zu Göttingen, Mathematisch-Physikalische Klasse, 235–257. In [Earman, Janssen, Norton 1993], 63–87.

Norton, John. 1989. How Einstein found his field equations. In [Howard and Stachel 1989], 101–159.

Padova, Ernesto. 1889. Sulle deformazioni infinitesime. *Rendiconti della Reale Accademia dei Lincei (IV)* 5(I): 174–178.

Pais, Abraham. 1982. *'Subtle is the Lord... ' The Science and the Life of Albert Einstein*. Oxford: Clarendon Press.

Pauli, Wolfgang. 1921. Relativitätstheorie. In *Encyklepädie der mathematischen Wissenschaften*, vol. 5, part 2, 539–775; Theory of relativity, G. Field, trans. London: Pergamon, 1958.

Reich, Karin. 1992. *Die Entwicklung des Tensorkalküls. Vom absoluten Differentialkalkül zur Relativitätstheorie*, Science networks, vol. 11. Basel: Birkhüuser.

Renn, J., and J. Stachel. 1999. Hilbert's foundation of physics: From a theory of everything to a constituent of general relativity, Max-Planck-Institut for Wissenschaftsgeschichte, Preprint 118.

Ricci, Gregorio, and Tullio Levi-Civita. 1901. Méthodes de calcul differentiel absolu et leurs applications. *Mathematische Annalen* 54: 125–201.

Rowe, David E. 1999. The Göttingen response to general relativity and Emmy Noether's Theorems, the symbolic universe. In *The Symbolic Universe Geometry and Physics, 1890–1930*, ed. Jeremy Gray. Oxford: Oxford University Press, 189–233.

Rowe, David E. 2001. Einstein meets Hilbert: At the crossroads of physics and mathematics. *Physics in Perspective* 3: 379–424.

Sauer, Tilman. 1999. The relativity of discovery: Hilbert's first note on the foundations of physics. *Archive for History of Exact Sciences* 53: 529–575.

Schouten, J. A., and Dirk Struik. 1924. Note on Mr. Harward's paper on the identical relations in Einstein's theory. *Philosophical Magazine* 47: 584–585.

Stachel, John. 1989. Einstein's search for general covariance, 1912–1915. In [Howard and Stachel 1989], 63–100.

Vermeil, Hermann. 1917. Notiz über das mittlere Krümmungsmass einer n-fach ausgedehnten Riemann'schen Mannigfaltigkeit, *Nachrichten der Königlichen Gesellschaft der Wissenschaften zu Göttingen. Mathematisch-physikalische Klasse. Nachrichten* 334–344.

von Laue, Max. 1921. *Die Relativitätstheorie, zweiter Band: Die allgemeine Relativitätstheorie und Einsteins Lehre von der Schwerkraft.* Braunschweig: Vieweg.

Voss, Aurel. 1880. Zur Theorie der Transformation quadratischer Differentialausdrücke und der Krümmung höherer Mannigfaltigketien. *Mathematische Annalen* 16: 129–178.

Weyl, Hermann. 1922. *Space-time-matter*, 4th ed., trans. Henry L. Brose. London: Methuen.

23 Puzzles and Paradoxes and Their (Sometimes) Profounder Implications

(*Mathematical Intelligencer* 33(1)(2011): 55–60)

In Remembrance of Martin Gardner

Martin Gardner had a special affinity for the enchantment of games, puzzles, and simple ideas that lead to unforeseen possibilities (Fig. 23.1). He was also clearly drawn to like-minded spirits of the present and past who shared his delight in such phenomena (a favorite of mine is (Gardner 2009)). By imbibing their lore and showcasing their findings in *Scientific American,* he celebrated creativity while making a signal contribution to our collective culture. For what better way could anyone convey the unlikely idea that doing mathematics can be fun? An avid reader, Gardner often stumbled upon many lost gems of the remote and not so distant past. In this essay, I endeavor to follow just a few of his leads.

Not long ago, I wrote about the largely forgotten German mathematician, Victor Schlegel, a leading disciple of a once obscure Gymnasium teacher named Hermann Grassmann (see Chap. 9). Schlegel's name, in fact, pops up in one of Gardner's delightful essays on the Fibonacci numbers (Gardner 2006), in which he shows how these are tied to an amusing series of dissection problems for plane figures. The standard paradox arises when one divides an 8 by 8 square into four pieces and then on reassembling these discovers that they form a 5 by 13 rectangle (Fig. 23.2):

As Gardner's Dr. Matrix reveals, one can produce any number of such paradoxes by making use of a property of the Fibonacci sequence: 1, 1, 2, 3, 5, 8, 13, 21, ..., namely that the square of any Fibonacci number differs by one from the product of the preceding and succeeding terms. The case illustrated above is just $8^2 + 1 = 5 \cdot 13$.

Now according to Gardner, Schlegel was the first "to generalize the square-rectangle paradox" by making use of this insight, a finding he published in 1879. Martin Gardner also mentions others who dabbled in this funny business, including one of his personal favorites, Lewis Carroll. But he does not indicate where and when this paradox seems to have first popped up. Schlegel published his version in the widely read *Zeitschrift für Mathematik und Physik,* edited by Otto Schlömilch. Thumbing back about 10 years in that journal, we encounter a little note by Schlömilch himself which shows that the original paradox was known at least 10 years earlier (Fig. 23.3).

So in all likelihood Schlegel knew about this oddity through Schlömilch's earlier note, which would also explain why he chose to publish his observations regarding the general square-rectangle paradox in the latter's journal. But moving forward in time, it is natural to wonder how this special dissection problem came to be known in wider circles, beginning with the German-speaking world. Indeed, I would like to show that there is much more to this tale than might first meet the eye, the reason being that some sharp-minded mathematicians realized the classical approach to the dissection of plane rectilinear figures invoked a general axiom in order to dispense with absurdities of the type we are discussing.

Implications for Foundations of Geometry

According to Euclid, when comparing magnitudes of the same type – those which have a ratio to one another – it is always the case that "the whole is greater than the part." Yet since Euclid's theory of plane areas is based on dissection into congruent figures – rather than a theory of measure for areas – a mathematician keen on rigor would clearly like to rely on purely geometrical principles, avoiding the general axioms for magnitudes needed for a general theory of ratios and proportions (Euclid's Book V). Let us now consider dissection paradoxes from this foundational point of view.

Twenty years after Schlegel's paper appeared, an algebraist by the name of Hilbert offered a lecture course on the foundations of Euclidean geometry in Göttingen. Hilbert's first doctoral student, Otto Blumenthal, recalled how he and others reacted with much surprise at this announcement; they had never heard Hilbert talk about anything other than algebraic number fields (Blumenthal 1935, 402). In 1898 projective and non-Euclidean geometry were still very much in vogue, of course, but Euclidean geometry? What, his

D.E. Rowe, *A Richer Picture of Mathematics*, https://doi.org/10.1007/978-3-319-67819-1_23

Fig. 23.1 Martin Gardner with some favorite gadgets (Photo by Elliot Erwitt).

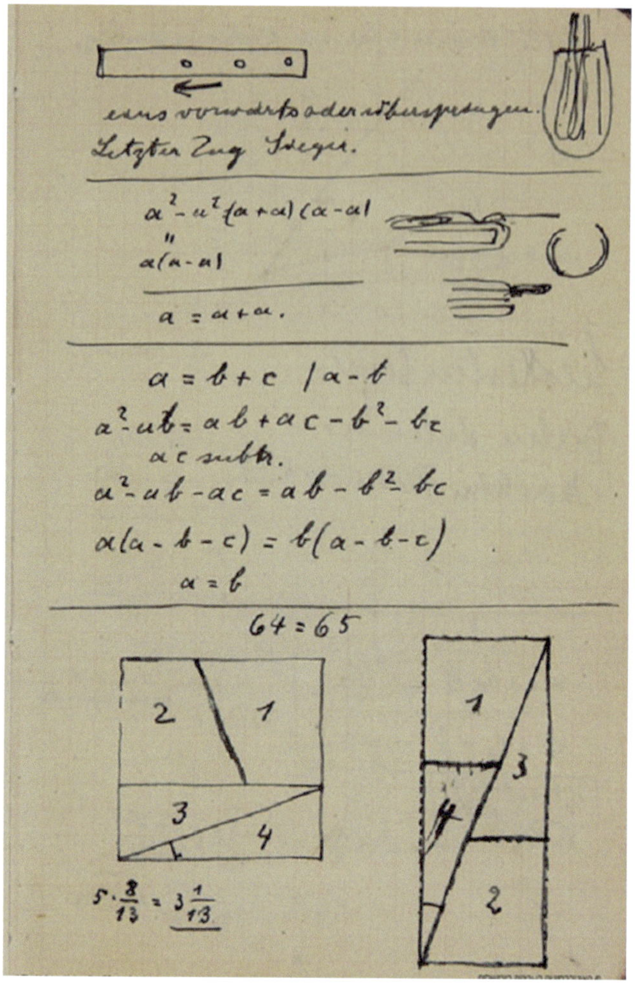

Fig. 23.2 Drawing made by a famous physicist around 1912–1913 showing the original dissection paradox (CPAE 3 1993, 584).

students wondered, could he possibly have up his sleeve? What they heard during the following winter semester was Hilbert's warm-up act for a *Festschrift* essay that would eventually lead to the many editions of his *Grundlagen der Geometrie.* We remember this classic today because with it Hilbert helped launch a movement to modernize the axiomatic method, an approach he extended to a wide range of disciplines that came to be called exact sciences. But if we look more closely at his immediate goals and the material he presented in his original lecture course, some surprising elements come into view (Hilbert 2004).

A central concern for Hilbert was to establish new foundations for a theory of area for plane rectilinear figures by recasting some key results found in the first two books of Euclid's *Elements.* This endeavor had much in common with efforts of several leading geometers who investigated the status of the continuum during the 1890s, though Hilbert hoped to finesse the deeper problems by showing that much of classical geometry can be done without appealing to continuity arguments. Thus his overall strategy aimed to recover standard results in what he called *elementary geometry* without any reliance on continuity assumptions or other axioms taken from a theory of magnitudes. Euclid had, in fact, made explicit use of the latter in the form of common notions, which included the axiom that the whole is always greater than the part. A special case of this had also been assumed by Wilhelm Killing in his book on foundations of geometry, a work Hilbert studied carefully and recommended to students at the beginning of the lecture course. Its importance in the present connection should not be overlooked, even though Hilbert did just that in the pages of his *Grundlagen der Geometrie* (Hilbert 1899).

In Chap. 4 in Hilbert's book he introduced a new theory of content for plane rectilinear figures. This was essentially modeled on Euclid's theory with the important exception that Hilbert based his on a purely geometrical theory of proportion. He developed the latter by introducing a segment arithmetic which gave him a number field whose properties he derived from Pappus' Theorem (or Pascal's Theorem for two lines), a result he proved without recourse to continuity axioms. Hilbert called this purely geometrical theory of content "one of the most remarkable applications of Pascal's theorem in elementary geometry." What he did not say was that this theory enabled him to circumvent Killing's postulate, which states that if a rectangle is decomposed into n rectilinear figures, then it is impossible to fill the rectangle

IX. Ein geometrisches Paradoxon. Um ad oculos zu demonstriren, dass das Schachbret nicht nur 64, sondern auch 65 Felder besitzt, schneide man dasselbe aus starkem Papier, zerlege es auf die in Fig. 1

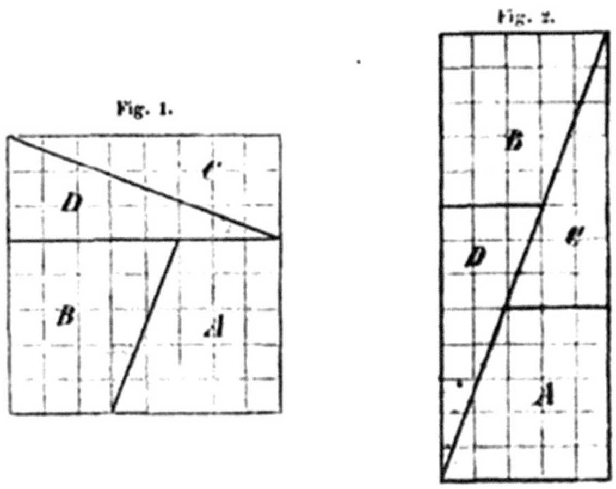

angegebene Weise in vier, zu je zweien congruente Stücke *A*, *B*, *C*, *D* und setze diese zu einem Rechtecke zusammen, welches, wie Fig. 2 zeigt, die Grundlinie 5 und die Höhe 13 besitzt also 65 Felder enthält. — Wir theilen diese kleine Neckerei mit, weil die Aufsuchung des begangenen Fehlers eine hübsche Schüleraufgabe bildet und weil sich an die Vermeidung des Fehlers die Lösung und Construction einer quadratischen Gleichung knüpfen lässt. Schl.

Fig. 23.3 Possibly the first published version of the Square-Rectangle Paradox, Otto Schlömilch, *Zeitschrift für Mathematik und Physik*, 1869.

with only $n - 1$ of these. Surely to complete a jig-saw puzzle we need all of its pieces, and just as surely for the pieces of a rectangle. But this, in effect, is the very situation that arises in the original rectangle-square paradox, for if we divide it into 65 squares, then allegedly only 64 are required to fill out the square formed from the four pieces of the original rectangle!

So did Hilbert know anything about this curious anomaly? Absolutely, he even referred to it explicitly in notes he wrote just prior to the time when he offered his lecture course on foundations of Euclidean geometry. What is more, he cited the book in which he found this figure: *Mathematische Mußestunden* (*Mathematical Pastimes*) by Hermann Schubert (1898), perhaps best known today as the inventor of the Schubert calculus in enumerative geometry. In his own time, though, Schubert's books on recreational mathematics were widely read, and some were also translated into English (Schubert 1903). Within the German-speaking world, Schubert's writings helped to popularize various games and puzzles, including the rectangle-square paradox.

Doodlings of a Physicist

A testimonial of sorts can be found in the pages of a notebook kept by a certain A. Einstein, better known for his physical thought experiments. By 1911 he had already speculated that light rays in the vicinity of the sun would be deflected slightly by the solar gravitational field. His notebook contains sketches from the period 1912–13, and sure enough we find drawings and calculations directly related to this phenomenon (see Fig. 23.4). It seems he even thought about the possibility of gravitational lensing back in those days (Renn et al. 1997).

Mixed together with these daring new speculations about the bending of light rays were some of Einstein's doodles on an array of mathematical curiosities, including those found in Fig. 23.2 above. Here we find, next to the dissection of a square "proving" that 64 = 65, allusions to several other samples from Schubert's garden of puzzles and problems:

(1) the Königsberg bridge problem, solved by Euler, which arose from the local challenge of trying to cross each of the seven bridges in the town of Königsberg just once. Euler treated this problem as a plane graph and demonstrated that it contains no cycle connecting all the vertices; hence the problem is unsolvable..
(2) the same graph problem posed for the 20 vertices of a pentagon dodecahedron. Einstein presumably read about this in Chap. 4 of Schubert's book, which deals with "Hamiltonian roundtrips," beginning with these remarks: "In the year 1859 two concentration games appeared in London, presented by the famous mathematician Hamilton, inventor of the quaternions. The first was called *The Traveler on the Dodecahedron or a Trip around*

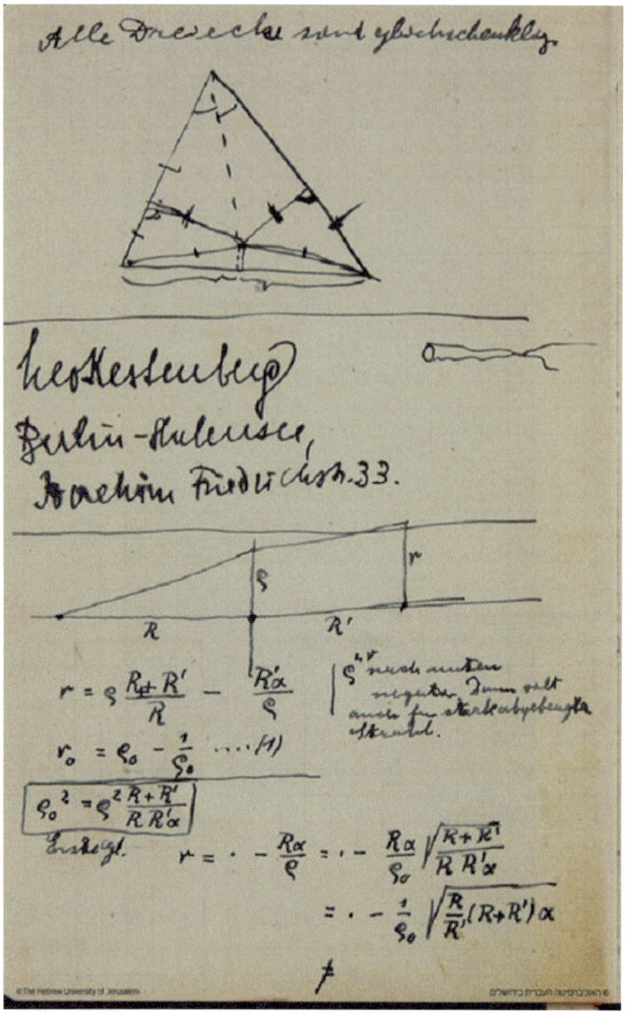

Fig. 23.4 Calculations of light deflection along with a curious absurdity in elementary geometry (CPAE 3 1993, 585).

the World, the other, *The Icosahedron Game*" (Schubert 1898, 68).

(3) a standard algebraic "proof" that a = 2a, followed by another similar one leading to the conclusion that for any numbers *a, b* one always can show that $a = b$.

At the top of Fig. 23.4 we find a dissected triangle above which Einstein has written: "Alle Dreiecke sind gleichschenklig" (all triangles are isosceles). The "proof" begins by bisecting the angle at the apex, erecting the perpendicular bisector of the base, and finding their point *P* of intersection. From *P* we drop perpendiculars to the remaining two sides and draw lines to the end points of the base. The triangle has now been dissected into six smaller ones, and these "clearly" come in three pairs of congruent triangles, revealing that the two sides are of equal length!

Fig. 23.5 Martin Gardner visiting Lewis Carroll's Alice in New York City's Central Park (Photo by Scot Morris).

This little geometrical jest was very well known years ago. In fact, Hilbert presented it at the very outset of his lecture course of 1898–99 on the foundations of Euclidean geometry (Toepell 1986). He wanted to make the point that rigorous geometric arguments should never draw conclusions merely by appeal to suggestive figures, however carefully drawn. Schubert drew his triangle very nicely indeed, making evident to the eye that the critically important point *P* can never fall inside the triangle!

Among Einstein's most famous ideas were his thought experiments with moving trains and elevators suspended in empty space, images he developed to explain key facets of relativity theory (Einstein 1917). He also introduced the thought experiments on time dilation that came to be widely known as the space-time paradoxes involving clocks and the twins who age at different rates. Given his predilection for whimsical humor and his keen eye for intellectual puzzles, it should perhaps come as no surprise that he took special delight in mathematical paradoxes, even those of a more trivial variety (Fig. 23.5).

On Machian Thought Experiments

One of Einstein's most controversial ideas stemmed from his reading of Ernst Mach, who famously criticized Newton's notion of absolute space as meaningless metaphysics. Newton was, indeed, a deeply metaphysical thinker, yet he also appealed to experimental evidence to support his belief in absolute space. His famous thought experiment with a rotating water bucket was meant to show that when a spinning object experiences inertial forces these stem from its motion with respect to space itself. Einstein, like Mach before him, thought otherwise, coining what he called the relativity of inertia, which he later reformulated as Mach's Principle. According to this, the metric properties of space, including its special inertial frames, are induced by global mass and energy.

Einstein was deeply attracted to this idea during the very years when he was groping for a generalized theory of relativity that could unite gravity and inertia. By 1916, after he had found what he was looking for, he tried to find a global solution for his gravitational field equations, one that would yield a static cosmology, but to no avail. (In these efforts he was supported by Jakob Grommer, one of Hilbert's many students.) Then Einstein hit upon the possibility of modifying his field equations by adding what became known as the "cosmological constant." By February 1917 he unveiled his cylindrical universe, a cosmological model whose underlying topology was $S^3 \times R$ (see Chap. 24). Riemann, in his celebrated habilitation lecture of 1854, had hinted that space might in fact be finite and closed, possessing the topology of a 3-sphere. Einstein now showed not only how, but why this was the case. He was much enamored by his new creation, in no little measure because it served to implement Mach's principle so nicely. Indeed, there was a simple correlation between the total mass M in the universe and the global curvature constant.

Yet not everyone found this pretty model so convincing, and some found the Machian arguments for it highly objectionable. Moreover, Einstein soon learned from the Dutch astronomer Willem de Sitter that his new field equations admitted another solution that led to another cosmological model, a 4-manifold of constant curvature. The problem for Einstein was that this de Sitter universe contained no matter at all! In fact, de Sitter invented it in the course of an ongoing debate with Einstein over Mach's principle, an idea that de Sitter took to be just as metaphysical as Newton's arguments for an absolute space. How, he asked, could an astronomer ever measure the influence of so-called distant matter?

Einstein was no astronomer, so he perhaps did not require an answer. Nor did this issue concern him much as the years went on. For by 1930, in the wake of Edwin Hubble's measurements of extra-galactic redshift effects, Einstein and de Sitter could easily reach a truce. Dropping the cosmological constant – which he famously called the "greatest mistake I ever made"[1] – Einstein left his static model behind. New cosmological models for an expanding universe were already on the drawing boards, while fresh data rendered the older ones (nearly) obsolete. Gone, too, was the initial motivation for Mach's principle, an idea that began to look more and more dated. By 1949, when he wrote an autobiographical sketch, Einstein noted that ". . . for a long time I considered [Mach's conception] as, in principle, the correct one. It presupposes implicitly, however, that the basic theory should be of the general type of Newton's mechanics. . . . The attempt at such a solution does not fit into a consistent field theory, as will be immediately recognized" (Einstein 1949, 29),

Nevertheless, Mach's Principle enjoyed a kind of renaissance after Einstein's death in 1955. During the latter half of the century one of its most foremost champions was the British cosmologist Dennis Sciama, who as a fellow at the Institute for Advanced Study once had the opportunity to speak with Einstein (Sciama 1978). Many years later, here's how he recalled their conversation (Fig. 23.6):

Fig. 23.6 Dennis Sciama, who sought to revive Machian cosmology.

[1] Mario Livio recently tracked this quote down and discovered that it was in all probability the invention of George Gamow (*Brilliant Blunders: From Darwin to Einstein*, 2014).

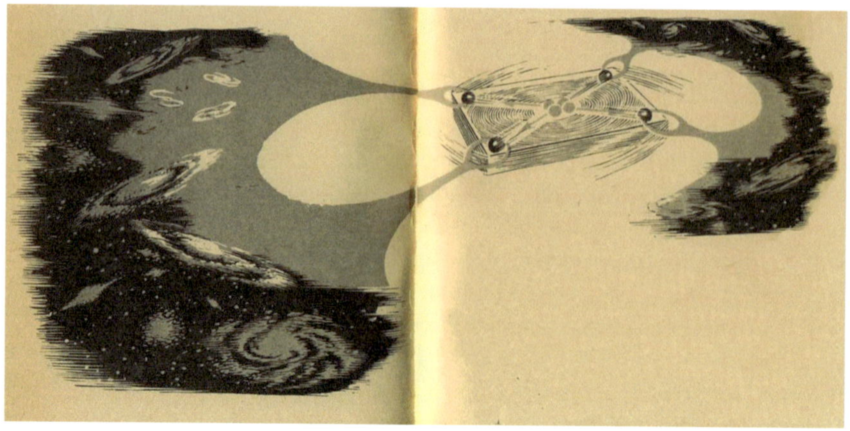

Fig. 23.7 Invisible hands of distant galaxies, as conceived by Martin Gardner and drawn by Anthony Ravielli (Gardner 1976, 124–125).

> . . . at the end of my year at the Princeton Institute, April 1955, I wanted to see him of course, and I plucked up courage only at the end of the year to go and see him, It was literally a week before he died, and I was with him for over an hour and a half. That was a great experience for me. Originally, of course, the very phrase Mach's Principle was Einstein's own phrase for that idea, and he'd used the principle as the guiding light for constructing general relativity. But he later came to feel that the principle wasn't so important, and in the autobiographical notes . . . he had said that he came to disown Mach's Principle.
>
> I started out a bit nervous of course. I'd read that he had a hearty laugh and a simple sense of humor, . . . So knowing that, I went to see him and I said, 'Professor Einstein, I've come to talk about Mach's Principle and I've come to defend your former self.' And it worked: he said, "Ho, ho, ho, that is gut, Ja!" Like that, really laughed. So that put me a bit at my ease. So then I talked about my way of doing Mach's Principle and he talked about his work and his doubts about quantum theory and so on. (Sciama Interview, Archive for History of Quantum Physics)

Dennis Sciama took the view that GR failed to attain what Einstein originally set out to achieve, namely to provide a physical theory of inertial effects. He sketched his ideas for such a theory n his 1953 paper, "On the Origin of Inertia," but unfortunately we don't know what Einstein thought about this approach, if anything.

What we do know is that Martin Gardner paid close attention to Sciama's speculations, which he described in his bestseller, *The Relativity Explosion* (Gardner 1976), the eighth chapter of which is entitled "Mach's Principle." According to Sciama's calculations, effects that arise from rotation and acceleration with respect to the celestial compass of inertia are induced by distant matter fields, which are almost entirely extragalactic. Mach and his contemporaries spoke about motion relative to the fixed stars, whereas Sciama estimated that all the stars in our galaxy contribute roughly one ten-millionth to the inertial effects measured in the vicinity of our planet. This gave Martin Gardner the inspiration for a wonderful thought experiment of his own, which he described as follows (Fig. 23.7):

> I once owned a small glass-topped puzzle, shaped like a square and containing four steel balls. Each ball rested on a groove that ran from the square's center to one of its corners. The problem was to get all four balls into the corners at the same time. The only way to solve it was by placing the puzzle flat on a table and spinning it. Centrifugal force did the trick. If Sciama is right, this puzzle could not be solved in this way if it were not for the existence of billions of galaxies at enormous distances from our own" (Gardner 1976, 125–26).

References

Blumenthal, Otto. 1935. Lebensgeschichte. In *David Hilbert Gesammelte Abhandlungen*, vol. 3, 388–429. Berlin: Springer.

CPAE 3. 1993. In *The Collected Papers of Albert Einstein*, ed. M. Klein, A. Kox, J. Renn, and R. Schulmann, vol. 3. Princeton: Princeton University Press.

Einstein, Albert. 1917. *Über die spezielle und die allgemeine Relativitätstheorie.* (Gemeinverständlich), Braunschweig: Vieweg.

———. 1949. Autobiographical Notes, in (Schilpp 1949, 1–95).

Gardner, Martin. 1976. *The Relativity Explosion*. New York: Vintage Books.

———. 2006. The Fibonacci Sequence. *Journal of Recreational Mathematics* 34: 183–190; (reprinted in (Gardner 2009, 106–113).

———. 2009. *When You Were a Tadpole and I Was a Fish*. New York: Hill and Wang.

Hilbert, David. 1899. Grundlagen der Geometrie, in (Hilbert 2004).

———. 2004. In *David Hilbert's Lectures on the Foundations of Geometry, 1891–1902*, ed. Michael Hallett and Ulrich Majer. Berlin/Heidelberg/New York: Springer.

Renn, Jürgen, Tilman Sauer, and John Stachel. 1997. The Origin of Gravitational Lensing: A Postscript to Einstein's 1936 Science Paper. *Science* 275: 184–186.

Schilpp, Paul Arthur, ed. 1949. *Albert Einstein: Philosopher-Scientist*. Vol. 1. New York: Harper Torchbooks.

Schubert, Hermann. 1898. *Mathematische Mußestunden*. Leipzig: Teubner.

———. 1903. *Mathematical Essays and Recreations*. Chicago: Open Court.

Sciama, Dennis. 1978. Interview with Dennis Sciama, with Spencer Weart, Middletown, Conn., 14 April 1978. *Archive for History of Quantum Physics*, http://www.aip.org/history/ohilist/4871.html.

Toepell, M.-M. 1986. *Über die Entstehung von David Hilberts Grundlagen der Geometrie*, Studien Zur Wissenschafts-, Sozial- und Bildungsgeschichte der Mathematik. Bd. 2 ed. Göttingen: Vandenhoeck & Ruprecht.

24 Debating Relativistic Cosmology, 1917–1924

(*Mathematical Intelligencer* 38(2)(2016): 46–58; 38(3)(2016): 52–60)

Introduction

Physical astronomy as we know it today matured during the latter half of the twentieth century. It was preceded by a period Jean Eisenstaedt has dubbed the "low water mark" in general relativity (GR), covering roughly the period 1925 to 1955 (Eisenstaedt 1988b).[1] Starting in the 1960s, however, a series of startling developments helped pave the way for what has since been called the "renaissance of general relativity," which suddenly took on great significance for astrophysics and cosmology. In the days of Einstein and Eddington, one could imagine a gravitational field so strong that it would produce a black hole, a true space–time singularity. People talked about such things, but hardly anyone believed they could actually occur (Thorne 1994). Yet after Penrose and Hawking proved the celebrated singularity theorems (Earman 1999; Hawking and Ellis 1973), experts began to look for evidence that might confirm the existence of black holes. This was only one of many unexpected developments in GR that helped to inaugurate a revolutionary shift in our understanding of the universe. Truly momentous discoveries soon followed, leading to findings that would eventually shatter the quaint universe inhabited by Albert Einstein at the time he unveiled his general theory of relativity.

In November 1915 Einstein came up with the "final" version of his gravitational field equations. This achievement was all the more impressive because he immediately used a linearized form of these equations to solve a problem that had plagued astronomers for over two decades. Einstein showed that with GR he could account for the missing 43 s of arc per century in the movement of Mercury's perihelion, a small deviation that could not be explained as a perturbation effect using standard methods of celestial mechanics. This dramatic breakthrough came during the midst of wartime, when scientific communications were difficult, if even possible at all. Yet despite these adverse circumstances a flurry of research activity in general relativity quickly ensued.[2]

In 1916 Johannes Droste and Karl Schwarzschild gave the first exact solutions of Einstein's field equations (Eisenstaedt 1988a); one year later Einstein and Willem de Sitter presented two different solutions of the modified equations with nonzero cosmological term. That same year, four mathematicians—Tullio Levi-Civita, Hermann Weyl, Jan Arnoldus Schouten, and Gerhard Hessenberg—independently pursued new ideas in differential geometry that generalized the semi-Riemannian metrics used by Einstein. Henceforth, the theory of affine connections would play a major role in efforts to find a yet more general field theory that could unite gravity and electromagnetism. In Leyden, Göttingen, Vienna, and Rome, scientists and mathematicians began to elaborate the groundwork for Einstein's new field-theoretic approach to gravitation, encountering numerous difficulties along the way.

In the meantime, de Sitter helped spread the new theory of gravitation to England, where Arthur Stanley Eddington quickly became a leading expert. Already by 1917, he and Frank Dyson, England's royal astronomer, were making plans to launch two eclipse expeditions that would test Einstein's prediction regarding the bending of light in the vicinity of the sun. They thus readied their instruments for the teams that would eventually journey to remote locations in Africa and South America. Eddington hoped that photographs taken at these two locations might reveal an apparent displacement of stars during the full solar eclipse that would take place two years hence, on 22 May 1919. What he was looking for would be hard to find: a truly tiny deviation, a mere 1.75″ of arc, predicted by Einstein as a consequence of the warping of space–time due to the sun's gravitational field. Measuring such a small angle presented astronomers with an enormous technical challenge; even

[1] Scott Walter pointed out to me that, on a purely quantitative basis, there is no clear evidence of any ebb in publication numbers with respect to other branches of physics, particularly if unification theories as taken into account.

[2] This flurry of activity also involved a good deal of conceptual confusion, as recently discussed in Darrigol (2015).

D.E. Rowe, *A Richer Picture of Mathematics*, https://doi.org/10.1007/978-3-319-67819-1_24

as late as 1955, the year of the Bern Jubilee Conference marking the 50th anniversary of the birth of relativity, experts still voiced disagreement over the data from this and other later eclipse expeditions (Mercier and Kervaire 1955, 106–113). Empirical support for Einstein's three classical tests for general relativity would long remain a delicate affair.

Given the miniscule size of such relativistic effects, it should also come as no surprise that GR was long ignored by most physicists and astronomers. Some scientists even scoffed when told that this new theory carried profound implications for our understanding of space and time, whereas others imagined that relativity stood for a new worldview that harmonized with the *Zeitgeist* of the post–war era (Rowe 2006). It mattered little that Einstein consistently explained why relativity was a fundamental, not a revolutionary theory. As a resident of Berlin, a city teeming with energy and celebrity life, he adapted to his new role instead of trying to escape from the limelight. Dubbed the "new Copernicus" by the German press, he felt obliged to satisfy public curiosity about relativity. So, whether at a dinner party or when speaking to a lay audience, Einstein often offered picturesque explanations of what he meant by a curved space. His earliest cosmological theory became an integral part of his fame, just as Einstein's "Weltbild" came to stand for the worldly wisdom he was believed to embody (Rowe 2012). Nearly all the popular books on relativity had something to say about Einstein's universe, though seldom did they mention that by 1930 he no longer believed in it. Scientists, after all, can and do change their minds; given the fertility of Einstein's scientific imagination, no one should be surprised that he changed his views rather often.

These circumstances should be born in mind when thinking about the early cosmological debates that began with Einstein's famous paper of 1917. Back then, when the European world was coming unhinged as a result of the Great War, scientific awareness of the celestial realm remained confined to phenomena within our own galaxy. The visible universe from a century ago was small and stable, much like the world conceived by Isaac Newton. His absolute space, on the other hand, extended without bound to infinity, a conception that inevitably led to metaphysical and theological speculations. From a purely physical standpoint, Newtonian cosmology presented a puzzling picture filled with anomalies that Einstein hoped relativity could resolve. As a theoretical physicist, he could also draw on his familiarity with thermodynamical systems. Thus, he began by picturing the physical universe as composed of stellar matter spread homogeneously throughout space and persisting in time in a state of thermal equilibrium. Like his contemporaries, Einstein thought only in terms of a static cosmology.

Long before this, astronomers had known, of course, that one could not speak literally of the "fixed stars" in the heavens. In fact, by the turn of the century the study of stellar motion had become a hot new specialty, challenging astronomers to determine whether such motion was merely random or, if not, what kind of systematic patterns it displayed (Smeenk 2014). The first to answer that challenge was Jacobus Kapteyn, de Sitter's mentor in Groningen. As a physical astronomer, Kapteyn drew on kinetic gas theory as an analogue for studying the dynamics of stellar systems. In 1904 he helped to launch modern statistical astronomy by demonstrating that stars could be divided into two streams that move in nearly opposite directions. This research led to a richer picture of the Milky Way galaxy, even though it had to be revised soon after his death in 1922. Kapteyn's universe was far too small; he was also mistaken about our place in it—our Sun lies closer to the outskirts of the galaxy rather than near its center. A decade later, astronomers also came to realize that Kapteyn's interpretation of star streaming was incorrect, but his data provided strong evidence for an important new celestial phenomenon, galactic rotation.

None of these discoveries, however, had any direct bearing on Einstein's cosmological reflections. So long as the velocities of stars remained small compared with the speed of light, he could afford to neglect all such motion and treat the universe as if it were in a quasi-static state of equilibrium. Newtonian gravitational theory, as Einstein well knew, could not account for a static cosmology, but could GR? That issue emerges particularly clearly in Einstein's early reflections on cosmology. By 1916 he had already turned his attention to the most puzzling problem in Newtonian celestial mechanics, a dilemma that had challenged experts for over 200 years: how to account for the stability of the universe?

During these years, Einstein was a steadfast defender of Ernst Mach's ideas regarding the relativity of inertia (Rowe 2015). Mach had advanced the notion that the inertial properties of matter were due to some kind of interaction with distant masses. This conjecture was closely connected with his famous critique of Newton's notions of absolute space and time, two of the cornerstones of Newtonian mechanics. Already in 1912 Einstein took up this Machian program in the context of a field–theoretic approach aimed at coupling gravity with inertia. Indeed, this Ansatz preceded his early work on general relativity, which only began in 1913 when he was collaborating with Marcel Grossmann. After the breakthrough that led Einstein to generally covariant field equations in November 1915, Machian notions regarding the origins of inertia remained a central motivating conviction. Indeed, they guided his search for a reasonable global solution of his field equations, one which could serve as a first approximation to the universe.

By this time Einstein had moved on to Berlin, where he held a special appointment as a distinguished member of the Prussian Academy of Sciences. Max Planck and his Berlin colleagues shared a deep admiration for Einstein's theoretical accomplishments, even though they could not quite fathom his passionate interest in this new gravitational theory. In fact, they were slow to appreciate what he had achieved. Elsewhere, on the other hand, general relativity caught fire. In Göttingen, where Einstein spoke about his new approach to gravitation during the early summer of 1915, GR was taken up with great interest by David Hilbert and Felix Klein. In Zurich, Grossmann's younger colleague, Hermann Weyl, pursued it with a passion. Still, the single most important outpost for the early reception and development of Einstein's new theory was Leyden, where the astronomer Willem de Sitter worked alongside two distinguished physicists, H. A. Lorentz and Paul Ehrenfest, both good friends of Einstein. In Part I of this survey we sketch the background leading up to the initial events and issues under debate, taking the story up to the middle of 1918. Part II then focuses on the ensuing discussions that came to a head during the waning months of the Great War. It was at this point that Weyl and Klein entered the fray. Weyl at first sided strongly with Einstein, but by the end of the period he adopted a position very close to de Sitter's views.

Cosmological Combatants

Einstein's paper from 1917 presented what was purported to be a picture of a new universe based on the principles of general relativity. To produce it, though, he had to alter his recently attained field equations. This celebrated work marks the beginning of relativistic cosmology, a field of inquiry born in an atmosphere of controversy and debate. The debates mainly focused on two competing models: the "cylindrical universe" of Einstein and the matter–free world of de Sitter. It should be noted, however, that the term cosmological model only became common parlance after around 1933, when H. P. Robertson employed it in an oft–cited review paper (Robertson 1933). Einstein and de Sitter were concerned with finding static solutions to Einstein's augmented field equations with "cosmological constant." In Einstein's case, his solution also aimed to implement what he called "Mach's Principle," a notion de Sitter rejected as pure speculation. From the standpoint of cosmology, de Sitter felt that to account for inertia by appealing to the influence of distant masses amounted to adopting a metaphysical principle in no way superior to Newton's original conception of absolute space. Einstein held firmly to his Machian convictions, even to the point of insisting that GR had to conform to this standpoint. De Sitter's matter–free universe thus flew in the face of his belief that inertial effects were solely due to the global matter–energy field. These differences of opinion were prominently featured in their famous debate, which unfolded during the period discussed here. Historical and mathematical details connected with these developments can be found in the references cited below.

Einstein and de Sitter first discussed the implications of general relativity for cosmology during the autumn of 1916, when Einstein was visiting Leyden. This marks the first of several friendly intellectual encounters, aspects of which have been examined in the scholarly literature (see, in particular, Kerszberg (1989), Janssen (2014), and Smeenk (2014)). As an early expert on and advocate of GR, de Sitter took a keen interest in astronomical phenomena that might be used to test Einstein's theory empirically. What interested him most, however, were its potential payoffs for astronomy as well as for understanding the nature of gravity (Martins 1999). On the latter subject, he offered these reflections in 1912:

> It is a remarkable fact that the law of gravitation, the simplest, longest known, and best investigated of all natural laws, has so far found no satisfactory physical explanation, and stands apart from all other physical phenomena, presenting no points of contact with any other physical theory. The explanation for this must be sought "in its extreme simplicity, its complete independence from everything that affects other phenomena of nature. Gravitation is not subject to absorption nor refraction, no velocity of propagation has been ascertained, it affects all bodies equally without any difference, always and everywhere we find it in the same simple and rigorous form, which defeats all attempts to penetrate into its inner mechanism." These considerations show the utmost importance of any attempt to discover a deviation from the rigorous law of Newton, or any influence of outside circumstances on its action. (De Sitter 1912, 387)[3]

Soon after his arrival in Leyden in 1908, de Sitter began a careful study of the perturbation effects that cause the precession of the perihelia of the inner planets (Röhle 2007). He later credited Lorentz for inspiring his interest in modern gravitational theories, especially those put forward by Poincaré and Minkowski, who independently developed scalar laws of gravitation compatible with SR. Based on these, de Sitter calculated an additional advance in the perihelion motion of Mercury amounting to 7.15″ (De Sitter 1911). This was much too small, which suggested the possibility of combining this with Hugo von Seeliger's zodiacal light hypothesis. He took up these issues again in 1913, this time calculating the perihelion precession for all four inner planets. From this he concluded that only Mercury would be suitable for testing a post-Newtonian theory of gravitation (De Sitter 1913) (Fig. 24.1).

One year later, having studied the Einstein–Grossmann theory, he applied it to calculate Mercury's perihelion motion (Röhle 2007, 197–200). His result was again much too small, 18″, which happened to agree with what Einstein and Michele Besso had found one year earlier. Neither they nor de Sitter published this result, however. In 1913 de Sitter conveyed it to Lorentz in a postcard, and a year later Lorentz informed his pupil, Johannes Droste. De Sitter's re-

[3]The passage in quotation marks was taken from de Sitter's 1908 inaugural lecture as professor of astronomy in Leyden.

Fig. 24.1 Willem de Sitter was a leading expert on the perturbation theory of planetary systems.

sult was then mentioned in print by Droste one year later: "As Prof. de Sitter has calculated from the equations of motion determined by Prof. Lorentz, it amounts for Mercurius to 18″ per century, the observed motion being 44″" (Röhle 2007, 198). Thus, when Einstein derived the full precessional value in November 1915, no one was in a better position to appreciate this accomplishment than de Sitter. In June 1916, he confirmed Einstein's result in the first of his several papers on GR, declaring therein that von Seeliger's hypothesis was now obsolete. At the same time he wrote a letter to Einstein explaining why Einstein's earlier choice of coordinate system led to unnecessary complications for solving the linearized form of the field equations. Einstein acknowledged the aptness of de Sitter's critique soon thereafter in Einstein (1916) (Fig. 24.2).

All this might seem as though it should have portended well for the discussions on cosmology that Einstein and de Sitter struck up later that year. This encounter, however, quickly brought out issues over which they sharply disagreed. In part this was a matter of differences in scientific temperament. For while de Sitter regarded Einstein's gravitational theory as a wonderful instrument for teasing out subtle astronomical effects, he had very little patience for speculative ideas. In particular, he took a skeptical view of Einstein's appeal to distant masses as a potential way to explain the origins of inertia. This Machian motif took on a new form in GR, since such effects would now be encoded in the metric tensor associated with the space–time geometry. De Sitter fully accepted the principles underlying Einstein's new theory, but he objected to his global speculations, which he considered to be mere metaphysics rather than ideas that could lead to a testable scientific hypothesis. He also recognized that, temperamentally, he and Einstein stood poles apart. He expressed this to him openly in these words:

Fig. 24.2 Einstein and Eddington in Leyden, September 1923, joined by Lorentz (front right), with Ehrenfest (middle) and de Sitter (back right). Leyden Archives.

"Our differences in belief amount to this: whereas you have a definite belief, I am a skeptic" (18 April, 1917).

In 1916 Einstein and the mathematician Jakob Grommer were trying to find global solutions of the original field equations. These can be written

$$G_{\mu\nu} = \kappa T_{\mu\nu},$$

where the left side is the Einstein-Tensor:

$$G_{\mu\nu} = R_{\mu\nu} - \frac{1}{2} g_{\mu\nu} R.$$

They hoped to find a global, everywhere regular solution of these equations that was centrally symmetric and assumed appropriate boundary values at spatial infinity. The resulting metric also had to be independent of time, corresponding to a static universe. Yet to accord with general relativity, the boundary conditions would also have to be valid in every possible coordinate system. In his conversations with Einstein, de Sitter criticized this approach to cosmology, taking the position that nothing could be said about the physical properties of the universe at distances infinitely far from the Earth. Einstein reacted to this stance in Einstein (1917), noting that de Sitter's pragmatic position was unassailable, but that this amounted to giving up all cosmological speculation just when GR had opened the way to new lines of inquiry.

Einstein's mathematical assistant at this time, Jakob Grommer, had studied in Göttingen, where he took his doctorate under Hilbert in 1914. A Russian orthodox Jew, Grommer had once hoped to become a rabbi in Brest-Litovsk, a dream shattered by a horrible illness. Grommer suffered from acromegaly, a disease that leads to enlargement of the bones, face, and jaw (Pais 1982, 487–488). Einstein first met him in Göttingen, during the summer of 1915 when he was invited to deliver a series of lectures on general relativity. Their collaboration began some time in 1916, the year during which GR took wing. One year later, having gained an appreciation of his abilities, Einstein wrote about him to his friend Paul Ehrenfest: "Herr Grommer, a marvelous mathematician whom you know from Göttingen (to aid your memory I remind you of the enormous proportions of his head and hands) would very much like to obtain a position in Russia ...." (22 July 1917) Nothing came of this idea, however, and Grommer had to eke out a living by assisting Einstein for more than a decade.

Unfortunately, the most ambitious goal that Einstein and Grommer tried to tackle together, the cosmological problem in GR, led them down a dead–end path. Having replaced Newtonian gravitation by a true field theory, Einstein hoped he could now resolve an old dilemma, first posed to Sir Isaac by the classicist and theologian Richard Bentley, later Master of Trinity College, Cambridge. Bentley inaugurated the Boyle Lectures in 1692, which meant that he was charged with the hardly simple task of reconciling science and religion. The estate of the deceased chemist Robert Boyle had set aside funding for this purpose, so that emminent scholars might have the opportunity to defend Christianity by explaining Holy Scripture in the light of modern natural philosophy. Little wonder that Bentley undertook this task by first studying Newton's *Principia*. He then turned to its author, the Lucasian Professor of Mathematics at Trinity, for personal advice. Newton signaled that he was delighted to help: "Sir, When I wrote my Treatise about our System, I had an Eye upon such Principles as might work with considering Men, for the Belief of a Deity; nothing can rejoice me more than to find it useful for that Purpose." (Holton 1960, 59).

Sharing the same conviction, Bentley felt free to query the great Newton about the cosmological implications of his theory. Could he explain how an infinite universe filled with an infinitude of particles could remain in perfect equilibrium given the force of gravitational attraction acting on these bodies? The reply he received surely fit his purpose, for Newton at once admitted how unlikely such a world would be. He then added:

> I reckon this as hard as to make, not one needle only, but an infinite number of them (so many as there are particles in an infinite space) stand accurately poised upon their points. Yet I grant it possible, at least by a divine power; and if they were once to be placed, I agree with you that they would continue in that posture without motion for ever, unless put into motion by the same power. (Norton 1999, 289)

Laplace later tried to prove the stability of the solar system in his magisterial five-volume study, *Mécanique Céleste*. Queried by Napoleon about the role of God in his work, he supposedly answered that he had no need for that particular hypothesis (presumably meaning that, unlike Newton, he could account for the enduring stability of the world without having to make room for occasional divine intervention). In Laplace's day, it would have been heretical to contemplate a world without a higher purpose that manifested God's plan. Still, Deists could point to the very perfection of that plan in order to argue that God's world was regulated alone by His laws of nature. Even in modern times, when theology and science largely went their separate ways, cosmological theorizing still required "ideological" assumptions (Hawking and Ellis 1973, 134), just as it does today. John North's survey of twentieth century cosmology up to 1965 shows how the field matured as its practitioners came to recognize and articulate certain metaphysical principles (North 1965). This, in turn, led to a new understanding of what went into proposing various types of cosmological models, a viewpoint that came well after the period under discussion here.

Einstein, along with most of his scientific contemporaries living in the year 1916, firmly believed in a quasi-static infinite universe, so for him the traditional cosmic problem remained. Moreover, on a cosmological scale under the assumption that space stretched to infinity, there seemed to be no simple way to adapt Newtonian concepts in order to achieve a model universe in stable equilibrium. Numerous astronomers had tried to do so, right up until the breakthrough with GR in 1915, but they failed (Norton 1999). Armed with his new field equations, Einstein took up this cosmological problem anew. His first attempt to introduce relativistic cosmology was based on a flat global space-time, a Minkowski space in which the gravitational potential took on infinite values at spatial infinity. Arguing against this, de Sitter thought such an approach could never be carried through independent of the choice of coordinates. It was simply incompatible with GR, a point Einstein eventually conceded in early 1917.

After encountering insuperable technical difficulties, Einstein eventually realized that he and Grommer had reached a dead end. But then he found an ingenious way to circumvent the main technical difficulty. Up until then they had been stymied by the problem of finding appropriate boundary conditions for the metric along the edges of an infinite space. That headache vanished, however, when Einstein realized he could alter his field equations slightly and thereby obtain a simple global solution for a static space that required no boundary conditions at all since it was both finite and without boundary.

In a letter written on the 2nd of February 1917, he informed de Sitter about his latest approach to the cosmological question: "Presently I am writing a paper on the boundary conditions in gravitation theory. I have completely abandoned my idea on the degeneration of the $g_{\mu\nu}$, which you rightly disputed. I am curious to see what you will say about the rather outlandish conception I have now set my sights on" (CPAE 8A. 1998a, 385). Two days later he reported to Paul Ehrenfest: "I have again perpetrated something in gravitational theory that puts me in a bit of danger of being committed to a madhouse. I hope you don't have any in Leyden so that I can visit you again safely." (CPAE 8A. 1998a, 386).

What Einstein had found was a new universe without any "edge at infinity." By simply adding a so-called cosmological term to his field equations, he could still maintain general covariance but now obtain a very simple global solution with startling consequences for cosmology. These new cosmological equations took the form:

$$G_{\mu\nu} - \lambda g_{\mu\nu} = T_{\mu\nu},$$

where the constant λ corresponds to a tiny repellent force that prevented matter from eventually collapsing. The metric tensor in this space-time allows for a global coordinatization that splits off the spatial components from a "cosmic time." So this space–time can be factored topologically as $\mathbf{S}^3 \times \mathbf{R}$, where the spatial cross-sections $\mathbf{S}^3$ are Riemannian 3-spheres of constant radius. Since the volume of such a 3-sphere is $2\pi^2 R^3$, a universe with mean density ρ will have $M = \rho \cdot 2\pi^2 R^3$ as its total mass. Felix Klein later dubbed this cosmology Einstein's "cylindrical world," a name that stuck.

Einstein speculated in private about the approximate radius of this new universe, but not in print. In (Einstein 1917) he avoided all discussion of possible astronomical evidence in support of this world picture. Instead he emphasized that the geometry of space could be curved even without the cosmological term. This addendum λ was merely needed to ensure the "quasi-static distribution of matter as required by the small velocities of stars." Writing to de Sitter, he stressed that "from the standpoint of astronomy ... I have erected but a lofty castle in the air. It was a burning question for me, however, whether the relativity thought can be carried all the way through or whether it leads to contradictions." (Einstein to de Sitter, 12 March 1917, (CPAE 8A. 1998a, 411)).[4] "Now I am no longer plagued by the problem," he wrote, "while previously it gave me no peace. Whether the scheme I formed for myself corresponds to reality is another question, about which we shall probably never gain information." A few days later de Sitter wrote back to say that he still needed time to absorb Einstein's findings, but that he had no objection to his conception, so long as he did not wish to impose it on reality, adding that "I have nothing against it as a contradiction-free chain of reasoning, and I even admire it" (15 March 1917, (CPAE 8A. 1998a, 413)).

De Sitter's Non-Machian Universe

De Sitter needed just five more days to study Einstein's paper before arriving at a startling conclusion. Einstein had altered his field equations in order to avoid the problem of stipulating boundary conditions at spatial infinity. The metric in his finite universe of constant curvature thus resulted directly from the matter it contained. Indeed, Einstein showed that the cosmological constant λ determined both the mean density ρ as well as the curvature of spherical space, according to the relation

$$\lambda = \frac{\kappa\rho}{2} = \frac{1}{R^2},$$

where $\kappa = \frac{8\pi G}{c^2}$ and G is the universal constant of gravitation from Newton's theory. These effects were, as he emphasized, far too small to be measurable, but nevertheless the result fit Einstein's expectations. So he was surely surprised to learn from de Sitter that his new gravitational equations with cosmological constant admitted another solution with an entirely unexpected property: it had no matter at all.

This news, imparted in a letter from 20 March, 1917, also came with what must have seemed like a stinging remark: "I do not know if it can be said that 'inertia is explained' in this way. I do not concern myself with explanations. If a single test particle existed in the world, where there was no

[4] Michel Janssen argues that this was the hidden agenda behind Einstein's cosmological speculations, namely to show that the cosmological problem could be tackled within the framework of general relativity without encountering some kind of contradiction (Janssen 2014, 207–208).

Fig. 24.3 Einstein and de Sitter renewing their cosmological conversations at Caltech in 1932.

sun and stars, it would have inertia" (CPAE 8A. 1998a, 414–416). So much for Einstein's search for an explanation of the origins of inertia: his new gravitational equations clearly admitted a non–Machian solution. Or so it seemed at first. As with the Einstein universe, de Sitter showed that this vacuum solution could be treated so that the spatial cross-sections were finite, homogeneous and of constant curvature. Thus, the only substantial difference between these two "models" was the presence of matter in Einstein's world, whereas de Sitter had created a world out of a yawning void. Could he have come up with such an idea in just five days (Fig. 24.3)?

As it happened, de Sitter had been talking to his colleague, Paul Ehrenfest. It seems that Ehrenfest was not altogether happy with Einstein's cosmic solution; in particular, he disliked the fact that its time coordinate could be distinguished from the three spatial coordinates. Einstein himself was at first surprised by this, as can be seen from a letter he wrote to Ehrenfest on 14 February, 1917:

> My solution may appear adventurous to you, but for the moment it seems to be the most natural one. From the measured stellar densities, a universe with radius of the order of magnitude 10^7 light years results, thus being unfortunately very large compared with the distances to observable stars. The odd thing is that now a quasi–absolute time and a preferred coordinate system do reappear in the end, while fully complying with the requirements of relativity (CPAE 8A. 1998a, 390).

When soon thereafter Ehrenfest discussed Einstein's solution with de Sitter, he planted the idea of a cosmology that would be homogeneous in all four dimensions of space-time rather than just the three spatial dimensions.

Initially this was the crucial issue under discussion when de Sitter came up with his alternative approach to the modified field equations. In his long letter to Einstein, written one month later, de Sitter brought out the differences between the two cosmological solutions by setting them side by side in neighboring columns, beginning with:

Three – dimensional	*Four – dimensional*
With super natural masses.	Without any masses.
$\lambda = \frac{1}{R^2}$	$\lambda = \frac{3}{R^2}$

By complexifying the time coordinate in his solution, de Sitter gave this striking comparison of the two metrics:

x_1, x_2, x_3, ct:	$x_1, x_2, x_3, x_4 = ict'$:
$x_1^2 + x_2^2 + x_3^2 \leq R^2$	$x_1^2 + x_2^2 + x_3^2 + x_4^2 \leq R^2$
$g_{44} = 1, \qquad g_{i4} = 0$	$g_{\mu\nu} = -\delta_{\mu\nu} - \frac{x_\mu x_\nu}{R^2 - (x_1^2 + x_2^2 + x_3^2 + x_4^2)}$
$g_{\mu\nu} = -\delta_{\mu\nu} - \frac{x_\mu x_\nu}{R^2 - (x_1^2 + x_2^2 + x_3^2)}$	$\mu, \nu = 1, 2, 3, 4$

De Sitter then wrote down the metrics in two other coordinate systems, the last of which he obtained by a stereographic projection that mapped a 4-D pseudo-sphere onto a hypersurface in Euclidean 5–space. This led to these two metrics in Cartesian coordinates:

$$ds^2 = c^2 dt^2 - \frac{dx^2 + dy^2 + dz^2}{[1 + \frac{1}{4R^2}(x^2 + y^2 + z^2)]^2}$$

for Einstein's universe, compared with

$$ds^2 = \frac{-dx^2 - dy^2 - dz^2 + c^2 dt^2}{[1 + \frac{1}{4R^2}(x^2 + y^2 + z^2 - c^2 t^2)]^2}$$

for de Sitter's solution. He then displayed the decisive difference between these two metrics at spatial infinity:

$$g_{\mu\nu} = \begin{Bmatrix} 0\,0\,0\,0 \\ 0\,0\,0\,0 \\ 0\,0\,0\,0 \\ 0\,0\,0\,1 \end{Bmatrix} \quad \Bigg| \quad g_{\mu\nu} = \begin{Bmatrix} 0\,0\,0\,0 \\ 0\,0\,0\,0 \\ 0\,0\,0\,0 \\ 0\,0\,0\,0 \end{Bmatrix}.$$

Only the second condition remains invariant under all coordinate transformations, whereas the first requires that $t' = t$. De Sitter further emphasized that the small value of λ within that portion of space-time accessible to measurement shows that the metric deviates only slightly from the standard values in a Minkowski space–time. "This is achieved," he wrote "*without* supernatural masses and alone through the introduction of the undetermined and *underterminable* constant λ in the field equation." He concluded his letter with these remarks:

> I am curious about whether you can agree with this approach and whether you prefer the three-dimensional or four-dimensional system. I personally much prefer the four-dimensional system, but even more so the original theory, without the underterminable λ, which is just philosophically and not physically desirable, and with the non–invariant $g_{\mu\nu}$'s at infinity. But if λ is only small, it makes no difference, and the choice is purely a matter of taste (CPAE 8A. 1998a, 416).

In his reply of March 24, 1917, Einstein raised various technical objections to de Sitter's solution, which he felt "does not correspond to a physical possibility." Most of all, he found it dissatisfying because it contained no matter producing the metric field, making it a non–Machian space–time. About this Einstein wrote: "It would be unsatisfactory, in my opinion, if a world without matter were conceivable. Rather, the $g^{\mu\nu}$–field *should be fully determined by matter and not be able to exist without the latter....* To me, as long as this requirement had not been fulfilled, the goal of general relativity was not yet completely achieved. This came about only with the λ–term (CPAE 8A. 1998a, 422)."

In the meantime, de Sitter was busy writing up his findings, which appeared in his paper "On the Relativity of Inertia. Remarks Concerning Einstein's Latest Hypothesis" (De Sitter 1917a). There he quoted the passage above from Einstein's letter, adding the comment that this might be called the "material postulate of inertia," which should be distinguished from the "mathematical postulate of inertia," according to which the metric tensor at infinity shall be invariant under all coordinate transformations. At this stage of the cosmological debate, de Sitter was mainly interested in clarifying matters of this kind. Although he now referred to the two solutions as *A* and *B*, nothing indicates that he was promoting his own solution (*B*) as an alternative to Einstein's (*A*). Indeed, he stated his own clear preference for dropping all speculation about boundary conditions at infinity and returning to the original field equations without the cosmological constant. But once he entered the fray of these discussions, there was no turning back. Indeed, he followed up this first paper with a second (De Sitter 1917b), and then summarized his findings for the British scientific community in De Sitter (1917c). Thus, by the end of 1917 relativistic cosmology had already come crashing onto the scene, though the debates had only just begun.

Geometric Motifs: Felix Klein

Among the more enlightened, Einstein's notion of a curved space–time universe brought to mind earlier discussions of non-Euclidean geometries and the fourth–dimension (North 1965, 56–63, 72–83). In 1900 the astronomer Karl Schwarzschild utilized contemporary measurements of stellar parallax in order to draw conclusions about the minimum size of the universe, assuming space were indeed a 3–manifold of constant curvature. Nearly a half–century earlier, in 1854, Riemann had discussed this possibility in his famous qualifying lecture to become a Dozent in Göttingen. And decades earlier still, Lobachevsky had written about the possibility that light rays follow paths in a space of constant negative curvature, a conjecture Riemann may not have known about. Still, for a long time these subtle ideas led a subterranean existence, while the truth status of Euclid's parallel postulate went largely unquestioned. Then, in the late 1860s, Eugenio Beltrami showed how the obscure non-Euclidean geometries of Janos Bolyai and Nicolai Lobachevsky could be realized by means of curves and surfaces in ordinary Euclidean space. By the 1870s, mathematicians were starting to become accustomed to these new theories, though many still rejected the possibility that physical space could have anything to do with non-Euclidean geometries.

In 1871 Felix Klein was able to throw new light on the purely mathematical status of these new geometries. Drawing on prior work of Arthur Cayley, Klein showed how to derive Euclidean as well as non-Euclidean geometries by using the so–called Cayley metric in projective geometry (Klein 1871). By this means he obtained the three principal cases for spaces of constant positive, negative, and null curvature. He then dubbed these elliptic, hyperbolic, and parabolic geometry, respectively, the last being the classical Euclidean case already dealt with by Cayley. Klein's approach allowed for an elegant global treatment of geometries with constant curvature, since it merely required the specification of a fixed quadric in a projective space. This was nothing other than Cayley's absolute figure. Still, Klein's viewpoint was far more liberal and led directly to the doctrine he set forth one year later in his famous "Erlangen Program" (Klein 1872). There he generated new geometries simply by taking any arbitrary figure F as the "absolute" which determines a subgroup G_F of the full projective group. The transformations in G_F are simply those that leave F fixed, and this, according to the principles of Klein's Erlangen Program, suffices in order to probe the underlying geometry.

Traditionally, geometrical investigations focused on special types of objects, beginning with the simplest figures: points, lines, and planes. By Klein's time, however, this object–oriented approach had given way to a variety of new methods, which he tried to systematize by means of the group concept, a notion he and the Norwegian geometer Sophus Lie borrowed from contemporary research in algebra. The notion of transformation groups was in the air when Lie and Klein arrived in Paris in the spring of 1870. There they met Camille Jordan, whose classic *Traité des substitutions et des équations algèbriques* was just rolling off the presses. Jordan's groups were finite, not continuous, but the analogy was clear enough, and so Klein and Lie began using infinite groups as a tool for studying the mutual relationships between complex geometrical objects: curves, surfaces, and envelopes of lines and spheres.

As Klein later emphasized in his Erlangen Program, one could start with virtually any object, no matter how exotic, in order to generate a higher–dimensional manifold induced by the action of an appropriate group of transformations. This manifold constituted just one possible realization of the underlying geometry, since the objects themselves were no longer of central importance. What mattered were their common underlying properties as reflected in the group of transformations. Thus seen, the new task for the geometer amounted to studying the invariants of the group and their mutual relations. Conceptually, this approach offered a kind of royal road to the global study of geometries of constant curvature. For geometries with variable curvature, on the other hand, one had to confront the local geometric structure. Since Einstein's gravitational theory was grounded on a space–time geometry in which masses affect the local geometry, this demanded a differential–geometric Ansatz à la Riemann.

What no doubt perplexed many non–mathematicians about Klein's projective methods was that one needed to distinguish real elements from those expressed using imaginary coordinates. Thus, for example, to introduce an elliptic geometry in the real projective plane, one fixes a non-degenerate conic F that contains no real points. The Cayley metric that arises in this context involves cross-ratios in which the imaginary points on F enter into the formulae. Hyperbolic geometry, on the other hand, corresponds to a real non-degenerate conic F, so that the coordinates of all four points in the cross-ratio formula are real. In this case, however, one must distinguish points that lie interior to F from exterior points. Only the former belong to the hyperbolic geometry per se, whereas the latter case leads directly to a projective model for de Sitter space, as Klein was the first to show (Klein 1918b). Much the same reasoning also applies in the parabolic case, where F degenerates to two points on a line (regarded as two pencils of lines). This case corresponds to the older treatment of Euclidean plane geometry as the invariant theory associated with $F = \{I, J\}$, where I and J are the imaginary circular points at infinity common to all circles in the plane.

Klein had occasion to recall these matters soon after the death of his Göttingen colleague, Hermann Minkowski, in January 1909 (Klein 1910). Minkowski's geometrization of special relativity sparked the flame that eventually came to be known as space–time physics. He introduced it in dramatic fashion to the larger scientific world in Cologne on 21 September, 1908. Speaking before a mixed audience of seventy-one mathematicians, physicists, and philosophers, Minkowski delivered what would become one of the most famous lectures in the history of mathematics, "Raum und Zeit." He began with these stirring words:

> The views of space and time which I wish to lay before you have sprung from the soil of experimental physics, and therein lies their strength. Their tendency is radical. From this moment onward, space by itself and time by itself will totally fade into shadows and only a kind of union of both will preserve independence. (Minkowski 1909, 75)

Klein had on occasion discussed this four–dimensional pseudo–Euclidean geometry with Minkowski, and both referred to it as the geometry of the Lorentz group. This meant, of course, that it fell within the framework of Klein's Erlangen Program (Klein 1872). Back then, Klein advocated an approach to geometrical research that was very close in spirit to Minkowski's new interpretation of SR. His central claim—that the study of a geometry amounted to the systematic investigation of the invariant theory associated with its transformation group—appeared particularly prescient in the light of Minkowski's achievement. Still, one should not overlook that Klein himself never anticipated anything remotely similar to space-time physics. As Minkowski emphasized, this conception arose from the fertile soil of experiment. Nevertheless, the invention of Minkowski space gave Klein a wonderful opportunity and vehicle for promoting his Erlangen Program once again. In fact, through the work of Sophus Lie and his followers the theory of continuous transformation groups had by now entered the mainstream of mathematical research, a trend that went hand in hand with growing interest in axiom systems for all kinds of geometrical systems. Physicists had begun to take notice of this, too.

Already in 1910 Bateman and Cunningham called attention to the fact that the Maxwell equations remain invariant under the conformal group, which contains the Lorentz transformations as a subgroup. Klein was immediately struck by this finding, for reasons not hard to appreciate. With Minkowski's breakthrough clearly in mind, he wrote to Einstein, asking for his opinion. The latter reacted only with a brief, altogether dismissive remark (Einstein to Klein, 21 April, 1917, (CPAE 8A. 1998a, 436)), but Klein was undeterred, thinking it better to leave no mathematical stone unturned. He therefore gave Erich Bessel-Hagen the task of working out the associated invariants using new variational methods that Emmy Noether had published in Noether (1918). Bessel-Hagen applied Noether's first theorem to derive the conserved quantities with respect to the conformal transformations in Bessel-Hagen (1921), (this, incidentally, was the first new application of Noether's results).

Soon after the publication of Einstein's paper on cosmology, Klein wrote him to point out an oversight in it. Einstein replied on 26 March, 1917, (CPAE 8A. 1998a, 425). Klein had noted the two possibilities for a space of constant positive curvature, whereas Einstein had considered only one: the Riemannian 3–sphere. So he should also have allowed for the possibility of an elliptic geometry. In fact, Einstein had already learned this lesson from the astronomer Erwin Freundlich. For now, he seemed quite happy with either option, and he assured himself that an elliptic space only meant that the universe would be half as large as in the spherical case. Over the next year, de Sitter showed a strong preference for a cosmology built on elliptic geometry, whereas for Einstein this remained largely a matter of secondary importance.

By 1918 Klein found himself thoroughly absorbed by the fast–breaking developments in GR. He now began an intense correspondence with Einstein in which they took up some of its most difficult problems, in particular those pertaining to energy conservation (CPAE 8B. 1998b, passim). This had posed a major stumbling block for GR almost from the beginning. Einstein and Hilbert both dealt with this matter in 1916, but their respective results left many questions unanswered. In their correspondence, Klein and Einstein expressed sharply different opinions with respect to the status of certain results that Klein insisted were devoid of physical significance. Hilbert also participated in the ensuing discussions, and in an epistolary exchange, published under Klein's name in Klein (1917), Hilbert asserted that energy conservation in GR should be sharply distinguished from the physical meanings ascribed to this law in both classical mechanics and special relativity. In order to clarify that distinction, Klein asked Emmy Noether—who had been working on Hilbert's theory all along—to explore the formal properties of quantities derived from various types of action integrals invariant under a given transformation group. Her analysis, which led to the famous Noether theorems (Noether 1918), not only answered Klein's request but also showed why general covariance would automatically lead to four mathematical identities among the Euler–Lagrange equations derived from an invariant variational problem (Rowe 1999).

Klein also hoped to make progress on the cosmological problem in GR, and thus wrote to Einstein about a preliminary calculation he had made using Schwarzschild's interior solution for a fluid sphere (Klein to Einstein, 20 March 1918, (CPAE 8B. 1998b, 685–688)). His idea was to apply this metric to all of elliptic space, but he then found that the value of λ came out twice as large as Einstein's. What, he wondered, might account for this discrepancy? Einstein answered a few days later, but only to say that the Schwarzschild solution dealt with a fundamentally different situation, since equilibrium for a liquid can only arise when its inner pressure varies, whereas the idealized matter in Einstein's homogeneous cosmos has no pressure at all; it is merely pressureless dust. Furthermore, Schwarzschild dealt with the original field equations, in which $\lambda = 0$, whereas no possibility exists for obtaining a static cosmological solution without a nonzero value for λ.

Soon after finishing his letter to Einstein, Klein wrote to de Sitter about his efforts to bring elliptic geometry into the cosmological discussions:

> When Einstein's Cosmological Considerations appeared in March 1917, I wrote to him immediately that elliptic geometry ... met all his requirements. What I did not see, and what I have learned in the meantime from your presentation ... is that Einstein's assumption (case A in your communication) does not after all conform with the elliptic assumption (case B). Right now the matter is particularly interesting to me, as I recently made clearer for myself the state of affairs in the elliptic case by placing it in Schwarzschild's for the interior of a homogeneous fluid sphere. ... In this way one arrives exactly at that which you emphasized in your paper in treating case B. I noticed, of course, that the details do not fully match Einstein's results, about which I just wrote to him yesterday. Your own reasoning, as well as Schrödinger's ... I still want to check more carefully. (Klein to de Sitter, 22 March 1918).

These remarks seem to indicate that Klein recognized the key geometric difference that separated Einstein's solution from de Sitter's: only the latter was a 4-manifold of constant curvature. This meant that solution *B* could be approached using the standard methods of projective non-Euclidean geometry so dear to Klein's heart. Only a few months later, he would present his first findings on the geometry of de Sitter space to the Göttingen mathematical community.

Einstein on the Counterattack

In the meantime, Einstein began to spell out more carefully what he understood by the key foundational principles underlying GR. His earlier explication of these had been criticized by Erich Kretschmann, a former student of Max Planck, who claimed that general covariance was a purely formal principle without physical significance (Rynasiewicz 1999). In response to Kretschmann's criticism, Einstein published a brief note in *Annalen der Physik* (Einstein 1918a), in which he articulated his three key assumptions:

(a) Principle of Relativity,
(b) Principle of Equivalence,
(c) Mach's Principle.

Einstein had little to say about (b), which had been central to his approach since 1907, whereas (c) was a different story. For although he had consistently invoked Mach's name when discussing the relativity of inertia, only now did he give an explicit formulation of what he meant by *Mach's Principle in GR*.

This states that the gravitional field (as expressed by the metric tensor $g_{\mu\nu}$) must be determined solely by the energy–matter field (given by the tensor $T_{\mu\nu}$). In a footnote, he added that until now he had not clearly separated (a) from (c), the latter being a generalization of Mach's ideas about the origin of inertia. Later in the text he went on to say that "the necessity to uphold [this principle] is by no means shared by all colleagues, but I myself feel it is absolutely necessary" (Einstein 1918a, 39). Clearly he had de Sitter especially in mind, since the latter had done his best to raise doubts about the legitimacy of Einstein's speculations on the origins of inertia throughout the previous year.

Although this paper is mainly remembered today for the response Einstein gave to Kretschmann's critique, a closer reading reveals that he had another even stronger motivation: he was clearly intent on using this occasion to bolster his case for Mach's principle and its relevance for cosmology. Thus, he went on to explain how (c) led him to modify the field equations by introducing the cosmological constant, etc. In fact, on 7 March, just one day after he submitted this note to the Prussian Academy, he wrote another aimed at showing why de Sitter's solution *B* to the cosmological equations was mathematically suspect (Einstein 1918b). Einstein now went over to launch a counter–attack, claiming that the de Sitter solution contained singularities at the boundary of the coordinate system. He interpreted the degeneration of the metric there as an indication that de Sitter's solution was not matter-free after all. Einstein was sure that masses lay hidden just over its horizon. For this reason he concluded that "the De Sitter system does not look at all like a world free of matter, but rather like a world whose matter is concentrated entirely on the [horizon] surface. This could possibly be demonstrated by means of a limiting process ..." (Einstein 1918b, 48). Not long afterward Hermann Weyl would help bolster this argument by working out some of the messy details.

Einstein's hunch was mainly grounded on his firm adherence to Mach's Principle in GR, but he also pinpointed the source of the mathematical problem as he saw it. For this purpose, he wrote the metric for de Sitter space using coordinates r, Ψ, θ, ct, so that

$$ds^2 = -dr^2 - R^2 sin^2\frac{r}{R}[d\psi^2 + sin^2\psi\, d\theta^2] + cos^2\frac{r}{R}c^2dt^2.$$

He then noted that the metric tensor $g_{\mu\nu}$, the contravariant $g^{\mu\nu}$, as well as their first derivatives had to be differentiable functions throughout the finite region in which they are defined. This implies that the determinant $g = \|g_{\mu\nu}\|$ cannot vanish in that region. This consideration leads to the problem that

$$g = det \begin{Bmatrix} -1 & 0 & 0 & 0 \\ 0 & -R^2 sin^2\frac{r}{R} & 0 & 0 \\ 0 & 0 & -R^2 sin^2\frac{r}{R} sin^2\psi & 0 \\ 0 & 0 & 0 & cos^2\frac{r}{R} \end{Bmatrix}$$

$$= -R^4 sin^2\frac{r}{R} sin^2\psi\, cos^2\frac{r}{R}$$

vanishes along the 2-dimensional boundary where $r = \frac{\pi}{2}R$.

Einstein added that these points on the boundary fall within the domain of the coordinates, since by holding ψ, θ, t constant, the integral

$$\int_P^{\frac{\pi}{2}R} ds$$

is finite. So "until the opposite is proven, we have to assume that the de Sitter solution has a genuine singularity along the surface $r = \frac{\pi}{2}R$ in the finite domain" (Einstein 1918b, 47). This he saw as an irreparable problem, but nevertheless went on to write:

> If the de Sitter solution were valid everywhere, it would show that the introduction of the 'λ–term' does not fulfill the purpose I intended. Because, in my opinion, the general theory of relativity is a satisfying system only if it shows that the physical qualities of space are completely determined by matter alone. Therefore, no $g_{\mu\nu}$–field can exist (that is, no spacetime continuum is possible) without matter that generates it. (Einstein 1918b, 47)

After de Sitter received a copy of Einstein's remarks, he sent him a short note containing a counter-argument (De Sitter to Einstein, 10 April, 1918, (CPAE 8B. 1998b, 712–713)). Calling Einstein's demand that the cosmological field equations be valid for all points within a finite domain a "philosophical requirement," he noted that his own analysis of the situation in De Sitter (1917c) regarding points on the equator $r = \frac{\pi}{2}R$ had shown that these points were physically inaccessible. To reach them, he concluded, would require an infinite time. He then contrasted their two positions, or mind-sets, as follows: "my solution satisfies the *physical* requirement, but not the *philosophical* one. You naturally have the right to lay down the philosophical condition and thus to reject my solution. I, on the other hand, also have the right to reject the philosophical requirement but not the physical one." (CPAE 8B. 1998b, 712–713). A cosmologist today would surely conclude that this disagreement came about because the distinction between an event horizon and a particle horizon was not yet clear (Earman 1995, 20–21). These notions would, in fact, remain unclear for a long time; not until the work of Wolfgang Rindler in the mid 1950s was real clarity achieved (Earman 1995, 130–133). In the context of research on black holes, these concepts have been of central importance for modern space-time physics.

In closing, de Sitter added that he would have to undertake the calculations before deciding whether Einstein was correct in asserting that solution *B* could be derived by assuming that all mass in de Sitter's universe was concentrated along its equator. A few days later, Einstein wrote back to assure him that this was, indeed, the case. Referring to Weyl's forthcoming book, he remarked that this showed how the de Sitter solution can be regarded as the limiting case for a fluid lying along the equator. "The calculation is very simple," he added, "and so it really involves a singularity-like surface completely analogous to that of a mass-point." (Einstein to de Sitter, 15 April 1918, (CPAE 8B. 1998b, 720)). With that, he consented to agree to disagree with de Sitter about whether GR allowed for more than one cosmic possibility. Both combatants were suffering from health problems at this time, which may also help explain why their correspondence petered out after this. Not until November 1920 did De Sitter resume these cosmological conversations, which then took on an interesting new turn.

Some months earlier, however, de Sitter wrote a letter to Wolgang Pauli expressing harsh criticism of Einstein's cosmological views. Pauli was then writing his classic report on relativity theory for the *Encyklopädie der mathematischen Wissenschaften* (Pauli 1921). His teacher in Munich, Arnold Sommerfeld, had at first assumed this daunting task after Einstein declined the offer. Sommerfeld surely breathed a great sigh of relief when the 20-year-old Pauli agreed to take over for him. This soon led to an intensive correspondence with Klein, who oversaw the whole *Encyklopädie* project, but Pauli also wrote to de Sitter, who answered him on 25 May, 1920:

> It seems that by now the λ-term introduced by Einstein is generally called the "cosmological term." That is incorrect. It has nothing whatsoever to do with cosmology. True, Einstein's paper Cosmological Observations [(Einstein 1917)] – in which he introduces it – begins by mentioning a cosmological problem; but, as I showed, that problem is not solved by λ, not even touched by it.... If one wants to give a name to λ, then one should call it the inertial-flow or the inertial-coefficient, since it serves to relativize inertia. But inertia has as little to do with cosmology as, for example, gravity, or light, or magnetism. (Pauli 1979, 17–18)

Confronting Space–Time Singularities

In the spring of 1918 Hermann Weyl (Fig. 24.4) was putting the finishing touches on his classic text on GR, *Raum–Zeit–Materie* (*RZM*) (Weyl 1918), which would eventually go through five editions. For this first edition, he had proofs sent to Einstein, who responded enthusiastically. "It is like a symphonic masterpiece," Einstein exclaimed, "every little word is related to the whole and the design of the work is grand" (Einstein to Weyl, 8 March 1918 (CPAE 8B. 1998b, 669–670)). What struck him even more, though, was something Weyl had announced to him a week earlier in a letter, namely that he had found a way to unite gravity and electromagnetism by means of a single action integral. Eager to learn how Weyl could have managed this, he confided that he had sought in vain to solve the same problem and urged Weyl to send him a paper for submission to the Berlin Academy. In a postscript, though, he added a brief remark pertaining to relativistic cosmology. One could stipulate that the components of the curvature tensor vanished at spatial infinity, he noted, but this hypothesis led to complications that can be avoided simply by assuming that space is closed.

Weyl was pleased to send Einstein a paper for the academy proceedings, but then both got entangled in a messy series of procedural problems (CPAE 8B. 1998b, passim). Einstein raised physical objections to Weyl's theory from the beginning, though he certainly found it fascinating and worthy of consideration. When he presented Weyl's sketch to his fellow academicians, however, Walther Nernst insisted that Einstein add a note stating why he found this daring new unified field theory physically untenable. Max Planck, as presiding member, then asked Einstein to resubmit Weyl's paper the following week, with or without such an appended note. Einstein took this to mean that the paper would be accepted one week later, but at this next meeting Nernst objected again, this time noting a precedent for such a policy.

Fig. 24.4 Hermann Weyl at first favored Einstein's cosmology, but later swung over to de Sitter's side.

After all this, Einstein had to write a sheepish letter to Weyl explaining what had gone wrong. In order to be published, the paper would have to contain a statement answering Einstein's principal criticism, namely that Weyl's theory appeared to make measurements of rods and clocks depend on their prehistories. To Einstein's great relief, Weyl responded to his concerns without the slightest sign of annoyance. At the same time, Weyl made clear that his approach was driven entirely by an underlying conceptual problem that Einstein's theory of gravitation had brought to the fore. For Weyl, field physics had to be grounded in a truly infinitesimal geometry based on general affine structures rather than treating these as aspects of a metric as in Riemannian geometry (Scholz 2001b). He would later elaborate on this theory in subsequent editions of *RZM*, but within the next two years he began to lose faith in this program for field physics, which faced the mounting difficulties of accounting for the properties of matter. By the time his first paper on his extended theory of relativity appeared in translation in Lorentz et al. (1922, 201–216), Weyl added a note alerting the reader that he had already abandoned it two years earlier.

While reading the proofs of *RZM* in April 1918, Einstein was delighted to see that Weyl had independently worked out a line of argument that supported his Machian position. In effect, Weyl showed that the singular points on the equatorial surface can be treated as mass points which induce the geometry of de Sitter space–time. This showed conclusively that the de Sitter metric could be interpreted as belonging to a universe in which all matter had been pushed out to the edge of space. Little wonder that Einstein now felt fully vindicated, so he was more than happy to inform de Sitter that his matter–free world was chimerical. But then he suddenly realized that something was wrong, and on 18 April, only a few days after he had written to de Sitter, he dashed off a frantic letter to Weyl, explaining that he had found a mistake in the argumentation. This led to a flurry of further exchanges, as Weyl struggled to put things right.[5] Since the war was still raging, Weyl had some reason to worry about communications between Zurich and Berlin, where Ferdinand Springer was patiently awaiting the final corrected proofs. Einstein thus kept the publisher informed while he and Weyl wrote back and forth. Both were relieved when *RZM* was released in early July, and by November Weyl could report to Einstein that the first edition had already sold out (Weyl to Einstein, 16 November, 1918, (CPAE 8B. 1998b, 948–949)).

In the meantime, Klein had been pondering certain topological difficulties posed by global solutions in GR, in particular de Sitter's. Like Einstein and de Sitter, he thought about both cases as static models, which presented no difficulties when it came to Einstein's solution A, whereas it seemed impossible to find a suitable coordinatization for the entire manifold associated with solution B. Since the latter was a 4-manifold of constant curvature, Klein could explore its properties by using projective methods, as discussed in Part I. From this point of view, the Cayley metric on induced on the spatial sections will autmotically yield elliptical 3-spaces, ruling out spherical geometries. But Klein now saw a perplexing new difficulty: how to give the worldlines a consistent direction in time? In describing this problem in a letter to de Sitter, he noted that a two-dimensional family of world lines forms an elliptical surface and is hence non-orientable. De Sitter answered these concerns in a letter from 25 April, 1918. He found Klein's argument sound mathematically, but he disputed its physical significance since the polar line of any point cannot be reached in finite time. The same being true for all the antipodal points associated with points on a worldline, there should never be any difficulty accounting for time orientation in a de Sitter space-time. These points were taken up in De Sitter (1918, 1310), which will be described in the section following. Klein also informed Einstein about the same in a letter from 25 April. He must have been amused when just two days later he received the physicist's reply, containing a nice little topological argument showing that only even–dimensional elliptic spaces are non-orientable (Einstein to Klein, 27 April, 1918, (CPAE 8B. 1998b, 738–740)). For this reason, he felt confident that Klein's considerations had no bearing on his own cosmology if one took its spatial-sections to be elliptical.

A few months later, in July, Klein delivered two lectures on relativistic cosmology before the Göttingen Mathematical Society (for concurrent events in Göttingen related to GR, see Rowe (1999)). He clearly saw this as a welcome opportunity to bring forth his favorite projective methods in order to discuss Einsteins solution in Einstein (1917) and especially de Sitter's solution B. As noted earlier, de Sitter had utilized three different coordinate systems in his very first paper De Sitter (1917a). He found it convenient to complexify the time coordinate in order to bring out analogies between his solution and Einstein's solution A. Reverting to real coordinates, he then obtained a hypersurface, analogous to a one–sheeted hyperboloid, which he mapped by stereographic projection into a flat 4–space M_4. In doing so, however, only a portion of the 4–space appears in the image, which is bounded by a 3–surface F_3 , itself a quadric with two sheets. This circumstance had immediately aroused Einstein's suspicions,

[5]These letters can be found in CPAE 8B. (1998b). Hubert Goenner explored Weyl's changing attitudes toward relativstic cosmology in his interesting paper Goenner (2001). Commenting on Weyl's earliest work, Goenner was struck by how a mathematician of his caliber could have been misled by these special types of coordinates systems, which only cover part of the manifold, into thinking that the boundary of a coordinate batch contained a real singularity. To a modern expert on GR, this seems especially odd since de Sitter space–time has constant curvature and so is homogeneous and isotropic.

though de Sitter tried to assuage him by arguing that this boundary F_3 was entirely inaccessible.

Klein realized that one could view F_3 as an absolute figure in M_4, thereby obtaining a metric geometry associated with the 10-parameter group of linear transformations G_{10} that leave F_3 fixed (Klein 1918b). He further noted that each pair of tangent hyperplanes fixes a two-dimensional axis A along with a double-wedge of 3–planes $P_3(A)$ none of which intersect F_3. Thus the geometry for each of these $P_3(A)$ will be elliptic with an identical metric. Letting these 3–planes run from one tangent through to the other, he could make use of the Cayley metric to introduce a time parameter, thereby obtaining a projective structure fully equivalent to de Sitter space in static coordinates. Klein called this a "de Sitter world" in which the axis A, the edge of the double-wedge, corresponds to the equator in de Sitter's hypersphere. Events in this space–time evolve as the pencil of hyperplanes $P_3(A)$ moves through time, stretching from one tangent of the quadric F_3 to the other at temporal infinity.

The upshot of this purely mathematical construct was simply to show that the "mass horizon" (Weyl's terminology in Weyl (1918, 226)) had no physical significance. What Einstein and Weyl took to be an essential singularity within the space-time manifold was merely an artefact of the coordinate system chosen. Thus, they were mistaken in assuming that hidden masses had to be present along the equator of de Sitter space, and that this matter was responsible for its curvature. Klein informed Einstein of this in a letter from 16 June, 1918: "I came to the conclusion that the singularity you noticed can simply be transformed away." (Klein to Einstein, 16 June, 1918, (CPAE 8B. 1998b, 805)).

What he meant by this was probably not completely transparent to Einstein, who was not well versed in projective geometry. Still, he would have understood the gist of Klein's argument, which indicated that each pair of tangent hyperplanes to the quadric F_3 determined a different "de Sitter world" with its own system for measuring time. All of these worlds are equivalent, however, in the sense that they can be mapped to one another by the transformations in G_{10}. A pair of tangent hyperplanes to the quadric F_3 will be fixed by giving their points of tangency, which requires fixing six coordinates. Thus the manifold of all possible "de Sitter worlds" will be six-dimensional. Clearly the group G_{10} acts transitively on this manifold, so any particular world can be mapped to any other. Since the axis A of a given double-wedge merely consists of points whose coordinates are indeterminate, one can transform these singularities away simply by choosing a different double-wedge that contains A. Klein likened this situation with the removal of the indeterminacy at the origin in a system of polar coordinates.

Whatever Einstein may have made of this, he was convinced that this argument had to be sound. Writing back to Klein, he admitted: "You are entirely right. De Sitter's world is, in and of itself, free of singularities and its spacetime points are all equivalent.... My critical remark about de Sitter's solution needs correction: a singularity-free solution for the gravitational field equations without matter does in fact exist. However, under no condition could this world come into consideration as a physical possibility." (Klein to Einstein, 20 June, 1918, (CPAE 8B. 1998b, 809). The final sentence was, of course, the decisive one, though it had the ring of dogmatism. Klein surely assumed that this was an allusion to the analysis that Weyl had given, and he later wrote Weyl asking for clarification. In his response from 7 February 1919 (Klein Nachlass XXII B), Weyl explained that he and Einstein rejected de Sitter's solution B because it did not correspond to a *static universe*. This in no way contradicted Klein's finding, which showed, however, that a system of static coordinates would only cover a portion of the space–time geometry (Fig. 24.5).

Most of the communications and discussions we have considered thus far took place behind the scenes, of course. So it is more than doubtful that contemporary readers—especially those who lacked inside information from the actors themselves—had anything like a clear picture of the underlying issues involved. Einstein certainly made no public statement, then or later, similar to the one he imparted privately to Klein when he conceded that de Sitter's space-time was free of singularities. In fact, Einstein completely ignored case B as a viable cosmology throughout his career. During the 1920s he also clung tenaciously to his defense of Mach's Principle, which he highlighted in his widely read Princeton lectures on *The Meaning of Relativity* (Einstein 1922).

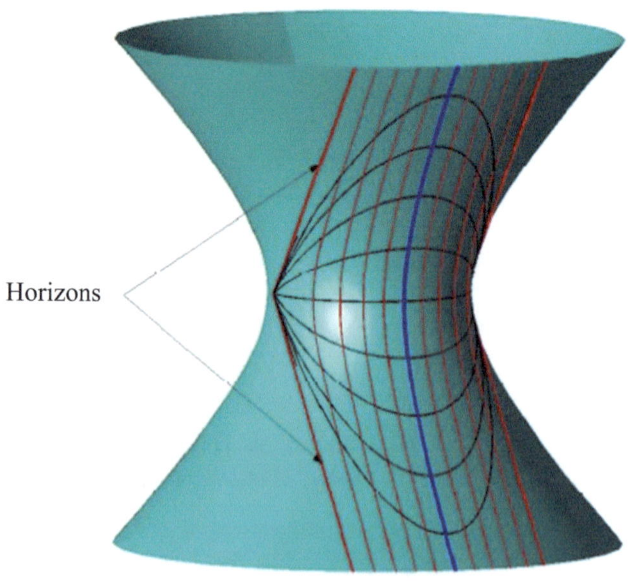

Fig. 24.5 A portion of de Sitter space in static coordinates as depicted in Moschella (2005, 7).

There, in his concluding lecture, Einstein argued just as he did in 1917, with no mention of de Sitter's alternative solution whatsoever. Furthermore, he presented the results of Thirring and Lense on the relativity of rotation in much the same way as when he discussed such Machian effects back in 1912. He also noted the purely theoretical character of such induction effects from frame dragging, since these remained much too small to be tested experimentally. In defending his finite, but unbounded cosmology, Einstein wrote: "If the universe were quasi-Euclidean, then Mach was wholly wrong in thinking that inertia, as well as gravitation, depends upon a kind of mutual interaction between bodies" (Einstein 1922, 64). So this would rule out solution C, Minkowski space, but not de Sitter's solution B, which was closed, finite, and matter–free. Given the intensity of the earlier discussions from 1917–1918, it would seem difficult to understand how Einstein could have simply ignored all mention of this, including the fact that his cosmological field equations admit a non-singular solution at odds with Machian principles.

On Losing Track of Time in de Sitter Space

When he wrote to Klein that the de Sitter world could not be taken seriously "as a physical possibility," Einstein might still have believed that it possessed a mass horizon, as Weyl had tried to show. If so, his statement could be interpreted as saying that it was physically unfeasible to imagine a universe in which virtually all the world's matter was removed to an unreachable boundary. Einstein, of course, favored a cosmology in which matter was distributed homogeneously throughout space. As a good Machian, on the other hand, he needed to explain how inertia could exist in a universe nearly devoid of matter. This led to what de Sitter called the *material postulate of relativity of inertia*, about which he gave this explanation: "Einstein originally supposed that the desired effect could be brought about by very large masses at very large distances. He has, however, now convinced himself that this is not possible. In the solution which he now proposes, the world matter is not accumulated at the boundary of the universe, but distributed over the whole world, which is finite, though unlimited." (De Sitter 1917c, 5).

Einstein could, on the other hand, just as well have pointed to Klein's own interpretation of the de Sitter solution as a strong argument against its physical feasibility. Klein's viewpoint, after all, led not to one, but to a vast manifold of worlds, each with its own coordinate system for telling time. As Klein himself emphasized, observers in these different worlds experience strange difficulties when they try to communicate with one another. Thus, he wrote: "It is amusing to picture how two observers living on the quasi-sphere with differing de Sitter clocks would squabble with each other. Each of them would assign finite ordinates to some of the events that for the other would be lying at infinity or that would even show imaginary time values." (Klein 1918a, 615). Surely this type of temporal confusion could have been cited as reason enough to cast aside de Sitter's cosmology as a physically senseless model. Yet apparently no one, aside from de Sitter himself, reacted to Klein's finding, leaving matters just as murky as before.[6]

De Sitter knew the passage cited above from Klein (1918a), of course, because he transmitted this short note for publication in the Proceedings of the Amsterdam Academy. He also cited a passage from a letter Klein sent him, dated 19 April 1918, in De Sitter (1918). This concerned the above-mentioned circumstance that even-dimensional elliptic spaces are non-orientable. Klein had already in 1890 pointed out the problem this raises for physical applications due to a possible reversal of the arrow of time. Recognizing the validity of this mathematical argument, de Sitter offered the following physical escape:

> ...we return to the starting point with the positive direction reversed ...only if we have travelled *along a straight line*, or at least along a line *which intersects the polar line of the starting point*. This "motion", though mathematically thinkable, is *physically impossible* .... (De Sitter 1918, 1310)

This might sound like another kind of "cosmic censorship", but at least de Sitter acknowledged the problem and took a position on its status for cosmology.

Within mathematical circles, Klein's projective approach to de Sitter space was probably little better known than among physicists (Röhle 2002). Largely for that reason, the geometer H. S. M. Coxeter wrote an interesting expository article about it during the Second World War (Coxeter 1943). Interestingly enough, he cites works by leading relativists—Eddington, Robertson, and A. A. Robb—but then remarks that he only learned about the projective derivation of de Sitter space from Prof. Patrick du Val in a conversation from around 1930. Coxeter added that he had not seen this in print Coxeter (1943, 224)), a rather curious admission since he could have easily found an exposition in Du Val (1924). However, neither Coxeter nor du Val made any reference to Klein (1918b), so this knowledge apparently only circulated in the geometers' "folk culture."

Using the 2-D figure below as a schematic for projective de Sitter space, Coxeter illustrated the temporal paradoxes that arise when different observers look into the past and future. The overlapping shaded region, for example, would lie in the future of A's world, but in the past of B's. Here the conic represents time–like infinity; each observer will have a different future and past horizon determined by the tangent cone enveloping the absolute, a 3-D quadric in projective 4–space. Coxeter likens this changing horizon with the visual

[6] For a survey of the connections between metrized projective geometries and their associated space–times, see Liebscher (2005).

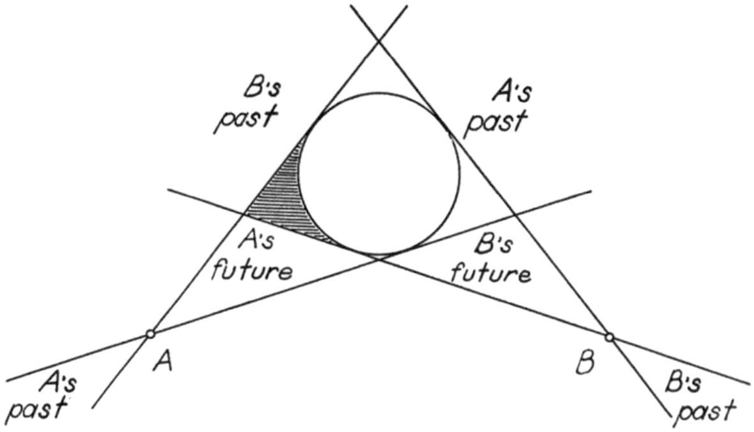

Fig. 24.6 The temporal anomalies of de Sitter space as portrayed in Coxeter (1943, 227).

effect when one tries to chase a rainbow. If the two observers move far apart the shaded region becomes larger, reaching a maximum when the distance $\overline{AB} = \pi R$, at which point A's future becomes identical with B's past, and vice–versa. Since the metric along a space–like line is elliptic, this maximal distance occurs when the points A and B coincide. As Coxeter puts it: "If you could travel all along a space–like line, you would return to the starting point with your past and future interchanged!" (Coxeter 1943, 227) (Fig. 24.6).

He saw two possible ways to evade this paradox: either one simply claimed that such a journey could never be completed on physcial grounds—as de Sitter had argued in De Sitter (1918)—or one could extend the space–time to its double covering space, which is an orientable manifold (a very modern-sounding idea). This can then be represented on a single-sheeted quadric in a Minkowskian space of 5 dimensions. Coxeter now asks: "do you not find it disturbing to envisage an exact replica of yourself at the 'antipodes' living backwards?" (Coxeter 1943, 227). This whole topsy–turvy universe reminded him of Lewis Carroll's *Through the Looking Glass*, a world in which cause and effect become inverted. As a prelude to his paper, he cited these words of the White Queen: "He's in prison now, being punished: and the trial doesn't even begin till next Wednesday: and of course the crime comes last of all."

As has long been known, Einstein worried a good deal about the loss of causality in quantum physics. But he also harbored some deep concerns about possible anomalous phenomena that might arise in general relativity, in particular when searching for global solutions to the gravitational field equations. Even before he found these equations in November 1915, he had expressed concern about the possibility that a space-time might admit closed time-like worldlines. Along such curves an observer could move continually forward into the future only to return back to his or her own past. In today's world of virtual realities one can find plenty of literature concerned with such exotic types of time travel in GR. Philosophically minded readers ought to ponder the reflections in Earman (1995), particularly its sixth chapter in which the author offers, among other things, some "therapies for time travel malaise." Clearly, problems associated with closed space–like and time–like curves played an important role in GR from the beginning, particularly in its most impressive field of application, relativistic cosmology. After Gödel's discovery in 1949 of a rotating universe with closed time–like geodesics, cosmologists began to investigate such anti–Machian models following this new lead.

By 1955, the year of the Bern Jubilee conference, such space–time anomalies were clearly on the minds of many experts. H. P. Robertson even expressed real dismay that GR admitted such exotic models. Responding to a question about whether Gödel's cosmological model did not constitute a proof that absolute rotation was, indeed, compatible with GR, Robertson replied: "I am afraid that is correct. The entire material field in his solution must be judged to be in rotation. I consider it a defect in the equations of the general theory of relativity that they allow such a solution".[7] Wolfgang Pauli, who presided at the conference, skirted over the problem of cosmological anomalies in his closing remarks. He also neglected to mention a short contribution by a 25-year-old Belgian mathematician named Jacques Tits (Mercier and Kervaire 1955, 46–47). Though little known at the time, Tits would go on to become a pre-eminent figure in modern algebra, particularly the theory of Lie groups and Lie algebras. He also happened to be well versed in classical geometry, which helps explain how he managed to promote the work of Coxeter among the mathematicians associated with the Bourbaki group in Paris.

Someone on the program committee for the Bern Jubilee must have heard about Tits's recent work and its relevance for cosmological modeling. Shortly before, Tits had classified all homogeneous and isotropic space–times, finding that these

[7](Mercier and Kervaire (1955, 145); the question was posed by the Hambrug cosmologist Otto Heckmann).

fell into five different classes.[8] The first two of these were the de Sitter and anti–de Sitter spaces and their covering spaces, which he defined projectively, following Klein's approach. He then went on to show which of these types of space–times had infinite extension in time in the sense that they did not contain closed time-like curves. These were: (1) de Sitter space and its double cover; (2) the infinite cover of anti-de Sitter space; (3) the special Robertson–Walker spaces with Euclidean space sections, and (4) the static spaces with cosmic time, where the space sections are the standard geometries of constant curvature for Euclidean, elliptic, spherical, and hyperbolic 3-space. In short, Tits answered a longstanding mathematical question of central importance for relativistic cosmology.

Einstein's Ether as Carrier of Inertia

During the immediate post–war period, Einstein and de Sitter remained in touch sporadically, though they apparently had no further occasion to debate cosmological issues. That, however, suddenly changed in the autumn of 1920. By this time life in Berlin had become both dreary and difficult for Einstein, making Leyden more attractive than ever. Thanks to his friends and colleagues there, he gained an appointment as an honorary guest professor. Thus, on 27 October 1920, he stepped to the podium to deliver his long-awaited inaugural lecture. As it turned out, this would be more than just a formal event as Einstein had some things of importance to say. His chosen theme, "Ether and Relativity Theory," served both to clarify his conception of space–without–matter in general relativity while paying tribute to H. A. Lorentz, Leyden's local genius, who had paved the way for a modern understanding of a non–material ether. Einstein also took this opportunity to elaborate on the Machian motifs that informed his own understanding of GR. It was to this end that he remarked:

> Mach's ideas find their full development in the ether of the general theory of relativity. According to this theory the metrical properties of the space–time continuum in the locality of individual space–time points differ and are simultaneously conditioned by matter outside the region in question. ... [This] has no doubt disposed of the notion that space is physically empty. But this has also once again given the ether notion a definite content—though one very different from that of the ether in the mechanical wave theory of light. The ether of the general theory of relativity is a medium devoid of all mechanical and kinematic properties, but it influences mechanical (and electromagnetic) phenomena. (Einstein 1920, 317)

Willem de Sitter, who suffered from tuberculosis, was at this time far from Leyden convalescing in the mountains of Switzerland. Though he could not attend Einstein's lecture, he received a copy of the text. After reading it, he sent Einstein an enthusiastic letter from the resort village of Arosa to express his delight in seeing how firmly Einstein rejected a mechanical explanation of nature. De Sitter then recalled how, as a student, he

> always bristled when someone explained matter by the ether or by electricity, only to turn around again and seek material explanations for the ether! ... Now you have decided to call the $g_{\mu\nu}$–field the "ether," and you show convincingly that this ether is just as good as "matter," if not better, as a primal physical substance. In my opinion, there is consequently *no* reason left to look for a *material* carrier of inertia. Mach's requirement also seems to me simply to be a residue of the quest for a *mechanical* explanation of nature (on the basis of action at a distance). The *ether* is the carrier of inertia. The material points are just discontinuities in the ether, i. e. in the $g_{\mu\nu}$–field; the field itself is what is real. (De Sitter to Einstein, 4 November, 1920, (CPAE 10. 2006, 477)).

Having sounded his main theme in this emphatic overture, de Sitter went on to argue for the superiority of his own ether-based world to the one Einstein brought forth in 1917. He began by recalling a well–known optical problem that arises when assuming that the geometry of space is that of a homogeneous 3-sphere, namely ghost images. In particular, an image of the sun ought to appear at night, formed by light rays coming from the antipodal location of its virtual image. In fact, on this hypothesis the heavens should be filled with these ghost images. Karl Schwarzschild had already discussed this problem 20 years earlier, arguing that if light were absorbed as it circled around space then these ghost stars would never appear (De Sitter 1917b). But de Sitter, citing Rayleigh's law, thought that the absorption would be far too small to dissipate the predicted effect. He also noted that the sun would display a gravitational ghost as well, though predicting its exact location would require complicated calculations. Here de Sitter could refer to his own investigations of gravitational absorption based on the moon's motion, from which he deduced that such an effect was so negligible it could be pronounced nonexistent (De Sitter 1912).

Still, it was not the "fear of ghosts" that bothered de Sitter most, but rather another feature of Einstein's universe: its reliance on an absolute, cosmic time, which "violates the principle of relativity." Realizing that Einstein was by now familiar with this complaint, he added a soothing remark: "We have often quarreled about this already, and it ultimately remains a matter of taste, which system one wants to consider more probable." (CPAE 10. 2006, 478). Finally, de Sitter called Einstein's attention to a recent note on cosmology that he had passed on to Lorentz for publication by the Amsterdam Academy (De Sitter 1920). This concerned a

[8]Isotropy here means with respect to the directions of light. Tits calls a Lorentzian manifold M isotropic at a point p if the isometries of M that fix p act transitively on the directions of light issuing from p, ($ds^2 = 0$). This is equivalent to requiring the existence of a space-like 3-plane ω containing p such that the isometries fixing p and ω act transitively on the lines of the light cone in ω.

matter that had some bearing on Einstein's argument in favor of a spatially closed universe, since his reasoning presumed that the mean density of matter, however small, was in a state of statistical equilibrium.

Einstein's reply to this long letter has not survived, but he clearly did react to several of the astronomical issues de Sitter had raised. In all likelihood he had not read the latter's recent note, which would explain why de Sitter provided information from it when he responded some weeks later (De Sitter to Einstein, 29 November, 1920, (CPAE 10. 2006, 500–501)). Writing without access to the literature, de Sitter reported that Eddington had worked on the stability of the Milky Way. He could only remember, however, that Eddington had found that the galaxy would probably be able to hold together on the basis of Newton's law of gravitation and the stellar velocities then available. What he could say with certainty, though, was that taking the mean mass in the galaxy to derive λ would lead to a universe far too small, whereas using the mean mass in the vicinity of the sun leads to a universe much too large. In short, there seemed to be no clear way to link astronomical data to the λ-term in the cosmological equations.

At this point, de Sitter repeated a query he received from Einstein: "Would it not be more satisfying to assume $\lambda = 0$ there, if one lays no significance on the existence of a mean density of matter anyway, nor on interpreting inertia as an interaction between bodies?" De Sitter reminded him that he had long held this very view, but then he added that "the repulsive force that results from my interpretation of the universe ... really does seem to exist!" He then summarized what he wrote in De Sitter (1920) in these words:

> When I wrote about this in 1917, the radial velocities of only 3 spiral nebulae had been measured by then. Today 25 have been, and with 3 exceptions, all are *positive*. The average, if one excludes the brightest, and thus probably the closest ones, is +631 km/sec. The largest observed velocity is 1200 km/sec. The velocities are *radial*; the nebuale are distributed unevenly throughout the whole sky. (CPAE 10. 2006, 501)

The upshot of these remarks was to throw fresh cold water on Einstein's universe, this time by questioning whether one could really assume that we live in a static cosmos. De Sitter surely thought there was far too little astronomical evidence to stake a serious claim for a dynamic cosmology based on GR, but he also felt that his solution B at least came closer to matching what was known. Einstein obviously found de Sitter's argument that solution A admitted ghost stars unconvincing, a reaction that led de Sitter to reply in kind:

> You say the universe is too inhomogeneous (optically opaque) to allow ghost stars to come into focus. But the world is unbelievably *empty*. It is an *observational fact* whose accuracy can scarcely be doubted, that outside of the Milky Way up to a distance of about 100,000 light–years there is still *no* sign of absorption or dispersion of light .... It is not *excluded*, of course, that over much larger distances (of 100,000,000 light–years) absorption and dispersion would become noticeable. But we still know nothing about that at present. (CPAE 10. 2006, 501)

Summarizing the content and tone of these two letters from November 1920, one is reminded of a famous line from *Hamlet* that might well have occurred to de Sitter:

> There are more things in heaven above, dear Einstein,
> Than are dreamt of in your philosophy.

A few years later, in fact, Weyl found a new interpretation of de Sitter's space-time, one that broke with the static views that had long dominated cosmological discussions (Bergia and Mazzoni 1999). He presented this in an amusing dialogue, "Massenträgheit und Kosmos" (Weyl 1924), which was published in *Die Naturwissenschaften*, the German counterpart to the British journal *Nature*. In this piece, Weyl re–enacts highlights from earlier debates in GR, treating these as quasi–theological disputes over the dogma of Mach's Principle as a condition for membership in the "church of relativity" (Rowe 2015). Turning to cosmology, he mentions current measurements of redshifts in order to argue that these show a dispersal of nebulae in space driven by a primal ether. To describe this, he uses a special foliation of de Sitter space (Figs. 24.7 and 24.8)[9] that serves to model this kinematic effect, while citing lines from the German poet Hölderlin in order to enhance the sense of Romantic wonder.

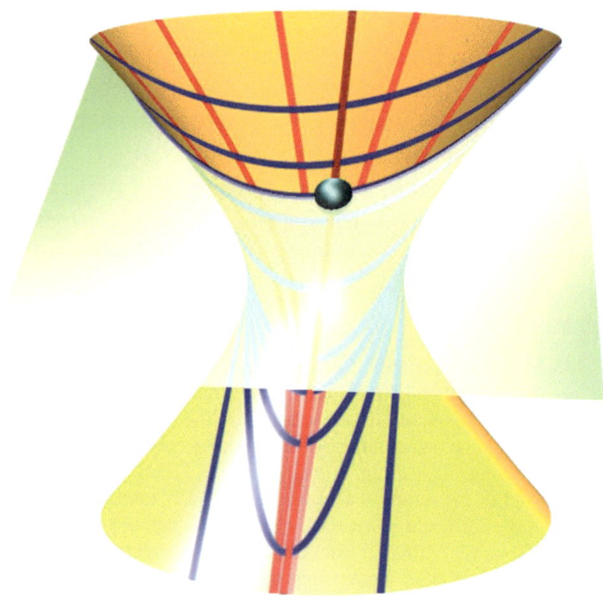

Fig. 24.7 Weyl's foliation of de Sitter space containing worldlines (in red) that spread out from a common source in the infinite past. Graphic by Oliver Labs. For an animation of this image showing the divergence of matter in de Sitter space–time, go to http://cosmology.MO-Labs.com.

[9]For a visual tour of de Sitter space and its various coordinatizatons and (partial) foliations, see Moschella (2005).

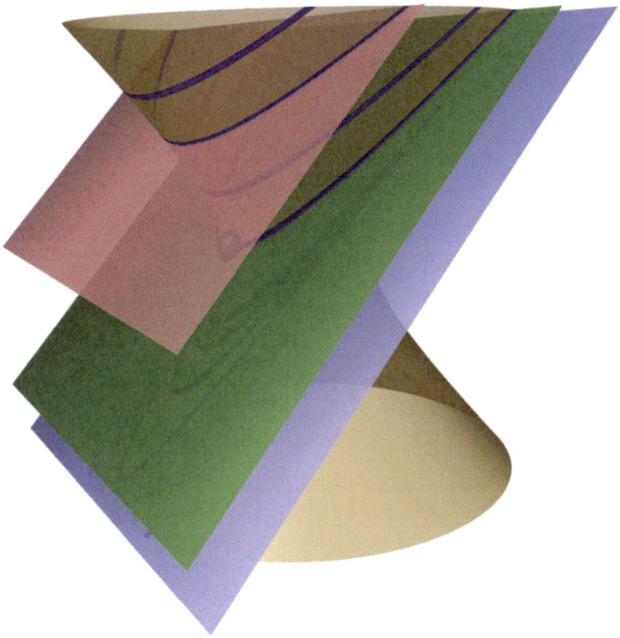

Fig. 24.8 The space-like sections are 3-planes that cover only the upper-half of the manifold. Graphic by Oliver Labs. For an animation go to http://cosmology.MO-Labs.com.

On the Persistence of Static Cosmologies

In the concluding remarks from his 1917 paper, Einstein emphasized that he had introduced the cosmological constant for one reason only: to obtain a global solution in static coordinates. Whether or not the universe was finite and closed, as he suggested, or space was infinite, as had long been assumed, that was another matter about which experts could disagree. There was no such disagreement, however, when it came to the quasi-static nature of the universe, which helps make clear why Einstein drew attention to that very point. His contemporaries, including de Sitter, could only nod their heads in agreement (Ellis 1988). And since there were very few static models to choose from, early relativistic cosmology was dominated by just two possibilities. Striking evidence of this is provided by the neglected work of the Russian mathematician Alexander Friedmann (North 1965, 113–117).

When Friedmann published a dynamical solution of Einstein's cosmological field equations in 1922, he showed how his approach delivered the static solutions of Einstein and de Sitter as special cases. Friedmann's work from the early 1920s became famous a decade later, well after his death in 1925. Yet during his lifetime it was either overlooked or, if read, then quickly forgotten. Einstein, who at first thought Friedmann's result was erroneous, sent a short note to *Zeitschrift für Physik* in order to make his opinion known. Afterward he received a lengthy letter from Friedmann clarifying the situation (A. Friedmann to Einstein, 6 December, 1922, (CPAE 13. 2012, 601–604)). This led to Einstein's public retraction 6 months later. "My criticism," he wrote, "was based on an error in my calculations. I consider Mr. Friedmann's results to be correct and to shed new light." In other words, this finding, while theoretically noteworthy, was without any physical significance considering the strong astronomical evidence in favor of a static universe.[10]

Einstein was hardly alone in holding this view, which was the consensus opinion in 1922, despite the fact that astronomers had by now detected large redshifts among a handful of stellar nebulae. As one of the era's leading astronomers, Willem de Sitter was, of course, well aware of these findings, though he took a cautious attitude when discussing the possibility of a dynamical approach to cosmology. De Sitter, Eddington, and Weyl had all drawn attention to these phenomena, but without connecting them to the possibility of an expanding universe. So the quasi–static nature of the universe went virtually unquestioned, while attention focused on the issue of size.

A century ago our universe was still a very small and tidy place. Or so many believed. Debates over its size would pick up after 1920, when two leading astronomers, Harlow Shapley and Heber Curtis, squared off over the status of remote spiral nebulae. Shapley believed these were relatively small celestial objects located at the far edges of our galaxy, whereas Curtis argued instead that some of these nebulae were exceedingly large and distant phenomena; he claimed these were whole new galaxies stretching out over vast empty realms of space. This encounter later came to be known as the "Great Debate," as it sparked arguments that ran both ways for nearly a decade. Ultimately, new data settled the question. Edwin Hubble's publications from the 1920s, based on observations made at the Mt. Wilson Observatory, gradually persuaded the astronomical community that Andromeda and other spiral nebulae were, indeed, extra-galactic. Some decades later, Carl Sagan and other astronomers helped to popularize the idea of "other worlds", the island universes once imagined by Immanuel Kant. Today our "visible universe" has grown so large as to contain some 200 billion galaxies rather than just our own Milky Way.

We have also come to think of our universe as a restless place, driven by powerful forces. Hubble's researches had something to do with that, too. In 1929 he announced what nowadays is called Hubble's law. This indicates a rough proportionality (Hubble's constant) between the distances of galaxies from the earth and the measurements of their redshifts. These findings provided dramatic evidence for the wild ideas of the cosmologist Georges Lemaître, who pioneered the notion of an expanding universe. Yet, to a consid-

[10]Einstein had actually written words to this effect, but then he struck the sentence out before sending off his note. See Frenkel (2002).

erable extent, Lemaître remained a lone prophet throughout the 1930s. Models for expanding universes became popular then, but without taking Lemaître's theory of the primeval atom (a kind of "big bang" theory) seriously into account. As Helge Kragh writes: "the notion of an expanding universe won general acceptance, and with it the notion of a world with a finite past. However, [these] world models ... do not necessarily qualify as big-bang models in a physical sense." (Kragh 1996, 22). General relativity had yet to forge strong links with astrophysics, itself a field of research in its infancy. The renaissance of relativity was still waiting to begin.

Acknowledgements I am grateful to Michel Janssen, Erhard Scholz, and Scott Walter for their comments on an earlier version of this paper. This being a snapshot of a complex story, I have tried to tell part of it here without taking in other contemporaraneous developments that would require serious attention in a more comprehensive study. Historical and mathematical details connected with the Einstein–de Sitter debates and other related matters can be found in several of the references cited below.

References

Bergia, Silvio, and Lucia Mazzoni. 1999. Genesis and Evolution of Weyl's Reections on de Sitter's Universe. In *The Expanding Worlds of General Relativity*, Einstein Studies, ed. H. Goenner, et al., vol. 7, 325–342. Boston: Birkhäuser.

Bessel-Hagen, Erich. 1921. Über die Erhaltungsätze der Elektrodynamik. *Mathematische Annalen* 84: 258–276.

Coxeter, H.S.M. 1943. A Geometrical Background for de Sitter's World. *American Mathematical Monthly* 50: 217–228.

CPAE 6. 1996. *Collected Papers of Albert Einstein, Vol. 6: The Berlin Years: Writings, 1914–1917*, ed. A.J. Kox et al. Princeton: Princeton University Press.

CPAE 7. 2002. *Collected Papers of Albert Einstein, Vol. 7: The Berlin Years: Writings, 1918–1921*, ed. Michel Janssen et al. Princeton: Princeton University Press.

CPAE 8A. 1998a. *Collected Papers of Albert Einstein, Vol. 8A: The Berlin Years: Correspondence, 1914–1917*, ed. Robert Schulmann et al. Princeton: Princeton University Press.

CPAE 8B. 1998b. *Collected Papers of Albert Einstein, Vol. 8A: The Berlin Years: Correspondence, 1918*, ed. Robert Schulmann et al. Princeton: Princeton University Press.

CPAE 10. 2006. *Collected Papers of Albert Einstein, Vol. 10: The Berlin Years: Correspondence, May-December 1920*, ed. Diana Kormos Buchwald et al. Princeton: Princeton University Press.

CPAE 13. 2012. *Collected Papers of Albert Einstein, Vol. 13: The Berlin Years: Writings & Correspondence, January 1922 to March 1923*, ed. Diana Kormos Buchwald et al. Princeton: Princeton University Press.

Darrigol, Olivier. 2015. Mesh and Measure in Early General Relativity. *Studies in History and Philosophy of Science Part B: Studies in History and Philosophy of Modern Physics* 52: 163–187.

De Sitter, Willem. 1911. On the Bearing of the Principle of Relativity on Gravitational Astronomy. *MNRAS* 71: 388–415.

De Sitter, Willem. 1912. Absorption of Gravitation. *Observatory* 35: 387–393.

De Sitter, Willem. 1913. Some Problems of Astronomy. VII The Secular Variation of the Elements of the Four Inner Planets. *Observatory* 36: 296–303.

De Sitter, Willem. 1917a. On the Relativity of Inertia. Remarks Concerning Einstein's Latest Hypothesis. *Koninklijke Akademie van Wetenschappen te Amsterdam, Proceedings* 19(1916–17): 1217–1225.

De Sitter, Willem. 1917b. On the Curvature of Space. *Koninklijke Akademie van Wetenschappen te Amsterdam, Proceedings* 20(1917–18): 229–243.

De Sitter, Willem. 1917c. On Einstein's Theory of Gravitation, and its Astronomical Consequences. Third Paper, *Royal Astronomical Society, Monthly Notices* 78: 3–28.

De Sitter, Willem. 1918. Further Remarks on the Solutions of the Field Equations of Einstein's Theory of Gravitation. *Koninklijke Akademie van Wetenschappen te Amsterdam, Proceedings* 20: 1309–1312.

De Sitter, Willem. 1920. On the Possibility of Statistical Equilibrium of the Universe. *Koninklijke Akademie van Wetenschappen te Amsterdam, Proceedings* 23(1920–22): 866–868.

Du Val, Patrick. 1924. Geometrical Note on de Sitter's World. *Philosophical Magazine* 47: 930–938.

Earman, John. 1995. *Bangs, Crunches, Whimpers, and Shrieks: Singularities and Acausalities in Relativistic Spacetimes*. Oxford: Oxford University Press.

Earman, John. 1999. The Penrose-Hawking Singularity Theorems: History and Implications. In *The Expanding Worlds of General Relativity*, Einstein Studies, ed. H. Goenner, et al., vol. 7, 235–270. Boston: Birkhäuser.

Einstein, Albert. 1916. *Näherungsweise Integration der Feldgleichungen der Gravitation*, Sitzungsberichte der Preussischen Akademie der Wissenschaften zu Berlin, physikalisch-math. Klasse, 1916: 688–696; Reprinted in (CPAE 6. 1996, 347–356).

Einstein, Albert. 1917. *Kosmologische Betrachtungen zur allgemeinen Relativitätstheorie*, Sitzungsberichte der Preussischen Akademie der Wissenschaften zu Berlin, physikalisch-math. Klasse, 1917, 142–152; Reprinted in (CPAE 6. 1996, 142–152).

Einstein, Albert. 1918a. Prinzipielles zur allgemeinen Relativitätstheorie. *Annalen der Physik* 55: 241–244; Reprinted in (CPAE 7. 2002, 38–41).

Einstein, Albert. 1918b. *Kritisches zu einer von Hrn. De Sitter gegebenen Lösung der Gravitationsgleichungen*, Sitzungsberichte der Preussischen Akademie der Wissenschaften zu Berlin, physikalisch-math. Klasse, 270–272; Reprinted in (CPAE 7. 2002, 46–48).

Einstein, Albert. 1920. *Aether und Relativitätstheorie*. Berlin: Julius Springer; Reprinted in (CPAE 7. 2002, 306–320).

Einstein, Albert. 1922. *Vier Vorlesungen über Relativitätstheorie gehalten im Mai 1921 an der Universität Princeton*. Braunschweig: Vieweg, 1922; Reprinted in (CPAE 7. 2002, 497–569). English trans., The Meaning of Relativity, London: Methuen.

Eisenstaedt, Jean. 1988a. The Early Interpretation of the Schwarzschild Solution. In *Einstein and the History of General Relativity*, Einstein Studies, ed. Don Howard and John Stachel, vol. 1, 213–233. Boston: Birkhäuser.

Eisenstaedt, Jean. 1988b. The Low Water Mark of General Relativity, 1925–1955, In *Einstein and the History of General Relativity*, Einstein Studies, ed. Don Howard and John Stachel, vol. 1, 277–292. Boston: Birkhäuser.

Ellis, George F.R. 1988. The Expanding Universe: A History of Cosmology from 1917 to 1960. In *Einstein and the History of General Relativity*, Einstein Studies, ed. Howard, Don, and John Stachel, vol. 1, 367–432. Boston: Birkhäuser.

Frenkel, Viktor. 2002. Einstein and Friedmann. In *Einstein Studies in Russia*, Einstein Studies, ed. Balashov, Yuri, and Vladimir Vizgin, vol. 10, 1–16. Boston: Birkhäuser.

Goenner, Hubert. 2001. Weyl's Contributions to Cosmology. In [Sch 2001a], pp. 105–137.

Hawking, S.W., and G.F.R. Ellis. 1973. *The Large Scale Structure of Space-Time*. Cambridge: Cambridge University Press.

Holton, Gerald. 1960. Notes on the Religious Orientation of Scientists. In *Science Ponders Religion*, ed. Harlow Shapley. New York: Appleton-Century-Crofts.

Janssen, Michel. 2014. 'No Success like Failure …': Einstein's Quest for General Relativity, 1907–1920. In [Jan and Leh 2014], 167–227.

Janssen, Michel, and Christoph Lehner eds. 2014. *The Cambridge Companion to Einstein*. Cambridge: Cambridge University Press.

Kerszberg, Pierre. 1989. *The Invented Universe: The Einstein-De Sitter Controversy (1916–1917) and the Rise of Relativistic Cosmology*. Oxford: Clarendon Press.

Klein, Felix. 1871. Über die sogenannte Nicht-Euklidische Geometrie. *Mathematische Annalen* 4: 573–625; Reprinted in [Kl-GMA], vol. 1, 254–306.

Klein, Felix. 1872. *Vergleichende Betrachtungen über neuere geometrische Forschungen*. Erlangen: Deichert; Reprinted in [Kl-GMA], vol. 1, 460–497.

Klein, Felix. 1910. Über die geometrischen Grundlagen der Lorentzgruppe. *Jahresbericht der Deutschen Mathematiker-Vereinigung*, 19: 281–300; Reprinted in [Kl-GMA], vol. 1, 533–552.

Klein, Felix. 1917. *Zu Hilberts erster Note über die Grundlagen der Physik*. Nachrichten der Königlichen Gesellschaft der Wissenschaften zu Göttingen 1917: 469–482; Reprinted in [Kl-GMA], vol. 1, pp. 553–567.

Klein, Felix. 1918a. Bemerkungen über die Beziehungen des de Sitter'schen Koordinatensystems B zu der allgemeinen Welt konstanter positiver Krümmung. *Koninklijke Akademie van Wetenschappen te Amsterdam. Proceedings* 20: 614–615.

Klein, Felix. 1918b. *Über die Integralform der Erhaltungssätze und die Theorie der räumlich-geschlossenen Welt*. Nachrichten der Königlichen Gesellschaft der Wissenschaften zu Göttingen 1918: 394–423; Reprinted in [Kl-GMA], vol. 1, 586–612.

Kragh, Helge. 1996. *Cosmology and Controversy*. Princeton: Princeton University Press.

Liebscher, D.-E. 2005. *The Geometry of Time*. Weinheim: Wiley-VCH.

Lorentz, H.A., et al. 1922. Das Relativitäsprinzip. Eine Sammlung von Abhandlungen, 4th ed. Leipzig: Teubner. English trans., The Principle of Relativity, London: Methuen, 1923.

Martins, Roberto de Andrade. 1999. The Search for Gravitational Absorption in the early Twentieth Century. In *The Expanding Worlds of General Relativity*, Einstein Studies, ed. H. Goenner, et al., vol. 7, 1–44. Boston: Birkhauser.

Mercier, André, and Michel Kervaire eds. 1956. *Fünfzig Jahre Relativitätstheorie/Cinquantenaire de la Théorie de la Relativité/Jubilee of Relativity Theory*, Helvetica Physica Acta, Supplementum IV. Basel: Birkhäuser.

Minkowski, Hermann. 1909. Raum und Zeit. *Physikalische Zeitschrift* 10: 104111; English trans., The Principle of Relativity, London: Methuen, 1923: 73–91.

Moschella, Ugo. 2005. The de Sitter and anti-de Sitter Sightseeing Tour. Seminar Poincaré 1. http://www.bourbaphy.fr/moschella.pdf.

Noether, Emmy. 1918. Invariante Variationsprobleme, Nachrichten von der Gesellschaft der Wissenschaften zu Göttingen. *Mathematisch-Physikalische Klasse* 235–257.

North, John. 1965. *The Measure of the Universe: a History of Modern Cosmology*. Oxford: Oxford University Press.

Norton, John. 1999. The Cosmological Woes of Newtonian Gravitation Theory. In *The Expanding Worlds of General Relativity*, Einstein Studies, ed. H. Goenner, et al., vol. 7, 271–324. Boston: Birkhauser.

Pais, Abraham. 1982. Subtle is the Lord. *The Science and the Life of Albert Einstein*. Oxford: Clarendon Press.

Pauli, Wolfgang. 1921. Relativitätstheorie, Encyklopädie der mathematischen Wissenschaften. 5: part 2 539–775; English trans., Theory of Relativity, London: Pergamon, 1958.

Pauli, Wolfgang. 1979. Wissenschaftlicher Briefwechsel mit Bohr, Einstein, Heisenberg u.a., ed. Karl von Meyenn, Armin Hermann, Victor F. Weisskopf, Bd. 1. Heidelberg: Springer.

Robertson, H.P. 1933. Relativistic Cosmology. *Review of Modern Physics* 5: 62–90.

Röhle, Stefan. 2002. Mathematische Probleme in der Einstein-de Sitter Kontroverse, Examensarbeit, JGU Mainz, 2000, Max-Planck-Institut für Wissenschaftsgeschichte, Preprint Nr. 210.

Röhle, Stefan. 2007. Willem de Sitter in Leiden: ein Kapitel in der Entwicklung der relativistischen Kosmologie. Dissertation, JGU Mainz.

Rowe, David E. 1999. The Göttingen Response to General Relativity and Emmy Noether's Theorems. In *The Symbolic Universe. Geometry and Physics, 1890–1930*, ed. Jeremy Gray, 189–233. Oxford: Oxford University Press.

Rowe, David E. 2006. Einstein's Allies and Enemies: Debating Relativity in Germany, 1916–1920. In *Interactions: Mathematics, Physics and Philosophy, 1860–1930*, Boston Studies in the Philosophy of Science, ed. Vincent F. Hendricks, et al., vol. 251, 231–280. Dordrecht: Springer.

Rowe, David E. 2012. Einstein and Relativity. What Price Fame? *Science in Context* 25(2): 197–246.

Rowe, David E. 2015. Einstein und die Anfänge der relativistischen Kosmologie, 1917–1924. *Mitteilungen der Mathematischen Gesellschaft Hamburg* 35: 93–136.

Rynasiewicz, Robert. 1999. Kretschmann's Analysis of Covariance and Relativity Principles. In *The Expanding Worlds of General Relativity*, Einstein Studies, ed. H. Goenner, et al., vol. 7, 431–462. Boston: Birkhauser.

Scholz, Erhard. 2001b. Weyls Infinitesimalgeometrie (1917–1925). In [Sch 2001a], 48–104.

Smeenk, Christopher. 2014. Einstein's Role in the Creation of Relativstic Cosmology. In [Jan and Leh 2014], 228–269.

Thorne, Kip S. 1994. *Black Holes and Time Warps: Einstein's Outrageous Legacy*. New York: W.W. Norton.

Weyl, Hermann. 1918. *Raum-Zeit-Materie. Vorlesungen über allgemeine Relativitätstheorie*. Berlin: Springer.

Weyl, Hermann. 1924. Massenträgheit und Kosmos. *Ein Dialog, Die Naturwissenschaften* 12: 197–204.

25 Remembering an Era: Roger Penrose's Paper on "Gravitational Collapse: The Role of General Relativity"

(*Mathematical Intelligencer* 30(1)(2008): 27–36)

Back in the 1960s, Einstein's theory of general relativity re-emerged as a field of important research activity. Much of the impetus behind this resurgence came from powerful new mathematical ideas that Roger Penrose and Stephen Hawking applied to prove general singularity theorems for global space-time structures. Their results stirred the imaginations of astrophysicists and gave relativistic cosmology an entirely new research agenda. A decade later, black holes and the big hang model were on the tongues of nearly everyone who followed recent trends in science. As popular expositions dealing with quasars, pulsars, and the geometry of black holes began to appear in magazines and textbooks, Stephen Hawking reached a wide audience in 1988 with a lucid little book called *A Brief History of Time.* Twenty years later, it has emerged as one of the greatest scientific best-sellers of all time with some 10 million copies in print.

Roger Penrose's talents as an expositor became widely known with the appearance of *The Emperor's New Mind* in 1989. More recently, he displayed the vast breadth of his interests in mathematics and physics in *The Road to Reality* (2004), a tour de force that bears the distinctive style of writing that Penrose has made all his own. Having long recognized that our contemporary scientific culture is not necessarily conducive to the pursuit of creative ideas when these ideas run counter to mainstream trends, Penrose makes a persuasive case for openness in scientific discourse (Fig. 25.1). His friend Michael Atiyah has given an apt description of how Penrose's intense pursuit of truth has made him a leading defender of unorthodox research agendas:

> These days most physicists follow the latest band-wagon, usually within microseconds. Roger steers his own path and eschews band-wagons. He may not always be right, but it is important that we have individuals who stick to their guns. Future progress with ideas, as in evolutionary genetics, depends on a sufficient stock so that some good ones will survive and prosper. Roger is one of those who are helping to diversify our "gene pool" of ideas.

Fig. 25.1 Roger Penrose (center) with James Eells (left) and Michael Atiyah (right) at the 1982 Durham conference on global Riemannian geometry (Photo by Dirk Ferus, Berlin; Courtesy of Archives of the Mathematisches Forschungsinstitut Oberwolfach).

While Penrose continues to forge ahead, we take this opportunity to recall where he was in 1969, the year that saw the publication of his expository paper on "Gravitational Collapse: The Role of General Relativity" in a special issue of *Revista de Nuovo Cimento.* Those familiar with his more recent popular writings will surely recognize in the excerpt reprinted here the familiar Penrose style that captures the excitement surrounding the new singularity theorems from the 1960s, In fact, this paper has since come to be regarded as a classic account, having been reprinted as a "Golden Oldie" in *General Relativity and Gravitation*, 34, 2002. The present excerpt, republished with the kind permission of the author, omits a few of the more technical passages in the original article. But the diagrams, which always enliven Penrose's writings, have been reprinted in their entirety. **D.E.R.**

D.E. Rowe, *A Richer Picture of Mathematics*, https://doi.org/10.1007/978-3-319-67819-1_25

Gravitational Collapse: the Role of General Relativity by Sir Roger Penrose Emeritus Rouse Ball Professor of Mathematics Oxford University

. . . I shall begin with what I think we may now call the "classical" collapse picture as presented by general relativity. Objections and modifications to this picture will be considered afterward. The main discussion is based on Schwarzschild's solution of the Einstein vacuum equations. This solution represents the gravitational field exterior lo a spherically symmetrical body. In the original Schwarzschild coordinates, the metric takes the familiar form

$$ds^2 = (1 - 2m/r)\,dt^2 - (1 - 2m/r)^{-1}dr^2 - r^2\left(d\theta^2 + \sin^2\theta\, d\phi^2\right). \tag{25.1}$$

Here θ and ϕ are the usual spherical polar angular coordinates. The radial coordinate r has been chosen so that each sphere r = const, t = const has intrinsic surface area $4\pi r^2$. The choice of time co-ordinate t is such that the metric form is invariant under.

$t \to t$ + const and also under $t \to -t$. The static nature of the space-time is thus made manifest in the formal expression for the metric. The quantity m is the mass of the body, where "general-relativistic units" are chosen, so that

$$c = G = 1$$

that is to say, we translate our units according to

$$1\text{s} = 3\cdot 10^{10}\text{ cm} = 4\cdot 10^{38}\text{ g}.$$

When $r = 2m$. the metric form (25.1) breaks down. The radius $r = 2m$ is referred to as the *Schwarzschild radius* of the body.

Let us imagine a situation in which the collapse of a spherically symmetrical (nonrotating) star takes place and continues until the surface of the star approaches the Schwarzschild radius. So long as the star remains spherically symmetrical, its *external field* remains that given by the Schwarzschild metric (25.1). The situation is depicted in Fig. 25.2.

Now the particles at the surface of the star must describe *timelike* lines. Thus, from the way that the "angle" of the light cones appears to be narrowing down near r = 2m, it would seem that the surface of the star can never cross to within the $r = 2m$ region. However, this is misleading. For suppose an observer were to follow the surface of the star in a rocket ship, down to $r = 2m$. He would find (assuming that the collapse does not differ significantly from free fall) that the total proper time that he would experience as elapsing, as he finds his way down to $r = 2m$, is in fact *finite*. This is despite the fact that the world line he follows has the *appearance* of an "infinite" line in Fig. 25.2. But what does the observer

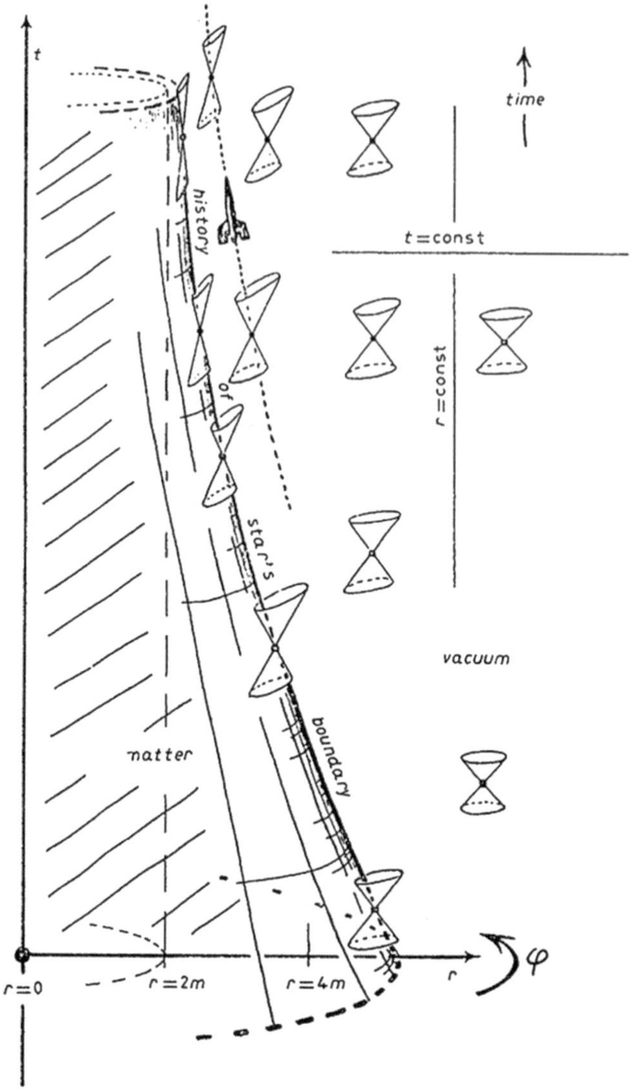

Fig. 25.2 Spherically symmetrical collapse in the usual Schwarzschiid co-ordinates.

experience after this finite proper time has elapsed? Two possibilities which suggest themselves are: (i) the observer encounters some form of space-time singularity – such as infinite tidal forces – which inevitably destroy him as he approaches $r = 2m$; (ii) the observer enters some region of space-time not covered by the (t, r, θ, ϕ) coordinate system used in (25.1). (It would be unreasonable to suppose that the observer's experiences could simply *cease* after some finite time, without his encountering some form of violent agency.)

In the present situation, in fact, it is possibility (ii) which occurs. The easiest way to see this is to replace the coordinate t by an advanced time parameter v, given by$v = t + r + 2m \log(r - 2m)$, whereby the metric (25.1) is transformed to the form (Eddington 1924; Finkelstein 1958)

$$ds^2 = (1 - 2m/r)\,dv^2 - 2dr\,dv - r^2\cdot d\theta^2 + \sin^2\theta\, d\phi^2 \tag{25.2}$$

This form of metric has the advantage that it does not become inapplicable at $r = 2m$. The whole range $0 < r < \infty$ is encompassed in a nonsingular fashion by (25.2). The part $r > 2m$ agrees with the part $r > 2m$ of the original expression (25.1). But now the region has been extended inwards in a perfectly regular way across $r = 2m$ and right down towards $r = 0$.

The situation is as depicted in Fig. 25.3. The light cones tip over more and more as we approach the centre. In a sense, we can say that the gravitational field has become so strong, within $r = 2m$, that even light cannot escape, and is dragged inwards towards the centre. The observer on the rocket ship, whom we considered above, crosses freely from the $r > 2m$ region into the $0 < r < 2m$ region. He encounters $r = 2m$ at a perfectly finite time, according to his own local clock, and he experiences nothing special at that point. The space-time there is locally Minkowskian, just as it is everywhere else ($r > 0$).

Let us consider another observer, however, who is situated far from the star. As we trace the light rays from his eye, back into the past towards the star, we find that they cannot cross into the $r < 2m$ region after the star has collapsed through. They can only intersect the star at a time *before* the star's surface crosses $r = 2m$. No matter how long the external observer waits, he can always (in principle) still see the surface of the star as it *was* just before it plunged through the Schwarzschild radius. In practice, however, he would soon see nothing of the star's surface – only a "black hole" – since the observed intensity would die off exponentially owing to an infinite red shift.

But what will be the fate of our original observer on the rocket ship? Alter crossing the Schwarzschild radius, he finds that he is compelled to enter regions of smaller and smaller *r*. This is clear from the way the light cones tip over towards $r = 0$ in Fig. 25.3, since the observer's world line must always remain a timelike line. As *r* decreases, the space-time curvature mounts (in proportion to $r - 3$), becoming theoretically infinite at $r = 0$. The physical effect of space-time curvature is experienced as a *tidal force:* objects become squashed in one direction and stretched in another. As this tidal effect mounts to infinity, our observer must eventually[1] be torn to pieces – indeed, the very atoms of which he is composed must ultimately individually share this same fate!

Thus, the true *space-time singularity,* resulting from a spherically symmetrical collapse, is located not at $r = 2m$. but at $r = 0$. Although the hypersurface $r = 2m$ has, in the past, itself been frequently referred to as the "Schwarzschild singularity", this is really a misleading terminology since $r = 2m$ is a singularity merely of the *t* coordinate used in

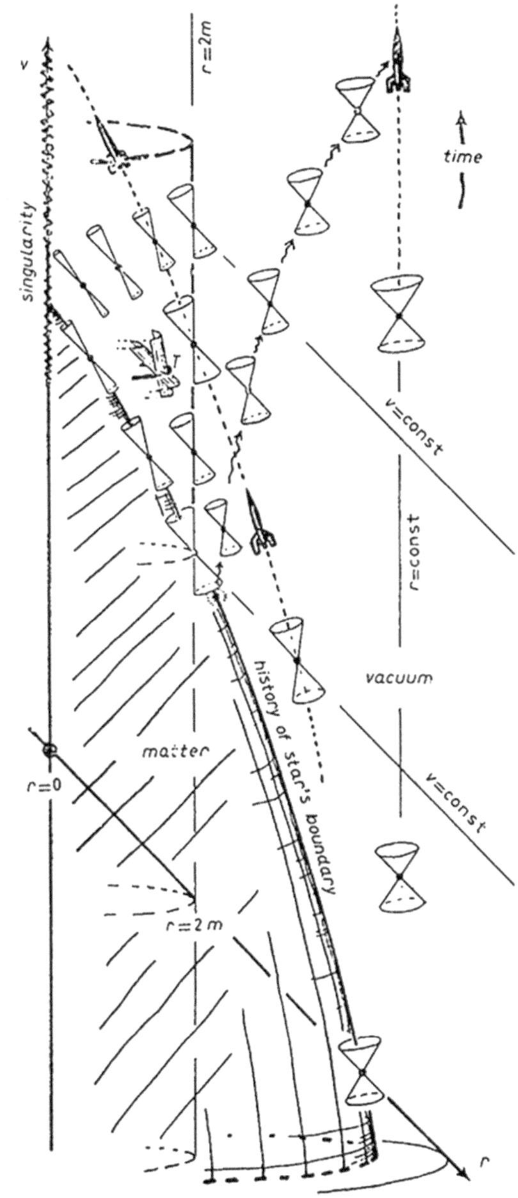

Fig. 25.3 Spherically symmetrical collapse in Eddington-Finkelstein co-ordinates.

(25.1) and not of the space-time geometry. More appropriate is the term "event horizon", since $r = 2m$ represents the absolute boundary of the set of all events which can be observed in principle by an external inertial observer. The term "event horizon" is used also in cosmology for essentially the same concept (cf. Rindler 1956). In the present case, the horizon is less observer-dependent than in the cosmological situations, so I shall tend lo refer to the hypersurface $r = 2m$ as the *absolute event horizon*[2] of the space-time (25.2).

[1] In fact, if *m* is of the order of a few solar masses, the tidal forces would already be easily large enough to kill a man in free fall, even at $r = 2m$. Bur for $m > 10^8 M_\odot$ the tidal effect at $r = 2m$ would be no greater than the tidal effect on a freely falling body near the Earth's surface.

[2] In a general space-time with a well-defined external future infinity, the absolute event horizon would be defined as the boundary of the union of all timelike curves which escape to this external future infinity. . . .

This, then, is the standard spherically symmetrical collapse picture presented by general relativity. But do we have good reason to trust this picture? Need we believe that it necessarily accords, even in its essentials, with physical reality? Let me consider a number of possible objections:

(a) densities in excess of nuclear densities inside,
(b) exact vacuum assumed outside,
(c) zero net charge and zero magnetic field assumed.
(d) rotation excluded,
(e) asymmetries excluded,
(f) possible λ-term not allowed for.
(g) quantum effects not considered.
(h) general relativity a largely untested theory,
(i) no apparent tie-up with observations.

As regards (a), it is true that for a body whose mass is of the order of $M\odot$ [mass of the sun], its surface would cross $r = 2m$ only after nuclear densities had been somewhat exceeded. It may be argued, then, that too little is understood about the nature of matter at such densities for us to be at all sure how the star would behave while still outside $r = 2m$. But this is not really a significant consideration for our general discussion. It could be of relevance only for the least massive collapsing bodies, if at all. For, the larger the mass involved, the smaller would be the density at which it would be expected to cross $r = 2m$. It could be that very large masses indeed may become involved in gravitational collapse. For $m > 10^{11}\ M_{\odot}$ (e.g., a good-sized galaxy), the averaged density at which $r = 2m$ is crossed would be less than that of air!

The objections (b), (c), (d), (e), and, to some extent, (f) can all be partially handled if we extract, from Fig. 25.3, only that essential qualitative piece of information which characterises the solution (25.2) as describing a collapse which has passed a "point of no return". I shall consider this in more detail shortly. The upshot will be that if a collapse situation develops in which deviations from (25.2) near $r = 2m$ at one time are not too great, then two consequences are to be inferred as to the subsequent behaviour. In the first instance, an absolute event horizon will arise. Anything which finds itself inside this event horizon will not be able to send signals to the outside worlds. Thus, in this respect at least, the qualitative nature of the "$r = 2m$" hypersurface in (25.2) will remain. Similarly, an analogue of the physical singularity at $r = 0$ in (25.2) will still develop in these more general situations. That is to say, we know from rigorous theorems in general-relativity theory that there must be *some* space-time singularity resulting inside the collapse region. However, we do not know anything about the detailed *nature* of this singularity. There is no reason to believe that it resembles the $r = 0$ singularity of the Schwarzschild solution very closely.

In regard to (c), (d), and (f) we can actually go further in that *exact* solutions are known which generalize the metric (25.2) to include angular momentum (Kerr 1963) and, in addition, charge and magnetic moment (Newman et al. 1965), where a cosmological constant may also be incorporated (Carter 1968). These solutions appear to be somewhat special in that, for example, the gravitational quadrupole moment is fixed in terms of the angular momentum *and* the mass, while the magnetic-dipole moment is fixed in terms of the angular momentum, charge, and mass. However, there are some reasons for believing that these solutions may actually represent the general exterior asymptotic limit resulting from the type of collapse we are considering. Any extra gravitational multipole moments of quadrupole type, or higher, can be radiated away by gravitational radiation; similarly, extra electromagnetic multipole moments of dipole type can be radiated away by electromagnetic radiation. (I shall discuss this a little more later.) If this supposition is correct, then (e) will to some extent also be covered by an analysis of these exact solutions. Furthermore, (b) would, in effect, be covered as well, provided we assume that all matter (with the exception of electromagnetic field – if we count that as "matter") in the neighbourhood of the "black hole", eventually falls into the hole. These exact solutions (for small enough angular momentum, charge, and cosmological constant) have absolute event horizons similar to the $r = 2m$ horizon in (25.2). They also possess space-time curvature singularities, although of a rather different structure from $r = 0$ in (25.2). However, we would not expect the detailed structure of these singularities to have relevance for a generically perturbed solution in any case.

It should be emphasized that the above discussion is concerned only with collapse situations which do not differ too much initially from the spherically symmetrical case we originally considered. It is not known whether a gravitational collapse of a *qualitatively different character* might not be possible according to general relativity. Also, even if an absolute event horizon *does* arise, there is the question of the "stability" of the horizon. An "unstable" horizon might he envisaged, which itself might develop into a curvature singularity. These, again, are questions I shall have to return to later.

As for the possible relevance of gravitational quantum effects, as suggested in (g), this depends, as far as I can see, on the existence of regions of space-time where there are extraordinary local conditions. If we assume the existence of an absolute event horizon along which curvatures and densities remain small, then it is very hard to believe that a classical discussion of the situation is not amply adequate. It may well be that quantum phenomena have a dominating influence on the physics of the deep interior regions. But whatever effects this might have, they would surely not be observable from the outside. We see from Fig. 25.3 that such effects would have to propagate outwards in *spacelike* directions over "classical" regions of space-lime. However, we must again bear in mind that these remarks might not apply in some qualitatively different type of collapse situation.

We now come to (h), namely the question of the validity of general relativity in general, and its application to this type of problem in particular. The inadequacy of the observational data has long been a frustration to theorists, but it may be that the situation will change somewhat in the future. There are several very relevant experiments now being performed, or about to be performed. In addition, since it has become increasingly apparent that "strong" gravitational fields probably play an important role in some astrophysical phenomena, there appears to be a whole new potential testing ground for the theory.

Among the recently performed experiments, designed to test general relativity, one of the most noteworthy has been that of Dicke and Goldenberg (1967) concerning the solar oblateness. Although the results have seemed to tell against the pure Einstein theory, the interpretations are not really clear-cut and the matter is still somewhat controversial. I do not wish to take sides on this issue. Probably one must wait for further observations before the matter can be settled. However, whatever the final outcome, the oblateness experiment had. For me, the importance of forcing me to examine once more the foundations of Einstein's theory, and to ask what parts of the theory are likely to be "here to stay" and what parts are most susceptible to possible modification. Since I feel that the "here to stay" parts include those which were most revolutionary when the theory was first put forward, I feel that it may be worthwhile, in a moment, just to run over the reasoning as I see it. The parts of the theory I am referring to are, in fact, the geometrical interpretation of gravity, the curvature of space-time geometry, and general-relativistic causality. These, rather than any particular field equations, are the aspects of the theory which give rise to what perhaps appears most immediately strange in the collapse phenomenon. They also provide the physical basis for the major part of the subsequent mathematical discussion....

So I want to admit the possibility that Einstein's field equations may be wrong, but *not* (that is, in the macroscopic realm, and where curvatures or densities are not fantastically large) that the general pseudo-Riemannian geometric framework may be wrong. Then the mathematical discussion of the collapse phenomenon can at least be applied. It is interesting that the general mathematical discussion of collapse actually uses very little of the details of Einstein's equations. All that is needed is a certain inequality related to positive-definiteness of energy. In fact, the adoption of the Brans-Dicke theory in place of Einstein's would make virtually no qualitative difference to the collapse discussion.

The final listed objection to the collapse picture is (h), namely the apparent lack of any tie-up with observed astronomical phenomena. Of course it could be argued that the prediction of the "black hole" picture is simply that we will not see anything – and this is precisely consistent with observations, since no "black holes" have been observed! But the real argument is really the other way around. Quasars *are* observed. And they apparently have such large masses and such small sizes that it would seem that gravitational collapse ought to have taken over. But quasars are also long-lived objects. The light they emit does not remotely resemble the exponential cut-off in intensity, with approach to infinite red shift that might be inferred from the spherically symmetrical discussion. This has led a number of astrophysicists to question the validity of Einstein's theory, at least in its applicability to these situations.

My personal view is that while it is certainly possible (as I have mentioned earlier) that Einstein's equations may be wrong, I feel it would be very premature indeed to dismiss these equations just on the basis of the quasar observations. For, the *theoretical* analysis of collapse, according to Einstein's theory, is still more or less in its infancy. We just do not know with much certainty what the consequences of the theory are. It would be a mistake to fasten attention just on those aspects of general relativistic collapse which *are* known and to assume that this gives us essentially the complete picture. (It is perhaps noteworthy that many general-relativity theorists have a tendency, themselves, to be a bit on the skeptical side regarding the "classical" collapse picture!) Since it seems to me that there are a number of intriguing, largely unexplored possibilities, I feel it may be worthwhile to present the "generic" general-relativistic collapse picture as I see it, not only as regards the known theorems, but also in relation to some of the more speculative and conjectural aspects of the situation.

To begin with, let us consider what the general theorems *do* tell us. In order to characterize the situation of collapse "past a point of no return", I shall first need the concept of a *trapped surface*. Let us return to Fig. 25.3. We ask what qualitative peculiarity of the region $r < 2m$ (after the star has collapsed through) is present. Can such peculiarities be related to the fact that everything appears to be forced inwards in the direction of the centre? It should be stressed again that *apart* from $r = 0$, the space-time at any individual point inside $r = 2m$ is perfectly regular, being as "locally Minkowskian" as any other point (outside $r = 0$). So the peculiarities of the $0 < r < 2m$ region must be of a partially "global" nature. Now consider any point T in the (v, r)-plane of Fig. 25.3 ($r < 2m$). Such a point actually represents a *spherical 2-surface* in space-time, this being traced out as the θ, ϕ coordinates vary. The surface area of this sphere is $4\pi r^2$. We imagine a flash of light emitted simultaneously over this spherical surface T. For an ordinary spacelike 2-sphere in flat space-time, this would result in an ingoing flash imploding towards the centre (surface area decreasing) together with an outgoing flash exploding outwards (surface area increasing). However, with the surface T, while we .still have an ingoing flash with decreasing surface area as before, the "outgoing flash, on the other hand, is in effect also falling inwards (though not as rapidly), and its surface area also decreases. The surface T (v = const, r = const $<2m$) of metric (25.2)) serves as the prototype of a trapped surface. If we perturb the metric (25.2) slightly, in the neighbourhood of an initial hypersurface, then we would still expect to get a surface T with the following property:

> T is a spacelike closed[3] surface such that the null geodesics which meet it orthogonally all *converge* initially at T.

This convergence is taken in the sense that the local surface area of cross-section *decreases,* in the neighbourhood of each point T, as we proceed into the future. (These null geodesics generate, near T, the boundary of the set of points lying causally to the future of the set T.) Such a T 'is called a *trapped surface.*

We may ask whether any connection is to be expected between the existence of a trapped surface and the presence of a physical space-time singularity such as that occurring at $r = 0$ in (25.2). The answer supplied by some general theorems (Penrose 1965, 1968; Hawking and Penrose 1969) is, in effect, that the presence of a trapped surface always *does* imply the presence of some form of space-time singularity.

There are similar theorems that can also be applied in cosmological situations. For example (Hawking and Penrose 1969; Hawking 1966a, b, 1967) if the universe is *spatially closed,* then (excluding exceptional limiting cases, and assuming $\lambda \leq 0$) the conclusion is that there must be a space-time singularity. This time we expect the singularity to reside in the past (the "big bang"). Other theorems (Hawking and Penrose 1969; Hawking 1966a, 1967, can be applied also to spatially open universes. For example, if there is any point (*e.g.* the Earth at the present epoch) whose past light cone starts "converging again" somewhere in the past (*i.e.* objects of given size start to have *larger* apparent angular diameters again when their distance from us exceeds some critical value), then, as before, the presence of space-time singularities is implied ($\lambda \leq 0$). According to Hawking and Ellis (1968, cf. also Hawking and Penrose 1969) the presence and isotropy of the 3°K radiation strongly indicates that the above condition on our past light cone is actually satisfied. So the problem of space-time singularities does seem to be very relevant to our universe, also on a large scale.

The main significance of theorems such as the above is that they show that the presence of space-time singularities in exact models is not just a feature of their high symmetry, but can be expected also in generically perturbed models. This is *not* to say that *all* general-relativistic curved space-times are singular – far from it. There are many exact models known which are complete and free from singularity. But those which resemble the standard Friedmann models or the Schwarzschild-collapse model sufficiently closely must be expected to be singular ($\lambda \leq 0$). The hope had often been expressed (cf. Lindquist and Wheeler 1957; Lifshitz and Khalatnikov 1963) that the actual space-time singularity occurring in a collapsing space-time model might have been a consequence more of the fact that the matter was all hurtling simultaneously towards one central point, than of .some intrinsic feature of general-relativistic space-time models. When perturbations are introduced into the collapse, so the argument could go, the particles coming from different directions might "miss" each other, so that an effective "bounce" might ensue. Thus, for example, one might envisage an "oscillating" universe which on a large scale resembles the cycloidal singular behaviour of an "oscillating" spatially closed Friedmann model; but the *detailed* behaviour, although perhaps involving enormous densities while at maximal contraction, might, by virtue of complicated asymmetries, contrive to avoid actual space-time singularities. However, the theorems.seem to have ruled out a singularity-free "bounce" of this kind. But the theorems do *not* say that the singularities need resemble those of the Friedmann or Schwarzschild solutions at all closely. There is some evidence (cf. Misner 1969, for example) that the "generic" singularities may be very elaborate and possess a qualitative structure very different from that of their smoothed-out counterparts. Very little is known about this, however...

A remark concerning the condition on the cosmological constant λ seems appropriate here. It is a weakness of the theorems that most of them do require $\lambda \leq 0$ for their strict applicability. However, it would appear that the condition $\lambda \leq 0$ is only really relevant to the initial setting of the global conditions on the space-time which are required for applicability of the theorem. If curvatures are to become large near a singularity, then (from dimensional considerations alone) the λ-term will become more and more insignificant. So it seems unlikely that a λ-term will really make much difference to the singularity structure in a collapse. The relevance of λ is really only at the cosmological scale.

Most of the theorems (but not all. cf. Hawking 1967) require, as an additional assumption, the nonexistence of closed, timelike curves. This is a very reasonable requirement, since a spacetime which possesses closed timelike curves would allow an observer to travel into his own past. This would lead to very serious interpretative difficulties! Even if it could be argued, say, that the accelerations involved might be such as to make the trip impossible in "practice"

[3]By a "closed" surface, hypersurface, or curve, I mean one that is "compact without boundary".

(cf. Gödel 1959). Equally serious difficulties would arise for the observer if he merely reflected some light signals into his own past! In addition, closed timelike curves can lead to unreasonable consistency conditions on the solutions of hyperbolic differential equations. In any case, it seems unlikely that closed timelike curves can substitute for a space-time singularity, except in special unstable models....

Finally, it should be remarked that none of the theorems *directly* establishes the existence of regions of approaching infinite curvature. Instead, all one obtains is that the space-time is not geodesically complete (in timelike or null directions) and furthermore, cannot be *extended* to a geodesically complete space-time. ("Geodesically complete" means that geodesics can be extended indefinitely to arbitrarily large values of their length or affine parameter – so that inertially moving particles or photons do not just "fall off the edge" of the space-time.) The most "reasonable" explanation for why the space-time is not extendible to a complete space-time seems to be (and I would myself believe this to be the most likely, in general) that the space-time is confronted with, in some sense, *infinite curvature* at its boundary. But the theorems do not quite say this. Other types of space-time singularity are possible, and theorems of a somewhat different nature would be required to decide which is the most likely type of singularity to occur.

We must now ask the question whether the theorems are actually likely to be relevant in the case of a collapsing star or superstar. Do we, in fact, have any reason to believe that trapped surfaces can ever arise in gravitational collapse? I think a very strong case can be made that at least *sometimes* a trapped surface must arise. I would not expect trapped surfaces necessarily always to arise in a collapse. It might depend on the details of the situation. But if we can establish that there can be nothing *in principle* against a trapped surface arising – even if in some very contrived and outlandish situation – then we must surely accept that trapped surfaces must at least occasionally arise in real collapse.

Rather than use the trapped-surface condition, however, it will actually be somewhat easier to use the alternative condition of the existence of a point whose light cone starts "converging again." From the point of view of the general theorems, it really makes no essential difference which of the two conditions is used. Space-time singularities are to be expected in either case. Since we are here interested in a collapse situation rather than in the "big bang," we shall be concerned with the *future* light cone C of some point p. What we have to show is that it is possible in principle for enough matter to cross to within C, so that the divergence of the null geodesics which generate C changes sign somewhere to the future of p. Once these null geodesics start to converge, then the "weak energy condition" will take over, with the implication that an absolute event horizon must develop (outside C).

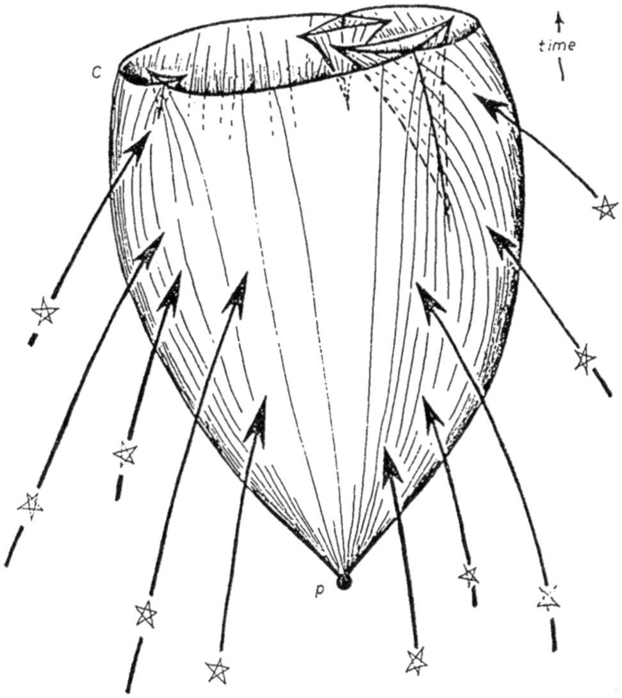

Fig. 25.4 The future light cone of p is caused to reconverge by the falling stars.

As a consequence of the stronger "energy condition" it will also follow that space-time singularities will occur.

Since we ask only that it be possible *in principle* to reconverge the null rays generating C, we can resort to an (admittedly far-fetched) "Gedanken experiment". Consider an elliptical galaxy containing, say, 10^{11} stars. Suppose, then, that we contrive to alter the motion of the stars slightly by eliminating the transverse component of their velocities. The stars will then fall inwards towards the centre. We may arrange to steer them, if we like, so as to ensure that they all reach the vicinity of the centre at about the same time without colliding with other stars. We only need to get them into a volume of diameter about fifty times that of the solar system, which gives us plenty of room for all the stars. The point p is now taken near the centre at about the time the stars enter this volume (Fig. 25.4). It is easily seen from the orders of magnitude involved that the relativistic light deflection (an *observed* effect of general relativity) will be sufficient to cause the null rays in C to re-converge, thus achieving our purpose.

Let us take it, then, that absolute event horizons can sometimes occur in a gravitational collapse. Can we say anything more detailed about the nature of the resulting situation? Hopeless as this problem may appear at first sight, I think there is actually a reasonable chance that it may find a large measure of solution in the not-too-distant future. This would depend on the validity of a certain result, which has been independently conjectured by a number of people. I shall refer to this as the *generalized*[4] *Israel conjecture* (abbreviated GIC). Essentially GIC would state: If an absolute event horizon develops in an asymptotically flat space-time, then the solution exterior to this horizon approaches a Kerr-Newman solution asymptotically with time. . . .

The following picture then suggests itself. A body, or collection of bodies, collapses down to a size comparable to its Schwarzschild radius, after which a trapped surface can be found in the region surrounding the matter. Some way outside the trapped surface region is a surface which will ultimately be the absolute event horizon. But at present, this surface is still expanding somewhat. Its exact location is a complicated affair, and it depends on how much more matter (or radiation) ultimately falls in. We assume only a finite amount falls in and that GIC is true. Then the expansion of the absolute event horizon gradually slows down to stationarity. Ultimately the field settles down to becoming a Kerr solution (in the vacuum case) or a Kerr-Newman solution (if a nonzero net charge is trapped in the "black hole") . . . .

But suppose GIC is not true, what then? Of course, it may be that there are just a lot more possible limiting solutions than that of Kerr-Newman. This would mean that much more work would have to be done to obtain the detailed picture, but it would not imply any qualitative change in the setup. On the other hand there is the more alarming possibility that the absolute event horizon may be *unstable*. By this I mean that instead of settling down to become a nice smooth solution, the space-time might gradually develop larger and larger curvatures in the neighbourhood of the absolute event horizon, ultimately to become effectively singular there. My personal opinion is that GIC is more likely than this, but various authors have expressed the contrary view.[5]

If such instabilities are present, then this would certainly have astrophysical implications. But even if GIC is true, the resulting "black hole" may by no means be so "dead" as has often been suggested. Let us examine the Kerr-Newman solutions, in the case $m^2 > a^2 + e^2$ in a little more detail. But before doing so, let us refer back to the Schwarzschild solution (25.2). In Fig. 25.5, I have drawn, what is in effect, a cross-section of the space-time, given by $v - r =$ const. The circles represent the location of a flash of light that had been emitted at the nearby point a moment earlier. Thus, they indicate the orientation of the light cones in the space-time. We note that for large *r*. the point lies inside the circle, which

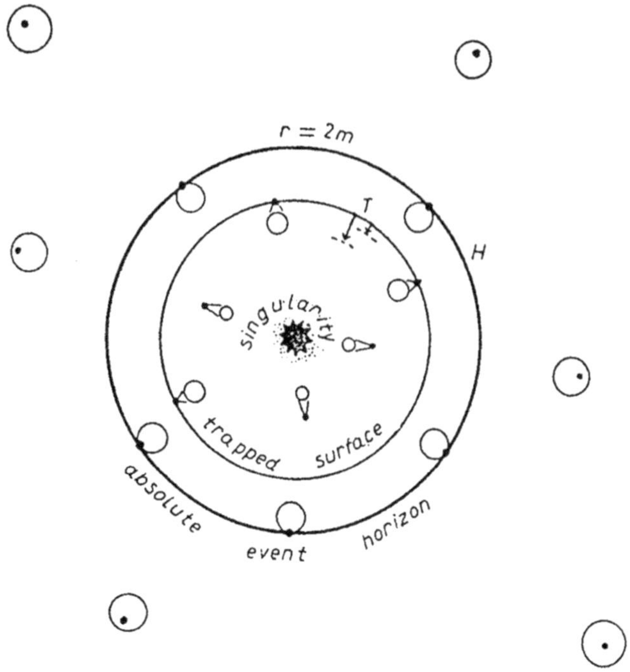

Fig. 25.5 Spatial view of spherical "black hole" (Schwarzschild solution).

is consistent with the static nature of the space-time (i.e., one can "stay in the same place" while retaining a timelike world line). On the other hand, for r < 2m, the point lies outside the circle, indicating that all matter must be dragged inwards if it is to remain moving in a timelike direction (so, to "stay in the same place" one would have to exceed the local speed of light). Let us now consider the corresponding picture for the Kerr-Newman solutions with $m^2 > a^2 + e^2$ (Fig. 25.6). I shall not be concerned here with the curious nature of the solution inside the absolute event horizon *H*, since this may not be relevant to GIC. The horizon *H* itself is represented as a surface which is tangential to the light cones at each of its points. Some distance outside *H* is the "stationary limit" *I*, at which one must travel with the local light velocity in order to "stay in the same place".

I want to consider the question of whether it is possible to extract energy out of a "black hole". One might imagine that, since the matter that has fallen through has been lost forever, so also is its energy content irrevocably trapped. However, it is not totally clear to me that this need be the case. There are at least two methods (neither of which is very practical) which might be constructed as mechanisms for extracting energy from a "black hole". The first is due to Misner (1968). This requires, in fact, a whole *galaxy* of 2^N "black holes", each of mass *m*. We first bring them together in pairs and allow them to spiral around one another ultimately to swallow each other up. During the spiraling, a certain fraction *K* of their mass-energy content is radiated

[4]Israel conjectured this result only in the stationary case, hence the qualification "generalized". In fact, Israel has expressed sentiments opposed to GIC. However, Israel's theorem (1967, 1968) represents an important step towards establishing of GIC, if the conjecture turns out to be true.

[5]Some recent work of Newman (1969) on the charged Robinson-Trautman solutions suggests that new features indicating instabilities may arise when an electromagnetic field is present.

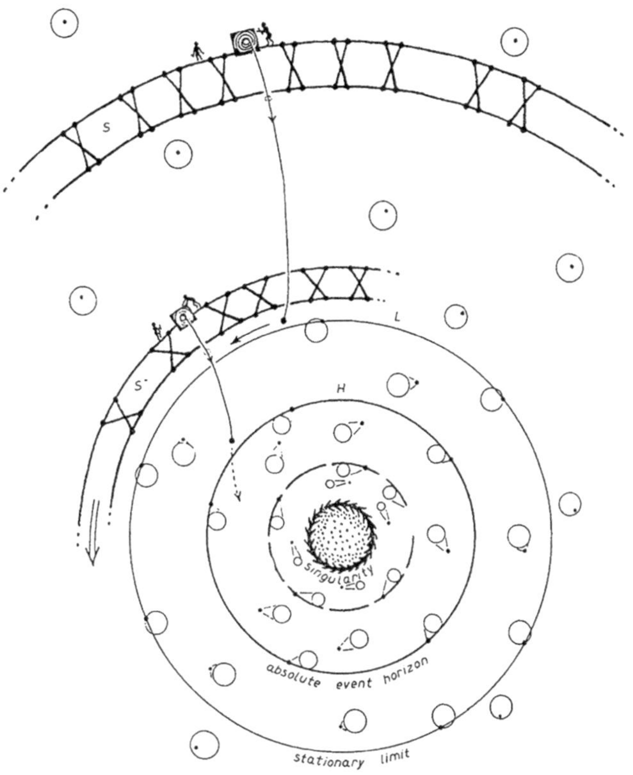

Fig. 25.6 Rotating "black hole" (Kerr-Newman solution with $m^2 > a^2 + e^2$). The inhabitants of the structures *S* and *S** are extracting rotational energy from the "black hole".

away as gravitational energy, so the mass of the resulting "black hole" is $2m(1 - K)$. The energy of the gravitational waves is collected and the process is repeated. Owing to the scale invariance of the gravitational vacuum equations, the same fraction of the mass-energy is collected in the form of gravitational waves at each stage. Finally we end up with a single "black hole" of mass $2^N m(1 - K)^N$. Now, the point is that *however small K* may in fact be, we can always choose N large enough so that $(1 - K)^N$ is as small as we please. Thus, in principle, we can extract an arbitrarily large fraction of the mass-energy content of Misner's galaxy.

But anyone at all familiar with the problems of detecting gravitational radiation will be aware of certain difficulties! Let me suggest another method, which actually tries to do something a little different, namely extract the "rotational energy" of a "rotating biack hole" (Kerr solution). Consider Fig. 25.6 again. We imagine a civilization which has built some form of stabilized structure *S* surrounding the "black hole". If they lower a mass slowly on a (light, inextensible, unbreakable) rope until it reaches *L*, they will be able to recover, at *S*, the entire energy content of the mass. If the mass is released as it reaches *L*, then they will simply have bartered the mass for its energy content. (This is the highest-grade energy, however, namely wound-up springs!) But they can do better than this! They also build another structure *S**, which rotates, to some extent, with the "black hole". The lowering process is continued, using *S**, to beyond *L*. Finally the mass is dropped through *H*, but in such a way that its energy content, as measured from *S*, is *negative*! Thus, the inhabitants of *S* are able, in effect, to lower masses into the "black hole" in such a way that they obtain *more* than the energy content of the mass, Thus they extract some of the energy content of the "black hole" itself in the process. If we examine this in detail, however, we find that the angular momentum of the "black hole" is also reduced.

Thus, in a sense, we have found a way of extracting *rotational energy* from the "black hole". Of course, this is hardly a practical method! Certain improvements may be possible, e.g., using a ballistic method.[6] But the real significance is to find out what can and what cannot be done *in principle,* since this may have some indirect relevance to astrophysical situations.

Let me conclude by making a few highly speculative remarks. In the first place, suppose we take what might be referred to, now, as the most "conservative" point of view available to us, namely that GIC is not only true, but it also represents the *only* type of situation that can result from a gravitational collapse. Does it follow, then, that nothing of very great astrophysical interest is likely to arise out of collapse? Do we merely deduce the existence of a few additional dark "objects", which do little else but contribute, slightly, to the overall mass density of the universe? Or might it be that such "objects", while themselves hidden from direct observation, could play some sort of catalytic role in producing observable effects on a much larger scale. The "seeding" of galaxies is one possibility which springs to mind. And if "black holes" are born of violent events, might they not occasionally be ejected with high velocities when such events occur? (The one thing we can be sure about is that they *would* hold together!) 1 do not really want to make any very specific suggestions here. I only wish to make a plea for "black holes" to be taken seriously and their consequences to be explored in full detail. For who is to say, without careful study, that they cannot play some important part in the shaping of observed phenomena?

But need we be so cautious as this? Even if GIC. or something like it, is true, have we any right to suggest that the *only* type of collapse which can occur is one in which the space-time singularities lie hidden, deep inside the protective shielding of an absolute event horizon? In this connection, it is worth examining the Kerr-Newman solutions for which $m^2 < a^2 + e^2$. The situation is depicted in Fig. 25.7. The absolute event-horizon has now complctcly disappeared! A

[6]Calculations show that this can indeed be done. A particle $p0$ is thrown from *S* into the region between *L* and *H*, at which point the particle splits into two particles $p1$ and $p2$. The particle $p2$ crosses *H*, but $p1$ escapes back to *S* possessing *more* mass-energy content than $p0$!

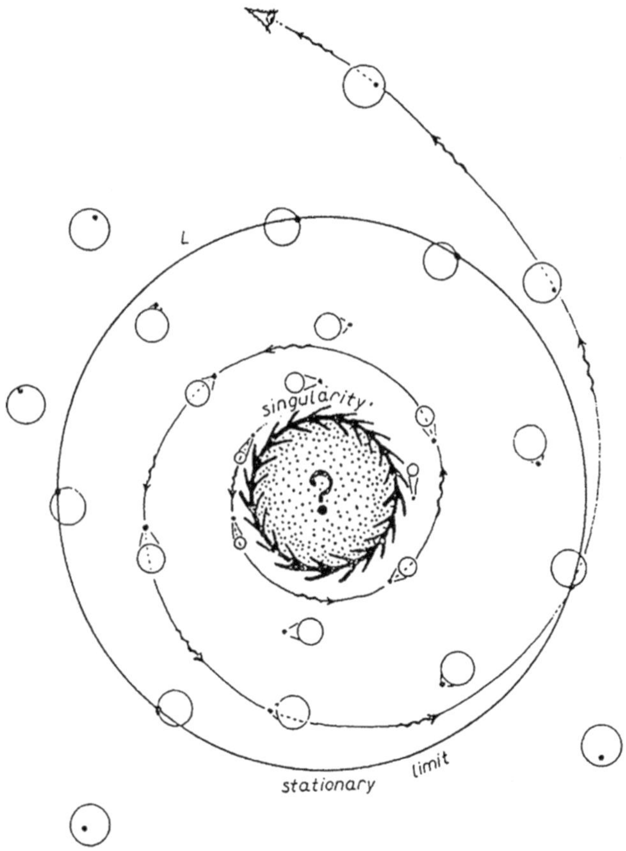

Fig. 25.7 A "naked singularity" (Kerr-Newman solution with $m^2 < a^2 + e^2$).

region of space-time singularity still exists in the vicinity of the centre, but now it is possible for information to escape from the singularity to the outside world, provided it spirals around sufficiently. In short, the singularity is *visible,* in all its nakedness, to the outside world!

However, there is an essential difference between the logical status of the singularity marked at the centre of Fig. 25.7 and that marked at the centres of Figs. 25.5 and 25.6. In the cases of Figs. 25.5 and 25.6, there are trapped surfaces present, so we have a *theorem* which tells us that even with generic perturbation, a singularity will still exist. In the situation of Fig. 25.7, however, we have no trapped surfaces, no known theorem guaranteeing singularities, and certainly no analogue of GIC. So it is really an open question whether a situation remotely resembling Fig. 25.7 is ever likely to arise.

We are thus presented with what is perhaps the most fundamental unanswered question of general-relativistic collapse theory, namely: does there exist a "cosmic censor" who forbids the appearance of naked singularities, clothing each one in an absolute event horizon? In one sense, a "cosmic censor" can be shown *not* to exist. For it follows from a theorem of Hawking (1967) that the "big bang" singularity is, in principle, observable. But it is not known whether singularities observable from outside will ever arise in a generic *collapse* which starts off from a perfectly reasonable nonsingular initial state.

If in fact naked singularities do arise, then there is a whole new realm opened up for wild speculations! Let me just make a few remarks. If we envisage an isolated naked singularity as a source of new matter in the universe, then we do not *quite* have unlimited freedom in this! For although in the neighbourhood of the singularity we have no equations, we still have normal physics holding in the space-time *surrounding* the singularity. From the mass-energy flux theorem of Bondi et al. (1962) and Sachs (1962), it follows that it is *not* possible for *more* mass to be ejected from a singularity than the original total mass of the system, *unless* we are allowed to be left with a singularity of *negative* total mass. (Such a singularity would *repel* all other bodies, but would still be attracted by them!)

While in the realm of speculation concerning matter production at singularities, perhaps one further speculative remark would not be entirely out of place. This is with respect to the manifest large-scale time asymmetry between matter and antimatter. It is often argued that small observed violations of T (and C) invariance in fundamental interactions can have no bearing on the cosmological asymmetry problem. But it is not all clear to me that this is necessarily so. It is a space-time singularity (i.e. presumably the "big bang") which appears to govern the production of matter in the universe. When curvatures are fantastically large – as they surely are at a singularity – the local physics will be drastically altered. Can one be sure that the asymmetries of local interactions will not have the effect of being as drastically magnified?

When so little is known about the geometrical nature of space-time singularities and even less about the nature of physics which takes place there, it is perhaps futile to speculate in this way about them. However, ultimately a theory will have to be found to cope with the situation. The question of the quantisation of general relativity is often brought up in this connection. My own feeling is that the purpose of correctly combining (quantum theory with general relativity is really somewhat different. It is simply a step in the direction of discovering how nature fits together as a whole. When eventually we have a better theory of nature, then perhaps we can try our hands, again, at understanding the extraordinary physics which must take place at a space-time singularity.

References

Bondi, H., M.G.J. van der Burg, and A.W.K. Metzner. 1962. *Proceedings of the Royal Society* A269: 21.
Carter, B. 1968. *Physical Review* 174: 1559.
Dicke, R.H., and H.M. Goldenberg. 1967. *Physical Review Letters* 18: 313.
Eddington, A.S. 1924. *Nature* 113: 192.
Finkelstein, D. 1958. *Physical Review* 110: 965.
Gödel, K. 1959. in: *Albert Einstein Philosopher Scientist,* edited by P. A. Schilpp (New York), 557.
Hawking, S.W. 1966a. *Proceedings of the Royal Society* A294: 511.
———. 1966b. *Proceedings of the Royal Society* A295: 490.
———. 1967. *Proceedings of the Royal Society* A300: 187.
Hawking, S.W., and G.F.R. Ellis. 1968. *Astrophysical Journal* 152: 25.
Hawking, S.W. and R. Penrose. 1969. *Proceedings of the Royal Society A* (in press)
Israel, W. 1967. *Physical Review* 164: 1776.
———. 1968. *Communications in Mathematical Physics* 8: 245.
Kerr, R.P. 1963. *Physical Review Letters* 11: 237.
Lifshitz, E.M., and I.M. Khalatnikov. 1963. *Advances in Physics* 12: 185.
Lindquist, R.W., and J.A. Wheeler. 1957. *Reviews of Modern Physics* 29: 432.
Misner, C.W. 1968. personal communication.
———. 1969. *Physical Review Letters* 22: 1071.
Newman, E.T. 1969. personal communication.
Newman, E.T., E. Couch, K. Chinnapared, A. Exton, A. Prakash, and R. Torrence. 1965. *Journal of Mathematical Physics* 6: 918.
Penrose, R. 1965. *Physical Review Letters* 14: 57.
———. 1968. In *Battelle Rencontres*, ed. C.M. De Witt and J.A. Wheeler. New York: Benjamin.
Rindler, W. 1956. *Monthly Notices of the Royal Astronomical Society* 116: 6.
Sachs, R.K. 1962. *Proceedings of the Royal Society* A270: 103.

Part V

Göttingen in the Era of Hilbert and Courant

26 Introduction to Part V

The shock of defeat at the end of the First World War left many German academics dumbfounded and numb. Even Hilbert, an outspoken internationalist, was deeply disillusioned by the chaos and instability that plagued the early Weimar years. Already during the war, political differences widened the gulf that had already formed within the Göttingen Philosophical Faculty, whose conservative members felt they were constantly being provoked by the "Hilbert faction." The controversy over Emmy Noether's candidacy to habilitate in 1915, mentioned in the introduction to Part IV, was only one of many such instances. Others were even more serious, as when Hilbert and his pacifist friends were accused of fomenting anti-German sentiment. These political conflicts reflect a quite general "two cultures" phenomenon found elsewhere in Germany. Einstein was struck by the same divide separating the more sober-minded scientists in the Berlin Academy from the rabid nationalists, who were typically philologists and philosophers. In Göttingen, the scientists who sided with the liberals in the "Hilbert clique" found themselves not only in a defensive position vis-à-vis their humanist colleagues but also increasingly isolated within the larger community as the Weimar era progressed.

A sense of crisis also loomed within the physics and mathematics communities at large, a mood that became palpable at the Bad Nauheim meeting of German *Naturforscher* in 1920 (Einstein-CPAE 7 2002). There Philipp Lenard clashed swords with Einstein in a dramatic confrontation over general relativity that overshadowed the entire scientific program (Rowe 2002). Few took notice when Weyl presented his brilliant new ideas for a unified field theory based on what he called *Nahgeometrie*. Fewer still paid heed to L.E.J. Brouwer when he spoke about his constructivist philosophy of mathematics, which came to be known as Brouwerian intuitionism. One who did, however, was Weyl himself, who was enamored as much by the Dutchman's persona as by his radical ideas. A year later, he declared Brouwer the revolutionary who offered not just a new program for the foundations of mathematics but a way out of its present state of crisis (see Chap. 27). At the same time, Lenard and Johannes Stark tried to mobilize a campaign against Einstein's general theory of relativity within the German physics community, arguing that it was purely mathematical and without relevance for physical theorizing (see Chap. 19).

Göttingen's mathematicians quickly closed ranks behind Einstein, who was appointed to the main editorial board of *Mathematische Annalen* in 1920. Einstein thereby joined the other three longstanding principal editors: Klein, Hilbert, and Blumenthal. This alliance was further strengthened by the appointment of Max Born to the Göttingen faculty shortly after Richard Courant took over as helmsman of the Mathematics Institute. These changes were part of a larger restructuring of the board undertaken when Julius Springer Verlag took over publication of the Annalen from the Leipzig firm of B. G. Teubner. Toward the end of the war, Ferdinand Springer had already taken advantage of his connections with Berlin mathematicians to launch a new journal, *Mathematische Zeitschrift*, edited by Leon Lichtenstein from the Technische Hochschule in Charlottenburg. Lichtenstein was supported by three other Berliners – Konrad Knopp, Erhard Schmidt, and Isaai Schur – who together comprised a four-man editorial team supported by a large board of distinguished advisors.[1] Already by 1918, this venture got off to a flying start, easily surpassing *Mathematische Annalen*, which had published very little during the war years. In 1914 the Annalen's managing editor, Otto Blumenthal, was called up for war service, after which the entire editorial operation slowed to a near standstill. By war's end, it even looked as though the Annalen might have to fold due to the weakness of the German economy. Academic publishing had suddenly become a much riskier affair, and the Teubner firm was no longer willing to publish at the same level as before.

This led to acrimonious negotiations between Teubner and Hilbert, who was intent on restoring the prestige of the

[1] This wissenschaftlicher Beirat was comprised of nine members: Wilhelm Blaschke, Leopold Féjer, Gustav Herglotz, Adolf Kneser, Edmund Landau, Oskar Perron, Friedrich Schur, Eduard Study, and Hermann Weyl.

D.E. Rowe, *A Richer Picture of Mathematics*, https://doi.org/10.1007/978-3-319-67819-1_26

Göttingen-based journal. Blumenthal thought this would still be possible, but only by initiating a propaganda campaign to promote the Annalen, something he was more than willing to do.[2] He had learned, however, from his colleague Theodor von Kármán that Springer was contemplating the idea of founding a second journal focused on applied mathematics. Blumenthal therefore urged Hilbert to open a new line of negotiations with Springer; he suggested that instead of trying to launch a new journal the Berlin firm should consider taking over *Mathematische Annalen* and merely restructuring its editorial board. This plan evidently appealed to all sides. Adding Einstein's name to the main board not only helped to accomplish this goal, it also gave the Annalen new luster just when the journal most needed it.[3]

The war years not only disrupted formerly cordial scientific relations within Europe, they also cut off many contacts with those working abroad. For Teiji Tagaki, who had solved Kronecker's *Jugendtraum* soon after spending three semesters in Göttingen, the Great War ironically spurred him on to follow his first love once again. As he later put it, the war "gave me a stimulus, rather a negative stimulus. No scientific information reached Europe for four years. Some said this would be the end of Japanese science, while newspaper articles wrote of their sympathy for Japanese professors losing their jobs. This made me realize the obvious truth that every researcher had to be independent. Possibly I would have done no research for myself had it not been for World War I."[4]

Takagi had benefitted a good deal from his studies in Göttingen, but his life story underscores how a true mathematical genius has to find his or her own way. A stimulating atmosphere can be very helpful, but sometimes isolation can be even more beneficial, particularly when it frees a person from the influential ideas of leading authorities. In Takagi's case, he later gave this impression of Hilbert's personality and abrupt manner when he first met him:

> He asked me: "You say you intend to study algebraic number theory. Is it really true?" In the world of that time, that theory was being studied almost exclusively in Göttingen. Thus it is no wonder that he never expected an Oriental to study it. I answered; "I intend to do so." "Then, by what object is an algebraic function determined?" he asked immediately. I could not answer at once. "It is determined by its Riemann surface," the professor himself gave the answer. Since it was true, I replied: "Oh yes, that's right." Perhaps he considered my mathematical ability to be not very reliable. Then he ordered me to accompany him, since he was on the point of going home. I told him that I intended to study *Kronecker's Dream of his Youth* in the case where the ground field was the Gaussian number field. "That's fine," said Hilbert. . . . (Honda 1975, 154–155).

Takagi's recollection of this encounter may have been highly selective, but it was significant in his own mind because he later realized he had been too quick to accept Hilbert's approach to class fields. Heinrich Weber had developed a more general theory, though it was also more complicated than Hilbert's approach (Edwards 1990). The latter was restricted to unramified class fields, where the factorizations of prime ideals of the ground field contain no repeated factors in the extensions fields. From the standpoint of the theory of algebraic functions, which are defined by Riemann surfaces, it was natural to limit consideration to this unramified case.

With the outbreak of the Great War and the ensuing breakdown of scientific exchanges between Japanese and European mathematicians, Takagi felt emboldened to take up Weber's more general theory instead of following Hilbert's approach. As he later recalled:

> I was freed from that idea and suspected that every abelian extension might be a class field if the latter is not limited to the unramified case. I thought at first that this could not be true. Were it false, the idea should contain an error and I tried my best to find this error. At that time I almost suffered from a nervous breakdown. I dreamt often that I had resolved the question. I woke up and tried to remember my reasoning but in vain. I tried my utmost to find a counterexample to the conjecture which seemed all too perfect. Finally I made my theory confirming this conjecture, but I could not rid myself of the doubt that it might contain an error which would invalidate the whole theory. I badly lacked colleagues who could check my work.[5]

After years of struggle, Takagi convinced himself that his theory was sound. He would next try to make his case to the mathematical world at large in two long papers, (Takagi 1920) and (Takagi 1922), which were written in German. In preparing the first of these, he also realized that he would have an excellent chance in the fall of 1920 to bring his work to the attention of leading mathematicians in Europe. The Japanese Ministry of Education, in fact, was financing a longer trip so that Takagi could report on academic conditions in various countries. At the same time, he planned to give an overview of his results in number theory at the forthcoming International Congress in Strasbourg, where out of necessity he would speak in French. Perhaps for that reason he spent a month in Paris before leaving for Strasbourg to take part in the congress.

---

[2]Blumenthal to Hilbert, 23 October 1919, Hilbert Nachlass, Niedersächsische Staats- und Universitätsbibliothek (SUB) Göttingen.

[3]The board of associate editors was consequently expanded to eleven persons: Ludwig Bieberbach, Harald Bohr, Max Born, L.E.J. Brouwer, Richard Courant, Constantin Carathéodory, Walther von Dyck, Otto Hölder, Theodor von Kármán, Carl Nuemann, and Arnold Sommerfeld.

[4]Teiji Takagi, Reminiscences and perspectives, in *Miscellaneous Notes on Mathematics* (Tokyo, 1935) as translated in the article on Takagi in MacTutor History of Mathematics Archive, http://www-history.mcs.st-and.ac.uk/

[5]Teiji Takagi, Reminiscences and perspectives, in *Miscellaneous Notes on Mathematics* (Tokyo, 1935) as translated in the article on Takagi in MacTutor History of Mathematics Archive, http://www-history.mcs.st-and.ac.uk/

The session in which Takagi spoke was chaired by the eminent number-theorist Leonard E. Dickson from the USA. Other leading experts who attended included Hilbert's former pupil, Rudolf Feuter from Switzerland, and Jacques Hadamard. No Germans were in the audience, however, as they as well as the German language were formally barred from the Strasbourg Congress. This decision followed stipulations set down by the International Mathematical Union, which adopted the strict guidelines of the newly established International Research Council headquartered in Brussels, about which more below. In any case, Takagi's talk left no lasting impression on the few mathematicians in the audience who would have been able to follow what he had to say.

Before returning to Japan, Takagi visited Hilbert in Göttingen, but apparently he missed seeing Carl Ludwig Siegel, who was then studying under Landau. Possibly he left Siegel an offprint of his paper (Takagi 1920), or perhaps he mailed it to him later that year. No one, though, seems to have realized what Takagi had achieved until Siegel loaned his copy of that paper to Emil Artin in 1922. Artin, who had come to Göttingen just that year, was the perfect person to read this daunting presentation of 133 pages. Many years later he remarked: "I felt strong admiration for it. It was not difficult to understand, since it was written very clearly" (Honda 1975, 161). Soon after this, Takagi's work began to receive the recognition it deserved, as Artin, Helmut Hasse, and others built on his ideas; this story is nicely summarized in (Yandell 2002, 230–257).

Hilbert was by now only vaguely aware of these fast-breaking developments in a field he had helped nurture more than two decades earlier. In the meantime, his influence made itself felt in areas far removed from pure mathematics. As noted in Chap. 19 in Part IV, Einstein's colleague Max von Laue made ample use of Hilbert's lecture notes in preparing his 1921 textbook, the first detailed presentation of general relativity theory. Hilbert continued to offer lectures on relativity, but his interests gradually shifted over to quantum theory, particularly after the summer of 1922 when Niels Bohr delivered a series of famous lectures in Göttingen during the "Bohr-Festspiele." By then, Hilbert may have given up his bold dreams for revolutionizing field physics, a vision that Einstein and Weyl had chosen to ignore (both took a skeptical view of Mie's electro-dynamic theory of matter). Most of Hilbert's time and energy, in fact, were by now taken up by his new proof theory aimed at securing the foundations of mathematics, a research program supported by the indefatigable Paul Bernays.

This work, too, unfolded in an atmosphere filled with a sense of crisis, as reflected in Hermann Weyl's polemical essay "The New Foundations Crisis" (Weyl 1921). This text, which effectively brought the core issues dividing Hilbert's formalism and Brouwer's intuitionism out into the open, is discussed in Chap. 27. Weyl's sudden leap over to Brouwer's intuitionist camp constituted a serious threat to Hilbert's foundational efforts. The latter saw this as nothing less than a frontal attack on his most cherished beliefs, indeed, as an attempt to undermine the central idea on which he had staked his intellectual capital ever since he pronounced it at the Paris Congress in 1900, namely, the decidability of every well-posed problem in mathematics. Brouwer claimed that this assertion was tantamount to the unrestricted validity of the logical law of *tertium non datur*, which in his view led to hollow claims of no epistemological value. Weyl's polemics brought this conflict to the attention of the larger mathematical community, but in such a dramatic way that readers must have sensed the personal tensions lurking behind his emotional language.

Indeed, the clash between Hilbert and Brouwer was symptomatic of the stresses and strains that the Göttingen community underwent during the 1920s. It also eventually reopened the old rift that had once divided mathematicians associated with Berlin and those who identified with Göttingen. The death of Frobenius in 1917 and the retirement of Schwarz in that same year put an end to this traditional rivalry, though not a lasting one. Schwarz was succeeded by Erhard Schmidt, one of Hilbert's star pupils and a good friend of Constantin Carathéodory, who had assumed Klein's chair in 1913. To the surprise of everyone, Schmidt enticed Carathéodory into leaving Göttingen in 1918 to take the position formerly held by Frobenius. Thus, for a brief time, these two regal personalities with strong links to Hilbert and Klein were the dominant figures in Berlin.

Carathéodory was not only a distinguished mathematician; he came from a Greek diplomatic family and spoke several European languages fluently (Georgiadou 2004). Soon after the war ended, he was tapped by Prime Minister Eleftherios Venizelos of Greece to submit plans for a new university to be founded at Smyrna in Asia Minor. After his appointment in 1920 as Dean of the University, he resigned his post in Berlin and dedicated himself to launching this new institution, the Ionian University of Smyrna, with a first-class library. Although scheduled to open in October 1922, it never did, as Turkish forces had meanwhile gained the upper hand in the Greco-Turkish War. During the ensuing battle of Smyrna, Carathéodory narrowly escaped the city as the Turkish army closed in. Still, he managed beforehand to get most of the library's books loaded onto cargo ships and sent to Athens. Four days after the Turks seized control of the city on 9 September, a fire broke out and raged out of control for 10 days, leaving tens of thousands of Greeks and Armenians dead. Without living quarters, the survivors from the Great Fire of Izmir (the city's Turkish name) were quickly evacuated. Carathéodory afterwards taught in Athens before accepting a professorship in 1923 at Munich University, where he succeeded Ferdinand Lindemann.

Carathéodory's departure from Berlin in 1920 came at a most unpropitious time for Erhard Schmidt and his colleagues. After Brouwer, Weyl, Herglotz, and Hecke all turned down offers to fill this position, the faculty eventually settled on Ludwig Bieberbach, who was then teaching in Frankfurt. Bieberbach had studied in Göttingen, taking his doctorate there in 1910 with a thesis on automorphic function written under Paul Koebe's supervision. In Berlin, he gained a reputation as a dynamic and multi-talented, but also vain and ambitious mathematician. Few noticed his ardent nationalism – in no sense unusual in this milieu – or suspected that his ambitions might tempt him to become a supreme opportunist.

The new rift between Berlin and Göttingen re-opened in the mid-1920s when Bieberbach became a political ally of Brouwer, the distinguished topologist turned intuitionist, and one of the key members of the editorial board of *Mathematische Annalen.* Tensions within the board grew as Brouwer insisted on taking a strong stance against French mathematicians, whose political views he deemed unacceptable. In the meantime, Carathéodory had taken Klein's place as one of the four principal editors, whereas Brouwer was one of the many associate editors, several of whom shared his anti-French views. Brouwer was also an effective propagandist who took full advantage of his relationship with Karl Kerkhof, author of *Der Krieg gegen die deutsche Wissenschaft* (Berlin 1922).[6]

When the Italian organizers of the forthcoming ICM in Bologna signaled their willingness to invite mathematicians from Germany, they broke with the policy set in place by the International Research Council in 1919. This called for a boycott of all scientists associated with the defeated Central Powers, who were thenceforth prohibited from attending international meetings, a policy implemented at the International Congress of Mathematicians held the following year in Strasbourg. There the mathematicians from the Entente Powers formed the *Union Mathématique Internationale* (UMI), which excluded representatives from former enemy states. By 1928, however, Salvatore Pincherle, who was then president of the UMI, and his Italian colleagues felt the time was at last ripe to restore civil relations in the European mathematical world, whereas Brouwer and his sympathizers in Germany saw matters differently.

Supported by Bieberbach, Brouwer called for a counter-boycott of the Bologna Congress, arguing that its organizers were still subservient to the UMI. Brouwer had long been on a mission to expose this organization as a political tool whose principal aim was the suppression of German science. Many others, including the internationalist G.H. Hardy, essentially shared that view. In June 1928, Bieberbach sent a letter to the Rector in Halle charging that the Italian invitation to the formerly ostracized Germans attempted to conceal the role of the UMI as sponsor of the Bologna congress. Copies of this letter were forwarded to other universities, including Göttingen, whose Rector made Hilbert aware of it. In response, Hilbert sent out a circular to all German universities protesting Bieberbach's action while disputing his claims (Hilbert Nachlass 494, 18). By this time, the political divisions separating Göttingen and Berlin were apparent for all to see.

Hilbert, who was fighting for his life due to a deficiency of Vitamin B12, decided he would personally lead a German delegation to the Bologna Congress in September. On that occasion, he was greeted with warm applause as he entered the congress hall, a moment he surely savored despite his poor health. He even planned to deliver a powerful political speech for this occasion (Siegmund-Schultze 2016b). In many respects, its message echoed the speech he delivered in 1909 at the opening of Poincaré Week, when he extolled the many bonds joining French and German mathematicians (see Chap. 16, the final essay in Part III). Twenty years later, he was intent on refuting contrary opinion: "Mathematics knows no races. When we look back on the history of our science, even only superficially, we see that all nations and peoples, great or small, have played their good and fair part" (Hilbert Nachlass 494, 19). Hilbert's speech was probably never delivered; otherwise, words like these would surely have caused quite a stir, had he actually spoken them. His text ends by giving thanks to the Italians for their steadfast support of international cooperation. Pincherle, in fact, had encountered strong pressure from the UMI and IRC, so that, in the end, the Bologna ICM was sponsored not by the UMI but by Bologna's famous university. It was also a great success, attended by 835 mathematicians, among whom the Germans formed the second largest national group. Thereafter, the UMI lost its political relevance and soon withered away (Curbera 2009).

This triumph notwithstanding, Brouwer's antics had seriously weakened the smaller liberal faction on the board of *Mathematische Annalen* (mainly Hilbert, Blumenthal, and Einstein), prompting Hilbert to take unilateral preemptive action against his fanatical foe. Around this time, he decided that Brouwer had to be removed from the editorial board due to his pernicious influence on its more conservative members. This decision set the stage for what Einstein would later dub "the battle of the frogs and mice," a story first told in detail by Dirk van Dalen in *The Mathematical Intelligencer* (van Dalen 1990). While Einstein was both shocked and bemused by the ruthlessness of this affair – since he had not imagined Hilbert capable of doing such a thing – Blumenthal and Carathéodory found themselves trapped between loyalty to their revered teacher and fairness to Brouwer. In a desperate move to avert a bloody power struggle, Carathéodory traveled to Holland to meet with Brouwer, hoping to intercept Hilbert's curt message informing him of his dismissal. He

[6]On the larger scope of this conflict, see Schroeder-Gudehus (1966).

Fig. 26.1 Otto Blumenthal and Egbertus Brouwer worked closely together as managing editor and associate editor of *Mathematische Annalen* (Courtesy of Dirk van Dalen, Brouwer Archive, Utrecht).

then tried to persuade his friend to step down voluntarily, a conversation that effectively marked the end of their friendship. Although a born diplomat, Carathéodory found he had no chance of persuading the implacable Brouwer, who was every bit as hardheaded as Hilbert. Knives having now been drawn, the power struggle could begin, though Hilbert let others do the fighting for him, principally Blumenthal, Courant, and Harald Bohr. The battle ended when Ferdinand Springer agreed to dissolve the entire board and give Hilbert *carte blanche* to form a new one. He did so, though this time he made it far smaller, while ensuring that it was comprised entirely of loyalists. As for Brouwer, he severed all ties with Göttingen and largely withdrew from the mathematical arena altogether, though he later founded his own journal, *Compositio Mathematica,* in 1934 (Fig. 26.1).

After Brouwer's ouster from the board of *Mathematische Annalen* in 1928, Bieberbach began mustering his own plans for countering the influence of Hilbert, Courant and co. Indeed, he emerged soon after 1933 as the leading exponent of a racist ideology that condoned the use of political pressure to remove Jewish mathematicians from their teaching posts. With both words and deeds, Bieberbach strongly supported the laws to "restore" the German civil service by expelling those who, according to those laws, were defined to be Jewish. In his eyes, these measures ensured that German youth would not be exposed to the corrupting influence of Jewish mathematicians. He explicitly pointed to the case of Göttingen's Edmund Landau, who allegedly inculcated a "foreign spirit," thereby thwarting the Nordic genius inherent in the German race. Bieberbach's easy adaptation of the Nazi ideology would seem difficult to square with his career prior to 1933, although Herbert Mehrtens was able to find evidence of an earlier shift in his thinking (Mehrtens 1987). Yet whatever the motivation, one can hardly trace his embrace of Nazism back to the ongoing foundational debates of the 1920s. In fact, Bieberbach never explicitly aligned himself with either Hilbert's formalism or Brouwer's intuitionism; indeed, his murky views linking mathematical styles with racial stereotypes had no real bearing on foundational issues in general.

The so-called foundations crisis of the 1920s has received a considerable amount of attention in the historical literature. Mehrtens, Paul Forman, and Reinhard Siegmund-Schultze have all dealt with it against the backdrop of cultural pessimism in the Weimar era, a mood reflected in the intellectual discourse of physicists as well as mathematicians. Dirk van Dalen's biography of Brouwer (van Dalen 2013) has shed much new light on the Dutchman's personal quirks as well as the philosophical views that motivated his intuitionist program. Complementing van Dalen's portrait of Brouwer, Dennis Hesseling undertook a broader study of the literature from the 1920s concerning the foundations crisis and the ongoing debates between formalists and intuitionists (Hesseling 2003). Since the number of prominent actors who participated in these debates was quite small, personal motives need to be taken into account in order to interpret what happened and why. I also think that Forman's original claim that the crises in physics and mathematics were linked should be examined more closely (Forman 1971).

Chapter 27 illustrates this confluence of mathematical and physical ideas, both of which colored the rivalry that developed between Hermann Weyl and his former teacher. Weyl was the most gifted and accomplished of Hilbert's many students; he was also the only one who dared to clash openly with the master. Weyl's entire intellectual outlook contrasted sharply with Hilbert's brand of rationalism, and though he held his erstwhile mentor's achievements as a pure mathematician in the highest regard, he grew increasingly skeptical of his program for axiomatizing physics and other fields of knowledge (Weyl 2009). This undercurrent of tension surfaces in Weyl's famous book, *Space-Time-Matter*. The far stronger and more familiar source of conflict, however, came at nearly the same time in the form of debates over the epistemological status of the real number continuum. In recounting the explosive events from these years, I offer an interpretation of Weyl's role as a "reluctant revolutionary" caught between two powerful personalities – Brouwer and Hilbert – who refused to concede anything to the other side.

Running alongside these intellectual battles were important institutional developments. These reflect the sudden and precarious decline in prestige of Germany's two leading universities in the wake of the country's wartime disaster. Weyl, who left Göttingen in 1913 to take a professorship in Zürich, was the recipient of numerous offers from German universities, including Göttingen and Berlin, but on each such occasion he chose to decline. These events and circumstances are taken up in Chap. 28, which describes how Richard Courant came to assume Klein's former chair in Göttingen. Courant was a gifted mathematician, though hardly comparable to Weyl. Yet in another respect, he clearly stood out when compared with other talented students of Hilbert from his generation: he was the most loyal of them all. Beyond that, he had an uncanny ability to tap new resources.

Before the war broke out, Klein had been making plans for a new mathematics building that would stand in the middle of the science complex located in the southwest corner of the city. That idea was postponed in 1914, and after the fall of the *Kaiserreich* it no longer looked feasible at all, though Klein clearly clung on to it. Soon after his death, Courant found a way forward by striking up negotiations with the International Education Board (IEB) of the Rockefeller Foundation. He realized that gaining their support would require considerable political acumen, but he also knew how to use the famous names of the past in appealing to higher principles. In making his case to the IEB, he wrote:

> The close association of mathematics and physics has at all times been a characteristic feature – and the strength – of the Göttingen tradition, in our special sphere. I need only recall the names of Gauss, Weber, Dirichlet, Riemann, H. Minkowski, Felix Klein. The last named entertained for decades the project of establishing a fixed home for mathematics and physics where both sciences would be cared for on the broadest possible basis, and in intimate mutual conjunction. In this way a series of new buildings and establishments has come to the front [leading to a] concentration of all university activities connected with our special domain. . . . (Siegmund-Schultze 2001, 146)

Courant was also the consummate insider, and he loved to tell stories about the good old days in Göttingen. Some of his tales became part of the standard lore found in Constance Reid's books (Reid 1970, 1976) and other works based on oral history. Not surprisingly, different variants of these anecdotes circulated over time, as Courant's memory was hardly the best.

In 1964, he was invited to speak at a colloquium hosted by the Department of History of Science and Medicine at Yale University. John Ewing later edited his manuscript and published it in an early issue of *The Mathematical Intelligencer* under the title "Reminiscences from Hilbert's Göttingen." On this occasion, Courant offered the following version of one of his favorite anecdotes:

> After Minkowski's death, Edmund Landau was called to Göttingen. I was an assistant of Hilbert at that time, but all the deep secrets of faculty policy were discussed between Hilbert, his friends, and his assistants. He liked to get advice from his wife and from his assistants, but not from his colleagues. There was a big question: who should one call to Göttingen as a successor of Minkowski? Minkowski died very tragically, very suddenly, and unnecessarily from the consequences of appendicitis which was not diagnosed. There were three candidates – it is interesting for people today to know how responsible scientists at that time acted – three candidates of first rank: one was Perron, the second was Hurwitz, and the third was Landau. Now, Hurwitz was rather ill and didn't want to move from Zürich. The question was between Perron and Landau. All their papers and everything that they had done were carefully scrutinized, not by one, but by quite a number of competent people of the faculty. It was a toss-up, and finally the decision went for Landau. Landau was called with the very explicit justification: of the two, Perron and Landau, Landau was less agreeable and less easy to handle, and it would be very important for the faculty not to have "yes" men, people who toe the line; and that is indeed how Landau came to Göttingen, and not Perron. It was very interesting and probably a very wise principle which could be very well used in many other cases today. (Courant 1981, 157–158)

A slight variant of this story appears in Reid's biography of Hilbert (Reid 1970, 118). According to this source, Klein made the decisive statement, thereby casting the decisive vote in favor of Landau. I once asked her about this anecdote, and she told me that Courant had repeated it to her on several occasions, and very emphatically, since it had made such a deep impression on him. The gist of Courant's story may well be correct, but significant parts of it are not. Oskar Perron was not one of the three candidates: the third was Otto Blumenthal. He, like Hurwitz, was a congenial personality, quite unlike Landau. So these differences in temperament were very likely taken into consideration during the faculty deliberations. Moreover, all three of these candidates, like Minkowski, were Jewish by birth, a circumstance so rare that one can only conclude that this was another factor taken heavily into account. Finally, Courant's picture of how Hilbert approached such important decision-making can hardly be taken seriously. In Chap. 28, I discuss the circumstances that led to Courant's appointment in Göttingen in 1920, partly in order to rectify his own rendition of how this came about.

Like Hilbert, Courant certainly knew how to instill loyalty among those closest to him. This is amply illustrated in Chap. 29, devoted to the academic career of the historian of mathematics, Otto Neugebauer, who came to Göttingen via Graz and Munich (Rowe 2016). Along the way his interests shifted from physics to mathematics, and then, quite suddenly, to the history of ancient mathematics and astronomy, the field in which he would stake his formidable reputation. Like Courant, Neugebauer left Germany shortly after the Nazis came to power and eventually immigrated to the United States (Siegmund-Schultze 2016a). Just as Courant built up the mathematics program at New York University, Neugebauer founded a distinguished, though short-lived graduate program for studies of the ancient exact sciences at Brown University.

Some of those who worked with him closely at Brown came to realize that Neugebauer's approach to historical studies was somehow deeply rooted in the training he received during his formative years. In fact, as I argue in Chap. 29, Neugebauer should be seen as a leading representative of the distinctive Göttingen mathematical culture he experienced while working with Courant. Indeed, he remained true to that mold throughout his career. His initial shift from physics to pure mathematics was clearly hastened by experiences gained in Courant's seminar on algebraic number fields and algebraic functions, which he attended during his first semester in Göttingen. As it happened, Kurt Otto Friedrichs, another new arrival, also took part in that seminar and remembered it well. Neugebauer and Friedrichs found themselves amidst a stellar gathering of ambitious young men. The group included Courant's *Assistent,* Hellmuth Kneser, who had written his dissertation on quantum theory under Hilbert, along with two other brilliant young post-docs, Carl Ludwig Siegel and Emil Artin. All three would soon go on to become world-class researchers.[7]

Chapter 31 comes from an interview with Friedrichs, whom I met only shortly before his death (Reid 1983). Although in poor health, he knew about my interest in Klein's life and he was eager to share his recollections of him. He met Klein just one day after Germany's foreign minister, Walther Rathenau, had been assassinated. Friedrichs had earlier provided Constance Reid with many insights for her biography of Courant, including some vivid memories of the atmosphere in that first seminar. Courant had a way of attracting talent and then coaxing them to take part in seminars, so "they all had to come and participate. Such a group of people, who knew everything about everything – it was very exciting to me" (Reid 1976, 91). Friedrichs also recalled how shocked he was at the level of difficulty. He had taken a course in Freiburg based on the first two chapters of the same text, whereas Courant simply skipped that material and started with chapter three. (The book he spoke about was a classic text: *Theorie der algebraischen Funktionen einer Variabeln und ihre Anwendung auf algebraische Kurven und abelsche Integrale* by Hensel and Landsberg, first published in 1902.) Neugebauer probably had even less knowledge of this subject than Friedrichs, but he dove straight in. After all, Courant was no expert in this field and did not pretend to be one. With people like Kneser, Artin, and Siegel in his midst, he clearly saw this as a splendid opportunity to learn about algebraic functions and algebraic number fields. So he twisted some arms to make sure plenty of bright young people got involved. The arm-twisting was perhaps new, but not the main impulse, which Courant had often experienced as a student in Hilbert's seminars. Indeed, he knew better than anyone how this type of spirit had animated nearly every memorable mathematics seminar in Göttingen going back to the days when Hilbert and Minkowski conducted these together. Neugebauer, too, found this type of communal learning environment much to his liking, and he soon came to realize that Courant was a mathematician of unusual breadth (Neugebauer 1963).

In subsequent semesters, Neugebauer continued to work closely with Courant as he gravitated toward pure mathematics. During his second and third semesters in Göttingen, he took courses with Edmund Landau, one of the world's leading experts on analytic number theory. Unlike Courant, Landau cultivated an uncompromising, rigorous formality in his books and lectures that was not to everyone's taste. Neugebauer, though, was predisposed to precision and order, and so felt drawn to Landau's teaching, the distinctive "Landau style": lean, precise, formal, and at times pedantic – theorem, lemma, proof, corollary. In many ways, he brought the Berlin tradition to Göttingen, thereby exemplifying the diversity of its mathematical community. Dirk Struik, who attended one of his courses a few years later, recalled how Landau sometimes offered teasers to his audience. "Occasionally he would present a well-known theorem in the usual way, and then while we sat there wondering what it was all about, he pontificated: 'But it is *false – ist aber falsch'* – and, indeed, there would be some kind of flaw in the conventional formulation" (see my interview with Struik, the final entry in Part V). Landau's blackboard technique was also famous and simply superb. While lecturing he would fill several boards with carefully numbered formulas. He then used these as back references as he unfolded his arguments so that everything was at hand to complete the proofs. Whereas Hilbert and Courant taught, Landau performed. Keeping up with his pace of delivery was next to impossible, so probably only a few of his auditors tried to take down everything he wrote, especially since much of this could be found anyway in his books.

[7]It was around this time that Siegel gave Artin an important tip, namely to read Takagi's classic 1920 paper of 133 pages on the theory of relative-Abelian number fields. Artin did, and thus began a new chapter in mathematics: modern class field theory; for a lively account of these developments, see Yandell (2002, 227–245).

Stories abound about all sorts of episodes that took place in Göttingen seminars, particularly those run by Hilbert. Göttingen was famous for its highly charged, competitive atmosphere, so it helped to have a thick skin. Dirk Struik remembered this very well when I interviewed him in 1987. Excerpts from that interview appear in Chap. 32, the final entry in Part V, providing a glimpse of the mathematical world he came to know in his youth. Struik's influence on subsequent generations of historians of mathematics was quite profound, so it is fitting that this volume ends with some personal memories of him.

During the mid-1920s, Struik visited Göttingen, arriving just after Klein had died. He and his wife were travelling on a Rockefeller stipend and had just spent nine enjoyable months in Rome with Levi-Civita in very cordial surroundings. What they encountered in Göttingen had little to do with cordiality. Courant was a very busy man, so Struik mainly saw him, if at all, from a distance. Still, Courant must have sensed that this Dutch visitor could help Otto Neugebauer and Stefan Cohn-Vossen with a special project that was dear to Courant's heart: the publication of Klein's wartime lectures on the development of mathematics in the nineteenth century (Klein 1926, 1927). Courant himself had assisted Klein in preparing these lectures before he was conscripted into the German army, and he knew that Klein had hoped in vain to edit them for publication during the last years of his life. Since the second volume focused on mathematical and physical theories connected with relativity, Struik was surely the perfect person to promote this effort given his background in tensor analysis.

Struik later made ample use of Klein's lectures when he wrote his popular *Concise History of Mathematics* (Struik 1948/1987). Neugebauer also came away from this undertaking deeply impressed by Klein's highly personal style as well as his overall approach to history. Clear evidence of this comes from one of his early papers, (Neugebauer 1927). Here he discusses the larger purpose of studies in the history of mathematics not only as a new type of scholarly discipline, but also especially as an antidote to specialization in mathematical research, about which he wrote:

> It is clear that the rigorous grounding of the newly formed sciences can be accomplished only by the greatest division of labor in careful individual investigations. As a consequence of this, however, there has arisen not only a sharp separation of the sciences from each other but also a crumbling of disciplines into subdivisions that are scarcely intelligible to one other and lack any common interests. No doubt a serious reaction must set in against this condition, and in part this has already taken place in a quite perceptible manner. Questions about the past and future course of the sciences, about their place in the broader sphere of our entire civilization, are being asked more and more frequently. In all fields we observe that only in the synthesis of modern research methods with the less hampered perspectives of a deeper intellectual substance can we hope to guarantee the restoration of unity among all the sciences.
>
> Like no other work, Felix Klein's *Lectures on the Development of Mathematics in the Nineteenth Century* shows what a historical view in this sense can mean for mathematics. Truly historical thinking combined with an intimate familiarity with research activity speaks to us here, beckoning us to see and to understand our own research inclinations as bound within a greater historical process. It will not be vouchsafed to many to write the history of a science in this sense. (Neugebauer 1927, 44–45)

As this passage shows, the young Otto Neugebauer saw himself as a proselytizer for a new approach to the history of mathematics, viewed as an integral part of the so-called exact sciences. This stressed the underlying unity of the mathematical sciences, long a watchword for Klein, Hilbert, and of course Courant. By this time, Neugebauer had made the transition to the field of ancient mathematics. His dissertation on *Die Grundlagen der ägyptischen Bruchrechnung* (Neugebauer 1926) was published in the same year he took his doctoral degree. It dealt with the famous *2/n* table in the Rhind Papyrus, which illustrates how the Egyptians represented such fractions as sums of unit fractions. Soon thereafter, however, he turned his attention to Babylonian mathematics, which led him into the vast terrain of Mesopotamian astronomy, the field in which he left a lasting legacy (Fig. 26.2).

Chapter 30, on the myriad traditions of ancient Greek mathematics, can be read as an outlier in Part V, as the text merely reflects my own views circa 2000. On the other hand, one might compare these with the discussion in the previous essay, which takes up Neugebauer's revisionist position with regard to ancient Greek mathematics. During the late 1920s, he made a careful study of the literature on Greek mathematics, the main topic of the history courses he taught in Göttingen (Rowe 2016). Much of the material he then covered was quite standard, but one can discern a clear trajectory of interests. Topics such as the confluence of "geometric and algebraic problems," early Greek "geometric algebra," and its later manifestations in the *Conica* of Apollonius were his dominant concerns. These would only converge into a new vision of ancient mathematics a few years later, but in Copenhagen rather than Göttingen. That particular physical transition followed the shattering of the Göttingen community, a painful period that witnessed the dismantling of the Mathematics Institute Courant had founded and that Neugebauer had helped to build and maintain (Figs. 26.3, 26.4, 26.5, and 26.6).

Although an unlikely successor to Felix Klein, Courant promoted the legacies of both Klein and Hilbert with great devotion. Beyond that, he was also a brilliant entrepreneur. As editor of Springer's "yellow series" he turned local oral knowledge – in the form of the edited lectures of famous mathematicians like Hurwitz, Klein, Hilbert, et al. – into internationally accessible knowledge in print form. Göttingen in its heyday had been an intensely interactive local culture,

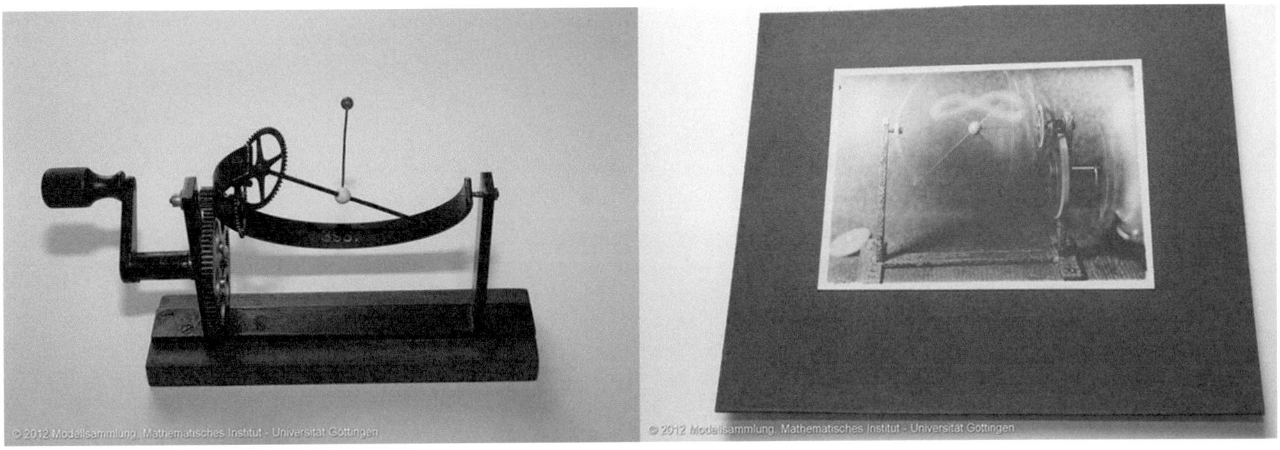

Fig. 26.2 Otto Neugebauer's kinematical model for generating a hippopede curve, used in the planetary system of Eudoxus to describe retrograde motion (Modellsammlung, Mathematisches Institut – Universität Göttingen).

but by Courant's time the community had lost much of its cohesion and communal character. In part, this was merely a natural consequence of having attained a far greater size and scope of operations as national funding organizations promoted scientific research throughout Germany. Thus, the locally based Göttingen *Vereinigung* that had funded its scientific institutes before the war was afterward swallowed up by the Helmholtz Foundation, which operated alongside the Notgemeinschaft (forerunner of today's DFG, the German Research Foundation). The Kaiser Wilhelm Gesellschaft had begun to found new institutes in Dahlem and elsewhere, while Courant coaxed the International Education Board of the Rockefeller Foundation into financing Göttingen's new mathematics institute, the building Neugebauer helped design (Siegmund-Schultze 2001). In short, Germany had entered the era of Big Science, and centers like Göttingen were the big beneficiaries. Stepping back from this scientific scene, on the other hand, reveals a quite different picture of life in Göttingen, about which a few words must be said.

Although its university was founded on Enlightenment principles back in 1737, the town of Göttingen had never been known as a bastion of liberalism. In fact, during the Weimar era it emerged as one of the strongest northern outposts of support for the National Socialist Workers Party. The Nazis founded a local organization in Göttingen in 1922, and by the end of 1923 they already had 200 *Sturmabteilung* men in uniform (Kühn 1983). The city's largest newspaper, the *Göttinger Tageblatt*, helped fuel the flames of resentment during the 1920s by praising Field Marshall Erich Ludendorff on his sixtieth birthday as a warrior against Jewry and calling the political satirist Kurt Tucholsky a Hebrew *Schmutzfink* (Wilhelm 1978, 38–39). By 1925 the *Tageblatt* had come out in support of the Nazi party; its readers were not long to follow. In national elections, the Nazis always fared far better in Göttingen than in the state of Prussia at large, where the Social Democrats fared much better. In 1930, when the Nazis garnered 20 percent of the vote across Germany, they won 38 percent in Göttingen; and in the great Nazi victory of July1932, when the party polled 37 percent of the nationwide vote, Göttingen gave them an absolute majority (Hasselhorn and Weinreis 1983, 47).

Nor was the student body any less right wing, as reflected by the fraternal organizations that dominated the scene. These were long known to be breeding nests for anti-Semitism and reactionary political views. When the Prussian Minister of Culture, Carl Heinrich Becker, proposed a constitution for a National Student Union that made discrimination by race and religion illegal, 86 percent of the Göttingen student body voted against it. Rather than accept a student union in which Jews and other undesirables would be granted free access, they evidently preferred to have none at all. Five years after the Nazis had founded their own student organization in Göttingen in 1926, they managed to attain an absolute majority in the student congress. This shift to the right largely met with the approval of the humanists within the Göttingen philosophical faculty, which was divided into two sections after the war. As noted in the introduction to Part IV (Chap. 17), sharp differences between a liberal faction, led by Hilbert, and the conservatives who dominated the humanities had exacerbated the already strained relations within the faculty. Although not as radical as the students, most professors who were politically active (roughly one-third) were affiliated with the two traditional parties of the right: the German National Peoples' Party (DNVP) and the German Peoples' Party (DVP) (Marshall 1972, 229–232, 117–118) (Fig. 26.7).

When Adolf Hitler's party finally seized power in January 1933, left-wing academics braced themselves for what was about to come. The Nazi purges led to three waves of dismissals that followed enactment of laws to "purify" the civil service (which included university professors). The first of these, which took effect in April 1933, sounded a virtual

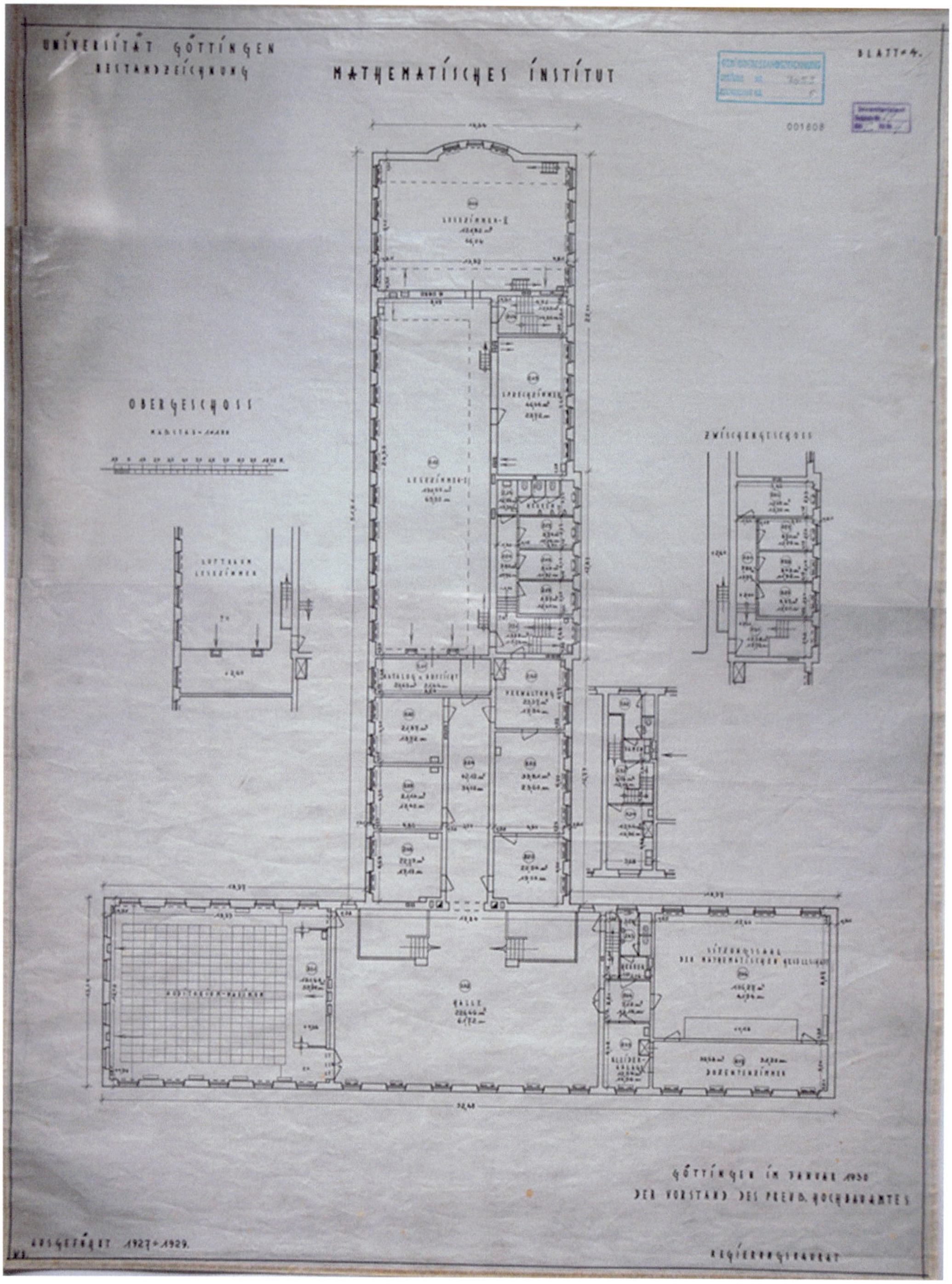

Fig. 26.3 Floor plan for the new Mathematics Institute in Göttingen as conceived by Otto Neugebauer (Photo by Bill Casselman, https://sites.google.com/site/neugebauerconference2010/web-exhibition-neugebauer).

death knell for mathematics in Göttingen; less than three weeks later Courant, Felix Bernstein, and Emmy Noether were informed by telegram that their services were no longer required. This sudden and wholly unanticipated action took many by surprise, especially Courant, who as a veteran of World War I should have been exempt from these regulations. Schappacher and Kneser, noting that Göttingen had long been a fortress for right-wing agitation, drew the plausible conclusion that the Ministry of Education did not want to risk letting grassroots leaders gain control over the situation (Schappacher and Kneser 1990). Rather than wait for the next blow to fall, Weyl accepted a position on the faculty of the newly founded Institute for Advanced Study. Proud and defiant, Landau decided he would continue teaching when

Fig. 26.4 Photo of the Mathematics Institute in Göttingen from the 1930s.

Fig. 26.5 View of the upstairs display of mathematical models and instruments (Reproduced with permission from Sattelmacher 2014).

the new semester opened in the fall of 1933. When he arrived at his lecture hall, he found SA-guards posted at the doorway and inside just one student awaited him. As a consequence, he "voluntarily" went into retirement. Oswald Teichmüller, the one truly brilliant mathematician among the Nazi rabble, led the boycott of Landau's class (Schappacher 1987, 1991).

In the process of decapitating Göttingen mathematics, brown-shirted activists like Teichmüller hoped to redirect the course of German mathematics along lines they deemed compatible with National Socialist ideology. What happened instead was that they alienated politically moderate mathematicians and forced them into a defensive posture. At the same time, their tactics may have contributed to the sense of discouragement that German students during the 1930s evidenced toward the subject. Between 1932 and 1939, the total number of students at the German universities declined from just under 100,000 to around 40,000. This decline, however, tells only a small part of the story, as this drop was far from uniform; some fields were affected far more than were others. For example, the fall in enrollments in medicine and chemistry, while significant, was not nearly so precipitous as in physics, which by 1939 had dropped to roughly a quarter of their former level. Still, no field felt the crunch as dramatically as mathematics. In the summer of 1932 there were well over 4000 mathematics students attending universities across Germany. Seven years later, that number had fallen to a mere 306, just 7.2% of the earlier figure (Schappacher and Kneser 1990, 17).

Fig. 26.6 Plaster models in the Göttingen collection (Reproduced with permission from Sattelmacher 2014).

Fig. 26.7 Academic gathering before the Göttingen Aula on 5 March 1933.

As mathematics enrollments began to plummet, a political battle of major dimensions within the German mathematical community loomed on the horizon. Now that the "Courant clique" had been driven from Göttingen, the scramble for power began in earnest, as Hilbert, on the verge of senility, failed to comprehend what was happening. Others, especially Ludwig Bieberbach, clearly did. Shortly after the Nazi takeover, he was appointed Dean of the Berlin Philosophical Faculty, although he would only become an official member of the party in 1937. His staunch supporter, Theodor Vahlen, had truly impeccable political credentials. He had already been fired during the Weimar era for agitation against the Republic. He became head of the Division of Higher Education in the Prussian Ministry of Culture after 1934. Vahlen and Bieberbach were publisher and editor, respectively, of the journal *Deutsche Mathematik,* an enterprise undertaken in the same spirit as the "Deutsche Physik" promulgated by Lenard and Stark (Beyerchen 1977).

Bieberbach had been a bit late in jumping on the Nazi bandwagon, but only months after the destruction of the Mathematics Institute in Göttingen, he made a bold move to politicize German mathematics. This began in April 1934 when he delivered a well-publicized lecture, entitled "Personality Structure and Mathematical Creativity" (Bieberbach 1934a), in which he held up the example of Felix Klein as a model for good healthy "Aryan" mathematics. To further clarify differences in mathematical styles by means of racial and national stereotypes, he pointed to Edmund Landau as a prime example of the "un-German type," while praising the Göttingen students who boycotted his classes for their "manly action" (*mannhaftes Auftreten*) (Schappacher and Kneser 1990, 58). When Harald Bohr, a close friend of Landau and Courant, criticized him in a Danish newspaper article, Bieberbach responded sharply. Acting as Schriftleiter of the DMV's official journal, he published an "Open Letter to Harald Bohr," in which he called him a "pestilence for all international cooperation" (Bieberbach 1934b, 3). Since Bieberbach had published this letter against the expressed wishes of two co-editors (Helmut Hasse and Konrad Knopp), his action set the stage for a dramatic confrontation that took place in September 1934 at the annual DMV meeting held in Bad Pyrmont. There, flanked by a throng of right-wing students whom he had invited, Bieberbach vied to implement the "Führerprinzip" so dear to Nazi ideologues. He did so by nominating Erhard Tornier, a second-rate mathematician with a first-rate Nazi pedigree, for the post of Führer of the DMV. When this coup attempt failed, he later tried to intimidate his rivals in the DMV by exploiting his connections with Vahlen. Having survived the showdown in Bad Pyrmont, however, Oskar Perron and Knopp were now prepared to call his bluff; in early 1935 they forced Bieberbach to step down from the executive committee of the DMV (Schappacher and Kneser 1990, 62–71).

A curious sidelight to this story concerns Klein's background and rumors that he came from a Jewish family. Similar stories circulated about Hilbert (see the comment by Oswald Veblen in the introduction to Part VI), but it seems these had no subsequent fallout during the Nazi era. Klein's case was different, no doubt owing to Bieberbach's efforts. In 1936, the *Göttinger Tageblatt* printed a short article with the headline: "Felix Klein was an Aryan. Which no one, in Göttingen at least, doubted." According to the *Tageblatt*, Klein's ancestry was cleared by his family after the *Völkischer Beobachter*, the official mouthpiece for the Nazi party, reported that he was of Jewish descent. The source of the error was said to be the *Jewish Encyclopedia*, but the writer added that it was well known that the Jews love to stamp famous men as Jewish in order to increase the prestige of their people (Fig. 26.8).

Fig. 26.8 Ludwig Bieberbach.

This newspaper article was almost surely based on a much more extensive piece published two years earlier in the official DMV journal (Manger 1934). Its author, Eva Manger, obviously worked closely with Bieberbach, since she cites not only his controversial lecture from April 1934 but also another – "Vom Wissenschaftsideal der Mathematiker" from 1925 – that was still unpublished. Moreover, Manger's extended defense of Bieberbach appears directly after his own open letter to Harald Bohr (Bieberbach 1934b). Her account ends with information about the ancestry on both sides of the family, information that was compiled by Felix Klein's younger brother Alfred in 1918. So clearly, Bieberbach wanted to take full advantage of his position as *Schriftleiter* of the *Jahresbericht* not only to vilify Bohr but also to publicize Klein's family history to the German mathematical community. This was a necessary expedient, of course, given that he had staked so much on making Klein, the great heir to the legacy of Gauss and Riemann, into a symbol for German mathematical genius.

Even after he was rebuffed by the DMV, Bieberbach continued to set forth his theory concerning the interconnection between race and mathematical style, despite the outrage his ideas provoked among foreign mathematicians. Once the "mystery" surrounding his former mentor's ethnic background had been settled, he could exploit Klein's immense prestige as a vehicle for legitimizing his own special role as a self-styled Nazi mathematician. In the process of doing so, Bieberbach conducted an apparently successful campaign to market Klein as a proto-Nazi thinker. In pursuing this radical course, Bieberbach believed he was merely following the lead of his former teacher, a claim he supported by citing an oft-quoted passage from Klein's Evanston Colloquium Lectures. This concerned Klein's conjecture that "a strong naïve space-intuition [was] an attribute pre-eminently of the Teutonic race [whereas] the critical, purely logical sense [was] more fully developed in the Latin and Hebrew races" (Klein 1893/1911, 46). This was by no means an isolated remark, as Klein often referred to racial stereotypes when discussing mathematical creativity (Rowe 1986).

Thus, Bieberbach not only called attention to Klein's Germanic mathematical style – with its emphasis on *Anschauung und Anwendungen* – he also emphasized that his hero had long ago pointed out that stylistic differences among mathematicians were grounded in racial typology (Lindner 1980). On the other hand, he avoided the rather obvious fact that Klein's Göttingen community was widely seen as a haven for Jewish mathematicians, not to mention that several of Klein's oldest allies and friends (Max Noether, Paul Gordan, et al.) were Jewish. Bieberbach singled out the "Landau style" as so imbued with the "Jewish spirit" that he even introduced the number π by means of an infinite series, but he conveniently forgot that Landau had been one of Klein's trusted colleagues for many years. One might also wonder whether Bieberbach had ever been close to Klein when the latter was still alive. Aside from a few perfunctory exchanges during the early 1920s, they seem to have little or no contact at all. Ludwig Bieberbach was without doubt a talented mathematician, but his accomplishments will forever be overshadowed by his misdeeds during the very darkest period in modern history.

The exodus of German-speaking mathematicians to the United States exerted a profound influence on the American mathematical community. Their story has been told often and was well documented by Reinhard Siegmund-Schultze in (Siegmund-Schultze 2009). Among the many sources he draws upon, perhaps the most important are the papers of Richard Courant (both publicly and privately held). Not surprisingly, Courant saw himself as the great protector and defender of the Göttingen legacy of Hilbert and Klein. At New York University, he succeeded in launching a novel research center, later named the Courant Center for the Mathematical Sciences. Kurt Friedrichs, one of its most distinguished members and a friend of Courant's since their days together in Göttingen, claimed that Courant, although a student of Hilbert, was intellectually far closer to Klein. My own impression is that Courant, as a mathematician, had little in common with either Klein or Hilbert. But, like Klein, he did have tremendous success as an innovator who used the full resources of the Göttingen tradition and its network of power. In this sense, he successfully transplanted an aggressive promotional approach to mathematics that he had cultivated with considerable flair in Göttingen.

After the defeat of the Nazi regime, Neugebauer never set foot in Germany again, whereas Courant returned to Göttingen nearly every summer for many years. The new postwar Director of the Mathematics Institute, Franz Rellich, happened to be his former student and a close friend (Courant 1957). Even after Rellich's premature death in 1955, Courant made regular trips to Göttingen. In the meantime, Reinhold Remmert had joined the faculty in 1963. Remmert later recalled how he and his colleagues found a way to make Courant feel welcome: by temporarily vacating the office of director so that Courant could (unofficially) assume his former position.

Given all that he accomplished at NYU, Courant's longing to hold on to past glory might make him seem like a nostalgic figure. Yet surely, this was not the way he thought of himself or how he was perceived by others who knew him well. Like Neugebauer, he identified strongly with the Göttingen tradition that he grew up with, and for both men this experience shaped their mathematical and historical sensibilities. Courant, after all, had heard Klein speak about the great figures of the nineteenth century, lectures that had inspired Neugebauer when he later read them. When he spoke to an audience of historians at Yale in 1964 about Hilbert's Göttingen, Courant began with remarks that clearly echoed the very same themes Klein had emphasized 50 years earlier:

> During the era since the French Revolution, a sizeable number of great scientific personalities have been largely responsible for the enormous development of science and technology in Europe, and by a chain reaction, outside Europe. It is true that the level of excellence and achievement in each case has invariably decayed after one or two generations. Also, it is remarkable that the centers of scientific activities have rapidly migrated from place to place and from country to country. Yet the overall progress from the time when the Ecole Polytechnique was established during the French Revolution to our own era of commercialization, public relations, and showmanship, has been very great indeed.
>
> If the young generation wants to resist the forces of decay, which are present in every civilization and must be resisted in every phase of civilization, then it seems to be vital that a sense of historical understanding and tradition be preserved and that some awareness of the role of leaders from the not too distant past should be kept alive (I think of leaders such as [Ernest] Rutherford, Niels Bohr, Harald Bohr, Arnold Sommerfeld, Ludwig Prandtl, and many others).
>
> I personally had the great fortune of a close personal and scientific association with some of these outstanding personalities. Since my days as a graduate student I belonged to the unique scientific center which had developed at the University of Göttingen. It had been initiated and guided by Felix Klein and, in my time, was filled with infinite energy and devoted enthusiasm by David Hilbert, until the moment when the Nazis broke the back of this very unique scientific center.
>
> The Mathematics Institute in Göttingen was not isolated at all; it was the organic center of a broad effort in the sciences reaching far beyond mathematics. It is perhaps fitting for me to describe unsystematically, informally, and personally some features of this old Göttingen as they come to my mind, and mostly attached to the name, to the personality of that great mathematician and leader of the younger generation, David Hilbert. (Courant 1981, 154–155)

In this brief survey, I hope to have conveyed some of the main features of the Göttingen community in the period of Hilbert and Courant. This Weimar-era community was both an intellectual construct – guided by Klein's overall vision of the mathematical sciences and reinforced by sophisticated modern mathematics – as well as an ongoing, largely improvised social experiment. Courant's later successes notwithstanding, this Göttingen experiment represented something unique und wholly incapable of replication. Just as the vast watershed of two world wars marked the end of a way of life for the privileged elites of German academia, so the very preconditions that had made the Göttingen phenomenon possible at all now ceased to exist. The Göttingen stories lived on, of course, as a collective memory whose harsher, less pleasant sides tended to be forgotten. Even for the broader mass of research mathematicians, Göttingen continued to survive as a potent symbol of a glorious past era, a time and place in which mathematics sparkled as never before. Göttingen and Hilbert, its charismatic leader, had the stuff from which legends are made, and their power was at once mythic and very real. To understand these, on the other hand, requires probing the larger conditions that shaped that specific, and in some ways quite unique intellectual environment.

References

Beyerchen, Alan D. 1977. *Scientists under Hitler: Politics and the Physics Community in the Third Reich*. New Haven: Yale University Press.

Bieberbach, Ludwig. 1934a. Die Kunst des Zitierens. Ein offener Brief an Herrn Harald Bohr on København. *Jahresbericht der Deutschen Mathematiker-Vereinigung* 44 2. Abteilung: 1–3.

———. 1934b. Persönlichkeitsstruktur und mathematisches Schaffen. *Unterrichtsblätter für Mathematik und Naturwissenschaften* 40: 236–243.

Brinkmann, Jens-Uwe und Hans-Georg Schmeling, Hrsg. 1983. *Göttingen unterm Hakenkreuz*, Göttingen: Stadt Göttingen Kulturdezernat.

Courant, Richard. 1957. Franz Rellich zum Gedächtnis. *Mathematische Annalen* 133: 185–190.

———. 1981. Reminiscences from Hilbert's Göttingen, John Ewing, ed., *Mathematical Intelligencer*, 3(4): 154–164.

Curbera, Guillermo. 2009. *Mathematicians of the World, Unite!: The International Congress of Mathematicians: A Human Endeavor*. Natick MA: AK Peters.

Edwards, Harold M. 1990. Takagi, Teiji. *Dictionary of Scientific Biography*, vol. 18, Supplement II, New York: Charles Scribner's Sons, 890–892.

Einstein-CPAE 7. 2002. *Collected Papers of Albert Einstein*, Vol. 7: *The Berlin Years: Writings, 1918–1921*, Michel Janssen, et al., eds., Princeton: Princeton University Press.

Forman, Paul. 1971. Weimar culture, causality, and quantum theory: adaptation by German physicists and mathematicians to a hostile environment. *Historical Studies in the Physical Sciences* 3: 1–115.

Georgiadou, Maria. 2004. *Constantin Carathéodory: Mathematics and Politics in Turbulent Times*. Berlin-Heidelberg: Springer.

Hasselhorn, Fritz, and Hermann Weinreis. 1983. Göttingens Weg in den Nationalsozialismus, dargestellt anhand der städtischen Wahlergebnisse, 1924–1933, in (Brinkmann u. Schmeling 1983, 47–58).

Hesseling, Dennis E. 2003. *Gnomes in the Fog. The Reception of Brouwer's Intuitionism in the 1920's*. Basel: Birkhäuser.

Honda, Kin-ya. 1975. Teiji Takagi: A Biography. *Commentarii mathematica Universitatis Sancti Pauli* 24: 141–167.

Jones, Alexander, Christine Proust, and John Steele, eds. 2016. *A Mathematician's Journeys: Otto Neugebauer and Modern Transformations of Ancient Science*. Archimedes, New York: Springer.

Klein, Felix. 1893/1911. *The Evanston Colloquium Lectures on Mathematics*, New York: Macmillan & Co., 1893. (Reprinted ed., New York: American Mathematical Society, 1911).

———. 1926. *Vorlesungen über die Entwicklung der Mathematik im 19. Jahrhundert*, Bd. 1, R. Courant u. O. Neugebauer, Hrsg., Berlin: Julius Springer.

———. 1927. *Vorlesungen über die Entwicklung der Mathematik im 19. Jahrhundert*, Bd. 2, R. Courant u. St. Cohn-Vossen, Hrsg., Berlin: Julius Springer.

Kühn, Helga-Maria. 1983. Die nationalsozialistische "Bewegung" in Göttingen von ihren Anfängen bis zur Machtergreifung (1922–1933), in (Brinkmann u. Schmeling 1983, 13–46).

Lindner, Helmut. 1980. "Deutsche" und "gegentypische" Mathematik. In *Naturwissenschaft, Technik und NS-Ideologie*, ed. Herbert Mehrtens and Steffen Richter, 88–115. Frankfurt am Main: Suhrkamp.

Manger, Eva. 1934. Felix Klein im Semi-Kürschner!. *Jahresbericht der Deutschen Mathematiker-Vereinigung*, 44, 2. Abteilung: 4–11.

Marshall, Barbara. 1972. The Political Development of German University Towns in the Weimar Republic, 1918–1930, Ph.D. dissertation, University of London.

Mehrtens, Herbert. 1987. Ludwig Bieberbach and "Deutsche Mathematik", *Studies in the History of Mathematics*, Esther R. Phillips, ed. MAA Studies in Mathematics, vol. 26, Washington: Mathematical Association of America, 195–241.

Neugebauer, Otto. 1926. *Die Grundlagen der ägyptischen Bruchrechnung*. Berlin: Springer.

———. 1927. Über Geschichte der Mathematik. *Mitteilungen des Universitätsbundes Göttingen* 9: 38–45.

———. 1963. Reminiscences on the Göttingen Mathematical Institute on the Occasion of R. Courant's 75th Birthday, Otto Neugebauer Papers, Box 14, publications vol. 11, Institute for Advanced Study, Princeton.

Reid, Constance. 1970. *Hilbert*. New York: Springer.

———. 1976. *Courant in Göttingen and New York: the Story of an Improbable Mathematician*. New York: Springer.

———. 1983. K. O. Friedrichs 1901–1982. *The Mathematical Intelligencer* 5 (3): 23–30.

Rowe, David E. 1986. "Jewish Mathematics" at Göttingen in the Era of Felix Klein. *Isis* 77: 422–449.

———. 2002. Einstein's Encounters with German Anti-Relativists, in (Einstein-CPAE 7 2002, 101–113).

———. 2016. From Graz to Göttingen: Neugebauer's Early Intellectual Journey, in (Jones, Proust, Steele 2016, 1–59).

Sattelmacher, Anja. 2014. Zwischen Ästhetisierung und Historisierung: Die Sammlung geometrischer Modelle des Göttinger mathematischen Instituts. *Mathematische Semesterberichte* 61 (2): 131–143.

Schappacher, Norbert. 1987. Das Mathematische Institut der Universität Gottingen 1929–1950. In *Die Universität Göttingen unter dem Nationalsozialismus*, ed. H. Becker, H.-J. Dahms, and C. Wegeler, 345–373. München: K.G. Saur.

———. 1991. Edmund Landau's Göttingen: From the Life and Death of a Great Mathematical Center. *Mathematical Intelligencer* 13 (4): 12–18.

Schappacher, Norbert, and Martin Kneser. 1990. In *Fachverband-Institut-Staat, in Ein Jahrhundert Mathematik, 1890–1990. Festschrift zum Jubiläum der DMV*, ed. Gerd Fischer et al., 1–82. Braunschweig: Vieweg.

Schröder-Gudehus, Brigitte. 1966. *Deutsche Wissenschaft und internationale Zusammenarbeit 1914–1928*. Geneva: Dumaret et Golay.

Siegmund-Schultze, Reinhard. 2001. *Rockefeller and the Internationalization of Mathematics between the Two World Wars: Documents and Studies for the Social History of Mathematics in the 20th Century, Science Networks, 25*. Basel/Boston/Berlin: Birkhäuser.

———. 2009. *Mathematicians fleeing from Nazi Germany: Individual Fates and Global Impact*. Princeton: Princeton University Press.

———. 2016a. "Not in Possession of any Weltanschauung": *Otto Neugebauer's Flight* from Nazi Germany and his Search for Objectivity in Mathematics, in Reviewing, and in History, in (Jones, Proust, Steele 2016, 61–106).

———. 2016b. "Mathematics Knows No Races": A Political Speech that David Hilbert Planned for the ICM in Bologna in 1928. *The Mathematical Intelligencer* 38 (1): 56–66.

Struik, Dirk J. 1948. *A Concise History of Mathematics*. New York: Dover; (4th rev. ed., 1987).

Takagi, Teiji. 1920. Über eine Theorie des relativ abelschen Zahlkörpers. *Journal of the College of Science, Imperial University of Tokyo* 41: 1–133.

———. 1922. Über das Reziprozitätsgesetz in einem beliebigen algebraischen Zahlkörper. *Journal of the College of Science, Imperial University of Tokyo* 44: 1–50.

van Dalen, Dirk. 1990. The War of the Frogs and the Mice, or the Crisis of the Mathematische Annalen. *The Mathematical Intelligencer* 12 (4): 17–31.

———. 2013. *L.E.J. Brouwer–Topologist, Intuitionist, Philosopher. How Mathematics is Rooted in Life*. London: Springer.

Weyl, Hermann. 1921. Über die *neue Grundlagenkrise* der Mathematik. *Mathematische Zeitschrift* 10: 39–79.

———. 2009. *Mind and Nature: Selected Writings in Philosophy, Mathematics, and Physics*, ed. Peter Pesic. Princeton: Princeton University Press.

Wilhelm, Peter. 1978. *Die Synagogengemeinde Göttingen, Rosdorf und Geismar, 1850–1942*, Studien zur Geschichte der Stadt Göttingen, 11, Göttingen: Vandenhoeck & Ruprecht.

Yandell, Ben H. 2002. *The Honors Class. Hilbert's Problems and their Solvers*. Natick, Mass.: A K Peters.

27 Hermann Weyl, The Reluctant Revolutionary

(*Mathematical Intelligencer* 25(1)(2003): 61–70)

"Brouwer – that is the revolution!" – with these words from his manifesto "On the New Foundations Crisis in Mathematics" (Weyl 1921) Hermann Weyl jumped headlong into ongoing debates concerning the foundations of set theory and analysis. His decision to do so was not taken lightly, knowing that this dramatic gesture was bound to have immense repercussions not only for him, but for many others within the fragile and politically fragmented European mathematical community. Weyl felt sure that modern mathematics was going to undergo massive changes in the near future. By proclaiming a "new" foundations crisis, he implicitly acknowledged that revolutions had transformed mathematics in the past, even uprooting the entire edifice of mathematical knowledge. At the same time he drew a parallel with the "ancient" foundations crisis commonly believed to have been occasioned by the discovery of incommensurable magnitudes, a finding that overturned the Pythagorean worldview that was based on the doctrine "all is Number." In the wake of the Great War that changed European life forever, the *Zeitgeist* appeared ripe for something similar, but even deeper and more pervasive.

Still, revolutions cannot occur without revolutionary leaders and ideologies, and these Weyl came to recognize in Egbertus Brouwer and his philosophy of mathematics, which Brouwer originally called "neo-intuitionism" (in deference to Poincaré's intuitionism, see (Dalen 1995)). Weyl had known Brouwer personally since 1912, and had studied his novel contributions to geometric topology as a prelude to writing *Die Idee der Riemannschen Fläche* (Weyl 1913). But the Brouwer he and most others knew back then was the brilliant topologist, not the mystic intuitionist Dirk van Dalen acquaints us with in his rich biography (Dalen 1999). Weyl simply had not known the whole Brouwer, and probably never did. True, he regarded him as a kindred philosophical spirit, but he seems never to have referred to Brouwer's *Leven, Kunst en Mystiek* (*Life, Art, and Mysticism*) or any of his other more general philosophical writings, presumably because he never read them (all were written in Dutch). If so, this surely precluded any chance of fully understanding the vision behind Brouwer's views. Nevertheless, he was swept off his feet both by Brouwer's personality and his revolutionary message for mathematics (Fig. 27.1).

Weyl had been teaching since 1913 at the ETH in Zurich (on his career there, see (Frei and Stammbach 1992)). His conversion experience took place in the summer of 1919 while vacationing in the Engadin, where Brouwer, too, was staying. Their encounter was brief, lasting only a few hours, but long enough for Weyl to see the light. Afterward, Brouwer lent him a copy of his 1913 lecture on "Formalism and Intuitionism," but Weyl returned it commenting that he already had "a copy... from the old days," presumably an allusion to the pre-revolutionary era. He further confessed that "at the time I did not pay attention to it or understand it..." (Weyl to Brouwer, 6 May 1920, quoted in (Dalen 1999, 320)), a remark befitting a new disciple of the faith.

Discipleship played a crucial role in the social relations among the mathematicians of this era, and no one felt this more keenly than Hermann Weyl when he studied under Hilbert in Göttingen. Hilbert's aura as a youth leader – the "Pied Piper of Mathematics" – was perhaps the most distinctive quality that separated him from all his contemporaries. Weyl must have felt a mixture of guilt and relief when, as he later described it, "during a short vacation spent together, I fell under the spell of Brouwer's personality and ideas and became an apostle of his intuitionism" (Weyl Nachlass, Hs 91a: 17). Even young Bertus Brouwer was strongly attracted by Hilbert's alluring persona. He spent a considerable amount of time with him during the summer of 1909 when Hilbert was vacationing in Scheveningen, a seaside-resort town near the Hague. This first personal encounter left a deep impression on Brouwer, as he related to his friend, the poet Adama van Scheltema: "This summer the first mathematician of the world was in Scheveningen; I was already in contact with him through my work, but now I have repeatedly made walks with him, and talked as a young apostle with a prophet. He is only 46 years old, but with a young soul and body; he swam vigorously and climbed walls and barbed wired gates with pleasure. It was a beautiful

D.E. Rowe, *A Richer Picture of Mathematics*, https://doi.org/10.1007/978-3-319-67819-1_27

Fig. 27.1 Brouwer with his two Russian friends, Pavel Alexandrov (left) and Pavel Urysohn (right) (Dirk van Dalen 2013).

new ray of light through my life." (Brouwer to Adama van Scheltema, 9 November 1909, quoted in (Dalen 1999, 128)).

Brouwer had already criticized Hilbert's axiomatic methods in his doctoral dissertation, submitted in 1907, where he concluded "that it has nowhere been shown that if a finite number has to satisfy a system of conditions of which it can be proved that they are not contradictory then the number indeed exists" (quoted in (Dalen 2000, 127)). For his part, Hilbert clearly recognized that the axiomatic method could never show more than consistency, but he emphatically asserted that this was all a mathematician needed to prove in order to assert that a mathematical object exists. As van Dalen has observed, it would not have been like Brouwer to pass up this golden opportunity to explain his foundational ideas to Hilbert firsthand. Unfortunately, neither apparently left any notes of what they talked about while strolling through the sand dunes of Scheveningen, but nearly 20 years later Brouwer did refer to these discussions while lamenting that Hilbert had in the meantime appropriated some of his key intuitionist principles (Brouwer 1928).

In May 1920, the ink from his "New Crisis" manuscript barely dry, Weyl sent it off to Brouwer along with the above-cited letter in which he explained his motives. "It should not be viewed as a scientific publication," he informed his new-found ally, "but rather as a propaganda pamphlet, thence the size. I hope that you will find it suitable for this purpose, and moreover suited to rouse the sleepers...." (Dalen 1999, 320). That it certainly did. Weyl's provocative broadside caused the long-bubbling cauldron of doubts about set theory and analysis to boil over into what came to be known as the modern "foundations crisis," a slogan taken directly from the title of this essay. Brouwer responded with almost gleeful delight: "your wholehearted assistance has given me an infinite pleasure. Reading your manuscript was a continuous delight and your exposition, it seems to me, will also be clear and convincing for the public..." (Dalen 1999, 321).

Among such delights was Weyl's use of politically inspired metaphor to convey a heightened sense of urgency. The antinomies of set theory, he wrote, had once been regarded as "border conflicts" in "the remotest provinces of the mathematical empire" (Weyl 1921, 143). But now they could be seen as symptomatic of a deep-seated problem, till now "hidden at the center of the superficially glittering and smooth activity," but which betrayed "the inner instability of the foundations upon which the structure of the empire rests" (*ibid.*). Weyl likened the ontological status of objects whose "existence" depends on proof by *reductio ad absurdum* to currency notes in a "paper economy," whereas true mathematical existence was surely a "real value, comparable to food products in the national economy." Nevertheless, "we mathematicians seldom think of cashing in this 'paper money.' The existence theorem is not the valuable thing, but rather the construction carried out in the proof. Mathematics, as Brouwer on occasion has said, is more an activity than a theory" (Weyl 1921, 157).

With Bismarck's mighty German Empire now in shambles, Weyl clearly thought that the empire of modern analysis would soon fall, too. Its mighty fortress in Göttingen, led by the fearless and often ferocious Hilbert, had weathered all assaults up till now, but Weyl saw its walls cracking and he prognosticized that they would soon come a crumblin' down, while the sage of Amsterdam stood ready to ride in and assume power. Weyl's defection to the intuitionist camp was clearly undertaken in order to tip the scales in the Dutchman's favor, thereby preparing the overthrow of the old regime. His manifesto, penned during the period of the abortive Kapp *Putsch* and its aftermath, reflected the mood of the times, when thoughts of revolution and counter-revolution abounded in Weimar Germany. Its principal target, of course, was his former mentor, Hilbert, who needed no rousing to see what was at stake. He struck back quickly, hard, and with plenty of polemical punch:

> What Weyl and Brouwer are doing amounts, in principle, to a walk along the same path that Kronecker once followed: they are attempting to establish the foundations of mathematics by throwing everything overboard that appears uncomfortable to them and erecting a dictatorship [*Verbotsdiktateur*] á la Kronecker. This amounts to dismembering our science, which runs the risk of losing a large part of our most valuable possessions. Weyl and Brouwer ban the general concept of irrational number, function . . ., the Cantorian numbers of higher number classes, etc.; the theorem that among infinitely many whole numbers there is always a smallest, and even the logical "Tertium non datur" . . . are examples of forbidden theorems and arguments. I believe, that just as earlier when Kronecker failed to do away with irrational numbers . . . so, too, today will Weyl and Brouwer not succeed; no: Brouwer is not, as Weyl contends, the revolution but rather only the repetition of a *Putsch* attempt with old means. If earlier it was carried out more sharply and still completely lost out, now, with the state so well armed and protected through Frege, Dedekind, and Cantor, it is doomed to failure from the outset (Hilbert 1922, 159–160).

This tense encounter, pitting the all-powerful Hilbert against his most gifted pupil, stands out as one of the more dramatic episodes of twentieth-century mathematics. Yet despite all its high drama, Hermann Weyl's commitment to Brouwer's intuitionist program soon lost its intensity. By the mid-twenties, Brouwer had put an immense amount of energy into his program for revolutionizing mathematics, while in Göttingen Hilbert and Bernays were just as busy developing proof theory as a bulwark of defense for classical analysis. By 1924, Brouwer had proved a series of results culminating in the theorem that every full function is uniformly continuous. Since these intuitionist findings had no counterparts in classical mathematics, Brouwer, who wasn't one to mince words, concluded that "classical mathematics is contradictory" (Dalen 1999, 376).

So where was Weyl? He largely stood by and watched this lively action from the sidelines, albeit with considerable interest. His flirtation with intuitionism seemed to many to have been just that, a fleeting affair doomed from the start to end in disappointment.[1] Weyl felt differently, but to understand why it will be helpful to glance back at his earlier interest not only in foundations of analysis but in mathematical physics as well. It was hardly an accident that these two fields coincided with Hilbert's principal research interests after 1910, as both men shared high hopes for breakthroughs in these two realms. In Weyl's case these took concrete form in 1918 with the nearly simultaneous publication of *Das Kontinuum* (Weyl 1918a) and his classic contribution to relativity theory, *Raum-Zeit-Materie* (Weyl 1918b).

On the Roots of Weyl's Ensuing Conflict with Hilbert

Hilbert set out his early foundational views on a number of prominent occasions, but for the most part he preferred to evade direct controversy (Rowe 2000). After 1904, however, when he delivered a highly polemical address on foundations issues at the Heidelberg ICM (Hilbert 1904), he remained virtually mute about these matters for over a decade. He did not return in earnest to research in this field until the late war years. Still, this hardly meant that he had in the meantime lost interest in foundational issues. As Volker Peckhaus has described, his long-standing efforts on behalf of Ernst Zermelo, who held a modest position in Göttingen teaching mathematical logic, as well as his support for the philosopher Leonard Nelson were part of Hilbert's long-term strategy aimed at providing institutional support for research in set theory, foundations, and mathematical logic (Peckhaus 1990, 4–22). Hilbert, by now at the height of his career, had emerged as Göttingen's second great empire-builder. He did so, however, not so much by building on the groundwork Felix Klein laid in various branches of applied mathematics (Rowe 2001), but rather by promoting research that extended the territorial claims he himself had already staked out in analysis, number theory, foundations of geometry, and mathematical physics. Compared with the Hilbertian production lines in these fields, Göttingen research efforts in set theory and foundations resembled a mere cottage industry. Presumably Hilbert hoped that by relegating this research to specialists he could turn to other matters, in particular the foundations of physics which dominated his attention after Minkowski's death in 1909 (Corry 1999).

Skúli Sigurdsson has addressed the theme of "creativity in the age of the machine" in connection with Weyl, who during his student days had close associations with Hilbert's "factory" for integral equation theory (Sigurdsson 2001, 21–29). Hermann Weyl clearly never wanted an ordinary job on this fast-moving assembly line, and he later downplayed the value of much that came off it. In one of his two obituaries for his mentor, he wrote that it had been due to Hilbert's influence that "the theory of integral equations became a world-wide fad in mathematics. .. producing an enormous literature of rather ephemeral value" (Weyl 1944, 126–127). Nevertheless, the young Weyl took a keen interest in the work of Erhard Schmidt, Ernst Hellinger, Otto Toeplitz, et al., and he also did a fair amount of mingling with his peers in the Göttingen mathematical community. This gave him ample opportunity to participate in discussions on set

[1] In an interview with B.L. van der Waerden from 1984, I asked him why he thought Weyl had retreated from intuitionism by the mid 1920s. His reply, no doubt based on logic rather than personal knowledge of Weyl's motives, offered a simple explanation. Weyl was essentially an analyst, he said, and if one accepts intuitionism then one cannot even prove the intermediate value theorem.

theory and foundations with Ernst Zermelo, whose proof of Cantor's well-ordering theorem in 1904 led to an intense debate regarding the admissibility of Zermelo's axiom of choice (Moore 1982). Four years later, in an effort to quell this controversy, Zermelo presented his well-known system of axioms for set theory (Zermelo 1908).

Soon thereafter, Weyl began to take a serious and active interest in set theory. Writing to his Dutch friend, Pieter Mulder, on 29 July 1910, he characterized his standpoint as closer to that of Borel and Poincaré than to Zermelo's views. But he also indicated that he would have to think these matters through very carefully, especially because he feared the controversy that typically ensued whenever issues in set theory and the foundations of analysis were addressed. Recalling these times, Weyl would later write: "I grew up a stern Cantorian dogmatist. Of Russell I had hardly heard when I broke away from Cantor's paradise; trained in a classical gymnasium, I could read Greek but not English" (Weyl Nachlass, Hs 91a: 17). Eight years later, Weyl alluded to the difficulties he encountered in trying to make sense of Zermelo's axiom system for set theory: "My investigations began with an examination of Zermelo's axioms for set theory.... Zermelo's explanation of the concept "definite set-theoretic predicate," which he employs in the crucial "Subset"-Axiom III, appeared unsatisfactory to me. And in my effort to fix this concept more precisely, I was led to the principles of definition of §2 [in *Das Kontinuum*]" (Weyl 1918a, 48). These principles were already enunciated in (Weyl 1910), a paper Solomon Feferman has discussed in connection with Tarski's ideas (Feferman 2000, 180) (Fig. 27.2).

Fig. 27.2 Hermann Weyl, *circa* 1910, around the time he first became sceptical of Zermelo's axioms for set theory.

Weyl described his initial orientation as similar to Dedekind's theory of chains, in that he sought to establish the principle of complete induction without recourse to the primitive notion of the natural numbers. This quest "drove me to a vast and ever more complicated formulation but, unfortunately, not to any satisfactory result." He finally abandoned this as a "scholastic pseudo-problem" after achieving "certain general philosophical insights," presumably derived from reading Edmund Husserl and distancing himself from Poincaré's conventionalism. Nevertheless, he concluded that Poincaré had been right regarding the status of the sequence of natural numbers as "an ultimate foundation of mathematical thought" (Weyl 1918a, 48).

What prompted Weyl to reenter this arena in 1918, a move that took his good friend, Erich Hecke, by surprise? Probably he had several motives, but he surely kept a keen eye on his mentor's activities about which he had firsthand knowledge. On 11 September 1917, Hilbert delivered a lecture on "Axiomatic Thought" (Hilbert 1918) before a meeting of the Swiss Mathematical Society in Zurich. This gave the first clear signs that he was about to take up the foundations of mathematics once again. Surely no one in Hilbert's audience listened more attentively than Hermann Weyl, who discussed this lecture in detail many years later. For Hilbert's talk offered a sweeping panorama of mathematical and physical ideas that stressed not only their mutual interdependence but the role of axiomatics in both realms (see (Corry 1997) on Hilbert's background interests) (Fig. 27.3).

Like many of his contemporaries, Hilbert regarded the growth of mathematical knowledge as an essentially teleological process in which thought obeys higher, transcendental laws. As such, his positivism had an added Hegelian flavor, only with the mathematician replacing the metaphysician as the highest human form of Reason. In his lecture, Hilbert thus described the manner in which axiomatization took place as part of a natural, organic process starting from an informal system of ideas ("*Fachwerk von Begriffen*"). However, these ideas, which arose spontaneously, as it were, in the course of the theory's development, were merely provisional in nature. Only during the next stage, when researchers attempted to provide deeper foundations for the theory, did the process of axiomatization actually begin. By invoking such architectonic imagery, Hilbert suggested how this process structured scientific thought:

Fig. 27.3 Hilbert in Zürich, where he delivered a lecture on "Axiomatisches Denken" in September 1917. The four men in the front row holding hats are, left to right, Constantin Carathéodory, Marcel Grossmann, David Hilbert, and Carl Geiser, followed by Hella (wearing hat) and Hermann Weyl (Alexanderson 1987).

> Thus arose the actual, present-day so-called *axioms* of geometry, arithmetic, statics, mechanics, radiation theory, and thermodynamics. These axioms build a deeper-lying layer of axioms than the axiom layer that was earlier characterized by the fundamental theorems of the individual fields. The process of the axiomatic method . . . thus amounts to a *deepening of the foundations* of the individual fields just as it becomes necessary to do with any building to the extent that one wants to make it secure as one builds it outward and upward (Hilbert 1918).

In his concluding remarks, Hilbert mentioned two particularly pressing problems confronting the foundations of mathematics: proving the consistency of the axioms for arithmetic (the second of Hilbert's 23 Paris problems), and showing the same for Zermelo's system of axioms for set theory. He emphasized that both of these problems were intimately wedded to a whole complex of deep and difficult epistemological questions of "specifically mathematical coloring": (1) the problem of the *solvability of every mathematical question in principle*, (2) the problem of the subsequent *verification* of the results of a mathematical investigation, (3) the question of a *criterion for the simplicity* for mathematical proofs, (4) the question of the relationship between *content and form* in mathematics and logic, and (5) the problem of the *decidability* of a mathematical question by means of a finite number of operations. Hilbert then summed up his position regarding all these complex issues as follows:

> All such fundamental questions . . . appear to me to form a newly opened field of research, and to conquer this field – this is my conviction – we must undertake an investigation of the concept itself of the specifically mathematical proof, just as the astronomer must take into account the movement of his position, the physicist must care for his apparatus, and the philosopher criticizes reason itself, (Hilbert 1918, 155).

While conceding that, for the present time, this program remained but a sketch for future research, Hilbert retained his optimistic outlook for this program:

> I believe that everything which can be the subject of scientific thought, as soon as it is ripe enough to constitute a theory, falls within the scope of the axiomatic method and thus directly to mathematics. By pursuing ever deeper-lying layers of axioms . . . we gain ever deeper insights into the essence of scientific thought itself, and we become ever more conscious of the unity of our knowledge. In the name of the axiomatic method, mathematics appears called upon to assume a leading role in all of science (Hilbert 1918, 156).

This *tour de force* performance clearly signaled Hilbert's intentions to take up once again the foundations program he had sketched 13 years earlier in his speech at the Heidelberg ICM. Indeed, his rhetorical flourishes clearly echoed themes Weyl and others would have recognized from Hilbert's even more famous address at the Paris ICM in 1900. Just as striking, however, were the parallels with his concluding remarks from his first contribution to the general theory of relativity, in which he made similarly sweeping claims regarding the strength and resilience of the axiomatic method:

> As one sees, the few simple assumptions expressed in Axioms I and II suffice by sensible interpretation for the development of the theory: through them not only are our conceptions of space, time, and motion fundamentally reformulated in the

> Einsteinian sense, but I am convinced that the most minute, till now hidden processes within the atom will become clarified through the herein exhibited fundamental equations, and that it must be possible in general to refer all physical constants back to mathematical constants—just as this leads to the approaching possibility, that out of physics in principle a science similar to geometry will arise: truly, the most glorious fame of the axiomatic method, while here, as we see, the mighty instruments of analysis, namely the calculus of variations and invariant theory, are taken into service (Hilbert 1915, 407).

Tilman Sauer has noted how Hilbert, quite ironically, made only a vague allusion to this vision for a unified field physics when he spoke in Zürich in 1917 (Sauer 2002). Weyl would have known, though, that Hilbert was never subdued about these prospects, even on this occasion. He also knew very well what Einstein thought of Hilbert's methodological approach. In a letter from November 1916, Einstein confessed:

> To me, Hilbert's *Ansatz* about matter appears to be childish, just like an infant who is unaware of the pitfalls of the real world . . .. In any case, one cannot accept the mixture of well-founded considerations arising from the postulate of general relativity and unfounded, risky hypotheses about the structure of the electron . . .. I am the first to admit that the discovery of the proper hypothesis, or the Hamilton function for the structure of the electron is one of the most important tasks of the current theory. The "axiomatic method," however, can be of little use in this (Einstein to Weyl, 23 November 1916, [Einstein 1998a, 366]).

Weyl took up these problems around this very time. In the summer semester of 1917 he offered a lecture course on general relativity and, on the advice of Einstein's close friend, Michele Besso, he decided to adapt his notes into a book on special and general relativity. This was published the following year by Julius Springer Verlag as the first edition of *Raum-Zeit-Materie* (Weyl 1918b); a second soon followed, and three substantially revised editions appeared between 1919 and 1923. A few months before the first edition came out Weyl had proofs sent to both Einstein and Hilbert. Their respective reactions reveal a good deal about both men.

Einstein was euphoric: "it's like a symphonic masterpiece. Every word stands in relation to the whole, and the design of the work is grand" (Einstein to Weyl, 8 March 1918, [Einstein 1998b, 669–670]). A week earlier, Hilbert wrote also, but he merely acknowledged receipt of the proofs (Hilbert to Weyl, 28 February 1918, Weyl Nachlass, Hs. 91: 604). Since he was on his way to Bucharest to attend a meeting on space and time in physics, he had no time to read them. Still, he expressed regret that he would not be able to meet Weyl in Switzerland over the semester break, but hoped to do so during the summer or early the following year. He then added some remarks about the professorship in Breslau recently offered to Weyl.

Hilbert had been consulted during the deliberations over potential candidates, and he had apparently recommended Weyl for the post. But he now counseled him against accepting the offer "*im Interesse des Reichsdeutschtum*," since if he left Zurich this would probably leave some worthy German mathematician without a job. Hilbert took a very active role in this game of musical mathematical chairs, and he apparently thought that Weyl should pass on this round. He also thought the Prussian government would be receptive to this argument, so that declining would not have unfavorable consequences for Weyl in the future. Apparently Weyl was not inclined to follow this advice; at any rate, in April he accepted the chair in Breslau (although he would later turn it down for health reasons). Hilbert, having returned from Bucharest, wrote him on 22 April, sending "congratulations on accepting the Breslau position," but quickly added "unfortunately this again creates a vacant mathematical position in Switzerland that will be difficult to fill and unlikely so with a suitable personality."

Miffed that Weyl had ignored his wishes, Hilbert added some curt praise for *Raum-Zeit-Materie*: "I have looked more carefully at the proofs of your book, which gave me great pleasure, especially also the refreshing and enthusiastic presentation. I noticed that you did not even mention my first Göttingen note from 1915. . .." He then proceeded to rattle off a litany of complaints bearing on this omission, a set of remarks that provide much insight into what Hilbert himself saw as the main achievements in his controversial paper on "Foundations of Physics" (Hilbert 1915). (For details, see the transcription below.)

Draft of letter from Hilbert to Weyl, NSUB, Hilbert Nachlass 457, 17.

22 April 1918.

Dear Herr Weyl,

above all my congratulations on accepting the Breslau position; unfortunately this again creates a vacant mathematical position in Switzerland that will be difficult to fill and unlikely so with a suitable personality.

Having returned from Bucharest I have looked more carefully at the proofs of your book which gave me great pleasure, especially also the refreshing and enthusiastic presentation. I noticed that you did not even mention my first Göttingen note from 1915 even though the foundations of the gravitational theory, in particular the use of the Riemannian curvature in the Hamiltonian integral which you present on page 191, stems from me alone, as does the separation of the Hamiltonian function in H - L, the derivation of the Maxwellian equations, etc. Also the whole presentation of Mie's theory is precisely that which I gave for the first time in my first note on the foundations of physics. For Einstein's earlier work on his definitive theory of gravitation appeared at the same time as mine (namely in November 1915); Einstein's other papers, in particular on electrodynamics and on Hamilton's principle appeared however much later. Naturally I am very gladly prepared to correct any of my mistaken assertions. It also appears odd to me that where you do take note of the fact that there are four fewer differential equations for gravitation than unknowns you only cite my second communication in the Göttinger Nachrichten, whereas precisely this circumstance served as the point of departure for my investigation and was especially enunciated as a theorem whose consequences were pursued there (Fig. 27.4) ....

With best greetings to you and your wife,
Your Hilbert

Weyl received this letter in time to make modifications in the text before it went off to press. He added a few citations and brief remarks on Hilbert's first note, but these remained shadowy features in his book compared with his own contributions and, of course, Einstein's. Hilbert could not have felt gratified by this, especially after reading the preface, which Weyl wrote while vacationing at his in-laws home in Mecklenburg. He could hardly have failed to notice his protégé's animus against his views on the axiomatization of physics. At the same time, Weyl announced that he had found a new avenue to a truly unified field theory:

Fig. 27.4 Einstein relaxing in his home office in Berlin, in which by 1918 he was carrying on an extensive scientific correspondence (Bildarchiv Preußischer Kulturbesitz, Berlin).

> The Einsteinian theory in its present form ends with the duality of electricity and gravitation, "field" and "ether"; these remain totally isolated and stand next to one another. Just now a promising path has opened to the author for the derivation of both realms of phenomena from a common source by an extension of the geometrical foundations. Evidently the development of the general theory of relativity had not yet been concluded. However, it was not the intention of this book to take the powerful, stirring life that stems from the field of physical knowledge to the point it has presently reached and transform it with axiomatic rigor into a dead mummy. (Weyl 1918b, vi)

Just as Weyl was drawing on Einstein's general relativity to open up new vistas for the geometrization of physics, he was also taking up the challenge Hilbert had laid down for the foundations of analysis. Indeed, immediately after Hilbert spoke about this in Zürich, Weyl began a lecture course in the winter semester of 1917–18 entitled "Logical Foundations of Mathematics." One of his potential auditors, Paul Bernays, was unable to attend since Hilbert had already snatched him up as his new assistant in Göttingen. Thus began the collaboration that enabled Hilbert to push forward proof theory in earnest (Sieg 2000). In the meantime, Weyl used his lectures as a platform for writing his booklet on *The Continuum* (Weyl 1918a). This aimed to salvage the foundations of analysis within the more modest scope of a constructivist program that avoided the complexities of a full-blown axiomatic set theory à la Zermelo. Its approach to the continuum thus clearly exposed the differences between Weyl's constructivist views and those of Hilbert. Weyl only gradually became aware of this insurmountable rift, but by 1918 he stood firmly on the other side of a deep abyss that separated his approach to foundations of analysis from conventional wisdom on this subject.

Even in Zurich, Weyl's ideas encountered strong opposition. His younger colleague, Georg Pólya, whom he deeply respected as an analyst, reacted incredulously, so much so that Weyl challenged him to a mathematical wager. He rounded up a dozen witnesses who validated a document stating that:

> Within 20 years Pólya and the majority of representative mathematicians will admit that the statements:
> *Every bounded set of reals has a precise supremum*
> *Every infinite set of numbers contains a denumerable subset*
> contain totally vague concepts, such as 'number', 'set', and 'denumerable', and therefore that their truth or falsity has the same status as that of the main propositions of Hegel's natural philosophy. However, under a natural interpretation 1) and 2) will be seen to be false. (Pólya 1972)

Thus, already by this time, Weyl's heretical views were well known within Zurich's mathematical circles. They would become even better known after he completed the manuscript of (Weyl 1918a) and followed it up in (Weyl 1919), a note on the "vicious circle" in conventional foundations of analysis that appeared in the widely read *Jahresbericht* of the German Mathematical Society.

In *Das Kontinuum*, he described the programmatic features of his new treatment of the continuum as follows:

> . . . in spite of Dedekind, Cantor and Weierstrass, the great task which has been facing us since the Pythagorean discovery of the irrationals remains today as unfinished as ever; that is, the *continuity* given to us immediately by intuition (in the flow of time and in motion) has yet to be grasped mathematically as a totality of discrete "stages" in accordance with that part of its content which can be interpreted in an "exact" way. (Weyl 1918a, 24–25)

Weyl insisted on the need to avoid the problem of impredicativity which arose when the set of real numbers was defined by means of arbitrary Dedekind cuts. Hilbert had tried to finesse this problem with his *Vollständigkeitsaxiom*, but in later editions of his *Grundlagen der Geometrie* he was forced to fall back on Dedekind cuts to prove that his axiom system characterized analytic geometry over the field of real numbers. Weyl's general philosophical orientation stood in sharp contrast to Hilbert's brand of rationalism, so in this respect it should not be surprising that he distanced himself from the fundamental tenets underlying his mentor's foundations research. Still, *Das Kontinuum* represented only a relatively mild break with Hilbertian precepts. Unlike Brouwer's approach, which took the continuum to be an irreducible given requiring the mathematician to master the new techniques of choice sequences, Weyl's goal in (Weyl 1918a) was far more pragmatic. As an analyst, he was less concerned with capturing the "essence" of the continuum than he was in extracting from it a sufficiently rich arithmeticized substructure which the "working mathematician" could use to recover most of the results of classical analysis (for more on this, see (Feferman 2000)).

Immediately after publishing *Das Kontinuum* and *Raum-Zeit-Materie*, Weyl began elaborating his gauge-invariant approach to a unified field theory. Einstein called it "a first-class stroke of genius," but nevertheless a useless one for physics (Einstein to Weyl, 6 April 1918, [Einstein 1998b, 710]). Arnold Sommerfeld had no such reservations, at least initially:

> What you are saying there is really wonderful. Just as Mie had grafted a gravitational theory onto his fundamental theory of electrodynamics that did not form an organic whole with it . . .Einstein grafted a theory of electrodynamics . . .onto his fundamental theory of gravitation that had little to do with that. You have worked out a true unity (Sommerfeld to Weyl, 7 July 1918, Weyl Nachlass, 91: 751).

Apparently Weyl had no difficulty persuading Hilbert, either. As he had promised, his former mentor visited him during September when he was vacationing in Switzerland. Whether or not they discussed foundational issues is unclear, but Weyl did report to Einstein that Hilbert "gave me his unqualified support" for Weyl's unified field theory (Weyl to Einstein, 18 September 1918, [Einstein 1998b, 879]). Einstein, on the other hand, continued to express his skepticism, which put Weyl under additional pressure, as he wrote to the physicist in December:

> I am now in a really difficult position; by nature so conciliatory that I'm almost incapable of discussion, I now must fight on all fronts: my attack on analysis and attempt to found it anew is encountering much more fierce opposition from the mathematicians who work on these logical things than my efforts in theoretical physics have received from you. (Weyl to Einstein, 10 December 1918, [Einstein 1998b, 966]).

By now Weyl was grappling not only with the foundations of classical analysis and mathematical physics but also with core problems that led to dramatic new developments in differential geometry. He described this research program in 1921 as follows:

> It concerns, on the one hand, impulses for a new foundation of the analysis of the infinite, the present foundation of which is in my opinion untenable; on the other hand, *in close connection with this* [my emphasis] and in connection with the general theory of relativity, a clarification of the relation of the two basic notions with which modern physics operates, that of *field of action and of matter, the theories of which can at present by no means be joined together continuously* [Weyl's emphasis]" (Dalen 2000, 311).

Erhard Scholz has emphasized the extent to which Weyl's mathematical work was intertwined with physical and epistemological problems (Scholz 2001b). All three of these fast-converging interests were clearly very much on his mind when he first learned about Brouwer's intuitionism.

Weyl's Emotional Attachment to Intuitionism

After meeting Brouwer in the summer of 1919, Weyl presented his provisional ideas on the "new crisis" the following December at three successive meetings of the mathematics colloquium, which he co-chaired with Georg Pólya. The topics he covered were identical with those eventually presented in his "propaganda pamphlet," and apparently he made no attempt to tone down his enthusiasm for intuitionism when he delivered these lectures. Not surprisingly, Pólya reacted just as vigorously in criticizing Weyl's philosophical claims in defense of intuitionism. According to notes taken by Ferdinand Gonseth, they exchanged words like these:

> Pólya: You say that mathematical theorems should not only be true, but also be meaningful. What is meaningful?
>
> Weyl: That is a matter of honesty.
>
> Pólya: It is erroneous to mix philosophical statements in science. Weyl's continuum is emotion.
>
> Weyl: What Pólya calls emotion and rhetoric, I call insight and truth; what he calls science, I call symbol-pushing (*Buchstabenreiterei*). Pólya's defense of set theory . . . is mysticism. To separate mathematics, as being formal, from spiritual life, kills it, turns it into a shell. To say that only the chess game is science, and that insight is not, *that* is a restriction. [Dalen 1999, 320].

Fig. 27.5 Hella and Hermann Weyl on a trip to the Alhambra in 1922 (From Hermann Weyl 2009).

Weyl was not vituperative by nature, but he clearly hoped his ideas would create a stir. He especially hoped it would arouse Hilbert from his dogmatic slumbers, something Brouwer had been unable to accomplish himself. Hard at work on his "New Crisis" paper, Weyl wrote Bernays in Göttingen (9 February 1920, Weyl Nachlass, 91: 10) that he had returned to foundations of analysis, and had "substantially modified [his] standpoint" thanks to Brouwer's work. Referring to their meeting the previous summer, he wrote "I was completely happy during the few hours we spent together." And he described Brouwer as a really "splendid fellow" (*Mordskerl*) and a wonderfully intuitive person, adding that "if you get him to come to Göttingen as Hecke's successor, then I envy you." The departures of Hecke and also Carathéodory from Berlin, two of Hilbert's most prominent protégés, reflected the fragile state of German academic life amid the surrounding political uncertainties (see Chap. 28). Brouwer turned down both chairs, but not before negotiating improved conditions in Amsterdam (Dalen 1999, 300–304) (Figs. 27.5 and 27.6).

Weyl's attraction to intuitionism and his feelings toward its founder were confirmed when he visited Brouwer at his home in Blaricum only a few months after completing his "New Crisis" essay. Writing to Felix Klein, he proclaimed that "Brouwer is a person I love with all my heart. I have now visited him in his home in Holland, and the simple, beautiful, pure life in which I took part there for a few days, completely and totally confirmed the impression I had made of him" [Dalen 1999, 298]. Even after this passionate phase had passed, Weyl always extolled Brouwer's positive achievement, which he saw as vital for understanding the

Fig. 27.6 A gathering of Zurich's mathematicians during the visit of József Kürschák (right front row), who came from Budapest along with Frédéric Riesz (third from left). Others pictured, from left to right, are Hermann Weyl, Louis Kollros, Riesz, Georg Pólya, Michel Plancherel, Jérome Franel, Mrs. Kürschák, (unknown), Andreas Speiser, Kürschák, and Rudolf Fueter (From Alexanderson 1987).

essential tension between form and content, a dichotomy he took to be characteristic of mathematical thought.

Yet Weyl was never an orthodox intuitionist, not even when he presented himself as a Brouwerian in his "New Foundations Crisis" (Scholz 2000, 199–200). What appealed to him most about Brouwer's philosophy was its honesty and its human dimension. Thus, Weyl shared Brouwer's aversion to Hilbert's overly sanguine, winner-take-all approach to mathematics that he felt bordered on intellectual dishonesty, just as he deplored the tendency to reduce the quest for mathematical meaning and truth to a meaningless formalist game. Brouwer refused to accept *carte blanche* the paper currency of pure existence proofs unless it could first be demonstrated that this money could actually be used to purchase a genuine mathematical article. Intuitionism thereby threatened the weakest flank in Hilbert's fortress by exposing the ontological assumptions of formalist dogma while advancing a vision for a new mathematics based on constructive principles. For Brouwer, Hilbert's formalist doctrine amounted to an impoverished view of mathematical activity that reduced it merely to showing that certain purely formal procedures could never produce a contradiction (Fig. 27.7).

Brouwer's views thus had a twofold significance for Weyl. First, they helped him to focus on certain conflicting issues he had long been struggling with but had never quite confronted, including his general dissatisfaction with Hilbert's claims regarding the axiomatic method. Second, the clarity of Brouwer's critique gave Weyl the courage to articulate his own views in strong, provocative language that could not be ignored. In his eyes, intuitionism was an ideology that dignified mathematical research as a human activity during a time of crisis. Still, it was only one element within the larger framework of ideas Weyl hoped to synthesize.

Many intellectuals – Weyl, Brouwer, Hilbert, and Einstein included – were acutely aware that they were living through revolutionary times when the whole "civilized" world was seeking new leaders, structures, and answers. But Weyl alone, working in the quiet seclusion of Zurich, felt the need to promote a new intellectual reorientation that would encompass both mathematics and physics. For Weyl, Einstein's general theory represented nothing less than a revolution in human thought, a view shared by Hilbert. Yet Weyl went much further; his pioneering work in differential geometry was motivated by the urge to explore the implications of space-time theories in order to reach a deeper conceptual understanding of the physical world. Moreover, he was convinced that this simultaneous revolution in mathematics and physics was already underway, having burst into the public eye well before the sudden collapse of Imperial Germany that ended the Great War.

Yet despite his sensitivity to these swirling intellectual currents, Weyl was not by temperament an iconoclastic or revolutionary thinker. Moreover, he lacked the hard-nosed, combative streak that made both Hilbert and Brouwer such forceful personalities. By announcing that the foundations of analysis was undergoing a foundational crisis, Weyl sought to expose the hollowness of Hilbert's foundational rhetoric and at the same time underscore his long-standing misgivings with regard to Cantorian set theory, which Hilbert and Zermelo had sought to retain and rigorize by means of axiomatics. He was, however, more than happy to accomplish this by carrying Brouwer's banner forward rather than holding up one of his own making.

Fig. 27.7 Weyl visiting Hilbert in the mid 1920s.

The political imagery in his "New Crisis" essay was all very apt, but the oedipal overtones in this conflict would seem even harder to ignore. For however strongly Weyl may have felt about Brouwer, he must have sooner or later realized that he had very little in common with him. In everything from their personalities and lifestyles to their philosophical views and mathematical tastes, he and Brouwer were polar opposites. Clearly, his feelings for Brouwer and his attachment to intuitionism were genuine, but just as clearly these helped him gain some emotional distance from his tyrannical *Doktorvater*, a man intent on directing the course of future mathematical research activity. Intellectually, he knew how much he owed Hilbert, and he largely shared his ambitions for the "new mathematics" as well as his vision for a unified field physics of gravitational and quantum phenomena. As the heir-apparent of the Göttingen master, Weyl must have felt almost predestined to compete with him on these two major fronts. When seen from this vantage point, his brief flirtation with Brouwer's intuitionism appears not only less perplexing, but also less Olympian, more human, and in some sense more believable.

References

Alexanderson, G.L. 1987. George Pólya: A Biographical Sketch. In *The Pólya Picture Album: Encounters of a Mathematician*. Boston: Birkhäuser.

Brouwer, L.E.J. 1928. Intuitionistische Betrachtungen über den Formalismus. *Sitzungsberichte der Preußischen Akademie der Wissenschaften*: 48–52.

Corry, Leo. 1997. David Hilbert and the Axiomatization of Physics (1894–1905). *Archive for History of Exact Sciences* 51: 83–198.

———. 1999. David Hilbert between Mechanical and Electromagnetic Reductionism (1910–1915). *Archive for History of Exact Sciences* 53: 489–527.

Einstein. 1998a. *Collected Papers of Albert Einstein (CPAE)*, Vol. 8A: *The Berlin Years: Correspondence, 1914–1917*, Robert Schulmann, et al., eds. Princeton: Princeton University Press.

———. 1998b *Collected Papers of Albert Einstein (CPAE)*, vol. 8B: *The Berlin Years: Correspondence, 1918*, Robert Schulmann, et al., eds. Princeton: Princeton University Press.

Feferman, Solomon. 2000. The Significance of Weyl's *Das Kontinuum*, in [Hendricks, Pederson, Jørgensen 2000], 179–193

Frei, Günther, and Urs Stammbach. 1992. *Hermann Weyl und die Mathematik an der ETH Zürich, 1913–1930*. Birkhäuser: Basel.

Hendricks, V.F., S.A. Pedersen, and K.F. Jørgensen, eds. 2000. *Proof Theory. History and Philosophical Significance*. Dordrecht: Kluwer.

Hilbert, David. 1904. Über die Grundlagen der Logik und der Arithmetik, *Verhandlungen des III. Internationalen Mathematiker-Kongresses in Heidelberg 1904*, 174–185.

———. 1915. Die Grundlagen der Physik. (Erste Mitteilung), *Königliche Gesellschaft der Wissenschaften zu Göttingen. Mathematisch-physikalische Klasse. Nachrichten*: 395–407.

———. 1918 Axiomatisches Denken, *Mathematische Annalen*, 79: 405–415. (Reprinted in [Hilbert 1932–35], vol. 3, 146–156.)

———. 1922. Neubegründung der Mathematik. Erste Mitteilung, *Abhandlungen aus dem Mathematischen Seminar der Hamburgischen Universität*, 1: 157–177. (Reprinted in [Hilbert 1932–35], vol. 3, 157–177.)

———. 1932–35. *Gesammelte Abhandlungen*. Vol. 3. Berlin: Julius Springer.

Moore, Gregory H. 1982. *Zermelo's Axiom of Choice: Its Origins, Development and Influence*. New York: Springer.

Peckhaus, Volker. 1990. *Hilbertprogramm und Kritische Philosophie*, Studien zur Wissenschafts-, Sozial- und Bildungsgeschichte der Mathematik, vol. 7, Göttingen; Vandenhoeck & Ruprecht.

Pólya, Georg. 1972. Eine Erinnerung an Hermann Weyl. *Mathematische Zeitschrift* 126: 296–298.

Rowe, David E. 2000. The Calm before the Storm: Hilbert's Early Views on Foundations, in [Hendricks, Pederson, Jørgensen 2000], 55–94.

———. 2001. Felix Klein as Wissenschaftspolitiker. In *Changing Images in Mathematics. From the French Revolution to the New Millennium*, ed. Umberto Bottazzini and Amy Dahan, 69–92. London: Routledge.

Sauer, Tilman. 2002. Hopes and Disappointments in Hilbert's Axiomatic 'Foundations of Physics. In *History of Philosophy of Science*, ed. M. Heidelberger and F. Stadler, 225–237. Dordrecht: Kluwer.

Scholz, Erhard. 2000. Hermann Weyl on the Concept of Continuum, in [Hendricks, Pederson, Jørgensen 2000], 195–220.

———. 2001a. *Hermann Weyl's Raum-Zeit-Materie and a General Introduction to his Scientific Work*, (DMV Seminar, 30.) Basel/Boston: Birkhäuser Verlag.

———. 2001b. Weyls Infinitesimalgeometrie, 1917–1925, in [Scholz 2001a], 48–104.

Sieg, Wilfried. 2000. Toward Finitist Proof Theory, in [Hendricks, Pederson, Jørgensen 2000], 95–115.

Sigurdsson, Skúli. 2001. Journeys in Spacetime, in [Scholz 2001a], 15–47.

van Dalen, Dirk. 1995. Hermann Weyl's Intuitionistic Mathematics. *The Bulletin of Symbolic Logic* 1: 145–169.

———. 1999. *Mystic, Geometer, and Intuitionist. The Life of L. E. J. Brouwer, vol. 1: The Dawning Revolution*. Oxford: Oxford Clarendon Press.

———. 2000. The Development of Brouwer's Intuitionism, in [Hendricks, Pederson, Jørgensen 2000], 117–152.

———. 2013. *L.E.J. Brouwer: Topologist, Intuitionist, Philosopher*, 419. Heidelberg: Springer-Verlag.

Weyl, Hermann. 1910. Über die Definitionen der mathematischen Grundbegriffe. *Mathematisch-naturwissenschaftliche Blätter* 7 (1910): 93–95, 109–113. (Reprinted in [Weyl 1968], vol. 1, 298–304.)

———. 1913. *Die Idee der Riemannschen Fläche*. Leipzig: Teubner.

———. 1918a. *Das Kontinuum*. Leipzig: Veit & Co., 1918. *The Continuum*. Trans. S. Pollard and T. Bole. New York: Dover.

———. 1918b. *Raum, Zeit, Materie*. 1st ed. Berlin: Springer.

———. 1919. Der circulus vitiosus in der heutigen Begründung der Analysis. *Jahresbericht der Deutschen Mathematiker-Vereinigung* 28: 85–92.

———. 1921. Über die neue Grundlagenkrise der Mathematik, *Mathematische Zeitschrift*, 10: 39–79. (Reprinted in [Weyl 1968], vol. 2, 143–180.)

———. 1944. Obituary: David Hilbert, 1862–1943, *Obituary Notices of Fellows of the Royal Society* 4: 547–553. Reprinted in [Weyl 1968], vol. 4, 120–129.

———. 2009. In *Mind and Nature: Selected Writings in Philosophy, Mathematics, and Physics*, ed. Peter Pesic. Princeton: Princeton University Press.

Zermelo, Ernst. 1908. Neuer Beweis für die Möglichkeit einer Wohlordnung. *Mathematische Annalen* 65: 107–128.

Transforming Tradition: Richard Courant in Göttingen

28

(*Mathematical Intelligencer* 37(1)(2015): 20–29)

Richard Courant had a knack for being at the right place at the right time. He came to Göttingen in 1907, just when Hilbert and Minkowski were delving into fast-breaking developments in electron theory. There he joined three other students who also came from Breslau: Otto Toeplitz, Ernst Hellinger, and Max Born, all three, like him, from a German Jewish background.[1] Toeplitz was their natural intellectual leader, in part because his father was an *Oberlehrer* at the Breslau Gymnasium (Müller-Stach 2014). Courant was 5 or 6 years younger than the others; he was sociable and ambitious, but also far poorer than they (Reid 1976, 8–13).

Max Born had been the first of these four Breslau Jews to enter the more intimate private sphere that made the Göttingen of Hilbert and Minkowski so special. But Courant was soon to follow, even though he got off to a rough start in their seminar (Reid 1976, 17–18). Born later suffered a similar fiasco when he gave his first talk in the *Mathematische Gesellschaft* (Born 1978, 134–135). In fact, this type of harrowing experience was so commonplace that it must have seemed like a kind of initiation rite for the young mathematicians who survived it. Even Emil Artin, who came to Göttingen in 1922 when Courant was director of the Mathematisches Institut, complained bitterly about the abuse he suffered from Hilbert, who of course set the tone.[2] Already from the time of Klein and Hilbert the mathematical atmosphere in Göttingen was fiercely competitive. You had to be a survivor.

Hilbert's special friendships with Hermann Minkowski and Adolf Hurwitz began during his student days in Königsberg (see Chap. 13). Both came from Jewish families and both figure prominently in our exhibition, "Transcending Tradition" (Bergmann/Epple/Ungar 2012). Some assumed that David Hilbert was also a Jew, if only because of his name (Rowe 1986, 422–423).[3] Later, when so many young Jews began to gravitate to Göttingen, the local atmosphere began to change. What Richard Courant experienced there was something new and exciting; indeed, Hilbert's Göttingen was very much a Weimar-culture phenomenon. Its community was uncharacteristically open, affording a young person like Courant opportunities that would have been unthinkable elsewhere. As the historian Peter Gay long ago pointed out, Weimar culture already took root during the Wilhelmian period (Gay 1968). Richard Courant was hardly the type of figure Gay had in mind, and yet the subtitle of his book fits him perfectly: "the outsider as insider".

In thinking about Courant's career, I am confronted by the difficulty of saying something about how he fits into the larger story suggested by the exhibition. Opinions about Courant varied sharply, and in some ways his character and approach to mathematics represents an anomaly – he was at once a daring innovator as well as a conservative with a deep belief in the vitality of older traditions. Courant identified very strongly with the Göttingen mathematical tradition he grew up with. If he was in some ways linked to Weimar culture, he had none of the oedipal urges that Gay saw in the German Expressionists, or that surfaced in Weyl's relationship with Hilbert (see Chap. 27 and Weyl 1944). Quite the contrary, he was the obedient son who honored his fathers. He lived in and for the Göttingen mathematical tradition, but not as a static relic of the past. He helped transform it into something that lived on right through the Weimar period, and in Courant's mind at least, he continued to carry this mission with him when he started all over again in New York. Courant's arrival at NYU in 1934 was his own personal *Stunde Null.*

[1] Reinhard Siegmund-Schultze points to the great importance of Breslau for German mathematics and culture in (Siegmund-Schultze 2009, xviii).

[2] "I have now given my lecture, but as far as Hilbert is concerned, I was not lucky. Landau and the number theorists liked it very much, as they also expressed, while Hilbert was interrupting me frequently. . . . I could not finish my talk and present the last results of my dissertation . . . Hilbert has spoiled my joy for work completely . . ." quoted in (Frei 2004, 270).

[3] The anti-Semitic philosopher Hugo Dingler harboured deep suspicions for three decades regarding Hilbert's favouritism toward Jews, see (Wolters 1992, 273) and (Rowe 1986, 422–424).

D.E. Rowe, *A Richer Picture of Mathematics*, https://doi.org/10.1007/978-3-319-67819-1_28

Richard Courant would surely have been surprised that he would be remembered today at NYU as a "Jewish mathematician from a German-speaking academic culture." Probably in later years he thought of himself as a "German mathematician in an academic world that spoke *amerikanisch*." Clearly, his own sense of Jewish identity was far less strong than that felt by other prominent contemporaries he knew, for example Einstein, but also Edmund Landau or Otto Toeplitz. He was one of those German Jews who suffered greatly when the Nazis rose to power precisely because he identified so strongly with German culture.[4] Still, there can be no doubt that he also recognized his outsider status; even as director of the Göttingen institute he lived in a kind of bubble. Town and gown were anything but philo-Semitic, and those in the so-called Courant clique engendered a great deal of resentment (Rowe 1986, 445–449). The fact that Courant was a war veteran, who nearly died fighting for Imperial Germany, made no difference; to those who never got over their country's defeat, he was just another symbol of what had gone wrong: too many Jews.[5]

After his appointment as Klein's successor in 1920, Courant continued to promote the interplay between pure and applied mathematics, especially by forging a close alliance with his colleagues in physics, Max Born and James Franck. All three were of Jewish background, which led to heightened tensions within the philosophical faculty. Hilbert and his allies had early on fought ferociously with their more conservative colleagues. Many in the latter camp felt that Courant, Born, and Franck owed their appointments to the turbulent situation in Germany immediately after the war, an attitude that afterward fuelled deep resentments and occasional open conflicts (Fig. 28.1).[6]

Fig. 28.1 Richard Courant in Göttingen (Reid 1976).

Anti-Semitism burst out into the open during the early years of the Weimar Republic. A prominent target then was Albert Einstein, who had close ties with the Göttingen community, especially with Hilbert. This was the period when Einstein discovered his own sense of Jewish identity (Rowe and Schulmann 2007, 136–171). In 1921 he came to New York, along with Chaim Weizmann, to raise money for the founding of a Hebrew University in Palestine. Asked by a journalist about the causes of recent anti-Semitism in Germany, Einstein offered these remarks: "To some degree, the phenomenon is based on the fact that Jews exert an influence on the intellectual life of the German people altogether out of proportion to their numbers. While in my opinion the economic position of the German Jews is vastly overestimated, Jewish influence on the press, literature, and science in Germany is very pronounced and obvious to even the casual observer." (Rowe and Schulmann 2007, 150).

Einstein had a keen eye for the tensions in this post-war climate. Having experienced politicized attacks by German physicists against his theory of relativity, he also realized that anti-Semitism was not merely confined to uneducated street thugs. Courant hoped it would fade away. Even when he and his friends were forced to step down in 1933, he clung on, hoping for another chance (Siegmund-Schultze 2009, 167–170). He was also most unhappy that Einstein, who was spending the winter of 1932–33 in Pasadena at Cal Tech and who would never again set foot in Germany, had issued public statements criticizing the policies of the new National Socialist government. Writing to his friend James Franck on 30 March, 1933, Courant vented his anger over this: "Even if Einstein does not regard himself as German, he has experienced a lot of good in Germany. So he should feel obligated to make amends for the trouble he has caused as far as he can" (quoted in (Siegmund-Schultze 2009, 84)).[7]

After the Nazis came to power, this so-called "Courant clique" was quickly singled out and dismissed as part of a policy to "purify" the German civil service. Courant and his friends tried to fight this dismissal, pointing not only to his patriotism during the Great War but also to his many achievements as director of the Göttingen Institute of Mathematics (Reid 1976, 143–152). To no avail, of course, a

[4]See (Siegmund-Schultze 2009, 167–170) and (Reid 1976, 142–163).

[5]On the destruction of Courant's institute in 1933 and the infighting thereafter, see (Schappacher 1991) and (Schappacher 1987).

[6]See (Rowe 1986, 438) and (Schappacher 1987, 346–349).

[7]Courant's attitude toward Einstein at this time was common among German Jews, as vividly revealed in a letter that Elsa Einstein wrote to her friend Antonina Vallentin in April 1933: "The greatest tragedy in my husband's life is that the German Jews make him responsible for all the horrors that happen to them over there. They believe he has provoked it all and in their resentment have announced their total dissociation from him. We get as many angry letters from the Jews as we do from the Nazis." (Vallentin 1954, 224).

turn of events that left their leader devastated and depressed. A few years later, his eyes now fully opened, he realized that he had been lucky that his next chance came while he was living in a country that would eventually go to war against Hitler's Germany.

Biography and Oral History

After Constance Reid completed her biography of Hilbert in 1969, Kurt Friedrichs asked her if she would be willing to help Courant write his memoirs. She had interviewed him earlier for her Hilbert book, which contains a number of favorite Courant stories about his fabled hero; the man Hermann Weyl called the "Pied Piper of Göttingen" (Weyl 1968, vol. 4, 132). After a brief meeting in New Rochelle, Courant reluctantly assented to this idea. He had been happy with the way she handled the story of Hilbert's life, so he no doubt found it hard to object to Friedrichs' plan (Reid 1976, 1–2). Reid soon discovered, however, that this was not going to be an easy task. In fact, when her Courant book came out in 1976, she admitted that it was something rather different than a conventional biography. Courant loved to talk about *certain parts* of his life, but when it came to other parts – such as his first 20 years, growing up in poverty in Breslau – he seemed to remember little (Reid 1976, 3–5).

Most of what Reid learned about those years came from written sources: Courant's father's unpublished memoir and the family chronicle written by Edith Stein, Richard's cousin (Stein 1965). Nina Courant née Runge tried to help, but she had to admit that Richard generally did not like to talk about his own life, least of all the many hardships he had endured; he was always looking forward, she said, not backward. Probably that applied to Nina, too.[8] Still, to understand Richard Courant's life one must take into account the hardships he endured before he became engaged to Nerina Runge.

Richard's problems began when his father, Siegmund, was blamed for his older brother's suicide following ill-fated business ventures that left the family bankrupt. Estranged from his siblings, Siegmund decided to leave Breslau and settle in Berlin. He took his wife and their two youngest sons with him, while Richard, who was only 16, remained behind. He lived alone, but occasionally visited the home of his paternal aunt, Auguste Stein, a widow who ran the family lumber business after her husband's death. Edith Stein was the youngest of her eleven children. Edith was also a close friend of Courant's first wife, Nelli Neumann, a talented mathematician who took her doctorate in Breslau in 1909, one year before him. She then went on to take courses that qualified her to become a secondary school teacher. But in 1912 she decided to marry Courant, who was teaching as a *Privatdozent* in Göttingen. Nelli grew up in Breslau as the only child of a judicial official, so Richard had married up: for the first time in his life he enjoyed some modest financial security. The marriage, however, went badly from the start. Nelli grew lonely, and so she asked her friend Edith to come to Göttingen. All this and more can be found in Edith Stein's *Aus dem Leben einer jüdischen Familie* (Stein 1965), which offers a vivid portrait of academic life in Göttingen before and during the war. She studied philosophy there under Edmund Husserl, eventually taking her doctorate *summa cum laude*.

When the Great War broke out, Courant was quickly called to serve in the army. Nelli then decided to leave Göttingen and went back to Breslau to live with her father. So Edith inherited the unhappy couple's apartment, thanks to which we have her fairly detailed description of its more than ample accommodations. Clearly, these had brought no joy, and so Nelli and Richard officially divorced in 1916. After the war she taught at a girls' school in Essen, but lost that position when the Nazis took power in 1933. Although a baptized Protestant, she was, according to Nazi law, a Jew. Nelli Neumann was later deported to Minsk where she was executed in 1942. Her friend Edith Stein, who converted to Catholicism in 1922 and afterward taught at a Catholic school in Münster, suffered the same fate. She died in Auschwitz. Richard Courant had some 30 cousins who lived in Germany in 1933. Of these, 19 left and went to live on five different continents, four managed to survive in Berlin, two committed suicide, and five died in gas chambers (Reid 1976, 247). He was a survivor, and he did not wish to dwell on the past. More I can't presume to say.

When I first came to New York in 1981 to study history of mathematics with Joe Dauben, I rented a room in New Rochelle that turned out to be right around the corner from the Courants' house on 142 Calton Road. So I introduced myself to Nina, who was still going strong at 90. I told her about my interest in Göttingen mathematics, especially Felix Klein, and she was happy to tell me about those days. She still had plenty of memories and I still have notes from one of our first chats. She dug around her house and offered to give me copies of whatever documents she could find. Best of all, she also invited me over when she and her daughter Lori were making music together with friends. So I soon got to meet a number of the women in this extended Courant family, including Nellie Friedrichs. She had just published her moving memoir of her early years in Braunschweig (Friedrichs 1988). This ends with the dramatic story of how she and her future husband – "Frieder" as she lovingly called him – plotted their successful escape from Nazi Germany.

The atmosphere in New Rochelle in the early 1980s was no longer so idyllic, but visiting Nina Courant's home was

[8]She told me in 1981 that she had been excited by the great adventure of going abroad and living in the US.

nevertheless like going back in time. A portrait of Hilbert hung in the dining room, and another Göttingen mathematician could be seen in the living room, near the grand piano: Nina's father, Carl Runge. She told me about how much she had enjoyed the adventure of coming to a new country: it was like starting life all over again. Her husband was always busy, of course, but he really loved it when he could get away from the city; their home was a real sanctuary for him, a place he could unwind, make music, and yes, even occasionally do mathematics.

I also got to interview Kurt Friedrichs back then (Chap. 30), a memorable occasion for me. He still had vivid memories of Göttingen, going back to 1922, the year he first arrived on the scene. Friedrichs had a keen sense for mathematical traditions and how they were transformed over the course of his career. His picture of Courant clearly dominates throughout Constance Reid's book. Richard Courant was a truly enigmatic figure, even for many who knew him well. His mathematical tastes spanned the gamut from classicism to romanticism, and while he admired number theorists like Edmund Landau and Carl Ludwig Siegel, his own work was eclectic, at times even sloppy. As Friedrichs once put it: Courant "always considered himself the mathematical son of Hilbert – and he always played down what he owed to Klein – but in fact he was the son of Klein" (Reid 1976, 241).[9] Friedrichs thought of him as "Hilbert in the spirit of Klein," an image that seems apt if we remember that Hilbert was fundamentally an algebraist, whereas Klein was a geometer, and yet both published heavily in analysis. So Courant, who was a *real true-blooded* analyst, could indeed see himself as carrying the heritage of Hilbert forward, even though his romantic style was far closer to Klein's.

Courant's research on conformal mapping, boundary-value problems, and Riemann surfaces combined methods and perspectives he learned from both of his mentors.[10] He admired the ideals these two giants stood for and later, after he assumed the reins of power, he strove to conserve the core values that lay at the heart of their Göttingen tradition. Yet his interests in applied mathematics reflected a broader outlook, a view consonant with the work of two other Göttingen figures, Ludwig Prandtl and Carl Runge. Their arrival in Göttingen in 1904 represents the crowning achievement in Klein's efforts to wed mathematics with modern scientific and technological developments, a move that eventually led to a longer-term transformation away from traditional pure research (Rowe 1989). Nevertheless, Courant remained open to both directions, pure and applied; thus he also took a deep interest in fields like analytic number theory, a discipline cultivated by Göttingen's Edmund Landau. Courant resembled Klein in another important respect: he liked to appeal to the vitality of an idealized Göttingen mathematical tradition. When Felix Klein attended the Chicago Congress in 1893 he spoke of wanting to return to the great tradition of Gauss and Wilhelm Weber. On other occasions, Klein held up Riemann as the key figure in the Göttingen tradition (Parshall and Rowe 1994, 310–312). For Courant, the great names were Riemann, Klein, and above all Hilbert. Courant never tired of telling stories about Hilbert, and of course a good story deserves to be exaggerated, at least a little.[11]

Counterfactual Courant Stories

I've long been interested in the oral dimension of Göttingen's mathematical culture, and Courant's story-telling clearly reflects part of that larger phenomenon. Courant happened to be Hilbert's assistant in January 1909 when Minkowski's sudden death shocked the Göttingen community. For Hilbert, the loss was devastating. Choosing Minkowski's successor also proved to be a delicate matter, and for a number of reasons. So Courant surely felt very privileged to learn about the faculty's confidential deliberations directly from Hilbert. After all, he was only 21, a mere student still without his doctorate.[12] Yet sitting in Hilbert's garden one day he learned that the decisive meeting had just taken place. The choice, Courant remembered, came down to two candidates: Edmund Landau and Oskar Perron. Landau was considered, in the words of Reid, "a brassy, rich 'Berlin Jew' while Perron was generally well liked" (Reid 1976, 25) (Fig. 28.2).

I once asked Constance Reid about this story, and she assured me that she had heard it from Courant on several occasions. Clearly, this was one of his favorite anecdotes, because he also told it to a group of historians of science at Yale when he spoke there in 1964 (Courant 1981). Edmund Landau was presumably unfamiliar with the finer details of the faculty deliberations, but if he ever heard Courant's rendition of what happened we can still easily imagine what he might have said: "ist aber falsch!"[13] Why? Because there were three candidates on the faculty's final list, and

[9] Courant's speech (Courant 1925) honoring Klein after his death reinforces what Friedrichs meant by this.

[10] For an overview of Courant's mathematical work, see (Lax 2003).

[11] For an impression of Courant as a story-teller, see (Courant 1981).

[12] From a letter Courant wrote to Otto Toeplitz, we know that he accompanied Hilbert on a trip to the Ministry of Culture in Berlin. He was allowed to be present at this meeting, during which Hilbert spoke in favor of four potential candidates for a professorship: Max Dehn, Ernst Hellinger, Issai Schur, and Toeplitz; all four were of Jewish background (Courant to Toeplitz, 27 August 1910, Toeplitz Teilnachlass, Universitäts- und Landesbibliothek Bonn Abteilung Handschriften und Rara).

[13] Dirk Struik recalled how Landau loved to embellish his lectures with mathematical truisms that turned out to be untrue in special cases; the explanation typically followed the proclamation: "ist aber falsch" (see Chap. 32).

Fig. 28.2 Edmund Landau, Göttingen's leading representative of pure mathematics.

Perron was not among them.[14] Still, Courant's story was really about Klein, not Perron, and here's what he said happened: At the decisive meeting – and to the surprise of his colleagues – Klein spoke in favor of Landau. "We being such a group as we are here, it is better if we have a man who is *not easy*" (Reid 1976, 25). Such a remark seems to ring true, especially if we remember what Otto Blumenthal wrote regarding a similar meeting in December 1894, the session in which Klein pushed through Hilbert's nomination. Klein told Blumenthal an anecdote about what transpired at that meeting when Klein's colleagues criticized him for wanting to appoint a younger man with whom he could get on comfortably. To this, Klein let them know something about Hilbert's character when he said, "ich berufe mir den allerunbequemsten" ("I want the most uncomfortable of them all!"), (Blumenthal 1935, 399).

[14]The faculty's list can be found in Universitätsarchiv Göttingen, UAG.Phil.II.36.d, Besetzung von Professorenstellen.

Filling Minkowski's position was, however, not a matter of choosing between two outsiders, Landau and Perron. Even more significantly, the three candidates under discussion were all Jews, at least by birth. Nor did the faculty even bother to rank them; all three were listed aequo loco. So there was no decisive meeting after all. Courant had been mistaken about that it seems. Still, there was a clear choice to be made between the outsider, Landau, and two insiders, Blumenthal and Adolf Hurwitz, both of whom had close ties with Hilbert and Klein. Both were also known to be warm, courteous, and genuinely likable personalities, unlike the abrasive and arrogant Landau. So Courant's story fits, except that Klein's *Sachlichkeit* had nothing to do with choosing a Jew over a non-Jew. Perhaps Courant never knew what was really at stake and why people were surprised by Klein's remarks.

A far more serious confusion, however, surrounds the circumstances that led to Courant's own professorial appointment in Göttingen in 1920, a story that has much to do with a former rival, Hermann Weyl. Among Hilbert's many distinguished students, Weyl stood in a special category all his own. His personal relationship with Hilbert, on the other hand, was highly ambivalent, in part because Weyl much preferred research to teaching (see Chap. 27). After joining the faculty at the ETH in Zürich in 1913, he afterward turned down a series of attractive offers from leading German universities, preferring to remain in Switzerland until 1930 (Frei and Stammbach 1992). In that year he was offered Hilbert's chair, an honor even Weyl could not refuse. Courant, of course, knew Hermann and Hella Weyl very well from their student days in Göttingen. Later in life, however, he may well have forgotten that his own career owed much to Weyl's reluctance to leave the beautiful surroundings of Zürich for the buzz-saw of mathematical activity in Göttingen. In this respect, he was the polar opposite of Courant, who loved to be at the center of action.

In Reid's biographies of Hilbert and Courant (Reid 1970, 1976), she tells a curious fable about a new professorship that Courant "apparently negotiated" in 1922; this was supposedly offered to Weyl, who then declined (Reid 1976, 90). One might naturally wonder how the Prussian Ministry of Education could have funded a new professorship in the midst of the hyperinflation and general scarcity of financial resources; moreover, if Weyl had in fact turned this position down, why was it then not offered to someone else? But, in fact, the true situation can be easily clarified and corrected: the year was 1920 and the position was Klein's former chair, the professorship Courant would ultimately obtain (Frei and Stammbach 1992, 33–48). Moreover, contrary to what one finds in Reid's book, Courant's call to Göttingen came about not through some carefully orchestrated plan hatched by Klein and Hilbert, but rather as the result of a complicated series of events that no one could have foreseen at the time. The actual course of ongoing negotiations in both Göttingen and Berlin can, in fact, be reconstructed from extant ministerial and faculty records, sources we can assume to be far more reliable than human memory. These documents not only clarify the chain of events that led to Courant's appointment but, even more, they throw fresh light on the surrounding circumstances, in particular, the truly abysmal living conditions in Germany at this time.

Given the prestige attached to these two vacant professorships, the faculties in Göttingen and Berlin naturally set their sights on the most accomplished mathematicians of the day. Both universities focused on three outstanding candidates: the Dutch topologist L.E.J. Brouwer, Leipzig's Gustav Herglotz, and Hermann Weyl. In Göttingen, these three were nominated in just that order, whereas the Berlin faculty placed Weyl after Brouwer and ahead of Herglotz.[15] Clearly, a strong consensus of opinion had been reached about these three men, but then something happened that would have been unthinkable in earlier times: all three candidates turned down *both offers*, preferring to remain in Amsterdam, Leipzig, and Zurich, respectively. In view of the ongoing political unrest in Berlin, which culminated with the unsuccessful Kapp Putsch in March 1920, one can easily understand their reluctance to reside in the Prussian capital. Weyl eventually dismissed this possibility, but not the idea of leaving Switzerland for Göttingen. It took him nearly 6 months before he finally declined, thereby opening the way for Courant's dark horse candidacy (Frei and Stammbach 1992, 46–48).

In the meantime, Courant's personal ties to Göttingen had become stronger than ever. Immediately after the war he was eking out a living as an *Assistent* to Carl Runge, Göttingen's Professor of Applied Mathematics. His relations with the Runge family grew even closer when in January 1919 he married their daughter, Nina. Housing being scarce, the newlyweds resided with her parents, and early the next year Nina gave birth to their son Ernst. Not long afterward, Courant was offered a professorship in Münster, the chair formerly occupied by Wilhelm Killing. He accepted, despite the drudgery of travelling back and forth from Göttingen. At this time he had no idea that he might be offered Klein's former position, though he probably knew that Hilbert and Klein were agitated over Weyl's inability to reach a decision. Still, there was no inside plan to recruit Courant from Münster, contrary to Reid's version of the ensuing events (Reid 1976, 78). In fact, the extant documentary evidence suggests a very different picture.

When Hecke left for Hamburg in the fall of 1919, Arthur Schoenflies wrote to Hilbert, offering him advice about potential candidates.[16] As a former protégé of Klein and Hilbert, Schoenflies was well aware of their general views regarding academic appointments. He thus left Courant's name off his list on the assumption that he could not be promoted from a mere titular professor, a status he acquired in 1918, to an *Ordinarius*. Even more to the point, Schoenflies explicitly noted that such a nomination would contravene the principle prohibiting *Hausberufungen* (in-house selections) since Courant had never held a position outside Göttingen. Schoenflies thus understood very well that Courant had strong support, but he also knew that his candidacy would have encountered great resistance, if only on purely formal grounds.

Schoenflies did not need to raise another inevitable hurdle, one that that he, as a Jew, knew all too well (Rowe 1986, 433–436). The philosophical faculty in Göttingen had long been open to accepting Jewish colleagues with the understanding that there should never be more than one in a given field. Thus when Minkowski suddenly died in 1909, he was succeeded by Landau, one of three Jews nominated for the position (the others were Blumenthal and Hurwitz). This suggests an implicit understanding that Minkowski's chair – which was created especially for him in 1902 to induce Hilbert to stay in Göttingen and turn down an offer from Berlin – was reserved for Jewish candidates. This situation clearly posed a potential obstacle for Courant's appointment to one of the other three chairs in mathematics. Thus, to gauge what was at stake here in 1920 one must also take into account the larger issue of the "Jewish question" as this relates to career opportunities in mathematics (see Chap. 15).

After the turn of the century, German universities had gradually drawn large numbers of talented Jewish students, many of whom excelled in fields like mathematics and theoretical physics. Institutions of higher education, however, were none too eager to employ them, particularly recent arrivals from the east. By the end of the war, several distinguished Jewish mathematicians still awaited their first regular academic appointment. A few had been passed over on several occasions in favor of Christian candidates; they were either stymied by resistance at the faculty level or occasionally at the higher level of the state ministries. Max Born, who had studied alongside Courant in Göttingen, managed to gain an appointment in Berlin during the war as an associate professor of theoretical physics (Greenspan 2005, 74–86). This was a fairly new field of research, spawned over the course of the preceding 30 years at several German universities. During the Weimar era, this field came to symbolize the ascendency of Jews in German science, spearheaded by figures like Born (Jungnickel and McCor-

[15]The complications in Berlin are described in (Biermann 1988, 192–194). The original list of candidates for Göttingen can be found in Rep. 76 Va Sekt. 6, Tit. IV, 1, Vol. XXVI, Bl. 423–424, Geheimes Staatsarchiv Preußischer Kulturbesitz.

[16]Schoenflies to Hilbert, 1919, Hilbert Nachlass 355, NSUB Göttingen.

mmach, 1990). As a side benefit of his appointment, Born got to strike up a useful friendship with Albert Einstein, since both were determined to do what they could to promote the careers of their kinsmen in mathematics and physics. Thus, in 1919 Einstein wrote to Felix Klein urging him to take steps with the Prussian Ministry of Education so that the brilliant female mathematician, Emmy Noether, would finally be given the title of *Privatdozent* in Göttingen.[17] Her lecture courses on modern algebra soon thereafter attracted a throng of enthusiastic young talent.

Born was no Einstein, but he nevertheless did his part to promote the cause of Jewish academics. In 1919 he wrote to Carl Heinrich Becker in the Prussian Ministry that the time had come to level the playing field, pointing to Courant's case as a glaring example of past injustices.[18] That same year a chair in mathematics came open in Halle. Hilbert was contacted by the physicist, Gustav Mie, who sought advice about prospective candidates. Hilbert named Isaai Schur, Paul Koebe, and Courant in that order.[19] Nevertheless, the chair instead went to Heinrich Wilhelm Jung, who had already been full professor in Kiel since 1913. Schur may well have been offered this position, however, for in 1919 he attained a long-sought promotion to full professor in Berlin, where he had been the star pupil of Frobenius. In 1917 the Berlin faculty had already placed him *aequo loco* with Carathéodory, hoping that the Ministry would appoint both. Instead the Greek was chosen over the Jew; as a sign of the times, Isaai Schur, a Russian-born Jew, was nominated no fewer than nine times to various chairs at German universities before he was finally appointed to a full professorship (Biermann 1988, 196).

A particularly striking instance illustrating the tensions aroused by this backlog of talented Jewish mathematicians can be seen from the private correspondence of Otto Toeplitz, who contacted several leading Göttingen mathematicians in early 1920 about a vacancy in Kiel. Like Courant and Born, Toeplitz came to Göttingen from Breslau, where his father taught mathematics at a local Gymnasium. He and Ernst Hellinger, both Jews from Silesia, became leading experts on Hilbert's theory of integral equations. In 1913 Toeplitz took a post as associate professor in Kiel, where he was promoted to full professor after the war. When his colleague, H. W. Jung, decided to accept the call to Halle, Toeplitz found himself in the unenviable position of serving on a commission charged with nominating candidates for this vacant chair in mathematics. As was surely expected, he began by seeking the advice of senior colleagues in Göttingen, including Felix Klein. Toeplitz wished to learn what Klein thought about five particular individuals – (1) Schur, (2) Steinitz, (3) Lichtenstein, (4) Hellinger, and (5) Bernstein – all of whom happened to be Jewish.[20] In responding, Klein added some remarks questioning the wisdom of this approach, particularly in the present political climate:

> . . . on the one hand we have not only the enormous advance of Jews as a result of their peculiar abilities but also through the rise of Jewish solidarity (where Jews seek in the first instance to help and support their clansmen in every way). On the other hand, in reaction to this, we have rigid anti-Semitism. This is a universal problem in which Germany plays only a secondary role, if we leave the new immigration from the East out of consideration. No one can say how things will develop (Klein to Toeplitz, 4 February, 1920, (Bergmann/Epple/Ungar 2012, 206)).

Klein, who was well aware of past injustices, commented further: "One could also almost argue that the flourishing of anti-Semitism at all universities has given Christian candidates such an advantage that only Jewish candidates are now available. But I ask you, please, to think about it again. We are potentially entering into conflicts that could become disastrous for our situation as a whole." (*ibid.*)

Toeplitz was more than a little surprised by the frankness of Klein's letter, but he responded in kind by explaining some of the special circumstances in Kiel. In the meantime Toeplitz had received a letter from Hilbert, who wrote in praise of Ernst Steinitz, also naming Felix Hausdorff, Ludwig Bieberbach, and Leon Lichtenstein as worthy of a second or third place on the list. Hilbert seems to have given little weight to the issue of racial background, though Bieberbach was "Aryan" of course.[21] His colleague, Edmund Landau, saw this as a potential problem, however, and thus advised Toeplitz to add one or two non-Jewish names, even if they were not likely to accept an offer from Kiel. Landau, who would later suffer the indignity of having his lectures boycotted by young Nazis, no doubt shared some of Klein's misgivings. Young Richard Courant felt otherwise. Like Born and Einstein, he thought the Prussian Ministry appreciated the predicament of young Jewish mathematicians and physicists. So Courant counseled Toeplitz to go ahead with his plan: he and his colleagues should submit a "racially pure" list of nominees, counting on Berlin to cooperate. Soon thereafter, Steinitz received the call to Kiel.

These background events at a provincial university would hardly be worth describing in such detail were they not symptomatic of much larger issues clearly reflected in the exchanges cited above. Ethnic and religious factors had always played a major role in academic appointments at the

[17]Einstein to Klein, 27 December, 1918, Klein Nachlass 22B, NSUB Göttingen; cited in (Rowe 1999, 198).

[18]Born to C. H. Becker, 1919, Geheimes Staatsarchiv Preußischer Kulturbesitz, I.HA.Rep.92. C.H.Becker, 7919.

[19]Hilbert to Mie, 1919, Hilbert Nachlass 487, NSUB Göttingen.

[20]Toeplitz to Klein, 11 February, 1920, Klein Nachlass, NSUB Göttingen.

[21]Hilbert to Toeplitz, 8 February, 1920. Universitäts- und Landesbibliothek Bonn, Toeplitz B: Dokument 47. Hilbert mistakenly thought that Hausdorff, too, was of non-Jewish background.

German universities, but in this new political climate the "Jewish question" took on a special urgency that strongly shaped and influenced concurrent deliberations over suitable candidates including the two positions that remained to be filled in Göttingen and Berlin.

In the meantime, the situation in Göttingen had become quite complicated due to the departure of the Dutch theoretical physicist Peter Debye, who chose to accept an attractive offer from the University of Zürich. Debye had worked quite closely with Hilbert, who was intent on finding a suitable successor. His first choice was Max Born, now teaching in Frankfurt. In mid February Hilbert wrote to Einstein, asking him to send a letter assessing Born's abilities as well as his suitability for the position in Göttingen. Einstein was happy to sing the praises of his friend, whom he once regarded as primarily a mathematical talent. Einstein now thought, however, that Born's more recent work showed a strong sense for physical reality.[22] This letter, written the very day the Philosophical Faculty convened, may well have given Hilbert the ammunition he needed. In any case, Born's name appeared second on the faculty's list, behind Arnold Sommerfeld's. No one imagined the latter would be tempted to leave Munich, as proved to be the case, so Born quickly emerged as the candidate of first choice (Greenspan 2005, 95–99).

Born afterward wrote Einstein for advice, but his friend only assured him that "theoretical physics will thrive wherever you are" (Born 1969, 47). He then plunged into a series of complex negotiations with the Berlin Ministry as well as the Göttingen faculty. Hilbert had already signaled to Born that he would have the opportunity to recommend an experimental physicist to fill another vacancy, so he already had a bargaining chip in hand. He played it forcefully by making plain that he would not leave Frankfurt unless the Göttingen faculty agreed to a double appointment; furthermore, he insisted that the second chair in experimental physics had to be offered to James Franck, then director of the physics division at the Kaiser Wilhelm Institute for Physical Chemistry in Berlin. Franck found this plan highly amenable, but various complications quickly ensued. Since both men were of Jewish background, this bold venture was bound to encounter resistance within the Philosophical Faculty, more, in fact, than Born had bargained with. As it turned out, the negotiations dragged on for several months. Some years later, Hilbert recalled how Born's appointment proved to be "the most ruthless and hardest fight [he] ever had to endure in the faculty."[23] There had been many such fights, in fact. Hilbert had a well-deserved reputation as a fearless warrior when it came to academic politics, a prime reason why he was much admired by those in the liberal camp and so loathed by his conservative colleagues.

This particular battle had not yet ended when, in early July, Weyl's letter finally arrived; after much soul-searching he decided to reject the Göttingen offer.[24] Now that the original list of candidates had been exhausted, the idea of calling Courant from Münster could at last come into play. Klein decided to lay all his cards on the table. He composed a letter to Courant, which he read in Hilbert's presence, setting forth the mutual understanding he assumed all three of them shared. This began:

> As you may have heard from other sources, I intend to advocate your appointment in Göttingen. It would be extremely helpful for me if you would confirm explicitly in writing that you are willing to promote energetically tasks which, in my opinion, have long been unduly neglected in our educational system as well as new demands which I can foresee as coming up" (Reid 1976, 83).

He then proceeded to enumerate which reforms he had in mind, and summed up by saying he was sure that none of these points would come as any surprise. Klein thereby obtained the proper assurances from Courant, who surely realized he would be assuming an awesome responsibility.

Klein and Hilbert now took their case to the faculty, but there they encountered a potential roadblock: Edmund Landau was not to be persuaded.[25] Landau saw no reason to doubt Courant's abilities, but he expressed strong reservations with regard to what he perceived as an unhealthy trend in Göttingen, one that was creating an imbalance between pure and applied mathematics. As a number theorist, Landau had long felt isolated in a community where analysis, mathematical physics, and applied mathematics dominated the scene, so he saw no reason to appoint yet another applied type like Courant. Instead he pushed for a pure mathematician, nominating Berlin's Isaai Schur in a strongly supportive letter. This went out to the Ministry on 12 July (just four days after Weyl had declined the offer) together with the counter-proposal, signed by Klein and Hilbert, with very different arguments in favor of Courant (including his bravery during the war).[26] Even now, no one could have been sure that the Ministry would agree to either of these two candidates, though soon thereafter Courant received the good news.

[22]See Hilbert to Einstein, 19 February 1920, and Einstein to Hilbert, 21 February 1920, (Einstein CP9 2004, 334, 440).

[23]Hilbert to Hermann Wagner, 1926, Cod. Ms. H. Wagner 27, Niedersächsische Staats- und Universitätsbibliothek (SUB) Göttingen.

[24]Weyl later recalled how he was still undecided the very day he sent a telegram to Göttingen declining the position (Weyl 1968, 650). Reid was under the mistaken impression that this took place two years later in 1922.

[25]Documentation on the faculty deliberations can be found in Universitätsarchiv Göttingen, UAG.Phil.II.36.d, Besetzung von Professorenstellen.

[26]The two faculty recommendations can be found in Ministry, Rep. 76 Va Sekt. 6, Tit. IV, 1, Vol. XXVI, Bl. 427–431 (signed by Landau) and Bl. 432–434 (signed by Klein and Hilbert), Geheimes Staatsarchiv Preußischer Kulturbesitz.

What transpired afterward in Berlin would also eventually have profound consequences for mathematics in Germany. Following the initial failure to fill Carathéodory's chair, the Prussian Ministry opened negotiations with Hamburg's Erich Hecke. However, he too declined, forcing the Berlin faculty to reconvene in order to start the search process all over again. It took until the end of 1920 before they could agree on a new list (Biermann 1988, 193–194). This time they named the Austrian geometer, Wilhelm Blaschke, Frankfurt's Ludwig Bieberbach, and the geometer Gerhard Hessenberg, who taught in Tübingen. After Blaschke declined the position, Bieberbach agreed to accept the post, one that accorded with his ambitions and inflated self-esteem (*ibid.,* 197–198). These personal attributes would become increasingly evident over time.[27] Bieberbach was, in many respects, the polar opposite of the more unassuming Courant, though they managed to stay out of each other's way for a long time to come. With Courant's return, followed by the double appointment of Born and Franck, Göttingen suddenly acquired an impressive trio of talent; they were not only gifted but, just as importantly, all three got along with each other exceptionally well. That they all happened to be secular Jews did not escape notice either; each got to know firsthand about various forms of local anti-Semitism.[28]

Courant as Innovator

Courant's sense of loyalty to Klein, Hilbert, and Runge ran very deep. No doubt his sincerity and sense of belonging was fully appreciated when they chose him. Diminutive and soft-spoken, Richard Courant must have appeared as the unlikeliest imaginable successor to Felix Klein, and yet he promoted the legacies of both Klein and Hilbert brilliantly. As a pupil of Hilbert, he took up classical analysis—variational methods, Dirichlet's principle and conformal mapping—a program that kept him busy all his life. What he accomplished in Göttingen was due in large part to his ability to build on the shoulders of Klein and Hilbert, the giants who dominated the scene during the pre-war years.

In choosing the title "transforming tradition," I have in mind a rather subtle process in which things clearly change and get transformed, but hardly in line with some great master plan. Courant was a brilliant innovator, but he was also in many ways a traditionalist, and this makes him an especially intriguing figure. His famous yellow series, for example, was largely drawn from or inspired by the lecture courses of former years. Even just the names of his co-editors—Blaschke, Runge, and Born—are enough to make one realize that this post-war project had a distinctly conservative character.

In the period from 1919 to 1925 mathematics publishing took on a vital new importance for Germany, both scientifically and economically. In an era of growing international contacts, German mathematicians and scientists were generally barred from attending congresses and meetings held in the countries of their wartime enemies. Many thought of German science as the last bastion of national prestige, yet this sphere of power, too, was clearly vulnerable, particularly if the products of German intellectual activity never found their way to the marketplace. Engineering and the applied sciences were hard pressed, but in the case of an ivory-tower field like mathematics the situation was particularly acute given the adverse political climate. A more aggressive approach to marketing the products of German mathematicians and scientists was needed, an approach embodied in the business practices of the firm of Julius Springer. Taking advantage of the vacuum created when B. G. Teubner pulled back from the mathematics market after the war, Springer soon emerged as a bold new player in this small niche within the publishing industry, promoting a surge in productivity that gave "mathematics made in Germany" an enduring allure (Remmert and Schneider 2010).[29]

Courant had already met Ferdinand Springer during the war; he was then temporarily stationed in Ilsenburg, a village in the Harz Mountains, working on terrestrial telegraphy (Sarkowski 1996, 262). This meeting, which took place on 28 September 1917, was facilitated by the editor of *Die Naturwissenschaften*, Arnold Berliner, whom Courant had known growing up in Breslau.[30] By the following year, plans for Courant's *Grundlehren der mathematischen Wissenschaften* – better known as the "yellow series" or, in Göttingen, as the "yellow peril" ("die gelbe Gefahr") – were already underway. Courant not only lined up Hilbert's support for this project, he also persuaded Hamburg's Wilhelm Blaschke and his father-in-law, Carl Runge, to join him as associate editors. By 1921 the first volume, Blaschke's *Vorlesungen über Differentialgeometrie, I*, was already in print with several more due to follow. That same year Springer opted to put Courant on his payroll as a consultant; he was

[27]For a portrait of Bieberbach's career, see (Mehrtens et al. 1987).

[28]These experiences would come to haunt them in the United States, too. In the winter of 1935–36 Courant received a disturbing letter from Franck, who urgently warned him not to become overly involved with helping European émigrés, adding that "we must not forget that we once deceived ourselves about the safety of the ground we were living on" (Franck to Courant, 1 November, 1935, quoted in (Siegmund-Schultze 2009, 212).

[29]The famous "yellow series" founded by Courant in 1920 continues to occupy a central niche in Springer's publishing program, though its character changed quite dramatically after 1945 when English became the dominant language for international publications in mathematics.

[30]Berliner's *Die Naturwissenschaften* often published articles celebrating Göttingen mathematics, such as those published in January 1922 for Hilbert's sixtieth birthday (see (Rowe 2012)).

paid the generous sum of 1500 marks (ca. 450 gold marks) quarterly (Sarkowski 1996, 264).

Courant's yellow series had just been launched when Otto Neugebauer showed up in Göttingen. Not surprisingly, he soon became an integral part of this local publishing project. Neugebauer was still only a student without a doctorate when Courant took him under his wing. Yet beginning already in the winter semester of 1923–24, he began to assume various administrative duties at the institute while helping Courant to write some of his books. Years later, he offered a vivid account of a typical scene during the end phase of this production process:

> A long table in Runge's old office was the battleground on which took place what Courant's assistants used to call the "Proof-Reading-Festivals" ("Korrekturfeste"). . . . During this period Courant wrote his first group of famous books, the second edition of the "Hurwitz-Courant," the first volume of the "Courant-Hilbert," and the "Calculus." All of his assistants during these years participated at one or the other time in the preparation of the manuscripts: [Kurt] Friedrichs, [Hans] Levy, [Willy] Feller, [Franz] Rellich, [B. L.] van der Waerden, and others; red ink, glue, and personal temperament were available in abundance. Courant had certainly no easy time in defending his position and reaching a generally accepted solution under the impact of simultaneously uttered and often widely divergent individual opinions about proofs, style, formulations, figures, and many other details. At the end of such a meeting he had to stuff into his briefcase galleys (or even page proofs) which can only be described as Riemann surfaces of high genus and it needed completely unshakeable faith in the correctness of the uniformisation theorems to believe that these proofs would ever be mapped on *schlicht* pages (Neugebauer 1963, 6–7).

One particular case deserves special attention here: Volume III by Hurwitz and Courant mentioned above. Its full title already hints at an unusual undertaking: *Vorlesungen über allgemeine Funktionentheorie und elliptische Funktionen* von Adolf Hurwitz, herausgegeben und ergänzt durch einen Abschnitt über *Geometrische Funktionentheorie* von Richard Courant. As Courant explained in the introduction to the first edition, Hurwitz had planned to publish these lectures before his death in 1919, so little by way of editing was actually needed. The contents of this first part of Hurwitz-Courant drew much of their inspiration from Weierstrass' lecture courses, offered during the late 1870s and early 1880s, which Hurwitz himself had attended.[31] Young Richard Courant had the opportunity to hear Hurwitz lecture on function theory before he came to Göttingen. So did another student from Breslau, Max Born, who called these "perhaps the most perfect [lectures] I have ever heard"[32] (see Chap. 15). Other *Ausarbeitungen* of Weierstrass' lectures were in circulation, of course, but it is surely ironic that Hurwitz, who was Felix Klein's star pupil, proved to be such an influential conduit for the ideas of the once revered Berlin *Meister*.

Courant, however, was not content to publish a volume that contained nothing but Hurwitz's version of Weierstrass' theory. As the new standard bearer for Göttingen mathematics, he felt compelled to add a dose of Riemannian function theory into the mix. In the preface to the first edition he wrote: "The viewpoint of the Weierstrassian theory can today no longer alone satisfy the student, despite the inner consistency with which it is erected."[33] Courant's supplementary text, however, did not meet with the same high critical acclaim as did Hurwitz's lectures. In fact, another Hilbert pupil, the American Oliver Kellogg, found it quite lacking in rigor. "It gives the impression," he wrote, "of being the work of a mind endowed with fine intuitive faculties, but lacking the self-discipline and critical sense which beget confidence.... The proofs offered often leave the reader unconvinced as to their validity and, at times, uncertain even as to whether they can be made valid" (Kellogg 1923, 416). In view of this criticism, Courant engaged Neugebauer to help him rewrite the Riemannian portion of the book, which came out in 1925 as the second edition. Several more editions of Hurwitz-Courant appeared after this, and the book grew thicker and thicker each time.

Courant's motivation in producing this work was thoroughly Kleinian[34]; he was guided by the notion that geometric function theory contains vital ideas that keep on giving life, whereas Weierstrassian complex analysis, while beautiful, was already complete and hence lifeless. Tributes to Klein abound in the yellow series, beginning with the very first volume in which Blaschke wrote: "May F. Klein's Erlanger Programm serve us as a guiding star" ("Als Leitstern möge uns F. Kleins Erlanger Programm dienen").[35] Courant prepared new editions of Klein's *Elementarmathematik vom höheren Standpunkte aus* (Bände XIV–XVI); he had Neugebauer and Stephan Cohn-Vossen edit Klein's wartime lectures on the mathematics of the nineteenth century (Bände XXIV–XXV); and he published authorized editions of several of the lecture courses that Klein had earlier circulated locally through mimeographed copies. All of this had a strikingly conservative, not to say

[31] Hurwitz's original *Ausarbeitungen* from that time can still be found among his scientific papers: they are numbers 112, 113, and 115 in his Nachlass at the ETH.

[32] (Born 1978, 72); Born also relates that he gave Courant his notebook for use in preparing the Hurwitz-Courant volume.

[33] „Bei aller inneren Konsequenz des so errichteten Gebäudes kann der Lernende sich heute mit den Gesichtspunkten der Weierstraßschen Theorie allein nicht mehr begnügen"(Hurwitz und Courant 1925, v).

[34] Courant's personal view of Klein's legacy can be seen in (Courant 1925).

[35] In part II on affine differential geometry (Band VII), Blaschke went even further writing: "Die erste, ehrfurchtsvolle Verbeugung Herrn F. Klein! Von ihm stammt die auf dem Begriff der stetigen Transformationsgruppen beruhende geometrische Denkart, die allem Folgenden zugrunde liegt."

nationalistic, tendency. Throughout his life, Courant saw himself as the great protector and defender of the Göttingen legacy associated with Klein and Hilbert, both of whom had far more mathematical breadth than did he (Rowe 1989). During the Weimar years Hilbert's star continued to shine on brightly, in no little part due to the reverence Courant held for him. Indeed, Hilbert's name and fame continued to grow long after his heyday in mathematical research had passed. Thanks to his assistants, Hilbert continued to pursue his research program in foundations of mathematics throughout the Weimar period. The legendary old man, who became increasingly eccentric with the years, remained a living symbol of past glory even after the demise of Göttingen as a world-class center in 1933.

Yet Courant was hardly a hidebound traditionalist, even if his mathematical tastes ran toward classicism. The single most famous volume in the yellow series, his Courant-Hilbert, attests to a vision that went far beyond the legacies of his teachers. In the preface to the first edition, Courant decried the tendency among analysts to focus undue attention on "refining their methods and finalizing their concepts" at the cost of forgetting that analysis has its roots in physical problems. At the same time, he emphasized that theoretical physicists had begun to lose touch with the mathematical techniques most relevant to their own research. As a result, two new disciplinary cultures had developed, each with its own language and methods, neither able to communicate in a meaningful way with the other. Courant, writing in February 1924 just after the country had nearly succumbed to runaway inflation, saw this not just as an unfruitful use of resources, to him this represented a familiar danger that both Klein and Hilbert had earlier tried to counteract: "Without doubt this tendency poses a threat to all science; the stream of scientific developments faces the danger of dissipating further and further, to seep away and dry up."[36] In preparing this volume, Courant relied on Hilbert's publications and *Vorlesungen* from the period 1902–1912. He also leaned heavily on the support of his own school of *Mitarbeiter*. These young men remained anonymous in 1924, but in the preface to the revised second edition from 1930 he gave credit to Kurt Friedrichs, Franz Rellich, and Rudolf Lüneburg, among others. He also alluded to the mathematical difficulties that had caused him to delay the publication of Courant-Hilbert II, which finally appeared in 1937.[37]

As principal editor of Springer's "yellow series" Courant turned local oral knowledge – the edited lectures of Hurwitz, Klein, Hilbert, et al. – into internationally accessible knowledge in print form. The scope of this undertaking eventually went far beyond the intellectual confines of the Göttingen tradition, and while its range was truly encyclopedic, Courant's brainchild exerted a far deeper and more lasting influence than Klein's massive *Encyklopädie der mathmatischen Wissenschaften*. The latter was a reference work, comprised of lengthy scholarly reports filled with footnotes that pointed to the vast specialized literature; it reflected Klein's penchant for detail rather than the needs of working mathematicians. The best volumes in the yellow series, on the other hand, were living mathematics of a kind that a younger generation of mathematicians could not only learn from but also build upon. That was precisely what Courant and his *Mitarbeiter* showed in producing the various new editions of Hurwitz-Courant and Courant-Hilbert, books that drew on research traditions with a long and rich history.

Courant's success as an institution builder had much to do with his unorthodox methods. Some people found him pushy, but those who were willing to be pushed got things done. He also had an uncanny ability to instill tremendous loyalty, a prime example being Otto Neugebauer, his *Oberassistent* in Göttingen (on Neugebauer's approach to the history of mathematics, see Chap. 29).

Neugebauer served not only as the real manager of Courant's old institute, he also designed the new building that opened in 1929 (Neugebauer 1930).[38] A far more direct, even blunt personality, Neugebauer came to share the very same values Courant stood for. Indeed, Neugebauer would ultimately devote himself to the study of the same nexus of mathematical sciences, though within the realm of ancient cultures. Once again, the Springer connection paved the way: its short-lived *Quellen und Studien* series, launched in 1929 and edited by Neugebauer, Julius Stenzel, and Otto Toeplitz, set a new standard for studies in the history of the ancient exact sciences. Like Courant, Neugebauer was a visionary (Rowe 2013), but neither man could have foreseen the explosion of interest in ancient as well as modern mathematics that would make this difficult decade a remarkably productive time for scholarly publications in Germany. Along the way to becoming a historian, Neugebauer gained an ever deeper respect for the unity of mathematical knowledge; much of that came through his interactions with Göttingen mathematicians. Regarding his former mentor's vision, Neugebauer later said this on the occasion of Courant's 75th birthday: "... the real core of his work [consisted] in the conscious continuation and ever widening development of the ideas of Riemann, Klein,

[36]Ohne Zweifel liegt in dieser Tendenz eine Bedrohung für die Wissenschaft überhaupt; der Strom der wissenschaftlichen Entwicklung ist in Gefahr, sich weiter und weiter zu verästeln, zu versickern und auszutrocknen." (Courant and Hilbert 1924, vi).

[37]Courant-Hilbert II was not listed in the bibliography of the Deutsche Bücherei. It was still listed in the Springer catalogues, however, in 1940. The *Sicherheitsamt* of the *Reichsführer* of the SS established a liaison office in the Deutsche Bücherei in 1934 to oversee the listing of books by Jewish authors. See (Sarkowski 1996, 353).

[38]On the planning and financing of this new institute, see (Siegmund-Schultze 2001).

Fig. 28.3 Courant and K. O. Friedrichs (right) in a meeting at NYU; on the far right is Cathleen Morowetz, who took her doctorate under Friedrichs in 1951.

and Hilbert, and in his insistence on demonstrating the fundamental unity of all mathematical disciplines. One must always remain aware of these basic motives if one wants to do justice to Courant's work and to realize its inner consistency" (Neugebauer 1963, 1) (see the preface to Neugebauer (1957)).

Like Courant, Neugebauer had been molded by his student days in Göttingen. His authoritarian manner complemented his boss's famous indecisiveness. Yet Courant had truly keen insight when it came to judging people, and his reticence, I would guess, was not just a personality quirk. More likely, it was a carefully learned social skill that served him well. Richard Courant had a certain ability to go with the flow of events. This, coupled with a fundamental optimism, gave him the strength to deal with a lifetime full of adversity. He faced very trying circumstances, both during the Weimar years and then in a totally different setting during the mid-1930s, when he began to build up the mathematics program at NYU in the midst of the Great Depression (Fig. 28.3). When he came to New York City, he found inspiration in Abraham Flexner's vision for higher education in the United States (Flexner 1930), especially the idea that New York contained a vast "reservoir of talent" (Reid 1976, 169). So Courant continued to innovate in the name of preserving past ideals. In both instances, however, a consistent theme remained: a longing to preserve the legacy of the Göttingen he had known in his youth.

Acknowledgments This essay was originally presented at a special symposium held at New York University in the fall of 2013. At that time the traveling exhibition "Transcending Tradition: Jewish Mathematicians in German-Speaking Academic Culture" was being shown at the nearby Leo Baeck Institute and Yeshiva University Museum. Three other talks from the symposium were published with this one in *The Mathematical Intelligencer* 37(2015). My thanks go to Moritz Epple, Michael Korey, and Marjorie Senechal for their efforts in preparing that special issue.

References

Bergmann, Birgit, Moritz Epple, and Ruti Ungar, eds. 2012. *Transcending Tradition: Jewish Mathematicians in German-Speaking Academic Culture*. Heidelberg: Springer.

Biermann, Kurt-R. 1988. *Die Mathematik und ihre Dozenten an der Berliner Universität, 1810–1933*. Berlin: Akademie Verlag.

Blumenthal, Otto. 1935. Lebensgeschichte, in (Hilbert 1932–1935, 3: 388–429).

Born, Max. 1978. *My Life. Recollections of a Nobel Laureate*. New York: Charles Scribner's Sons.

Courant, Richard. 1925. *Felix Klein*. *Die Naturwissenschaften* 37: 765–772.

———. 1981. *Reminiscences* from *Hilbert's Göttingen*. *Mathematical Intelligencer* 3 (4): 154–164.

Courant, Richard und David Hilbert. 1924. *Methoden der mathematischen Physik,* Berlin: Julius Springer.

Einstein CP9. 2004. *Collected Papers of Albert Einstein*, Vol. 9: *The Berlin Years: Correspondence, January 1919-April 1920*, Diana Kormos Buchwald, et al., eds. Princeton: Princeton University Press.

Flexner, Abraham. 1930. *Universities: American, English, German*. Oxford: Oxford University Press.

Frei, Günther. 2004. On the History of the Artin Reciprocity Law in Abelian Extensions of Algebraic Number Fields: How Artin was led to his Reciprocity Law, in (Laudal & Piene 2004, 267–294).

Frei, Günther, and Urs Stammbach. 1992. *Hermann Weyl und die Mathematik an der ETH Zürich, 1913–1930*. Basel: Birkhäuser.

Friedrichs, Nellie H. 1988. *Erinnerungen aus meinem Leben in Braunschweig 1912–1937*. 2nd ed. *Braunschweig*: Stadtarchiv und Stadtbibliothek.

Gay, Peter. 1968. *Weimar Culture: The Outsider as Insider*. New York: Harper & Row.

Greenspan, Nancy Thorndike. 2005. *The End of the Certain World: The Life and Science of Max Born: The Nobel Physicist Who Ignited the Quantum Revolution*. New York: Basic Books.

Hilbert, David. 1932–1935. *Gesammelte Abhandlungen*. 3 Bde., Berlin: Springer.

Jungnickel, Christa, and Russell McCormmach. 1990. *Intellectual Mastery of Nature. Theoretical Physics from Ohm to Einstein*. Vol. 2. Chicago: University of Chicago Press.

Kellogg, O.D. 1923. Review of Hurwitz and Courant, *Vorlesungen über allgemeine Funktionentheorie und elliptische Funktionen*, *Bulletin of the American Mathematical Society* 29: 415–417.

Laudal, Olav Arnfinn, and Ragni Piene. 2004. *The Legacy of Niels Henrik Abel*. Berlin: Springer.

Lax, Peter. 2003. Richard Courant, *Biographical Memoirs. National Academy of Sciences*, 82: 78–97.

Mehrtens, Herbert, Ludwig Bieberbach and Deutsche Mathematik. 1987. *Studies in the History of Mathematics*, ed. Esther R. Phillips, *MAA Studies in Mathematics*, vol. 26, Washington: Mathematical Association of America.

Müller-Stach, Stefan. 2014. Otto Toeplitz: Algebraiker der unendlichen Matrizen. *Mathematische Semesterberichte* 61 (1): 53–77.

Neugebauer, Otto. 1930. Das Mathematische Institut der Universität Göttingen. *Die Naturwissenschaften* 18: 1–4.

———. 1957. *The Exact Sciences in Antiquity*, 2nd ed., Providence: Brown University Press; first edition 1951, Munksgaard, Copenhagen; reprinted 1969, New York: Dover.

———. 1963. Reminiscences on the Göttingen Mathematical Institute on the Occasion of R. Courant's 75th Birthday, Otto Neugebauer Papers, Institute for Advanced Study, Princeton, Box 14, publications vol. 11.

Parshall, Karen H., and David E. Rowe. 1994. *The Emergence of the American Mathematical Research Community, 1876–1900. J.J. Sylvester, Felix Klein, and E.H. Moore*, History of Mathematics. Vol. 8. Providence: American Mathematical Society.

Reid, Constance. 1970. *Hilbert*. New York: Springer.

———. 1976. *Courant in Göttingen and New York: the Story of an Improbably Mathematician*. New York: Springer.

Remmert, Volker, and Ute Schneider. 2010. *Eine Disziplin und ihre Verleger. Disziplinenkultur und Publikationswesen der Mathematik in Deutschland, 1871–1949*. Bielefeld: Transkript.

Rowe, David E. 1986. "Jewish Mathematics" at Göttingen in the Era of Felix Klein. *Isis* 77: 422–449.

———. 1989. Klein, Hilbert, and the Göttingen Mathematical Tradition, *Science in Germany: The Intersection of Institutional and Intellectual Issues*, ed. Kathryn M. Olesko (Osiris, 5, 1989), 189–213.

———. 1999. The Göttingen Response to General Relativity and Emmy Noether's Theorems. In *The Symbolic Universe. Geometry and Physics, 1890–1930*, ed. Jeremy Gray, 189–234. Oxford: Oxford University Press.

———. 2012. On Stage and Behind the Scenes in Göttingen: Otto Blumenthal, Richard Courant, Emmy Noether, and Paul Bernays, in [Bergmann/Epple/Ungar 2012], 79–87.

———. 2013. Otto Neugebauer's Vision for Rewriting the History of Ancient Mathematics. *Anabases – Traditions et réceptions de l'Antiquité* 18: 175–196.

Rowe, David E., and Robert Schulmann, eds. 2007. *Einstein on Politics: His Private Thoughts and Public Stands on Nationalism, Zionism, War, Peace, and the Bomb*. Princeton: Princeton University Press.

Sarkowski, Heinz. 1996. *Springer-Verlag. History of a Scientific Publishing House, Part I: 1842–1945*. Heidelberg: Springer.

Schappacher, Norbert. 1987. Das Mathematische Institut der Universität Gottingen, 1929–1950, (Becker/ Dahms/ Wegeler 1987), 345–373.

———. 1991. Edmund Landau's Göttingen: From the Life and Death of a Great Mathematical Center. *The Mathematical Intelligencer* 13 (4): 12–18.

Siegmund-Schultze, Reinhard. 2001. *Rockefeller and the Internationalization of Mathematics between the Two World Wars: Documents and Studies for the Social History of Mathematics in the 20th Century*, Science Networks, 25, Basel, Boston and Berlin: Birkhäuser.

———. 2009. *Mathematicians Fleeing from Nazi Germany: Individual Fates and Global Impact*. Princeton: Princeton University Press.

Stein, Edith. 1965. *Aus dem Leben einer jüdischen Familie*. Freiburg: Verlag Herder.

Vallentin, Antonina. 1954. *The Drama of Albert Einstein*. New York: Doubleday.

Weyl, Hermann. 1944. David Hilbert and his Mathematical Work. *Bulletin of the American Mathematical Society* 50: 612–654; (Reprinted in [Weyl 1968, vol. 4, 130–172].)

———. 1968. In *Gesammelte Abhandlungen*, ed. K. Chandrasekharan, vol. 4. Berlin: Springer.

Wolters, Gereon. 1992. Opportunismus als Naturanlage: Hugo Dingler und das 'Dritte Reich', *Entwicklungen der methodischen Philosophie,* hrsg. Peter Janich, Frankfurt a. M., 257–327

29 Otto Neugebauer and the Göttingen Approach to History of the Exact Sciences

(*Mathematical Intelligencer* 34(2)(2012): 29–37)

The common belief that we gain "historical perspective" with increasing distance seems to me utterly to misrepresent the actual situation. What we gain is merely confidence in generalizations which we would never dare make if we had access to the real wealth of contemporary evidence.

—Orro Neugebauer, *The Exact Sciences in Antiquity* (Neugebauer 1969, viii)

Otto Neugebauer (1899–1990) was, for many, an enigmatic personality. Trained as a mathematician in Graz, Munich, and Göttingen, he had not yet completed his doctoral research when in 1924 Harald Bohr, brother of the famous physicist, invited him to Copenhagen to work together on Bohr's new theory of almost periodic functions. Quite by chance, Bohr asked Neugebauer to write a review of T. Eric Peet's recently published edition of the Rhind Papyrus (Neugebauer 1925). In the course of doing so, Neugebauer became utterly intrigued by Egyptian methods for calculating fractions as sums of unit fractions (e.g. $3/5 = 1/3 + 1/5 + 1/15$). When he returned to Göttingen, he wrote his dissertation on this very topic. In 1927 he published the first of many studies on Babylonian mathematics and astronomy, a pioneering study on the evolution of the sexagesimal (base 60) number system (Neugebauer 1927). These works received high praise from leading Egyptologists and Assyriologists, helping to launch Neugebauer's career as a historian of ancient mathematics and exact sciences. Indeed, he would go on to revolutionize research in these areas, leaving a deep imprint on our understanding of these ancient scientific cultures to this very day.[1]

Yet Neugebauer's general orientation as a historian seems strangely remote from today's perspective, so much so that even scholars who know his work well and respect it highly have great difficulty identifying with his methodological views. One who worked closely with him during his later career at Brown University, Noel Swerdlow, gave a most apt description of the "zwei Seelen" that dwelled within Otto Neugebauer and that colored all his work:

> At once a mathematician and cultural historian, Neugebauer was from the beginning aware of both interpretations and of the contradiction between them. Indeed, a notable tension between the analysis of culturally specific documents, whether the contents of a single clay tablet or scrap of papyrus or an entire Greek treatise, and the continuity and evolution of mathematical methods regardless of ages and cultures, is characteristic of all his work. And it was precisely out of this tension that was born the detailed and technical cross-cultural approach, in no way adequately described as the study of "transmission," that he applied more or less consistently to the history of the exact sciences from the ancient Near East to the European Renaissance. But if the truth be told, on a deeper level Neugebauer was always a mathematician first and foremost, who selected the subjects of his study and passed judgment on them, sometimes quite strongly, according to their mathematical interest (Swerdlow 1993, 141–142).

Taking up this last point, one can easily appreciate why Neugebauer's approach to history persuaded few, while provoking some of his detractors to take a firm stand against his methodological views and what they felt was a deleterious influence on studies of the ancient sciences. Neugebauer firmly believed in the immutable character of mathematical knowledge, which meant that his field of historical inquiry, the exact sciences, differed from all other forms of human endeavor in one fundamental respect: in this realm there was no room for historical contingency. The methodological implications Neugebauer drew from this were simple and clear: once an investigator had cracked the linguistic or hieroglyphic codes that serve to express a culture's scientific knowledge he or she then suddenly held the keys to deciphering ancient sources. And since the content of these sources pertained to mathematical matters, one could, in principle, argue inductively in order to reconstruct what they originally contained, namely a fixed and determinable pattern of scientific results. Clearly, this type of puzzle solving held great fascination for Neugebauer, and he practiced it with considerable success in his research on Mesopotamian astronomy, beginning in the mid-1930s.

Neugebauer's work on Greek mathematics during these politically turbulent times was far scantier. Nevertheless, his views on Greek mathematics formed a central component of his overall view of the ancient mathematical sciences. When it came to purely human affairs, Neugebauer professed that he held no *Weltanschauung*, and he took pains to make this known to those who, like Oskar Becker, mingled ideology with science (see Siegmund-Schultze 2009, 163). Regarding

[1] For recent reassessments of his career and impact, see the essays in Jones et al. (2016), in particular Brack-Bernsen (2016), Høyrup (2016), Ritter (2016), Rowe (2016), and Siegmund-Schultze (2016).

D.E. Rowe, *A Richer Picture of Mathematics*, https://doi.org/10.1007/978-3-319-67819-1_29

historiography, on the other hand, Neugebauer adopted a rigorously empirical approach that worked well in some cases, but often led him to make sweeping claims based on little more than hunches. Not surprisingly, his views on historiography had much to do with the special context in which he first experienced higher mathematics.

Neugebauer's Cornell Lectures

In 1949, when Otto Neugebauer delivered six lectures on ancient sciences at Cornell University, he was the first historian of mathematics to be given the honor of speaking in its distinguished Messenger lecture series. He did not waste this opportunity. Afterward, he went over his notes and gave the text its final, carefully sculpted form that we find today in the six chapters of Neugebauer's *The Exact Sciences in Antiquity*, published in 1951 with high-quality plates (see Fig. 29.1). The text begins by describing a famous work in the history of art:

> When in 1416 Jean de France, Duc de Berry, died, the work on his "Book of the Hours" was suspended. The brothers Limbourg, who were entrusted with the illuminations of this book, left the court, never to complete what is now considered one of the most magnificent of late medieval manuscripts which have come down to us. A "Book of Hours" is a prayer book which is based on the religious calendar of saints and festivals throughout the year. Consequently we find in the book of the Duc of Berry twelve folios, representing each one of the months. As an example we may consider the illustration for the month of September. As the work of the season the vintage is shown in the foreground. In the background we see the Château de Saumur, depicted with the greatest accuracy of architectural detail. For us, however, it is the semicircular field on top of the picture, where we find numbers and astronomical symbols, which will give us some impression of the scientific background of this calendar. Already a superficial discussion of these representations will demonstrate close relations between the astronomy of the late Middle Ages and antiquity. (Neugebauer 1969, 3).

Fig. 29.1 Septembre from the Très Riches Heures du Duc de Berry, one of the most famous works in the French Gothic tradition. The manuscript first gained public attention after 1856 when it was acquired by the Duc d'Aumale, founder of the Musée Condé Musée in Chantilly.

Neugebauer went on to note four different types of writing for the numbers that appear in the *Book of Hours:* Hindu-Arabic as well as Roman numerals, number words (September through December for the seventh to the tenth months of the Roman calendar), and alphabetic numbers, here calculated modulo 19, the system used in connection with the Metonic lunar cycle. Regarding the latter, he noted that for a given year, the associated number between 1 and 18 was called the "golden number" in the late Middle Ages, after a thirteenth-century scholar wrote that this lunar cycle excels all others "as gold excels all other metals."

He then comments as follows about the state of scientific progress in the Latin West when seen against the backdrop of earlier developments: "In the twelfth century this very primitive method [for calculating the date of a new moon] was considered by scholars in Western Europe as a miracle of accuracy, though incomparably better results had been reached by Babylonian and Greek methods since the fourth century B.C. and though these methods were ably handled by contemporary Islamic and Jewish astronomers" (Neugebauer 1969, 8). Clearly, Neugebauer wanted his audience to realize that it was one thing to appreciate a magnificent work of art, quite another to think of it as a canvas for clues about the state of mathematical and astronomical knowledge in the culture within which it was produced.

For the second edition, he updated the material and added two technical appendices, but he still hoped to have "avoided. . . converting my lectures into a textbook" (Neugebauer 1969, ix). Evidently, he valued the less formal form of exposition associated with oral exposition, a hallmark of the Göttingen tradition (Rowe 2004). Still, Neugebauer grew up in Austria, not Prussia, which may help account for his playful sense of humor. A typical example comes in a passage where he comments on how astronomers took delight in harmonizing their science with anthropocentric religious views, whereas modern celestial mechanics teaches us to be humble creatures living in a solar system conditioned by accidental circumstances:

> The structure of our planetary system is indeed such that Rheticus [an early champion of the Copernican theory] could say "the planets show again and again all the phenomena which God desired to be seen from the earth." The investigations of Hill and Poincaré have demonstrated that only slightly different initial conditions would have caused the moon to travel around the earth in a curve [with small loops]. . .. Nobody would have had the idea that the moon could rotate on a circle around the earth and all philosophers would have declared it as a logical necessity that a moon shows six half moons between two full moons. And what could have happened with our concepts of time if we were members of a double-star system (perhaps with some uneven distribution of mass in our little satellite) is something that may be left to the imagination (Neugebauer 1969, 152–153).

Neugebauer and Courant in Göttingen

Significantly, Neugebauer dedicated this now classic book to "Richard Courant, in Friendship and Gratitude." Elaborating on that dedication in the preface, he wrote that it was Courant who enabled him to pursue graduate studies in ancient mathematics, and he went on to remark: "more than that I owe him the experience of being introduced to modern mathematics and physics as a part of intellectual endeavour, never isolated from each other nor from any other field of our civilization" (Neugebauer 1969, vii). Neugebauer was a man who chose his words carefully, and so we may be sure that this public acknowledgement of his debt to Courant was far more than just a friendly gesture. He wrote further that Courant's vision saw mathematics and physics as fields of intellectual endeavor "never isolated from each other nor from any other field of our civilization." This brief remark comes very close to capturing the essence of Neugebauer's own understanding of what it meant to study the history of mathematics as an integral part of human cultural life. Regarding Courant's personal outlook, he described this in connection with the Göttingen mathematical tradition they both shared and valued:

> . . . the real core of his work [consisted] in the conscious continuation and ever widening development of the ideas of Riemann, Klein, and Hilbert, and in his insistence on demonstrating the fundamental unity of all mathematical disciplines. One must always remain aware of these basic motives if one wants to do justice to Courant's work and to realize its inner consistency (Neugebauer 1963, 1).

As a close ally of Courant, Neugebauer shared a positivist vision of mathematics as an integral part of scientific culture. In particular, both men were deeply influenced by the universalism advocated by Göttingen's two aging sages, Felix Klein and David Hilbert, who broke with an older German tradition in which mathematical research was largely isolated from developments in neighboring disciplines, like astronomy and physics. Hilbert's strong epistemic claims for mathematics had also deeply alienated conservative humanists on the Göttingen faculty, many of whom feared a realignment of traditional disciplinary boundaries (Rowe 1986). Neugebauer's personal relationship with Richard Courant reflects many of the broader mathematical and scientific interests the two men shared (Fig. 29.2).

Fig. 29.2 Richard Courant, Director of the Göttingen Mathematics Institute, 1922–1933.

As director of the Göttingen Mathematics Institute during the Weimar years, Courant was faced with numerous challenges as he struggled to uphold its international scientific reputation. Part of his strategy was conservative in nature. Through his connections with Ferdinand Springer, Courant launched the famed "yellow series," one of several initiatives that enabled Springer to attain a pre-eminent position as a publisher in the fields of mathematics and theoretical physics (Remmert and Schneider 2010). Courant was an innovator with a deep belief in the vitality of older traditions. His yellow series looked backward as well as forward; in fact, surprisingly few of its volumes betray a commitment to what came to be identified as modern, abstract mathemat-

Fig. 29.3 Plaque honoring Otto Neugebauer erected on the Mathematical Institute in Göttingen, the building he helped to design.

ics. Far more evident was the way in which Courant and his co-editors built on the tradition of Klein and Hilbert, and with the yellow series he found a way to make local knowledge accessible well beyond the borders of Germany. Neugebauer would ultimately devote himself to the study of the same nexus of mathematical sciences in antiquity. For the history of the ancient exact sciences, Springer's short-lived *Quellen und Studien* series, launched in 1929 and edited by Neugebauer, Julius Stenzel, and Otto Toeplitz, created a new standard for studies in this fast-breaking field.

Soon after Neugebauer arrived in Göttingen in 1922, Courant gave him various special duties to perform at the hub of operations, located on the third floor of the *Auditorienhaus*. There one found the famous *Lesezimmer* together with an impressive collection of mathematical models, long cared for by Felix Klein's assistants. Now Neugebauer stood guard while Klein received nearly daily reports through those who were busy helping him prepare his collected works. Neugebauer's new interest in Egyptian mathematics also came to Klein's attention, along with a complaint that he had stuffed all the books on mathematics education tightly together on a high shelf, making them nearly inaccessible. By now Klein was an infirm old man who rarely left his home, which overlooked the botanical garden immediately behind the *Auditorienhaus*, but he still kept up a busy and tightly organized schedule. Neugebauer remembered how Klein called him over to be gently scolded. When he arrived, Klein greeted him by saying: "there came a new Moses into Egypt and he knew not Pharaoh!" (Reid 1976, 100) (a play on: "Now there arose up a new king over Egypt, which knew not Joseph", Exodus I.8). The young Neugebauer surely realized that watching over the *Lesezimmer* was no trifling matter (Fig. 29.3).

Neugebauer's Revisionist Approach to Greek Mathematics

Neugebauer saw himself as a "scientific historian"; he had no patience for those who simply wanted to chronicle the great names and works of the past. George Sarton, who did little else, saw the history of science as a humanistic endeavor; nevertheless, he had the highest respect for Neugebauer's achievements. Sarton's views emerge clearly from correspondence during September 1933 with Abraham Flexner. At the time, Flexner was contemplating the possibility of founding a school for studies of science and culture at the Institute for Advanced Study. Sarton thought that Neugebauer was just the man for such an enterprise, a point he made by humbly contrasting the nature of their work: "As compared with Neugebauer I am only a dilettante. He works in the *front trenches* while I amuse myself way back in the rear – praising the ones, blaming the others; saying this ought to be done, etc. – & doing very little myself. What Neugebauer does is fundamental, what I do, secondary" (Pyenson 1995, 268). Neugebauer certainly did view Sarton as a dilettante through and through. When I interviewed him in 1982, he made a point of telling me this by lumping him together with Moritz Cantor, another encyclopedist of great breadth and little depth.

Although plans to bring Neugebauer to Princeton came to naught, Harald Bohr managed to arrange a three-year appointment for him in Copenhagen beginning in January 1934. Neugebauer managed to get most of his property out of Germany, but had to abandon a house with a partially paid mortgage. In Copenhagen, his research was supported in part by the Rockefeller Foundation. Almost immediately he began preparing a series of lectures on Egyptian and

Babylonian mathematics that he would publish in Courant's yellow series as *Vorgriechische Mathematik* (Neugebauer 1934). According to Swerdlow this volume was "as much a cultural as a technical history of mathematics" and represents "Neugebauer's most thorough and successful union of the two interpretations" (Swerdlow 1993, 145) More striking still is the unfinished character of this work, which represents the first and final volume in a projected trilogy that remained incomplete. Neugebauer had planned to tackle Greek mathematics proper in the second volume, whereas the third would have dealt with mathematical astronomy, both in the Greek tradition culminating with Ptolemy as well as the largely unknown work of late Babylonian astronomers. Thus, his original aim, as spelled out in the foreword to the first volume, was to achieve a first overview of the ancient mathematical sciences in their entirety, something that had never before been attempted.

Swerdlow has offered compelling reasons to explain why Neugebauer dropped this project, one being that he simply found the rich textual sources for Mesopotamian mathematical astronomy far more important than anything he could ever have written about Greek mathematics. Nevertheless, we can trace a fairly clear picture of the line of argument Neugebauer originally had in mind by examining the summary remarks at the conclusion of his *Vorgriechische Mathematik* as well as some of his other publications from the 1930s. Neugebauer's writings from the 1920s contain few hints that his understanding of ancient mathematics was fundamentally opposed to older views. By the early 1930s, however, his analyses of Babylonian texts led him to a new conception, namely that the Greek penchant for geometrization represented a retrograde step in the natural development of the exact sciences. This did not mean, of course, that he held a low opinion of Euclid's *Elements*; he simply thought that historians and philosophers had distorted its true place in the history of mathematics. Thus, he once imagined how scholars in some future civilization might easily form a deceptive picture of mathematical knowledge *circa* 1900 if the only important text that happened to survived were Hilbert's *Grundlagen der Geometrie*.

In the course of this transition, Neugebauer's assertions about the character of ancient mathematics often took on a strident tone. Particularly suggestive is an essay entitled "Zur geometrischen Algebra," published in 1936 in *Quellen und Studien* (Neugebauer 1936). Significantly, Neugebauer takes as his motto a famous fragment from the late Pythagorean Archytas of Tarentum, which reads: "It seems that logistic far excels the other arts in regard to wisdom, and in particular in treating more clearly what it wishes than geometry. And where geometry fails, logistic brings about proofs." (Neugebauer 1936, 245) Much has been written about this passage, in particular about what might be meant by the term "logistic", a matter Jakob Klein discussed at great length in his study "Die griechische Logistik und die Entstehung der Algebra" (Klein 1936), which appeared alongside Neugebauer's article (it was later translated into English by Eva Brann (Klein 1968)). In fact, both scholars were chasing after the same elusive goal, though there the similarity ends.

Jakob Klein was a classical philologist who later became a master teacher of the "Great Books" curriculum at St. Johns College in Annapolis Maryland. Not surprisingly, he was intent on squeezing as much out of Plato as he possibly could. Thus he distinguished carefully between practical and theoretical logistic, offering a new interpretation of Diophantus' *Arithmetica* that placed it within the latter tradition. Neugebauer had no patience for the nuances of meaning classicists liked to pull out of their texts. Indeed, he had an entirely different agenda. His point was that rigorous axiomatic reasoning in the style of Euclid arose rather late, and that Archytas, a contemporary of Plato, was bearing witness to the primacy of algebraic content over the geometrical form in which the Greeks dressed their mathematics. With that, we can take another step forward toward attaining a closer understanding Neugebauer's *Weltanschauung*.

Decades earlier, the Danish historian of mathematics H. G. Zeuthen already advanced the idea that the Greeks had found it necessary to geometrize their purely algebraic results after the discovery of incommensurable magnitudes (Zeuthen 1896).[2] Neugebauer took up this by now standard interpretation, adopted by Heath and nearly everyone else, but he then went much further, arguing that the algebraic content – found not only in Book II of Euclid but throughout the entire corpus of Apollonius' *Conica* (Neugebauer 1932) – could be traced back to results and methods of the Babylonians:

> The answer to the question what were the origins of the fundamental problem in all of geometrical algebra [meaning the application of areas, as given by Euclid's propositions I.44 and VI.27–29][3] can today be given completely: they lie, on the one hand, in the demands of the Greeks to secure the general validity of their mathematics in the wake of the emergence of irrational magnitudes, on the other, in the resulting necessity *to translate the results of the pre-Greek "algebraic" algebra as well.* Once one has formulated the problem in this way, everything else is completely trivial [!] and provides *the smooth connection between Babylonian algebra and the formulations of Euclid* (Neugebauer 1936, 250, my translation, his italics).

[2]Ancient sources only hint at the circumstances surrounding this discovery, which probably took place during the latter half of the fifth century. Before this time, it was presumed that magnitudes of the same kind, for example two lengths, could always be measured by a third, hence commensurable. This is equivalent to saying that their ratio will be equal to the ratio of two natural numbers. This theory had to be discarded when it was realized that even simple magnitudes, like the diagonal and side of a square, have an irrational ratio because their lengths are incommensurable lengths. The discovery of such irrational objects in geometry had profound consequences for the practice of Greek geometry in the fourth century (see Fowler 1999).

[3]See the discussion below.

The mathematical concepts underlying this argument are by no means difficult. It must be emphasized, however, that what may seem mathematically trivial (*i.e.* obvious) should hardly be thought of as historically self-evident. Since Zeuthen's time, it had been customary to interpret Greek problem-solving methods as manipulations closely related to techniques like "completing the square", used to solve quadratic equations. These Greek methods, called applications of areas, occupy a prominent place in Euclid's *Elements* as well as in his *Data,* a kind of handbook for problem solving. Neugebauer was struck by the parallelism between certain standard Babylonian problems and the Greek methods for solving very similar problems geometrically (Neugebauer 1969, 40–41, 149–150) (Fig. 29.4).

A typical algebra problem found in several cuneiform tablets from the Old Babylonian period requires that one find two numbers whose sum (or difference) and product are both given (Neugebauer called this the "normal form" leading to a single quadratic equation). This pair of problems, depending on whether the sum or difference is given, can also be found as Propositions 84 and 85 in Euclid's *Data.* Moreover, according to the neo-Platonic commentator Proclus – on the authority of Aristotle's student, Eudemus, author of a lost *History of Geometry* written just before Euclid's time – the three types of applications of areas (later used by Apollonius to distinguish the three types of conic sections: ellipse, parabola, and hyperbola) were discovered long before Euclid: "These things, say Eudemus, are ancient and are discoveries of the Muse of the Pythagoreans" (Heath 1956, 343).

Fig. 29.4 Otto Neugebauer: a pioneer for the study of ancient mathematical cultures (Jones et al. 2016, xiii).

Neugebauer would have been the last to argue that the Pythagoreans had anything to do with this ancient knowledge; for him, the key fact was merely that the original ideas were old, hence likely to have roots in still older cultures from which the Greeks borrowed freely. Having established that the mathematical content of the Babylonian texts was fundamentally algebraic, he now claimed that Mesopotamia was the original source of the algebra underlying the "geometric algebra" uncovered by Zeuthen at the end of the nineteenth century. Neugebauer was fully aware, of course, that his interpretation required a really bold leap of the historical imagination, since making a claim for the transmission of such knowledge over such a vast span of time meant accepting that this mathematical linkage sufficed to fill a gap devoid of any substantive documentary evidence. Summarizing his position, he offered these remarks: "Every attempt to connect Greek thought with the pre-Greek meets with intense resistance. The possibility of having to modify the usual picture of the Greeks is always undesirable, despite all shifts of view, . . . [and yet] the Greeks stand in the middle and no longer at the beginning" (Neugebauer 1936, 259).

When we try to square this with Neugebauer's stated belief that we should be wary of generalizations about the distant past—the position quoted in the motto to this essay—the problems with such an argument only become more acute. Perhaps these evident difficulties help explain the intensely passionate language in the concluding parts of his text. The tone in *The Exact Sciences in Antiquity* is far milder, and yet his arguments remain substantively the same (Neugebauer 1969, 146–151). There is even brief mention of the same quotation from Archytas, and one senses what Swerdlow might have meant when he wrote that Neugebauer grew bored with Greek mathematics (Swerdlow 1993, 146) (Fig. 29.5).

Neugebauer's research represented part of a large-scale intrusion by mathematicians into a field that was formerly dominated by classicists. Before he entered the field the history of Greek mathematics was traditionally seen as strongly linked with the works and influence of Plato and Aristotle, a view that would later be contested by the influential American historian Wilbur Knorr (see, for example, the

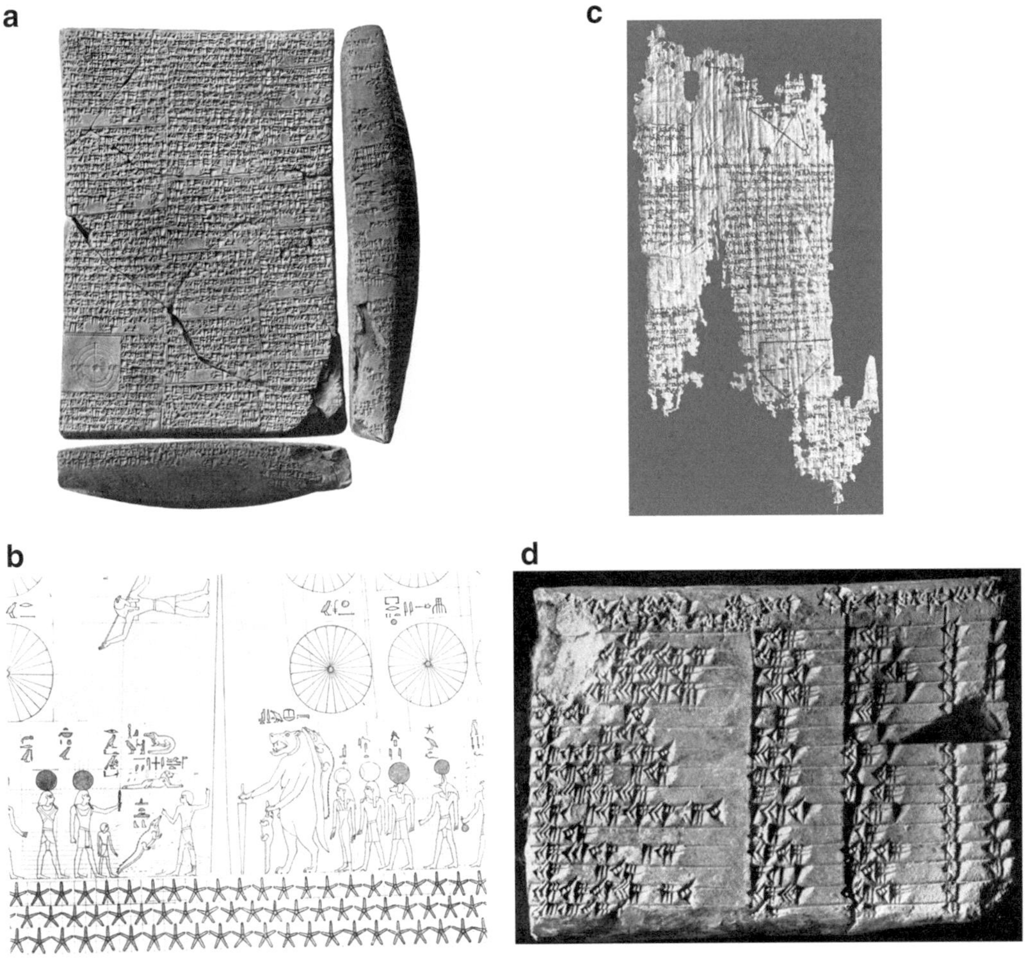

Fig. 29.5 (**a**) BM85194, a cuneiform tablet from the "Old Babylonian" period, ca. 1600 B.C.E. (**b**) Drawing of the Ceiling in the Tomb of Senmut made by Expedition of the Metropolitan Museum of Art, New York. (**c**) Geometrical Problems in the Style of Heron, Greek Papyrus in the Cornell Collection. (**d**) Plimpton 322, the most famous of all Babylonian mathematical tablets, discussed by Neugebauer in *The Exact Sciences in Antiquity,* pp. 36–40.

essays by Knorr in Christianidis (2004)). Neugebauer's work thus struck a sympathetic chord among a younger generation of experts on Greek mathematics, even though he had left the field by the mid 1930s. Ever the anti-philosopher, he wanted to undermine the special German fascination with Greek philosophy, most particularly the Platonic tradition. In this respect, his work stood poles apart from that of Oskar Becker, or for that matter, Otto Toeplitz, both of whom, like Neugebauer, published regularly in *Quellen und Studien.* These two older contemporaries combined fine-tuned mathematical analyses with careful philological readings of classical Greek texts. Neugebauer, on the other hand, showed very little interest in studies of this kind. Furthermore, he had an entirely different agenda than they: he aimed to overthrow the standard historiography that made mathematics look like the handmaiden of Greek philosophy.

Neugebauer's original vision thus entailed a radical rewriting of the history of ancient mathematics and exact sciences. One of his central theses was that rigorous axiomatic reasoning in the style of Euclid arose rather late. At the same time he liked to call on the testimony of Archytas, a much earlier mathematician contemporaneous with Plato, who – according to Neugebauer's reading – tells us that the Greeks of that era understood the primacy of algebraic content over geometrical form. If one probed the later Greek sources with a mathematically trained eye – as Neugebauer tried to show in his study of Apollonius' *Conica* – what one found was a fundamentally algebraic style of thought. His revisionist stance also aimed to debunk the notion of a "Greek miracle" that sprang up during the sixth century from the shores of Ionia. Neugebauer was convinced that most of the sources that reported on the legendary feats of ancient heroes – Thales, Pythagoras, and their intellectual progeny – were just that: legends that had grown with the passing of time. So his constant watchword remained skepticism with regard to the accomplishments of the early Greeks, whereas Toeplitz, Becker, and others began to analyze extant sources with a critical eye toward their standards of exactness.[4]

Greek Mathematics Reconsidered

One can well imagine that for some experts on ancient Greek philosophy and early science, Neugebauer's views regarding the historical development of Greek mathematics were simply anathema. On the other hand, he published almost nothing that dealt with early Greek mathematics *per se*, partly no doubt in order to avoid controversy. Still, he had a number of notable allies in classics who shared his general skepticism. In fact, a debate was then underway in which these skeptics questioned the level of truly scientific activity among the followers of sixth-*century physiologoi,* particularly the early Pythagoreans. German classical philology had witnessed a very different type of debate when Nietzsche published his *Birth of Tragedy,* but in a sense the parallel holds true. Leading classicists saw themselves as *Kulturträger*, which meant that they were quite accustomed to playing for "high stakes" (or at least imagining they were). Owing to their spiritual affinity with the ancient Greeks, they did not think of themselves as mere scholars: their discipline and special expertise carried with it an implicit social responsibility, namely to explain the deeper meaning of Greek ideals to that special class of German society, its *Bildungsbürgertum,* who perhaps alone could appreciate the true mission of the German people, especially when faced with momentous "world-historical" events like the Great War.

After Imperial Germany collapsed following that calamitous struggle, it should come as no surprise that fresh fissures developed within the humanities and, in particular, the discipline of classical philology. This suggests that by the time Neugebauer brought forth his new vision for understanding the history of the exact sciences a quite general reorientation had long been underway among experts who specialized in classical Greek science and philosophy. At any rate, Neugebauer had plenty of good company. He could thus cite the work of classical scholars like Eva Sachs and Erich Frank – dubbed by their opponents as "hyper-critical" philologists – while defending his case for recasting the early history of Greek mathematics.

Thus, in a synopsis of van der Waerden (1940) for *Mathematical Reviews*, he wrote:

> In the first paragraph the author shows that the famous paradoxa of Zeno (for example, of the tortoise and Achilles) are not at all directed against the infinite divisibility of geometrical magnitudes, but that their aim is simply to support the assumption of Parmenides that all movement is only a human fiction. The second part points out that *in Zeno's time no mathematical theory of importance existed* in which infinitesimal methods played a role. This fits in with the general concept of the development of Greek mathematics, which is familiar, at least since E. Frank's book "Plato und die sogenannten Pythagoreer" [Halle, 1923]. The last paragraph emphasizes that the so-called "crisis" of the foundation of Greek mathematics did not originate in the problem of infinite divisibility but from the discovery of irrationals (Neugebauer 1940).[5]

B. L. van der Waerden (1903–1996) was a distinguished Dutch mathematician who had taken a course on ancient mathematics with Neugebauer in Göttingen. They remained good friends and corresponded regularly about historical matters, but they also often disagreed. Only a year after he wrote the above, Neugebauer came back to the same issue while reporting on (van der Waerden 1941), a paper on Pythagorean astronomy: "The author gives an outline of the development of Greek astronomy in its earlier phases. He seems to have overlooked the book of E. Frank, *Plato und die sogenannten Pythagoreer* [Niemeyer, Halle, 1923], where essential points of his theory are already published" (Neugebauer 1941).

Neugebauer's persistent references to Frank's book appear to have made no impression on van der Waerden, who remained in Leipzig after the Nazis rose to power. This makes it highly unlikely, of course, that he knew of Neugebauer's printed remarks from 1940–41, at least not until some time after the war had ended. Yet when he brought out the original Dutch edition of *Science Awakening* in 1950 – a more popular account of the exact sciences in antiquity that drew heavily on Neugebauer's researches – van der Waerden presented the legendary Pythagoras as the founder of a scientific school, one in which the sage's teachings had a profoundly mathematical character as opposed to the doctrines of a religious sect that practiced number mysticism.

Neugebauer, who was not Jewish, could have stayed on in Göttingen. After Courant's dismissal, however, he chose instead to leave for Copenhagen, where Harald Bohr arranged a three-year appointment beginning in January 1934. From this new outpost he continued editing Springer's *Zentralblatt* until 1938, at which point he resigned in protest of Nazi racial policies that had led to the removal of Jewish colleagues from its board. These events then paved the way for the founding

[4]See Christianidis (2004) for a recent account of older as well as newer historiography on Greek mathematics.

[5]Neugebauer here alludes to the so-called "foundations crisis" that supposedly ensued with the discovery of incommensurable magnitudes. This interpretation became popular during the 1920s, but later fell out of favour (Christianidis 2004).

of *Mathematical Reviews,* which Neugebauer co-managed beginning in 1940, after his arrival at Brown University. Courant, who was now teaching at New York University, had by this time severed his publishing connections with Springer. Ten days after the devastating blow to Jewish property and life during the *Reichskristallnacht,* he wrote to Ferdinand Springer informing him that he wished to resign as editor of the "yellow series" (Reid 1976, 208–209). Still, Courant continued to maintain his former contacts in Göttingen after the war. He often visited the Mathematics Institute, whose new director Franz Rellich had earlier been part of the "Courant clique" that was forced to leave in 1933. Neugebauer, by contrast, refused ever to set foot in Germany again (he did visit Austria once or twice however).

Despite his loathing for the Nazis, Neugebauer steered clear of politics when commenting on the work of scholars whom he surely knew to be faithful followers of Hitler's brand of fanatical German nationalism. A striking example of this can be seen in his review of the German translation of the well-known *Commentary on Book I of Euclid's Elements*, written by the neo-Platonic philosopher Proclus in the fifth century (Steck 1945). One should note that this rather large volume with extensive commentary by Max Steck, a hardcore Nazi from Munich, managed to get published in the year 1945. Neugebauer praised the work of the translator and then wrote this about Steck's contribution:

> The introduction [33 pp.] contains many words which fortunately have no English equivalent, e.g., "deutscher Geistraum," "Geistschau," "in- und ausstrahlen," etc. By means of this "denkanschauend" method Proclus is made a founder of the German Idealismus for which Cusanus, Copernicus, Kepler, Hegel, Gauss (!) and many others are quoted. On the other hand, Proclus is considered as the culmination of Greek mathematics. The author here follows [Andreas] Speiser with whom he shares the tendency to consider the last phase of Greek metaphysics as representative of Greek mathematics. The subsequent commentary on Proclus shows the same contempt for the chronological element of history. There is hardly a combination of any pair of famous names missing, however great their distance may be (Neugebauer 1945).

Once he was located in the United States Neugebauer published regularly in English in the Danish journal *Centaurus* as well as in numerous American publications, including George Sarton's *Isis,* the official journal of the History of Science Society. By the early 1950s, however, a first wave of negative reaction began to swell up among émigré scholars now residing in the United States. In 1951 Neugebauer's revisionist interpretation came under strong attack in *Isis* in an article entitled "Philolaos in Limbo, or: What Happened to the Pythagoreans?", written by Georgio de Santillana and Walter Pitts. The first author, well-known for his book *The Crime of Galileo,* had fled fascist Italy to take up a post at the Massachusetts Institute of Technology. Thus, it was fitting that the authors began their essay by citing these famous words: "Several years ago there was published in Rome a salutary edict which, in order to obviate the dangerous tendencies of our present age, imposed a seasonable silence upon the Pythagorean opinion that the earth moves...". They then proceeded to explain their present purpose:

> These are the opening words of Galileo's preface to his Dialogue on the World Systems. One would be tempted to repeat them almost word for word today, apropos [sic] of certain contemporary philological research. The invisible edict or "trend" to which we refer has decreed that the whole development of Greek mathematics and astronomy must be condensed into a rather short interval of time around 400 B.C., so that almost all the mathematics, astronomy, and music theory of the "so-called Pythagoreans" becomes contemporary with Plato and his successors (Santillana and Pitts 1951, 112).

Three different groups are then identified as being responsible for this trend. The first of these is only vaguely named by alluding to "the massed power of Platonic and Aristotelian scholarship." Far more important for their critique is the role played by the aforementioned "hyper-critical philologists", especially Sachs and Frank, but also the American, W. A. Heidel, author of "The Pythagoreans and Greek Mathematics" (Heidel 1940). Erich Frank, who had succeeded Heidegger in 1928 as professor of philosophy in Marburg, had been forced to flee Germany after losing this chair in 1935; he eventually came to Harvard as a Rockefeller Fellow. Unable to secure regular employment in the United States, he died in Amsterdam in 1949. His older study Frank (1923) argued that when Aristotle spoke about "so-called Pythagoreans" he was referring to the circle around Archytas of Tarentum, who was a friend of Plato as well as a gifted mathematician. This argument supported Frank's larger thesis, according to which the early Pythagoreans were merely a religious sect and played no substantive role in early Greek science.

The third group of trend setters was "the recent school of scientific historians which has attempted to trace the connection between Babylonian and Greek mathematics." Several works are cited by three authors: Neugebauer, van der Waerden, and the mathematician Kurt Reidemeister. "Relying on Frank," it is charged, "these authors have dismissed the entire tradition about early Greek mathematics, and supplanted it either with a most improbably late transference of Babylonian mathematics to Greece in the Vth century, or else have tried to fill the gap with speculations, conceived certainly in a true and subtle mathematician's spirit, derived from conjectural traces in Euclid and Plato" (*ibid.*). Having identified Erich Frank as the key culprit responsible for this hyper-critical treatment of sources on the Pre-Socratics – in the present case the authenticity of fragments attributed to the Pythagorean Philolaos form the principal matter under dispute – de Santillana and Pitts proceed to demolish the arguments in his book. Since Frank was no longer among the living, there was small chance of a rebuttal, although they also chided the distinguished classicist Harold Cherniss for having been duped by Frank's arguments regarding the authenticity of the Philolaos fragments (Cherniss 1935, 386).

The year 1951 also saw the publication of the original Copenhagen edition of *The Exact Sciences in Antiquity*. It was reviewed at length in *Isis* by George Sarton, who noted that no one but Neugebauer could have written such a book. Sarton also paid tribute to Cornell University for its role in helping the author produce this idiosyncratic *synthesis* based on his six Messenger Lectures from 1949. This opportunity, Sarton felt sure, gave Neugebauer just the incentive he needed to address a broader set of historical issues, something he was otherwise loathe to do. In his review, Sarton put the matter this way: "as he does not like synthetic work and even affects to despise it, he would probably not have written this book without that flattering invitation, and we, his readers, would have been the losers" (Sarton 1952, 69).

One can easily read between the lines here, since Sarton, the doyen of American historians of science, certainly saw himself as a leading representative of that very genre of scholarship to which he here alluded. Nor was this review altogether positive. Sarton voiced skepticism, for example, when it came to Neugebauer's claims regarding the historical impact of Babylonian mathematics and astronomy. Noting that neither Hipparchus nor Ptolemy made mention of earlier Babylonian theoretical contributions, he wondered how historians could ever know that these Greek astronomers drew on such sources? As for Babylonian algebra, why should we assume that this knowledge survived long after the period of Hammurabi when there is no extant evidence *for a continuous tradition* of high mathematical culture in Mesopotamia? And if such mathematical knowledge persisted, how was it transmitted? After all, the complexity of the Babylonian algebraic and astronomical techniques required an expertise similar to Neugebauer's own. Sarton also took sharp issue with Neugebauer over the "centrality of Hellenistic science," especially his claim that this melting pot of ancient science later spread to India before entering Western Europe, where it held sway until the time of Newton. In Sarton's view, the Hellenistic period marked the final phase of Babylonian science, though he admitted some slight influences on both the Indian and Islamic cultural spheres. For the most part, however, Sarton contrasted the larger long-term impact of Greek science with the relatively meager legacy of the Babylonian tradition. For him, this was the gravest shortcoming of all; how could Neugebauer write a book called *The Exact Sciences in Antiquity* and virtually ignore the achievements of the Greeks? Doing that was comparable to writing a play entitled *Hamlet* while leaving out the figure of Hamlet himself. With that quip Sarton could chide Neugebauer's Danish editors – identified as Zeuthen's countrymen – for allowing their distinguished friend to make such a blunder.

Sarton's criticisms reflect the views of a generalist who clearly found Neugebauer's overall framework far from convincing (Sarton 1936a). He had the highest respect for the author's specialized contributions to research on the ancient exact sciences – work that required not only formidable mathematical abilities but also immense discipline – but this review makes plain that he saw Neugebauer's book as the product of a remarkable specialist. Sarton's overall verdict – seen from his personal vantage point as someone who hoped to open inroads for the history of science within the curriculum of American higher education – echoed Neugebauer's own forthright opinion that he "did not like synthetic work." *Exact Sciences*, Sarton opined, was of limited value for introductory courses; it should not and could not be taken as a model for teaching the history of ancient science. Though full of nicely chosen anecdotes and a good deal of general information, it simply did not pass muster as a global account of the history of the exact sciences in ancient cultures. Noel Swerdlow later expressed a very different opinion when he wrote: "Neugebauer here allowed himself the freedom to comment on subjects from antiquity to the Renaissance. The expert can learn something from it, and from its notes, every time it is read, and for the general reader it is, in my opinion, the finest book ever written on any aspect of ancient science." (Swerdlow 1993, 156).

George Sarton saw himself as a champion of what he called a synthetic approach to the history of mathematics (Sarton 1936b, 11). What Neugebauer thought about this can well be surmised from the preface to the first edition of *Exact Sciences in Antiquity:* "I am exceedingly skeptical of any attempt to reach a "synthesis" – whatever this term may mean – and I am convinced that specialization is the only basis of sound knowledge" (Neugebauer 1969, vii–viii). Paging through Sarton's booklet, *The Study of the History of Mathematics,* one can easily understand Neugebauer's dismissive attitude. There one reads that:

> The main reason for studying the history of mathematics, or the history of any science, is purely humanistic. Being men, we are interested in other men, and especially in such men as have helped us to fulfill our highest destiny. As soon as we realize the great part played by individual men in mathematical discoveries – for, however these may be determined, they cannot be brought about except by means of human brains –,we are anxious to know all their circumstances (Sarton 1936b, 12).

Sarton's humanistic approach to the history of mathematics thus derives from simple human curiosity, which he admits is the same instinct that feeds public fascination with murderers. Whereas newspapers skillfully exploit this "insatiable desire to know every detail of a murder case, those who are more thoughtful wish to investigate every detail of scientific discoveries or other creative achievements" (*ibid.*). This loftier interest apparently has much to do with Sarton's sympathy for hero worship: "One soon realizes that mathematicians are much like other men, except in the single respect of their special genius, and that genius itself has many shapes and aspects" (*ibid.*).

Not surprisingly, Neugebauer drew a sharp line between his work and that of dabblers like Sarton, though he never launched a frontal attack on the latter's own works. He did, however, occasionally publish critical responses to Sarton's opinions in *Isis*, one of which sheds much light on the intellectual fault lines that divided them. In a review of B. L. van der Waerden's *Science Awakening,* Sarton expressed dismay over the author's "shocking ingratitude" towards Moritz Cantor, whom he called "one of the greatest scholars of [the] last century, a man to whom every historian of mathematics owes deep gratitude." After citing this passage, Neugebauer went on to explain why he was writing this "Notice of Ingratitude" (Neugebauer 1956):

> Since I must conclude that this statement in its generality would also apply to myself, I should like to point out that I never felt a trace of indebtedness to Cantor's voluminous production. I do not deny, of course, the fact that it had a great influence, though in a direction quite opposite to what Professor Sarton's statement implies. I always felt that its total lack of mathematical competence as well as its moralizing and anecdotal attitude seriously discredited the history of mathematics in the eyes of mathematicians, for whom, after all, the history of mathematics has to be written. In methodological respects, Cantor's work might be of some value for historians of science since it contains so many drastic examples of how one should not approach a problem.. .. If Cantor had not philosophized about a goose counting her young or about oriental mathematics, which was equally inaccessible to him, but instead had studied the texts themselves, he would have avoided countless misinterpretations and inaccuracies which have become commonplace. It was with good reasons that the *Bibliotheca Mathematica* for years ran a special column devoted to corrections of errors in Cantor's *Geschichte der Mathematik*. But no amount of corrections can ever remedy consistent mediocrity (Neugebauer 1956, 58).

Given that Neugebauer's academic career was decisively shaped by his training and background as a mathematician, one can easily understand his aversion to the writings of Cantor and Sarton. He was most definitely not a "synthetic" historian in the sense of Sarton, but we can say just as assuredly that his work was guided by a larger view of the history of mathematics. His was an approach to history deeply grounded in the mathematical culture he grew up in, and his sensibilities as a historian were from the very beginning guided by a grandiose vision. Neugebauer worked on details, but always with a larger landscape in mind. His attitude toward his own work seems to have also contained elements of playful irony. When he came to the end of his Messenger lectures on the exact sciences in antiquity, he offered a simile to describe the historian's craft:

> In the Cloisters of the Metropolitan Museum in New York there hangs a magnificent tapestry which tells the tale of the Unicorn [Fig. 29.6]. At the end we see the miraculous animal captured, gracefully resigned to his fate, standing in an enclosure surrounded by a neat little fence. This picture may serve as a simile for what we have attempted here. We have artfully erected from small bits of evidence the fence inside which we hope to have enclosed what may appear as a possible, living creature. Reality, however, may be vastly different from the product of our imagination; perhaps it is vain to hope for anything more than a picture which is pleasing to the constructive mind when we try to restore the past (Neugebauer 1969, 177).

Fig. 29.6 The Unicorn in Captivity, one of seven tapestries dating from 1495–1505 located in The Cloisters in New York. In the pagan tradition, the unicorn was a one-horned creature that could only be tamed by a virgin; whereas Christians made this into an allegory for Christ's relationship with the Virgin Mary.

References

Brack-Bernsen, Lis. 2016. Otto Neugebauer's Visits to Copenhagen and his Connections with Denmark, in (Jones, Proust, Steele 2016, 107–126).

Cherniss, Harold. 1935. *Aristotle's Criticism of the Presocratics*. Baltimore: Johns Hopkins Press.

Christianidis, Jean, ed. 2004. *Classics in the History of Greek Mathematics*. Dordrecht: Kluwer.

Fowler, David H. 1999. *The Mathematics of Plato's Academy. A New Reconstruction*. 2nd ed. Oxford: Oxford University Press.

Frank, Erich. 1923. *Plato und die sogenannten Pythagoreer : ein Kapitel aus der Geschichte des griechischen Geistes*. Halle: Niemeyer.

Heath, Thomas L., ed. 1956. *The Thirteen Books of Euclid's Elements*. Vol. 1. New York: Dover.

Heidel, W.A. 1940. The *Pythagoreans and Greek Mathematics*. *American Journal of Philosophy* 61: 1–33.

Høyrup, Jens. 2016. As the Outsider walked in the Historiography of Mesopotamian Mathematics until Neugebaruer, in (Jones, Proust, Steele 2016, 165–196).

Jones, Alexander, Christine Proust, and John Steele, eds. 2016. *A Mathematician's Journeys: Otto Neugebauer and Modern Transformations of Ancient Science*. Archimedes: Springer.

Klein, Jakob. 1936. Die griechische Logistik und die Entstehung der Algebra. *Quellen und Studien zur Geschichte der Mathematik, Astronomie und Physik, B: Studien* 3 (18-105): 122–235.

———. 1968. *Greek* Mathematical Thought and the *Origin of Algebra*. Trans. Eva Brann. Cambridge, MA: MIT Press.

Neugebauer, Otto. 1925. Litteraturanmeldelse, (T. E. Peet, The Rhind Mathematical Papyrus). *Matematisk Tidsskrift A*: 66–70.

———. 1927. *Zur Entstehung des Sexagesimal*systems. *Abhandlungen der Akadamie der Wissenshaften zu Göttingen* 13: 1–55.

———. 1932. Apollonius-Studien. *Quellen und Studien zur Geschichte der Mathematik, Astronomie und Physik, B: Studien* 2: 215–254.

———. 1934. *Vorlesungen über Geschichte der antiken mathematischen Wissenschaften, Erster Band: Vorgriechische Mathematik. Berlin: Verlag Julius Springer*.

———. 1936. Zur geometrischen Algebra (Studien zur Geschichte der Algebra III). *Quellen und Studien zur Geschichte der Mathematik, Astronomie und Physik, B: Studien* 3: 245–259.

———. 1940. Review of (van der Waerden 1940), *Mathematical Reviews*.

———. 1941. Review of (van der Waerden 1941), *Mathematical Reviews*.

———. 1945. Review of (Steck 1945), *Mathematical Reviews*.

———. 1956. A Notice of Ingratitude. *Isis* 47: 58.

———. 1963. Reminiscences on the Göttingen Mathematical Institute on the Occasion of R. Courant's 75th Birthday, Otto Neugebauer Papers, Box 14, publications vol. 11.

———. 1969. *The Exact Sciences in Antiquity*. 2nd rev. ed. New York: Dover.

Pyenson, Lewis. 1995. Inventory as a Route to Understanding: Sarton, Neugebauer, and Sources. *History of Science* 33 (3): 253–282.

Reid, Constance. 1976. *Courant in Göttingen and New York: the Story of an Improbably Mathematician*. New York: Springer.

Remmert, Volker, and Ute Schneider. 2010. *Eine Disziplin und ihre Verleger*. In *Disziplinenkultur und Publikationswesen der Mathematik in Deutschland, 1871–1949*. Bielefeld: Transkript.

Ritter, Jim. 2016. Otto Neugebauer and Ancient Egypt, in (Jones, Proust, Steele 2016, 127–164).

Rowe, David E. 1986. "Jewish Mathematics" at Gottingen in the Era of Felix Klein. *Isis* 77: 422–449.

———. 2004. Making Mathematics in an Oral Culture: Göttingen in the Era of Klein and Hilbert. *Science in Context* 17 (1/2): 85–129.

———. 2016. From Graz to Göttingen: Neugebauer's Early Intellectual Journey, in (Jones, Proust, Steele 2016, 1–59).

de Santillana, George, and Walter Pitts. 1951. Philolaos in Limbo, or: What Happened to the Pythagoreans? *Isis* 42 (2): 112–120.

Sarton, George. 1936a. *The Study of the History of Science, with an Introductory Bibliography*. Cambridge: Harvard University Press.

———. 1936b. *The Study of the History of Mathematics*. Cambridge: Harvard University Press.

———. 1952. Review of Otto Neugebauer, *The Exact Sciences in Antiquity*. *Isis* 43 (1): 69–73.

Siegmund-Schultze, Reinhard. 2009. *Mathematicians Fleeing from Nazi Germany: Individual Fates and Global Impact*. Princeton: Princeton University Press.

———. 2016. "Not in Possession of any Weltanschauung": *Otto Neugebauer's Flight* from Nazi Germany and his Search for Objectivity in Mathematics, in Reviewing, and in History, in (Jones, Proust, Steele 2016, 61–106).

Steck, Max, Hrsg. 1945. *Proklus Diadochus, Kommentar zum ersten Buch von Euklids "Elementen"*, Halle.

Swerdlow, Noel M. 1993. Otto E. Neugebauer (26 May 1899–19 February 1990). *Proceedings of the American Philosophical Society* 137 (1): 138–165.

van der Waerden, B. L. 1940. Zenon und die Grundlagenkrise der griechischen Mathematik, *Mathematische Annalen*.

———. 1941. Die Astronomie der Pythagoreer und die Entstehung des geozentrischen Weltbildes, *Himmelswelt*.

Zeuthen, H.G. 1896. *Geschichte der Mathematik im Altertum und Mittelalter*. Kopenhagen: Verlag A. F. Hoest.

30 On the Myriad Mathematical Traditions of Ancient Greece

(*Mathematical Intelligencer* 24(1)(2002): 61–64)

To exert one's historical imagination is to plunge into delicate deliberations that involve personal judgments and tastes. Historians can and do argue like lawyers, but their arguments are often made on behalf of a picture of the past, and these historical images obviously change over time. Why should the history of mathematics be any different? When we imagine the world of ancient Greek mathematics, the works of Euclid (Heath 1926), Archimedes (Heath 1897b), and Apollonius (Heath 1897a) easily spring to mind. Throughout most of the twentieth century, our dominant image of Greek mathematical traditions has been shaped by the high standards of rigor and creative achievement that are purportedly found in extant texts presumed to have been written by these three famous authors. Thanks to the efforts of Thomas Little Heath, the English-speaking world has long enjoyed easy access to this trio's major works and much else besides (Heath 1921). Yet despite this plentiful source material, our conventional picture of Greek mathematics has largely been sustained by a far smaller corpus of knowledge. Our image of Greek geometry, as conveyed in countless mathematical texts and most books on the history of mathematics, has tended to stress the formal structure and methodological sophistication found in a handful of canonical works or, more accurately, selected portions of the same. Even the first two books of Euclid's *Elements* – which concern the congruence properties of rectilineal figures and culminate in theorem II-14 showing how to square such a figure – have often been trivialized. Many writers have distilled their content down to a few definitions, postulates, and elementary propositions intended merely to illustrate the axiomatic-deductive method in classical geometry.

When considering the origins of Greek mathematics, we find similar tendencies toward selectivity. A likely picture that readily comes to mind shows a group of Athenian mathematicians gathered over a diagram, a scene that conjures up Plato's Academy during the early fourth century. Some of these geometers are famous names that appear in Plato's Dialogues, which contain several vivid scenes and vital clues. For David Fowler, these and other sources raised, but did not answer, a pressing historical question: how did the geometers of Plato's time (427?–347?) represent ratios of incommensurable magnitudes? Fowler was by no means the first to ask this question, but what interests us here is the way he went about answering it.

The discovery of incommensurables, though shrouded in mystery, presumably took place around the time of Plato's birth. Two younger contemporaries, Theaetetus and Eudoxus, both of whom had ties with the Academy, are credited with having developed the mature theories found in Euclid's *Elements* that bear on this problem. Theaetetus's classification scheme for ratios of lines appears in Book X, the longest and most technically demanding of the 13 books, whereas Book V presents Eudoxus's general theory of proportions, which elegantly skirts the problem of representing ratios of incommensurable magnitudes by providing a general criterion for determining when two ratios are equal (Chap. 31). Fowler's query led him, naturally enough, to reexamine the sources that shed light on the relevant prehistory, a task undertaken before him by Wilbur Knorr in Knorr (1975). But inquisitive minds have a way of turning over new stones before all the old ones can be found, and Fowler's inquiry eventually became a quest to grasp the central problems that preoccupied the mathematicians in Plato's Academy. Since this lost world has left only a few faint traces, but quite a few tempting mathematical clues, Fowler makes the most of the latter in an imaginative attempt to restore the historical setting. In *The Mathematics of Plato's Academy* (Fowler 1999), he offers an unabashed rational reconstruction of mathematical life in ancient Athens, replete with fictional dialogues.

Ironically, we seem to know more about the activities of the mathematicians affiliated with Plato's Academy than we do about those of any other time or place in the Greek world, including the museum and library of Alexandria, where many of the mathematical texts that have survived the rise and fall of civilizations and empires were first written. The Alexandrian mathematicians, first and foremost among them Euclid, dedicated themselves to assimilating and systematizing the

D.E. Rowe, *A Richer Picture of Mathematics*, https://doi.org/10.1007/978-3-319-67819-1_30

work of their intellectual ancestors. But we know next to nothing about their lives and how they went about their work. Even the famous author of the 13 books known today as Euclid's *Elements* remains a shadowy figure. Was he a gifted creative mathematician or a mere codifier of the works of his predecessors? Is it even plausible that a single human being could have written all the numerous works that Pappus of Alexandria later attributed to Euclid? On the basis of internal evidence alone, it seems unlikely that the author of the *Data* and the *Elements* were one and the same person. But what about all the other mostly nameless scholars who surely must have mingled with Euclid in Alexandria shortly after Alexander's death? Perhaps our Euclid was actually a gifted administrator who worked at the library and was charged with the task of heading a research group to produce standard texts of ancient mathematical works. Is it too farfetched to imagine Euclid as the ancient Greek counterpart to the twentieth century's Bourbaki?

But leaving these biographical speculations aside, we can easily agree that the *Elements* established a dominant paradigm for classical Greek geometry, or what came to be known as ruler-and-compass geometry. Indeed, synthetic geometry in the style of Euclid's *Elements* still served as the centerpiece of the English mathematical curriculum until well into the nineteenth century. For Anglo-American gentlemen steeped in the classics, no formal education was complete without a sprinkling of Euclidean geometry. This mainly meant mimicking an old-fashioned style of deductive reasoning that many believed disciplined the mind and prepared the soul to understand and appreciate Reason and Truth. With Hilbert's *Grundlagen der Geometrie* (Hilbert 1899) the Euclidean style may be said to have made its peace with mathematical modernity. Hilbert upgraded its structure and redesigned its packaging, but most of all he gave it a new modernized system of axioms. Within this universe of "pure thought" Greek mathematics could still retain its honored place. Enshrined in the language of modern axiomatics, it took on new form in countless English-language texts that presented Greek geometry as a watered-down version of Heath's Euclid.

The history of mathematics abounds with examples of this kind: a good theorem, so the adage goes, is always worth proving twice (or thrice), just as a good theory is one worthy enough to be renovated. In the case of an old warhorse like Euclidean geometry, we take this for granted. But if mathematicians will never tire of modernizing older theories, we might still do well to ask about the consequences this activity has for historical understanding. In this respect, Euclid's *Elements*, a work that has gone through more shifts of meaning and context than any other, represents a fascinating source for reflections on historical inquiry. Reading Euclid (carefully) had profound consequences for Isaac Newton, who soon thereafter immersed himself in the lesser known works of ancient Greek geometers. He emerged a different mathematician, set on defending the Ancients against Moderns like Descartes, who claimed to have found a methodology superior to Greek analysis (Newton 1976). We need not puzzle over why Newton wrote his *Principia* in the language of geometry once we understand his strong identification with what he understood by the problem solving tradition of the ancient Greeks. Nothing rankled him more than Cartesian boasting about how this tradition had been supplanted by modern analysis.

If we consider Greek mathematics from a more recent, post-Hilbertian perspective, the question of consequences for historical understanding can be posed in something like the following terms. If our dominant image of Greek mathematics was shaped by ideas from the more recent past, and if we continue to view it through the prism of Euclid's *Elements*, which itself is seen as an early model of axiomatic rigor, what effect has this had on our conception of the far more remote past in which Greek mathematics thrived? One of the more obvious consequences has been the time-honored glorification of the intellectual heritage of the ancient Greeks at the expense of other ancient cultures. This theme has been the subject of much bickering ever since the publication of Martin Bernal's *Black Athena* (Bernal 1994), a debate we need not enter here. This fracas does suggest, however, that our pictures of ancient mathematics are in the process of undergoing significant change, and this applies equally to the indigenous mathematical traditions of ancient Greece as well as those that were transmuted and transformed through interaction with other cultures. By accenting the plural in traditions, I mean to emphasize that several different currents of Greek mathematical thought continued to flourish in the Hellenistic world and beyond: we should not imagine Greek mathematics monolithically as if a single mathematical style dominated all others.

Nor should we overestimate the unity of Greek mathematics even within the highbrow tradition of Euclid, Archimedes, and Apollonius. In his *Conica* and other lesser known works, Apollonius systematically exploits an impressive repertoire of geometrical operations and techniques in order to derive a series of complex metrical theorems whose significance is often obscure. In this respect, his style contrasts sharply with the one we find in Euclid's *Elements*. When we compare their works with those of Archimedes, whose inventiveness is far more striking than any stylistic element, the contrasts only widen. Unlike Apollonius, Archimedes apparently had little interest in showcasing all possible variants of his results merely to demonstrate the power of his arsenal of techniques. He was first and foremost a problem-solver, not a systematizer, and many of the problems he tackled were inspired by ancient mechanics. Ivo Schneider has suggested that Archimedes's early career in Syracuse was probably closer to what we would today call mechanical engineering than to mathematics (Schneider 2016). Not that this was unusual; practical and applied mathematics flourished in ancient Greece and again in early modern Europe when Galileo taught these subjects as professor of mathematics at the University of Padua, which belonged to the Venetian Republic (Remmert 1998). Like Venice, Syracuse had an impressive navy, and we can be fairly sure that Archimedes spent a considerable amount of time around ships and the machines used to build them. From these, he must have learned the principles behind the various mechanical devices that Heron and Pappus of Alexandria would later describe and classify under the five classical types of machines for generating power.

Archimedes was neither an atomist nor a follower of Democritus. Nevertheless, the parallels between these two bold thinkers are both striking and suggestive. In one of his flights of fancy, Archimedes devised a number system capable of expressing the number of "atoms" in the universe. For this purpose he took a sandgrain as the prototype for these tiny, indivisible corpuscles. Obviously this doesn't mean he was an atomist, but he surely saw Democritus's atomic theory as a powerful heuristic device in mathematics. Democritus had introduced infinitesimals in geometry, and by so doing had found the volume of a cone, presumably arguing along lines similar to the ideas that led Cavalieri to announce his general principle for finding the volumes of solids of known cross-sectional area.

As is well known, Eudoxus is credited with having introduced the "method of exhaustion" in order to rigorously demonstrate theorems involving areas and volumes of curvilinear figures, including the results obtained earlier by Democritus. Archimedes used this Eudoxian method with impressive virtuosity, but since this technique could only be applied when one knew the correct result in advance he had to rely first on ingenious thought experiments to obtain provisional results. His inspiration came from mechanics. By performing sophisticated thought experiments with a fictitious balance, Archimedes could "weigh" various kinds of geometrical objects by treating these as if they were composed of "geometrical atoms" – indivisible slivers of lower dimension than the objects themselves. As he clearly realized, this mechanical method was a definite no-no for a Eudoxian geometer, but he also knew that there was "method" to this madness, since it enabled him to "guess" the areas and volumes of curvilinear figures such as the segment of a parabola, cylinders, and spheres. As Heath once put it, here we gain a glimpse of Archimedes in his workshop, forging the tools he would need before he could proceed to the next stage, namely formal demonstration (Heath 1921).

Going one step further, he carried out thought experiments inspired by a problem of major importance to the economic and political welfare of Syracuse: the stability of ships. Archimedes' idealized vessels had hulls whose cross-sections were parabolic in shape, enabling him to determine the location of their centers of gravity precisely. Had he performed a similar service in seventeenth-century Sweden for King Gustav Adolfus, the latter might have been spared the horror of witnessing one of the great blunders in maritime history: the disaster that befell his warship, the Vassa, when it flipped over and sank in the harbor on her maiden voyage. (If you've ever visited the Vassa Museum in Stockholm you'll notice that it wouldn't have taken an Archimedes to realize that this magnificent vessel was likely to keel over as soon as it caught its first strong gust of wind.) (Fig. 30.1).

Fig. 30.1 King Gustav Adolf's Vasa was the pride of the Swedish fleet, but she sank on her maiden voyage in 1628 without even leaving Stockholm's harbor.

Archimedes's work was presumably related to his other duties as an advisor to the Syracusan court, which later called upon him when the city was besieged by Marcellus's Roman armies. Plutarch immortalized the story of how Archimedes single-handedly held back the Roman legions with all manner of strange, terrifying war machines. These legendary exploits inspired Italian Renaissance writers to elaborate on Archimedes's feats of prowess as a military engineer. No longer content with mechanical contraptions, this new-age Archimedes devises a system of mirrors that could focus the sun's rays on the sails of Roman ships, setting them all ablaze. These mythic elements reflect the imaginative reception of Archimedes during the Renaissance as a symbol of the power of human genius, a central motif in Italian humanism. Within the narrower confines of scientific thought, the reception of Archimedes's works underwent a long, convoluted journey during the Middle Ages, so that by Galileo's time they had begun to exert a deep influence on a new style of mathematics. By the seventeenth century, the

Archimedean tradition had become strongly interwoven with the Euclidean tradition, but these two currents were by no means identical from their inception.

Another major tradition within Greek mathematics can be traced back to Pythagorean idealism, which continued to live on side-by-side with the rationalism represented by Euclid's *Elements*. If the Pythagorean dogma that "all is number" could no longer hold sway, this does not mean that all traces of it simply vanished. Far from it: we have every reason to believe that the Pythagorean and Euclidean traditions interpenetrated one another, influencing both over a long period of time. Euclid's approach to number theory in Books VII–IX differs markedly from that found in the *Arithmetica* of Nicomachus of Gerasa (Nicomachus 1926), who continued to give expression to the Pythagorean tradition during the first century A.D. Still, the distinctive Pythagorean doctrine of number types (even and odd, perfect, etc.) can be found in both Euclid and Nicomachus, albeit in very different guises. Thabit ibn Quarra knew both works and assimilated these arithmetical traditions into Islamic mathematics. Finding Nicomachus's treatment of amicable numbers inadequate (Euclid ignores it completely), Thabit developed this topic further. Al-Kindi later translated the *Arithmetica* into Arabic and applied it to medicine. These two writers thus helped perpetuate and transform the Pythagorean mathematical tradition within the world of Islamic learning.

Taking Pythagorean cosmological thought into account, we see an even deeper interpenetration of mythic elements within the Euclidean tradition. For Plutarch, a writer whose imagination often outran his critical judgment, Euclid's *Elements* was itself imbued with Pythagorean lore (Plutarch 1961). He linked Euclid's beautiful Proposition VI.25 with the creation myth in Plato's *Timaeus*, a work rife with Pythagorean symbolism. Plato's Demirurge, the Craftsman of the universe, fashions his cosmos out of chaos following a metaphysical principle, one that Plutarch identified with this theorem: given two rectilinear figures, to construct a third equal in area to the first figure and similar to the second. In other words, Euclid's geometrical craftsman must transform a given quantity of matter into a desired form.

But we need no Plutarchian wings of imagination to see that Euclid's *Elements* contains numerous and striking allusions to Pythagorean/Platonic cosmological thought as noted by Proclus and other commentators. The theories of constructible regular polygons and polyhedra appear in Books IV and XIII, respectively, thereby culminating the first and last major structural divisions in the *Elements* (Books I–IV on the congruence properties of plane figures; Books XI–XIII on solid geometry). In both cases, the figures are constructed as inscribed figures in circles or spheres, the perfect celestial objects that pervade all of Greek astronomy and cosmology. Perhaps most striking of all, in Book XIII, which ends by proving that the five Platonic solids are the only regular polyhedra, Euclid determines the ratio of the side length to the radius of the circumscribed sphere according to the classification scheme presented in Book X for incommensurable lines. It takes no flight of the imagination to see this as an allusion to a doctrine of celestial harmonies, an idea whose origins are obscure, but which undoubtedly stems from Pythagorean lore.

This doctrine of celestial harmonies, according to which a sublime astronomical music was produced by the movements of the invisible heavenly spheres that carry the stars and planets, continued to ring forth in the works of Plato and Cicero. Kepler went further, proclaiming in *Harmonice Mundi* (Kepler 1967) the underlying musical, astrological, and cosmological significance of Euclid's *Elements* (Field 1988). For him, Book 4 on the theory of constructible polygons, contained the keys to the planetary aspects, the cornerstone of his "scientific" astrology. Historians of science have long overlooked Kepler's self-acknowledged *magnum opus* from 1619, preferring instead to emphasize his "positive contributions" to the history of astronomy, namely Kepler's three laws. Few seem to have been puzzled about the connection between these laws and Kepler's cosmological views as first set forth in *Mysterium Cosmographicum*, where he tries to account for the distances between the planets by means of a famous system of nested Plantonic solids (Fig. 30.2).

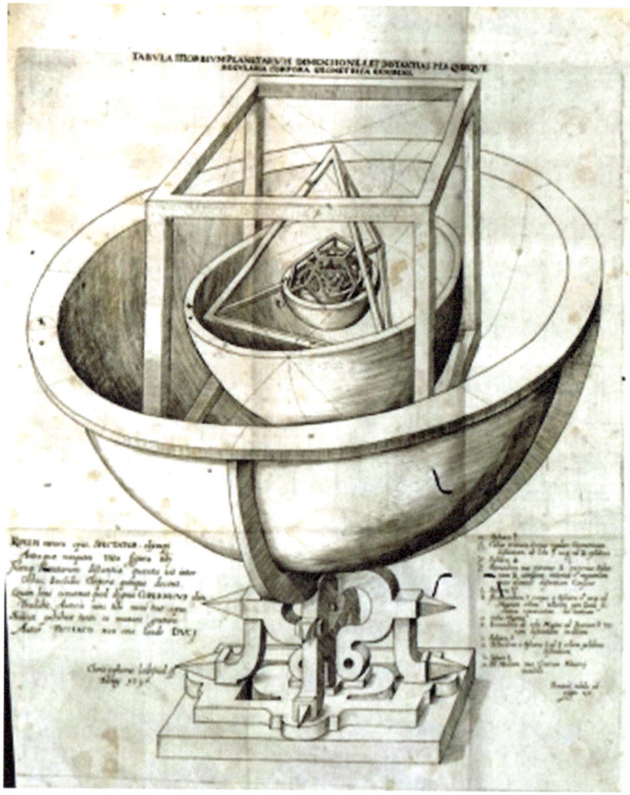

Fig. 30.2 Kepler's nested Platonic solids from the *Mysterium Cosmographicum* (1596): five solids for six planets.

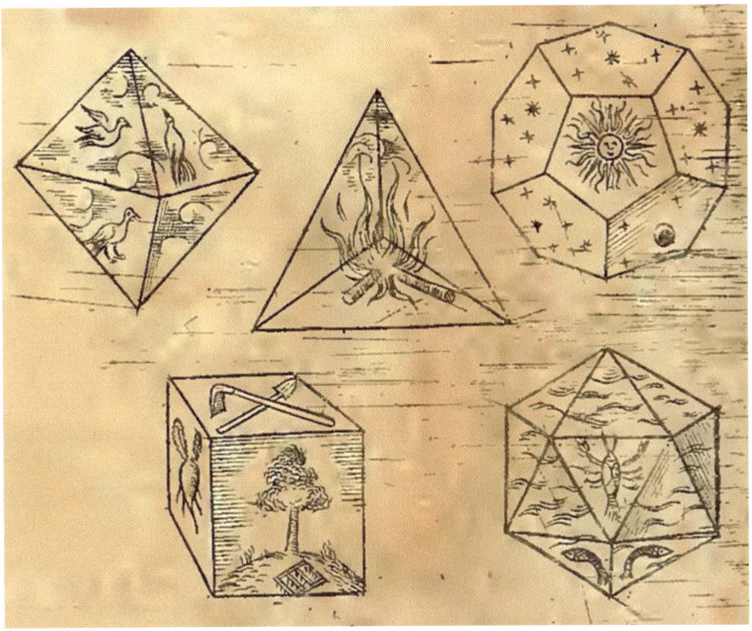

Fig. 30.3 The Platonic solids in the cosmogony of Plato's *Timaeus* as depicted in Kepler's *Harmonices Mundi*.

Kepler published his first two laws (that the planets move around the sun in elliptical orbits, and that from the sun's position they sweep out equal areas in equal times) 13 years later in *Astronomia Nova*, which presented the astronomical results of his long struggle to grasp the motion of Mars. The third law (that for all planets the ratio of the square of their mean distance to the sun to the cube of their period is a fixed constant) only appeared another 10 years later in *Harmonice Mundi*. Unlike the first two astronomical laws, the third had a deeper cosmic significance for Kepler, who never abandoned the cosmological views he advanced in 1596. Indeed, for him the third law vindicated his cosmology of nested Platonic solids by revealing the divine cosmic harmonies that God conceived for this system as elaborated by Kepler in Book V of *Harmonice Mundi*.

Kepler knew Euclid's *Elements* perhaps better than any of his contemporaries, and his imagination ran wild with it in *Harmonices Mundi*. Like so many early moderns, he saw his work as the continuation of a quest first undertaken by the ancient Greeks. Kepler believed that the Ancients had already discovered deep and immortal truths, none more important than those found in the 13 books of the *Elements*. And since truth, for Kepler, meant Divine Truth, he saw his quest as inextricably interwoven with theirs. His historical sensibilities were shaped by a profoundly felt religious faith that led him to identify his Christian God with the Deity that pagan Greeks described in the mythic language of Pythagorean symbols. We gasp at the gulf that separates our post-historicist world from Kepler's naïve belief in a transcendent realm of bare truth. We can only marvel in the realization that it was Kepler's sense of a shared past that enabled him to compose his *Harmonice Mundi* while contemplating the truths he thought he saw in the works of ancient Greek writers (Fig. 30.3).

These brief reflections suggest some broader conclusions for the history of mathematics: that mathematical knowledge, as a general rule, is related to various other types of knowledge, that its sources are varied, and that the form and content of its results are affected by the cultures within which it is produced. Those who have produced mathematics historically have done so in a number of quite different societies within which these producers have had quite varied functions. Western mathematics owes much, of course, to ancient Greek mathematicians, but even within the scope of their traditions we encounter considerable variance in the styles and even the content and purposes of their mathematics. For this reason, we should avoid the temptation to identify Greek mathematics with one dominant paradigm or style.

References

Bernal, Martin. 1994. *Black Athena: the Afroasiatic Roots of Classical Civilization*. Vol. 1. New Brunswick: Rutgers Univ. Press.

Field, J.V. 1988. *Kepler's Geometrical Cosmology*. Chicago: University of Chicago Press.

Fowler, David H. 1999. *The Mathematics of Plato's Academy*. 2nd ed. Oxford: Clarendon Press.

Heath, Thomas L. 1897a. *Apollonius of Perga: Treatise on Conic Sections*. Cambridge: Cambridge Univ. Press.

———. 1897b. *The Works of Archimedes*. Cambridge: Cambridge Univ. Press. (Reprint, New York: Dover, 1912.)

———. 1921. *A History of Greek Mathematics*. 2 vols. Oxford: Clarendon Press.

———., Trans. 1926. *The Thirteen Books of Euclid's Elements*, vol. 3. Cambridge: Cambridge Univ. Press, (Reprint, New York: Dover, 1926).

Hilbert, David. 1899. Grundlagen der Geometrie. In *Festschrift zur Einweihung des Göttinger Gauss-Weber Denkmals*. Leipzig: Teubner.

Kepler, Johannes. 1967. *Weltharmonik*. Trans. Max Caspar. München: Oldenbourg.

Knorr, Wilbur R. 1975. *The Evolution of the Euclidean Elements: A Study of the Theory of Incommensurable Magnitudes and Its Significance for Early Greek Geometry*. Dordrecht: Reidel.

Newton, Isaac. 1976. Geometry: the first Book 1. In *The Mathematical Papers of Isaac Newton*, ed. D.T. Whiteside, vol. 7, 305–309.

Nicomachus, Gerasenus. 1926. *Nicomachus of Gerasa: Introduction to Arithmetic*. Trans. M. L. D'Ooge, F. E. Robbins, and L. C. Karpinski. New York: Macmillan.

Plutarch. 1961. *Plutarch's Moralia*, vol. 9, *Table Talk*, VIII.2, 720A. Cambridge, Mass.: Harvard University Press.

Remmert, Volker. 1998. *Ariadnefäden im Wissenschaftslabyrinth. Studien zu Galilei: Historiographie—Mathematik—Wirkung, in Freiburger Studien zur Frühen Neuzeit*. Bd. 2 ed. Berlin: Peter Lang.

Schneider, Ivo. 2016. *Archimedes: Ingenieur, Naturwissenschaftler und Mathematiker. Mathematik im Kontext*. Heidelberg: Springer.

The Old Guard Under a New Order: K. O. Friedrichs Meets Felix Klein

31

(*Mathematical Intelligencer* 6(2)(1984): 74–76)

Constance Reid's recent tribute to K. O. Friedrichs (Reid 1983) undoubtedly brought back fond memories to those who knew the man and his many achievements (see also Reid 1986). It was a great pleasure for me to interview Friedrichs in January 1982, only about a year before his death. He was already in very delicate health. His wife Nellie (see Biegel 2012) was kind enough to arrange the interview, but warned me beforehand that her husband tired rather easily and was somewhat hard of hearing. Nevertheless, he was extremely forthcoming in discussing his early career with me, and quick to dismiss some of my faulty misconceptions regarding Göttingen mathematics in the 1920s, which was the main topic of conversation. Most of what Friedrichs spoke about can be found either in Reid's eulogistic article or in her book on Richard Courant. One particular incident he related to me, however, was only alluded to in her article, and because of its personal interest and the larger themes it suggests, it seems appropriate to repeat it here, in a somewhat embellished form. This is the story Friedrichs told me about the day he met Felix Klein (Figs. 31.1, 31.2).

The meeting took place in June 1922, well into the twilight of Klein's career. The aged Olympian rarely went out anymore and could only get about in a wheelchair. His main exercise was occasional excursions through the botanical garden that was located immediately between his home and the *Auditorienhaus*, where he had taught for some 30 years. The winter before he did manage to attend the banquet honoring Hilbert on his 60th birthday, on which occasion he presented his colleague with a copy of the *Vortrag* Hilbert had delivered in Klein's Leipzig seminar of 1885. But at the birthday party that evening, Klein was noticeably absent, as his health could only withstand a certain amount of activity and excitement in one day.

By this time, Klein also had greatly curtailed his administrative and organizational activities, most of which, in any event, had been disrupted by the War. At the end of March 1922, he stepped down from his position as head of the subcommittee for mathematics, astronomy, and geodesy of the *Notgemeinschaft der Deutschen Wissenschaften.* In the meantime, the second volume of his *Gesammelte Mathematische Abhandlungen* was published and, with the assistance of Erich Bessel-Hagen, considerable progress already had been made in preparing the third and final volume for publication. The *Notgemeinschaft* originally authorized an expenditure of 50,000 Marks to support the publication of Klein's Collected Works, but spiraling inflation eventually forced the organization to lay out an additional 400,000 Marks in order to complete the project. This third volume, the culminating achievement of his career, presented for the first time Klein's correspondence of 1881–82 with Henri Poincaré, which documents their famous rivalry over the early development of the theory of automorphic functions (see Chap. 11). It was only a few months before meeting Friedrichs that Klein finally recovered Poincaré's half of this correspondence, which had been lost for over 10 years. Ever since October 1914, when he and 92 other German scholars signed the chauvinistic appeal "To the Cultural World," Klein's international reputation had taken a precipitous slide downward, and losing the Poincaré letters certainly had not helped. Numerous parties prodded him to produce them: Mittag-Leffler asked for the letters before the War broke out, as he was planning to dedicate a volume of *Acta Mathematica* to Poincaré who died in 1912. He, for one, was greatly miffed that Klein could have lost such important correspondence, and he even suspected foul play on Klein's part. It was not until 1923 that this volume (no. 39) honoring Poincaré finally appeared, containing the long lost Klein-Poincaré correspondence.

Probably little or none of this background was known to the 21-year old Friedrichs, as he approached his first and only meeting with Felix Klein with a mixture of excitement and trepidation. Certainly he had arranged for it carefully enough: his father was indirectly acquainted with Klein's brother Alfred, as they were both lawyers in Düsseldorf, and

D.E. Rowe, *A Richer Picture of Mathematics*, https://doi.org/10.1007/978-3-319-67819-1_31

Fig. 31.1 Kurt Otto Friedrichs as a student.

Fig. 31.2 Felix Klein as portrayed in a painting by Max Liebermann from 1912 (Mathematisches Institut, Universität Göttingen).

the latter agreed to write a short letter of introduction paving the way for the meeting. Still, the shy young mathematician was not at all sure how he would be received once he arrived at Wilhelm Weber Strasse 3 (Fig. 31.3).

But whatever his misgivings, the reception he was accorded – and Friedrichs spoke about it vividly and with great animation – exceeded all his hopes and expectations.

"I was amazed," he told me, "simply swept off my feet by Klein's grace and charm. I was just a student and completely unknown, but he treated me like I was some kind of big-shot from Paris." Klein inquired about Friedrichs' study plans in Göttingen, offered his considered advice, and gave his complete attention to what the aspiring young student had to say. Friedrichs was "simply astounded by his performance," not that he was unaware of the other side of Klein's personality: "He could be very charming and gentlemanly when all went his way, but with anyone who crossed him, he was a tyrant!" And to reinforce this point he mentioned some of the familiar anecdotes about Klein's run-ins with Max Born, and the unpleasant time Carl Ludwig Siegel had living in Klein's home.

Nor did Friedrichs have much respect for the circle around Klein at the end of his career: "Most were good second-rate mathematicians, not so much research-oriented as they were educators. Klein wanted to place them in the schools in order to upgrade educational standards, but his people were not really effective enough." This negative judgment extended to Klein's nephew and close collaborator, Robert Fricke, with whom Klein co-authored the four large volumes devoted to elliptic modular functions and automorphic functions. Friedrichs succeeded him at the *Technische Hochschule* in Braunschweig, a few days after his death in 1930. Although Fricke's work has been regarded with esteem by other contemporary mathematicians, it is easy to see why it appeared old-fashioned to Friedrichs, who grew up reading the latest works of people like Hermann Weyl and John von Neumann.

The exact date of this meeting – June 25, 1922 (not during the autumn as reported by Reid) – is known because of another recollection Friedrichs had from his conversation with Klein: their discussion took place just one day after the assassination of the German Foreign Minister, Walther Rathenau.. It was Klein's reaction to this event that Friedrichs remembered most vividly: "He was very upset about it; you could tell it was really bothering him that this had happened. He felt it was a terrible blow to the Republic,

Fig. 31.3 The home of Anna and Felix Klein at Wilhelm Weber Strasse 3 in Göttingen.

and he was deeply concerned and worried about Germany's future." Rathenau has since become something of a symbol of the fragility of the Weimar Republic. A Jewish intellectual who had studied mathematics, physics, and chemistry before assuming leadership of the *Allgemeine Elektrizitäts-Gesellschaft* (AEG) founded by his father, Walther Rathenau later went on to organize and direct the Wartime Raw Materials Department, which managed to keep the German army supplied during the critical first stages of World War I. An expert on economic questions, he later came to be seen as one of the strongest advocates for fulfillment of the terms of the Versailles Treaty. His assassination at the hands of right-wing terrorists took place less than five months after he was appointed Foreign Minister.

Felix Klein, on the other hand, was one of the most influential figures in educational affairs (which had especially strong social and political overtones) throughout the Wilhelmian era. Moreover, his politics were marked by much the same kind of ambivalence that was characteristic of that period: he was certainly progressive regarding technical education and opportunities for women, foreigners, and Jews, views that were consonant with modernist thinking. At the same time he was an ardent nationalist with imperialist sympathies and a strong predilection for autocratic rather than democratic decision-making structures.

How did Klein react to the whirlwind events of the Weimar era? Friedrichs at least gives a glimpse of his views at a critical point in German history when little else seems to be known. It is true that Courant was enjoined by Klein to run for office (which he successfully did) as a member of the Social Democratic Party, but this was most likely a purely pragmatic decision, as Klein wanted someone at the University who could stay in touch with the local political scene. Probably his own views were closer to the liberal-oriented German Democratic Party, in which his daughter Elisabeth was an active member. It may also well be that Klein eventually succumbed to the fatalism that was so endemic during this period in Germany. In his Autobiographical Sketch, written in early 1923, he ends with this reflection: "In the course of time it has become ever clearer to me, that it is only possible in a very limited sense for human beings to control their own fate, as outer circumstances independent of our will are in large measure always stepping in between." And Norbert Wiener, who visited Klein only a few months before his death, said that when he spoke "the great names of the past ceased to be mere shadowy [figures] . . . [while] . . . there was a timelessness about him which became a man to whom time no longer had a meaning."

References

Biegel, Gerd, u.a., Hrsg. 2012. *Jüdisches Leben und akademisches Milieu in Braunschweig: Nellie und Kurt Otto Friedrichs wissenschaftliche Leistungen und illegale Liebe in bewegter Zeit,* Braunschweiger Beiträge zur Kulturgeschichte, 2.

Reid, Constance. 1983. K O Friedrichs 1901–1982. *The Mathematical Intelligencer* 5(3): 23–30.

———. 1986. The Life of Kurt Otto Friedrichs. In *Kurt Otto Friedrichs Selecta*, vol. 1. Boston: Birkhäuser Boston.

32 An Enchanted Era Remembered: Interview with Dirk Jan Struik

(*Mathematical Intelligencer* 11(1)(1989): 14–26)

Dirk J. Struik was born in Rotterdam in 1894, where he attended the Hogere Burger School from 1906–1911 before entering Leiden University. At Leiden he studied algebra and analysis with J. C. Kluyver, geometry with P. Zeeman, and physics under Paul Ehrenfest. After a brief stint as a high school teacher at Alkmaar, he spent seven years at Delft as the assistant to J. A. Schouten, one of the founders of tensor analysis. Their collaboration led to Struik's dissertation, *Grundzüge der mehrdimensionalen Differentialgeometrie in direkter Darstellung,* published by Springer in 1922, and numerous other works in the years to follow.

From 1923 to 1925 Struik was on a Rockefeller Fellowship while studying in Rome and Göttingen (Fig. 32.1). During these years he and his wife Ruth, who took her degree under Gerhard Kowalewski at Prague, met many of the leading mathematicians of the era—Levi-Civita, Volterra, Hilbert, Landau, et al. After befriending Norbert Wiener in Göttingen, Struik was invited to become his colleague at M.I.T., an offer he accepted in 1926. He taught at M.I.T. until his retirement, except for a five-year period during the McCarthy era when he was accused of having engaged in subversive activities. He has also been a guest professor at universities in Mexico, Costa Rica, Puerto Rico, and Brazil.

Beyond his work in differential geometry and tensor analysis, Professor Struik is widely known for his accomplishments as a historian of mathematics and science. His *Concise History of Mathematics* (recently reissued with a new chapter on twentieth century mathematics) has gone through several printings and has been translated into at least 16 languages. His *Yankee Science in the Making,* a classic account of science and technology in colonial New England, is considered by many to be a model study of the economic and social underpinnings of a scientific culture. As one of the founding editors of the journal *Science and Society,* Professor Struik has been one of the foremost exponents of a Marxist approach to the historical analysis of mathematics and science. At the present time he is completing a study on the history of tensor analysis while working on his autobiography. He is a passionate devotee of Sherlock Holmes.

This interview was based on a December 1987 conversation.

Rowe *You entered the University of Leiden in 1912 with the intention of becoming a high school mathematics teacher. What made you change your mind and how did you manage to break into the academic world?*

Struik The man who enabled me to enter academic life was Paul Ehrenfest. Ehrenfest was born in Vienna and studied in St. Petersburg and Göttingen. He and his Russian wife, Tatiana (Tanja), had made a name for themselves with their book on statistical mechanics. It was the first work to take into account the achievements of Boltzmann and Gibbs, a great step forward at the time. In 1912 Ehrenfest was appointed professor of mathematical physics at Leiden, succeeding the great H. A. Lorentz. Ehrenfest felt greatly honored to serve as Lorentz's successor, but he was dismayed by the stiff formality of the Leiden academic world where students only saw their professors in class and half the student body disappeared by train before sundown. Having come from Göttingen, he was greatly influenced by the atmosphere there, and so he implemented some of the same reforms that Felix Klein had introduced. One of these was the mathematical-physical library, the *Leeskammer,* which on Ehrenfest's instigation was housed in Kamerlingh Onnes's laboratory. There students could browse through a wide variety of books rather than being confined to picking out a few at a time from the dusty university library. Just like in Göttingen, the *Leeskammer* proved to be a central meeting spot for students and faculty alike, and before long there was considerable intermingling between them.

Rowe *Were physics and mathematics closely allied fields at Leiden?*

D.E. Rowe, *A Richer Picture of Mathematics*, https://doi.org/10.1007/978-3-319-67819-1_32

Fig. 32.1 Dirk Struik during his student days.

Struik To a considerable degree, although perhaps no more than elsewhere at this time. This was of course a period in which revolutionary changes were taking place in physics, and Leiden was one of the leading centers in the world with Lorentz, Ehrenfest, and Kamerlingh Onnes. Lorentz and Kamerlingh Onnes were Nobel Prize winners. The latter presided over his cryogenic lab where he ran experiments on the liquification of gases under low temperatures; only shortly before this he had discovered superconductivity. Lorentz was by now curator of Teyler's Museum in Haarlem, but he came to Leiden once a week to lecture on a variety of subjects from statistical mechanics to electrodynamics, all in his serene and masterful way. It was often said that his lectures were full of pitfalls for the unwary, as he had a way of making even the most difficult things look easy. We heard other esteemed visitors from around the world – Madame Curie, Rutherford, Einstein – so it is little wonder that the most talented students were attracted to physics. I never felt quite at home in this field; I was always more adept at thinking in terms of mathematical, and especially geometrical concepts.

Rowe *Did Ehrenfest's ideas influence you in any definite way beyond the impact of his personality?*

Struik Oh indeed, he himself had been influenced by Felix Klein's views, which stressed the underlying unity of ideas that were historically unrelated, like group theory, relativity theory, and projective and non-Euclidean geometry. The way he taught statistical mechanics and electromagnetic theory, you got the feeling of a growing science that emerged out of conflict and debate. It was alive, like his lectures, which were full of personal references to men like Boltzmann, Klein, Ritz, Abraham, and Einstein. He told us at the beginning that we should teach ourselves vector analysis in a fortnight–no babying. Ehrenfest's students all acknowledge how much his method of exposition has influenced their own teaching. I remember a digression he once entered into on integral equations that I later used in my own course. He also recommended extracurricular studies; in my case he advised me to study group theory (again Klein's influence) together with a fellow pupil. I once asked Ehrenfest what was then one of the difficult questions of that day: whether or not matter exists. He proceeded to explain not only the status of matter as of 1915 (when $E = mc^2$ had just been put on the map), but also how the facts of sound and electricity tie in with the three dimensions of space, noting that if the Battle of Waterloo had been fought in a two-dimensional space we would be able to detect the sound of its cannon fire even today.

Rowe *Who were some of the other students you got to know in Leiden?*

Struik There were several whom I got to know quite well, especially through our scientific circle "Christiaan Huygens." One of the most outstanding was Hans Kramers. He, too, came from Rotterdam, but he attended the *Gymnasium,* so we did not know one another in high school. He later took his Ph.D. under Niels Bohr at Copenhagen and eventually succeeded Ehrenfest at Leiden. Another was Dirk Coster who also studied under Bohr and co-discovered a new element (Hafnium – Hafniae is the Latin for Copenhagen). He later returned to the Netherlands and became professor of physics in Groningen.

Rowe *What did you do after graduation?*

Struik My stipend had run out so I had to look for work, which was not difficult to find in the summer of 1917 with so many young fellows tucked away in garrison towns. I took a job as a teacher of mathematics at the H.B.S. (high school) in Alkmaar, 20 miles north of Amsterdam. But in

Fig. 32.2 Jan Arnoldus Schouten.

November I received a letter from Professor J. A. Schouten in Delft inviting me to join him as his assistant there. Schouten was by training an engineer, but he eventually succumbed to his love of mathematics. His doctoral dissertation, which was published by Teubner, dealt with the construction and classification of vector and affinor (tensor) systems on the basis of Felix Klein's Erlangen Program. After some soul-searching, I decided to accept his offer, and I ended up spending the next seven years in Delft. The salary was less than at Alkmaar, but it gave me a wide-open window on the academic world (Fig. 32.2).

Rowe *It must have been an exciting period to work on tensor calculus.*

Struik It surely was. Schouten had shown that an application of the ideas in Klein's Erlangen Program could lead to an enumeration not only of the rotational groups underlying ordinary vector analysis, but others like the projective and conformal groups for any number of dimensions. Schouten's formal apparatus was algebraic, but it was accompanied by suggestive geometric constructs. We now know, of course, that Elie Cartan was working on related problems from a different point of view. With his great mastery of Lie group theory and Darboux's *trièdre mobile,* Cartan was able to dig deeper and obtain his own results with an almost deceptive elegance. But none of us knew of Cartan's work in 1918; his fame came much later. Schouten's work appealed to me first because of its close ties with Klein's Program, which was already familiar to me through Ehrenfest, and secondly because of its close connection with Einstein's general theory of relativity. It was not just the formal apparatus of tensors that interested me, it was the dialectics involved. For Klein, these were the interplay between complex functions, Euclidean and non-Euclidean geometry, continuous and discontinuous groups, Galois theory and the properties of the Platonic solids, et al.

For Einstein, his field theory established connections between geometry, gravitation, and electrodynamics.

Rowe *To what extent was Schouten's mathematics related to recent developments in Einstein's theory?*

Struik At the time I joined him in Delft he was busy applying his ideas to general relativity theory, i.e., the direct analysis of a Riemannian space of four dimensions. The algebra involved was fairly simple, but the differentiation required new concepts because the curvature is non-zero. Schouten was able to introduce covariant differentiation on such a space by considering what he called geodesically-moving coordinate systems. This enabled him to introduce new structure into the already existing tensor calculus utilized by Einstein. It was top-heavy with formalism, but Lorentz took an interest in it and helped to see that it was published by the Dutch Academy of Sciences. One day in 1918 Schouten came bursting into my office waving a paper he had just received from Levi-Civita in Rome. "He also has my geodesically-moving systems," he said, "only he calls them parallel." This paper had in fact already been published in 1917, but the war had prevented it from arriving sooner. As it turned out, Levi-Civita's approach was much easier to read, and of course he had priority of publication. But few people realize that Schouten barely missed getting credit for the most important discovery in tensor calculus since its invention by Ricci in the 1880s.

Rowe *You must have had a good working relationship with him.*

Struik Yes, though Schouten was neither an easy chap to work with nor to work for, but we had few difficulties, especially after I outgrew the position of being merely his assistant and became his collaborator and friend. I certainly learned a great deal from him; especially the combination of algebraic and geometric thinking typical of Klein and Darboux. Our first common publication appeared in 1918; it investigated the connection between geometry and mechanics in static problems of general relativity. Thus it accounted for the perihelion movement of Mercury, then a

Fig. 32.3 Levi-Civita and his wife (right) visiting with Vito Volterra and his family.

crucial test for Einstein's theory, by a change of the metric corresponding to a corrective force.

Rowe *When did you complete your doctoral thesis?*

Struik Originally, I planned to write my dissertation with Kluyver at Leiden on a subject in algebraic geometry, either on the application of elliptic functions to curves and surfaces, or a topic related to the Riemann-Roch theorem in the spirit of the Italian and German schools. De Rham's work appeared shortly afterward, revealing that there was a future in this field of research, especially since he showed how one could utilize concepts from tensor analysis. But in 1919 I was not aware of these possibilities, and anyway I had become increasingly occupied with tensor calculus through Schouten. I therefore arranged to have W. van der Woude, the Leiden geometer, as my thesis advisor, although the actual work grew out of my collaboration with Schouten on the application of tensor methods to Riemannian manifolds. I finally completed my thesis in 1922 and received my Ph.D. in July of that year. It was written in German and published by Springer in Berlin. The title was *Grundzüge der mehrdimensionalen Differentialgeometrie in direkter Darstellung*. Following a time-honored tradition, I paid for the book myself, which was an easy proposition in 1922. The inflation in Germany was such that it is entirely possible that the little party I threw afterwards for family and friends cost me more in guilders than the whole dissertation of 192 pages.

Rowe *I believe it was around this time that you and your wife first met.*

Struik Yes, Ruth and I met at a German mathematical congress in 1922 and were married in the ancient Town Hall of Prague in July of the following year. She had a Ph.D. from the University of Prague, where she had studied under Georg Pick and Gerhard Kowalewski. Her thesis was a demonstration of the use of affine reflections in building the structure of affine geometry, a new subject at the time. After our marriage we settled in Delft for a brief time before travelling to Rome on a Rockefeller Fellowship. We spent 9 months there while I worked with Tullio Levi-Civita.

Rowe *What sort of a man was Levi-Civita?*

Struik He was short and vivacious; his manner combined great personal gentleness and charm with tremendous will power and self-discipline. He was then about 50 years old and at the height of his fame as a pure and applied mathematician. His internationalist outlook derived from the ideals of the Risorgimento. His wife was a tall blonde woman of the Lombard type who was equally charming and graceful. She had been a pupil of his and was now his faithful friend and devoted companion; they had no children (Fig. 32.3).

Rowe *Did you learn a lot from him about tensor calculus?*

Struik No, not really. In Rome he suggested that I should take up a new field. He showed me a paper he had recently published on the shape of irrational periodic waves in a canal of infinite depth and asked if I would like to tackle the same problem for canals of finite depth. It involved complex mapping in connection with a non-linear integro-differential equation to be solved by a series expansion and a proof of its convergence. Even though Levi-Civita's methods appeared applicable to this case, the problem was far from trivial. It also appealed to me, as I liked to test my strength in an unfamiliar field.

Rowe *Did he give you any further guidance with this problem, or was he too busy with his own affairs?*

Struik I had the benefit of seeing him often, either at his apartment in the Via Sardegna or at the University near the Church of San Pietro in Vincoli where I often had a look at Michelangelo's Moses, which I greatly admired. The Leiden philosopher Bolland once said that the Moses remains gigantic even in the smallest reproduction. Yes, Levi-Civita was one of those persons who in spite of a busy and creative career always seemed to find time for other people. He was remarkably well organized. I can still hear him saying, after I asked him to write a letter for me, "Scriveró immediatemente" – and he did.

Rowe *How did your work on canal waves come out?*

Struik It went well, and I was able to bring it to a successful conclusion. Abstracts of it were published in the *Atti* of the Accademia dei Lincei, and later the full text came out in *Mathematische Annalen.* It was evidently read and studied, and later the theoretical results were experimentally verified by a physicist in California.

Rowe *Who were some of the other interesting figures you met during your year in Italy?*

Struik On the floor above Levi-Civita's apartment lived Federigo Enriques, who was known for his research in algebraic geometry and the philosophy of science. When he heard that Ruth had graduated with a thesis in geometry, he invited her to prepare an Italian edition of the tenth book of Euclid's *Elements.* She accepted, and spent much of her time preparing the text with modern commentary. Maria Zapelloni, a pupil of Enriques, corrected her Italian. It was published along with the other books in the Italian edition of the *Elements.* Besides Enriques, there were a number of other prominent mathematicians whom I met either at the university or at small dinner parties thrown by Levi-Civita and his wife. There was gentle Hugo Amaldi, who was then writing a book on rational mechanics with Levi-Civita. Then there was Guido Castelnuovo with his strong Venetian accent, and (but only at the university) the grand old man, Senatore Vito Volterra, President of the Accademia dei Lincei. I followed his lectures on functional analysis, which were largely based on his own researches. His delivery was impeccable, a style that reminded me of Lorentz's lectures. Volterra was a senator, as was Luigi Bianchi, who came to Rome from Pisa for sessions of the Senate, and whose books and papers had been among my principal guides in differential geometry. On a day excursion with the Levi-Civitas we also met Enrico Fermi, but since he was a physicist we saw little of him thereafter. Little did we imagine that he would one day be a man of destiny, a real one, not like the fascist braggart known as "Il Duce."

Rowe *It seems that Italian mathematicians took a fairly active role in politics from the time of Cremona and Brioschi.*

Struik Indeed, political involvement was not uncommon among Italian scientists ever since the Risorgimento. The ones I knew were all anti-fascists with the sole exception of Francesco Severi, another outstanding algebraic geometer. On the other hand, their antipathy towards the Mussolini regime was not a militant one, so far as I could see. Volterra was an exception in this regard. He and Benedetto Croce actively attacked the regime from their seats in the Senate. After 1930 Volterra was dismissed from the University and stripped of his membership in all Italian scientific societies. The same thing later happened to Levi-Civita. To the honor of the *Santa Sede,* he and Volterra (both of whom were Jews) were soon thereafter appointed by Pope Pius XI to his Pontifical Academy.

Rowe *What was the political atmosphere like during your stay in Rome?*

Struik One could not help being aware of fascism with all the blackshirts strutting through the streets of the city, but the political climate was relatively mild in those days, at any rate compared with what came later. The murder of Giacomo Matteotti, the Socialist opposition leader in the parliament, was still fresh in everyone's mind, and the resulting crisis in the government was very much unresolved. Mussolini tried to disavow the murder and tighten police control, but his dictatorship was off to a shaky start. Opposition papers could still appear, even the Communist *Unità.* I was able to establish a contact with one of their contributors, and I used the information he passed on to me to write an occasional article for the Dutch party paper, *De Tribune.* I do not believe that I ever saw Mussolini in person, despite his high visibility. He used to parade around on horseback in the Villa Borghese, but I had too much contempt for the sawdust Caesar to go out of my way to see this spectacle.

Rowe *I guess Rome had plenty to offer mathematically in those days. Were there many other foreigners who came to study or visit?*

Struik Yes, there were other Rockefeller fellows in Rome, and we struck up an amiable acquaintance with Mandelbrojt and Zariski, both of whom went on to become famous in their respective fields of research, Mandelbrojt at the Sorbonne and Zariski at Harvard. Mandelbrojt had worked on problems in analysis with Hadamard, and Zariski was studying algebraic geometry. Paul Alexandroff, the Russian topologist, also spent some time in Rome. At that time he especially enjoyed the relative luxury of Italy after enduring the many privations in his homeland, which was just recovering from hunger and civil war. He told us that to do topology in Russia at that time you had to convince the authorities that it was useful for economic recovery. So the topologists told them that their field could be of service to the textile industry. Alexandroff admired my winter coat, and when he learned that I had bought it with money from my stipend he dubbed it the "paletot Rockefeller."

Rowe *It sounds as though all in all you had a splendid time in Italy.*

Struik Yes, we grew very fond of life there and enjoyed many memorable experiences. I remember visiting the Vatican on Christmas Eve to witness the opening of the Anno Santo, the Holy Year 1925, in which one could receive special indulgences. The enormous basilica was filled with throngs of worshipers who had come to see the pope. He entered through a special door, the Porta Santa, in an ornately decorated chair carried high on the shoulders of selected members of the papal nobility, while the crowd shouted: "Viva il Papa!" Another occasion I recall occurred at a meeting of the Accademia dei Lincei. Levi-Civita often took us to these sessions held in its ancient palace on the left bank of the Tiber. On this particularly ceremonious occasion the Academy was visited by *Il Re* Vittorio Emanuele and his still handsome Queen, Helena of Montenegro, once a famous beauty. He had short *legs – eine Sitzgrösse* as the Germans would say – but both majesties did very well on their decorous chairs, he with a bored face listening to the speeches. After the ceremony one was permitted to go up and shake hands with the monarchs, and I was amused to see how the Americans in the audience crowded around them. I preferred to sample the pastry and sherry instead, and I was pleased to see that Levi-Civita also kept his distance from the Presence. It is said that when someone once asked Einstein what he liked about Italy, he answered, "Spaghetti and Levi-Civita." I felt pretty much the same way. I also learned to like, with some amusement, that curious blend of Catholicism and anticlericalism found among many intellectuals and socialistically-inclined workers. I had not encountered this attitude in the Netherlands, where Catholics (at any rate in public) faithfully followed their clergy in matters of morals and politics. Catholic anticlericalism in Italy, on the other hand, dates back at least as far as the Risorgimento, when the Pope was an obstacle to reform and unity, but may in fact have had its roots in the Renaissance. Galileo is a good example. As my colleague Giorgio De Santillana once told me, Galileo's attitude can only be understood if one is aware of the phenomenon of anticlericalism among Italian Catholics. Giordano Bruno's statue on the Campo di Fiore in Rome is a typical example of this challenge to the papacy.

Rowe *Where did your ventures take you after Italy?*

Struik My fellowship from the Rockefeller Foundation was renewed for another year, but on the condition that we continue our studies in Göttingen. In and of itself this was fine: Göttingen was after all the mecca of mathematicians. But we had grown fond of Italy, its people (except for the blackshirts), its history, art, and science. And we had come to take its atmosphere of courtesy among mathematicians somewhat for granted.

Rowe *What was the atmosphere in Gottingen like?*

Struik Mathematically it was very stimulating, of course, but you had to have a thick skin to survive; the Göttingen mathematicians were known for their sarcastic humor. Emmy Noether, who was shy and rather clumsy, was often the butt of some joke, as was the good-natured Erich Bessel-Hagen. In von Kerekjarto's topology book there is a reference to Bessel-Hagen in the index, but when you turn the page cited there is no reference to him in the text, only a topological figure that looks like a funny face with two big ears. That was the way they could treat you at Göttingen, where ironical jokes about one's colleagues were always in vogue. It was a world apart from the courteous atmosphere in Italy.

Rowe *Did you have any contact with the older generation of mathematicians in Göttingen?*

Struik Yes, although I never met Felix Klein, who died in the summer of 1925. Ruth and I attended his funeral, which was attended by most of the academic community in Göttingen. There were a few short speeches, one by Hilbert, and I joined the group of those who threw a spade of earth over the grave. I felt as though I had lost one of my teachers. Ehrenfest had always emphasized the importance of Klein's lectures to his students, and we read many of those that circulated in lithograph form. They are full of sweeping insights that reveal the interconnections between different mathematical fields: geometry, function theory, number theory, mechanics, and the internal dialectics of mathematics that manifest themselves through the concept of a group. During my stay in Göttingen, Courant invited me to help prepare Klein's lectures on the history of nineteenth and early twentieth century mathematics for publication, which I did. These first appeared in Springer's well-known "yellow series," and they remain, with all their personal recollections, the most vivid account of the mathematics of this period.

Rowe *What about Hilbert?*

Struik I saw a fair amount of him in those days, although he was quite old by then. His main interest was foundations questions, as he was still in the thick of his famous controversy with L. E. J. Brouwer. Hilbert was very good at reinforcing his own enormous power and authority by making use of clever assistants whose time and brains he ruthlessly exploited, but not withholding credit where credit was due. Emmy Noether had been his assistant during the war years when he worked on general relativity. Hilbert was an East Prussian, and there was a distinctly Prussian quality about him that was reflected in his relationships with his assistants. Ruth and I once asked Hilbert's assistant, Paul Bernays, to join us on a Sunday morning walk. Bernays was then in his mid-thirties and already a well-known mathematician, but he actually had to ask the *Herr Geheimrat* (which was the title one used in addressing Hilbert) whether he could spare him for a few hours.

I often attended Hilbert's seminar, which generally had anywhere from 40 to 70 participants. Often the speaker was a visitor who had come to talk about his research. It was a daunting experience to speak before such a critical audience, and many who came were justifiably apprehensive. Afterward came the chairman's *judicium,* and his verdict, usually to the point, could help or harm a young mathematician's standing considerably, at least in the eyes of his colleagues. I once spoke about my work on irrotational waves and was happy that it received a friendly reception. Others were not so fortunate. Young Norbert Wiener, for example, was too insecure and nervous to do justice to his excellent research in harmonic analysis and Brownian motion.

Rowe *Are there any particular Hilbert anecdotes that come to mind?*

Struik Oh sure, but a good Hilbert anecdote has to be told with an East Prussian accent, which he never quite lost. Once a young chap, lecturing before Hilbert's seminar, made use of a theorem that drew Hilbert's attention. He sat up and interrupted the speaker to ask: "That is really a beautiful theorem, yes, a beautiful theorem, but who discovered *it? – wer hat das erdacht?*" The young man paused for a moment in astonishment and then replied: *"Aber, Herr Geheimrat, das haben Sie selbst erdacht! –* But, Lord Privy Councilor, you discovered that yourself!" That is a true story – I witnessed it myself. Another episode I remember took place in one of Hilbert's lectures on number theory, which I followed during my stay in Göttingen.[1] The previous day he had written the prime numbers less than 100 on the blackboard, and now he came rushing into class to tell us: "*Ach,* I made a slip, a bad slip. I forgot the number 61. That should not have happened. These prime numbers are beautiful; they should be treated well – *man muss sie gut behandeln.*" On another occasion we were waiting for him in the seminar room. He finally came rushing in only to berate us: "Oh, you smug people, here you are sitting around talking about your petty problems. I have just come from the physics seminar where they are playing with ideas that will turn physics upside down!" That was Max Born's seminar, which week after week was attracting a hundred or more physicists, mostly younger men. Heisenberg and Pauli were then discussing the new matrix theory approach they were developing as an alternative to Schrödinger's wave theory.

Rowe *Did you ever get invited to Hilbert' s home?*

Struik Yes, he and his wife occasionally invited us to an evening party at their home, usually to meet some visiting celebrity. I have a better recollection of the parties at the Landaus. He was a stocky fellow and looked more like a butcher than a scientist. Having married the daughter of well-to-do Professor Paul Ehrlich, the famous chemist who found the first effective remedy against syphilis, Landau lived in upper middle-class comfort in a large and splendid home on the outskirts of town. After a sumptuous dinner our host led us to his study, a large room whose walls were covered with books, all of them mathematical. There were complete runs of important journals, collected works of famous figures, and nearly every imaginable work in number theory and analysis. No frivolous stuff here. There was nothing frivolous about his writing either. He presented his ideas as precisely as possible, in the unemotional style of Euclid: theorem, lemma, proof, corollary. He lectured the same way: precise, some of us thought pedantically precise. Occasionally he would present a well-known theorem in the usual way, and then while we sat there wondering what it was all about, he pontificated: "But it is *false – ist aber falsch*" – and, indeed, there would be some kind of flaw in the conventional formulation. Once the guests were assembled with refreshments, Landau started organizing mathematical games. One of them I still like to play once in a while. Suppose you define "A meets B" to mean that at some time A shook hands with B, or at any rate A and B touched each other. Now construct the shortest line of mathematicians connecting say Euler with Hilbert. Can you shorten it by admitting non-mathematicians in your chain, like royalty or persons who circulated widely and reached old age, like Alexander von Humboldt? All kinds

[1] Struik's notes from this lecture course, partly written in Dutch, can be found under item 570a in Hilbert's Nachlass, SUB Göttingen.

of variations are possible. Can you forge a link to Benjamin Franklin? to Eleanor of Aquitaine?

Rowe *That sounds a little like the present pastime of constructing a mathematician's ancestral tree or determining one's "Erdös number."*

Struik Yes, only the possibilities are much more open-ended. I can't resist telling one more Landau story that my former M.I.T. colleague Jesse Douglas liked to recall. One day at Göttingen Landau was speaking about the so-called Gibbs phenomenon in Fourier series, and remarked: *"Dieses Phänomen ist von dem englischen Mathematiker Gibbs (pronounced Jibbs) in Yale (pronounced Jail) entdeckt."* Only my respect for the great mathematician, said Jesse, withheld me from saying: "Herr Professor, what you say is absolutely correct. Only he was not English, but American, he was not a mathematician, but a physicist, he was not Jibbs, but Gibbs, he was not in jail, but at Yale, and finally, he was not the first to discover it."

Rowe *Who else did you meet in Göttingen?*

Struik There were many mathematicians from all over the world: Harald Bohr, Leopold Fejér, Serge Bernstein, Norbert Wiener, Øystein Ore, and of course, B. L. van der Waerden. I had met him already at the Mathematical Society in Amsterdam. He and Heinrich Grell could often be seen strolling down the Weender Strasse on either side of Emmy Noether. They were sometimes called her *Unterdeterminanten* (minor determinants). I had some contact with Courant when I first arrived. I had met him in Delft a year earlier, and he and Levi-Civita had both supported my application for a Rockefeller fellowship. Courant was then working on existence questions connected with the Dirichlet problem as these bore on potential theory and solutions to partial differential equations. These ideas were at that time elaborated in the famous Courant-Hilbert text *Methoden der mathematischen Physik*. I was very interested in this field, and was already somewhat familiar with it through Ehrenfest's lectures at Leiden. Courant's assistant, Dr. Alwin Walther, took the time to introduce me to the latest developments, which was fortunate considering that Courant was burdened with his many academic obligations. Courant was of course a brilliant man, but to me he seemed then to lack Levi-Civita's talent for organizing his time.

Rowe *When did you first begin to take a serious interest in the history of mathematics?*

Struik It was on the historic soil of Italy that I met two historians of mathematics, Ettore Bortolotti from Bologna and Giovanni Vacca, and from this point on my interest in the field has grown steadily. I also met Gino Loria, who like Castelnuovo, Enriques, Bianchi, and Severi, was a geometer, though on a more modest scale. We talked about the desirability of having more ancient texts published with commentary. Vacca was a professor in Rome. I remember when we went to visit him and were looking for his apartment along the narrow street that he lived on. Some girls were playing outside, and we asked them where Professor Vacca lived. "*Mamma mia, siamo tutte vacche*" ("We are all cows"), they giggled. But we found the house, and talked among other things about ancient Chinese mathematics, a subject that was then hardly touched. "Learning enough Chinese characters for mathematical purposes is not difficult when you try," he said; but I never tried.

Later I met the director of the Dutch archeological institute in Rome, G. J. Hoogewerff, who was then working on Dutch Renaissance painters, the *Zwerfvogels* (wander-birds). When he heard of my interest in the history of mathematics, he suggested that I take a look at a Dutch Renaissance mathematician who had become an Italian bishop and advisor on calendar reform at the Fifth Lateran Council of 1512–1517. His name was Paul van Middelburg – Paolo di Middelborgo. It was more work than I anticipated, as it required reading Latin texts in incunabula and post-incunabula, but it was a nice occasional break from my work on hydrodynamics and function theory. For once I could profit from the Latin I learned in preparation for my entrance at the university. And so I persevered, my research leading me to a number of Rome's antiquarian libraries, like the Alessandrina and the Vatican. To get permission to enter the Vatican archives I had to go through the office of the Netherlands' ambassador, but at least it was no longer necessary to "prostrate oneself before the feet of his Holiness," which, as I was told, had been the case not long before. These libraries are only heated on cold days by a brazier with smoldering charcoal, so that you had to study with your coat on; luckily such ancient palaces had thick walls. In some of them you had to overcome the inertia of custodians who resented the intrusion of readers as an attack on their privacy.

I discovered some interesting things about the mathematical bishop who left his native Zeeland because, as he wrote, the people there considered intoxication the *summum* of virtue. An abstract of my findings was published in the *Atti* of the Accademia dei Lincei, and the full text appeared later in the *Bulletin of the Netherlands Historic Institute*. Only a few people have taken the time to glance at it, but let us say that the work was good for my soul.

Rowe *When did you begin taking a wider view of the history of mathematics and science, taking into account the social context that shaped them?*

Struik This question interested me from quite early on, and I followed the role played by science, and particularly mathematics, in the wake of the Russian Revolution. In fact, I saw this question of mathematics in society as a testing ground for my newly acquired Marxist views. Did "external" factors actually influence the "internal" structure of science, its growth or stagnation? Until fairly recently, it seems that everyone assumed this was not the case, that mathematics was a purely intellectual undertaking whose development is best understood by analyzing ideas and theories independent of the social system that produced them. But Marxist scholars had already shown that almost equally exalted fields like literature and biology could be successfully tackled using the tools of historical materialism. So what about mathematics? Around the turn of the century mathematics flourished in a state of blissful innocence. One could do geometry, algebra, analysis, and number theory in a delightful social vacuum, undisturbed by any extraneous pressure other than that exerted by one's immediate academic and social milieu. Even as late as 1940 G. H. Hardy could maintain that the "real" mathematics of the great mathematicians had, thank goodness, no useful applications. Yet 50 years earlier Steinmetz in the USA and Heaviside in England were already applying advanced mathematical concepts in electrical engineering, and probability and statistics were being utilized in biology, the social sciences, and industry. None of these developments, however, seemed to influence the mathematicians' purist outlook on the field.

When I assisted in editing Klein's lectures on nineteenth century mathematics during my stay in Göttingen, I learned how profoundly the French Revolution had influenced both the form and content of the exact sciences and engineering, as well as the way in which they were taught. This was especially due to the impact of the newly-founded Ecole Polytechnique in Paris, headed by Gaspard Monge. Quite clearly the educational reforms of this period were intended to benefit the middle classes and not the *sans culottes.* This realization gave me more confidence in the potential efficacy of historical materialism as an approach to the development of mathematics.

This confidence was strengthened a few years later when I read Boris Hessen's landmark paper on seventeenth-century English science. Hessen emphasized that even an Olympian figure like Newton was a man of his times who was inspired by problems that were central to the expanding British mercantile economy – problems posed by mining, hydrostatics, ballistics, and navigation. The British Social Relations in Science Movement, which included such figures as J. D. Bernal, J. B. S. Haldane, J. Needham, L. Hogben, and Hyman Levy, followed the trail blazed by Hessen, producing a number of germinal ideas for the history of science. These writers were a strong source of inspiration to me in thinking about the historical relationship between mathematics and society, and my views were strengthened by conversations with Levy and J. G. Crowther who were visiting the Boston area from England. Such an attitude also implies concern for the social responsibility of the scientist. In 1936 1 helped to launch the quarterly *Science and Society,* which for 50 years now has been bringing this message of responsibility to the academic world. Some of my contributions to early issues of *S&S* deal with the sociology of mathematics.

Rowe *I understand that it was through Norbert Wiener that you first came to the United States.*

Struik Yes, Wiener was one of those Americans who had come to Göttingen in the mid-twenties, and he and I took to each other from the beginning. We talked a good deal of shop, as was wont in Göttingen and with Wiener. I became acquainted with his work in harmonic analysis and Brownian motion, which made it clear to me that I had met an exceptionally strong mathematician. But in matters of the world, such as politics, he was rather naive. He then seemed to think that the main problem in the world was overpopulation. But at the same time he was fiercely internationalist and detested the way many scientists from the allied countries still snubbed the Germans.

Anyway, we drank beer together and took walks through the woods in the Hainberg overlooking the town. He asked me about my future plans and I admitted that they were rather vague and unpromising. I had spent seven years as an assistant in Delft, which was a very nice job but with no future prospects. Academic openings in those days were few and far between in the Netherlands. Wiener then suggested that I come to the United States. He told me about New England and M.I.T., where he was an assistant professor; they were looking for new blood and he thought I might fit in.

Rowe *Were you attracted by the prospect of joining the M.I.T. faculty?*

Struik Yes, I knew of M.I.T. through the *Journal of Mathematics and Physics* that it issued, where papers by C. L. E. Moore and H. B. Phillips on projective and differential geometry had appeared. So I knew there were congenial spirits in the mathematics department there. Wiener also made it all sound very attractive by describing the natural beauties of New England, his father's farm in the country, and the mountain climbing he and his sister Constance had been doing. Of course, I was foot-loose at the time and this would have been a step up the academic ladder, which was particularly important as it was then, as I said, quite difficult to land a promising job in mathematics. I told him that I might well take him up on this proposition if an offer came my way, and my wife Ruth also liked the idea.

Rowe *So you were interested in coming to the United States, but perhaps open to other offers as well.*

Struik Yes, and by the time I heard from M.I.T. I received another tempting offer from the Soviet Union, where my brother Anton had been working as an engineer.

Otto Schmidt, a mathematician and academician in Moscow, sent me an invitation to give lectures there. My work in differential geometry was not unknown in Russia, as I discovered in 1924 when I was invited to join the committee that was preparing the collected works of Lobachevsky. Shortly before I left for the United States, Kazan University also bestowed on me its seventh Lobachevsky prize. I sometimes wonder what might have happened had I accepted Otto Schmidt's offer and gone to work as one of his collaborators. Schmidt was not only a gifted scientist he was also a first-rate organizer. Not long after I heard from him the conquest of the Arctic became an important part of the Socialist program, leading to the famous airplane expeditions of 1936–37 to the North Pole and the scientific expedition that spent 274 days on an ice floe. Schmidt was one of the leaders of these expeditions and the research that led to settlements in the huge wastelands of Northern Russia and Siberia. Under him I might have turned my attention to soil mechanics, for which my work on hydromechanics could have served as a preparation. Or perhaps I would have gone in for Polar exploration ... On the other hand, my natural Dutch obstinacy, also in politics, might have gotten in the way and brought me into conflict with the trend toward conformity typical of the later Stalin years. At any rate, I weighed this decision very carefully, including the factor of Ruth's health, which was not good at the time. We both agreed that life in the United States would be an easier adjustment, both in terms of the economic circumstances and the language and culture. And so I accepted the offer from M.I.T., with the idea that I might consider accepting the offer from the Soviet Union at a later date (Fig. 32.4).

Rowe *Was your choice by any chance influenced by an attraction to the culture of New England?*

Struik Not at all. I really had no idea of New England and Yankees and the whole variety of American cultures at this point. As a matter of fact when I received the invitation from President Samuel Stratton of M.I.T. in September 1926, I had to take out my atlas to see where Massachusetts was located. I was surprised to learn that it was in the northeast and not on the Mississippi – perhaps I confused it with Missouri. Since that time I have always been very tolerant of those Americans who think that Hamburg is in Bavaria, or that Pisa and not Padua is near Venice.

Fig. 32.4 Dirk Struik lecturing on tensor calculus in 1948, the year he published his *Concise History of Mathematics* and *Yankee Science in the Making*.

Rowe *You were a good friend of Norbert Wiener. What qualities did you admire most in him?*

Struik I would say his courage and his sensitivity. He was a man of enormous scientific vitality which the years did not seen to diminish, but this was complemented by extreme sensitivity; he saw and felt things for which most of us were blind and unfeeling. I think this was partly due to the overly strict upbringing he had as a child prodigy. Wiener was a man of many moods, and these were reflected in his lectures, which ranged from among the worst to the very best I have ever heard (Fig. 32.5). Sometimes he would lull his audience to sleep or get lost in his own computations – his performance in Göttingen was notoriously bad. But on other occasions I have seen him hold a group of colleagues and executives at breathless attention while he set forth his ideas in truly brilliant fashion. Wiener was among those scientists who recognized the full implications of the scientist's unique role in modern society and his responsibilities to it in the age of electronic computers and nuclear weapons. I well remember how upset he was the day after Hiroshima was bombed. When I remarked that because of Hiroshima the war against Japan should now come to a speedy close without much further bloodshed – a common sentiment at the time

Fig. 32.5 Norbert Wiener (center) and Dirk Struik (right) in the centennial procession at M.I.T., April 1961.

and the official justification still heard today – he replied that the explosion signified the beginning of a new and terrifying period in human history, in which the great powers might prove bound to push nuclear research to a destructive potential never dreamed of before. He also recognized and detested the racism and arrogance displayed in using the bomb against Asians. He just saw further than the rest of us. In Wiener's day robots were largely the stuff of fiction. His favorite parables concerned such robots or similar devices with the capability of turning against those who built them: Rabbi Loew's *Golem,* for example, or Goethe's *Sorcerer's Apprentice,* the Genie of the *Arabian Nights,* and W. W. Jacobs' *Monkey's Paw.* Today we all know that cybernetics, the science of self-controlling mechanisms, has an increasing impact on industry and employment, on warfare and the welfare of human beings.

Rowe *You have continued to combine scholarship with political activism since you came to this country. Tell me something about your political activities.*

Struik During the Second World War I stayed at M.I.T. and taught mathematics to the "boys in blue" sent to us by the navy. For some time I also spent weekends in Washington working at one of the Netherlands' desks in connection with the war effort, and I participated in the activities of the Queen Wilhelmina Fund, the Russian War Relief Fund, and the Massachusetts Council of American-Soviet Friendship. This latter work, which was a logical consequence of the anti-fascist campaign for collective security waged in the late thirties, together with my support for the Indonesians in their fight for independence and for the 1948 campaign of the Progressive Party, attracted the attention of sundry cold and hot warriors of the postwar period. I was called before the witch-hunting committees and an ambitious district attorney had me and my friend Harry Winner indicted on three counts of "subversion." That was in 1951, the beginning of the McCarthy era. There were wild newspaper headlines, M.I.T. suspended me (but luckily not my salary), and I was let out on heavy bail. Bertrand Russell was then lecturing at Harvard. When told that I was accused of attempting to overthrow the governments of Massachusetts and the United States, he murmured gravely, "Oh, what a powerful man he must be!"

Rowe *What became of the charges against you?*

Struik The case never came to trial, but it was not until 1955 that the indictment was finally quashed and I regained my position at M.I.T. It might have taken even longer if it had not been for the strong community support that Winner and I received, the dedication of our lawyers, and the Supreme Court ruling in the Pennsylvania case of Steve Nelson, which declared that subversion was a federal offense – Steve and I had been indicted under state law. During the five years of my suspension, I lectured all over the country on the right of free speech, and at home I worked on editing the mathematical works of Simon Stevin.

Rowe *You continued to collaborate with Schouten throughout the 1930s. When did you give up doing differential geometry and concentrate on history?*

Struik In the late thirties Schouten and I co-authored a two-volume work entitled *Einführung in die neueren Methoden der Differentialgeometrie.* This gave the first systematic introduction of the kernel-index method and incorporated a number of new techniques – exterior forms, Lie derivatives, etc. – that had since been developed. My last major mathematical publication was *Lectures on Classical Differential Geometry,* which appeared in 1950. After I became an emeritus in 1960 I gradually gave up following the course of new mathematical developments. I felt a little too old for that. My goal instead has been to learn as much as I can about mathematics up to about 1940. That's a big enough field for one human being, I think: the history of mathematics from the Stone Age to the outbreak of World War II! (Fig. 32.6)

Fig. 32.6 Dirk and Ruth Struik in 1987.

Selected Publications of Dirk J. Struik

(With J.A. Schouten), On Some Properties of General Manifolds Relating to Einstein's Theory of Gravitation, *American Journal of Mathematics*, 43(1921), 213–216.

———, Über Krümmungseigenschaften einer n–dimensionalen Mannigfaltigkeit, die in einer n—dimensionalen Mannigfaltigkeit eingebettet ist, *Rendiconti del Circolo Matematico di Palermo*, 46(1922).

Grundzüge der mehrdimensionalen Differentialgeometrie in direkter Darstellung, Berlin: Springer, 1922 (Doctoral Dissertation, University of Leiden, 1922).

(With J.A. Schouten), Einführung in die neueren Methoden der Differentialgeometrie, in Christiaan Huygens} 1(1922): 333–353; 2(1923): 1–24, 155–171, 291–306; Groningen: Noordhoff, 1924.

Sull' opera matematica di Paolo di Middelburg, *Rendiconti della Accademia dei Lincei* (6), 1(1925): 305–308.

Paolo di Middelburg e il suo posto nella storia delle scienze esatte, *Periodico di Matematiche* (4), 5(1925): 337-347.

Sur les ondes irrationelles das les canaux, *Rendiconti della Accademia dei Lincei* (6), 1(1925): 373–377.

Über die Entwicklung der Differentialgeometrie, *Jahresbericht der Deutschen Mathematiker-Vereinigung*, 34(1925): 14–25.

Paulus van Middelburg, Meded. Ned. Hist. Inst. Rome, 5(1925): 79–118.

Determination rigoureuse des ondes irrotationelles périodiques dans un canal à profondeur finie, *Mathematische Annalen*, 95(1926): 595–634.

(With N. Wiener), A Relativistic Theory or Quanta, *Journal of Mathematics and Physics*, 7(1927a): 1–23.

———, Sur la théorie rélativiste des quanta, I, II, *Comptes Rendus Academie des Sciences Paris*, 185(1927b) I: 42–44 , II: 184–185.

———, Quantum Theory and Gravitational Relativity, *Nature*, 118(1927c): 852–854.

On the Geometry of Linear Displacement, *Bulletin of the American Mathematical Society*, 33(5)(1927), 523–564.

(With N. Wiener), The Fifth Dimension in Relativistic Quantum Theory, *Proceedings of the National Academy of Sciences*, 14(1928): 262–268.

(With R. Struik) Cauchy and Bolzano in Prague, *Isis*, 11 (1928): 364–366.

Kepler as a Mathematician, in *Johann Kepler 1571–1630*, (Tercentenary Commemoration, History of Science Society), Baltimore: Williams and Wilkins 1931, 39–57.

Outline of a History of Differential Geometry, I. II, *Isis*, 19(1933): 92-120; 20(1933): 161–191.

On the foundations of the theory of probability, *Philosophy of Science*, 1(1934): 50–67.

Concerning Mathematics. *Science & Society*, 1(1936–37): 81–101.

Mathematics in the Netherlands during the First Half of the XVIth Century, *Isis*, 25(1936): 46–56.

(With J. A. Schouten), *Einführung in die neueren Methoden der Differentialgeometrie*, vol. 2. Groningen: Noordhoff, 1935 , 1938.

On the Sociology of Mathematics, *Science & Society*, 6(1942): 58–70.

Marx and Mathematics, *Science & Society*, 12(1948): 118–196.

Yankee Science in the Making, Boston: Little, Brown & Co., 1948.

Lectures on Classical Differential Geometry, Cambridge, Mass.: Addison- Wesley 1950 ; (2nd ed. Reading, Mass.: Addison-Wesley, 1961.)

A Concise History of Mathematics, New York: Dover, 1948, 2nd rev. ed. 1951, (3rd rev. ed. 1967, 4th rev. ed. 1987)

Lectures on Analytic and Projective Geometry, Cambridge, Mass.: Addison-Wesley, 1953.

Julian Lowell Coolidge, *American Mathematical Monthly*, 62(1955): 669–682.

Het land van Stevin en Huygens, Amsterdam: Pegasus 1958; English trans., *The Land of Stevin and Huygens*, Dordrecht: Reidel, 1981.

Editor, *The Principal Works of Simon Stevin II, Mathematics*, vol. 2, IIA and IIB, Amsterdam: Swets and Zeitlinger, 1958.

———, *Economic and Philosophic Manuscript of 1844 by Karl Marx*, New York: International Publ., 1964.

The Determination on Longitude at Sea, *Actes du XIe congrès international d'histoire des sciences Varsovie-Cravie 1965*, 4(1968): 262–272.

Editor, *A Source Book in Mathematics, 1200–1800*, Cambridge, Mass.: Harvard University Press, 1969.

———, *Birth of the Communist Manifesto*, New York: International Publ., 1971.

J.A. Schouten and the Tensor Calculus, *Nieuw Archief voor Wiskunde*, 3(26)(1978): 96–107.

The History of Mathematics from Proklos to Cantor, *NTM,* 17(1980): 1–22.

Why Study the History of Mathematics?, *UMAP (Undergraduate Mathematics and Applications) Journal*, 1(1980): 3–28.

Mathematics in the Early Part of the Nineteenth Century, in *Social History of Nineteenth-Century Mathematics*, ed. H. Mehrtens, et al., Boston: Birkhäuser, 1981, 6–20.

The Sociology of Mathematics Revisited, *Science & Society*, 50(1986): 280–299.

Schouten, Levi-Civita, and the Emergence of Tensor Calculus, in *The History of Modern Mathematics*, vol. 2, eds. D.E. Rowe and J. McCleary, Boston: Academic Press, 1989: 99–105.

The M.I.T. Department of Mathematics during its First Seventy-Five Years, in *A Century of Mathematics in America*, ed. Peter Duren, Providence: American Mathematical Society, 1989, vol. 3: 163–177.

Further Thoughts on Merton in Context, *Science in Context*, 8(1)(1989): 227–238.

Marx and Engels on the History of Science and Technology, in *Amphora. Festschrift für Hans Wussing zu seinem 65. Geburtstag*, ed. S. Demidov et al., Basel: Birkhäuser, 1992, 739–749.

Part VI

People and Legacies

33 Introduction to Part VI

This last set of essays leaves the terrain of Göttingen, even though its tradition still looms large in the background. Here I present an assortment of mathematical people, some of whom are likely to be known to many readers. The scene now shifts somewhat abruptly to figures in North America during the last century, though usually with an eye cast toward their links with Europe. Curiosity about the roots of research mathematics in the United States – an important chapter in the larger story of American higher education – was a major factor that influenced my early interest in the German universities, particularly Göttingen's Georgia Augusta. One can hardly exaggerate the strength of that university's attraction on the nascent American community in the time of Klein and Hilbert, a short period during which foreign mathematicians flocked to Göttingen in droves (Parshall and Rowe 1994). Considering the tiny number of graduate students worldwide at that time, it is simply astounding to realize that between 1898 and 1914 over 60 of them took doctoral degrees under Hilbert's supervision! Not a few of his students later went on to enjoy distinguished careers as researchers and teachers both in Germany as well as abroad.

Glancing across the ocean, we notice a parallel development: several of the most distinguished figures in the United States also took their degrees from a single institution: the University of Chicago. Among this group, the two most prominent graduates were George David Birkhoff and Oswald Veblen, the former long associated with Harvard, the latter with Princeton. After studying together under E. H. Moore in Chicago, they taught together at Princeton before Birkhoff joined the Harvard faculty in 1912. Unlike nearly all their predecessors, though, Birkhoff and Veblen were homegrown products, mid-Westerners who remained in the U.S. for their graduate and post-doctoral education. Both, however, eventually became well known to European mathematicians: in Veblen's case by the early 1920s, in Birkhoff's even earlier. G. D. Birkhoff's fame as a mathematician has often been dated to the year 1912, a story that ought to be recalled here.

It was in 1912 that Poincaré announced his famous last theorem, published without proof in the *Rendiconti del Circolo Matematica di Palermo* just two months before his death (Gray 2013, 297–299). This theorem arose from Poincaré's investigations on the restricted three-body problem, which led him to conjecture that a continuous area-preserving map on an annular region that moves the inner and outer rings in opposite directions must necessarily have two invariant points. Birkhoff had long been studying similar dynamical systems, and so he was able to prove this result just a few months later, thereby winning fame in Europe as the first American to solve a mathematical problem of great importance. Such, at least, is the standard story found in most accounts of Birkhoff's life.

Readers of *The Mathematical Intelligencer* were recently offered a somewhat different picture of all this, as seen through Veblen's own eyes, in Barrow-Green (2011). During the academic year 1913–1914, he travelled to Europe with his wife and wrote three letters to Birkhoff about his experiences. These letters, published by June Barrow-Green, cast the events from this time in a new and quite interesting light. In September, Veblen had the opportunity to attend the Third Scandinavian Congress of Mathematicians, which was held in Christiania (Oslo). There he made the acquaintance of a number of mathematicians, including the elderly Gösta Mittag-Leffler, who greatly impressed the young American, even though Veblen probably could not understand a word of Swedish. He described him as "a fairly tall, erect and slender old man with a white moustache and long grey hair. At the first session he wore a grey frock coat evidently designed to match the hair flowing over his collar. He spoke on several occasions and, so far as I could judge did it very

D.E. Rowe, *A Richer Picture of Mathematics*, https://doi.org/10.1007/978-3-319-67819-1_33

Fig. 33.1 Oswald Veblen and George David Birkhoff.

gracefully, with some humor and a great deal of historical reference" (Barrow-Green 2011, 40). Veblen felt flattered when Mittag-Leffler cordially invited him to stop over for a visit in Stockholm, but he also sensed that "some of the men find him too condescending." (Fig. 33.1).

Following their Scandinavian travels, the Veblens went on to Germany, where they spent 10 weeks in Göttingen and another four in Berlin. In summing up his impressions for Birkhoff, Veblen wrote, "mathematically, even more than politically, [the country] is a monarchy." In affirming that opinion, he mentioned an interesting retort made by Landau, when asked whether any mathematicians were working on Abelian functions and such like things. Rather than answering this question directly, the acerbic number-theorist merely stated a general principle, from which Veblen could easily deduce the answer himself. If Hilbert has not worked in some particular mathematical field, he opined, no one in Germany will be interested in it.

Veblen noted further that Landau had nothing very flattering to say about the attitudes of mathematicians in Germany in general. According to him, they evinced little interest in what went on elsewhere, except in those cases when this bore directly on their own work. As a true representative of the Berlin tradition, Edmund Landau never wavered in his commitment to pure mathematics in Göttingen. He surely saw this as an uphill struggle, though, since Hilbert's research interests were increasingly influenced by developments in mathematical physics. Thus, the Göttingen community itself reflected the ongoing struggle across the spectrum of pure and applied images of mathematical knowledge discussed in Chap. 34, which asks whether mathematics is an art or a science. Landau thought he knew the answer, so he opposed Courant's appointment in 1920, arguing that this would only tilt the scales even further toward applications (see Chap. 28). As it turned out, Emmy Noether helped restore the balance by launching a dynamic new school in modern algebra, a turn of events no one could have foreseen (Corry 2004, 220–244).

As for Hilbert, Veblen only encountered Germany's mathematical monarch from afar, but he was duly impressed by what he saw:

> Hilbert came fully up to my expectations. He has a much better style of exposition than I had been led to expect and one could not fail to realize his extreme intelligence after watching him a few minutes. He also struck me as being both urbane and magnanimous, although the stories one hears do not bear this out—for example, the stories told from the German point of view about Poincaré's visit to Göttingen put Hilbert and the others in rather a bad light. (Barrow-Green 2011, 43)

Klein had recently retired, so Hilbert was now conducting the weekly meetings of the Mathematical Society. Veblen's brief description of the way he ran these sessions shows that Hilbert more or less followed the format long established by his older colleague:

> He opens the sessions with remarks about his own work or what he has heard of interest from the physicists, then he calls for voluntary reports from anyone who has a new theorem or result

> which may be stated briefly. Meanwhile the literature which has come in during the week is circulated about, and if Hilbert is specially interested he may make a few remarks. Then there is a literature report on definite papers usually made by a professor, then they have the principal paper. In this I was pleased to see that the hearers, especially Hilbert, constantly interrupted with questions and comments, much as you and I used to do in Princeton. About once in two weeks the meeting is followed by a supper in a beer restaurant. (*ibid.*)

Veblen heard a great deal of gossip in Göttingen, but also in Berlin, where he got to meet H. A. Schwarz, among others. No doubt this penchant for storytelling was enhanced by the raucous beer-drinking culture, particularly prominent in Berlin. Even though Schwarz was getting on in years, he could still hold his own with the younger crowd in a *Bierstube*. There were, of course, plenty of titillating stories in Göttingen, too, including rumors about Hilbert. To Veblen, his manner in conducting the society's meetings recalled that of E. H. Moore, who usually showed his best side in public, though he was also famous for behaving very badly at times. Hilbert's unusual voice made Veblen think of another American, Saul Epsteen, who had spent some time in Göttingen around the turn of the century. Presumably, Epsteen had not been there long enough to pick up Hilbert's striking East Prussian accent, but Veblen thought he heard something else. The similarity he adduced led him to conclude that Hilbert must be Jewish, a rumor that was very much in the air at this time. He asked several people, only to receive shrugs: no one really knew.

As the essays in Part V reveal, Göttingen was too small a place to easily accommodate so many big egos, which helps to explain why acrimonious relations were nothing unusual. Oswald Veblen was more than a little annoyed by the indifference people in Göttingen showed toward his own work as well as Birkhoff's. Although some voiced respect for the latter's recent proof of Poincaré's last theorem, Veblen felt this was only grudging recognition based largely on hearsay knowledge. Thus, he related an incident when he was present just as Carathéodory was opening a packet of offprints Birkhoff had sent him. Veblen had imagined Carathéodory, as an expert on variational methods, would have followed Birkhoff's works very avidly. But his only comment was that "he seems to cover a great many subjects" (Barrow-Green 2011, 42). Worse still was a lecture given by Felix Bernstein that Veblen heard; this allegedly dealt with one of Birkhoff's recent papers, but instead mainly emphasized related work by Hilbert that Birkhoff had cited. Veblen was incensed over Bernstein's performance: "when it came to explaining what your work was about, he showed that he had not the least idea. He was not able to answer the simplest questions" (Barrow-Green 2011, 43).

Birkhoff and Veblen were good friends; as Barrow-Green points out, Veblen's letters from Europe reflect their "natural solidarity" as leading representatives of the first generation of American mathematicians to study exclusively in the US. Birkhoff had a keen interest in promoting research mathematics in the United States, particularly in fields that were weakly represented, like mathematical physics. In 1926 he made his first trip to Europe serving as an advisor to the Rockefeller Foundation's International Educational Board, part of a larger story told in (Siegmund-Schultze 2001). Birkhoff spent much of that year living in Paris, from whence he made several side trips to the leading mathematical centers in Europe. Felix Klein had undertaken a similar fact-finding mission when he visited the United States back in 1893, but Birkhoff's travels were far more extensive and of far greater consequence. At the end of his stay, he wrote a 12-page report summarizing his impressions of the relative standing and levels of support for mathematical research in the countries he visited. None of his findings was particularly surprising, but the very fact that an American mathematician felt competent to judge the relative standing of the various Western European nations – his survey only extended to Hungary in the east – clearly reflects how much had changed since the days when Americans looked to Klein and Hilbert for guidance in their studies.

In Birkhoff's comparative assessment, support for research was far higher in Italy, France, Germany, and Sweden than in England, Holland, Belgium, or Denmark. He also named some of the leading figures in these countries: Volterra and Levi-Civita in Italy; Picard, Hadamard, Lebesgue and Borel in France; and Hilbert, Landau, Hecke, and Carathéodory in Germany. Outside the main centers, he mentioned several other outstanding researchers: Brouwer in Holland; Weyl in Switzerland; Harald Bohr in Denmark, and Hardy and Whittaker in Great Britain. Birkhoff got to meet them all and many others, too. In fact, he was bombarded with so many invitations from European colleagues that he had to turn down several requests to deliver lectures. What a contrast with the attitude Veblen encountered little more than a decade earlier.

Throughout his stay abroad, Birkhoff consulted closely with the physicist Augustus Trowbridge, who was also stationed in Paris as head of the IEB's European office (Siegmund-Schultze 2015). Both were engaged in ongoing discussions with Émile Borel, who hoped to gain support from the Rockefeller Foundation for a new mathematics building to be named the Institut Henri Poincaré. At the same time, Trowbridge had opened negotiations with Richard Courant, who had long been seeking a way to obtain funding for a similar structure that would house the Mathematisches Institut in Göttingen (Fig. 33.2). In early July 1926, Birkhoff and Trowbridge visited Göttingen, where they were given the red-carpet treatment. As visiting members of the National Academy of Sciences, they sat in on a meeting of the Göttingen Academy, a session that featured papers by the physicists Max Born and Robert Pohl. They

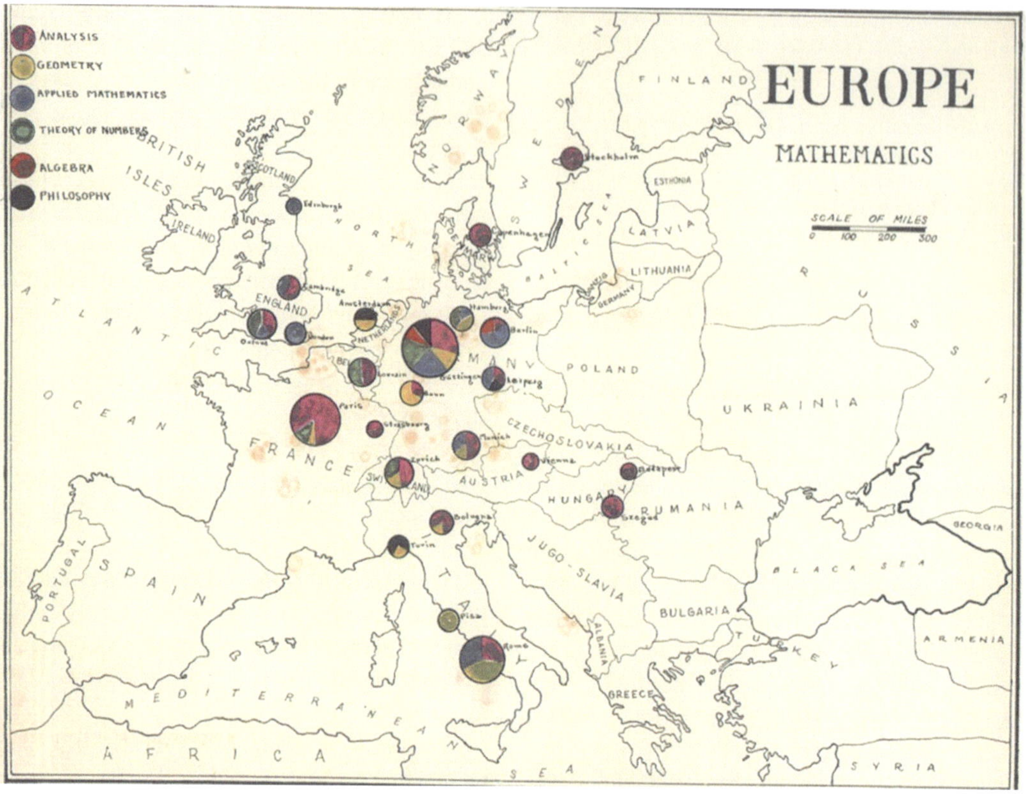

Fig. 33.2 European Centers in the Mathematical World as seen by G. D. Birkhoff in 1927 (Reproduced with permission from *Rockefeller and the Internationalization of Mathematics between the Two World Wars* Siegmund-Schultze 2001, 44).

were then invited to an evening supper at the home of the third Göttingen physicist, James Franck. Afterward, a whole group of Göttingen faculty came to greet them, including Hilbert, Landau, Courant, and Runge.

Trowbridge recorded some of his impressions of these personalities:

> Hilbert has been in precarious health, but seems very alert and is distinctively the scientific leader of the mathematical group, though in the matter of the proposed institute, Courant is the driving force. Hilbert is said to be 64 years of age, though he looks much older. Landau is . . . a rather aggressive person. Courant is the persuasive type, very tactful, etc. and the exact antithesis of Landau. Runge who began as a physicist, looks to be in the late sixties, a very charming personality, man of the world, well known and much liked as well as respected outside of Germany. (Siegmund-Schultze 2001, Appendix 6)

Trowbridge was also quick to take note of the tensions between Landau and his colleagues, a matter that came up when he asked about who was likely to succeed Hilbert when he retired. Both Courant and Franck were quick to recognize that they needed to give Trowbridge strong assurances that the Göttingen faculty would never choose Landau when that time came. For his part, their American visitor assured them that the IEB could tolerate personal differences so long as these did not pose a danger to scientific productivity. In any case, the Rockefeller program was deeply interested in the kind of integrative research that had been a Göttingen hallmark since Klein's days. Birkhoff, too, thought the Göttingen group was strongest in applied research, even though he held Landau in high esteem.

A curious factor in these deliberations, as was noted by Siegmund-Schultze, stems from the view the Americans took of the elderly Hilbert, whose importance for research activity in Göttingen was consistently overrated. Naturally, he continued to play his traditional role as mathematical monarch, but by the mid-1920s his former powers were rapidly dwindling. That his reputation continued to shine on so brightly surely had much to do with Courant's devotion to him. As described in Part V, Courant continued to uphold Hilbert's name and fame long after he had ceased to be actively involved in mathematical research. Indeed, he became a living symbol of Göttingen's past glory for all those who identified with its magical atmosphere from before the Great War – a sentiment that runs through Max Born's evocative autobiography in a striking way (Born 1978).

Some who knew Hilbert and Göttingen better, among them Hermann Weyl, were surely less inclined to view him as the heroic figure who championed mathematical truth in the face of the naysayers. In Weyl's famous image, he was the Pied Piper of Mathematics, whose sweet tune proved irresistible to the many young rats who followed him into the currents of that swiftly moving river (Weyl 1944, 614). His persuasive powers and almost hypnotic effect on those in his circle were clearly grounded in deeply held beliefs about the unique status of mathematical knowledge. For Weyl, Hilbert's outlook brought to mind the contrast with that of Königsberg's most famous thinker, Immanuel Kant:

> If Kant through critique and analysis arrived at the principle of the supremacy of practical reason, Hilbert incorporated, as it were, the supremacy of pure reason – sometimes with laughing arrogance (*arrogancia* in the Spanish sense), sometimes with the ingratiating smile of intellect's spoiled child, but most of the time with the seriousness of a man who believes and must believe in what is the essence of his own life. (*ibid.*)

Hilbert, like almost no other mathematician of his day, was seen as an advocate of the highest standards of rigor. On numerous highly visible occasions, he proclaimed, in fact, that mathematics lived or died by its status as a rigorous body of knowledge. The underlying irony here lies in the glaring gap between what he promised to deliver (or claimed to have shown) and what he actually produced. In 1900 he intimated that a direct proof for the consistency of the real number continuum was just around the corner; three decades later, Kurt Gödel proved that Hilbert's methods were inadequate even for proving the consistency of the natural numbers. Clearly, Hilbert was a rigorous thinker, but his authority and stature derived from other qualities, and these tell us much about the mathematical atmosphere in Göttingen both before and after the Great War.

This brings to mind the famous foundations crisis and the personal conflict with Brouwer. In light of what I wrote in the introduction to Part V about Hilbert's probable motive for dismissing Brouwer from the editorial board of *Mathematische Annalen*, the following passage from the first essay below stands in need of clarification:

> To some in Göttingen circles, it looked as though Hilbert had defeated the mystic from Amsterdam, but their victory celebration was unearned. Formalism never faced intuitionism on the playing field of the foundations debates. Rather, the Dutchman had merely been shown the door, ostracized from the Göttingen community that had once offered him Felix Klein's former chair.

There surely were many in Göttingen who interpreted Hilbert's role in thwarting the effort to boycott the 1928 ICM in Bologna – his dramatic triumph there directly preceded Brouwer's removal from the board – as a heroic defense of classical mathematics. Ever since be delivered his famous speech on mathematical problems in Paris, but even more so after Poincaré's death in 1912, Hilbert grew accustomed to the idea that he alone spoke for mathematics. Landau's remark to Veblen one year later clearly confirms just how dominant Hilbert's views had become. Nevertheless, formalism did not satisfy everyone. Zermelo's axiomatization of set theory in 1908, incorporating the controversial axiom of choice, was sharply criticized by those outside – and in Poincaré's case, even inside Göttingen.

Well before this, in correspondence with Hilbert, Gottlob Frege sharply criticized the ontological implications of his work on foundations of geometry (Frege 1980). Frege was an arch-realist: he took the position that, while an axiom system might faithfully describe the properties of a real mathematical entity, one could never establish the existence of a mathematical concept merely by showing that a system of axioms pertaining to it is logically consistent. This clash of opinion later attracted considerable attention among philosophers, but it barely caused a stir during the lifetime of these two combatants. In fact, Frege challenged Hilbert to publish their correspondence, but the latter refused. That refusal prompted Frege to take up his side of the argument in the pages of the DMV's official journal (Frege 1903). Frege's public critique of formalism surely was widely noticed at the time, but it failed to spark serious discussion of Hilbert's axiomatic methodology (Rowe 2000).

Hilbert was also well aware that Brouwer's intuitionism took dead aim at central tenets of his formalist philosophy, though it took over a decade before Weyl openly challenged formalism. Hilbert responded merely with *arrogancia*; indeed, throughout the 1920s he adopted the monarchial principle by pouring scorn on the upstart princes of pessimism who dared to attack his citadel. Yet, setting all of Hilbert's rhetorical flourishes aside, one should not overlook that Brouwer had long been hoping to draw him into an open debate over foundational issues. He saw an excellent chance for this in 1920 when he chose to speak about intuitionism at the Bad Nauheim conference, since Hilbert initially planned to present a lecture on his new proof theory there, too. Otto Blumenthal was in touch with both men shortly before that meeting took place, and he informed Hilbert that Brouwer was deeply disappointed after learning that Hilbert had withdrawn his talk and would not attend.[1] This was only shortly before Weyl published his famous propaganda piece on the foundations crisis (Weyl 1921), discussed in Chap. 27. Brouwer was at first elated to have gained such an important ally, but then saw with dismay how Weyl gradually drifted toward a more neutral position. Yet Hilbert's grand optimism and swift dismissal of all criticism seemed to carry the day. In fact, throughout the 1920s Brouwer's radical prescription for reforming set theory and foundations found only a few adherents (Hesseling 2003).

[1] On August 20, 1920 Blumenthal wrote to Hilbert: "Ich hoffe doch sehr, dass Sie nach Nauheim kommen werden. Von allem anderen abgesehen, halte ich die Auseinandersetzung mit Brouwers ‚Intuitionismus"für so wichtig, dass Sie dabei nicht schweigen sollten."(Nachlass Hilbert, SUB Göttingen, 403, Nr. 31, Beilage 6). After learning that Hilbert would not attend, he wrote him again, on September 16: ‚[Brouwer] ist auch schon enttäuscht, dass Sie nicht nach Nauheim kommen und dort Ihren angekündigten Vortrag halten."(Nachlass Hilbert, SUB Göttingen, 403, Nr. 31, Beilage 7).

By 1927 Brouwer had lost all patience with Hilbert, who one year before had published his grandiose lecture "Über das Unendliche" (Hilbert 1926) which culminated with a sketch for a proof of the continuum hypothesis (the proof turned out to be faulty). Brouwer decided to throw down the gauntlet in a review of the literature entitled "Intuitionistic observations on formalism" (Brouwer 1928). His main thesis – a barely concealed charge of plagiarism – asserted that the entire debate rested on four insights, all central principles of intuitionism that Hilbert was gradually coming to adopt in his work. Brouwer then predicted that the foundations debate would end once Hilbert adopted, explicitly or not, all four insights originally due to Brouwer's intuitionism. Unluckily for Brouwer, events played out differently, and the issues he hoped to make central concerns for foundational research were largely shunted aside once his student, Arend Heyting, began promoting intuitionistic logic (van Dalen 2013).

Hilbert and Brouwer were both incredibly strong-willed personalities, so they could not possibly agree just to disagree. The essays that follow, though, deal mainly with other sorts of mathematical people. Two of the more colorful characters from the last century were Donald Coxeter (Chap. 35) and Clifford Truesdell (Chap. 36), both of whom had quite distinctive historical interests and tastes. Coxeter's ran toward the playful and artistic end of the spectrum of mathematical experience. In Chap. 35, I touch on these interests, mainly by way of his best-known work, *Regular Polytopes*, first published in 1948. But I might have mentioned his work on the eleventh edition of *Mathematical Recreations and Essays*, which was originally published by Rouse Ball in 1892. Beyond these passions, Siobhan Roberts writes about Coxeter's lifelong interest in music, among other things; he was a composer as well as an accomplished pianist. He also felt that mathematics and music were intimately related and he wrote an article about this in the *Canadian Music Journal* (Roberts 2006).

At the 1954 ICM in Amsterdam, Coxeter met and befriended the Dutch artist M. C. Escher, who learned about hyperbolic tessellations from him. Escher took this idea up later in his *Circle Limit* series. Soon thereafter, Martin Gardner brought Escher's work to the attention of a wider public in one of his columns in *Scientific American* (see the April 1966 issue). In "Snapshot of Debates on Relativistic Cosmology" (Chap. 24), I had occasion to describe Coxeter's delightful take on the space-time paradoxes that confront those living in a de Sitter universe. These reminded him of Lewis Carroll's *Through the Looking Glass*, which describes a world in which cause and effect are inverted. So he cited the words of Carroll's White Queen: "He's in prison now, being punished: and the trial doesn't even begin till next Wednesday: and of course the crime comes last of all."

Truesdell's mathematical tastes were, if anything, even more exotic than Coxeter's, though with more nastiness than playfulness. As the founding editor of two Springer journals, *Archive for Rational Mechanics and Analysis* as well as *Archive for History of Exact Sciences*, one might be inclined to place him at the applied end of the spectrum. Chapter 36, however, suggests why doing so might be misleading; in fact, placing Truesdell anywhere in the landscape of twentieth-century opinion would seem pointless. He was his own man: an admirer of Euler, Hilbert, and G. D. Birkhoff, but also a harsh critic of those who sought short cuts in order to promote new applications of mathematical ideas. His wartime experiences at a summer school on applied mathematics held at Brown University form the centerpiece in that essay. Phil Davis discovered the typescript of Truesdell's personal impressions of the various Europeans who taught in that program, and he thankfully passed it on to his namesake, Chandler Davis, who was then editor of *The Mathematical Intelligencer.* It took me a while to realize the importance of this document for understanding Truesdell's later views.

One can easily trace the influence of Hilbert's sixth Paris problem on the course of Truesdell's thinking, and he often wrote about its broader impact. After the war, he wrote to Hilbert's former student, Georg Hamel, who had taken up Hilbert's lead in his work on the axiomatization of classical mechanics. Through Hamel, he got in touch with a young German, Walter Noll, who later came to work with Truesdell in Indiana. In Chap. 36, I also mention the influence Paul Neményi had on the young Clifford Truesdell. A decade after Neményi's death in 1952, Truesdell wrote about their relationship in an editorial note published in the second volume of *Archive for History of Exact Sciences*. This accompanied Neményi's essay on the historical development of fluid mechanics. Truesdell probably never knew that Paul Neményi was the biological father of Bobby Fischer, whose career both on and off the chessboard would make history. But he surely did know about his legitimate son, Peter, who studied mathematics in Princeton and later taught at the University of Wisconsin in Madison.

A certain confusion regarding father and son, Paul and Peter, ought to be clarified here, since this pertains to the English translation of the classic volume *Anschauliche Geometrie* (Hilbert and Cohn-Vossen 1932), briefly discussed in the introduction to Part III. The translation was published 20 years later by Chelsea under the certainly apt title *Geometry and the Imagination*. Although Paul Neményi is often cited on websites as the translator, the title page merely gives P. Neményi. This, together with the fact that Paul was an expert on continuum mechanics, already makes this attribution seem rather unlikely.

Peter, on the other hand, studied under Max Dehn at Black Mountain College before going on to Princeton. Dehn, in fact, asked Emil Artin to visit the college and serve as his student's outside examiner, which probably explains how Peter Neményi got into Princeton's graduate program.

Fig. 33.3 Max and Toni Dehn at Black Mountain College in 1948 (Photo by Nancy Newhall).

Max Dehn, who began his career as one of Hilbert's most gifted students before going on to do pioneering work in low-dimensional topology, was almost certainly instrumental in arranging this translation; in fact, he recommended the freshly translated text to Dorothea Rockburne, who was then studying art at Black Mountain. Dehn, too, died quite suddenly in 1952, the year *Geometry and the Imagination* was released, so he clearly knew about this translation before it ever appeared in print. Soon after his father's death, Peter Neményi wrote to Regina Fischer, Bobby's mother, who was seeking his help. He provided none, other than the name of someone he knew at Chelsea Publishing, who might provide her with translation work. Those who have never held the Hilbert & Cohn-Vossen volume in their hands might wish to simply page through it, if only to admire the extraordinary quality of the drawings (there are 330 of them).

Max Dehn, who came from a Jewish family in Hamburg, taught in Frankfurt until he was dismissed by the Nazi regime. He and his wife, Toni, had to flee the country soon after the reign of terror that began on 9 November 1938, the "Night of Broken Glass." They managed to reach Copenhagen, and from there Dehn was able to gain an appointment at the Institute of Technology in Trondheim, Norway. Barely settled in, they had to escape a second time when the Germans occupied the country in 1940. Their harrowing adventures afterward eventually brought them to the famous experimental college in Black Mountain, North Carolina (Yandell 2002, 129–135).

In many ways, it was a perfect fit for Dehn, who loved to commune with nature and contemplate beautiful things. In Frankfurt, he stood at the center of a vibrant mathematical community that was nearly the antithesis of Göttingen's competitive factory for knowledge production. Dehn and his colleagues – Carl Ludwig Siegel, Ernst Hellinger, and Paul Epstein – practiced a different model inspired by communal ideals and sensibilities. This background helps account for Dehn's remarkable ability to adapt to the atmosphere he found at Black Mountain College (Fig. 33.3). There he taught courses in philosophy and foreign languages, but also "mathematics for artists," a course in which students learned how to draw figures using principles from projective geometry, like the theorems of Pascal and Desargues. This approach complemented the teaching of the influential art instructor, Josef Albers, who came to BMC when the college opened in 1933. Albers had been teaching until then at the Bauhaus in Dessau before the Nazis forced it to close (Fig. 33.4).

Dehn was a multi-talented mathematician with strong aesthetic sensibilities. His career serves as a reminder that some of the most influential figures of the twentieth century have often operated far from the established power centers that have figured so prominently in this book. Birkhoff's map of "mathematical Europe" may be illuminating, but part of what it shows us is the "view from Harvard" in the 1920s. One should not imagine that the typical mathematician of that era was as ambitious as a Klein, Mittag-Leffler, or Birkhoff, though different forms of organized competition have clearly played an important role in shaping modern research activity and the attitudes of practitioners.

Still, rivalries between competing centers are hardly unique to mathematics; one finds them in practically all disciplines and cultures. Competition for resources has also had a strong effect on relations within mathematics faculties, a topic touched on in Chap. 34, which offers a thumbnail sketch of tensions between pure and applied mathematicians, both at Göttingen and beyond. In posing the naïve question – is (was) mathematics an art or a science? – I might have cited the opinion of Emil Artin, (Fig. 33.5) who once wrote:

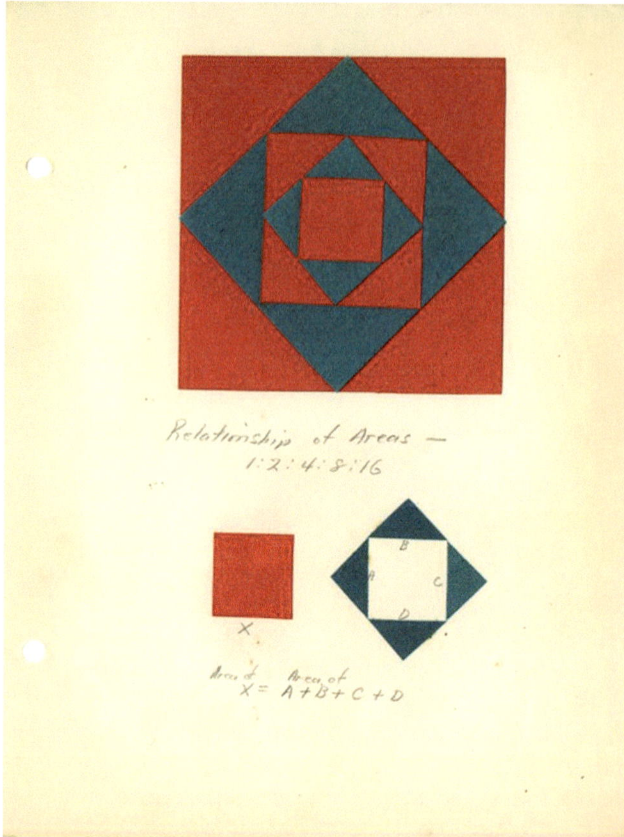

Fig. 33.4 Drawing from an art course taught by Josef Albers at Black Mountain College (Courtesy of Western North Carolina Archives).

> We all believe that mathematics is an art. The author of a book, the lecturer in a classroom tries to convey the structural beauty of mathematics to his readers, to his listeners. In this attempt, he must always fail. Mathematics is logical to be sure: each conclusion is drawn from previously derived statements. Yet the whole of it, the real piece of art, is not linear; worse than that, its perception should be instantaneous. We have all experienced on some rare occasion the feeling of elation in realizing that we have enabled our listeners to see at a moment's glance the whole architecture and all its ramifications. (Artin 1953, 475)

Many will recognize this quotation, with its ironic claim that "we have all experienced" – as lecturers and not just as listeners! – a certain elation that was even rare for an Emil Artin. These words have been cited countless times, though almost never with reference to the context. Artin was reviewing the latest installment of Bourbaki's monumental *Éléments de mathématique*, Book II: *Algebra*, and he was obviously delighted by what he had read. Clearly this work spoke to him, even if it could never quite live up to his high aesthetic standards, and indeed he went on to make a number of concrete criticisms. But he also wrote that Bourbaki was quite aware of the larger dilemma, namely that "clinging stubbornly to the logical sequence inhibits the visualization of the whole, and yet this logical structure must predominate or chaos would result." As for the larger goal, here is what he wrote at the very outset of his review:

> Our time is witnessing the creation of a monumental work: an exposition of the whole of present day mathematics. Moreover this exposition is done in such a way that the common bond between the various branches of mathematics becomes clearly visible, that the framework which supports the whole structure is not apt to become obsolete in a very short time, and that it can easily absorb new ideas. Bourbaki achieves this aim by trying to present each concept in the greatest possible generality and abstraction. The terminology and notations are carefully planned and are being accepted by an increasing number of mathematicians. Upon completion of the work a standard encyclopedia will be at our disposal. (Artin 1953, 474)

Here Artin even called to mind the giant project Klein long oversaw, *Die Encyklopädie der mathematischen Wissenschaften*, but only to contrast this with the organic structural approach of Bourbaki. Artin was also right when he predicted that this structure would stand the test of time, at least for many mathematicians. As for prestige, clearly the "artists" gained the upper hand throughout most of the last century. One need only glance at the list of names of Fields Medalists to recognize the dominance of pure mathematics over work done in more applied fields.

Artin joined the Princeton faculty in 1946, the year of its Bicentennial Conference (Chap. 37; he is pictured there in the back row; see Fig. 37.1). Harvard and Princeton by this time had developed their own special rivalry, vaguely

Fig. 33.5 This group photo from a *Göttinger Ausflug* in the early 1930s was possibly taken by Natascha Artin, whose husband Emil is standing behind Emmy Noether. Others include, from l. to r., Ernst Witt, Paul Bernays, Hella, Hermann, and Joachim Weyl. The picture was first published in *Emmy Noether: A Tribute to Her Life and Work*, James Brewer and Martha Smith (eds.), 1981. Clark Kimberling contacted many people before publication and decided that the photo was probably taken in 1932 (early that year Artin lectured in Göttingen). More recently Christophe Eckes and Norbert Schappacher reviewed the evidence in Kimberling's papers and found it more likely that the photo was made in July 1933 – thus after Bernays and Noether had been dismissed by the Nazi government (see their detailed analysis in the Oberwolfach Photo Collection, https://owpdb.mfo.de/detail?photo_id=9265).

reminiscent of the earlier one between Göttingen and Berlin. Two years after G. D. Birkhoff's death in 1944, his son Garrett represented Harvard at the Princeton conference, an important event from the immediate post-war period that I recall in Chap. 37. There the younger Birkhoff clashed with Princeton's Solomon Lefschetz, an encounter symptomatic of the tensions that had grown between these two leading centers during the 1930s and the ensuing war years.

It was also on this occasion that Hermann Weyl gave a memorable after-dinner speech in which he recalled numerous highlights from the recent past.[2] Not surprisingly, he spoke of Hilbert and Brouwer as the two greatest geniuses he had ever known, while emphasizing some of their achievements before they clashed over foundational issues. He remembered the excitement in Göttingen when Hilbert presented his solution of Waring's problem – that for any integer k there is a positive integer $s(k)$ such that any integer n can be written as a sum of no more than $s(k)$ k-th powers. In the case $k = 2$, Lagrange had proved that $s(k) = 4$, but no one had really tackled the general problem. Hilbert made this dramatic breakthrough at the very time that his dear friend Minkowski passed away.

Of more immediate significance for Weyl were Brouwer's fundamental papers in topology. These were published rapid-fire in *Mathematische Annalen* in the years 1911–12, and Weyl rightly regarded them, together with Poincaré's six memoirs from 1895 to 1904, as the works that led to the flowering of modern topological research. He called Brouwer and Paul Koebe the godfather's of his book on Riemann surfaces, which came out in 1913. "A strange couple," he went on, "Koebe the rustic, and Brouwer the mystic. Koebe at that time used to define a Riemann surface by a peculiar gesture of his hands; when I lectured on the subject I felt the need for a more dignified definition. I used the idea of cohomology for establishing the invariance of genus. Topology was in an innocent stage, then."

In his reflections on foundational problems, Weyl alluded to the brief period when he became an apostle of Brouwer and began preaching intuitionism. He also spoke of Hilbert's heroic effort to prove the consistency of classical mathematics and thereby end the foundational debates once and for all. But then came "Gödel's great discoveries. Move and countermove [with] no final solution … in sight." The philosophically minded Weyl expressed dismay over the

[2]There are two manuscripts for this speech in Weyl's Nachlass at the ETH. Peter Pesic used both in preparing the version found in Weyl, Hermann. 2009. In *Mind and Nature: Selected Writings in Philosophy, Mathematics, and Physics*, ed. Peter Pesic, 162–174. Princeton: Princeton University Press.

complacency of the typical mathematician in the face of this "serious crisis," asking, "what makes him so confident? How does he know that he builds on solid rock and does not merely pile clouds on clouds?" These queries were linked to Weyl's general misgivings. He voiced concern about how far mathematicians had drifted away from concrete concerns in favor of lofty abstractions. To a certain degree, he still found himself sympathizing with Brouwer's position. "In ... solving an algebraic equation one can see that the demands of an intuitionistic or constructive approach coincide completely with the demands of the computer," he noted. The task "is to compute the roots – with an accuracy that can be increased indefinitely when the coefficients become known with ever greater accuracy." Kronecker would certainly have nodded his head in agreement.

The final essay, Chap. 38, concerns two people who deeply influenced my own career as an historian of mathematics: Joe Dauben and Dirk Struik. When I first came to New York in the fall of 1981, I had read Joe's biography of Georg Cantor (Dauben 1979) and Dirk's *Concise History of Mathematics* (Struik 1948). But I had no idea that the two of them were on very friendly terms going back to the time when Joe Dauben was researching his book as a graduate student at Harvard. He recalls those days very vividly in Chap. 38, which I prefaced with some remarks about Struik's unique place in our field as well as his lasting influence on it. The photo below was taken during the "intermission" between the two parts of the International Congress on the History of Science. This took place in August 1989, the first half in Hamburg, the second in Munich. Instead of taking the special train reserved for participants, we drove down to Munich with Dirk, spending the first night in Göttingen, where he and I went out for a beer at the *Ratskeller*. Our waiter quickly noticed that he had a special guest in his midst. So he brought us our drinks, quickly took this photo, dashed back into the kitchen, and returned with two copies of it, mounted and framed. We were very touched by this kind gesture (Fig. 33.6).

Struik had not set foot in Göttingen for 63 years, which is to say, not since the period he recalled in the interview reprinted in Chap. 32. I should say a word about how that interview came about. In November 1984, I had the opportunity to spend a week in Leipzig as a guest of the Karl Sudhoff Institut, whose co-director was the historian of mathematics Hans Wussing. We met only briefly then, but the following summer Wussing attended the International Congress on the History of Science, held that year in Berkeley. We met there and decided to visit San Francisco together, which gave Hans the chance to tell me about the awful heat and humidity in Washington, D.C.; he had just spent several weeks there working in the Library of Congress. I remember how excited he was about two Americans in particular: Benjamin Franklin and Dirk Struik. So he told me that I absolutely must get up to the Boston area to meet the latter, which of course I did fairly soon thereafter. That first visit with him in Belmont is still a vivid memory for me.

Struik gave up doing original research in mathematics by around age 50; this decision came to him quite easily, since his work with Jan Arnoldus Schouten made no use of modern developments in topology and the theory of differential forms created by Élie Cartan. Counting ahead, this meant he still had well over a half-century to set his sights on the history of mathematics. During that time, he wrote about all manner of mathematical things, drawing on his vast knowledge of the primary as well as secondary literature. Historians of my generation will remember the new wave of interest during the 1970s in social history of mathematics, a trend already presaged in many of Struik's works. His viewpoint from the beginning was global, looking beyond the ramparts of Eurocentrism decades before it became popular to assail them. In a 1963 article on "Ancient Chinese Mathematics," he introduced mathematics teachers to the work of Yoshio Mikami, Joseph Needham and others. Ever since his first visit to Mexico in 1935, Struik has taken a passionate interest in Latin American history, and he published several articles on the history of mathematics in the Americas. He also took a keen interest in the work of Paulus Gerdes and others in the new field of ethnomathematics.

Others might show disdain or disinterest in older historical studies that now appear dated, not so Dirk Struik. He had a deep appreciation for past historiographic traditions, and he felt these have much to teach us. To see what I mean, one need only read his survey of "The Historiography of Mathematics from Proklos to Cantor," which can be warmly recommended to every historian of mathematics, but particularly to those just starting out in the field. For most of us who consider ourselves professionals, questions like "why?" or "what for?" rarely receive more than just nodding attention. Struik's reflections on this topic can be found in his essay "Why Study the History of Mathematics?" Finally, I should not neglect to add an article that deals with his own favored approach to the subject, "The Sociology of Mathematics Revisited: A Personal Note". These and other writings are described with full bibliographic information in Rowe (1994). As should be more than apparent by now, the present volume reflects to a considerable extent the inspiration I gained not only from reading Dirk Struik's works but also especially from having known him personally.

Fig. 33.6 Dirk Struik with David Rowe toasting the good life in the Göttingen Ratskeller, August 1989. The occasion marks Struik's return visit 63 years after he came to the town as a Rockefeller Fellow (see the interview in Chap. 32).

References

Artin, Emil. 1953. Review of N. Bourbaki, *Éléments de mathématique*, Book II: *Algebra*, chapters I–VII. *Bulletin of the American Mathematical Society* 59 (5): 474–479.

Barrow-Green, June. 2011. An American goes to Europe: three letters from Oswald Veblen to George Birkhoff in 1913/1914. *The Mathematical Intelligencer* 33 (4): 37–47.

Born, Max. 1978. *My Life: Recollections of a Nobel Laureate*. New York: Charles Scribner's Sons.

Brouwer, L.E.J. 1928. Intuitionistische Betrachtungen über den Formalismus, *Koninklijke Akademie van Wetenschappen te Amsterdam. Section of Sciences. Proceedings*, 31: 374–379. English translation in (Mancosu 1998, 40–44).

Corry, Leo. 2004. *Modern Algebra and the Rise of Mathematical Structures*. Basel: Birkhäuser.

Dauben, Joseph W. 1979. *Georg Cantor. His Mathematics and Philosophy of the Infinite*. Cambridge, Mass.: Harvard University Press.

Frege, Gottlob. 1903. Über die Grundlagen der Geometrie. *Jahresbericht der Deutschen Mathematiker-Vereinigung* 12: 319–324.

Frege. 1980. In *Gottlob Frege: Philosophical and Mathematical Correspondence*, ed. Gottfried Gabriel et al., 33–51. Oxford: Blackwell.

Gray, Jeremy. 2013. *Henri Poincaré: A Scientific Biography*. Princeton: Princeton University Press.

Hesseling, Dennis E. 2003. *Gnomes in the Fog. The Reception of Brouwer's Intuitionism in the 1920's*. Birkhäuser: Basel.

Hilbert, David. 1926. Über das Unendliche. *Mathematische Annalen* 95: 161–190.

Hilbert, David, und Stefan Cohn-Vossen. 1932/1952. *Anschauliche Geometrie*, Berlin: Springer, 1932; *Geometry and the Imagination*. Trans. Peter Nemenyi. New York: Chelsea, 1952.

Mancosu, Paolo, ed. 1998. *From Hilbert to Brouwer*. In *The Debate on the Foundations of Mathematics in the 1920s*, Oxford: Oxford University Press.

Parshall, Karen H. and David E. Rowe. 1994. *The Emergence of the American Mathematical Research Community 1876–1900: J. J. Sylvester, Felix Klein and E. H. Moore*, AMS/LMS History of Mathematics 8, Providence/London.

Roberts, Siobhan. 2006. *King of Infinite Space: Donald Coxeter, the Man Who Saved Geometry. New York: Walker & Company.*

Rowe, David E. 1994. Dirk Jan Struik and his Contributions to the History of Mathematics, (Introductory Essay to the Festschrift celebrating Struik's 100th Birthday). *Historia Mathematica* 21 (3): 245–273.

———. 2000. The Calm before the Storm: Hilbert's early Views on Foundations. In *Proof Theory: History and Philosophical Significance*, ed. Vincent Hendricks, 55–94. Dordrecht: Kluwer.

Siegmund-Schultze, Reinhard. 2001. *Rockefeller and the Internationalization of Mathematics between the Two World Wars: Documents and Studies for the Social History of Mathematics in the 20th Century, Science Networks, 25*. Basel/Boston/Berlin: Birkhäuser.

———. 2015. Rockefeller Philanthropy and Mathematical Emigration between World Wars. *The Mathematical Intelligencer* 37 (1): 10–19.

Struik, Dirk J. 1948/1987. *A Concise History of Mathematics*, New York: Dover, 1948. (4th rev. ed., 1987.)

van Dalen, Dirk. 2013. *L.E.J. Brouwer–Topologist, Intuitionist, Philosopher. How Mathematics is Rooted in Life*. London: Springer.

Weyl, Hermann. 1921. Über die *neue Grundlagenkrise* der Mathematik. *Mathematische Zeitschrift* 10: 39–79.

———. 1944. David Hilbert and his Mathematical Work. *Bulletin of the American Mathematical Society* 50: 612–654.

Yandell, Ben H. 2002. *The Honors Class. Hilbert's Problems and their Solvers*. Natick, Mass.: AK Peters.

34 Is (Was) Mathematics an Art or a Science?

(*Mathematical Intelligencer* 24(3)(2002): 59–64)

If you teach in a department like mine, the answer to this timeless question may actually carry consequences that seriously affect the resources your program will have available to teach mathematics in the future. In Mainz, no one is likely to protest that mathematics has long been counted as part of the *Naturwissenschaften* (natural sciences). If it were part of the *Geisteswissenschaften* (humanities), this would probably have serious budgetary implications. Of course most mathematics departments are now facing a far more immediate and pressing issue, one that can perhaps be boiled down to a different question: is mathematics closer to (a) an art form or (b) a form of computer science? If your students think the answer is certainly (b), then you can dismiss the above query as irrelevant for higher education in the twenty-first century. But since I'm mainly concerned with historical matters, let me turn to the loftier issue raised by the (parenthetical) question in the title above.

Looking into the recent past, we might wonder to what degree leading mathematicians of the twentieth century saw their work as rooted in the exact sciences, as opposed to the purist ideology espoused by G. H. Hardy in *A Mathematician's Apology*? Not surprisingly, then as now, opinions about what mathematics is (or what it ought to be) differed. For every Hardy, so it would seem, there was a Poincaré, advocating a realist approach, and vice-versa. About a century ago, when the prolific number-theorist Edmund Landau learned that young Arnold Sommerfeld – who started his career as a mathematician – was expending his talents on an analysis of machine lubricants, he summed up what he thought about this dirty business with a single sneering word (pronounced with a disdainful Berlin accent): *Schmieröl.* What could have been more distasteful to a "real" mathematician like Landau than this stuff – even the word itself sounded *schmutzig*. And so *Schmieröl* became standard Göttingen jargon, a term of derision that summed up what many felt: applied mathematics was inferior mathematics; or was it even worthy of the name? Sommerfeld himself may have grown tired of hearing about "monkey grease." In 1906 he left the field of engineering mathematics to become a theoretical physicist, one of the most successful career transitions ever made.

Of course such squabbles between pure and applied mathematicians hardly come as a surprise. Still, even within pure mathematics there has been plenty of room for hefty disputes about what mathematics ought to be. Foundational issues that had been smoldering throughout the nineteenth century stoked the brush-fire debates that spread after 1900. By the 1920s the foundations of mathematics were all ablaze, while Hilbert battled Brouwer in the center of the inferno. Their power struggle culminated with Hilbert's triumphal speech at the ICM in Bologna in 1928, followed shortly thereafter by his unilateral decision to dismiss Brouwer from the editorial board of *Mathematische Annalen* (see Chap. 26 and van Dalen 1990). To some in Göttingen circles, it looked as though Hilbert had defeated the mystic from Amsterdam, but their victory celebration was unearned. Formalism never faced intuitionism on the playing field of the foundations debates. Rather, the Dutchman had merely been shown the door, ostracized from the Göttingen community that had once offered him Felix Klein's former chair. By the time Kurt Gödel pinpointed central weaknesses in Hilbert's program in 1930, the personal animosities that had fueled these fires ceased to play a major role any longer. The foundations crisis proclaimed by Hermann Weyl in 1921 was thus already over by the time Gödel proved his incompleteness theorem. Never really extinguished, the fire had just blown out, enabling the foundations experts to go on with their work in a far more peaceful atmosphere (for an overview, see the essays in Hendricks et al. 2000).

Herbert Mehrtens has suggested that fundamental differences regarding the issue of mathematical existence reflected a broader cultural conflict that divided modernists and anti-modernists (Mehrtens 1990). No doubt philosophical disputes over existential difficulties cut deeply, but Mehrtens emphasizes that the intense foundational debates during the early twentieth century took place against a background

D.E. Rowe, *A Richer Picture of Mathematics*, https://doi.org/10.1007/978-3-319-67819-1_34

of rapid modernization, and this had a major impact on mathematical research. The impact of modernity on higher education, in general, and on mathematics, in particular, is easy enough to discern, and yet the effects on mathematical practice depended heavily on how higher mathematics was already situated in various countries. Thus, Hardy's purism can best be appreciated by remembering that nineteenth-century Cambridge had long upheld applied mathematics in the grand tradition of its famous Wranglers and physicists. During roughly the same time just the opposite tendency prevailed in Germany. There, mathematical purism held sway, reaching a high water mark in Berlin during the 1870s and 80s, the heyday of Kummer, Weierstrass, and Kronecker (see Chap. 4 and Rowe 1998).

Modernization at the German universities elevated the status of the natural sciences, which had long been overshadowed by traditional humanistic disciplines. As part of this trend, mathematicians began to pay closer attention to scientific and technological problems. Felix Klein took this as his principal agenda in building a new kind of mathematical research community in Göttingen, where Karl Schwarzschild, Ludwig Prandtl, and Carl Runge promoted various facets of applied research. Ironically, this community has often come to be remembered as "Hilbert's Göttingen," whereas Hilbert himself has often been seen through the lens of his later "philosophical" work, the formalist program of the 1920s (for reassessments of his approach to foundations, see Corry 2000, Rowe 2000, and Sieg 2000). Clearly, Richard Courant had a very different image of Hilbert in mind when he wrote the first volume of Courant-Hilbert, *Mathematische Methoden der Physik* in 1924. Just as clearly, Hilbert himself saw mathematics in very broad terms, a vision sustained by strong views about the nature of mathematical thought. The same can be said of his leading rival, Henri Poincaré, whose ideas had a lasting impact on philosophers of science.

Poincaré's work often took its inspiration from physical problems, and he made numerous important contributions to celestial mechanics and electrodynamics (see Barrow-Green 1997 and Darrigol 1995). In *Science and Hypothesis*, Poincaré examined the role played by hypotheses in both physical and mathematical research, arguing against many of the standard views about mathematical knowledge that had prevailed a century earlier. In particular, he sought to demonstrate that it was fallacious to believe "mathematical truths are derived from a few self-evident propositions, by a chain of flawless reasonings," that they are "imposed not only on us, but on Nature itself." (Poincaré 1905, p. xxi). Poincaré's alternative view, a doctrine that came to be known as conventionalism, was supported by a trenchant analysis of the geometry of physical space, then a matter of considerable controversy (for an analysis of Poincaré's views, see Ben-Menahem 2001).

Hermann von Helmholtz had attacked the Kantian doctrine according to which our intuitions of space and time have the status of synthetic a priori knowledge. This position had become central to Neo-Kantian philosophers, who insisted that Euclidean geometry alone was compatible with human cognition. Helmholtz argued, on the contrary, that the roots of our space perception are empirical, so that in principle a person could learn to perceive spatial relations in a different geometry, either spherical or hyperbolic. Like Helmholtz, Poincaré rejected the Kantian claim that the structure of space was necessarily Euclidean, but he stopped short of adopting the empiricist view, which implied that the question as to which space we actually live in could be put to a direct test. Poincaré noted that any such test would first require finding a physical criterion that would enable one to distinguish between the candidate geometries. This, however, amounted to laying down conventions for the affine and metric structures of physical space in advance, which effectively undermined any attempt to determine the geometrical structure of space without the aid of physical principles.

Poincaré's conventionalism reflected his refusal to separate geometry from its roots in the natural sciences, a position diametrically opposed to Hilbert's approach in *Grundlagen der Geometrie* (1899). Hilbert would have been the last to deny the empirical roots of geometrical knowledge, but these ceased to be relevant the moment the subject became formalized in a system of axioms. By packaging his axioms into five groups (axioms for incidence, order, congruence, parallelism, and continuity), Hilbert revealed that these intuitive notions from classical geometry continued to play a central role in structuring the system of axioms employed by the modern geometer. Nevertheless, these groupings played no direct role within the body of knowledge, since they played no role in the proofs of individual theorems. Thus, for Hilbert, the form and content of geometry could be strictly separated. In contrast with Poincaré's position, he regarded the foundations of geometry as constituting a pure science whose arguments retain their validity without any reliance on intuition or empirical support.

Hilbert originally conceived his famous lecture on "Mathematical Problems" as a reply to Poincaré's lecture at the inaugural ICM in Zurich (see Chap. 15). Stressing foundations, axiomatics, and number theory, he set forth a vision of mathematics that was at once universal and purist. Aside from the sixth of his twenty-three Paris problems, he gave only faint hints of links with other fields. The central theme of Hilbert's address, in fact, was based on the claim that mathematics, as a purely rigorous science, was fundamentally different from astronomy, physics, and all other exact sciences. Taking up a theme popularized by the physiologist Emil du Bois-Reymond, who maintained that some of mankind's most perplexing questions could never be answered by science, Hilbert turned the tables the other way around. For him, this was the quintessential difference between mathematics and the natural sciences: in mathematics alone there *could be no ignorabimus* because every well-posed mathematical question had an answer; moreover, with enough effort, that answer could be found. But Hilbert went further. This seemingly bold claim, he maintained, was actually an article of faith that every mathematician shared. Of course this was Hilbert in 1900; he hadn't yet met Brouwer!

For much of the twentieth century, young North American mathematicians were taught to believe that doing mathematics meant proving theorems (rigorously). This ethos gained a great deal of legitimacy from the explosion of interest after 1900 in foundations, axiomatics, and mathematical logic, fields which emerged along with the first generation of homegrown research mathematicians in the United States. During the 1880s and 90s, a number of young Americans came to Göttingen to study under Felix Klein, who gladly took on the role of training those who became mentors to that first generation. But by the mid 1890s, Hilbert gradually took over this formidable task (Parshall and Rowe 1994, pp. 189–234, 439–445). Modernity was sweeping through the German universities, and throughout the two decades preceding the outbreak of World War I enrollments in science and mathematics courses in Göttingen grew dramatically, as did the number of foreigners attending them.

Hilbert's ideas exerted a major impact on American mathematics, but not just on those who studied under him in Göttingen (Gray 2000). Among those who responded to his message, none did so with more enthusiasm than Eliakim Hastings Moore, who helped launch research mathematics at the University of Chicago during the 1890s. Moore's school owed much to Hilbert's research agenda, particularly the latter's axiomatic approach to the foundations of geometry. Oswald Veblen pursued this same program, first as a doctoral student at Chicago and later at Princeton, but the leading proponent of this style was another Chicago product, the Texan Robert Lee Moore.

Like his namesake and mentor, R. L. Moore served as a "founding father" for a distinctively American style of mathematics (Wilder 1982). He and his followers acted on their belief in a fundamentally egalitarian approach to their subject based on the (unspoken) principle that "all theorems are created equal" (so long as you can prove them!). Moore's students at the University of Texas spread this gospel, making point-set topology one of the most popular subjects in the mathematics programs of American graduate schools. True, this ethos in its purer form remained largely confined to colleges and universities in the heart of the country. General topology made only modest inroads at the older elite institutions on the East coast, where the Princeton school of J. W. Alexander, Solomon Lefschetz, and Norman Steenrod emerged as the nation's leading center for algebraic topology.

R. L. Moore's Socratic teaching style, the so-called "Moore method," played an integral part in his philosophy of mathematics, which evinced the rugged individualism typical for mathematicians from the prairie states. Book-learning had little appeal for them: this was mathematics for the self-made man who didn't need to rely on anyone except perhaps a friendly neighbor. Over on the West coast, Stanford's George Pólya gave mathematical pedagogy a new Hungarian twist aimed at fostering mathematical creativity. Whereas advocates of the "Moore method" taught that doing mathematics was synonymous with proving theorems and finding counterexamples, Pólya stressed the importance of inductive thinking in solving mathematical problems. His *How to Solve It* sold over a million copies and was translated into at least 17 languages (Alexanderson 1987, p. 13). Not to be outdone, Richard Courant enlisted Herbert Robbins to help him write another popular text: *What is Mathematics?* Presumably Courant thought he had the answer, but then so did Pólya, R. L. Moore, and Bourbaki!

Back in Göttingen during the Great War, physics and mathematics had become ever more closely intertwined. Einstein's general theory of relativity captivated the attention of Hilbert and his circle, and this wave of interest in the subtleties of gravitation soon traveled across the Atlantic. Columbia's Edward Kasner was the first American mathematician to take up the challenge, but he was soon followed by two of E. H. Moore's star students, G. D. Birkhoff and O. Veblen. Harvard's Birkhoff had already begun to depart from the abstract style of his Chicago mentor. Inspired by the achievements of Poincaré, he tackled some of the toughest problems that physics had cast upon the mathematicians' shore. His monograph *Relativity and Modern Physics* appeared in 1923; although nearly forgotten today it contains a result of major significance for modern cosmology:

> (Birkhoff's Theorem) any spherically symmetric solution of Einstein's empty space field equations is equivalent to the Schwarzschild solution, i.e. the static gravitational field determined by a homogeneous spherical mass. (see Hawking and Ellis 1973, Appendix B, for a modern statement and proof of this theorem).

Both Birkhoff and Veblen got to know Einstein in 1921, when he delivered a series of lectures in Princeton. Einstein afterward adapted these into book form, and they were published the following year under the title *The Meaning of Relativity*. Around this time, Veblen took up differential geometry, joining his colleague Luther Eisenhart's quest to build new tools adapted to the needs of general relativity. This research explored the virgin territory of spaces with semi-Riemannian metrics, non-degenerate quadratic differential forms that need not be positive definite. Following Weyl's lead, the Princeton trio of Eisenhart, Veblen, and Tracy Thomas spearheaded research on the projective space of paths, which led to a new foundation for general relativity closely connected to the theory of Lorentzian manifolds (for a survey of their work, see Thomas 1938).

General relativity and cosmology remained major playing fields for mathematicians throughout the 1930s. By the time Einstein joined the faculty at Princeton's new Institute for Advanced Study in 1933, however, quantum mechanics had long since emerged as the dominant field of interest among theoretical physicists. Led by John von Neumann, a new wave of activity took place aimed at developing operator theory and other mathematical methods that became the central tools for quantum theorists. In the meantime, after 15 years of intense efforts to formulate a field theory that could unite gravity and electromagnetism, a lull of activity in this area began to set in (for an overview, see Goldstein and Ritter 2003). Einstein, of course, remained in the arena until his death in 1955, surrounded by a small group of younger men who provided him with technical mathematical assistance.

Back in Berlin, the first of Einstein's many assistants had been Jakob Grommer, whom he apparently met in Göttingen through Hilbert in the summer of 1915. An orthodox Jew from Brest-Litovsk, Grommer had gravitated to Göttingen, where he was "discovered" in a seminar run by Otto Toeplitz. Beginning in 1917, he worked off and on as Einstein's assistant for some 10 years, longer than anyone else (see Pais 1982, pp. 483–501). Thereafter, Einstein was never without similar technical assistance in his quest for a unified field theory, an effort that took on a more purely mathematical character the longer he pursued this goal. Just as Einstein's theory of gravitation transformed differential geometry, so he hoped that mathematics would some day return the favor to physics, if only by showing the physicists the kind of theory they needed in order to explore the outermost and innermost regions of the universe. In Princeton, most of Einstein's assistants were recent European émigrés who had managed to flee before the full force of Nazi racial policies took hold.

Due to his seniority, Edmund Landau was not among those who lost their jobs in the Nazi's initial effort to purify the German civil service (Schappacher 1991). His exodus from the scene was both more poignant and chilling, especially in light of recent discussions of how "ordinary Germans" behaved during the events leading up to the Holocaust (Landau escaped its jaws when he died in Berlin in 1938). Landau's lectures on number theory and analysis at Göttingen were delivered in the grand style of the *artiste*, and his personal tastes and idiosyncrasies at the blackboard came to be known as the "Landau style." Some found his passionate dedication to rigor overly pedantic, while others resented his ostentatious lifestyle. By November of 1933, the Göttingen student body was convinced that Landau's mathematical art could no longer tolerated. Posting brown-shirted SA troopers at the doors of his lecture hall, they organized a successful boycott of his classes. This effort, however, was not led by the usual Nazi-rabble but rather by one of Germany's most talented young mathematicians, Oswald Teichmüller (see Schappacher and Scholz 1992). Afterward, Ludwig Bieberbach, the new spokesman for Aryan mathematics, praised the Göttingen students for their "manly actions," which showed their steadfast refusal to be taught in such an "un-German spirit" (see Mehrtens 1987). For NS ideologues, Landau's work, like that of the famous Berlin portraitist Max Liebermann, was just "decadent art" (*entartete Kunst*) and treated as such. Symptomatic of what was to follow throughout Germany, nearly all of the more talented Göttingen mathematicians were gone by the mid 1930s, their services no longer needed or desired (for an overview, see Siegmund-Schultze 1998).

In the earlier era of Klein and Hilbert, "art for art's sake" had always played a prominent part in the Göttingen milieu (Rowe 1989). Tastes differed, but style mattered, and mathematical creativity found various forms of personal expression. To Hermann Weyl, Hilbert's most gifted student and a masterful writer, the preface to his mentor's *Zahlbericht* was a literary masterpiece. Still, in many Göttingen circles, the spoken word, uttered in lecture halls and seminar rooms, carried an even higher premium. Some preferred Klein's sweeping overviews, coupled with vivid illustrations, while others favored Hilbert's systematic approach, aimed at reducing a problem to its bare essentials.

The European émigrés realized, of course, that the Hilbertian legacy comprised far more than just axiomatics; nor was Hilbert's style exclusively designed for the pure end of the mathematical spectrum. After Richard Courant arrived at New York University, he continued to work in the tradition of his 1924 classic, Courant-Hilbert, eventually producing its long-awaited second volume, with the help of K. O. Friedrichs. Courant, who went on to become one of the foremost advocates of applied mathematics in the United States, always imagined that the spirit of "Hilbert's Göttingen" lived on at NYU's Courant Institute. Meanwhile, in the quieter environs of Princeton's Institute for Advanced Study, Einstein, Gödel, and Weyl cultivated their respective arts while contemplating the significance of mathematics for science, philosophy, and the human condition.

References

Alexanderson, G.L. 1987. George Pólya: A Biographical Sketch. In *The Pólya Picture Album: Encounters of a Mathematician*. Boston: Birkhäuser.

Barrow-Green, June E. 1997. *Poincaré and the Three-Body Problem*. Providence: American and London Mathematical Societies.

Ben-Menahem, Yemima. 2001. Convention: Poincaré and Some of His Critics. *British Journal for the Philosophy of Science* 52: 471–513.

Corry, Leo. 2000. The Empiricist Roots of Hilbert's Axiomatic Approach. In *Proof Theory. History and Philosophical Significance*, Hendricks, V. F., S. A. Pederson, and K. F. Jorgensen, eds. Synthese Library, vol. 292, Dordrecht: Kluwer, 35–54.

Darrigol, Olivier. 1995. Henri Poincaré's Criticism of fin de siècle Electrodynamics. *Studies in the History of Modern Physics* 26 (1): 1–44.

Goldstein, Catherine, and Jim Ritter. 2003. The Varieties of Unity: Sounding Unified Theories, 1920–1930. In *Revisiting the Foundations of Relativistic Physics (Festschrift for John Stachel)*, ed. A. Ashtekar et al., 93–149. Dordrecht: Kluwer.

Gray, Jeremy J. 2000. *The Hilbert Challenge*. Oxford: Oxford University Press.

Hawking, S.W., and G.F.R. Ellis. 1973. *The Large Scale Structure of Space-Time*. Cambridge: Cambridge University Press.

Hendricks, V. F., S. A. Pederson, and K. F. Jorgensen, eds. 2000. *Proof Theory. History and Philosophical Significance*, Synthese Library, vol. 292, Dordrecht: Kluwer.

Mehrtens, Herbert. 1987. Ludwig Bieberbach and "Deutsche Mathematik". In *Studies in the History of Mathematics*, ed. Esther R. Phillips, 195–241. Washington, D. C.: Mathematical Association of America.

———. 1990. *Moderne-Sprache-Mathematik. Eine Geschichte des Streits um die Grundlagen der Disziplin und des Subjekts formaler Systeme*. Suhrkamp Verlag: Frankfurt am Main.

Pais, Abraham. 1982. *'Subtle is the Lord...' The Science and the Life of Albert Einstein*. Oxford: Clarendon Press.

Parshall, Karen, and David E. Rowe. 1994. *The Emergence of the American Mathematical Research Community, 1876–1900. J.J. Sylvester, Felix Klein, and E.H. Moore*, AMS/LMS History of Mathematics Series. Vol. 8. Providence: American Mathematical Society.

Poincaré, Henri. 1905. *Science and Hypothesis*. London: Walter Scott; repr. New York: Dover, 1952.

Rowe, David E.. 1989. Felix Klein, David Hilbert, and the Göttingen Mathematical Tradition. In *Science in Germany: The Intersection of Institutional and Intellectual Issues*, Kathryn Olesko, ed., *Osiris*, 5, 186–213.

———. 1998. Mathematics in Berlin, 1810–1933. In *Mathematics in Berlin*, ed. H.G.W. Begehr, H. Koch, J. Kramer, N. Schappacher, and E.-J. Thiele, 9–26. Birkhäuser: Basel.

———. 2000. The Calm before the Storm: Hilbert's early Views on Foundations. In *Proof Theory. History and Philosophical Significance*, Hendricks, V. F., S. A. Pederson, and K. F. Jorgensen, eds. Synthese Library, vol. 292, Dordrecht: Kluwer, 55–93.

Schappacher, Norbert. 1991. Edmund Landau's Göttingen. From the Life and Death of a Great Mathematical Center. *Mathematical Intelligencer* 13: 12–18.

Schappacher, Norbert and Erhard, Scholz, eds. 1992. Oswald Teichmüller. Leben und Werk, *Jahresbericht der Deutschen Mathematiker-Vereinigung* 94: 1–39.

Sieg, Wilfried. 2000. Towards Finitist Proof Theory. In *Proof Theory. History and Philosophical Significance*, Hendricks, V. F., S. A. Pederson, and K. F. Jorgensen, eds. Synthese Library, vol. 292, Dordrecht: Kluwer, 95–114.

Siegmund-Schultze, Reinhard. 1998. *Mathematiker auf der Flucht vor Hitler,Dokumente zur Geschichte der Mathematik*. Bd. 10 ed. Braunschweig/Wiesbaden: Vieweg.

Thomas, T.Y. 1938. Recent Trends in Geometry. In *Semicentennial Addresses of the American Mathematical Society*, 98–135. New York: American Mathematical Society.

van Dalen, Dirk. 1990. The War of the Mice and Frogs, or the Crisis of the *Mathematische Annalen*. *The Mathematical Intelligencer* 12 (4): 17–31.

Wilder, Raymond L. 1982. The Mathematical Work of R. L. Moore: Its Background, Nature, and Influence. *Archive for History of Exact Sciences* 26: 73–97.

35 Coxeter on People and Polytopes

(*Mathematical Intelligencer* 26(3)(2004): 26–30)

H. S. M. Coxeter, known to his friends as Donald, was not only a remarkable mathematician. He also enriched our historical understanding of how classical geometry helped inspire what has sometimes been called the nineteenth-century's non-Euclidean revolution (Fig. 35.1). Coxeter was no revolutionary, and the non-Euclidean revolution was already part of history by the time he arrived on the scene. What he did experience was the dramatic aftershock in physics. Countless popular and semi-popular books were written during the early 1920s expounding the new theory of space and time propounded in Einstein's general theory of relativity. General relativity and subsequent efforts to unite gravitation with electromagnetism in a global field theory gave research in differential geometry a tremendous new impetus. Geometry became entwined with physics as never before, and higher-dimensional geometric spaces soon abounded as mathematicians grew accustomed not just to four-dimensional space-times but to the mysteries of Hilbert space and its infinite-dimensional progeny.

Seen from this perspective, Coxeter's work on polytopes surely must have looked quaint to many contemporary observers. Still, his research was by no means a lonely adventure; he found plenty of others who shared his fascination with the symmetry properties of geometrical configurations. Aided by the wonders of computer graphics, this classical geometric style of mathematics has enjoyed a tremendous resurgence. Donald Coxeter surely saw this not so much as a personal triumph but as a triumph for geometry and its practitioners. One has only to read Coxeter with an historian's eyes to appreciate this part of his legacy. The most noteworthy example is his *Regular Polytopes*, which first appeared in 1948, marking the culmination of 24 years of loving labors in a field that was considered exotic and well outside the mainstream of mathematical research. Coxeter seems to have relished his role as the consummate outsider. He was an old-fashioned geometer who thoroughly grasped the significance of modern methods, some of which he invented (Coxeter groups, Coxeter diagrams). Yet he identified with gifted amateurs and artists all his life. Indeed, he spent as much effort resurrecting the work of forgotten heroes as he did rehabilitating the theory of polytopes, his main field of research.

In *Regular Polytopes* Coxeter aimed to reach a broad audience and by all accounts he succeeded. His text offers a reasonably self-contained introduction to the classical background, much of which would have been familiar to Kepler. It begins with a tour of polyhedra and their symmetry groups, tessellations and honey combs, kaleidoscopes, and star-polyhehra (Fig. 35.2). Gently guiding the reader through the first six chapters, Coxeter hoped to provide the novice with adequate preparation for the adventurous journey that follows. Clearly he knew that the land of polytopes was fraught with difficulties, and so he suggested for all those who might feel "at all distressed by the multi-dimensional character of the rest of the book" that they consult the textbooks (Manning 1914; Sommerville 1929). Since both of these were seriously dated in 1948, this advice betrays how unfashionable higher geometry had become since the heyday of Italian projective geometry when hyperspace constructions abounded.

At some early point in his career, Coxeter clearly was drawn to the whole lore surrounding this branch of geometry. In *Regular Polytopes* he reaches across the boundaries of time, embracing kindred spirits, both living and dead, who shared his geometrical and aesthetic vision. However skeptical one might be of this bow toward a fictive community of polytope afficianados, one cannot help but admire Coxeter's enthusiasm and the generosity he showed toward those who shared in the enterprise. This liberal attitude was a natural one for Coxeter, who came from a Quaker family. He saw himself as an internationalist and felt that mathematical knowledge enriches humankind as a whole. A particularly striking comment reflecting his views can be found in the preface of *Regular Polytopes*:

D.E. Rowe, *A Richer Picture of Mathematics*, https://doi.org/10.1007/978-3-319-67819-1_35

Fig. 35.1 Coxeter as a student at Trinity College, Cambridge.

Fig. 35.2 Coxeter admiring a spherical tiling.

> The history of polytope theory provides an instance of the essential unity of our western civilization, and the consequent absurdity of international strife. The Bibliography lists the names of thirty German mathematicians, twenty-seven British, twelve American, eleven French, seven Dutch, eight Swiss, four Italian, two Austrian, two Hungarian, two Polish, two Russian, one Norwegian, one Danish, and one Belgian (Coxeter 1973, p. vii).

Coxeter's book is strewn with historical summaries that give us a glimpse of the human side of the polytope industry. His remarks are always informative, often warm, witty, and erudite, but never stuffy. In several cases they also provide us with important autobiographical clues as well, a few of which will be mentioned below.

Coxeter's Heroes

The preface of *Regular Polytopes* opens in typical Coxeter style with some interesting historical remarks. He notes that the term polytope was coined by Reinhold Hoppe in 1882 and introduced into the English-language literature around the turn of the century by Alicia Boole Stott (1860–1940), a granddaughter of George Boole of Boolean logic fame. Coxeter reminds us that the subject had its roots in Greek mathematics. Before Euclid's time the regular polygons and polyhedra were shrouded in Pythagorean lore, as seen from the prominent role played by the Platonic solids in Plato's *Timaeus*. Euclid took up the construction of these perfect bodies, first by studying the constructible polygons in Book IV and then showing how to construct the five regular polyhedra in Book XIII. By making this the culminating topic of the Elements—which ends with the observation that the tetrahedron, cube, octahedron, dodecahedron, and icosahedron are the only such polyhedra—Euclid helped preserve their exalted status within classical geometry.

Coxeter clearly identified with this tradition; he flatly stated that the main motivation for studying the Platonic solids was aesthetic. We are drawn to these figures, just like the ancient Pythagoreans, because "their symmetrical shapes appeal to our artistic sense." True, Felix Klein's *Lectures on the Icosahedron* cast the theory of the quintic equation in a fresh new light, but Coxeter saw this as wholly unnecessary: "if Klein had not been an artist he might have expressed his results in purely algebraic terms" (Coxeter 1973, p. vi). Klein surely would have disagreed: Kronecker had done just that, but Klein insisted that one needed geometry in order to do Galois theory properly!

Coxeter's much-sung hero in *Regular Polytopes* is the Swiss mathematician Ludwig Schläfli (1814–1895), whose work on this topic only appeared posthumously in (Schläfli 1901). During his lifetime, Schläfli was best known for his investigation of the "double-six" configuration of twelve lines which bears an intimate connection with the 27 lines on a cubic surface (see the introduction to Part II and (Hilbert and Cohn-Vossen 1932, 146–151), whereas his monograph on polytopes remained virtually unnoticed for some 50 years. As Coxeter tells it:

> The French and English abstracts of this work, which were published in 1855 and 1858, attracted no attention. This may have been because their dry-sounding titles tended to hide the geometrical treasures that they contain, or perhaps it was just because they were ahead of their time, like the art of van Gogh (Coxeter 1973, p. 143).

Schläfli studied the general polytope (which he called a "Polyschem") and developed a criterion for determining regular polytopes in dimensions four and higher. For this purpose he introduced the now standard Schläfli symbol, which contains all the information needed to characterize a polytope. Let's consider the case of a regular polyhedron. If n denotes the number of vertices or edges on a face and p the number of edges or faces that pass through a point, then the Schläfli symbol $\{n, p\}$ can take on just five values satisfying the condition $1/n + 1/q < 1$. The general Schläfli symbol is defined analogously. Proceeding to dimension four, we define r as the number of faces or solids that meet at an edge. Schläfli could then show that the only admissible values of $\{p,q,r\}$ are precisely $\{3,3,3\}$, $\{4,3,3\}$, $\{3,3,4\}$, $\{3,4,3\}$ $\{5,3,3\}$, and $\{3,3,5\}$. In a laborious fashion, he also showed that it was possible to construct these six 4-dimensional figures. Finally, Schläfli proved that only three types of regular polytope exist in dimensions five and higher. These are the n-simplex, the general measure polytope and its dual figure, all of which are easily constructed as the analogues of the tetrahedron, cube, and octahedron in ordinary 3-space. Thus, Schläfli was the first to recognize that dimensions three and four are unique in that they contain "exotic polytopes."

Coxeter's book gives a lucid account of these matters and much more. He provides a brief synopsis of Schläfli's lengthy monograph along with a few biographical remarks on his career. He also notes that ignorance of Schläfli's work meant that it had virtually no impact on other researchers, including an American named Washington Irving Stringham. A student of J. J. Sylvester's at Johns Hopkins University during the late 1870s, Stringham wrote his dissertation on 4-dimensional regular polytopes and published his main results in [Stringham 1880], which was published in the *American Journal of Mathematics*. After graduation he went to Leipzig where he presented these latest findings in Felix Klein's seminar.

Stringham's approach was based on an analysis of the number of regular polyhedra that can meet together at a vertex point without filling up the spatial region surrounding it (Fig. 35.3). This gave his work a visual appeal, and numerous others began to experiment with solids that bound a 4-D cell. Stringham was also able to show that there are three "exotic polytopes" in dimension four—the self-dual 24-cell and the dual 120- and 600 cells. Although partisan to Schläfli's pioneering achievement, Coxeter readily admitted that "[Stringham's] treatment was far more elementary and perspicuous, being enlivened by photographs of models and by drawings. The result was that many people imagined Stringham to be the discoverer of the regular polytopes" (Coxeter 1973, p. 143). Coxeter further noted that another seven authors independently rediscovered the six 4-dimensional regular polytopes between 1881 and 1900, that is, before the publication of (Schläfli 1901). Only two of them, however, came up with a notation as elegant as the Schläfli symbol.

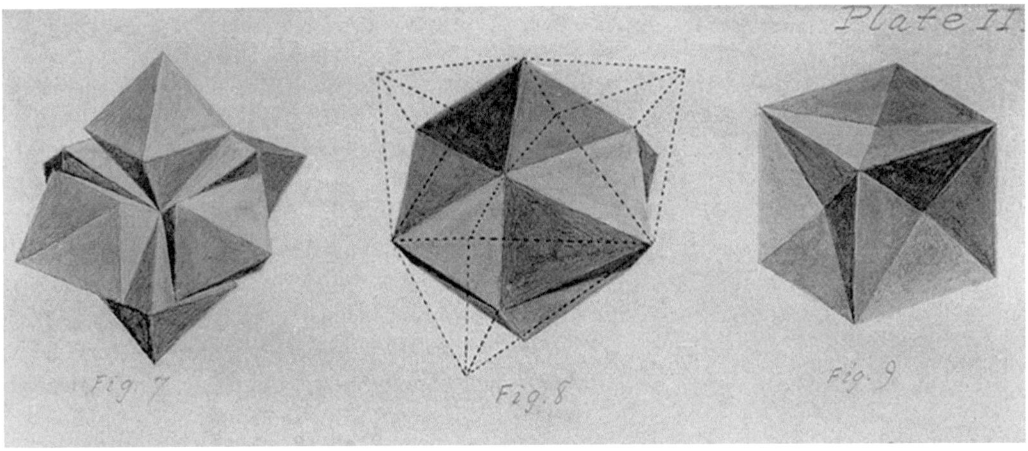

Fig. 35.3 Constructing a 24-cell (Reproduced with permission from Stringham 1880).

This research activity on polytopes went hand in hand with a surge of interest in non-Euclidean and higher-dimensional geometries during the last two decades of the nineteenth century. Stringham's drawings captured the imagination of numerous amateurs, some of whom were artists. As Linda Dalrymple Henderson has shown, these new geometrical ideas had a profound impact on modern art that was independent of the discussions on relativity theory and space-time geometries (Henderson 1983). In 1909 *Scientific American* opened a prize competition for the best popular essay explaining the fourth dimension. The hefty prize of $500 helped attract 245 entries, four of which were published in the magazine. In the wake of the relativity revolution, the editors opted to produce a full book of essays from the contest. Henry Parker Manning from Brown University, author of (Manning 1914), was asked to choose a suitable set of essays from the original entries and to write an introduction to the volume (Manning 1921). In the publisher's promotional preface *Scientific American* noted that the subject of the fourth dimension "has unfortunately been classed with such geometrical absurdities as the squaring of a circle and the trisection of an angle" (Manning 1914, p. 4). Yet in one important respect those who sought to plumb the mysteries of the fourth dimension were quite unlike ordinary circle squarers and angle trisectors. For the latter activities have always been attractive to amateur puzzle-solvers, whereas the former theme has a mysterious quality about it that appeals to a different kind of mindset.

Coxeter on the Intuitive Approach to the Fourth Dimension

Coxeter had a very matter-of-fact attitude about the fourth dimension. He asserted that we can approach higher-dimensional Euclidean spaces in three different ways: axiomatically, algebraically, or intuitively. The first two methods pose no real difficulties, whereas the third relies on dimensional analogy, which can easily lead one astray. He nevertheless took the intuitive approach seriously since it can be "very fruitful in suggesting what results should be expected" p. 119]. He even cited Edwin Abbott's *Flatland* approvingly in this connection, but he also issued these words of warning to the mystically minded:

> Many advocates of the intuitive method fall into an . . . insidious error. They assume that, because the fourth dimension is perpendicular to every other direction known through our senses, there must be something mystical about it. Unless we accept Houdini's exploits at their face value, there is no evidence that a fourth dimension of space exists in any physical or metaphysical sense (Coxeter 1973, p. 119).

Coxeter surely had many encounters with those who believed otherwise. His friendship with the aforementioned Alicia Boole Stott, which began in 1930, no doubt gave him additional insights into the role of the fourth dimension in Victorian culture. Alicia was only four years old when her father died, and she grew up with her four sisters in poverty. Coxeter gave this vivid description of her early life:

> She spent her early years, repressed and unhappy, with her maternal grandmother and great uncle in Cork. When Alice was about thirteen the five girls were reunited with her mother (whose books reveal her as one of the pioneers of modern pedagogy) in a poor, dark, dirty, and uncomfortable lodging in London. There was no possibility of education in the ordinary sense, but Mrs. Boole's friendship with James Hinton attracted to the house a continual stream of social crusaders and cranks. It was during those years that Hinton's son Howard brought a lot of small wooden cubes and set the youngest three girls the task of memorizing the arbitrary list of Latin words by which he named them, and piling them into shapes. To Ethel, and possibly Lucy too, this was a meaningless bore; but it inspired Alice (at the age of about eighteen) to an extraordinary intimate grasp of four-dimensional geometry. Howard Hinton wrote several books on higher space, including a considerable amount of mystical interpretation. His disciple did not care to follow him along these other lines of thought, but soon surpassed him in geometrical knowledge. Her methods remained purely synthetic, for the simple reason that she had never learnt analytical geometry (Coxeter 1973, p. 258) (Fig. 35.4).

Coxeter presumably obtained much of this information firsthand from Alicia Boole Stott. Alicia's mother was Mary Everest Boole (1832–1916), whose father was a minister (see Mihalowicz 1996). The Everest family is mainly remembered, however, for the exploits of Mary's uncle, Colonel Sir George Everest, who worked for many years as the Surveyor General of India. In 1841 he charted the Himalayas and determined the position and height of its largest peak, known today as Mount Everest. Mary lived in France until she was eleven, during which time she took mathematics lessons from a private tutor. After returning to England she worked as her father's assistant, teaching Sunday school classes and helping him prepare his sermons. She kept her interest in mathematics alive by reading books in her father's library.

Another of her uncles, John Ryall, happened to be a professor of Greek at the newly established Queens College in Cork, West Ireland. When Mary Everest went to visit him, he introduced her to his colleague, the mathematician George Boole, who found she had a burning desire to learn higher mathematics. Boole went to England two years later to give Mary Everest private lessons and she eventually assisted him in writing his *Investigation of the Laws of Thought* (1854). Mary's father died around the time it appeared, and in 1855 she and Boole married. Their marriage was a happy one, but it lasted only nine years: Boole died of pneumonia in 1864 leaving her with five children to care for. Soon afterward, Mary was offered a job as a librarian at Queens College. She also began writing books on various topics around this time. One of her interests was psychic phenomena and the spirit world, which led to local controversy at the college when

Fig. 35.4 Alicia Boole Stott with Pieter Hendrik Schoute.

she tried to publish a book called *The Message of Psychic Science for Mothers and Nurses*. As a consequence, she was forced to give up her position at Queens College, but her father's friend, James Hinton, hired her as his secretary.

As Coxeter duly noted, the Hintons, father and son, made life in the Boole household, well, different than it had been. James Hinton drew Mary's attention to evolution and the art of thinking, ideas she began to develop in a series of articles and books. Charles Howard Hinton (1853–1907) was at least as flamboyant as his father. His interests in four-dimensional geometry no doubt deeply influenced Alicia Boole, but he married Mary's eldest daughter, Mary Ellen, instead. Perhaps his preoccupation with relationships in four dimensions caused him to be somewhat disoriented when it came to sorting things out in ordinary three-space. At any rate, when it was discovered that he was also married to a second woman named Maud Wheldon, Howard was put on trial for bigamy. He and Mary Ellen thereafter fled to Japan, but eventually he found his way to Princeton. A prolific writer, C. H. Hinton's ideas about the fourth dimension influenced contemporaries as diverse as Edwin Abbott and the Theosophist Rudolf Steiner. Some of his writings are still available in (Hinton 1980).

Among other works, Mary Everest Boole wrote a book entitled *Philosophy and Fun of Algebra*, described by Ivars Peterson in (Peterson 2000). She also invented what she called "curve stitching" to help children learn basic geometry. Her daughter Alicia shared a similar interest in geometrical visualization, creating various models for projections of four-dimensional constructs to satisfy her own curiosity. She had little time for this after 1890, however, when she married Walter Stott, an actuary. As Coxeter described it "for some years she had a life of drudgery, rearing two children on a very small income" (Coxeter 1973, p. 258). Her return to mathematics came about through her friendship with Pieter Hendrik Schoute (1846–1913), the leading Dutch expert on polytopes. Citing Coxeter again:

Mr. Stott drew his wife's attention to Schoute's published work; so she wrote to say that she had already determined the whole sequence of [middle] sections . . . for each polytope agreeing with Schoute's result. In an enthusiastic reply, he asked when he might come over to England and work with her. He arranged for the publication of her discoveries in 1900, and a friendly collaboration continued for the rest of his life. Her cousin, Ethel Everest, used to invite them to her house in Hever, Kent, where they spent many happy summer holidays. Mrs. Stott's power of geometrical visualization supplemented Schoute's more orthodox methods, so they were an ideal team. After his death in 1913 she attended the tercentenary celebration of his university of Groningen, which conferred upon her an honorary degree and exhibited her models (Coxeter 1973, pp. 258–259) (Fig. 35.5).

Alicia Boole Stott remained mathematically inactive following Schoute's death, but that changed when she met Donald Coxeter in 1930. He was then working on his doctorate in Cambridge under the supervision of H. F. Baker. Their acquaintance was facilitated by Alicia's nephew, G. I. Taylor, whose mother Margaret was the second of George and Mary Boole's five daughters. Coxeter and Stott collaborated on an investigation of Thorold Gossett's semi-regular polytope {3, 4, 3}, which Coxeter had recently rediscovered. She showed that its vertices lie on the edges of the regular polytope {3, 4, 3}, dividing them in the ratio of the golden section.

Fig. 35.5 Models by Alicia Boole Stott on display at Cambridge University.

Coxeter as Promoter of Geometrical Art

Geometry and art were intimately connected in Coxeter's mind, and so it was natural that he should befriend the Dutch artist M. C. Escher. Nearly all the obituaries of Coxeter mention his relationship with Escher, whom he met at the Amsterdam ICM in 1954. Readers of *MI* were twice treated to articles by Coxeter dealing with Escher and the mathematics underlying his art (Coxeter 1985, 1996). Their "collaboration" was wonderfully recounted in (Coxeter 1979), which explains how Escher came to produce his woodcut "Circle Limit III" after Coxeter explained the general procedure for constructions in Poincaré's model for the hyperbolic plane. (For a more recent account of the mathematics involved, illustrated with beautiful color pictures, see Dunham 2003.)

Four years after their first meeting, Escher wrote to thank Coxeter for having sent him a booklet from a symposium on symmetry that contained Coxeter's article "Crystal Symmetry and its Generalizations." Escher found it "much too learned for a simple, self-made plane pattern-man like me," but he was intrigued by one of the diagrams, which he said "gave me quite a shock." He then explained why:

> Since a long time I am interested in patterns with "motives" getting smaller and smaller till they reach the limit of infinite smallness. The question is relatively simple if the limit point is a point in the centre of a pattern. Also a limit line is not new to me, but I was never able to make a pattern in which each "blot" is getting smaller gradually from a center towards the outside limit circle, as shows your picture. I tried to find out how this figure was geometrically constructed, but I succeeded only in finding the centers of the largest inner-circles. If you could give me a simple explanation [for] how to construct the following circles, whose centers approach gradually from the outside till they reach the limit, I should be immensely pleased and very thankful to you! (Coxeter 1979, p. 19)

Escher also sent Coxeter a copy of his woodcut "Circle Limit I," his first attempt to exploit the geometry of the Poincaré model to achieve his vision. He later wrote that this woodcut ". . . displays all sorts of shortcomings. Not only the shape of the fish, still hardly developed from rectilinear abstractions into rudimentary animals, but also their arrangement and relative position leave much to be desired. There is no continuity, no 'traffic flow,' no unity of colour in each row!" (Coxeter 1979, p. 20).

Fig. 35.6 M. C. Escher, Circle Limit III (1959) (Wikipedia).

In his response to Escher's letter, Coxeter explained that there are infinitely many patterns that lead to a tessellation of the hyperbolic plane. He also explained how "scaffolding" outside the boundary circle can be used to determine the centers of circles with arcs passing through the interior and perpendicular to the boundary. Escher took full advantage of these tips in constructing three new woodcuts based on similar patterns. The best of these, Circle Limit III, was carefully analyzed by Coxeter 20 years later in (Coxeter 1979) (Fig. 35.6).

Shortly before his death, Coxeter was honored by the Fields Institute at Toronto University where a complex geometric sculpture by Marc Pelletier was unveiled. For the trained eye its aesthetic appeal was enhanced by the virtuosity of the achievement: for this sculpture shows an orthogonal projection of the 120-cell into three-space, a model much appreciated by Coxeter, John Conway, and the other geometers present. Fittingly enough, the Pelletier was present and spoke about the work of another artist, whose work both he and Coxeter greatly admired, Paul Donchian (see the tribute to Donchian in Coxeter 1979, p. 260).

Born in Hartford, Connecticut in 1895 into a family of Armenian descent, Donchian took over the rug business established by his father. Many of his ancestors had been jewelers and craftsmen, a tradition that came to fruition in his wire models showing three-dimensional projections of the four-dimensional regular polytopes. Coxeter had this to say about the artist's techniques:

> [Donchian's models use] straight pieces of wire for the edges and globules of solder for the vertices. The vertices are distributed on a set of concentric spheres (not appearing in the model), one for each par of opposite sections. Donchian did not attempt to indicate the faces, because any kind of substantial faces would hide other parts (so that the model could only be apprehended by a four-dimensional being). The cells appear as "skeletons," usually somewhat flattened by foreshortening but still recognizable. Parts that would fall into coincidence hve been artificially separated by slightly altering the direction of projection, or introducing a trace of perspective (Coxeter 1979, p. 242).

Donchian's models were displayed at the 1934 Century of Progress Exposition in Chicago. At the Coxeter celebration, Marc Pelletier pointed out that many of Donchian's works are held in storage at the Franklin Institute in Philadelphia, where they were last displayed in 1967, the year of his death. Perhaps they should be dusted off and put back on display. For those willing to settle for two-dimensional images, it should not be forgotten that Coxeter adorned his classic book with photographs of the Donchian models, another form of tribute to the artist who brought them to life. These are only a few of thelovely things to be found in Donald Coxeter's *Regular Polytopes*, one of the great books of twentieth-century mathematics. One must hope it will continue to find an appreciative audience for many years to come.

References

Coxeter, H.S.M. 1973. *Regular Polytopes*. 3rd ed. New York: Dover.

———. 1979. The non-Euclidean symmetry of Escher's Picture 'Circle Limit III'. *Leonardo* 12: 19–25, 32.

———. 1985. Review of M. C. Escher: His Life and Complete Graphic Work. *Mathematical Intelligencer* 7 (1): 59–69.

———. 1996. The Trigonometry of Escher's Woodcut "Circle Limit III". *Mathematical Intelligencer* 18 (4): 42–46.

Dunham, Douglas. 2003. Hyperbolic Art ad the Poster Pattern, *Math Awareness Month – April 2003*, http://mathfoum.org/mam/03/essay1.html.

Henderson, Linda Dalrymple. 1983. *The Fourth Dimension and Non-Euclidean Geometry in Modern Art*. Princeton: Princeton University Press.

Hilbert, David, and Stephen Cohn-Vossen. 1932. *Anschauliche Geometrie*. Berlin: Springer.

Hinton, Charles Howard. 1980. In *Speculations on the Fourth Dimension: Selected Writings of C. H. Hinton*, ed. Rudolf Rucker. New York: Dover.

Manning, Henry Parker. 1914. *Geometry of Four Dimensions*. New York: Dover.

———, ed. 1921. *The Fourth Dimension Simply Explained*. New York: Scientific American Publishing.

Mihalowicz, Karen Dee. 1996. Mary Everest Boole (1832–1916): An Erstwhile Pedagogist for Contemporary Times. In *Vita Mathematica: Historical Research and Integration with Teaching*, ed. Ronald Calinger. Washington, D.C.: Mathematical Association of America.

Peterson, Ivars. 2000. Algebra, Philosophy, and Fun, www.ma.org/\penalty\z@mathland/\penalty\z@mathtrek_1_17_00.html.

Schläfli, Ludwig. 1901. Theorie der vielfachen Kontinuität. *Denkschriften der Schweizerischen naturforschenden Gesellschaft* 38: 1–237.

Sommerville, Duncan M.Y. 1929. *An Introduction to Geometry of N Dimensions*. New York: Dover.

Stringham, W.I. 1880. Regular figures in n-dimensional space. *American Journal of Mathematics* 3: 1–15.

36 Mathematics in Wartime: Private Reflections of Clifford Truesdell

(*Mathematical Intelligencer* 34(4)(2012): 29–39)

Our featured guest writer in this column of *Years Ago*, Clifford Ambrose Truesdell III (1919–2000), was a cantankerous mathematical physicist turned historian of mathematics, who taught at the Midwestern universities of Michigan and Indiana before going on to distinguished career at Johns Hopkins University. Truesdell was a man of many parts. Those who knew him, whether admirers or not, would all attest that he stood apart as an American thoroughly in love with European culture. Yet Truesdell's Europe was not that of leftists like the writer Jean-Paul Sartre or the filmmaker Rainer Werner Fassbinder; nor was he enamored of the music of modernist composers, not even Igor Stravinsky's, much less the electronic experimentalism of a Karlheinz Stockhausen. What moved him instead were the achievements of Europeans of the eighteenth century. Truesdell loved baroque music, insisting however that it be played if at all possible on instruments from the era. All his tastes—in science as well as art—ran decidedly toward classicism. And since he took his interests in the arts and sciences seriously, he chose to spend his last decades in Baltimore living the life of a cultivated European surrounded by memorabilia from the baroque age (Figs. 36.1 and 36.2).[1]

In twentieth century America, many sought their fortune and fame in California. But for Truesdell, who was born and raised there, still loftier American dreams remained to be discovered on a different shore, though in another time and place. His vivid imagination, coupled with personal discipline and a thirst for knowledge, pointed the way. For him and his wife, Charlotte, Italy held a special fascination, and they travelled there often. At home, both strove to re-enact a courtly lifestyle that had vanished long ago, one in which aristocrats mingled with intellectuals and artists, promoting their common love of language, literature, and art. As self-made aristocrats, the Truesdells loved to flaunt their wealth and refined tastes at their redesigned Renaissance-style home, a small mansion they named *Il Palazzetto* (the little palace). Here they entertained their guests, while dressed in lavish attire from a bygone era, replete with frilly collars and cuffs. The internal décor splendidly befit such occasions, showcasing a vast collection of European artwork, furniture, and sculpture. One could gaze in wonderment at the ornate bronze statuary, gilded baroque furniture, and exquisite silverware – simply out of this world! Or, on further reflection, one might wonder why these gracious and convivial hosts chose to live in a museum. Needless to say, many at Johns Hopkins found this lifestyle more than a little exotic; some surely thought their illustrious colleague must have been clinically insane. For, along with his other eccentricities, Clifford Truesdell was exceptionally extroverted, hence anything but shy when it came to expressing his views about the failings of modern society in general and American higher education, in particular.

Even as a mathematician, Truesdell could only look backward with longing for the ideals of the past. His great hero in the world of mathematics was Leonhard Euler, truly an excellent choice. He identified so strongly with the Swiss-born genius that his own research soon began to blend with detailed studies of Euler's contributions to mechanics and hydrodynamics. As a mathematical physicist, Truesdell faced an uphill battle. Recognition for his work on Euler, on the other hand, came quickly; in 1958 he was awarded the Euler Medal by the Soviet Academy of Sciences (he would receive this distinction a second time in 1983). In 1957, on the occasion of the 250th anniversary of Euler's birth, Truesdell was invited to deliver a major commemorative address in Basel. He took this opportunity to give a highly personal testimonial that revealed perhaps as much about the speaker as it did his subject, which was "Eulers Leistungen in der Mechanik."[2] The following translated passages help to capture the flavor of this performance.

[1]For a brief, yet highly illuminating look back on his life, see the obituary (Buchwald and Cohen 2001). Autobiographical vignettes are strewn throughout the essays (Truesdell 1983). See further (Giusti 2003) and (Ignatieff and Willig 1999).

[2]The following is my *Rückübersetzung* from the published German text (Truesdell 1957). Sandro Caparinni pointed out to me that the original version, written in English, can be found among Truesdell's papers.

D.E. Rowe, *A Richer Picture of Mathematics*, https://doi.org/10.1007/978-3-319-67819-1_36

Fig. 36.1 Among the Truesdells' friends was the famous Dutch keyboard artist, Gustav Leonhardt, who often performed works of Bach on their beautiful harpsichord. Here the historian of mathematics, Jesper Lützen, gets to tickle its keys (Courtesy of Jed Buchwald).

Fig. 36.2 Truesdell mounted this momento mori above his desk to remind him of his mortality (Courtesy of Jed Buchwald).

> Euler was the greatest mathematician of all time; he was, however, only a mathematician, not a magician, a prophet, or a savior. Still he pursued mathematics in the conviction that the world is ruled by the best possible laws, created and ordered by a benevolent God. . . . he believed that mankind had a special God-given task, through an understanding of these sacred laws, to discover and develop improvements for humanity within the scope of human possibilities. . . . The great Book of Nature lies open for us, though it is written by God in a language that we cannot immediately understand. It is written in the language of mathematics. The Book of Nature also contains certain God-given tasks. . . . In striving to clarify and answer these eternal questions, humanity tries to grasp in the end the best possibilities that can be realized in accordance with the fixed order of the world.

Truesdell here invokes a famous metaphor from Galileo's *Assayer* but with a new Eulerian twist. For when the Florentine physicist wrote that the Book of Nature was written in the language of mathematics, he went on to say that "its characters are triangles, circles, and other geometric figures, without which it is humanly impossible to understand a single word." Galileo's mathematics, after all, differed little from that of the ancient Greeks; indeed, he wrote as a leading representative of the neo-Archimedean tradition during the Renaissance. Euler's mathematics, on the other hand, represented a truly new language for investigating the laws of nature, something Galileo probably never dreamed of.

Given this elective affinity with Euler's scientific outlook, it should come as no surprise that Truesdell made explicit reference here to rational mechanics, the field of research that came to be associated with his name. After noting that Euler's works rely in no sense on empirical data, nor do they compare theoretical results with experimental findings, he emphasized that "they belong even less . . . to so-called applied mathematics, in which already known mathematical methods are used to solve physical problems." Quite the contrary, Euler "practiced mechanics as an independent mathematical science, as rational mechanics." (Fig. 36.3).

That same year, Truesdell launched his new journal with the Springer-Verlag, *Archive for Rational Mechanics and Analysis.* Three years later, he inaugurated another Springer periodical, *Archive for History of Exact Sciences.* Both have become monuments to his intellectual legacy. What follows is only an episode in Truesdell's rich life, though one that carried reverberations throughout his later career, as will be seen. We focus our attention on the candid impressions of a thoughtful young mathematician during wartime, when in the summer of 1942 he found himself at Brown University amidst an entourage of exotic European émigrés (one of whom would decisively influence his future research interests). First, though, a few words about the mood in the country during this time.

Fig. 36.3 Clifford Truesdell in the late 1990s (Courtesy of Jed Buchwald).

Building Applied Mathematics at Brown

The smoke had barely lifted from the fleet of ships destroyed by the Japanese air force in their surprise attack on Pearl Harbor when American public opinion about the ongoing war began to shift abruptly. Isolationism, a view of the world long prevalent in the vast heartland, no longer seemed a viable option. Support for it collapsed completely after President Roosevelt went to Congress the next day to ask for a formal declaration of war. He spoke for just 6 min, delivering a rhetorical masterpiece that began by calling December 7, 1941 a "date which will live in infamy." Less than an hour after FDR stepped down from the podium, Congress declared war on Japan, invoking a power stipulated by Article 1, Section 8 of the Constitution. It would be the last time that an American president sought a formal declaration of war from Congress to authorize committing the country's armed forces to engage in military combat.

Only one member of Congress, the Republican Representative from Montana, Jeannette Rankin, voted against the motion to declare war on Japan. Her reasoning was simple: "As a woman, I can't go to war and I refuse to send anyone else." A lifelong pacifist and advocate of women's rights, Rankin had every reason to regard warlike behavior as a male domain. Indeed, during her first tenure in Congress – she was the first woman ever to be elected when she arrived in 1917, three years before women gained the right to vote – she was among the far larger contingent that voted against the declaration of war against Germany and its allies. Her pacifism knew limits, however, and when Hitler upped the ante by declaring war on the United States, she decided to abstain during the vote to reciprocate. By taking such a stand, she became extremely unpopular, so much so that she decided not to run for re-election in 1942. After the war, she travelled to India, where she studied methods of civil disobedience introduced by Gandhi. Never one to shrink from her political convictions, she became a familiar figure in pacifist circles, expressing staunch opposition to U.S. military engagements in Korea and Vietnam.

Pearl Harbor clinched a longstanding debate within the United States that pitted isolationists against interventionists. Yet well before, political pundits on the right and left of the spectrum helped promote a shift in public opinion that left Congresswoman Rankin standing virtually alone. In the pages of *Life* magazine, Walter Lippmann prophesied in 1939, "what Rome was to the ancient world, what Great Britain has been to the modern world, America is to be to the world of tomorrow." One year later, as Nazi armies stood poised to enter a prostrate Paris, *Life's* editor-in-chief, Henry Luce, ran another special issue on "America and the World." In it, he extolled the wisdom of the "prophet Lipmann," who wrote a sequel declaring, "The hour of our destiny has come."

Born and raised in China as the son of Christian missionaries, Luce helped promote this quasi-religious view of what he called the "American century." In the same issue of *Life,* dated 3 June 1940, he offered his own reflections of the crisis facing the world taken from a speech entitled "America and Armageddon." Having just returned from the ravaged countries of Western Europe portrayed by photographs in his magazine, Luce, a stalwart Republican, called Franklin Roosevelt "a very great leader," while lamenting his propensity to cuddle up to mediocre minds while giving America's talented business elites the cold shoulder. No doubt anticipating FDR's re-election, he stopped short of endorsing him in order to take the high ground of self-proclaimed American patriot. As such, he made clear that he would throw his steadfast support behind any of the country's elected leaders. What he expected from them, however, was moral leadership, the kind needed to undertake the colossal effort of arming the nation, not just materially but spiritually, preparing the country for war by giving its people a sense of what they were fighting for and why the sacrifices were needed. Taking up Lipmann's theme, Luce wrote, "The arming of America must in itself be the first practical test of our ability to act as a united people.

For many years we have been anything but a united people. We have been a very expensively divided nation. The arming of America must be our first great act of national unity." Luce did not have long to wait; American military might played a decisive role during the war. Afterward, it helped fill the vacuum left by the collapse of the British Empire, and during the Cold War it became a permanent feature of twentieth-century politics.

Scientific research for military purposes had a long prehistory, of course, but the character of that relationship changed dramatically during the course of World War II.[3] Inevitably, the vast U.S. army undertaking, code-named the Manhattan Project, comes immediately to mind. Yet there were many far less spectacular examples of the growing entanglement that drew the world of academia and the military closer together, and in virtually all cases, European immigrants played key roles. Some of these projects had a major impact on mathematical research in the United States both during and after the war. Such was the case at Brown University, where Roland G. D. Richardson promoted applied mathematics as Dean of the Graduate School.

A native of Nova Scotia, Richardson had studied under Klein and Hilbert in Göttingen before his appointment as chair of the mathematics department at Brown in 1914 (Archibald 1950). After the First World War, he began to build up the graduate program there, while cultivating numerous ties within the national community through his activities as Secretary of the American Mathematical Society, an office he held from 1921 to 1940. As an institution builder, Richardson had considerable success at Brown, much of it based on his recruitment of European émigrés. That policy, strongly supported by Brown President Henry Wriston, began with the appointment of the Russian analyst Jacob D. Tamarkin, who first came to the U.S. in 1925 (Hille 1947). Tamarkin took a position at Dartmouth College before joining the Brown faculty in 1927, remaining there until his retirement in 1946. Soon after Hitler assumed power, Richardson invited the German Hans Lewy and the Hungarian Otto Szász to teach at Brown; both were supported by the Emergency Committee in Aid of Displaced German Scholars. Another Hungarian, Georg Pólya, long affiliated with the ETH in Zürich, was appointed as visiting professor from 1940 to 1942; he also taught at Smith College during this period before accepting a permanent professorship at Stanford (Dresden 1942, 426).

A major turning point for mathematics at Brown took place in 1939 when Otto Neugebauer visited Providence. Soon afterward, he decided that the university library and other resources were adequate for his historical researches on ancient mathematics and astronomy. Neugebauer, a close associate of NYU's Richard Courant since their days together in Göttingen, had already left Germany in January 1934. Afterward he found a new home in Copenhagen under the aegis of Harald Bohr, from which station he edited two major Springer journals: *Zentralblatt für Mathematik* as well as *Quellen und Studien zur Geschichte der Mathematik, Astronomie und Physik.* As the political situation in Germany continued to deteriorate, culminating in the violent events of November 9–10, 1938—the "Night of the Broken Glass"—Neugebauer grew increasingly alarmed. When in that year Tullio Levi-Civita was removed from the editorial board of *Zentralblatt*, he decided he could no longer serve as editor of these German-language publications.

These circumstances led to the new arrangements at Brown, unveiled by President Wriston in his annual report from June 1939:

> Professor Neugebauer paid a visit to this country this spring and found rich material for his researches. At the same time the American Mathematical Society decided to establish an international journal of mathematical abstracts, to be known as *Mathematical Reviews,* and located that journal at Brown, appointing Professors Neugebauer and Tamarkin as its editors. The Society also made available the necessary funds to bring Dr. Willy Feller as assistant to the editors and as lecturer in Brown University.

Neugebauer's salary costs, along with those for his Danish assistant, Olaf Schmidt, were to be supported by grants from the Carnegie Corporation, Rockefeller Foundation, and the Emergency Committee in Aid of Displaced Foreign Scholars. Furthermore, to promote access to the work described in *Mathematical Reviews,* a photographic service was to be made available, supported by funding from the Rockefeller Foundation. This would disseminate copies of journal articles and extracts from books at a nominal cost to all those at institutions whose libraries were too small or underfunded to carry specialized mathematical literature.

Dean Richardson and President Wriston were clearly elated by these new developments, as the latter's report made plain: "In the course of the last few years the center of research in mathematics has shifted to the United States. The library at Brown is already one of the most distinguished collections in the world. The recent developments bid fair to make it even more so, and to make this institution a center for mathematical research known the world around."

Richardson was a firm advocate of the view that the United States would eventually be compelled to enter the war. Nor was he alone. Already by 1940, the AMS and the MAA had established a War Preparedness Committee that brought together such leading figures as John von Neumann, Norbert Wiener, and Thornton Fry from Bell Labs. In that same year, the National Resources Planning Board, a body within the National Research Council, submitted a report to Congress entitled, "Research – A National Resource."

[3]For a recent look at mathematical activities in relation to the military, see the essays in Booß-Bavnbek and Høyrup (2003) and, in particular, Siegmund-Schultze (2003b).

This contained an influential 38-page chapter on industrial mathematics written by Fry, long a proponent of applied research.[4] Richardson would later refer to this document often, claiming that its message helped launch Brown's program in applied mathematics (Richardson 1943b, 422).

Fry pushed hard for university programs that could train a new type of "industrial mathematician," noting how rare such people were. "The typical mathematician," in his view, was "not the sort of man to carry on an industrial project. He is a dreamer" (Fry 1941, 4). What the country needed instead was a new breed of mathematician: versatile problem-solvers able to tackle the diverse technical challenges posed by a fast-growing industrial society. As the country stood on the brink of war, Thornton Fry estimated that no more than 150 individuals had the requisite skills and training for this kind of work (Fry 1941, 9). At the same time, he predicted that graduate programs in applied mathematics would offer significant economic benefits, for example through the employment of such individuals to carry out experiments for the military.

Thus, by 1940 alarm bells had begun to ring, alerting American leaders to take stock of the country's intellectual resources, looking for new possibilities to use this untapped talent in preparation for fighting what would soon become the first truly global war. Among these resources were the mathematicians, physicists, and engineers at the nation's universities and research laboratories. These institutions, as it happened, had recently been enriched by the acquisition of numerous émigrés from Europe, some of whom were struggling to eke out an existence at less stellar American institutions of higher education.[5] A number of these colorful individuals from disparate backgrounds found their way to Brown University, where they taught in a special program that aimed to give talented young men a deeper exposure to topics in applied mathematics. Richardson obtained financial support for this venture from the U.S. Office of Education and the Carnegie Foundation, and in the summer of 1941 he was able to launch his Program of Advanced Instruction and Research in Mechanics as a pilot project. This program, closely tied to the war effort and the perceived needs of industry, would eventually molt into the Division of Applied Mathematics within Brown's Graduate School.

Brown's first summer program was a modest affair, consisting of four courses taught by members of the regular faculty with the support of three visitors.[6] Nevertheless, it managed to attract 63 enrollees. They could attend lectures on differential and integral equations in mathematical physics, offered by Tamarkin and Feller, as well as another course on advanced partial differential equations given by Stefan Bergmann, recently arrived from Paris, after spending three years in the Soviet Union. Bergmann was a protégé of Richard von Mises, with whom he studied in Berlin.[7] Both left Germany soon after the Nazis came to power. Mises accepted a chair in Istanbul, where he was joined one year later by Hilda Geiringer, whom he would later marry (Binder 1992). By 1939, events in Europe had reached such a stage that he prudently decided to leave for the U.S. As one of the most prominent applied mathematicians of the era, he was offered a lectureship at Harvard University; he rose to become McKay Professor in 1945. In 1941 Richardson enticed him to teach a course on fluid dynamics at his first summer school; he was assisted by Kurt Friedrichs from NYU. Rounding out the curriculum, Ivan Sokolnikoff taught a course on elasticity theory, his special field of expertise.

Sokolnikoff came to the U.S. the hard way, after an ill-fated military career in which he fought on the losing side during the Russian Revolution. While still a teenager, he was wounded in battle off the Kurile Islands. Making his way to China, he found employment with a subsidiary of an American electrical firm. Then, in 1921, with the assistance of the American Consul in Harbin, he was able to immigrate to Seattle. The following year, he entered the University of Idaho to study electrical engineering. After completing his undergraduate studies there, he then went on to do graduate work in mathematics at the University of Wisconsin, taking his doctorate in 1930. This led to an appointment on the Wisconsin faculty, where he taught until 1946, attaining full professorship in 1941. After the war, he joined the mathematics faculty at UCLA.

These six instructors brought to Brown considerable specialized knowledge of topics in applied mathematics, but all of them worked within academic environments. Richardson surely recognized that the type of mathematics they represented was still a far cry from the kind that Fry had described in his report on the needs of industry. Perhaps in order to blunt potential criticism, the Dean supplemented the regular courses with a series of guest lectures, several of them delivered by "industrial mathematicians." Four of these guest speakers even spoke twice. Dr. Arpad Ludwig Natai from Westinghouse gave talks on properties of metals; Dr. Hillel Poritsky, from General Electric, lectured on numerical and graphical methods for solving PDEs; Dr. Theodore Theodorsen, of the National Advisory Committee on Aeronautics (predecessor of NASA) delivered two lectures on wing flutter; and Ronald Foster from Bell Labs gave talks on electric circuits and their analogy with dynamical systems.

[4]Many openly acknowledged that the U.S. lacked expertise in applied mathematics; for historical reflections on these developments, see Siegmund-Schultze (2003a).

[5]For an overview of those mathematicians who immigrated to the United States, see Dresden (1942). The most complete historical study of this phenomenon to date is found in Siegmund-Schultze (2009).

[6]Information on Brown's special program in applied mathematics during the years 1941–1943 can be found in PAIRM (1943).

[7]On von Mises's career in Berlin, see Siegmund-Schultze (1993).

Toward the end of this summer session, Thornton C. Fry also came to make a more general presentation on "Mathematical Research in the Industries." Prominent academic mathematicians were also invited as guest speakers, including Norbert Wiener from MIT, Stephan Timoschenko (Stanford), and NYU's Richard Courant.

In all likelihood, the financial outlays for these guest speakers equaled or even exceeded the costs incurred from the four regular courses. Yet Richardson surely knew what he was doing, calculating that the public relations payoff would make this a worthwhile investment. Brown's department, though still a fledgling operation at this time, stood on the cusp of a great new opportunity. Recent appointments, coupled with the growing likelihood that the nation would go to war, suddenly placed it in a new situation. Taking his cue from Fry, Richardson now hoped that funding for applied mathematics would enable Brown to play a new part on a much larger stage.

As a former secretary of the AMS, he was also banking on strong support within the American mathematical community, knowing that his dreams for Brown might go up in smoke without it. He and President Wriston therefore devised a simple plan to obtain a seal of approval from a number of influential figures: they would appoint a committee charged with evaluating the success and future promise of Richardson's first summer school. Its five members were carefully chosen as "movers and shakers" in the world of pure and applied mathematics: Marston Morse of the Institute for Advanced Study and President of the American Mathematical Society; Mervin J. Kelly, Research Director of the Bell Telephone Laboratories; George B. Pegram, Dean of the Graduate School at Columbia University; Theodore von Kármán, Director of the Aeronautics Laboratory at the California Institute of Technology; and Warren Weaver, Director for Natural Sciences at the Rockefeller Foundation.[8]

This committee duly deliberated immediately following the conclusion of Brown's first summer school session. In fact, its members convened only a short distance away from the Brown campus. Nor was their verdict ever really in doubt, though it was also clear that the other four would have to defer to von Kármán's opinion, he being one of the foremost representatives of scientific engineering in the U.S. (von Kármán came to Caltech in the 1920s after having studied under Ludwig Prandtl in Göttingen). Richardson later published some key passages from the committee's report, beginning with its assessment of the overall situation as seen from this vantage point in time:

> . . . there has been in American mathematics since 1900 a marked tendency to emphasize pure mathematics. The success of this development is a great source of national strength and should be a cause for national pride. But it is highly unfortunate that in our enthusiasm for pure mathematics, we have foolishly assumed that applied mathematics is something less attractive and less worthy. . . . It is obvious that, in the long run, pure mathematics and applied mathematics ought to be in the closest relationship of mutual respect, parallel development, and continuously stimulating interaction. (PAIRM 1943, 4)

Regarding appropriate means to achieve this long-term goal, the committee emphasized the need for specific programs that would upgrade the quality of mathematics instructions in scientific and engineering schools. Furthermore, it called for programs specifically designed to train mathematicians who could do effective work for government agencies and in private industry. Richardson and Wriston had played their hand wisely; with their handpicked committee's nod of approval, they could grin and prepare for the next round of action – "full speed ahead!"

Only a few months later, the Japanese bombed Pearl Harbor. Richardson apparently now felt he had history on his side, and so he drew the obvious lessons. In an article that appeared in a journal for physics teachers, he wrote:

> Is it coincidence that the founding of the École Polytechnique just preceded the beginning of Napoleon's successful army campaigns? Is it coincidence that fundamental research in ship construction was assiduously prosecuted in Britain during the period just before 1900, when the maritime commerce of that great nation held an assured position of world leadership? Is it coincidence that over the last quarter century airplane research at Göttingen and other German centers was heavily subsidized and vigorously pursued and that in this war German aviation has come spectacularly to the forefront? Is there a lesson to be learned here in America from the consideration of such concurrence? (Richardson 1943a).

Still, Brown lacked the kind of faculty that would be needed to launch a serious program in applied mathematics. Indeed, other institutions seemed far better situated in this regard. After all, MIT had Norbert Wiener, Harvard had acquired von Mises, and Richard Courant was already building up a strong group of analysts at NYU. Brown could boast of having Jacob Tamarkin and Willy Feller, of course, but it lacked anyone in mathematics with a strong engineering background. Richardson realized this. Presumably on the advice of von Mises, he decided to take a chance on a displaced scholar, one Willy Prager, professor of mechanics in Istanbul.

Prager had formerly taught at the Institute of Technology in Karlsruhe and had served as a consultant to the Fiesler Aircraft Company in Kassel. His research interests centered on engineering design, especially problems involving vibrations, plasticity, and the theory of structures. Richardson first had to persuade President Wriston to hire Prager sight unseen, no easy task.[9] Next came a serious logistical

[8]On Weaver's subsequent role in promoting mathematical research for the military, see Owens (1989).

[9]What follows is based on Martha Mitchell, "William Prager," Encyclopedia Brunoniana, 1993; http://www.brown.edu/Administration/News_Bureau/Databases/Encyclopedia/search.php?serial=P0350.

problem: how was Prager going to travel from Turkey to the U.S. in wartime? As a German citizen of military age, he obviously had to avoid moving through German-held territories. This meant planning a lengthy trip across the Pacific, via Odessa and Japan, and on to San Francisco, a plan that came to naught once Germany attacked Russia in June 1941.

Prager's sense of desperation can be gleaned from a telegram he cabled to his friend Otto Neugebauer in Providence: "US Visa Cancelled/Cabled Details Dean/Furniture Sold/Position Resigned/Situation Here Expected Deteriorate Soon/Impossible Stay for Czechoslovakians/Implore Help/ Willy." Finally, Prager decided to leave Istanbul with his wife and their 12-year-old son on a train that took them to Bagdad. From there they boarded a plane bound for Karachi, India, making their way to Bombay. Their ship, the *President Monroe,* sailed around Cape Town to New York, a 40-day journey that brought Prager to Providence in November 1941. He began his teaching career at Brown the following spring term with a course on engineering mechanics.

For the summer of 1942, Richardson and Prager now unveiled a truly impressive set of course offerings in applied mathematics, a program with courses for beginning graduate students as well as young post-docs. Instead of just four lecture courses, there were now 15, including elementary reviews of algebra (taught by Hilda Geiringer) and function theory (given by Stefan Bergmann). Tamarkin and Feller repeated their course on mathematical physics; Sokolnikoff was recruited again from Wisconsin to teach elasticity theory, and von Mises returned from Harvard to launch a new course on the theory of flight. Perhaps coincidentally, Brown acquired the services of Hans Reissner, a true pioneering figure in the field of aerodynamics. Reissner barely escaped from Germany in 1938 to join the faculty at the Illinois Institute of Technology. At Brown, he offered two courses, one on graphical statics and another on the hydrodynamics of propellers.

A Russian-born expert on communications technology, Sergei Alexander Schelkunoff, came from Bell Labs to teach a course on electromagnetic waves. Schelkunoff's life story closely paralleled that of Ivan Sokolnikoff. A student at Moscow University when the revolution broke out, he was drafted into the Czar's army. Under obscure circumstances, he eventually escaped, crossing Siberia to reach Japan, and then finding a crossing to Seattle. He arrived there in 1921, the same year as Sokolnikoff, but unlike him, Schelkunoff stayed to study at the University of Washington. He later took his Ph.D. from Columbia in 1928, after which he joined Bell Labs. Prager, the most recent arrival to the U.S., taught a course on plasticity, his special field of expertise. He also arranged the appointment of the distinguished physicist, Léon Brillouin, who managed to leave Vichy France for the U.S. in late 1940. From then until 1942 Brillouin held an appointment a visiting professor at the University of Wisconsin in Madison. During the latter years of the war, he worked in the field of RADAR as a research scientist with the National Defense Research Committee at Columbia. After the war, he taught from 1947 to 1949 as professor of applied mathematics at Harvard.

With such a team of experts sounding forth on so many topics in applied mathematics all at once, one can easily imagine that the atmosphere at Brown was full of inventive energy that summer. Luckily, a first-hand account of this historic gathering, though brief and highly subjective, has been preserved. Moreover, unlike the programmatic accounts of the politically minded Richardson, this personal account emphasizes the *quality* of the teaching that took place on this occasion. As a recent graduate from Caltech, where he had taken his master's degree under Harry Bateman, Clifford Truesdell was one among the some 115 young participants enrolled at this second summer school. Through Bateman, who was trained in Britain, he had already acquired certain distinct tastes about how to do mathematical physics, and so his brief encounter with these Continental Europeans, many of whom he found rather exotic, was bound to illicit an interesting reaction. In reading his private, off-the-cuff remarks about them, it is well to remember the author's age (he was 23 in the summer of 1942) and exuberant personality.

Truesdell's Private Reflections

Tamarkin is a powerful built man of about 250 pounds, slightly under 6 feet in height. His clumsy gait, swaying from side to side either from an imparity in the length of his legs or from excessive girth of his thighs, combines with his ruddy countenance, squinty eyes, and polished bald head to give a perhaps false impression of joviality, which is not at first destroyed by his high petulant voice that seems so exaggerated as of necessity to be a pretense masking high good humor. This impression of clownishness is a little strengthened by a marked but pleasant accent, a sharp barking voice, and an almost complete éschéance of the articles. There is in him a touch of old Fyodor-Pavlovitch in The Brothers Karamazov. Always wearing a well-pressed suit or sporting costume in good style, usually a different one from the previous days, his entrance into the lecture hall never ceases to be impressive. He surveys the domain like a monarch, rolling along with his coat unbuttoned, so that its ample yardage and great pockets give a particular air of affluence; before saying a word he lays out half a dozen new pieces of chalk in the trough already filled with several hundreds of all sizes, even some unbroken ones left from the day before, and although he always breaks them before writing, he seems to prefer virgin pieces.

His lectures are fluent and polished, although occasionally he is stopped by some point, which he always resolves with difficulty by consulting his notes, written on small cards, which extreme short-sightedness forces him to bring almost into contact with his nearly squinted shut left eye. He complains that the students do not ever give any indication of comprehension or non-comprehension of his remarks, and when he has made some mistake at the board he always triumphantly observes that he could say anything and make every possible error before anyone would stop him. This little game seems to give him real pleasure, but whenever anyone has the temerity to ask something, he replies by shouting over again in a loud angry voice what he has just said, along with other remarks implying politely that the questioner is a little stupid.

Schelkunoff is a Mephistophelian with a pointed goatee. His darting black eyes, high nervous voice and laugh, and hasty walk, together with much external and apparently genuine good humor, give him a likeness to Porfiry Petrovitch in Crime and Punishment, a man who was very far from being a sot.

Prager is the embodiment of elegance in every detail; were it not for his bald head, he might be a fashion plate in Esquire. His accent, while unmistakable, is markedly refined, and his platform manner is almost ostentatiously free from peculiarities.

Mrs. Geiringer is a fine looking blond woman of about 40 with an unfortunately solid figure, principally in the middle parts, and a face full of European sophistication. Her lectures are said to be difficult for her elementary students, but judging from several peeps through the crack of the door I should say she took extreme pains to make everything as clear as possible, speaking with great energy and enthusiasm. She gave the second of the colloquium lectures, which turned out to be quite an event, attended not only by all the students but also by all the mathematical faculty, a regular command performance, in fact, at the order of Dean Richardson, who was annoyed that the first in the series had been attended by no one at all.

Under these circumstances it was not surprising that Mrs. Geiringer was nervous, so nervous in fact that nine tenths of her European sophistication vanished, only the one tenth that was permanently engraved in her countenance remaining. She spoke for one and a half hours at top speed about a sort of six-dimensional vector variously called a screw, motor, or dynam; the subject was of interest to few, and because of her speed and her nervous confusion, comprehensible to none except those already familiar with the subject. She wore a suit, very unbecoming to a woman of her girth, the coat of which she took off from the heat of her discourse, and after that she frequently tucked in her shirt-tail in back, like a little boy. After the lecture during the question period she, Prager, von Mises, and Saunders MacLane had a slightly embarrassing discussion of something or other, mumbling softly so as hardly to be heard by the audience, which was uncertain whether to go or stay.

Feller is a singularly open-faced man with soft brown eyes. He lectures vehemently and tries to enforce the comprehension of the listener. At the close of his lecture on matrices, during the question period, there was a heated triangular discussion in which Feller bellowed, Tarmarkin barked, and Geiringer pleaded; their words were largely inaudible, but one heard "trivial" and "obvious" in irascible flight from one to another.

Sokolnikoff is a tall, thin picture of summer elegance. With his small moustache and short sensual mouth he gives an impression of Russian hauteur, although he is, I think, at heart more cordial than Tamarkin. During his lectures, which are fluent and completely without errors, delivered in flawless English with only a slight trace of accent, he darts up to the board and back again with long, jerky steps, and often falls backward off the small platform. This clumsiness and seeming too large for his surroundings somewhat dampens the elegance of his appearance. He starts at the upper left board and covers it methodically and completely. When illness forced him to be absent for some days he was replaced by a student, equally long and thin but lacking in sartorial elegance, whose platform manners were identical, even to the falling off backward; the similarity was so marked one was occasionally brought up short by the substitute's pure American accent.

Brillouin is a grey, academic figure. Years of study and walking probably in small chambers have shortened his step. His right shoulder blade protrudes slightly, the left side of his face receded a little, and he speaks as if his mouth were full of warm rice, due either to some defect or to his in all probability having observed that English-speaking people keep their mouths nearly shut at all times. He has a complete disregard of ictus, saying sometimes "rela̲tivity" and sometime "relativi̲ty." His lectures are masterpieces of slow dullness. At first he got his two courses mixed, and started out in "Introduction to Part. Diff. Eqns." with a review of tensor analysis, in "Advanced Dynamics" with the most elementary vector concepts, and although his error was pointed out to him after two weeks, he seemed to persist in it for some time afterward.

As he is the highest paid of the staff, Dean Richardson is disturbed about this trouble, and he had instructed Miss Stokes, Brillouin's assistant, to discover the opinions of the students. This Miss Stokes, by the way, a plump grey-haired virgin of 40, whom I suspect to be a gratis volunteer, is head of the mathematics department at the South Carolina College for women, and is inordinately and childishly proud of being both a woman and a mathematician; it is said she is striving manfully and succeeding in keeping a few pages ahead of the students whom she is coaching, and from some questions she asked me I should say she is largely unfamiliar with Fourier's series, although she has a doctor's degree from Duke. Her findings are that the elementary students find the work terribly difficult, and that the advanced students are bored at their enforced return to immaturity. Brillouin seems deeply struck with the successes of modern physics, for in his course in dynamics he has more to say about the limitations of classical mechanics than about the mechanics itself. I recently saw a female attached to him, presumably his wife, whose face, surmounted by much very pretty golden brown hair, was decorated with the little sour smile characteristic of French sophistication, and whose clothes suggested she had consulted Queen Mary's dressmaker.

Bergman is a small man with a mincing step, stooped over so that his head protrudes, and his short cropped black hair makes him the caricature of a little bear. He is pathetically eager to impart knowledge, of which he seems to have a good deal. His lectures are nerve-racking experiences which reduce him to a limp perspiring rag and his students to a state of somnolence, nervous itching, or amusement, depending on their constitutions. This effect is due partly to his unfamiliarity with the language, his horrible grimaces, and his peculiar accent which causes him to articulate some words with the greatest of care and quite incorrectly and to mumble other words in a jumbled hoarse whisper, and partly to his complete inefficiency. He keeps the air filled with sound waves by repeating words sometimes five or six times and by filling in all pauses with miscellaneous vowel sounds. Starting at the centre of the board he scatters his marks at random over the outer parts. He seems an exceptionally good-hearted and well-meaning man, but one soon learns to have as little as possible to do with him, for he is so inefficient that he wastes hours mulling over the details of various arrangements, never has anything really planned in advance, and is always lacking some necessary person or implement to carry out his work.

Thus, for example, when he assigned some problems, he had to have three special meetings of 1 h each, in which he partially explained a very simple procedure, before he got us started, and when we attempted to do the problems, there was always some piece of equipment, such as thumb tacks or a T-square, that was not to be found; his assistant, one Dr. Vaszyoni, is seldom present, and when he is, his aid is of doubtful value. After about a week of lectures Bergman got down to the flow around an airfoil, where I suspect he will stay indefinitely, as he is passionately fond of airplanes. Each day he struggles in with a six-foot skeleton model of a plane, in a dilapidated state, which he triumphantly places before the class, saying "Each day I bring zee plane, so that you have constantly before you the idea of zee plane." The model of course is of no use in illustrating his lectures, although Bergman sometimes picks it up and gestures with it, as he does with erasers or chalk or paper, to illustrate particles of fluid in motion, and sometimes kicks it or steps on it a little by accident. He once asked me, as we were struggling with it together to get it around a corner and up a narrow flight of stairs, "How you like the plane?" and I was at a loss what to answer.

The courses are lamentably mismanaged, disregarding the preparation and ability of the students. The elementary students seem very poorly prepared and immature, but they are driven so hard as to make them complain a good deal. In their courses there are lectures separate from the text, readings in the text, and problems on both, as, I should say, an advanced course should be run. In the advanced courses, however, there are mimeographed notes followed exactly by the lecturers and no problems at all, so that there is next to no work at all. Some of the courses are also misnamed so that one cannot find out what they are in advance. Advanced dynamics has turned out to be elementary special relativity and Lagrange's equations, and Review of Function Theory is concerned with the general theory of orthogonal functions for mapping complex domains, a subject most of the advanced students are unfamiliar with. Of the courses I have observed, I should say Tamarkin's in partial differential equations was the best, and Sokolnikoff's elasticity next. Compared with other books on elasticity, Sokolnikoff's notes are much clearer and more direct, although they are perhaps unnecessarily elementary; they contain also some elegant treatments due, I think, to Mushelisvili. Bergman's fluid dynamics and Brillouin's dynamics are a waste of time. Prager's plasticity, which makes a pass at tensor analysis, promises to be pretty good. (This expectation was deceived.)[10]

A Participant's Perspective

These colorful remarks provide a vivid snapshot of what the Brown summer school of 1942 looked like from the vantage point of an attendee, that is, a potential beneficiary of the program. In an era predating the vogue for student evaluations, we are fortunate still to have this scrapbook entry from Truesdell's personal memorabilia, a document that sheds fresh light on Richardson's ambitious operation. Clearly, the latter was not a "hands-on" administrator in the style of a Richard Courant. For when Truesdell writes that the whole summer school was "lamentably organized," he backs this up, not only with funny anecdotes about the teaching staff but also with general remarks about why the two-tiered course offerings were bound to be ineffective.

To gain a sense of the gulf separating Richardson's realm and "events on the ground," one should contrast Truesdell's remarks with Brown's highly promotional report, "Program of Advanced Instruction and Research in Mechanics" (PAIRM 1943). That document showcases an impressive quasi-military structure, headed by five "Officers of Admin-

[10]The final remark was pencilled in and was presumably written somewhat later than the text preceding.

istration" (including Richardson); an Advisory Committee that included Thornton Fry and Harvard's Marshall Stone; a liaison officer, the electrical engineer Frederick Neale Tompkins, representing the Engineering, Science, and Management War Training Division of the U.S. Department of Education; and 19 "Officers of Instruction." The latter taught designated courses (all carrying numbers to look like a carefully constructed curriculum) in the applied mathematics program either during the regular academic year and/or during the three summers from 1941 to 1943. On paper, very impressive indeed, but in reality – unless Willy Prager was somehow able to reshape the Brown program almost immediately after his arrival – no one was running the show. In all likelihood, something like Clifford Truesdell's description of a truly chaotic scene continued to prevail throughout the war years.

Still, for Truesdell himself, the Brown experience would prove a major turning point in his mathematical career. But to account for this, we must add a brief note to his private reflections. Richardson chose to cut back the number of guest speakers for the summer of 1942, although several of the regular "officers of instruction" presented special lectures. As noted by Truesdell, Geiringer spoke on geometrical mechanics and Feller on matrix computations; Prager lectured on plasticity, Schelkunoff on the antenna problem, and Brillouin talked about the megatron vacuum tube. Yet only two visitors were invited to speak that summer. The first was Isaac Opatowski, an instructor at the University of Minnesota who spoke about applications of Laplace transformations. Originally from Italy, where he taught at the University of Turin and then worked for Fiat, Opatowski left for the United States in 1938. The second guest speaker was a Hungarian expert on fluid mechanics, Paul Felix Neményi, who also left Europe in 1938 and eventually gained an appointment in 1941 at Colorado State College. He visited Brown during the last week in July, when he gave two lectures on "the fundamental phenomena of fluid motion and the various theories based upon them." Truesdell listened in rapt attention and soon struck up a close friendship with Neményi, who would afterward exert a strong influence on the young mathematician during his formative years.

From Brown, Truesdell went on to Princeton, where he took his doctorate under Solomon Lefschetz in 1943. The topic of his dissertation work, however, was far removed from his mentor's chosen field of algebraic topology. Already that year, in fact, Truesdell and Neményi coauthored a short article on "The Stress Function for the Membrane Theory of Shells of Revolution" that was published in the Proceedings of the National Academy of Sciences. At Princeton, Truesdell's dissertation research was simply a direct continuation of this collaboration with Paul Neményi. At a meeting of the AMS held at Rutgers in September 1943, he spoke about his work on the membrane theory, for which he received his doctorate that same year. The principal results from his dissertation were later published in the AMS Transactions (Truesdell 1945).

Thirty-five years later, he would have other reasons to recall that particular meeting, at which his mentor, Lefschetz, introduced him to Harvard's George David Birkhoff, then in the twilight of his stellar career. Truesdell also had reason to reflect back on the significance of his early contact with Neményi:

> My early research did not fall into any professional category. The only encouragement I received, and it was neither much nor steady, was from older men, . . . from my teachers, who were Bateman, Neményi, Lefschetz, and Bochner, and from Hadamard, Villet, Bouligand, Synge, Hamel, Picone, and Finzl. I doubt that any but Neményi had gone into what I was doing, and so perhaps what seemed to me encouragement from the rest was only the gentlemanly courtesy of an older generation. I take this occasion to express once again my gratitude to Neményi, who taught me that mechanics was something deep and beautiful, beyond the ken of schools of "applied mathematics" and "applied mechanics" in the 1940s (Truesdell 1983, vii).

Paul Neményi died unexpectedly in 1952, forgotten by nearly everyone except Truesdell, who wrote a short obituary for *Science*. Ten years later, in an editorial note in *Archive for History of Exact Sciences* that accompanied the posthumous publication of Neményi's essay on the historical development of fluid mechanics (Neményi 1962), Truesdell again paid homage to his real mentor. There he recalled how Neményi and he had collaborated on research in fluid mechanics after the war, when both were employed by the Naval Ordnance Laboratory. They also planned to publish a jointly authored essay on the history of fluid mechanics, a task that Truesdell tried to complete after his friend's death, but eventually abandoned after he realized the complexities of the task. His decision to publish Neményi's portion of the manuscript, despite its gaps and flaws, reflects a deeply personal reverence for his memory.

Truesdell knew Paul Neményi well, but presumably he did not know that much about his personal life. Ironically, he had already gone to his own grave when his long deceased friend's name made minor headlines. Early in the new millennium the world learned about Neményi's activities during the war years when he befriended a young woman named Regina Fischer, she a student at the University of Denver. Fischer's husband, Hans Gerhardt, happened to be a German-born communist – nothing unusual at the time, except that the FBI suspected he was an active Soviet agent. As it turned out, J. Edgar Hoover's boys followed Regina's trail closely as well. This took them to Chicago, where she gave birth to their second child, a son, in 1943. The boy grew up fatherless after the Fischers divorced in 1945. Eventually the family settled in Brooklyn, where Regina's young lad learned to play chess. Thus began the tragi-comic life of Robert James "Bobby" Fischer, whose exploits on and off the chessboard would become the stuff of legends

(Edmonds and Eidinow 2005). Acting in compliance with the Freedom of Information Act, in 2002 the FBI released its extensive files on Regina Fischer, who had died in 1997, long since estranged from her reclusive son. Thirty years of documentary evidence then revealed that Bobby Fischer's real father was not the man named on his birth certificate as Hans Gerhardt Fischer, but rather Truesdell's own fatherly mentor, Paul Neményi.

Truesdell Looks Back

In 1978, Truesdell was named as one of three recipients of the George David Birkhoff Prize in applied mathematics, an award given jointly by the AMS and SIAM for outstanding contributions to "applied mathematics in the highest and broadest sense." The other two recipients were Harvard's Garrett Birkhoff and Mark Kac of Rockefeller University. All three honorees were invited to speak at the January meeting of the AMS that year (Birkhoff et al. 1978). On that occasion, Birkhoff described some of his research in applied algebra, whereas Kac's talk had a lighter quality, merely repeating some remarks he made at a symposium on "The Future of Applied Mathematics" held at Brown in 1972. Thus, Kac recalled a wartime cartoon that showed two chemists staring at what looked like a small pile of sand. They were surrounded by test tubes and impressive laboratory apparatus, the caption underneath reading: "Nobody really wanted a dehydrated elephant but it is nice to see what can be done." To which Kac commented, "I am sure that we can all agree that applying mathematics should not lead to the creation of 'dehydrated elephants.'"

Truesdell delivered an entirely different kind of speech, full of highly personal opinions, in which he began by paying homage to G. D. Birkhoff, while making plain his disdain for Richardson and Brown's program in applied mathematics:

> I met Birkhoff just once. It was at the summer meeting of the Society on September 12, 1943, at Rutgers. In those days there were about 2,000 members, of whom some 200 were active mathematicians; at a meeting in the East you could usually encounter about half of those. My teacher, Lefschetz, who was often kind to his students, introduced me to Birkhoff and to many other senior men. After dinner there were two speeches. The report I published in the *Bulletin* is brief: "Dean R. G. D. Richardson of Brown University spoke of the importance of applied mathematics in the war effort. Professor G. D. Birkhoff of Harvard University urged mathematicians and scientists to maintain a proper balance of values during the emergency." Having survived a summer school of applied mathematics at Brown, I knew what to expect of Richardson; the *Bulletin* sums his lecture. Not so Birkhoff's. With commanding dignity and in superb, native English, Birkhoff lashed the universities for their subservience to government and warfare. He warned us that by giving junior men heavy teaching loads, as much as eighteen hours, with no assistance, and by admitting unselected and mainly unqualified undergraduates, the universities were destroying what it was their prime duty, above nations and above emergencies, to foster. The many incompetents pressed into instruction were unable to teach, for they did not know, while the competent few were unable to learn because they were left no time. (Truesdell 1978, 99).

Five years before this, Birkhoff had made similar remarks in a speech that caused considerable stir within the American mathematical community (see Siegmund-Schultze 2009, 225–227). Citing the recent influx of first-rate European mathematicians to the United States, he expressed concern that those born and trained in this country might find themselves relegated to serving as "hewers of wood and drawers of water" (a phrase taken from Joshua 9:23 in the King James Bible). This situation, in Birkhoff's opinion, had already reached "the point of saturation," a view that helps explain why Harvard, under Birkhoff's leadership, took a conservative stance when it came to academic appointments. When the elder Birkhoff passed from the scene in 1944, his friend Oswald Veblen took note of this obliquely in his obituary. Veblen also distanced himself from Birkhoff's views regarding the dangers of a foreign invasion, writing, "the American mathematical community has at least been healthy enough to absorb a pretty substantial number of European mathematicians without serious complaint" (Veblen 1944, 52).

Truesdell, however, remembered a quite different message form Birkhoff's later speech from 1943, namely the higher duty of universities that was "above nations and above emergencies." Thirty-five years later, he was intent on pounding this point home:

> I wonder what Birkhoff would have said, had he lived to see them, about our mass universities today, universities which in a time of peace and ease have forgotten that the task of higher learning is not only to sow, dung, and harvest, but also and above all to winnow. I wonder what he would have said about the pollution by social sciences, changing values, team work, computing, sponsored research, involvement in the community, and even soft mathematics, to the point that mathematics of his kind is today decried as being "esoteric." I wonder what he would have said about the iron rule of mediocrity the Government today imposes as the price of the manna called "overhead" it scatters to the voracious and insatiable education mongeries. I wonder what he would have said about our management by regular corporate administrations which lack, however, the profit motive and in perfect Parkinsonian policy have but a single real interest alike in students and in employees (some of whom are still called "faculty"): Get money out of them directly or for them through subsidy, in quantity sufficient to sustain exponential growth in number and power of administrators - mad pursuit of bigness, until, one day, every boy and girl will be conscripted to serve a term of unstructured play garnished with social indoctrination called "higher learning," and all universities will be branch offices of one Government bureau. (Truesdell 1978, 99).

Following this torrent, Truesdell managed to add a few words of admiration for Birkhoff's orientation and achievements as a research mathematician, likening his style with that of Poincaré (see further (Birkhoff 1950)). He then turned to make some remarks about Hilbert's sixth Paris problem on the axiomatization of physics, adding these personal comments:

> Hilbert's influence on recent rational mechanics is not so widely known as it should be. While he published scantly in that field, years ago I found in the libraries of Purdue University and the University of Illinois notes taken by two Mid-Westerners in Hilbert's course on continuum mechanics, 1906/7. After studying those notes, in my own expositions from 1952 onward I followed Hilbert's lead, not in detail but in standpoint and program; basic laws are formulated in terms of integrals, and the science is treated as a branch of pure mathematics, by systematic, rigorous development motivated by its logical and conceptual structure, descending from the general to the particular.
>
> The word "axiom" may confuse, since on the one hand all of mathematics is essentially axiomatic, while on the other hand "axiomatics" is a term often used pejoratively to suggest a fruitless rigorization of what everybody believed already. A mathematician must know that good axioms spring from intrinsic need—details scatter, concept is not clear, or paradox stands unresolved. Formal axioms represent only one response to the call for conceptual analysis before significant new problems can be stated. You cannot solve a problem that has not yet been set. Formal axiomatics make an important part of modern rational mechanics; I refer in particular to Walter Noll's solution of the relevant portion of Hilbert's sixth problem. But by far the greater part of modern mechanics is neither more nor less axiomatic than is any other informally stated branch of mathematics. *The essence is conceptual analysis,* analysis not in the sense of the technical term but in the root meaning: logical criticism, dissection, and creative scrutiny. After that comes the poetry of statement and proof, the ornament of illuminating examples. A mathematical discipline is made by mathematicians. However much they may begin from, digest, clarify, and build upon the ideas of physicists and engineers, it is the mathematicians who make sense out of them. (Truesdell 1978, 101)

Truesdell's penchant for poetry slips into the picture here, and yet such aesthetic concerns, very close to Hilbert's in their way, form a central feature in his understanding of mathematical physics. A passage from his lecture in honor of Euler makes this point most emphatically: "Mechanics is beautiful because the mathematicians of the Baroque period, among them especially Euler, made it beautiful . . . To him, above all others, we owe the heavenly clear structure of the entire theory of mechanics and the divinely simple expression of its fundamental laws." Yet, returning to his speech from 1978, Truesdell had no wish to dwell on such gushy matters. Instead, he stayed on message, true to his role as academy's Cassandra. He ended by repeating a word of warning issued by Richard Courant in the *Vorwort* to the first edition of Courant-Hilbert, *Mathematische Methoden der Physik,* a book Truesdell had studied with Bateman as an undergraduate at Caltech:

> Now, I fear, many people accept a picture too narrow of mathematical activity. They forget that the great theories which enable us to understand in part the world about us and on which engineers and computers base their applications were created by great mathematicians; they forget also that the problems suggested by new mathematical theories of nature often fail to fall within any of the previously existing fields of "applied mathematics", which fields, all of them, grew out of older theories of nature. They disregard a prophetic warning published in 1924 by that canny and shrewd observer, Richard Courant:. . . . many analysts are no longer aware that their science and physics . . . . belong to each other, while . . . . often physicists no longer understand the problems and methods of mathematicians, indeed, even their sphere of interest and their language. Obviously this trend threatens the whole of science: the danger is that the stream of scientific development may ramify, ooze away, and dry up. (Truesdell 1978, 101)

Twenty years after Luce and Lipmann proclaimed the "American century," Dwight D. Eisenhower, the war hero who oversaw the D-Day invasion, issued a famous warning to his nation. In his farewell address as its outgoing president, Ike warned his fellow compatriots about the encroaching power of what he dubbed the military-industrial complex:

> In the councils of government, we must guard against the acquisition of unwarranted influence, whether sought or unsought, by the military-industrial complex. The potential for the disastrous rise of misplaced power exists and will persist. We must never let the weight of this combination endanger our liberties or democratic processes. We should take nothing for granted. Only an alert and knowledgeable citizenry can compel the proper meshing of the huge industrial and military machinery of defense with our peaceful methods and goals, so that security and liberty may prosper together.

Eisenhower's words went largely unheeded, even if the term "military-industrial complex" gained widespread currency within New Left circles and beyond. During the anti-war demonstrations at Columbia and Berkeley in the 1960s, its leaders also aimed to expose the complicity of American institutions of higher education in "selling out" to these powerful special interests. Clifford Truesdell surely loathed the protesters who disrupted lecture halls during the height of the Vietnam War. Ironically, however, this died-in-the-wool conservative shared with them a deep belief that the leaders of American universities had been corrupted by the money and power of the Establishment, which in his view included government agencies such as the National Science Foundation. In recalling his experiences during the war years, he painted his own grim picture of how the events of that period had forever reshaped American academic life.

Acknowledgments This essay owes much to the helpful contributions of various colleagues. My thanks go to Philip Davis for calling attention to Truesdell's notes on the Brown summer school, to Reinhard Siegmund-Schultze for his careful reading and advice on sources, to Jed Buchwald for personal memories of music making at Il Palazzetto, to Sandro Caparinni for sharing his knowledge of Truesdell's work and career, and to Marjorie Senechal, whose thoughtful editing helped smooth the text. Thanks also must go to Clifford Truesdell, who certainly had a way with words.

References

Archibald, Raymond Clare. 1950. R. G. D. Richardson, 1878–1949. *Bulletin of the American Mathematical Society* 56 (3): 256–265.

Binder, Christa. 1992. Hilda Geiringer: Ihre ersten Jahre in Amerika. In *Amphora, Festschrift for Hans Wußing*, ed. S.S. Demidov et al., 25–53. Basel: Birkhäuser.

Birkhoff, G.D. 1950. Intuition, Reason and Faith in Science, *Collected Mathematical Papers*. Vol. 3, 652–659. New York: American Mathematical Society.

Birkhoff, Garrett, Mark Kac, and Clifford Truesdell. 1978. The Birkhoff Prize Talks, Atlanta 1978. *Mathematical Intelligencer* 1: 93–101.

Booß-Bavnbek, Bernhelm, and Jens Høyrup, eds. 2003. *Mathematics and War*. Basel: Birkhäuser.

Buchwald, Jed, and I. Bernard Cohen. 2001. Eloge: Clifford Truesdell, 1919–2000. *Isis* 92(1): 123–125.

Dresden, Arnold. 1942. The Migration of Mathematicians. *The American Mathematical Monthly* 49 (7): 415–429.

Edmonds, David, and Eidinow, John. 2005. *Bobby Fischer goes to War.* New York: HarperCollins.

Fry, Thornton. 1941. Industrial Mathematics. *American Mathematical Monthly* 58 (6): 1–38.

Giusti, Enrico. 2003. Clifford Truesdell (1919–2000), Historian of Mathematics. *Journal of Elasticity* 70: 15–22.

Hille, Einar. 1947. Jacob David Tamarkin – His Life and Work. *Bulletin of the American Mathematical Society* 53 (7): 440–457.

Ignatieff, Yurie, and Heike Willig. 1999. *Clifford Truesdell: Eine wissenschaftliche Biographie des Dichters, Mathematikers und Naturphilosophen*. Aachen: Shaker Verlag.

Neményi, Paul. 1962. The Main Concepts and Ideas of Fluid Dynamics in their Historical Development, C. Truesdell (ed.), *Archive for History of Exact Sciences* 2: 52–86.

Owens, Larry. 1989. Mathematicians at War: Warren Weaver and the Applied Mathematics Panel, 1942–1945. In *The History of Modern Mathematics*, ed. David E. Rowe and John McCleary, vol. 2, 287–305. Boston: Academic Press.

PAIRM. 1943. Program of Advanced Instruction and Research in Mechanics. *Bulletin of Brown University* XL (8): 1–32.

Richardson, R.G.D. 1943a. Advanced instruction and research in mathematics. *American Journal of Physics* 11 (2): 67–73.

———. 1943b. Applied Mathematics and the Present Crisis. *American Mathematical Monthly* 50 (7): 415–423.

Siegmund-Schultze, Reinhard. 1993. Hilda Geiringer-von Mises, Charlier Series, Ideology, and the Human Side of the Emancipation of Applied Mathematics at the University of Berlin during the 1920s. *Historia Mathematica* 20: 364–381.

———. 2003a. The late Arrival of Academic Applied Mathematics in the United States: a Paradox, Theses, and Literaure. *NTM* 11 (2): 116–127.

———. 2003b. Military Work in Mathematics, 1914–1945: an Attempt at an International Perspective, in (Booß-Bavnbek & Hoyrup), 23–82.

———. 2009. *Mathematicians Fleeing from Nazi Germany: Individual Fates and Global Impact*. Princeton: Princeton University Press.

Truesdell, Clifford. 1945. The Membrane Theory of Shells of Revolution. *Transactions of the American Mathematical Society* 58: 96–166.

———. 1957. Eulers Leistungen in der Mechanik. *L'enseignement mathematique* 3 (4): 251–262.

———. 1978. The Birkhoff Prize Talks, Atlanta 1978, *Mathematical Intelligencer* 1: 93–101, pp. 99–101.

———. 1983. *An Idiot's Fugitive Essays on Science: Methods, Criticism, Training, Circumstances*. Berlin: Springer.

Veblen, Oswald. 1944. George David Birkhoff, 1884–1944, *Biographical Memoirs. National Academy of Sciences* 80: 44–56.

Hilbert's Legacy: Projecting the Future and Assessing the Past at the 1946 Princeton Bicentennial Conference

37

(*Mathematical Intelligencer* 25(4)(2003): 8–15)

Saunders Mac Lane on Solomon Lefschetz (Mac Lane 1989, 220): In 1940 when he was writing his second book on topology, [Lefschetz] sent drafts of one section up to Whitney and Mac Lane at Harvard. The drafts were incorrect, we wrote back saying so – and every day for the next seven or eight days we received a new message from Lefschetz, with a new proposed version. It is no wonder that the local ditty about Lefschetz ran as follows:

—Here's to Lefschetz, Solomon L,
Ir-re-pres-si-ble as hell
When he's at last beneath the sod
He'll then begin to heckle God.

While working on this essay, I found myself thinking about some general questions raised by discussions that took place in Princeton several decades ago. For example, does it make sense to talk about "progress" in mathematics in a global sense, and, if so, what are its hallmarks and how do mathematicians recognize such improvements? Or does mathematics merely progress at the local level through conceptual innovations and technical refinements made and appreciated only by the practitioners of specialized subdisciplines? Of course, specialists in modern mathematical communities are regularly called upon to assess the quality of work undertaken in their chosen field. But what criteria do mathematicians apply when they express opinions about the depth and importance of contemporary research fairly far removed from their own expertise? Those in leadership positions surely expect their general opinions to carry real weight and sometimes even to have significant practical consequences. So how do opinion leaders justify their views when trying to assess the importance of past research or guide it into the future? How do they determine the relative merits of work undertaken in distinct disciplines, and on what basis do they reach their conclusions? Clearly various kinds of external forces – money springs to mind – influence mathematical research and channel the talent and energy in a community. Yet as every researcher knows, even under optimal working conditions and without external constraints, success can be highly elusive. Small research groups are often more effective than isolated individuals, but projects undertaken on a larger scale can also pose many unforeseeable difficulties. So to what extent can mathematicians really direct the course of future investigations? How important are clearly conceived research programs? Or do such preconceived ideas tend to hamper rather than promote creative work? And if "true progress" can only be assessed in retrospect or within the context of specialized fields of research, shouldn't opinionated mathematicians think twice before making sweeping pronouncements about the significance of contemporary developments?

These kinds of questions are, of course, by no means new; their relevance has long been recognized, even if mathematicians have usually tried to sweep them under their collective rugs. More recently, historians and sociologists have cast their eyes on such questions just as mathematicians have become increasingly sensitized to the contingent nature of most mathematical activity (see Rowe 2003). Until recent decades, however, conventional wisdom regarded mathematical knowledge as not just highly stable, but akin to a stockpile of eternal truths. If since *The Mathematical Experience* (Davis and Hersh 1981) this classical Platonic image of mathematics has begun to look tired and antiquated, we might begin to wonder how this could have happened. Those who eventually turned their backs on conventional Platonism surely realized that doing so carried normative implications for mathematical research (as well as for historians of mathematics, see Rowe 1996). So long as doing mathematics was equated with finding eternal truths, practitioners could ply their craft as a high art and appeal to the ideology of "art for art's sake," like the fictive expert on "Riemannian hypersquares" in *The Mathematical Experience.* But once deprived of this traditional Platonist crutch, mathematicians have had difficulty finding a substitute prop to support their work. When Davis and Hersh poked fun at the inept responses of their expert on Riemannian hypersquares who was unable to explain what he did (never mind why), this didn't mean that these kinds of questions are easy to answer. How, after all, *do* leading authorities form judgments about the quality or promise of a fellow mathematician's work?

D.E. Rowe, *A Richer Picture of Mathematics*, https://doi.org/10.1007/978-3-319-67819-1_37

What criteria are used to assess the relative importance of work undertaken in two different, but related fields?

Mathematics may well be likened with a high art, but then artists are normally exposed to public criticism by non-artists, such as professional critics. Clearly, mathematicians seldom find themselves in a similar position; their work is far too esoteric to elicit comment other than in the form of peer review. So what is good mathematics, and who decides whether it is really good or merely "fashionable"? If research interests shift with the fashions of the day, to what extent do fashionable ideas reflect ongoing developments in other fields? And who, then, are the fashion moguls of a given mathematical era or culture and how do they make their influence felt? Can *anyone* really predict the future course of mathematical events or at least sense which areas are likely to catch fire?

Hilbert's Inspirations

No doubt plenty of people have tried, most famously David Hilbert, the leading trendsetter of the early twentieth-century. In 1900 he captured the attention of a generation of mathematicians who subsequently took up the challenge of solving what came to be known as the 23 "Hilbert problems" (see Browder 1976; Gray 2000). Some of these had been kicking around long before Hilbert stepped to the podium at the Paris ICM in 1900 to speak about "Mathematische Probleme" (Hilbert 1935, 290–329). Moreover, a few of the fabled 23 (numbers 6 and 23 come readily to mind) were not really problems at all but rather broadly conceived research programs. In fact, the whole idea behind Hilbert's address was to suggest fertile territory for the researchers of the early twentieth century rather than merely to enumerate a list of enticing problems. Indeed, his main message emphatically asserted that mathematical progress – signified by the solution of difficult problems – leads to simplification and unification rather than baroque complexity. "Every real advance," he concluded, "goes hand in hand with the invention of sharper tools and simpler methods which at the same time assist in understanding earlier theories and cast aside older mathematical developments. . . The organic unity of mathematics is inherent in the nature of this science, for mathematics is the foundation of all exact knowledge of natural phenomena." .

Hilbert badly wanted to make a splash at the Paris ICM. Initially he thought he could do so by challenging the views of the era's leading figure, Henri Poincaré, who stressed that the vitality of mathematical thought was derived from physical theories. This vision rubbed against Hilbert's deeply engrained purism, so he sought the advice of his friend, Hermann Minkowski. The latter dowsed cold water on Hilbert's plans to counter Poincaré's physicalism, but then gave him an enticing idea for a different kind of lecture. "Most alluring," Minkowski wrote, "would be the attempt to look into the future, in other words, a characterization of the problems to which the mathematicians should turn in the future. With this, you might conceivably have people talking about your speech even decades from now. Of course, prophecy is indeed a difficult thing" (Minkowski to Hilbert, 5 January 1900, (Minkowski 1973, 119–120)).

Hilbert, who now stood at the height of his powers, rose to Minkowski's challenge. He never doubted his vision for mathematics, and his success story – indeed, the whole Hilbert legend – took off with the publication of the Paris lecture and its full list of 23 problems (at the Paris ICM he presented only ten of them). Now that more than a century has elapsed, we realize that Hilbert's views on foundations, as adumbrated in his 1900 speech, were hopelessly naïve and far too optimistic. Even his younger contemporaries – most notably Brouwer and Weyl – sensed they were inadequate, though Hilbert continued to fight for them bravely as an old man (Weyl 1944). When Kurt Gödel dealt the formalist program a mortal blow in 1930, Hilbert's vision of a simple, harmonious Cantorian paradise died with it. Still, his reputation as the "Göttingen sage" lived on, making Minkowski's prediction – that mathematicians might still be "talking about your speech even decades from now" – the most prophetic insight of all.

Hilbert died on a bleak day in mid-February 1943, just after the German army surrendered at Stalingrad setting the stage for the final phase of the Nazi era (Reid 1970, 213–215). As the SS and Gestapo intensified their efforts to round up and exterminate European Jews, Hilbert's first student, Otto Blumenthal, was caught in their web; he died a year later in the concentration camp in Theresienstadt. In the meantime, several of Hilbert's numerous students had found their way to safer havens (see Siegmund-Schultze 1998 for a detailed account of the exodus). Two of them, Hermann Weyl and Richard Courant, met again in Princeton nearly four years after Hilbert's death to take part in an event that brought to mind their former mentor's famous Paris lecture. There, on the morning of 17 December 1946, Luther Eisenhart opened Princeton's Bicentennial Conference on "Problems of Mathematics," a three-day event that brought together some one hundred distinguished mathematicians (Duren et al. 1989, 309–359) (Fig. 37.1).

Princeton Agendas

The stated purpose of this event was "to help mathematics to swing again for a time toward unification" after a long period during which a "unified viewpoint in mathematics" had been neglected. Its program was both broad and ambitious, but as a practical consideration the Conference Committee decided to omit applied mathematics, even though significant connections between pure mathematics and its applications

Fig. 37.1 (**a**) Participants at the 1946 Princeton Bicentennial Conference. (**b**) Participants identified by number (see following page).

were discussed. The larger vision set forth by its organizers also carried distinctly Hilbertian overtones:

> The forward march of science has been marked by the repeated opening-up of new fields and by increasing specialization. This has been balanced by interludes of common activity among related fields and the development in common of broad general ideas. Just as for science as a whole, so in mathematics. As many historical instances show, the balanced development of mathematics requires both specialization and generalization, each in its proper measure. Some schools of mathematics have prided themselves on digging deep wells, others on excavation over a broad area. Progress comes most easily by doing both. The increasing tempo of modern research makes these interludes of common concern and assessment come more and more frequently, yet it has been nearly 50 years since much thought has been broadly given to a unified viewpoint in mathematics. It has seemed to us that our conference offered a unique opportunity to help mathematics to swing again for a time toward unification (Lefschetz et al. 1947, p. 309).

These pronouncements make clear that the Princeton Conference on "Problems of Mathematics" was no ordinary meeting of mathematical minds. As the editors of *A Century of Mathematics in America* noted, "the world war had just ended, mathematicians had returned to their university positions, and large numbers of veterans were beginning or resuming graduate work. It was a good time to take stock of open problems and to try to chart the future course of research" (Duren et al. 1989, p. ix). The Conference Committee, chaired by Solomon Lefschetz, reflected the pool of talent that had been drawn to Princeton as a result of the flight from European fascism, listing such stellar names as Emil Artin, Valentin Bargmann, Solomon Bochner, Claude Chevalley, and Eugene Wigner. Thus, the Princeton Bicentennial came at a propitious time for such a meeting, though the scars of the Second World War were still fresh and the threat of nuclear holocaust a looming new danger. The tensions of this political atmosphere, but above all the Princeton mathematicians' hopes for the future were echoed in their conference report:

> Owing to the spiritual and intellectual ravage caused by the war years, it seemed exceedingly desirable to have as many participants from abroad as possible. As the list of members shows, considerable success was attained in this. Our conference became, as it were, the first international gathering of mathematicians in a long and terrible decade. The manifold contacts and friendships renewed on this occasion will, we all hope, in the words of the Bicentennial announcement, "contribute to the advancement of the comity of all nations and to the building of a free and peaceful world" (Lefschetz et al. 1947, p. 310).

Just over a decade had passed since the last International Congress of Mathematicians was held in Oslo, and several who were present at that 1936 event also attended the Princeton conference, including Oswald Veblen, Norbert Wiener, Hermann Weyl, Garrett Birkhoff, Lars Ahlfors, and Marcel Riesz. Among the distinguished foreigners who attended the Princeton Bicentennial were Paul Dirac and William Hodge from England, Zurich's Heinz Hopf, and China's L. K. Hua. Of the 93 mathematicians – all of them males – pictured in the group photo above, 11 (all seated in the front row) came from overseas. A large percentage of the others, however, were European émigrés, many of whom had come to North America during the previous 10 years.

The Problems of Mathematics

1. Morse, M., Institute for Advanced Study
2. Ancocbea, G., University of Salamanaca, Spain
3. Borsuk, K., University of Warsaw, Poland
4. Cramér, H., University of Stockholm, Sweden
5. Hlavaty, V., University of Prague, Czechoslovakia
6. Whitehead, J. H. C., University of Oxford, England
7. Garding, L. J., Princeton
8. Riesz, M., University of Lund, Sweden
9. Lefachetz, S., Princeton
10. Veblen, O., Institute for Advanced Study
11. Hopf, H., Federal Technical School, Switzerland
12. Newman, M. H. A., University of Manchester, England
13. Hodge, W. V. D., Cambridge, England
14. Dirac, P. A. M., Cambridge University, England
15. Hua, L. K., Tsing Hua University, China
16. Tukey, J. W., Princeton
17. Harrold, O. G., Princeton
18. Mayer, W., Institute for Advanced Study
19. Mautner, F. I., Institute for Advanced Study
20. Gödel, K., Institute for Advanced Study
21. Levinson, N., Massachusetts Institute of Technology
22. Cohen, I. S., University of Pennsylvania
23. Seidenberg, A., University of California
24. Kline, J. R., University of Pennsylvania
25. Ellenberg, S., Indiana University
26. Fox, R. H., Princeton
27. Wiener, N., Massachusetts Institute of Technology
28. Rademacher, H., University of Pennnsylvania
29. Salem, R., Massachusetts Institute of Technology
30. Tarski, A., University of California
31. Bargmann, V., Princeton
32. Jacobson, N., The Johns Hopkins University
33. Kac, M., Cornell University
34. Stone, M. H., University of Chicago
35. von Neumann, J., Institute for Advanced Study
36. Hedlund, G. A., University of Virginia
37. Zariski, O., University of Illinois
38. Whyburn, G. T., University of Virginia
39. McShane, E. J., University of Virginia
40. Quine, W. V., Harvard
41. Wilder, R. L., University of Michigan
42. Kaplansky, I., Institute for Advanced Study
43. Bochner, S., Princeton
44. Leibler, R. A., Institute for Advanced Study
45. Hildebrandt, T. H., University of Michigan
46. Evans, G. C., University of California
47. Widder, D. V., Harvard
48. Hotelling, H., University of North Carolina
49. Peck, L. G., Institute for Advanced Study
50. Synge, J. L., Carnegie Institute of Technology
51. Rosser, J. B., Cornell
52. Murnaghan, F. D., The Johns Hopkins University
53. Mac Lane, S., Harvard
54. Cairns, S. S., Syracuse University
55. Brauer, R., University of Toronto, Canada
56. Schoenberg, I. J., University of Pennsylvania
57. Shiffman, M., New York University
58. Milgram, A. N., Institute for Advanced Study
59. Walker, R. J., Cornell
60. Hurewicz, W., Massachusetts Institute of Technology
61. McKinsey, J. C. C., Oklahoma Agricultural and Mechanical
62. Church, A., Princeton
63. Robertson, H. D., Princeton
64. Bullitt, W. M., Bullitt and Middleton, Louisville, Ky.
65. Hille, E., Yale University
66. Albert, A. A., University of Chicago
67. Rado, T., The Ohio State University
68. Whitney, H., Harvard
69. Ahlfors, L. V., Harvard
70. Thomas, T. Y., Indiana University
71. Crosby, D. R., Princeton
72. Weyl, H., Institute for Advanced Study
73. Walsh, J. L., Harvard
74. Dunford, N., Yale
75. Spenser, D. C., Stanford University
76. Montgomery, D., Yale
77. Birkhoff, G., Harvard
78. Kleene, S. C., University of Wisconsin
79. Smith, P. A., Columbia University
80. Youngs, J. W. T., Indiana University
81. Steenrod, N. E., University of Michigan
82. Wilks, S. S., Princeton
83. Boas, R. P., Mathematical Reviews, Brown University
84. Doob, J. L., University of Illinois
85. Feller, W., Cornell University
86. Zygmund, A., University of Pennsylvania
87. Artin, E., Princeton
88. Bohnenblust, H. F., California Institute of Technology
89. Allendoerfer, C. B., Haverford College
90. Robinson, R. M., Princeton
91. Bellman, R., Princeton
92. Begle, E. G., Yale
93. Tucker, A. W., Princeton

Fig. 37.1 (continued)

Nativism Versus Internationalism in American Mathematics

Yet if internationalism had a nice ring, this theme played a secondary role at the Princeton Bicentennial, which had little in common with the ICMs of the past. On the contrary, as the first large-scale gathering of America's mathematical elite at the onset of the post-war era, domestic conflicts strongly colored this meeting. Intent on laying the groundwork for their own vision of a "new mathematical world order," the Princetonians seized on their university's bicentennial as an opportunity to place themselves at the fulcrum of a now dynamic, highly Europeanized American mathematical community. Princeton's Veblen, quite unlike Harvard's G. D. Birkhoff, had played a major part in helping displaced European mathematicians find jobs in the United States. Given these circumstances, Princeton could legitimately host an intellectual event with the explicitly stated moral agenda of aiming to promote harmonious relations among the world's mathematicians. But the Princeton community was, in this respect, almost singular in the United States.

Harvard's reputation as a bastion of conservatism placed it in natural opposition to Princeton, thereby heightening tensions within the American mathematical community. G. D. Birkhoff had long despised Lefschetz in an era when anti-Semitism at Ivy League universities was pervasive (Reingold 1981, 182–184). As the first native-trained American to compete on equal terms with Europe's elite mathematicians, Birkhoff sought to bring the United States to the forefront of the world scene. Coming from E. H. Moore's ambitious Chicago school, he embodied the Midwestern ideals of Americans determined to demonstrate their own capabilities and talent through incessant hard work. During the 1920s, he molded Harvard into the strongest department in the U.S., particularly in his own field, analysis and dynamical systems. Like other Harvard departments, it was not a model of ethnic diversity, a fact appreciated by M.I.T.'s Norbert Wiener and, somewhat later, New York University's Richard Courant (see Siegmund-Schultze 1998, 181–185). Five years after the Nazi takeover, Birkhoff offered a survey of the first 50 years of American mathematics as part of the AMS Semicentennial celebrations. This lecture caused a major stir because of certain oft-repeated remarks about the influx of first-class foreign mathematicians to the United States. The latter, Birkhoff felt, threatened to reduce the chances of native Americans, who could become "hewers of wood and drawers of water" within their own community. He then added, "I believe we have reached the point of saturation. We must definitely avoid the danger." (Birkhoff 1938, 276–277).

During the final years of Birkhoff's career – he died in 1944 – he tangled with Princeton's Hermann Weyl in a dispute over gravitational theories. Birkhoff had set forth an alternative to the general theory of relativity, which dispensed with the equivalence principle, the very cornerstone of Einstein's theory. After some rather petty exchanges, Birkhoff and Weyl broke off the debate, agreeing that they should disagree. Veblen, who enjoyed having both Einstein and Weyl as colleagues, took a rather dismissive view of Birkhoff's ideas about gravitation. He also distanced himself from the Harvard mathematician's rather provincial views about the "dangers" posed by foreigners within the American mathematical community. In his necrology of Birkhoff, he wrote:

> . . . a sort of religious devotion to American mathematics as a "cause" was characteristic of a good many of [Birkhoff's] predecessors and contemporaries. It undoubtedly helped the growth of the science during this period. By now [1944] mathematics is perhaps strong enough to be less nationalistic. The American mathematical community has at least been healthy enough to absorb a pretty substantial number of European mathematicians without serious complaint. (Veblen 1946)

After the war, the senior Birkhoff having passed from the scene, Lefschetz no longer had to contend with his former nemesis. During the Princeton Bicentennial Conference, he emerged in his full glory as the new gray eminence of American mathematics (Fig. 37.2). Princeton's 12-man organizing committee clearly set its sights high in preparing for this memorable event. The conference dealt with developments in nine fields, some venerable (algebra, algebraic geometry, and analysis), others more modern (mathematical logic, topology), and a few of even more recent vintage (analysis in the large, and "new fields"). Each of the nine sessions was chaired by a distinguished figure in the field, who made some opening remarks that were followed by extensive discussions, usually led by one or more experts.[1] This format was designed to promote informal exchanges, rather than forcing the participants to spend most of their time listening to a series of formal presentations. The results were carefully recorded by specially chosen reporters, who summarized the main points discussed.

A Rivalry Lives on

Judging from these conference reports, which the organizers characterized as giving "much of the flavor and spirit of the conference," they must have found many of the sessions disappointing (assuming they took the stated agenda of the conference seriously). Still, the Russian-born Lefschetz must have felt a deep satisfaction in hosting an event that demonstrated the dominance of Princeton's Europeanized community over its traditional rival, the Harvard department once led by G. D. Birkhoff. This rivalry nevertheless lived on and was manifest throughout the meeting. In the opening session on algebra, chaired by Emil Artin, Harvard's Garrett Birkhoff began by noting the contrast between the discussion format adopted for the Princeton meeting and the more conventional one adopted at the Harvard Tercentenary meeting 10 years earlier (though he apparently did not state any preference). The younger Birkhoff then proceeded to make some rather pompous pronouncements about the state of his discipline. He characterized algebra as "dealing only with operations involving a finite number of elements," noting that this led to

[1] The nine sessions were (1) algebra (chair (C): E. Artin, discussion leader(s) (D): G. Birkhoff, R. Brauer, N. Jacobson; (2) algebraic geometry (S. Lefschetz (C), W. V. D. Hodge, O. Zariski (D)); (3) differential geometry (O. Veblen (C), V. Hlavatý, T. Y. Thomas (D)); (4) mathematical logic (A. Church (C), A. Tarski (D)); (5) topology (A. W. Tucker (C), H. Hopf, D. Montgomery, N. E. Steenrod, J. H. C. Whitehead (D)); (6) new fields (J. von Neumann (C), G. C. Evans, F. D. Murnaghan, J. L. Synge, N. Wiener (D)); (7) mathematical probability (S. S. Wilks (C), H. Cramér, J. L. Doob, W. Feller (D)); (8) analysis (S. Bochner (C), L. V. Ahlfors, E. Hille, M. Riesz, A. Zygmund (D)); (9) analysis in the large (H. P. Robertson (C), S. Mac Lane, M. H. Stone, H. Weyl (D)).

Fig. 37.2 Solomon Lefschetz was impulsive, frank and opinionated, enough so that many found him obnoxious. He loved to argue and never openly admitted his mistakes, however glaring. But his student Albert W. Tucker was convinced that Lefschetz's bark was worse than his bite. On a train ride from Princeton to New York he overheard a conversation between Lefschetz and Oscar Zariski, who were both discussing an important new paper in algebraic geometry. Lefschetz wasn't sure whether to classify the author's techniques as topological or algebraic, which led Zariski to ask: "How do you draw the line between algebra and topology?" Lefschetz answered in a flash: "Well, if it's just turning the crank, it's algebra, but if it's got an idea in it, it's topology!" (Albers and Alexanderson 1985, p. 350).

three distinct types of algebraic research: (1) "trivial" results; (2) those which also employ the axiom of choice, which he felt were "becoming trivial"; and (3) general results, like his own work relating to the Jordan-Hölder theorem.

Artin had only recently arrived from Indiana, so he had not yet fully emerged as the "cult figure" of the Princeton department later described by Gian-Carlo Rota (Rota 1989). Still, he had known Garrett Birkhoff for some time, as the latter had twice stopped off in Hamburg during the mid-1930s to visit him on the way to European conferences (Birkhoff 1989, p. 46). Predictably, he brushed aside Birkhoff's definition of algebra based on systems to which finitely many operations are applied. "What about limits," he fired back, noting that these are indispensable for valuation theory? Birkhoff merely replied that he didn't consider this part of algebra, but added, "this doesn't mean that algebraists can't do it." After this, a number of others chimed in – Mac Lane, Dunford, Stone, Radó, and Albert – mainly adding special remarks that seem to have contributed little toward clarifying major trends in algebraic research. One senses a number of different competing agendas here, particularly in the exchange between Artin and Birkhoff. As Rota later recalled, at Princeton Artin made no secret of his loathing for the whole Anglo-American algebraic tradition "associated with the names Boole, C.S. Peirce, Dickson, the later British invariant theorists, . . ., and Garrett Birkhoff's universal algebra (the word "lattice" was strictly forbidden, as were several other words)." Birkhoff presumably had more than an inkling of this attitude, which must have grated on him, since Artin's arrogance was almost in a class by itself (Fig. 37.3).

This particular rivalry may be seen as part of the ongoing conflict between "nativists" and "internationalists" within the American mathematical community, Garrett Birkhoff having been a leading representative of the former group. The year 1936 was undoubtedly still very vivid in Birkhoff's mind when he attended the Princeton conference a decade later. Many years later he recalled how he was "dazzled by the depth and erudition of the invited speakers" at the 1936 ICM in Oslo. He was pleased that the two Fields medalists – Lars Ahlfors and Jesse Douglas – "were both from Cambridge, Massachusetts, and delighted that the 1940 International Congress was scheduled to be held at Harvard, with my father as Honorary President!" (Birkhoff 1989, 46). He remembered the "serene atmosphere of Harvard's Tercentenary celebration" which took place the following September in conjunction with the summer meeting of the American Mathematical Society. The event attracted more than one thousand persons, including 443 members of the AMS. He admitted that the invited lectures were over his head, but knew that only very few in the large audiences that attended could follow the presentations.

Ten years later, much had changed now that Birkhoff had become a leading figure in the American mathematical community. With his famous name and rising reputation, he clearly saw himself as carrying Harvard's banner into the rival Princeton camp, and he probably missed the kind of serene pleasures he associated with his *alma mater*. He may well have been unhappy about the format of the conference, given that Princeton mathematicians chaired all nine sessions. Just a glance at their names would have been enough to bring home that Cambridge, even with the combined resources of both Harvard and M.I.T., was no match for the mathematical community in Princeton with Artin, Lefschetz, Veblen, Alonzo Church, A. W. Tucker, John von Neumann, S. S. Wilks, Bochner, Marston Morse, and H. P. Robertson.

Fig. 37.3 Garrett Birkhoff had numerous opportunities to witness the traditional rivalry between Harvard and Princeton during G. D. Birkhoff's heyday. He later recalled this incident: one day Lefschetz came to Harvard—this must have been around 1942—to give a colloquium talk. After the talk my father asked him, "What's new down at Princeton?" Lefschetz gave him a mischievous smile and replied, "Well, one of our visitors solved the four color problem the other day." My father said: "I doubt it, but if it's true I'll go on my hands and knees from the railroad station to Fine Hall." He never had to do this; the number of fallacious proofs of the four color problem is, of course, legion (Albers and Alexanderson 1985, pp. 12–13).

With its university and the Institute for Advanced Study, Princeton had drawn together an unprecedented pool of mathematical talent, all of it on full display at this celebratory meeting. Even Einstein was in the audience, at least briefly.

More sparks flew in the session on algebraic geometry, chaired by Lefschetz, in which William Hodge and Oscar Zariski served as discussion leaders. The latter two made illuminating remarks on the Hodge conjecture, one of the Clay prize problems for the twenty-first century, and on minimal models in birational geometry. Lefschetz, commenting on Zariski's presentation, remarked, "To me algebraic geometry is algebra with a kick. All too often algebra seems to lack direction to specific problems." To this, Birkhoff countered: "If the algebraic geometers are so ambitious, why don't they do something about the real field?" Lefschetz answered by suggesting that the geometry of real curves was analogous to number theory before the utilization of analytic methods, when one had only scattered results without a unifying theory. He pointed to Hilbert's still unsolved sixteenth problem on the nesting configurations for the components of real curves as an illustration of the lack of suitable general methods.

In the session on mathematical logic, chaired by Alonzo Church, most of the discussions centered on decision problems. Oddly enough, Hilbert's tenth Paris problem, the decision problem for Diophantine equations (proved unsolvable by Yuri Matijacevic in 1970) was not even mentioned, though it was only in the 1930s that the notion of a computable algorithm became tractable. Church called attention to the recent theorem of Emil Post, who proved that the word problem for semi-groups is unsolvable. This prompted him to suggest that the word problem for groups and the problem of giving a complete set of knot invariants ought to be tackled next. J. H. C. Whitehead expressed a different opinion about these problems in the topology session, where he mentioned the word problem in the same breath as the Poincaré conjecture, noting, "our knowledge of these matters is practically nil."

Alfred Tarski, who conducted a survey of the decision problem in various logical domains, led further discussions on mathematical logic. An interesting argument ensued when Kurt Gödel proposed an expansion of the countable formalized systems that he had investigated on the way to his famous incompleteness theorem of 1931. Church apparently took issue with Gödel's claim that "the set of all things of which we can think" is probably denumerable. A philosophical debate then ensued over what it meant to have a "proof" and when a purported proof could be "reasonably" doubted. These reflections appear to have enriched rather than deflected the general discussions in this session, which were both focused and informative. Unlike some of the discussions between participants in the algebra and algebraic geometry sessions, the logicians avoided the temptation to grandstand or make sweeping pronouncements about the status of a particular area of research. The contrast was reflected by the organizers characterizations of the logic session, which showed "the liveliness of mathematical logic and its insistent pressing on toward the problems of the general mathematician," as opposed to the discussion about general algebra. Should limits and topological methods, which were required for many vital results, be *defined* out of algebra?

Clearly, Artin and Lefschetz did not think they should, as otherwise "algebra would lose much power."

Taking Stock

Having touched upon the overall atmosphere at this meeting shortly before Christmas 1946 as well as some of the specific exchanges during these three days of discussions, let us now jump ahead to the year 1988 when the AMS was celebrating its own centenary. The following year saw the publication of the second volume of *A Century of Mathematics in America* in which the proceedings of the Princeton meeting were reprinted (Duren et al. 1989, 309–334). The editors also asked several experts in relevant fields to comment on the discussions that had taken place in 1946 as recorded for these proceedings.[2] In view of the purpose of the Princeton meeting—which aimed to cast its eye on what the future held in store—one might have thought that such a retrospective analysis would have proven useful in order to take stock of the progress made during the intervening period. If so, the editors were forced to conclude that these commentaries underscored "how different mathematics was in 1946."

Almost all of the experts noted the immense gap that separated state of the art research in their field ca. 1988 and the interests of leading practitioners 40 years earlier. Several noted that some highly significant work already published before 1946 received no attention at the Princeton conference. Thus, Robert Osserman was astonished that names like Cartan, Chern, and Weyl did not appear in the report on recent work on differential geometry. Chern's intrinsic proof of the generalized Gauss-Bonnet theorem was presented "in Princeton's own backyard at the Institute for Advanced Study" in 1943! J. L. Doob's comments on the probability session are particularly illuminating given that he had participated in the conference as a discussant:

> The basic difference between the roles of mathematical probability in 1946 and 1988 is that the subject is now accepted as mathematics whereas in 1946 to most mathematicians mathematical probability was to mathematics as black marketing to marketing; that is, probability was a source of interesting mathematics but examination of the background context was undesirable. And the fact that probability was intrinsically related to statistics did not improve either subject's standing in the eyes of pure mathematicians. In fact the relationship between the two subjects inspired heated fruitless discussions of "What is probability?", and thereby encouraged the confusion between probability and the phenomena to which it is applied (Doob 1989, 353).

Doob went on to note that Kolmogorov's program for the foundations of probability had already been set forth back in 1933. It nevertheless took several decades before the idea of treating random variables as measurable functions gained acceptance. As Doob put it, "some mathematicians sneered that probability should not bury its spice in the bland soup of measure theory, that perhaps probability needed rigor, but surely not *rigor mortis*."

Two commentators, William Browder and Karen Uhlenbeck, were struck by some general remarks that Hermann Weyl made in his after-dinner speech at the close of the meeting. As one of the last great representatives of the Göttingen mathematical tradition, it was surely fitting that Weyl was asked to speak at the closing ceremonies. And it was equally fitting that Weyl mentioned Minkowski's 1905 speech honoring Dirichlet, in which Weyl's former teacher stated that the "true Dirichlet principle" was to solve mathematical problems "with a minimum of blind calculation and a maximum of seeing thought." Hilbert had been a leading advocate of this philosophy, but even in his youth Weyl had deep reservations about this whole approach to mathematical knowledge (see Chap. 27). These misgivings had evidently not abated during the twilight of his career, and in Princeton he went so far as to formulate a counter-principle. "I find the present state of mathematics, that has arisen by going full steam ahead under this slogan (the "true Dirichlet principle"), so alarming that I propose another principle: *Whenever you can settle a question by explicit construction, be not satisfied with purely existential arguments.*"

Although he had long since parted company with Brouwer's brand of intuitionism, Weyl continued to believe that pure mathematics could only thrive when its tendency toward abstraction is sustained by ideas of a non-formal nature. He made the point in Princeton by quoting himself in 1931 when he offered these remarks at a conference in Bern: "Before one can generalize, formalize, or axiomatize, there must be a mathematical substance. I am afraid that the mathematical substance in the formalization of which we have exercised our powers in the last two decades shows signs of exhaustion. Thus I foresee that the coming generation will have a hard lot in mathematics." Despite the tumultuous political events that had interceded, Weyl's views in 1946 reflected much the same opinion:

> The challenge, I am afraid, has only partially been met in the intervening 15 years. There were plenty of encouraging signs in this conference. But the deeper one drives the spade the harder the digging gets; maybe it has become too hard for us unless we are given some outside help, be it even by such devilish devices as high-speed computing machines.

No doubt, John von Neumann, who had chaired the session entitled simply "New Fields," was smiling in approval. Neither he nor Weyl knew what the future held in store, but they probably sensed that they were standing on the brink of a new era.

[2]The commentaries covered eight of the nine sessions: algebra (J. Tate and B. Gross), algebraic geometry (H. Clemens), differential geometry (R. Osserman), mathematical logic (Y. N. Moschovakis), topology (W. Browder), mathematical probability (J. L. Doob), analysis (E. M. Stein), and analysis in the large (K. Uhlenbeck).

References

Albers, Donald J., and G.L. Alexanderson, eds. 1985. *Mathematical People. Profiles and Interviews*. Boston: Birkhäuser.

Birkhoff, G.D. 1938. Fifty Years of American Mathematics. In *Semicentennial Addresses of the American Mathematical Society*, vol. 2, 270–315. Providence: American Mathematical Society.

Birkhoff, Garrett. 1989. Mathematics at Harvard, 1836–1944, in [Duren 1989, 3–58].

Browder, Felix, ed. 1976. *Mathematical Developments Arising from Hilbert's Problems*, Symposia in Pure Mathematics, vol. 28, Providence: American Mathematical Society.

Davis, P.J., and Reuben Hersh, eds. 1981. *The Mathematical Experience*. Birkhäuser: Boston.

Doob, J. L. 1989. Commentary on Probability, in [Duren 1989, 353–354].

Duren, Peter, et al., eds. 1988. *A Century of Mathematics in America*. Vol. 1. Providence: American Mathematical Society.

———., eds. 1989. *A Century of Mathematics in America*. Vol. 2. Providence: American Mathematical Society.

Gray, Jeremy J. 2000. *The Hilbert Challenge*. Oxford: Oxford University Press.

Hilbert, David. 1935. *Gesammelte Abhandlungen*. Vol. 3. Berlin: Springer.

Lefschetz, Solomon, et al., eds. 1947. *Problems of Mathematics*, Princeton University Bicentennial Conferences, Series 2, Conference 2, reprinted in [Duren 1989, 309–334].

Mac Lane, Saunders. 1989. Topology and Logic at Princeton, in [Duren 1989, 217–221].]

Minkowski, Hermann. 1973. *Briefe an David Hilbert*. Hg. L. Rüdenberg und H. Zassenhaus, New York: Springer.

Nye, Mary Jo, ed. 2003. *The Cambridge History of Science. Volume 5: The Modern Physical and Mathematical Sciences*. Cambridge: Cambridge University Press.

Reid, Constance. 1970. *Hilbert*. New York: Springer.

Reingold, Nathan. 1981. Refugee Mathematicians in the United States of America 1933–1941. *Annals of Science* 38: 313–338. (Reprinted in [Duren 1988, 175–200]).

Rota, Gian-Carlo. 1989. Fine Hall in its Golden Age: Remembrances of Princeton in the Early Fifties, in [Duren 1989, 223–236].

Rowe, David E. 1996. New Trends and Old Images in the History of Mathematics. In *Vita Mathematica. Historical Research and Integration with Teaching*, MAA Notes Series, ed. Ronald Calinger, vol. 40, 3–16. Washington, D.C.: Mathematical Association of America.

———. 2003. Mathematical Schools, Communities, and Networks, in [Nye 2003, 113–132].

Siegmund-Schultze, Reinhard. 1998. *Mathematiker auf der Flucht vor Hitler, Dokumente zur Geschichte der Mathematik*. Bd. 10 ed. Braunschweig: Vieweg.

Veblen, Oswald. 1946. George David Birkhoff, 1884–1944, Proceedings of the American Philosophical Society, Yearbook, 279–285.

Weyl, Hermann. 1944. David Hilbert and his Mathematical Work. *Bulletin of the American Mathematical Society* 50: 612–654.

Personal Reflections on Dirk Jan Struik By Joseph W. Dauben

38

(*Mathematical Intelligencer* 33(2)(2011): 36–41)

Introduction: Dirk Struik and the History of Mathematics

Dirk Jan Struik, who taught for many years at the Massachusetts Institute of Technology and died on 21 October 2000 at the age of 106, was a distinguished mathematician and influential teacher. He was also widely known as a leading Marxist scholar and social activist. His early work on vector and tensor analysis, undertaken together with Jan Arnoldus Schouten, helped impart new mathematical techniques needed to master Einstein's general theory of relativity. This collaboration lasted for over 20 years, but by the end of the 1930s, Struik came to realize that the heyday of the Ricci calculus had passed. After the Second World War, having now entered his 50s, he gave up mathematical research in order to focus his attention on the history of mathematics and science. It was through his work as an historian that he left a truly lasting mark, not only as a writer but also as a mentor to those who had the pleasure of knowing him personally.

I had the great pleasure of knowing Dirk during the last 15 years of his life. Initially our mutual interest in the history of mathematics brought us together. But like so many young people – Dirk thought anyone under 70 was young – I quickly became fascinated with his whole life. One of my favorite memories came in 1994 when Dirk spoke to a general audience Mainz. This was right after the wonderful centennial celebration in Amsterdam, which already had him in high spirits (Koetsier 1996). We drove down to Mainz where I introduced him to a packed audience, easily 300 people. He liked it when I introduced him not as a member of the '68 generation but the generation of 1917. His topic that day was simply "Some Mathematicians I have known." The whole while this amazing centenarian stood at the podium, a few note cards in hand, and proceeded to tell a string of fascinating anecdotes. He began in English, but kept slipping into German. Then he would catch himself and start speaking in English again. It was a wonderful example of how naturally he connected with people in so many languages and cultures, always with an immense sense of good will. He took lots of questions after his talk; people were sitting in the aisles, and I didn't notice anybody leaving. That evening he kept a small group of my students spellbound with his stories. A few of the young ladies present got extra attention, and when they told him about their own ongoing work in the history of mathematics he got that twinkle in his eyes that made all of us feel special and glad to be around him.

As he grew older, Dirk was often asked, of course, about the secret to his longevity (Fig. 38.1). Those who knew him realized it had nothing to do with abstention from life's little vices (a pipe of tobacco and a glass of sherry were part of his daily regimen). Characteristically, he answered by attributing his good health and zest for life to the three pillars of his spiritual strength, "the three M's": Mathematics, Marxism, and Marriage. He shared these passions with his wife of some 70 years, Ruth Struik, née Ramler, herself a mathematician and a native of Prague; she died in 1993 at the age of 99. Dirk and Ruth Struik were political activists, deeply moved by the struggle against fascism in Europe and filled with high hopes that the Soviet Union's socialist experiment would eventually triumph as a new model for human society (Alberts 1994).

In 1934, after eight years in the United States, they became naturalized American citizens, after which time they began taking a more active part in supporting various political causes. In the wake of the Nazi racial laws aimed at "purifying" the civil service, they both tried to help several European mathematicians find refuge in the United States. Dirk was also involved in the often-bitter disputes among American leftists regarding efforts to support the Spanish Loyalists in their battle against Franco's insurgent Fascists during the 1930s. During the war years, he worked for the Council of American-Soviet Friendship, and in 1944 he helped found the Samuel Adams School in Boston, a short-lived experiment that nevertheless came under the eye of J. Edgar Hoover's F.B.I. Dirk Struik had a boundless faith in the capacity of human beings to build a just society. He saw science and mathematics as liberating forces within

D.E. Rowe, *A Richer Picture of Mathematics*, https://doi.org/10.1007/978-3-319-67819-1_38

Fig. 38.1 Dirk Jan Struik (1894–2000).

society, but he also realized that the modern scientist has a responsibility to consider the social consequences of scientific research. In this regard, he was strongly influenced by his MIT colleague and friend Norbert Wiener, who refused to place his fertile mind at the disposal of government technocrats.

As an historian, Struik saw culture, science, and society as tightly intertwined (Rowe 1994). Mathematics, his first love, was deeply embedded in the cauldron of cultures, not as something freely imported from without, but rather built from within as a product of human intellectual and social activity. Taking this approach, he tried to ferret out the links between scientific "high culture" and the work of artisans, technicians, and the myriad other practitioners who represent the "applied science sector" within a society's workforce. In *Yankee Science in the Making* (1948) Struik analyzed the local social, geographical, and economic forces that shaped the lives of those inventors and amateurs who contributed to the emergence of a new scientific culture in Colonial New England (Stapleton 1997). A similar focus on local conditions animates his book on early modern Dutch science, *The Land of Stevin and Huygens* (Engl. Trans. 1981, *Het land van Stevin en Huygens*, 1958). Struik made no secret of his Marxism, but neither did it dictate his historical analyses, the best of which were guided by an intuitive grasp of the salient features that led to the formation of distinctive scientific cultures. Like his friend Jan Romein, he sought global patterns of development, but he also avoided reifying Marxist ideas or taking a reductionist approach to historical materialism. Inspired by the work of figures like J. D. Bernal, J. B. S. Haldane, Lancelot Hogben, and others in the British tradition of the Social Relations in Science Movement, he became increasingly interested in the social underpinnings of scientific knowledge. Multiculturalism would only become a buzzword decades later, but Struik's scholarship already reflected a deep awareness of how ideas are molded and transported in a complex mix of cultural contexts. The best-known example is his *Concise History of Mathematics*, first published in 1948, which went through four English editions in as many decades. Eventually it was translated into at least 18 different languages. It would be fair to say that no historical survey has done more to promote interest in the rich diversity of mathematical ideas and cultures.

Teaching always played an integral part in Struik's academic life. During the 1940s, he began offering an informal seminar on historical materialism and Marxism, a topic then outside the official curricula of universities throughout the country. George Mosse, who came to Harvard as a graduate student in the early 1940s and was eager to learn some Marxist theory, quickly found his way to Struik's seminar (George L. Mosse, *Confronting* History, pp. 121, 136). He went on to become a distinguished historian at the University of Wisconsin and a leading expert on the intellectual origins of fascism in Germany. Back in the 1930s, George Sarton pioneered studies of the history of science in the United States at Harvard. He regularly invited Dirk Struik to teach special session of his Harvard seminar on the topic of history of mathematics. I. B. Cohen, then a graduate student at Harvard, later recalled how Sarton and Struik used to argue over the role of social factors in the history of mathematics (Sarton's favorite counterexample was magic squares). Yet their differences notwithstanding, both men shared a cultural approach to the history of science of universal scope.

A generation later, a young man named Joseph Dauben came to Harvard to study history of science, *en route* to writing a doctoral dissertation on the mathematics and philosophy of Georg Cantor. Along the way, he too fell under the sway of Struik's influence, a story he vividly recalls here. It gives me special pleasure to see his recollections appear in *Years Ago*, as I well remember the delight I took in reading his wonderful biography of Cantor when I was a graduate student in Oklahoma. Soon afterward, I wrote to the author and I was even more delighted to receive a warm letter from him inviting me to study with him at the CUNY Graduate Center. So began my own journey into the field of inquiry that owes so much to the inspiration of Dirk Struik.

D.E.R.

***

Personal Reflections on Dirk Jan Struik By Joseph W. Dauben

I first encountered Dirk Struik through his *Concise History of Mathematics*, which I read when I was a senior in High School. At the time I was taking an advanced calculus class at Pasadena High School in Southern California, and through a quirk of fate, the teacher had an evening class at the local Junior College and I was invited to give a lecture on the derivative, which it was suggested I might want to motivate with some historical background. Thus rather than start my introduction to history of mathematics with one of E.T. Bell's questionable anecdotes, I was introduced at the outset to the subject through one of the adepts, someone with a true feeling for both historical and political sensibilities. Of course, I did not really appreciate what Struik was up to in his book at the time, but then I was only in high school.

Nevertheless, this says something about the readership Struik's work enjoyed—reaching even to impressionable history-school students. I next read Struik's book for a second time, and more thoroughly, in college (Claremont McKenna College, one of the Claremont Colleges) where I was taking a course on history of science taught by Granville Henry of the mathematics department. I was also writing my senior thesis on nonstandard analysis, overseen by Janet Myhre, and I found Struik's book again helpful in explaining the early concerns for infinitesimals in the seventeenth century and the rise of *epsilon-delta* techniques to avoid them entirely in the nineteenth century. But again, I still had no real appreciation for what the *Concise History* had achieved for the history of mathematics generally, or what a revolutionary, some might have even said how radical, a book it was.

The Marxist approach Struik had adopted only became apparent and important to me just before I was about to meet Struik for the first time, when I was a graduate student in the newly-founded Department for History of Science at Harvard. I was fortunate at Harvard to be part of a cohort of new graduate students, one of whom was Wilbur Knorr, and the two of us could not have been better prepared for our general examinations than by John Murdoch in history of ancient and medieval mathematics, and in modern mathematics than by Judith Grabiner, who had just finished her Ph.D. at Harvard with a thesis on the mathematics of Lagrange. In 1967–68, as I was studying for these exams, I was again reading Struik, but with a much more sensitive eye to the political and social contexts that he made clear were sometimes as important for understanding the development of mathematics as were its purely internal developments.

Thus I knew Dirk Struik quite well on paper, from his writings, but I had not as yet met him, and I was in fact unprepared when we did meet for the first time, at one of the departmental Christmas parties held every year at the Harvard Faculty Club. I recall his tall, gangly figure discussing something with I. Bernard Cohen, who introduced us and then went off, leaving us to talk about history of mathematics. Struik had a way of making whomever he was with seem like the center of attention, at least his center of attention, and at that moment he wanted to know how I came to study history of mathematics, and to be at Harvard. After a brief answer, I asked Struik about the *Concise History*, into how many languages had it been translated? I also confessed that his treatment of the nineteenth century in particular had given me a much broader panorama of interests to consider as I thought about the subject of my dissertation. At that point I had decided to write about Bernhard Bolzano, and through John Murdoch had been in touch with Luboš Nový at the archives of the Academy of Sciences in Prague about my working with him there. I had actually decided to divide my time between Prague and Vienna, to work on Bolzano's logic and his paradoxes of the infinite. But in August of 1968, as I was enjoying a brief vacation in Southern California to see my family before setting off for Prague, news came of the Russian invasion of Czechoslovakia. That changed things considerably, and over the next year I rethought the subject of my dissertation. Back at Harvard, in all of my reading I realized I had never come across a detailed biography of Georg Cantor, founder of transfinite set theory. I had always been interested in set theory, and it was a subject that would at least force me to learn German, a language I could read but by no means use with any fluency.

This also seemed like a good choice since Erwin Hiebert had just come to Harvard from the University of Wisconsin. Hiebert was a specialist in the history of modern German physics, and that would be a good match with what I wanted to do in mathematics. Hiebert also suggested that since Struik was emeritus from MIT, but still affiliated with the Department at Harvard for just such purposes as directing the odd thesis that might turn up, I should ask him about working with me on Cantor, and so I approached him about serving as one of my thesis advisors. I am forever grateful that when I first brought up this possibility, he was enthusiastic and agreed to help however he could.

He began by suggesting that I get in touch with Christoph Scriba, who was then Director of the History of Science group at the Technische Universität in Berlin, and that I should also arrange to meet the Director of the Alexander von Humboldt Forschungsstelle at the German Akademie der Wissenschaften in East Berlin. I did, and subsequently spent a very productive year, 1970–1971, living what I imagined to be the life of a double-agent, stationed in West Berlin while doing most of my archival work in East Berlin. From time to time I would write to Struik to inform him of my progress, and he would usually write back with a reference or two he thought I should read as background or as a foil to how I might otherwise have been thinking about a particular aspect of Cantor's life and work.

When I got back to Harvard after my year in Berlin, Struik invited me to his home in Belmont to provide a "full report" of the year in Berlin, and as I arrived on his doorstep he asked me with a wink in his eye, "Is it still '*eine Reise wert*'?" This was one of the slogans often used to refer to Berlin, and I assured him that it was definitely "worth a visit." Actually, I considered the year I had spent in Berlin as my "Bertolt Brecht Zeit," given that I was living in a cold-water flat with an octogenarian former singer from the German Opera over what can best be described as a bar that served as a local conduit to the nearest brothel. Struik was delighted to hear of all this, and we spent that afternoon, a lazy day in late September, swapping stories about European capitals, his stories better than mine, and full of names well known to every mathematician. But he was genuinely interested in Berlin and how it was faring with the wall up and the city divided.

I then spent regularly at least one afternoon each month visiting Struik as I delivered, chapter-by-chapter, my thesis as it began to unfold. I can remember in particular one afternoon in the early spring – I had started delivering chapters to Struik in October, and by March I had gotten to Chap. 6. My writing was accelerating, and I had reached the point of Cantor's major mid-career work, the *Grundlagen*, his first large-scale introduction to set theory and transfinite numbers, although at that point only the transfinite ordinal numbers had been worked out and the transfinite *alephs* were **another decade in coming. Struik was surprised by that – and** wanted to know why Cantor had chosen *alephs* when he came to introduce the transfinite cardinal numbers in the 1890s. The answer to that question was one that amused Struik to no end, because the answer was social, political, and in a sense economic as well, a nice Marxian trio, as he put it to me at the time, and I had come to see it that way as well. As Struik also said, he had always thought Cantor had chosen the *alephs* for his transfinite numbers because he was *Jewish*. That too turned out to be another fable. In fact, Cantor's mother was Roman Catholic (his father's side of the family most likely had its roots in the Jewish community of Copenhagen). Cantor, who was raised as a Lutheran and seems to have been very comfortable in his correspondence with Catholic theologians (a part of my thesis in which Struik was also particularly interested), seems not to have been a practicing Lutheran or a follower of any particular faith. E.T. Bell's famous characterization of Cantor and Kronecker as the epitome of two antagonistic Jewish professors who were enemies to the death was even further from the mark. Their views on foundations were certainly at odds with each other, but they did their best to get on with one another, and at the end of both their careers, as Cantor was actively working to establish the German Union of Mathematicians, he invited Kronecker to be Union's first keynote speaker in 1891.

In any case, the true story about the alephs, which Struik particularly enjoyed, was basically pragmatic; Cantor knew that his transfinite numbers were special, and he wanted a special notation for the transfinite cardinals. As he told Giulio Vivanti (December 13, 1893), all of the usual alphabets were taken, and letters from the Roman and Greek alphabets were too common in mathematics, whereas it would have been costly to design an entirely new symbol that most printers would not have on hand. But in Germany, virtually all printers had the Hebrew alphabet at their disposal. It occurred to Cantor that since the Hebrew aleph also represented the number one, it was the perfect choice for the first of his transfinite cardinal numbers.

By March of 1972, despite his interest in what I had been writing, Struik was getting worried that I wasn't going to finish in time for a June degree. I assured him that I was writing at full speed and was certain I could finish the last two chapters (as I then envisioned them at the time) in April, mid-May at the latest. One was on Cantor's "*Beiträge*," his last major presentation of his set theory, including both the ordinal and cardinal transfinite numbers, the other a concluding chapter about his philosophy of mathematics, the paradoxes of set theory, and the slow but eventual acceptance of set theory and the new mathematics. Struik apparently wasn't convinced, because within the week I got a phone call from my mentor at Harvard, I. Bernard Cohen. I was also the head TA for Cohen's Scientific Revolution course at Harvard, for which he was justly famous. I thought he was calling about the next meeting of the course when I heard him say: "Joe, I've heard something rather disturbing about your thesis." This did not sound like good news, and so I asked him what the problem seemed to be. "I understand," said Cohen, with a suitable pause for dramatic effect, "that it is getting rather long, and I've spoken with Hiebert and Struik and they both assure me that what you've written is plenty for your Ph.D. We think you should stop. Where are you?" "I'm at home," I said, without really thinking. "No – I mean where are you in the *thesis*?" "Oh," I replied, "more than half-way – I've only got two more chapters to go and I'll be done."

Nevertheless, I did indeed stop where I was, more or less, and my thesis, instead of being a history of Cantor's entire work, became instead "The Early Development of Cantorian Set Theory." And so, thanks possibly to Dirk's behind-the-scenes intervention, I received my degree in June and began teaching at Herbert H. Lehman College of the City University of New York in the fall of 1972. But before leaving Cambridge for New York, I had one last afternoon on the porch with Dirk at his home in Belmont. With a cold bottle of Riesling, we sat and watched the sunset and talked about his many travels. Among the most nostalgic, he mentioned Rome, knowing that I had spent a part of the past summer at the American Academy in Rome, where I had visited the mathematician Lucio Lombardo-Radice. Struik knew Lombardo-Radice and had instructed me to visit him in hopes of getting permission to see letters between Cantor and Vivanti that were still in the hands of Vivanti's family. Struik reminisced about the time he had spent at the University of Rome with his wife Ruth in the 1920s. His memory was as accurate about the university and the friends he had made while there as if he had only been away a few weeks, rather than decades, and he spoke of Rome as if it were an old friend. He could still see, smell and feel its pulse as he talked about walks along the Tiber or coffee in the Piazza Navona.

As I remember, I next saw Dirk not in Belmont, but in Hamburg, Germany, in the summer of 1989. The occasion was the XVIIIth International Congress of the History of Science in Hamburg, where Struik was to receive the first award of the Kenneth O. May medal for outstanding contributions to the history of mathematics, an award he was pleased to share with his old friend and colleague from the Soviet Union, the Russian historian of mathematics Adolf P. Yushkevich. As the Chairman of the International Commission on History of Mathematics, which had established the prize, it was my pleasure to make the actual presentation of the medals, which thanks to Christoph J. Scriba, former Chair of the ICHM and one of the co-organizers of the Congress in Hamburg, were presented on boats in the "Binnen Alster," the inner harbor in Hamburg. At the appointed time, three boats carrying members and guests of the ICHM pulled alongside each other and were tethered together while the presentations and speeches were made (Figs. 38.2 and 38.3).

After accepting his award, Struik came to the microphone and talked about the importance of the history of mathematics and how the field had grown from the time he started working, when he was something of a lone figure. Indeed, when he wrote his *Concise History* nothing like it had really ever been written before, given its level of political and social consciousness that sat very comfortably with the technical mathematical history he also had to tell.

Over the years, whenever I was in Cambridge with time to spare, I would make a point of visiting Dirk Struik in Belmont. The last time I saw him was shortly after his 100th birthday. This was in the fall of 1995, just after publication of my biography of Abraham Robinson, and I wanted to give Struik a copy. I made the familiar trip out to Belmont, and congratulated him personally for having passed his 100th birthday. I pointed out that the phrase with which he had ended his remarks on the occasion of the *Festschrift* Robert Cohen had published in his honor in 1974. There, after noting that he and his wife had just celebrated their 50th wedding anniversary and had three daughters and ten grandchildren, he simply added: "Wish me luck." I remarked about this, that it seemed to have worked, to which he replied in his usual laconic way: "much better than expected!" (Fig. 38.4).

Once again, as so many times before, we were sitting out on his porch, enjoying a late September afternoon in Belmont, and along with the book I had brought a bottle of Riesling. As Struik poured out two glasses, he asked me what I had been working on since Robinson, and I told him about a paper I was reworking about Charles Sanders Peirce and a Tiffany watch the Coast and Geodetic Survey had provided him. Once, on a trip to New York, it had been stolen, and Peirce, in accounting for how he managed to track it down, used this as an example of his theory of abductive reasoning, which he compared to the methods of Sherlock Holmes. The case of his finding the Tiffany watch by a process of elimination and intuition, as Peirce put it, made

Fig. 38.2 Dirk Struik, Hamburg, August 1989 (Photo courtesy the International Commission on History of Mathematics).

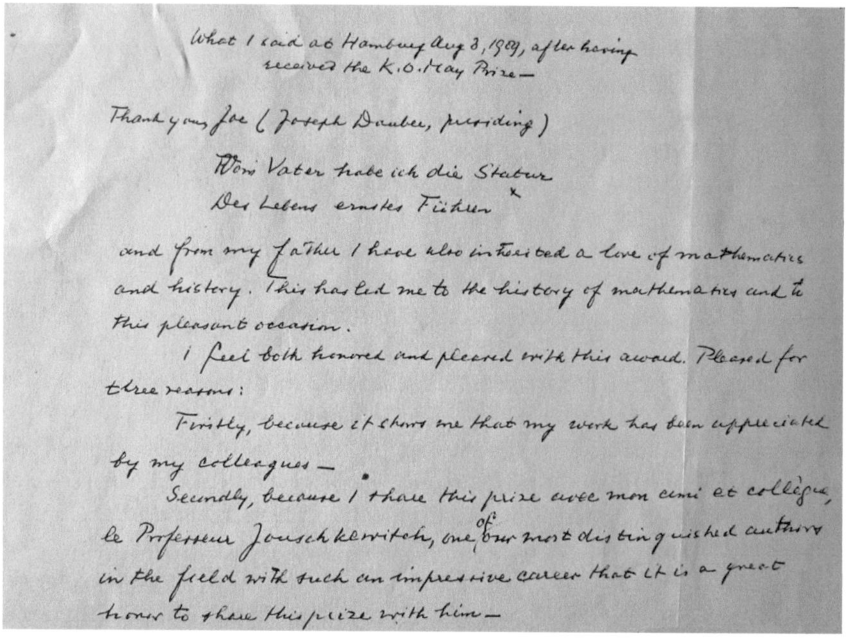
What I said at Hamburg Aug 3, 1989, after having received the K.O. May Prize—

Thank you, Joe (Joseph Dauben, presiding)

Vom Vater habe ich die Statur
Des Lebens ernstes Führen ×

and from my father I have also inherited a love of mathematics and history. This has led me to the history of mathematics and to this pleasant occasion.

I feel both honored and pleased with this award. Pleased for three reasons:

Firstly, because it shows me that my work has been appreciated by my colleagues —

Secondly, because I share this prize avec mon ami et collègue, le Professeur Jouschkewitsch, one of our most distinguished authors in the field with such an impressive career that it is a great honor to share this prize with him—

Fig. 38.3 From Struik's handwritten remarks, beginning with a passage from Goethe, on accepting the K. O. May Medal, Hamburg, 3 August, 1989.

Fig. 38.4 Dirk Struik speaking at the Amsterdam symposium in celebration of his 100th birthday.

clear that "when all other possibilities have been excluded, what remains, however improbable, must be true."

Whereupon Struik immediately launched into a detailed account of Holmes and the connection between Holmes and Watson, recalling details of an article he had written that, he said, compared Watson to Zeno of Elea—in that both are known only through the writings of others, in this case, what we know of Holmes, Struik said, came mainly from Watson. He said he thought he had copies of the article and would be happy to give me one. He suggested we go upstairs to his study to find it. As he made his way to the second floor, he observed wryly that at his age he no longer went upstairs as quickly as he used to, with the added footnote that "you know, I'm nearly a hundred-and-one." It took him a few minutes, but he indeed found the article in question—written in 1947!

Back on the front porch, as he sipped his glass of Riesling, Struik told me about his active membership in the Boston Holmes Society, the Speckled Band of Boston, and pointed out that, in fact, he was in very good company: Holmesian devotees included Ellery Queen, Basil Rathbone, Isaac Asimov, Franklin Roosevelt, and T.S. Eliot, among many others.

He didn't mention that his article, which later appeared in a collection edited by Philip Shreffler, *Sherlock Holmes by Gas Lamp* (1989), not only suggested that Watson's Holmes was akin to Aristotle's Zeno or Plato's Socrates, but also covered a vast terrain of cultural information from Mortimer Snerd to the motets of Orlando de Lassus. But this was typical of Struik – he was interested in everything, and that is a large measure, I am sure, of his longevity. It is what kept him young – being constantly interested in everything and everyone around him. How remarkable, to have literally spanned a century, to have lived through the entire twentieth century and even into the twenty-first. As historians of mathematics, we must be grateful that Dirk Struik made clear that our subject is not only of highly abstract, theoretical interest, but has very real, significant social roots. These – as he made apparent in one of the most widely-read books on the subject – profoundly affect the societies in which mathematics today plays a pervasive role in virtually every aspect of the world in which we all live.

References

Alberts, Gerard. 1994. On connecting socialism and mathematics: Dirk Struik, Jan Burgers, and Jan Tinbergen. *Historia Mathematica* 21 (3): 280–305.

Koetsier, Teun. 1996. Dirk Struik's autumn 1994 visit to Europe: Including "My European extravaganza of October, 1994" by Struik, *Nieuw Archief voor Wiskunde* (4) (1) 14: 167–176.

Rowe, David E. 1994. Dirk Jan Struik and his contributions to the history of mathematics. *Historia Mathematica* 21 (3): 245–273.

Stapleton, Darwin H. 1997. Dirk J. Struik's *Yankee Science in the Making*: A Half-Century Retrospective. *Isis* 88 (3): 505–511.

Struik, Dirk J. 1948. *Yankee Science in the Making*, Boston: Little, Brown & Co.

Struik, Dirk J. 1958. *Het land van Stevin en Huygens*, Amsterdam: Pegasus.

Struik, Dirk J. 1981. *The Land of Stevin and Huygens*, English trans. of (Struik 1958), Dordrecht: Reidel, 1981.

Struik, Dirk J. 1989. The Real Watson. In *Sherlock Holmes by Gas-Lamp. Highlights from the First Four Decades of The Baker Street Journal*, ed. Philip A. Shreffler, 174–178. New York: Fordham University Press.

Name Index

D.E. Rowe, *A Richer Picture of Mathematics*, https://doi.org/10.1007/978-3-319-67819-1

H

I

J

K

Q

R

S

MIX
Papier aus verantwortungsvollen Quellen
Paper from responsible sources
FSC® C105338

If you have any concerns about our products,
you can contact us on
ProductSafety@springernature.com

In case Publisher is established outside the EU,
the EU authorized representative is:
Springer Nature Customer Service Center GmbH
Europaplatz 3, 69115 Heidelberg, Germany

Printed by Libri Plureos GmbH
in Hamburg, Germany